Coarse-Graining of Condensed Phase and Biomolecular Systems

Coarse-Graining of Condensed Phase and Biomolecular Systems

Edited by

Gregory A. Voth

CRC Press
Taylor & Francis Group
Boca Raton London New York

CRC Press is an imprint of the
Taylor & Francis Group, an **informa** business

CRC Press
Taylor & Francis Group
6000 Broken Sound Parkway NW, Suite 300
Boca Raton, FL 33487-2742

Printed in the United States of America on acid-free paper
10 9 8 7 6 5 4 3 2 1

International Standard Book Number-13: 978-1-4200-5955-7 (Hardcover)

Library of Congress Cataloging-in-Publication Data

Coarse-graining of condensed phase and biomolecular systems / editor, Gregory A. Voth.
p. ; cm.
Includes bibliographical references and index.
ISBN 978-1-4200-5955-7 (hardcover : alk. paper)
1. Molecular dynamics--Computer simulation. 2. Biomolecules--Computer simulation. 3. Condensed matter--Computer simulation. I. Voth, Gregory A. II. Title.
[DNLM: 1. Computer Simulation. 2. Computational Biology--methods. 3. Models, Molecular. 4. Models, Statistical. 5. Molecular Biology--methods. QA 76.9.C65 C652 2009]

QP517.M65C63 2009
541'.394--dc22 2008027690

Visit the Taylor & Francis Web site at
http://www.taylorandfrancis.com

and the CRC Press Web site at
http://www.crcpress.com

Table of Contents

Acknowledgments

My own research contributions to this book would not have been possible if it were not for the remarkable dedication, talent, and hard work of the members of my research group, both past and present. I thank my assistant, Shawna Derry, for her indispensable help and patience in the preparation of this book, and Lance Wobus of CRC Press/Taylor & Francis for his help in formulating the concept of the book and for his advice and guidance during its preparation. Most of all, I thank my two children Michael and Carolyn, for supporting me through the many long hours I have worked in my career, my two brothers and mother for putting up with someone who tries to think "outside the box" a bit too much, and my father who, while he was living, taught me the value of loyalty, courage, and perseverance.

Editor

Gregory A. Voth is a distinguished professor of chemistry and the director of the Center for Biophysical Modeling and Simulation at the University of Utah. He received a PhD in theoretical chemistry from the California Institute of Technology in 1987.

Selected honors and awards include: John Simon Guggenheim Memorial, Fellowship, 2004–2005; Miller Professorship, University of California, Berkeley, 2003; Elected Fellow of the American Association for the Advancement of Science, 1999; Elected Fellow of the American Physical Society, 1998; IBM Faculty Research Award, 1997–99; Camille Dreyfus Teacher-Scholar Award, 1994–99; Alfred P. Sloan Foundation Research Fellow, 1992–94; National Science Foundation Presidential Young Investigator Award, 1991–96; David and Lucile Packard Foundation Fellowship in Science and Engineering, 1990–95; Camille and Henry Dreyfus Distinguished New Faculty Award, 1989; IBM Postdoctoral Fellowship, University of California, Berkeley, 1987–88; The Francis and Milton Clauser Doctoral Prize, California Institute of Technology, 1987; The Herbert Newby McCoy Award, California Institute of Technology, 1986; The Procter and Gamble Award for Outstanding Research in Physical Chemistry, American Chemical Society, 1985. Current professional affiliations include American Chemical Society (ACS), American Physical Society (APS), the Biophysical Society (BPS), and the American Association for the Advancement of Science (AAAS).

Professor Voth is the author or co-author of more than 300 peer-reviewed scientific articles and mentor to more than 100 postdoctoral fellows, graduate students, and undergraduate research assistants. His research interests include multiscale simulation and theoretical modeling of biomolecular systems; proton transport processes in biological, material, and solution phase systems; computer simulation and modeling of soft materials; room-temperature ionic liquids; theory and simulation of solvation phenomena; structure and dynamics of interfaces; theory and simulation of condensed-phase quantum dynamical processes; and high-performance computing.

Contributors

Anton Arkhipov
Department of Physics
University of Illinois at Urbana-Champaign
Urbana, Illinois

Gary S. Ayton
Center for Biophysical Modeling and Simulation and Department of Chemistry
University of Utah
Salt Lake City, Utah

Kurt Binder
Institut für Physik
Johannes Gutenberg-Universität Mainz
Mainz, Germany

Grace Brannigan
Center for Molecular Modeling
Department of Chemistry
University of Pennsylvania
Philadelphia, Pennsylvania

Frank L. H. Brown
Department of Chemistry and Biochemistry and Department of Physics
University of California at Santa Barbara
Santa Barbara, California

N.-V. Buchete
School of Physics
University College Dublin
Dublin, Ireland

M. G. Burke
Institute for Mathematical Sciences, and Department of Chemistry
Imperial College London
London, England, U.K.

Elaine R. Chan
Electronics and Electrical Engineering Laboratory
National Institute of Standards and Technology
Gaithersburg, Maryland

Zhongzhou Chen
Department of Physics
University of Illinois at Urbana-Champaign
Urbana, Illinois

Cecilia Clementi
Department of Chemistry
Rice University
Houston, Texas

Andrew V. Colasanti
Department of Chemistry & Chemical Biology
BioMaPS Institute for Quantitative Biology
Rutgers, The State University of New Jersey
Piscataway, New Jersey

Peter T. Cummings
Center for Nanophase Materials Sciences
Oak Ridge National Laboratory
Oak Ridge, Tennessee

Luke Czapla
Department of Chemistry & Chemical Biology
BioMaPS Institute for Quantitative Biology
Rutgers, The State University of New Jersey
Piscataway, New Jersey

Cezary Czaplewski
Baker Laboratory of Chemistry and Chemical Biology
Cornell University
Ithaca, New York
and
Faculty of Chemistry
University of Gdańsk
Gdańsk, Poland

K. Ch. Daoulas
Institut für Theoretische Physik
Georg-August Universität
Göttingen, Germany

J. J. de Pablo
Department of Chemical and Biological Engineering
University of Wisconsin-Madison
Madison, Wisconsin

F. A. Detcheverry
Department of Chemical and Biological Engineering
University of Wisconsin-Madison
Madison, Wisconsin

Russell DeVane
The Laboratory for Research on the Structure of Matter
University of Pennsylvania
Philadelphia, Pennsylvania

Batsal Devkota
School of Biology
Georgia Institute of Technology
Atlanta, Georgia

Pemra Doruker
Department of Chemical Engineering and Polymer Research Center
Bogazici University
Bebek, Istanbul, Turkey

Ron Elber
Department of Computer Science
Cornell University
Ithaca, New York

Roland Faller
Department of Chemical Engineering and Materials Science
University of California, Davis
Davis, California

Peter L. Freddolino
Center for Biophysics and Computational Biology
University of Illinois at Urbana-Champaign
Urbana, Illinois

Marc Fuhrmans
Groningen Biomolecular Sciences and Biotechnology Institute and
Zernike Institute for Advanced Materials
University of Groningen
Groningen, The Netherlands

Jayeeta Ghosh
Department of Chemical Engineering and Materials Science
University of California, Davis
Davis, California

Sharon C. Glotzer
Department of Chemical Engineering and
Department of Materials Science and Engineering
University of Michigan
Ann Arbor, Michigan

Stephen C. Harvey
School of Biology
Georgia Institute of Technology
Atlanta, Georgia

Changbong Hyeon
Center for Theoretical Biological Physics
University of California, San Diego
La Jolla, California

Sergei Izvekov
Center for Biophysical Modeling and Simulation and
Department of Chemistry
University of Utah
Salt Lake City, Utah

Robert L. Jernigan
LH Baker Center for Bioinformatics and Biological Statistics
Department of Biochemistry, Biophysics, and Molecular Biology
Iowa State University
Ames, Iowa

Mikko Karttunen
Department of Applied Mathematics
The University of Western Ontario
London, Ontario, Canada

Rajmund Kaźmierkiewicz
Baker Laboratory of Chemistry and Chemical Biology
Cornell University
Ithaca, New York
and
University of Gdańsk
Gdańsk, Poland

Michael L. Klein
The Laboratory for Research on the Structure of Matter
University of Pennsylvania
Materials Research Science and Engineering Center
Philadelphia, Pennsylvania

Kurt Kremer
Max Planck Institute for Polymer Research
Mainz, Germany

Ozge Kurkcuoglu
Department of Chemical Engineering and Polymer Research Center
Bogazici University
Bebek, Istanbul, Turkey

Richard Lavery
Institute de Biologie et Chimie des Protéines
Université de Lyon
Lyon, France

Adam Liwo
Baker Laboratory of Chemistry and Chemical Biology
Cornell University
Ithaca, New York
and
Faculty of Chemistry
University of Gdańsk
Gdańsk, Poland

Jianpeng Ma
Baylor College of Medicine
Verna and Marrs Mclean Department of Biochemistry and Molecular Biology
Houston, Texas

Luis González MacDowell
Departamento de Quimica Fisica
Universidad Compluteuse de Madrid
Madrid, Spain

Peter Májek
Department of Computer Science
Cornell University
Ithaca, New York

Mariusz Makowski
Baker Laboratory of Chemistry and Chemical Biology
Cornell University
Ithaca, New York
and
Faculty of Chemistry
University of Gdańsk
Gdańsk, Poland

Siewert J. Marrink
Groningen Biomolecular Sciences and Biotechnology Institute and
Zernike Institute for Advanced Materials
University of Groningen
Groningen, The Netherlands

Silvina Matysiak
Institute for Computational Engineering and Science
The University of Texas at Austin
Austin, Texas

Clare McCabe
Department of Chemical Engineering
Vanderbilt University
Nashville, Tennessee

J. Andrew McCammon
Department of Chemistry and Biochemistry
Center for Theoretical Biological Physics
Howard Hughes Medical Institute
University of California, San Diego
La Jolla, California

Osamu Miyashita
Department of Biochemistry and Molecular Biophysics
The University of Arizona
Tucson, Arizona

Marcus Müller
Institut für Theoretische Physik
Georg-August Universität
Göttingen, Germany

Rajesh K. Murarka
Baker Laboratory of Chemistry and Chemical Biology
Cornell University
Ithaca, New York

Teemu Murtola
Laboratory of Physics and Helsinki Institute of Physics
Helsinki University of Technology
Espoo, Finland

P. F. Nealey
Department of Chemical and Biological Engineering
University of Wisconsin-Madison
Madison, Wisconsin

W. G. Noid
Department of Chemistry
Pennsylvania State University
University Park, Pennsylvania

Stanisław Ołdziej
Baker Laboratory of Chemistry and Chemical Biology
Cornell University
Ithaca, New York
and
Faculty of Chemistry
University of Gdańsk
Gdańsk, Poland

Wilma K. Olson
Department of Chemistry & Chemical Biology
BioMaPS Institute for Quantitative Biology
Rutgers, The State University of New Jersey
Piscataway, New Jersey

José N. Onuchic
Center for Theoretical Biological Physics
University of California, San Diego
La Jolla, California

L. Paramonov
Institute for Mathematical Sciences, and Department of Chemistry
Imperial College London
London, England, U.K.

Wolfgang Paul
Institut für Physik
Johannes Gutenberg-Universität Mainz
Mainz, Germany

Xavier Periole
Groningen Biomolecular Sciences and Biotechnology Institute and
Zernike Institute for Advanced Materials
University of Groningen
Groningen, The Netherlands

Christine Peter
Max Planck Institute for Polymer Research
Mainz, Germany

Anton S. Petrov
School of Biology
Georgia Institute of Technology
Atlanta, Georgia

H. Jelger Risselada
Groningen Biomolecular Sciences and Biotechnology Institute and
Zernike Institute for Advanced Materials
University of Groningen
Groningen, The Netherlands

Ana V. Rojas
Baker Laboratory of Chemistry and Chemical Biology
Cornell University
Ithaca, New York
and
Department of Physics and Astronomy
Louisiana State University
Baton Rouge, Louisiana
and
Center for Computation and Technology
Louisiana State University
Baton Rouge, Louisiana

Sophie Sacquin-Mora
Laboratoire de Biochimie Théorique
Institut de Biologie Physico-Chimique
Paris, France

Harold A. Scheraga
Baker Laboratory of Chemistry and Chemical Biology
Cornell University
Ithaca, New York

Alexander Schug
Center for Theoretical Biological Physics
University of California, San Diego
La Jolla, California

Klaus Schulten
Department of Physics
University of Illinois at Urbana-Champaign
Urbana, Illinois

Amy Y. Shih
Center for Biophysics and Computational Biology
University of Illinois at Urbana-Champaign
Urbana, Illinois

Wataru Shinoda
Research Institute of Computational Science
National Institute of Advanced Industrial Science and Technology
Philadelphia, Pennsylvania

Guang Song
LH Baker Center for Bioinformatics and Biological Statistics
Department of Computer Science
Iowa State University
Ames, Iowa

Mark J. Stevens
Sandia National Laboratories
Albuquerque, New Mexico

J. E. Straub
Chemistry Department
Boston University
Boston, Massachusetts

Alberto Striolo
School of Chemical, Biological and Materials Engineering
The University of Oklahoma
Norman, Oklahoma

Qi Sun
Department of Chemical Engineering and Materials Science
University of California, Davis
Davis, California

Florence Tama
Department of Biochemistry and Molecular Biophysics
The University of Arizona
Tucson, Arizona

Robert K.-Z. Tan
School of Biology
Georgia Institute of Technology
Atlanta, Georgia

D. Thirumalai
Biophysics Program
Institute for Physical Science and Technology
University of Maryland
College Park, Maryland

Valentina Tozzini
NEST-CNR-INFM Scuola Normale Superiore
Pisa, Italy

Nico F. A. van der Vegt
Max Planck Institute for Polymer Research
Mainz, Germany

Ilpo Vattulainen
Department of Physics
Tampere University of Technology
Tampere, Finland

Angela Violi
Department of Mechanical Engineering
University of Michigan
Ann Arbor, Michigan

Peter Virnau
Institut für Physik
Johannes Gutenberg-Universität Mainz
Mainz, Germany

Gregory A. Voth
Center for Biophysical Modeling and Simulation and
Department of Chemistry
University of Utah
Salt Lake City, Utah

Harel Weinstein
Department of Physiology and Biophysics
and the HRH Prince Alwaleed Bin Talal Bin Abdulaziz Alsaud Institute for Computational Biomedicine
Weill Medical College of Cornell University
New York, New York

S. N. Yaliraki
Institute for Mathematical Sciences, and Department of Chemistry
Imperial College London
London, England, U.K.

Lei Yang
LH Baker Center for Bioinformatics and Biological Statistics
Department of Biochemistry, Biophysics, and Molecular Biology
Iowa State University
Ames, Iowa

Leonid Yelash
Institut für Physik
Johannes Gutenberg-Universität Mainz
Mainz, Germany

Ying Ying
Department of Physics
University of Illinois at Urbana-Champaign
Urbana, Illinois

Guohui Zheng
Department of Chemistry & Chemical Biology
BioMaPS Institute for Quantitative Biology
Rutgers, The State University of New Jersey
Piscataway, New Jersey

Daniel M. Zuckerman
Department of Computational Biology
University of Pittsburgh School of Medicine
Pittsburgh, Pennsylvania

1 Introduction

Gregory A. Voth
Department of Chemistry, University of Utah

The computer simulation of condensed phases and biomolecular systems has resulted in profound new insight into the molecular-scale phenomena that occur in these complex systems. However, many processes that occur in liquids, soft materials, and biomolecular systems occur over length and time scales that are well beyond the current capabilities of atomic-level simulation. As such, new and novel approaches continue to be developed that can access longer time and length scale phenomena. One such approach is coarse-grained (CG) simulation, the topic of this book. In coarse-graining, groups of atoms are clustered into new CG "sites". These CG sites then interact through more computationally efficient effective interactions. The combination of these efficient interactions with the reduction in the total number of degrees of freedom of the system allows for a significant jump in the accessible spatial/temporal scales. As such, coarse-graining is the reduction of molecular-scale information (structural and interactions) into lower-resolution models that seek to retain the key physical features of the system of interest but are also simplified in their form (sometimes even greatly simplified). Such CG models are then most often used in a molecular simulation context, usually molecular dynamics (MD) or Monte Carlo (MC) simulation, to obtain the target properties of the system of interest.

The key motivation for CG molecular modeling and simulation thus primarily derives from the need to bridge the atomistic and mesoscopic scales. Typically speaking, there are two to three orders-of-magnitude in length and time separating these regimes. At the mesoscopic scale, one sees the emergence of critically important phenomena (e.g., self-assembly in biomolecular or soft matter systems). CG simulations, especially as they seek to make increasing contact with experimental results on complex systems, can therefore play a crucial role in the exploration of mesoscopic phenomena and, in turn, of the behavior of real biomolecular and materials systems. Coarse-graining promises to provide a revolutionary advance for the scientific community, especially in the field of computer simulation.

However, new challenges emerge when the CG approach is employed. These challenges are described in more detail below and in the chapters of this book. One of the challenges involves the establishment of a proper formal connection between the behavior of the CG representation of the system and the underlying all-atom (full atomic resolution) model. Additional challenges involve the degree of "believable" predictive power of CG models and their transferability between dissimilar systems.

The main current approaches to coarse-graining are represented in this book. These include highly "minimalist" CG models that are intended to reveal the essential physics of a given class of system. These models are usually very computationally efficient and qualitatively informative, but they do not necessarily provide quantitatively accurate predictions. Another approach is to develop CG models using experimental, thermodynamic, and/or average structural properties. This can be called the "inversion" approach to coarse-graining. Yet a third approach is to bridge atomistic information upward in scale to the CG level in a "multiscale" fashion. All of these approaches have their strengths and weaknesses, and they are certainly complementary to each other.

However, at some level coarse-graining must ultimately be understood within the context of statistical mechanics. This venerable and remarkable theoretical framework provides us with connections between the macroscopic world of thermodynamics and the atomistic world of molecular interactions. In that vein, most CG methods are best cast within the context of the following formula:

$$\exp(-F / k_B T) = (\text{const.}) \int d\mathbf{x} \exp[-V(\mathbf{x}) / k_B T] \tag{1.1}$$
$$\approx (\text{const.}') \int d\mathbf{x}_{CG} \exp[-V_{CG}(\mathbf{x}_{CG}) / k_B T]$$

where in the first line F is the (Helmholtz) free energy of the system, $V(\mathbf{x})$ is the system potential energy as a function of the coordinates $\mathbf{x}$ of all of the atoms of the system, T is the thermodynamic temperature, k_B is Boltzmann's constant, and "const." is a normalization constant (the prime in the second line being a different constant). Importantly, in the second line of Equation 1.1, the expression for the free energy is rewritten in terms of the CG variables $\mathbf{x}_{CG}$ and the CG effective potential $V_{CG}(\mathbf{x}_{CG})$. The CG variables, by virtue of the definition of coarse-graining, are fewer in number than the atomistic variables such that the number of these variables satisfies $N_{\mathbf{x}_{CG}} < N_{\mathbf{x}}$.

It should be noted that Equation 1.1 is rarely solved directly. However, its underlying structure forms the basis for various distribution functions, equilibrium averages, and properties, etc. Moreover, the equation clearly illustrates the principle of coarse-graining. It is well known that the evaluation of the integral in the first line of Equation 1.1 (and all integrals like it), using either MD or MC methods, is a great challenge for rugged multidimensional potential energy functions such as those for biomolecular systems. The promise of CG modeling is therefore to substantially reduce this computational challenge through a combination of fewer CG degrees of freedom and also the likely fact that the CG effective potential $V_{CG}(\mathbf{x}_{CG})$ will be "smoother" than the full all-atom resolution one, $V(\mathbf{x})$. However, in this concept lie two of the main challenges in coarse-graining. First, one may not know beforehand the optimal choices of the CG sites since one does not know the solution to Equation 1.1. Second, one does not know the CG effective potential, $V_{CG}(\mathbf{x}_{CG})$, so it must somehow be determined or modeled. On this latter point, Equation 1.1 also reveals just how difficult this latter task may be, because the equation indicates that the CG effective "potential" should actually be a *free energy surface* (i.e., the so-called many-body potential of mean force) for the CG variables. This is because, in a formal sense, certain degrees of freedom have been integrated out in going from the first to the second line of Equation 1.1. As such, the effective CG potential must contain these "missing entropy" effects arising from the degrees of freedom that have been integrated over when transforming the equation to the CG variables. These entropic effects can be ill defined and hard to predict in their behavior, wherein lies the origin of one of the key challenges in coarse-graining.

At one level or another, most current coarse-graining schemes attempt to solve the problem embodied in Equation 1.1. Some methods may seek to only approximately satisfy this equation for a particular system and thermodynamic state, so that at the same time the CG model is transferable over a wider range of systems and conditions. Other coarse-graining methods, such as those being developed in my own research group, seek to provide a precise and systematic route to Equation 1.1 so that the approximation sign in the second line of the equation is as close to an equality as possible. This approach may, however, come at the expense of complete transferability of the CG model between disparate systems and thermodynamics conditions, so that additional formal methodology will need to be developed to enhance the model transferability.

Following in the spirit of the above discussion, the individual chapters of this book describe most of the important current developments in the field of CG simulation and modeling, with a focus on approaches that provide CG representations of complex systems such as liquids, polymers, lipid bilayers, peptides, proteins, nucleic acids, and protein complexes. Each chapter focuses on specific examples of evolving coarse-graining methodology and presents results for a variety of these complex systems. Each author was asked to carefully describe their own CG approach, its motivation, strengths, and weaknesses, and to give one or two important example applications. These individual contributions contain an excellent cross-section of much of the important work being undertaken at the present time. For the reader the book represents the first time that most of the various current coarse-graining researchers have collated their work in such a fashion. Indeed, the field of coarse-graining is so new and so fluid at the present time that the format of the present book seems optimal, as it is difficult to imagine how a single-author book could capture the full diversity of this rapidly emerging field.

For scientists interested in CG modeling, and also for those researchers interested in implementing such methods, the various chapters therefore provide a good overview of the current state of the art from a variety of different perspectives. For example, Chapters 2 and 3 provide two of the most successful current coarse-graining schemes. These two approaches are in fact quite different and complementary to one another. The work of Marrink and co-workers in Chapter 2 is a good example of the "inverse" approach to coarse-graining, wherein thermodynamic and other properties are used to parameterize CG force fields. Our own contribution in Chapter 3 presents the multiscale coarse-graining (MS-CG) approach in which atomistic force information is utilized within a variational framework to systematically develop CG models from the "bottom up". In this sense, the MS-CG method adheres closely and strictly to the concept of coarse-graining embodied in Equation 1.1, while the work of Marrink and co-workers is a looser interpretation of that equation. However, it has the benefit of significant transferability between a variety of systems. Chapters 4 through 7 go on to present various other coarse-graining schemes, especially for lipid bilayers as a key example. Several of these schemes rely heavily on the so-called "reverse Monte Carlo" approach, and further develop it to help define the effective CG interactions based on an inversion of equilibrium structural (radial distribution function) data.

Chapters 8 through 11 discuss current CG model development for peptides and proteins at the amino acid level (i.e., amino acids in the primary sequence are coarse-grained into a single or a few CG sites). Here, these systems are very complex, so one can rightfully expect significant diversity in the coarse-graining approaches. There is presently no single "best way" to coarse-grain such systems, and there may never be. In addition, Chapters 12 and 13 describe special methods for "mixed-resolution" studies and for characterizing conformational transitions, respectively.

At larger length scales, one typically utilizes more "aggressive" (lower-resolution) coarse-graining schemes. Here individual amino acids or base pairs in nucleic acids may not even be completely resolved at the CG level. Chapters 14 and 15 describe coarse-graining of nucleic acids (DNA) along these lines, while Chapters 16 through 21 provide various aggressive coarse-graining schemes for proteins, including elastic network models and normal mode-based approaches.

The book concludes with Chapters 22 through 28, which present important coarse-graining (and multiscale) methods and applications in soft-matter materials science (polymers, surfactants, etc.) and in nanoscience. While there is a significant overlap in methodology with the earlier chapters, the materials science problems described in these chapters also present challenges and opportunities of their own for CG modeling. Chapter 28 concludes the book by describing an approach to the issue of "coarse-graining in time", in which full atomic resolution is retained but the coarse-graining occurs in the dynamics so as to significantly extend the effective time scale of the simulation. The issue of time scale and realistic dynamics in CG modeling is clearly an important topic for the future.

Despite its promising future, coarse-graining faces a number of significant challenges before it can become widely utilized by the research community, especially by experimental researchers as a tool to help interpret their experiments. In order for such a broad degree of acceptance to occur, coarse-graining must become a systematic, fully predictive technique in molecular simulation. For example, at present it often seems there is a risk that such models could have bias built into them because one sometimes "knows" (or has an idea) of the answer one wants when building a CG model to study a particular system or class of systems. It is therefore absolutely essential that a clear set of standards be developed (albeit with an appropriate degree of latitude) so that one can fully trust the predictions of a CG model or simulation. Along these lines, it is critical that CG simulation researchers "push" their models and methods into unknown territory and not be afraid to report their failures along with the successes. We must also make our procedures, both their strengths and weaknesses, clearly known to our audience, both in our written papers and in our oral presentations. Generating beautiful graphics and CG simulations of systems for which the end result is already largely known will not serve to advance the field and, in fact, could well undermine it. We can be rightfully optimistic, however, that this will not happen, but we should also be realistic that there are various impediments that must be surmounted.

In addition to providing a true predictive capability for CG modeling, there are also various immediate challenges faced by all CG methods. One essential challenge is the degree of transferability of CG models between various systems and from one set of thermodynamic conditions to another. In principle, a CG model cannot be completely transferable because it is a simplified (reduced degree of freedom) picture of a complex system and certain information has been effectively averaged away for those given conditions. On the other hand, many aspects of the CG model must certainly be transferable. A key goal then is both to define and to *understand* what is and what is not transferable in a given CG model and why. This is more than a technical issue. It is actually a very significant problem deeply rooted in the foundations of statistical mechanics, and a problem that has not yet been completely solved.

There is also the question of CG dynamics (i.e., time-dependent behavior), because CG models do not have the same dynamics as the real underlying atomistic MD. CG dynamics are often significantly faster than the real dynamics, and this is in fact a desirable feature of CG models if statistical sampling is their primary goal (i.e., the sampling is faster and probably more extensive). Some progress has been made on the CG dynamics problem. However, it also presents a paradox because if one were to develop a CG model with the correct (slower) dynamics, it would in turn undermine the efficient statistical sampling of the CG model. Thus, such a dynamically correct CG model would need to be extremely efficient computationally in order to simultaneously achieve both objectives. This is clearly a challenge for the future.

Another important question to consider is whether coarse-graining will stand the test of time. As of this writing, it has become an explosively growing methodology in the field of molecular simulation. In addition to the fertile intellectual environment feeding this growth, the primary current driving force for coarse-graining is a desire among researchers to access the length and time scales in biomolecular and soft-matter systems that cannot be reached by present-day all-atom MD or MC methods. However—and this is an important point—one can certainly expect these all-atom simulation methods (especially biomolecular MD) to increase their power significantly for the foreseeable future, including new MD algorithms, CPU speeds, and parallel execution on very large computing clusters. If this is the case, will the need for CG modeling and simulation then become obsolete? While the relevant CG methods and problems studied by CG modeling will surely evolve with time, all facts suggest that the answer is clearly "no". There are many orders of magnitude in length and time scale that must be bridged for molecular-inspired simulation to make contact with numerous real biological and materials phenomena. Moreover, and this is perhaps a key issue, it seems clear that coarse-graining will always remain a vital methodology for the *interpretation* of the behavior of complex systems, simply because the all-atom description is often "overkill" in that it contains too much detailed information and hence a reduced CG picture offers great advantages as an interpretive tool. This aspect of coarse-graining will always be an important and valuable asset to many scientific researchers.

To sum up this introduction, it is thus very clear that coarse-graining is an exciting conceptual and algorithmic challenge in the field of computer simulation and statistical mechanics. It is an approach that is providing a great step forward in the molecular modeling and simulation of real, complex systems. This research effort continues to be very rewarding for all of the contributors to this book, and they will all certainly make an ongoing contribution to finding the solutions to the critical challenges facing CG modeling. Only time can tell if we have succeeded, but there is every reason to be optimistic about the future growth and impact of this revolutionary advance in molecular simulation methodology.

2 The MARTINI Force Field

Siewert J. Marrink, Marc Fuhrmans,
H. Jelger Risselada, and Xavier Periole
Groningen Biomolecular Sciences and Biotechnology
Institute and Zernike Institute for Advanced Materials,
University of Groningen, The Netherlands

CONTENTS

2.1 INTRODUCTION

The use of coarse-grained (CG) models in a variety of simulation techniques has proven to be a valuable tool to probe the time and length scales of systems beyond what is feasible with traditional all-atom (AA) models. Applications to lipid systems in particular, pioneered by Smit et al.,[1] have become widely used. A large diversity of coarse-graining approaches is available; they range from qualitative, solvent-free models, via more realistic models with explicit water, to models including chemical specificity (for recent reviews see Refs. 2–4). Models within this latter category are typically parameterized based on comparison to atomistic simulations, using inverted Monte Carlo schemes[5–7] or force matching[8] approaches. Our own model,[9,10] coined the MARTINI force field, has also been developed in close connection with atomistic models; however, the philosophy of our coarse-graining approach is different. Instead of focusing on an accurate reproduction of structural details at a particular state point for a specific system, we aim for a broader range of applications without the need to reparameterize the model each time. We do so by extensive calibration of the chemical building blocks of the CG force field against thermodynamic data, in particular oil/water partitioning coefficients. This is similar in spirit to the recent development of the GROMOS force field.[11] Processes such as lipid self-assembly, peptide membrane binding, and protein–protein recognition depend critically on the degree to which the constituents partition between polar and nonpolar environments. The use of a consistent strategy for the development of compatible CG and atomic-level force fields is of additional importance for its intended use in multiscale applications.[12] The overall aim of our coarse-graining approach is to provide a simple model that is computationally fast and easy to use, yet flexible enough to be applicable to a large range of biomolecular systems.

Currently, the MARTINI force field provides parameters for a variety of biomolecules, including many different lipids, cholesterol, and all amino acids. A protocol for simulating peptides and proteins is also available. Extensive comparison of the performance of the MARTINI model with respect to a variety of experimental properties has revealed that the model performs generally quite well ("semi-quantitatively") for a broad range of systems and state points. Properties accurately reproduced include structural (e.g., liquid densities,[9] area/lipid for many different lipid types,[9] accessible lipid conformations,[13] or the tilt angle of membrane spanning helices[14]), elastic (e.g., bilayer bending modulus,[9] rupture tension[10]), dynamic (e.g., lipid lateral diffusion rates,[9] water transmembrane (TM) permeation rate,[9] time scales for lipid aggregation[9,15]), and thermodynamic (e.g., bilayer phase transition temperatures,[16,17] propensity for interfacial versus TM peptide orientation,[14] lipid desorption free energy[10]) data.

The remainder of this chapter is organized as follows. A detailed description of the CG methodology is presented in the next section, discussing both its abilities and its limitations. Subsequently, examples of three applications are given, namely the fusion of vesicles, the formation of membrane domains, and the aggregation of membrane proteins. A short look at the future prospects of the MARTINI force field concludes this chapter.

2.2 METHOD

2.2.1 BASIC PARAMETERIZATION

The mapping: The MARTINI model is based on a four-to-one mapping;[10] that is, on average four heavy atoms are represented by a single interaction center, with an exception for ring-like molecules. To map the geometric specificity of small ring-like fragments or molecules (e.g., benzene, cholesterol, and several of the amino acids), the general four-to-one mapping rule is insufficient. Ring-like molecules are therefore mapped with higher resolution (up to two-to-one). The model considers four main types of interaction sites: polar (P), nonpolar (N), apolar (C), and charged (Q). Within a main type, subtypes are distinguished either by a letter denoting the hydrogen-bonding capabilities (d = donor, a = acceptor, da = both, 0 = none) or by a number indicating the degree of polarity (from 1 = low polarity to 5 = high polarity). The mapping of representative biomolecules is shown in Figure 2.1.

Nonbonded interactions: All particle pairs i and j at distance r_{ij} interact via a Lennard–Jones (LJ) potential:

$$V_{\mathrm{LJ}} = 4\varepsilon_{ij}[(\sigma/r_{ij})^{12} - (\sigma/r_{ij})^{6}]. \tag{2.1}$$

The strength of the interaction, determined by the value of the well-depth ε_{ij}, depends on the interacting particle types. The value of ε ranges from $\varepsilon_{ij} = 5.6$ kJ/mol for interactions between strongly polar groups to $\varepsilon_{ij} = 2.0$ kJ/mol for interactions between polar and apolar groups mimicking the hydrophobic effect. The effective size of the particles is governed by the LJ parameter $\sigma = 0.47$ nm for all normal particle types. For the special class of particles used for ring-like molecules, slightly reduced parameters are defined to model ring–ring interactions; $\sigma = 0.43$ nm, and ε_{ij} is scaled to 75% of the standard value. The full interaction matrix can be found in the original publication.[10] In addition to the LJ interaction, charged groups (type Q) bearing a charge q interact via a Coulombic energy function with a relative dielectric constant $\varepsilon_{\mathrm{rel}} = 15$ for explicit screening:

$$V_{\mathrm{el}} = q_i q_j / 4\pi\varepsilon_0\varepsilon_{\mathrm{rel}} r_{ij}. \tag{2.2}$$

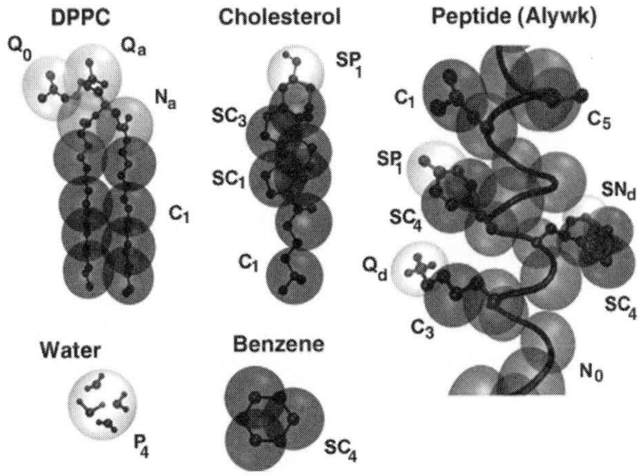

FIGURE 2.1 Mapping between the chemical structure and the coarse-grained model for DPPC, cholesterol, water, benzene, and a peptide fragment (with five amino acids highlighted). The coarse-grained bead types that determine their relative hydrophilicity are indicated, with more polar groups shown in lighter shades. The prefix "S" denotes a special class of CG sites used to model rings.

Note that the nonbonded potential energy functions are used in their shifted form. The nonbonded interactions are cut off at a distance $r_{cut}=1.2$ nm. The LJ potential is shifted from $r_{shift}=0.9$ nm to r_{cut}. The electrostatic potential is shifted from $r_{shift}=0.0$ nm to r_{cut}. Shifting of the electrostatic potential in this manner mimics the effect of a distance-dependent screening.

Bonded interactions: Bonded interactions are described by the following set of potential energy functions:

$$V_b = \frac{1}{2} K_b (d_{ij} - d_b)^2, \tag{2.3}$$

$$V_a = \frac{1}{2} K_a [\cos(\phi_{ijk}) - \cos(\phi_a)]^2, \tag{2.4}$$

$$V_d = K_d [1 + \cos(\theta_{ijkl} - \theta_d)], \tag{2.5}$$

$$V_{id} = K_{id} (\theta_{ijkl} - \theta_{id})^2, \tag{2.6}$$

acting between bonded sites i, j, k, l with equilibrium distance d_b, angle ϕ_a, and dihedral angles θ_d and θ_{id}. The force constants K are generally weak, inducing flexibility of the molecule at the CG level resulting from the collective motions at the fine-grained level. The bonded potential V_b is used for chemically bonded sites, and the angle potential V_a to represent chain stiffness. Proper dihedrals V_d are presently only used to impose secondary structure of the peptide backbone, and the improper dihedral angle potential V_{id} is used to prevent out-of-plane distortions of planar groups. LJ interactions between nearest neighbors are excluded.

Implementation: The functional form of the CG force field was originally developed for convenient use with the GROMACS simulation software.[15] Example input files for many systems can be downloaded from http://md.chem.rug.nl/~marrink/coarsegrain.html. The general form of the potential energy functions has allowed other groups to implement our CG model (with small modifications) also into other major simulation packages such as NAMD[20] and GROMOS.[13]

Effective time scale: For reasons of computational efficiency, the mass of the CG beads is set to 72 amu (corresponding to four water molecules) for all beads, except for beads in ring structures, for which the mass is set to 45 amu. Using this setup, the systems described in this paper can be simulated with an integration time step of 30–40 fs, which corresponds to an effective time of 120–160 fs. In the remainder of the paper, we will use an effective time rather than the actual simulation time unless specifically stated. The CG dynamics is faster than the AA dynamics because the CG interactions are much smoother compared to atomistic interactions. The effective friction caused by the fine-grained degrees of freedom is missing. Based on comparison of diffusion constants in the CG model and in atomistic models, the effective time sampled using CG interactions is 3–8-fold longer. When interpreting the simulation results with the CG model, a standard conversion factor of 4 is used, which is the effective speed-up factor in the diffusion dynamics of CG water compared to real water. The same order of acceleration of the overall dynamics is also observed for a number of other processes, including the permeation rate of water across a membrane,[9] the sampling of the local configurational space of a lipid,[13] and the aggregation rate of lipids into bilayers[9] or vesicles.[15] However, the speed-up factor might be quite different in other systems or for other processes. Particularly for protein systems, no extensive testing of the actual speed-up due to the CG dynamics has been performed, although protein translational and rotational diffusion was found to be in good agreement with experimental data in simulations of CG rhodopsin.[26] In general, however, the time scale of the simulations has to be interpreted with care.

2.2.2 Reproducing Thermodynamic Data: Optimizing Nonbonded Parameters

In order to parameterize the nonbonded interactions of the CG model, a systematic comparison to experimental thermodynamic data has been performed. Specifically, the free energy of hydration, the free energy of vaporization, and the partitioning free energies between water and a number of organic phases were calculated for each of the 18 different CG particle types. Concerning the free energies of hydration and vaporization, the CG model reproduces the correct trend.[10] The actual values are systematically too high, however, implying that the CG condensed phase is not as stable with respect to the vapor phase as it should be. The same is true with respect to the solid phase. This is a known consequence of using a LJ 12-6 interaction potential, which has a limited fluid range. Switching to a different nonbonded interaction potential could, in principle, improve the relative stability of the fluid phase. As long as its applications are aimed at studying the condensed phase and not at reproducing gas/fluid or solid/fluid coexistence regions, the most important thermodynamic property is the partitioning free energy. Importantly, the water/oil partitioning behavior of a wide variety of compounds can be accurately reproduced with the current parameterization of the CG model. Table 2.1 shows results obtained for the partitioning between water and a range of organic phases of increasing polarity (hexadecane, chloroform, and octanol) for a selection of the 18 CG particle types. The free energy of partitioning between organic and aqueous phases, $\Delta G^{\mathrm{oil/aq}}$, was obtained from the equilibrium densities ρ of CG particles in both phases:

$$\Delta G^{\mathrm{oil/aq}} = kT \ln(\rho_{\mathrm{oil}}/\rho_{\mathrm{aq}}). \tag{2.7}$$

The equilibrium densities can be obtained directly from a long MD simulation of the two-phase system in which small amounts (around 0.01 mol fraction proved sufficient to be in the limit of infinite dilution) of the target substance are dissolved. With the CG model, simulations can easily be extended into the multimicrosecond range, enough to obtain statistically reliable results to within 1 kJ/mol for most particle types. As can be judged from Table 2.1, comparison to experimental data for small molecules containing four heavy atoms (the basic mapping of the CG model) reveals a close agreement to within 2 kT for almost all compounds and phases; indeed, agreement is within 1 kT for many of them. Expecting more accuracy of a CG model might be unrealistic. Note that the multiple nonbonded interaction levels allow for discrimination between chemically similar building blocks, such as saturated versus unsaturated alkanes or propanol versus butanol (which would be modeled as N_{da}) or ethanol (P_2). A more extensive table including all particle types and many more building blocks can be found in the original publication.[10]

TABLE 2.1
Oil, Chloroform, and Octanol/Water Partitioning Free Energies for a Selection of the 18 CG Particle Types, Compared to Experimental Values of the Corresponding Chemical Building Blocks

		Hexadecane/Water		Chloroform/Water		Octanol/Water	
Building Block	**Type**	**CG**	**Exp**	**CG**	**Exp**	**CG**	**Exp**
Acetamide	P_5	−28	−27	−18	−20	−10	−8
Water	P_4	−23	−25	−14	–	−9	−8
Propanol	P_1	−11	−10	−2	−2	−1	0
Propylamine	N_d	−7	−6	0	1	3	3
Methylformate	N_a	−7	−6	0	4	3	0
Methoxyethane	N_0	−2	1	6	–	5	3
Butadiene	C_4	9	11	13	–	9	11
Chloropropane	C_3	13	12	13	–	14	12
Butane	C_1	18	18	18	–	17	16

The experimental data are compiled from various sources (see Ref. 10); the simulation data are obtained using Equation 2.7. All values are expressed in kilojoules per mole and obtained at $T = 300$ K.

To select particle types for the amino acids, systematic comparison to experimental partitioning free energies is also used. Table 2.2 shows the resulting assignment of the amino acid side chains and the associated partitioning free energies. The simulation data are calculated from equilibrium densities of low concentrations of CG beads dissolved in a water/butane two-phase system, using Equation 2.7. The experimental data refer to partitioning of side-chain analogues between water and cyclohexane.[18] Both the simulation and the experimental data are obtained at 300 K. Where available, the experimental values are reproduced to within 2 kT, a level of accuracy that is difficult to obtain even with atomistic models. Most amino acids are mapped onto single standard particle types, similarly to the recent work of other groups.[19,20] Figure 2.1 shows the mapping of a few of them. The apolar amino acids (Leu, Pro, Ile, Val, Cys, and Met) are represented as C-type particles, the polar uncharged amino acids (Thr, Ser, Asn, Gln) by the class of P-type particles, and the small negatively charged side chains (Glu, Asp) as Q type. The positively charged amino acids (Arg, Lys) are modeled by a combination of a Q-type and an N- or C-type particle. The bulkier ring-based side chains are modeled by three (His, Phe, Tyr) or four (Trp) beads of the special class of ring particles. The Gly and Ala residues are only represented by the backbone particle. The type of the backbone particle depends on its secondary structure; when free in solution or in a coil or bend, the backbone has a strong polar character (P type), while as part of a helix or beta strand the interbackbone hydrogen bonds reduce the polar character significantly (N type). Details of the parameterization of the amino acids can be found elsewhere.[14]

2.2.3 Reproducing Structural Data: Optimizing Bonded Parameters

To parameterize the bonded interactions, we use structural data that are either directly derived from the underlying atomistic structure (such as bond lengths of rigid structures) or obtained from comparison to fine-grained simulations. In the latter procedure, the fine-grained simulations are first converted into a "mapped" CG (MCG) simulation by identifying the center of mass of the corresponding atoms as the MCG bead. Second, the distribution functions are calculated for the mapped simulation and compared to those obtained from a true CG simulation. Subsequently the CG parameters are systematically changed until satisfactory overlap of the distribution functions is

TABLE 2.2
Free Energy Based Mapping of the Amino Acids

Side Chain	Type	Oil/Water CG	Oil/Water Exp
Leu	C1	22	22
Ile	C1	22	22
Val	C2	20	17
Pro	C2	20	–
Met	C5	9	10
Cys	C5	9	5
Ser	P1	−11	−14
Thr	P1	−11	−11
Asn	P5	<−25	−28
Gln	P4	−23	−25
Asp	Qa	<−25	–
Asp (uncharged)	P3	−18	−19
Glu	Qa	<−25	–
Glu (uncharged)	P1	−11	−11
Arg	N0-Qd	<−25	–
Arg (uncharged)	N0-P4	−22	−25
Lys	C3-Qd	<−25	–
Lys (uncharged)	C3-P1	0	−2
His	SC4-SP1-SP1	−19	−20
Phe	SC4-SC4-SC4	19	17
Tyr	SC4-SC4-SP1	−1	−2
Trp	SC4-SP1-SC4-SC4	12	9

The experimental partitioning free energies are obtained for cyclohexane/water,[18] the simulation results for butane/water are obtained using Equation 2.7. All values are expressed in kilojoules per mole and obtained at $T = 300$ K.

obtained. Using this procedure, simulations of bulk alkanes have been used to determine the optimal values of the "standard" equilibrium bond distance of 0.47 nm and force constant of $K_b = 1250$ kJ/mol/nm^2, and equilibrium angle of 180° with force constant of $K_a = 25$ kJ/mol. Likewise, standard bonded parameters have been derived for unsaturated alkanes,[10] the phospholipid headgroup,[9] and for cholesterol.[10]

For the bonded interactions involving amino acid side chains and for the peptide backbone, we used a similar procedure. However, instead of deriving target distributions from fine-grained simulations we derived distributions directly from the protein databank, using the same mapping protocol. The amount of data is so large that statistically very accurate distributions can be obtained for all the required bonded interactions. These distributions reflect all possible configurations for a large number of different systems under a variety of conditions. Keeping the aim of our CG model in mind, namely to be able to simulate many biomolecules with a single set of parameters, this is the least biased information. Using this procedure, bonded parameters were derived for the backbone (BB) potentials, namely the BB–BB bonded potential, the BB–BB–BB angle potential, and the BB–BB–BB–BB dihedral potential. The last two terms are used to enforce the secondary structure of the backbone, which is therefore an input parameter in our CG model. In the current version, different dihedral and angle parameters are used to distinguish a helix, a strand, or a random coil. It is therefore not possible to study realistic folding–unfolding events at this stage. Furthermore, for each amino acid, side chain (SC) distributions were obtained for the BB–SC bonded potential, the BB–BB–SC angle potential, and for the intra-SC potentials for amino acids containing more than one CG particle. The complete set of bonded parameters for proteins will be published elsewhere.[14]

2.2.4 Coarse-Graining Recipe

Here we present a simple three-step recipe, or guide, to how to proceed in parameterizing new molecules using the CG model. The first step consists of mapping the chemical structure to the CG representation, the second step is the selection of appropriate bonded terms, and the third step is the optimization of the model by comparing to AA level simulations and/or experimental data.

Step I, mapping onto CG representation: The first step consists of dividing the molecule into small chemical building blocks, ideally of four heavy atoms each. The mapping of CG particle types to chemical building blocks, examples of which are presented in Table 2.1, subsequently serves as a guide towards the assignment of CG particle types. Because most molecules cannot be entirely mapped onto groups of four heavy atoms, some groups will represent a smaller or larger number of atoms. In fact, there is no reason to map onto an integer number of atoms, for example, a pentadecane mapped onto four C_1 particles implies that each CG bead represents 3¾ methyl(ene) groups. In case of more substantial deviations from the standard mapping scheme, small adjustments can be made to the standard assignment. For instance, a group of three methyl(ene)s is more accurately modeled by a C_2 particle (propane) than the standard C_1 particle for saturated alkanes. The same effect is illustrated by the alcohols: whereas the standard alcohol group is modeled by a P_1 particle (propanol), a group representing one less carbon is more polar (P_2, ethanol), whereas adding a carbon has the opposite effect (N_{da}, butanol). Similar strategies can be used for modulation of other building blocks. To model compounds containing rings, a more fine-grained mapping procedure can be used. In those cases, the special class of S-particles is appropriate.

Step II, selecting bonded interactions: For most molecules the use of a standard bond length (0.47 nm) and force constant of $K_b = 1250$ kJ/mol/nm^2 appears to work well. In cases where the underlying chemical structure is better represented by using different values, there is no restriction in adjusting these values. Especially for ring structures, much smaller bond lengths are required. For rigid rings, the harmonic bond and angle potentials are replaced by constraints, as was done for benzene and cholesterol. For linear chain-like molecules, a standard force constant of $K_a =$ 25 kJ/mol with an equilibrium bond angle $\phi_a = 180°$ best mimics distributions obtained from fine-grained simulations. The angle may be set to smaller values to model unsaturated cis-bonds (for a single cis-unsaturated bond $K_a = 45$ kJ/mol and $\phi_a = 120°$) or to mimic the underlying equilibrium structure more closely in general. In order to keep ring structures planar, improper dihedral angles should be added. For more complex molecules (e.g., cholesterol), multiple ways exist for defining the bonded interactions. Not all of the possible ways are likely to be stable with the preferred time step of 30–40 fs (actual simulation time). Some trial-and-error testing is required to select the optimal set.

Step III, optimization: The coarse-graining procedure does not have to lead to a unique assignment of particle types and bonded interactions. A powerful way to improve the model is by comparison to AA level simulations, analogous to the use of quantum calculations to improve atomistic models. Structural comparison is especially useful for optimization of the bonded interactions. For instance, the angle distribution function for a CG triplet can be directly compared to the distribution function obtained from the AA simulation, using the mapping procedure described earlier. The optimal value for the equilibrium angle and force constant can thus be extracted. Comparison of thermodynamic behavior is a crucial test for the assignment of particle types. Both AA level simulations (e.g., preferred position of a probe inside a membrane) and experimental data (e.g., the partitioning free energy of the molecule between different phases) are useful for a good assessment of the quality of the model. The balance of forces determining the partitioning behavior can be very subtle. A slightly alternative assignment of particle types may significantly improve the model. Once more, it is important to stress that Table 2.1 serves as a guide only; ultimately the comparison to AA simulations and experimental data should be the deciding factor in choosing parameters.

2.2.5 Limitations

The potential range of applications of the CG model is very broad. There are, however, certain important limitations that should be kept in mind. First of all, the model has been parameterized for the fluid phase. Properties of solids, such as crystal packing, are not expected to be accurate. Both the gas and the solid phase appear somewhat too stable with respect to the fluid phase. The thermodynamic behavior of solid/fluid and gas/fluid interfaces should therefore be interpreted with care, at least at the quantitative level. In applications where such interfaces are formed (especially the water/vapor interface in, for example, rupture of lipid monolayers) these limitations have to be kept in mind.

Furthermore, the parameterization is based on free energies. The inherent entropy loss on coarse-graining is necessarily compensated for by a reduced enthalpy term.[13] The enthalpy/entropy balance of many processes is therefore biased when modeled at the CG level. Consequently, the temperature dependence is affected, although not necessarily weaker. For instance, the temperature-dependent hydration free energy for linear alkanes was found to be more pronounced in the CG representation compared to an AA representation.[13] As is true for any force field, applications outside the temperature range used for parameterization ($\sim$270–330 K) have to be considered with care.

Another difficulty encountered in our CG model, and perhaps in most coarse-graining approaches, is the correct modeling of the partitioning of polar and charged compounds into a low dielectric medium. Because of the implicit screening, the interaction strength of polar substances is underestimated in nonpolarizable solvents. Applications involving the formation of polar/charged complexes in a nonpolar environment are especially prone to be affected. The inability to form a TM water pore upon dragging a lipid across the membrane is an example.[10] Apart from the implicit screening in the CG model, the neglect of long-range electrostatic forces poses a further limitation. Pairwise interactions beyond 1.2 nm (between two and three CG beads away) are not taken into account. In principle long-range electrostatic interactions could be added to the CG model, in ways similar to those used in atomistic simulations. One has to realize that a modification of the electrostatic interaction scheme will affect other system properties.

Finally, in applications of peptides and proteins one has to be aware that secondary structure transformations are not modeled in the current parameterization. The secondary structure is essentially fixed by the use of a dihedral potential energy function, allowing discrimination between various secondary structure elements but preventing realistic transitions between them. Processes in which folding and unfolding are playing a substantial role are therefore not suitable for modeling with our current CG force field. Movements of secondary structure elements with respect to each other are possible, however, and were shown to be quite realistic in a recent application of the gating of a membrane-embedded mechanosensitive channel.[21]

2.3 APPLICATIONS

The list of applications of our CG model to date is broad, reflecting the flexibility underlying our coarse-graining protocol. Some important recent applications include, among others, vesicle fusion,[22,23] lamellar phase transformations,[16,17,24,25] the structure and dynamics of membrane–protein assemblies,[19,20,21,26] the phase behavior of bicelles[27] and monolayers,[28] nanotube–lipid interactions,[29] and confinement of copolymers.[30] In most of these studies the CG model, sometimes with small changes or extensions from the published version, performed well when compared to either experimental data or to more detailed atomistic models. Here we selected three applications of ongoing studies performed in our own lab, namely vesicle fusion, domain formation, and membrane-embedded protein self-assembly. All simulations were performed with the GROMACS simulation software,[15] versions 3.0–3.3.

2.3.1 Vesicle Fusion

Membrane fusion lies at the heart of important biological processes such as endo- and exocytosis, intracellular trafficking, and viral infection. In vivo, membrane fusion is tightly regulated by proteins. The basic mechanism, however, is primarily determined by the physics of lipid-lipid interactions.[31] Previously various CG models have been used to simulate the fusion between small unilamellar vesicles.[2,4] Simulations using our CG model[22,23] were conducted on vesicles of 15 nm diameter, containing common phospholipids such as phosphatidylcholine (PC), phosphatidylethanolamine (PE), lysoPC, and mixtures of the above lipids. The fusion intermediates found are in general agreement with the stalk-pore mechanism.[32] Transient pores sometimes form adjacent to the stalk, however, resulting in the mixing of lipids from the outer and inner monolayers. The rate of stalk formation and the opening of the fusion pore can be modulated by altering the lipid composition in qualitative agreement with experimental observations.

Recently, we performed simulations using somewhat bigger vesicle sizes of approximately 20 nm diameter (Fuhrmans, unpublished results). The vesicles consisted of a total of 2528 lipids with different mixtures of DPPE and DPPC (3:1, 1:1, 1:3). The simulations were performed at 323 K. In order to cut down computational costs, the vesicles were fused with their own periodic images, using semi-isotropic pressure coupling to overcome the repulsive forces acting between the opposing bilayer leaflets. Already small pressure differences of ~5 bar proved sufficient to induce adhesion and subsequent stalk formation. It was observed that the minimum pressure to form a stalk within a given time interval decreased with the amount of DPPE in the lipid mixture. This can be understood from the relatively small effective size of the PE headgroup resulting in a negative spontaneous curvature, favoring the formation of inverted phases. The stalk is also characterized by a dominant negative curvature, and therefore stabilized by the presence of PE lipids. To reach the next step of the fusion process, the hemifused state, the stalk must expand. As is illustrated in Figure 2.2, this can happen either in a two-dimensional fashion, by a circular expansion, or in a one-dimensional linear expansion. The first constitutes the more direct route, but would create an empty space in the center of the stalk that has to be filled by the tails of the lipids of the inner leaflet, which are still separated at this point. The linear expansion avoids the energetic cost associated with the required stretching of the lipid tails. Here, the expansion proceeds in a bent ("banana-shaped") manner until both ends of the linear expanded stalk meet, at which point the two ends fuse and form a closed circular stalk. This gives rise to an intermediate structure in which an inverted micelle is trapped in between the two vesicles. The formation of a pore (through rupture of one of the membranes separating the vesicle interiors from the water bubble) then leads to the hemifused state. This type of pore formation is observed to happen either before or after the closure of the circular stalk. Previous studies involving smaller vesicles indicated that both routes may be followed for a single composition.[22] However, the new set of simulations, using vesicles of increased size, behave differently. A clear bias towards the bent expansion route for each of the mixtures studied is observed, pointing to an interesting curvature effect.

It should be noted that the simulations described here, as well as those obtained earlier, were obtained with the original version of the CG model.[9] This model overestimates the negative curvature of both PC and PE lipids, as shown by a variety of studies,[10,27] and therefore the stalk-forming propensity is too large. The recent reparameterization of the model reduced the negative curvature of the lipids. Consequently, in the current MARTINI model,[10] stalk formation between small juxtaposed PC/PE vesicles no longer happens spontaneously on a microsecond time scale. In fact,

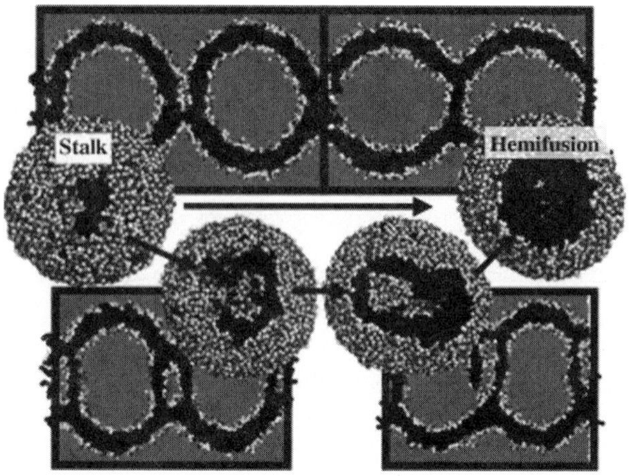

FIGURE 2.2 Two possible alternative transition pathways from the stalk to the hemifused state during the fusion process of small unilamellar vesicles (DPPC:DPPE 3:1) at $T = 323$ K. The upper arrow shows the direct route in which the stalk expands radially; the lower pathway involves a number of steps. First the stalk expands in a curved way, second, upon closing an inverted micelle becomes trapped in between the fusing vesicles, and finally, by opening a pore in one of the vesicles, the pathway to the hemifused state is completed. Lipid headgroups are shown in light shades, lipid tails in black, and the solvent in gray.

this might be the more realistic situation. In vivo, proteins play an important role in regulating the fusion process. Apart from inducing proximity and curvature, another role of the proteins (e.g., by means of fusion peptides) might be to induce local destabilization of the lipid matrix to facilitate stalk formation.

2.3.2 Domain Formation

Biological membranes contain a large number of different lipid species. Rather than forming a homogeneous phase, such a multicomponent mixture exhibits a complex phase behavior with regions of structural and compositional heterogeneities. Especially the domains formed by ternary mixtures including cholesterol have received a lot of attention as these so-called rafts are presumably linked to biological activity. In order to understand the complexity of multicomponent lipid mixtures, much effort is also directed to study the phase behavior of simpler, two-component systems including mixtures of saturated/unsaturated lipids, short/long-tail lipids, and lipids with different headgroups. In such systems, macroscopic domains (on a micrometer length scale) are visualized with confocal fluorescence microscopy on giant unilamellar vesicles, whereas much smaller domains (down to tens of nanometers) are seen by AFM measurements. The geometry of the domains can change from striped patterns to quasicircular and dendritic, or may resemble percolated networks. Whether or not these structures are true equilibrium structures is not clear. The kinetics of the phase separation process can become extremely slow, with correlation times reported in the order of hours or more. The presence of curvature (in vesicular studies), the sample type (supported versus unsupported bilayers), and detection technique may further contribute to the large variety of domain sizes and geometries observed experimentally.

The use of simplified models makes it computationally feasible to study phase separation processes in binary lipid membranes.[33,34] Recent simulations of lipid mixtures consisting of short- and long-tail lipids, using our CG model, indicated the existence of a small two-phase region consisting of a fluid domain enriched in the short-tail lipid surrounded by a gel-phase matrix of the long-tail lipid.[24] The limited system sizes (mostly 256–512 lipids) combined with the small difference in phase-transition temperature of the pure compounds made interpretation of the results not straightforward.

In this chapter we show simulations of the lateral phase separation in binary mixtures of a saturated lipid, diC16-PC (dipalmitoyl-phosphatidylcholine, DPPC), and a polyunsaturated one, diC16:2-PC (dihexadecadienyl-PC, DHPC). By extending the length scale (up to 4000 lipids) and choosing a mixture of two components exhibiting a very different main phase-transition temperature, we show that clear phase separation can be observed in simulations. At temperatures below the main phase-transition temperature of DPPC (estimated to be 295 K for this CG model,[17] somewhat lower than the experimental temperature 315 K), the mixture separates into a gel and a fluid domain. A snapshot of the equilibrium structure of the two-phase region of a DHPC/DPPC 1:3 mixture at 273 K is presented in Figure 2.3. Note the difference in bilayer thickness of the gel phase with respect to the fluid phase. Due to the large gap between the respective phase-transition temperatures of both lipids (the pure unsaturated component remains fluid even at temperatures around 250 K, in line with experimental data for double unsaturated lipids), the gel phase is formed almost exclusively by the DPPC lipids. Naturally, the remaining fluid phase is enriched by the unsaturated component. Interestingly, the domain borders are enriched in the unsaturated lipid. Similar stabilization of domain edges by the non-gel-forming lipid has previously been predicted based on mean field lattice models.[33] Apparently the line tension arising from the packing differences between the gel and fluid phases is lowered by the presence of the more flexible unsaturated lipids. The temperature of phase separation in the simulations depends on the fraction of DHPC. Larger fractions require further cooling. For mixtures containing 50% or more of the unsaturated lipid, phase separation at the accessible temperatures (i.e., not limited by freezing of water) is no longer observed. Mixtures with only a very small amount of the unsaturated lipid form a homogeneous gel phase.

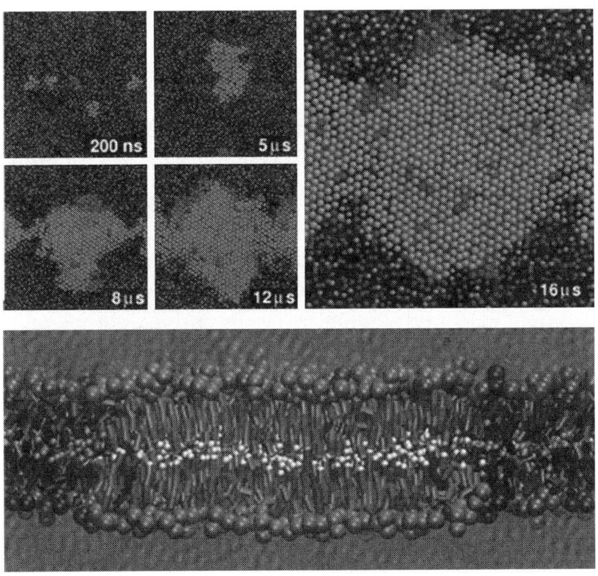

FIGURE 2.3 (See color insert following page 238.) Coarse-grained simulation of domain formation in an unsaturated/saturated (diC16:2/diC16 PC 1:3) phospholipid mixture. The saturated lipid is colored green, the unsaturated lipid purple. Upper snapshots: top view of the membrane showing only the second tail beads of the lipids. Large/small beads are used for tails in the gel/fluid phase. A darker shade is used for the underlying monolayer. At $t=0$ the system was quenched from the fluid phase at $T=295$ K into the gel–liquid coexistence region at $T=273$ K. During the first microseconds the system remains in a metastable fluid phase. Small gel-clusters composed of mainly the saturated component are seen to form and disappear. Only when a critical threshold size is exceeded, a gel-cluster remains stable (snapshot at 5 μs) and starts growing, until the system boundaries are reached ($t=8$ μs). The gel cluster slowly optimizes its interface with the remaining fluid phase, which is enriched in the unsaturated component. Lower figure: final snapshot of the coexisting gel and fluid domain. Light/dark shades distinguish gel/fluid phase lipids. The terminal tail beads are depicted as white/gray.

A major advantage of CG MD simulations is that the kinetics of the phase separation process can be followed at near-atomic resolution. The upper part of Figure 2.3 shows graphical images of the phase-separation process upon quenching a DHPC/DPPC 1:3 mixture. Initially the temperature of the system is $T=295$ K, above the main phase transition temperature of both components, and the system remains in a fluid phase. At $t=0$ the system is quenched to a temperature of 273 K, just below the fluidus line. Within hundreds of nanoseconds, small gel patches are formed consisting mainly of the saturated lipid. These patches, containing between 5 and 30 lipids, are only meta-stable. They melt again on a time scale of a few nanoseconds for the smallest domains to several hundred of nanoseconds for the largest domains. In this particular simulation, at $t=5$ μs a gel domain of ~30 DPPC lipids appeared suddenly, which no longer disappears. Instead, the domain starts growing. Initially, the domain forms in one monolayer only. After 500 ns the gel domain extends to the opposite monolayer and expands to match the size of the domain first formed. The monolayer-spanning domain keeps growing until the boundaries of the simulation cell are reached. At this stage ($t=8$ μs) the domain fuses with its periodic image. The growth of the domain continues for a while in the perpendicular direction, but after 10 μs it appears to have reached its equilibrium size. The final stage is characterized by a slow optimization of the domain wall into a more regular structure. This process is not completely finished by the end of the simulation ($t=16$ μs). The nucleation and growth mechanism observed here is similar to that observed recently for gel formation in pure DPPC membranes,[17] including the coupling between the two monolayer leaflets. In the one-component system the gel cluster grows until the entire system is converted, whereas in the two-component system gel–fluid phase coexistence is the final state. Details of this work will appear elsewhere.[35]

2.3.3 Protein Aggregation

The classical fluid mosaic model of cell membranes describes a two-dimensional liquid-like solution of membrane proteins. In many cases, integral membrane proteins, including channels and receptors for TM signaling, assemble into oligomeric structures during biogenesis or in response to ligand binding. In addition, self-assembly or self-organization of membrane proteins into dimers and higher-order structures seem to be involved in sorting and compartmentalization of membrane components. Understanding the interplay between specific lipid–protein interactions and continuum elastic membrane properties is one of the major challenges in biophysical chemistry.

The CG model is ideally suited to address the question of how the physicochemical properties of the membrane lipids might affect self-assembly of integral membrane proteins. In particular the role of lipid diversity in the constitutive self-assembly process of the visual receptor rhodopsin, a 7-TM helical G protein-coupled receptor, is poorly understood.[36,37] We carried out multicopy rhodopsin simulations (16 proteins per unit cell) at a protein-to-lipid ratio of 1:100 in different bilayer environments at $T=300$ K.[26] From an initial condition out-of-equilibrium, where the proteins were fully dispersed and ordered in the bilayer cell, we followed the relaxation of the system and analyzed the simulations with the aim of characterizing the mechanism by which rhodopsins progressively self-assemble into aggregates and ordered linear arrays. We found that localized adaptation of the membrane bilayer to the protein and surface complementarity are key factors defining the rate, extent, and orientational preference of protein–protein association.

Figure 2.4 shows snapshots of the aggregation process of rhodopsin proteins embedded in a diC16:1-PC bilayer. Initially, rhodopsin monomers predominate. However, the formation of protein–protein contacts rapidly increases, with most monomers recruited within 1–2 μs. The system rapidly rearranges into growing clusters with a strong bias towards higher order. The final snapshots show extended string-like clusters of eight monomers. Furthermore, from results obtained with other bilayers, a clear dependency on the lipid chain length can be appreciated. With increasing bilayer thickness in the series diC12:0-PC, diC16:1-PC, and diC20:1-PC, the kinetics of surface burial slows down progressively. However, in the thickest bilayer studied, diC20:0-PC, rhodopsin association again accelerates. These observations are in line with recent fluorescence resonance

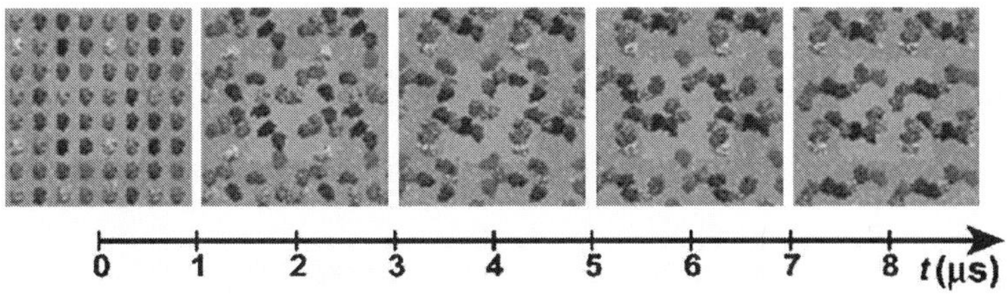

FIGURE 2.4 Rhodopsin oligomerization in an explicit diC16:1-PC lipid bilayer. Snapshots of instantaneous configurations of the proteins in the membrane plane after 0, 2, 4, 6, and 8 μs of CG simulation. The central box in light gray indicates the simulation cell surrounded by periodic images in darker gray. The 16 proteins are shown in different shades of gray. Starting from a homogeneous initial placement (4 × 4 array), the rhodopsins are seen to aggregate in response to a subtle interplay of protein–lipid hydrophobic mismatch and specific protein–protein interactions.

energy transfer (FRET) experiments.[37] The results for the homologous series of unsaturated lipids diCn:1-PC (at 20°C and ~ 1:100 protein-to-lipid ratio) gave clear evidence for nonrandom (nonideal) dispersion of rhodopsin in membranes for bilayer thickness deviating from an optimal acyl chain length *n* 20–22, in good agreement with the CG simulations.

2.4 OUTLOOK

Especially during the past decade, CG simulations have proven to be a powerful method to fill the gap between more fine-grained simulations on the one hand, and mean-field methods on the other. The power of CG simulation methods lies in the fact that mesoscale processes can be probed without losing sight of the chemical details from which the collective interactions arise. The applications described in this chapter are illustrative of what CG simulations are capable of achieving nowadays. Complicated collective processes such as fusion, domain formation, and protein self-assembly can be probed on the microsecond time scale, something unachievable with more detailed AA models. At the same time the role of individual lipids or amino acids can be revealed, specificity that is lacking with coarser methods. With the current development of multiscale simulation techniques,[12] even more progress is to be expected over the next years. Hybrid simulations in which a small, central part is simulated at full detail, surrounded by a part treated at the CG level, allow the advantages of fine- and coarse-grained models to be combined. Alternatively, CG simulations might serve to sample phase space of a particular system more globally, after which fine-grained models are used to sample the areas of interest in more detail. Direct coupling of the approaches is also possible using exchange simulation protocols. Various groups, including our own, are currently working to optimize the use of such multiscale techniques. Furthermore, improvement and refinement of the CG parameterization is possible. For our MARTINI force field we want to include an effective local dielectric constant instead of the unrealistic use of a uniform screening factor throughout the entire system. The use of softer nonbonded interactions might enhance the stability of the fluid phase. Extension of the model to include all kind of biomolecules (e.g., sugars, nucleotides) is anticipated. Perhaps the potential energy landscape for proteins can be more closely mimicked to even allow realistic simulations of secondary structure transitions. Taking for granted the ongoing increase in computational power, we might conclude that a very fascinating era of biomolecular simulations lies ahead of us.

ACKNOWLEDGMENTS

SJM acknowledges the input of numerous people into the ongoing development of the MARTINI force field, and in particular that of Senthil Kandasamy and Luca Monticelli, for the parameterization of the CG peptide force field. The work on rhodopsin aggregation was performed in close collaboration with Thomas Huber and Thomas Sakmar. Alex de Vries is acknowledged for proofreading of the manuscript.

REFERENCES

1. Smit, B., Hilbers, P. A. J., Esselink, K., Rupert, L. A. M., van Os, N. M., Schlijper, A.G. 1990. Computer-simulations of a water oil interface in the presence of micelles. *Nature* 348:624–25.
2. Shillcock, J. C., Lipowsky, R. 2006. The computational route from bilayer membranes to vesicle fusion. *J. Phys.: Condens. Matter* 18:S1191–219.
3. Venturoli, M., Sperotto, M. M., Kranenburg, M., Smit, B. 2006. Mesoscopic models of biological membranes. *Phys. Lett.* 437:1–54.
4. Müller, M., Katsov, K., Schick, M. 2006. Biological and synthetic membranes: What can be learned from a coarse-grained description? *Phys. Lett.* 434:113–76.
5. Shelley, J. C., Shelley, M., Reeder, R., Bandyopadhyay, S., Klein, M. L. 2001. Simulations of phospholipids using a coarse-grained model. *J. Phys. Chem. B* 105:4464–70.
6. Lyubartsev, A. P. 2005. Multiscale modeling of lipids and lipid bilayers. *Eur. Biophys. J. Biophys. Lett.* 35:53–61.
7. Elezgaray, J., Laguerre, M. 2006. A systematic method to derive force fields for coarse-grained simulations of phospholipids. *Comput. Phys. Commun.* 175:264–68.
8. Izvekov, S., Voth, G. A. 2005. Systematic coarse-graining of nanoparticle interactions in molecular dynamics simulation. *J. Phys. Chem. B* 109:2469–73.
9. Marrink, S. J., de Vries, A. H., Mark, A. E. 2004. Coarse-grained model for semiquantitative lipid simulations. *J. Phys. Chem. B* 108:750–60.
10. Marrink, S. J., Risselada, H. J., Yefimov, S., Tieleman, D. P., de Vries, A. H. 2007. The MARTINI forcefield: Coarse grained model for biomolecular simulations. *J. Phys. Chem. B* 111:7812–24.
11. Oostenbrink, C., Villa, A., Mark, A. E., van Gunsteren, W. F. 2004. A biomolecular force field based on the free enthalpy of hydration and solvation: The GROMOS force-field parameter sets 53A5 and 53A6. *J. Comput. Chem.* 25:1656–76.
12. Chu, J. W., Ayton, G. S., Izvekov, S., Voth, G. A. 2007. Emerging methods for multiscale simulation of biomolecular systems. *Mol. Phys.* 105:167–75.
13. Baron, R., Trzesniak, D., de Vries, A. H., Elsener, A., Marrink, S. J., van Gunsteren, W. F. 2007. Comparison of thermodynamic properties of coarse-grained and atomic-level simulation models. *Chem. Phys. Chem.* 8:452–61.
14. Monticelli, L., Kandasamy, S., Periole, X., Larson, R. G., Tieleman, D. P., Marrink, S. J. 2008. The MARTINI coarse-grained force field: Extension to proteins. *J. Chem. Theory Comp.* In press.
15. van der Spoel, D., Lindahl, E., Hess, B., Groenhof, G., Mark, A. E., Berendsen, H. J. C. 2005. GROMACS: Fast, flexible, and free. *J. Comput. Chem.* 26:1701–18.
16. Marrink, S. J., Mark, A. E. 2004. Molecular view of hexagonal phase formation. *Biophys. J.* 87:3894–900.
17. Marrink, S. J., Risselada, J., Mark, A. E. 2005. Simulation of gel phase formation and melting in lipid bilayers using a coarse grained model. *Chem. Phys. Lipids* 135:223–44.
18. Radzicka, A., Wolfenden, R. 1988. Comparing the polarities of the amino-acids-side-chain distribution coefficients between the vapor-phase, cyclohexane, 1-octanol, and neutral aqueous-solution. *Biochemistry* 27:1664–70.
19. Bond, P. J., Sansom, M. S. P. 2006. Insertion and assembly of membrane proteins via simulation. *J. Am. Chem. Soc.* 128:2697–704.
20. Shih, A. Y., Arkhipov, A., Freddolino, P. L., Schulten, K. 2006. Coarse grained protein–lipid model with application to lipoprotein particles. *J. Phys. Chem. B* 110:3674–84.
21. Yefimov, S., van der Giessen, E., Onck, P. R., Marrink, S. J. 2008. Mechanosensitive membrane channels in action. *Biophys. J.* 94:2994–3002.
22. Marrink, S. J., Mark, A. E. 2003. The mechanism of vesicle fusion as revealed by molecular dynamics simulations. *J. Am. Chem. Soc.* 125:11144–45.
23. Kasson, P. M., Kelley, N. W., Singhal, N., Vrljic, M., Brunger, A. T., Pande, V. S. 2006. Ensemble molecular dynamics yields submillisecond kinetics and intermediates of membrane fusion. *Proc. Natl. Acad. Sci. USA* 103:11916–21.

24. Faller, R., Marrink, S. J. 2004. Simulation of domain formation in DLPC-DSPC mixed bilayers. *Langmuir* 20:7686–93.
25. Dahlberg, M. 2007. Polymorphic phase behavior of cardiolipin derivatives studied by coarse-grained molecular dynamics. *J. Phys. Chem. B* 111:7194–200.
26. Periole, X., Huber, T., Marrink, S. J., Sakmar, T. P. 2007. G protein-coupled receptors self-assemble in dynamics simulations of model bilayers. *J. Am. Chem. Soc.* 129:10126–32.
27. de Joannis, J., Jiang, F. Y., Kindt, J. T. 2006. Coarse-grained model simulations of mixed-lipid systems: Composition and line tension of a stabilized bilayer edge. *Langmuir* 22:998–1005.
28. Baoukina, S., Monticelli, L., Marrink, S. J., Tieleman, D. P. 2007. The pressure–area isotherm of a lipid monolayer from molecular dynamics simulations. *Langmuir* 23:12618–23.
29. Wallace, E. J., Sansom, M. S. P. 2007. Carbon nanotube/detergent interactions via coarse-grained molecular dynamics. *Nano Lett.* 7:1923–28.
30. Hatakeyama, M., Faller, R. 2007. Coarse-grained simulations of ABA amphiphilic triblock copolymer solutions in thin films. *Phys. Chem. Chem. Phys.* 9:4662–72.
31. Jahn, R., Grubmüller, H. 2002. Membrane fusion. *Curr. Opin. Cell Biol.* 14:488–95.
32. Kozlov, M. M., Markin, V. S. 1983. Possible mechanism of membrane fusion. *Biofizika* 28:242–47.
33. Jorgensen, K., Klinger, A., Biltonen, R. L. 2000. Nonequilibrium lipid domain growth in the gel–fluid two-phase region of a DC16PC-DC22PC lipid mixture investigated by Monte Carlo computer simulation, FT-IR, and fluorescence spectroscopy. *J. Phys. Chem. B* 104:11763–73.
34. Liam McWhirter, J., Ayton, G., Voth, G. A. 2004. Coupling field theory with mesoscopic dynamical simulations of multicomponent lipid bilayers. *Biophys. J.* 87:3242–63.
35. Marrink, S. J., Risselada, H. J. 2008. in preparation.
36. Fotiadis, D., Liang, Y., Filipek, S., Saperstein, D. A., Engel, A., Palczewski, K. 2003. Atomic-force microscopy: Rhodopsin dimers in native disc membranes. *Nature* 421:127–28.
37. Botelho, A. V., Huber, T., Sakmar, T. P., Brown, M. F. 2006. Curvature and hydrophobic forces drive oligomerization and modulate activity of rhodopsin in membranes. *Biophys. J.* 91:4464–77.

3 The Multiscale Coarse-Graining Method: A Systematic Approach to Coarse-Graining

W.G. Noid
Department of Chemistry, The Pennsylvania State University

Gary S. Ayton, Sergei Izvekov, and Gregory A. Voth
Center for Biophysical Modeling and Simulation and
Department of Chemistry, University of Utah

CONTENTS

3.1 INTRODUCTION

Atomistic molecular dynamics (MD) simulations [Allen and Tildesley 1987; Frenkel and Smit 2002] remain one of the most powerful tools for computationally modeling biomolecular and condensed phase systems. By providing at each timestep in the simulation the coordinates, velocities, and forces for each atom within a system, atomistic MD simulations allow the direct investigation of molecular structure and dynamics with Angstrom-level detail and femtosecond resolution. Although conventional MD methods cannot readily describe phenomena such as bond cleavage that require explicit treatment of quantum mechanical effects [although see, e.g., Hammes-Schiffer 2006; Voth 2006; Warshel 2002], classical atomistic force fields have been carefully parameterized to accurately reproduce many experimental observables [Gnanakaran et al. 2003; Jorgensen, Maxwell, and Tirado-Rives 1996; Lindahl and Edholm 2001; MacKerell et al. 1998]. With available computational resources [Mervis 2001; Reed 2003] and software [Brooks et al. 1983; Case et al. 2005; Phillips et al. 2005; Smith and Forester 1996; Van der Spoel et al. 2005], classical MD simulations routinely model the equilibrium fluctuations of biomolecular or condensed phase systems over length scales of tens of nanometers for time scales on the order of 100 ns. Moreover, in some exceptional cases, MD simulations have modeled processes such as protein fluctuations [Freddolino et al. 2006; Sanbonmatsu, Joseph, and Tung 2005; van Gunsteren et al. 2001] or equilibrium bilayer undulations [Blood and Voth 2006; Lindahl and Edholm 2000], on somewhat longer length and time scales. Consequently it is indisputable that classical atomically detailed MD simulations have tremendously impacted our present understanding of physical chemistry, structural biology and biochemistry [Karplus and McCammon 2002], and materials science.

Nevertheless, many important biomolecular and condensed phase processes evolve on length scales and time scales that extend far beyond the range of conventional atomistic MD methodologies. For example, relatively small proteins fold on the microsecond time scale or longer [Snow et al. 2002] and signal transduction cascades involve the interaction of many proteins on even longer time scales over relatively large length scales [Kusumi and Suzuki 2005; Simons and Toomre 2000]. An adequate investigation of the mechanisms and relevant transitions involved in these processes remains well beyond the present capabilities of conventional atomistic MD simulation methods. Moreover, an atomically detailed model of such complex processes may be neither necessary nor even desirable. In complex processes involving slowly evolving global structural changes, complete atomistic detail may actually obscure the fundamental physical or biological mechanisms underlying the process of interest. In general, an ideal computational model should be tailored for a specific process of interest, providing the temporal and spatial resolution necessary for describing the key transitions and structures but "coarse-graining" over high-resolution details that are not necessary to understand the particular process. Furthermore, such coarse-grained (CG) models are particularly natural when addressing experimental data that is inherently limited in spatial and temporal resolution, such as in cryoelectron microscopy [Chacon, Tama, and Wriggers 2003; Tama et al. 2003; Wu et al. 2003].

These considerations have recently motivated significant interest in low-resolution CG models of biological and condensed phase systems [see, e.g., Ayton, Noid, and Voth 2007a; Bahar and Rader 2005; Tozzini 2005 for several recent reviews, as well as the chapters within this book]. Although many disparate types of CG models have been developed, the present chapter considers only CG models that are a systematically defined low-resolution mechanical model of a particular molecular system. In this case, the fundamental interacting "particles" are not atoms but rather CG "sites" that each represent groupings of one or more atoms. Multiple or mixed resolution models are also possible. CG models of a slightly different form have been considered elsewhere, for example, works by Brannigan, Lin, and Brown (2006), Buchete, Straub, and Thirumalai (2004), Groot and Warren (1997), Liwo et al. (2001), Villa et al. (2004), and Wriggers (2004). All CG computational models are anticipated to be highly efficient because within a given spatial volume there are far fewer interacting particles than in an atomically detailed model of the same system so that fewer force calculations are necessary per integration timestep in the CG model. Moreover, because the interactions between CG sites should represent averages over atomic interactions, these interactions are expected to be "softer" than the interactions within the atomistic model, thus allowing for a larger integration timestep [Marrink, de Vries, and Mark 2004].

In principle, CG models can be designed to provide the appropriate resolution for a specific system or process of interest. Consequently, CG models allow a tremendously simplified picture and interpretation of complex molecular phenomena by averaging over unessential atomic details. In fact, as indicated in the early work of Levitt and Warshel [Levitt 1976; Levitt and Warshel 1975], the promise of CG modeling is that an appropriately constructed low-resolution model can capture the key physical interactions and biologically relevant transitions for a complex process such as protein folding. When this is possible, CG models provide a tremendously powerful computational tool yielding fundamental insight into complex processes through simplified models that only include the critical features of the system. However, although CG models may provide a highly efficient computational tool for rapidly investigating with a desired resolution the expansive configuration space that is important for slowly evolving complex processes, the conclusions drawn from such low-resolution models may be misleading unless the CG model is consistent with an accurate high-resolution model. An outstanding challenge in the development of CG models is therefore the construction of a formal and methodological multiscale bridge [Ayton, Noid, and Voth 2007a, 2007b] between accurate atomistic models and computationally efficient low-resolution CG models for a given system.

With the goal of providing a systematic multiscale approach to coarse-graining, Izvekov and Voth have introduced the multiscale coarse-graining (MS-CG) method [Izvekov and Voth 2005a, 2005b]. This method constructs a rigorous multiscale bridge between atomistic and CG models by incorporating force data obtained from atomistic MD simulations [Ayton, Noid, and Voth 2007a, 2007b; Chu, Izvekov, and Voth 2006; Chu et al. 2007] within a systematic variational procedure [Noid et al. 2007, 2008a, 2008b]. The MS-CG method has been applied as of this writing to develop accurate CG models for peptides [Liu and Voth 2007; Zhou et al. 2007], pure bilayers [Izvekov and Voth 2005a], mixed bilayers [Izvekov and Voth 2006b], carbohydrates [Liu, Izvekov, and Voth 2007], simple fluids [Izvekov and Voth 2005b, 2006a], ionic liquids [Wang et al. 2006; Wang and Voth 2005], soot nanoparticles [Izvekov, Violi, and Voth 2005], and even mixed-resolution models of transmembrane proteins [Shi, Izvekov, and Voth 2006]. Figure 3.1 provides an overall perspective on the MS-CG methodology for biomolecular systems. Panel (a) depicts how the MS-CG method may be applied to model a wide variety of lipid bilayer systems that incorporate interactions between lipids, cholesterol, and membrane proteins, such as the N-BAR domain [Gallop et al. 2006; Masuda et al. 2006], in either an implicit or an explicit solvent model. In an MS-CG implicit solvent model, the interactions with explicit solvent molecules in the atomistic model have been incorporated into effective interactions between the remaining CG sites in the solvent-free bilayer model. Also illustrated in Figure 3.1 are applications of the MS-CG method for modeling peptides [panel (b)] and monosaccharide solutions [panel (c)].

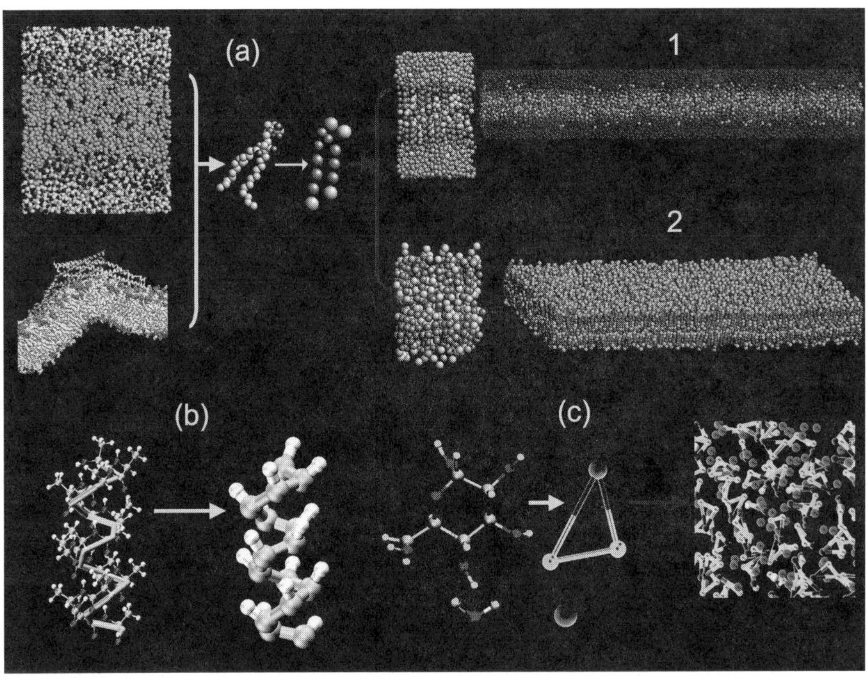

FIGURE 3.1 (See color insert following page 238.) A schematic depicting various aspects of the MS-CG methodology for biomolecular systems. (a) The MS-CG method being applied for modeling interactions between lipids, cholesterol, and membrane-bound proteins, such as the N-BAR domain. The MS-CG method has been implemented with both explicit solvent (1) and solvent-free (2) CG models. The MS-CG method also applied for modeling peptides (b) and monosaccharide solutions (c).

In each instance illustrated in Figure 3.1, a general formal statistical mechanical framework provides a rigorous foundation for the particular application [Ayton, Noid, and Voth 2007b; Noid et al. 2007, 2008a, 2008b]. Recent analysis has demonstrated that, if no approximations are introduced into the method, the MS-CG variational principle determines the exact many-body CG potential of mean force (PMF) [Noid et al. 2007, 2008a]. Consequently, the MS-CG method may be employed to develop CG models that are consistent with a given underlying atomistic model [Noid et al. 2007, 2008a]. Moreover, it has been demonstrated that the MS-CG method variationally determines the optimal approximation of a given form to the exact many-body PMF by considering critical information regarding many-body structural correlations between CG sites that is determined from the atomistic model [Noid et al. 2007, 2008b]. For simple systems governed by central pair potentials, the MS-CG method determines an optimal pair potential between CG sites by treating three-body effects according to the well-known Yvon–Born–Green (YBG) equation [Hansen and McDonald 1986] for the liquid state [Noid et al. 2007]. For more complex molecular systems with both bonded and nonbonded interactions, the MS-CG method incorporates many-body correlations not only between nonbonded sites but also between sites in distinct bonded interactions and between sites in bonded and nonbonded interactions [Noid et al. 2008b].

The present chapter surveys the theory and application of the MS-CG method. The following methods section discusses the rigorous statistical mechanical foundations of the MS-CG method and outlines the key formal results mentioned above. The applications section describes the development of accurate MS-CG models for two short peptides [Zhou et al. 2007] and also for a mixed lipid–cholesterol bilayer that has been modeled both with [Izvekov and Voth 2006b] and without explicit solvent. The concluding section provides closing remarks regarding the present successes and future directions for the MS-CG method.

3.2 METHODS

3.2.1 CONSISTENT CG MODELS

Consider a classical atomistic model of a particular system containing n atoms. These atoms interact as point particles according to a classical n-particle potential energy function, $u(\mathbf{r}^n)$, that depends upon the Cartesian coordinates for the n atoms, $\mathbf{r}^n = \{\mathbf{r}_1, \ldots, \mathbf{r}_n\}$. The canonical equilibrium coordinate distribution function [Hansen and McDonald 1986] for the atomistic model at the given volume, V, and temperature, T, is

$$p_r(\mathbf{r}^n) = \frac{1}{z_n} e^{-u(\mathbf{r}^n)/k_B T}, \tag{3.1}$$

where k_B is Boltzmann's constant, the classical canonical configuration integral is $z_n = z(n,V,T) = \int d\mathbf{r}^n e^{-u(\mathbf{r}^n)/k_B T}$, and each integral over atomic coordinates ranges over the entire system volume, V, in this and in all ensuing integrals. Note that the present discussion considers only flexible systems without rigid intramolecular constraints. A more general discussion for atomistic models with rigid intramolecular constraints has been presented [Noid et al. 2008a].

A given atomically detailed configuration $\mathbf{r}^n$ may also be described by a CG representation with $N<n$ structureless CG "sites," each of which corresponds to one or more atoms. The Cartesian coordinates for the CG site I, $\mathbf{R}_I$, may be determined from the coordinates of the atoms "involved" in the site by applying a mapping function, $\mathbf{M}_{\mathbf{R}I}(\mathbf{r}^n)$, that is defined as a linear function of atomic Cartesian coordinates [Noid et al. 2008a]

$$\mathbf{M}_{\mathbf{R}I}(\mathbf{r}^n) = \sum_{i=1}^{n} c_{Ii}\mathbf{r}_i \tag{3.2}$$

for $I=1,\ldots,N$. The set, I_I, of atoms i that are "involved" in a given CG site I is defined as the set of atoms for which $c_{Ii}\neq 0$. The Cartesian coordinates for each CG site are therefore determined as a linear combination of the Cartesian coordinates for the atoms that are involved in the site. If all of the atoms involved in a CG site are displaced by a given vector, **t**, it is clear that the coordinates of the CG site must also be displaced by the same vector. Consequently, it follows that $\sum_{i=1}^{n} c_{Ii}=1$ for each $I=1,\ldots,N$. A particularly convenient mapping for many CG models is the center of mass mapping, for which $c_{Ii}=m_i/M_I$, where $M_I=\sum_{i\in I_I} m_i$ is the mass of CG site I and is defined as the total mass of all the atoms involved in the site.

A CG configuration, $\mathbf{R}^N$, (i.e., a low-resolution representation of an atomistic configuration) can be determined from a given atomic configuration, $\mathbf{r}^n$, by applying the CG mapping, $\mathbf{M}_{\mathbf{R}I}(\mathbf{r}^n)$, for each CG site $I=1,\ldots,N$, to the given atomic configuration. The atomistic equilibrium coordinate distribution function, $p_r(\mathbf{r}^n)$, and the set of N CG mapping operators, $\mathbf{M}_{\mathbf{R}}^N(\mathbf{r}^n)=\{\mathbf{M}_{\mathbf{R}1}(\mathbf{r}^n),\ldots,\mathbf{M}_{\mathbf{R}N}(\mathbf{r}^n)\}$, then determine the equilibrium probability for observing within the atomistic model an atomically detailed configuration $\mathbf{r}^n$ that maps onto a particular CG configuration, $\mathbf{R}^N$,

$$p_R(\mathbf{R}^N)=\int d\mathbf{r}^n p_r(\mathbf{r}^n)\,\delta[\mathbf{R}^N-\mathbf{M}_{\mathbf{R}}^N(\mathbf{r}^n)], \tag{3.3}$$

where

$$\delta[\mathbf{R}^N-\mathbf{M}_{\mathbf{R}}^N(\mathbf{r}^n)]=\prod_{I=1}^{N}\delta[\mathbf{R}_I-\mathbf{M}_{\mathbf{R}I}(\mathbf{r}^n)].$$

A CG model for the same system may be developed in terms of the same N structureless CG sites. Within the CG model, these sites interact according to an N- site "potential energy function", $U(\mathbf{R}^N)$, which depends upon the CG configuration, $\mathbf{R}^N=\{\mathbf{R}_1,\ldots,\mathbf{R}_N\}$. (Note that for consistency and clarity within the present discussion, CG quantities are represented by capitalized symbols, while atomistic quantities are represented by lower-case symbols.) The canonical equilibrium coordinate distribution function at the given volume, V, and temperature, T, for the CG model is

$$P_R(\mathbf{R}^N)=\frac{1}{Z_N}e^{-U(\mathbf{R}^N)/k_BT}, \tag{3.4}$$

where $Z_N=Z(N,V,T)=\int d\mathbf{R}^N e^{-U(\mathbf{R}^N)/k_BT}$ is the classical configuration integral for the CG model (Hansen and McDonald 1986). For the purposes of the present chapter, a CG model is "consistent" with a particular atomistic model when the equilibrium probability of observing in the CG model a given CG configuration, $\mathbf{R}^N$, is equal to the equilibrium probability of observing in the atomistic model an atomically detailed configuration, $\mathbf{r}^n$, that maps onto the same CG configuration [Noid et al. 2008a]. Therefore, in a consistent CG model

$$P_R(\mathbf{R}^N)=p_R(\mathbf{R}^N). \tag{3.5}$$

Consequently, the CG potential energy function appropriate for a consistent CG model is uniquely determined (to within an additive constant) by the atomistic potential energy, $u(\mathbf{r}^n)$, and the set of N CG mapping functions, $\mathbf{M}_{\mathbf{R}}^N(\mathbf{r}^n)$, such that

$$e^{-U(\mathbf{R}^N)/k_BT}=\frac{Z_N}{z_n}\int d\mathbf{r}^n e^{-u(\mathbf{r}^n)/k_BT}\delta[\mathbf{R}^N-\mathbf{M}_{\mathbf{R}}^N(\mathbf{r}^n)]. \tag{3.6}$$

From Equation 3.6 it is clear that the appropriate interaction potential $U(\mathbf{R}^N)$ for a consistent CG model is not a conventional potential energy function, but rather a many-body configurational free energy function, or PMF, that incorporates both energetic and entropic effects and depends upon the thermodynamic state point [Noid et al. 2007, 2008a].

Simulations of a CG model employing a CG force field determined from the gradients of the many body PMF $U(\mathbf{R}^N)$ will generate an equilibrium distribution of CG configurations that is consistent with the underlying atomistic model [Noid et al. 2008a]. A general procedure for obtaining this force field has recently been developed by Voth and coworkers [Noid et al. 2008a]. As will be shown shortly, in principle this force field may be numerically determined by the MS-CG procedure introduced by Izvekov and Voth (2005a, 2005b).

For CG mappings in which no atom is involved in the definition of more than one CG site, it can be shown that in a consistent CG model the appropriate force on site I in a given CG configuration, $\mathbf{R}^N$, is the conditional expectation value of the "atomistic force experienced by the same site," $\mathbf{f}_I(\mathbf{r}^n)$, averaged over all atomistic configurations, $\mathbf{r}^n$, that map onto the given CG configuration [Noid et al. 2007, 2008a]; that is,

$$\begin{aligned} \mathbf{F}_I(\mathbf{R}^N) &= -\frac{\partial}{\partial \mathbf{R}_I} U(\mathbf{R}^N) \\ &= E[\mathbf{f}_I(\mathbf{r}^n) | \mathbf{M}_{\mathbf{R}}^N(\mathbf{r}^n) = \mathbf{R}^N] \\ &= \frac{1}{Z(\mathbf{R}^N)} \left\langle \mathbf{f}_I(\mathbf{r}^n) \delta[\mathbf{M}_{\mathbf{R}}^N(\mathbf{r}^n) - \mathbf{R}^N] \right\rangle, \end{aligned} \tag{3.7}$$

where $Z(\mathbf{R}^N) = \langle \delta[\mathbf{M}_{\mathbf{R}}^N(\mathbf{r}^n) - \mathbf{R}^N] \rangle$, $\mathbf{f}_I(\mathbf{r}^n) = \sum_{i \in \mathrm{I}_I} \mathbf{f}_i(\mathbf{r}^n)$ is the net force on the atoms involved in site I, $\mathbf{f}_i(\mathbf{r}^n) = -\partial u(\mathbf{r}^n)/\partial \mathbf{r}_i$ is the net force on atom i in configuration $\mathbf{r}^n$, and the angular brackets in the previous expressions indicate canonical averages evaluated for the atomically detailed model according to the atomistic coordinate distribution function defined in Equation 3.1. Equation 3.7 expresses the CG force field determined from gradients of the many-body PMF as an average of atomistic forces evaluated in the canonical ensemble for the atomistic model. Although the atomistic force field may often be readily decomposed into simple additive interactions, the CG force field derived from gradients of the many-body CG PMF may not necessarily be simply decomposed. As a result of the correlations between the CG sites that are incorporated into the canonical averaging, the CG force field defined in Equation 3.7 in principle depends upon the entire CG configuration. The interested reader is directed to work by Noid et al. (2007, 2008a) for further details and a more general analysis.

3.2.2 The MS-CG Method for Deriving Consistent CG Models

The MS-CG method introduced by Izvekov and Voth (2005a, 2005b) determines an optimized force field describing the interactions between CG sites from atomistic force information. This is accomplished by variationally minimizing the residual functional [Noid et al. 2007, 2008a] of N vector valued functions

$$\chi^2[\tilde{\mathbf{F}}] = \frac{1}{3N} \left\langle \sum_{I=1}^{N} \left| \tilde{\mathbf{F}}_I[\mathbf{M}_{\mathbf{R}}^N(\mathbf{r}^n)] - \mathbf{f}_I(\mathbf{r}^n) \right|^2 \right\rangle \tag{3.8}$$

where $\tilde{\mathbf{F}} = \{\tilde{\mathbf{F}}_1(\mathbf{R}^N), \ldots, \tilde{\mathbf{F}}_N(\mathbf{R}^N)\}$ is an arbitrary set of N vector valued functions of the CG configuration, $\mathbf{R}^N$, that describe the force on each CG site $I = 1, \ldots, N$, and the angular brackets indicate an equilibrium canonical ensemble average evaluated with the atomistic model. The MS-CG variational principle asserts that the force field determined by gradients of the many-body CG PMF in Equation 3.7 may be determined as the unique minima of the residual functional in Equation 3.8. The proof of this result has been previously provided [Noid et al. 2007, 2008a] and is summarized below.

It has been demonstrated [Noid et al. 2008a] that the MS-CG residual functional may be expressed as

$$\chi^2[\tilde{\mathbf{F}}] = \chi^2[\mathbf{F}] + \frac{1}{3N}\left\langle \sum_{I=1}^{N} \left| \tilde{\mathbf{F}}_I[\mathbf{M}_{\mathbf{R}}^N(\mathbf{r}^n)] - \mathbf{F}_I[\mathbf{M}_{\mathbf{R}}^N(\mathbf{r}^n)] \right|^2 \right\rangle, \tag{3.9}$$

where $\mathbf{F} = \{\mathbf{F}_1(\mathbf{R}^N), \ldots, \mathbf{F}_N(\mathbf{R}^N)\}$ is the CG force field that is defined in Equation 3.7 by the gradients of the CG many-body PMF and that is appropriate for simulations of a consistent CG model; that is, $\mathbf{F}_I(\mathbf{R}^N) = -\partial U(\mathbf{R}^N)/\partial \mathbf{R}_I$. Thus the value of the MS-CG residual functional for a given trial CG force field, $\tilde{\mathbf{F}}$, is the sum of two nonnegative quantities: (1) the magnitude of the MS-CG residual functional evaluated for the CG force field determined by the many body CG PMF and (2) a term that describes the difference between this CG force field, $\mathbf{F}$, and the given trial CG force field, $\tilde{\mathbf{F}}$. This latter term vanishes only if $\mathbf{F}_I[\mathbf{M}_{\mathbf{R}}^N(\mathbf{r}^n)] - \tilde{\mathbf{F}}_I[\mathbf{M}_{\mathbf{R}}^N(\mathbf{r}^n)] = \mathbf{0}$ for all sites $I = 1, \ldots, N$ in all atomistic configurations $\mathbf{r}^n$. Therefore, $\mathbf{F}$ determines the unique minimum of the MS-CG residual functional defined in Equation 3.8. Consequently, if no approximations are introduced into the method, the MS-CG variational principle provides a prescription for determining the exact many-body CG PMF, $U(\mathbf{R}^N)$, to within an arbitrary additive constant. Moreover, the resulting CG model will be consistent with the atomistic model according to the definition in Equation 3.5.

In practice, two approximations are typically introduced into the MS-CG method [Izvekov and Voth 2005a, 2005b]. The first approximation replaces the canonical average over the atomistic configuration space in the definition of the MS-CG residual functional with an average over n_t configurations, $\mathbf{r}_t^n$, sampled from a finite length atomistic MD simulation, $\langle A(\mathbf{r}^n)\rangle \approx \langle A(\mathbf{r}_t^n)\rangle_t \equiv (1/n_t)\sum_{t=1}^{n_t} A(\mathbf{r}_t^n)$. If the MD simulation has appropriately sampled the canonical ensemble for the atomistic model, then this approximation becomes exact. The following analysis will assume that this approximation is valid. More importantly, in general it is not practical to either numerically determine or even to represent a many-body CG force field that depends upon the entire CG configuration, $\mathbf{R}^N$. Rather, in practice the MS-CG force field approximates the many-body CG force field determined by the exact CG PMF with a linear combination of N_d basis vectors,

$$\mathbf{F}_I^{MS}(\mathbf{R}^N;\phi) = \sum_{d=1}^{N_d} \phi_d \boldsymbol{G}_{I;d}(\mathbf{R}^N) \approx \mathbf{F}_I(\mathbf{R}^N), \tag{3.10}$$

where the basis vectors, $\boldsymbol{G}_{I;d}(\mathbf{R}^N)$, are functions of the CG configuration that span a vector subspace of the vector space of N vector valued functions of the CG configuration [Noid et al. 2007, 2008a]. The precise form of these basis vectors depends upon both the interactions included in the approximate decomposition of the many-body CG force field (e.g., approximate decomposition of the CG PMF into additive pair potentials) and also on the functions used in representing those interactions. The coefficients of these basis vectors, ϕ_d, are then determined by variationally minimizing the MS-CG residual function,

$$\chi^2(\phi) = \frac{1}{3N}\left\langle \sum_{I=1}^{N} \left| \mathbf{F}_I^{MS}[\mathbf{M}_{\mathbf{R}}^N(\mathbf{r}_t^n);\phi] - \mathbf{f}_I(\mathbf{r}_t^n) \right|^2 \right\rangle_t. \tag{3.11}$$

The derivation of Equation 3.9 from Equation 3.8 generalizes within any vector space of trial CG force fields [Noid et al. 2008a]. In particular, it follows that within the vector space of trial CG force fields spanned by the basis vectors of Equation 3.10, the CG force field determined by the unique set of parameters that minimize the residual function defined in Equation 3.11 is the closest approximation within that space to the many-body CG force field determined by the exact CG PMF [Noid et al. 2008b]. (The relevant inner product is determined by the correlation function in Equation 3.13.) In other words, the MS-CG force field in Equation 3.10 that is parameterized by minimizing the MS-CG residual function defined in Equation 3.11 with respect to the force field parameters, ϕ_d, is the best approximation to the many-body CG force field $\mathbf{F}$ among all possible CG force fields of the form determined by the approximate decomposition in Equation 3.10.

Because the MS-CG force field has been represented as a linear combination of basis vectors, the numerical task of determining the optimal approximation to the many-body CG PMF has been reduced [Izvekov and Voth 2005a, 2005b] to a linear least-squares problem [Demmel 1997]. The MS-CG residual may be re-expressed in a form explicitly manifesting this simplicity:

$$\chi^2(\phi) = \frac{1}{3N}\left\langle \sum_{I=1}^{N} \left| \sum_{d=1}^{N_d} \phi_d \boldsymbol{G}_{I;d}[\mathbf{M}_{\mathbf{R}}^{N}(\mathbf{r}_t^n)] - \mathbf{f}_I(\mathbf{r}_t^n) \right|^2 \right\rangle_t$$

$$= \sum_{d,d'=1}^{N_d} G_{dd'}\phi_d\phi_{d'} - 2\sum_{d=1}^{N_d} b_d\phi_d + \chi^2(0) \tag{3.12}$$

where

$$b_d = \frac{1}{3N}\left\langle \sum_{I=1}^{N} \mathbf{f}_I(\mathbf{r}_t^n)\cdot \boldsymbol{G}_{I;d}[\mathbf{M}_{\mathbf{R}}^{N}(\mathbf{r}_t^n)] \right\rangle_t$$

$$G_{dd'} = \frac{1}{3N}\left\langle \sum_{I=1}^{N} \boldsymbol{G}_{I;d}[\mathbf{M}_{\mathbf{R}}^{N}(\mathbf{r}_t^n)]\cdot \boldsymbol{G}_{I;d'}[\mathbf{M}_{\mathbf{R}}^{N}(\mathbf{r}_t^n)] \right\rangle_t . \tag{3.13}$$

The MS-CG force field may be numerically determined by directly solving (within a least-squares sense) the overdetermined system of $3n_tN$ equations for N_d parameters given by

$$\mathbf{F}_I^{MS}\left(\mathbf{M}_{\mathbf{R}}^{N}(\mathbf{r}_t^n);\phi\right) = \sum_{d=1}^{N_d} \phi_d \boldsymbol{G}_{I;d}[\mathbf{M}_{\mathbf{R}}^{N}(\mathbf{r}_t^n)] = \mathbf{f}_I(\mathbf{r}_t^n) \tag{3.14}$$

for each of the N CG sites in each of the n_t sampled configurations $\mathbf{r}_t^n$ (Demmel 1997). Intuitively the MS-CG method determines the optimized CG force field, $\mathbf{F}^{MS}$, that on average provides the best "match" to the instantaneous forces [Izvekov et al. 2004] on the CG sites, $\mathbf{f}$, in the configurations sampled from an atomistic MD simulation. Alternatively, if the MS-CG force field is determined by minimizing the second expression for the MS-CG residual function [Noid et al. 2007, 2008a] in Equation 3.12, the optimal set of parameters are obtained by solving the normal system (Demmel 1997) of N_d linear equations for N_d parameters:

$$\sum_{d'=1}^{N_d} G_{dd'}\phi_{d'} = b_d . \tag{3.15}$$

The two systems of equations for determining the MS-CG force field are equivalent in principle (Demmel 1997). In practice, one or the other system of equations may be more convenient for a particular application. These numerical aspects have been previously discussed [Noid et al. 2008b]. However, the normal system of equations expressed in Equation 3.15 are particularly convenient for analysis because both $G_{dd'}$ and b_d have been expressed in terms of structural correlation functions that are calculated from atomistic MD simulations [Noid et al. 2007, 2008a]. In particular, the quantity b_d may be considered the projection of the atomistic force field onto the basis vector $\boldsymbol{G}_{I;d}$ and the quantity $G_{dd'}$ describes correlations between the d and d' basis vectors. If the basis vectors were orthonormal, then $G_{dd'} = \delta_{dd'}$ and $\phi_d = b_d$. In general, though, these basis vectors describe correlated molecular interactions and are not orthogonal.

The preceding analysis [Noid et al. 2007, 2008a, 2008b] is quite general and applies for complex molecular systems as well as simple models without bonded topology. However, the physical significance of these many-body structural correlations and the role of the basis vectors is conveniently illustrated by considering a CG model for a simple one-component system in which each molecule is represented by a single CG site [Noid et al. 2007]. If it is further assumed that the many-body CG force field determined by the exact CG PMF can be approximately represented by an MS-CG potential consisting only of central pair potentials, then the total force on CG site I in a given CG configuration, $\mathbf{R}^N$, may be expressed

$$\mathbf{F}_I^{MS}(\mathbf{R}^N) = \sum_{J(\neq I)} \hat{\mathbf{R}}_{IJ} F(R_{IJ}) \tag{3.16}$$

where $\mathbf{R}_{IJ} = \mathbf{R}_I - \mathbf{R}_J$ is the vector displacement from site J to site I, $R_{IJ} = |\mathbf{R}_{IJ}|$ is the magnitude of this vector, $\hat{\mathbf{R}}_{IJ} = \mathbf{R}_{IJ}/R_{IJ}$ is the associated unit vector, and $F(R_{IJ})$ is the magnitude of the force on CG site I arising from CG site J and depends only upon the distance between the two CG sites. If this force function were tabulated on a discrete grid at a set of N_d points, $\{R_d\}$, that are evenly spaced by δR, then

$$F(R) = \sum_{d=1}^{N_d} F_d \delta_D(R - R_d) \tag{3.17}$$

where $\delta_D(R - R_d)$ is a "discrete delta function" with a value of 1 when $|R - R_d| < \delta R/2$ and 0 otherwise, so that $F(R) = F_d = const$ over the range of distances $R_d - \delta R/2 < R < R_d + \delta R/2$. It then follows that $\phi_d = F_d = F(R_d)$ and the basis vector associated with this parameter is

$$\boldsymbol{G}_{I;d}(\mathbf{R}^N) = \sum_{J(\neq I)} \hat{\mathbf{R}}_{IJ} \delta_D(\mathrm{R}_{IJ} - \mathrm{R}_d), \tag{3.18}$$

i.e., $\boldsymbol{G}_{I;d}(\mathbf{R}^N)$ is the sum of unit vectors directed towards site I from all sites J that are a distance R_{IJ} from I such that $|R_{IJ} - R_d| < \delta R/2$. In the CG configuration $\mathbf{R}^N$, the net force on CG site I from all CG sites that are within this distance range from site I is then $F_d \boldsymbol{G}_{I;d}(\mathbf{R}^N) = \phi_d \boldsymbol{G}_{I;d}(\mathbf{R}^N)$, since the magnitude of the force from all CG sites J in that distance range is the same. For this simple example it then follows that

$$\begin{aligned} b_d &= \frac{1}{3N}\left\langle \sum_{I=1}^{N} \sum_{J(\neq I)} [\mathbf{f}_I(\mathbf{r}^n) \cdot \hat{\mathbf{R}}_{IJ}] \delta_D(R_{IJ} - R_d) \right\rangle \\ G_{dd'} &= \frac{1}{3N}\left\langle \sum_{I=1}^{N} \sum_{J(\neq I)} \sum_{K(\neq I)} (\hat{\mathbf{R}}_{IK} \cdot \hat{\mathbf{R}}_{IJ}) \delta_D(R_{IJ} - R_d) \delta_D(R_{IK} - R_{d'}) \right\rangle. \end{aligned} \tag{3.19}$$

In Equation 3.19, all CG quantities are determined by applying the CG mapping to generate CG configurations from atomistic configurations sampled by the MD trajectory and the averages correspond to the time averages over that atomistic trajectory. (For example, $R_{IJ}=|\mathbf{M}_{\mathbf{R}I}(\mathbf{r}^n)-\mathbf{M}_{\mathbf{R}J}(\mathbf{r}^n)|$.) The quantity b_d is proportional to $k_BTR_d^2\, g'(R_d)$, where $g(R)$ is the radial distribution function (RDF) for the CG sites calculated from the CG representation of atomistic configurations sampled from the atomistic trajectory. Consequently, b_d is related to the conventional two-body PMF. In contrast, the quantity $G_{dd'}$ describes both two- $(J=K)$ and three- $(J\neq K)$ particle correlations. It has been demonstrated [Noid et al. 2007] that in this case, the normal MS-CG equations are formally equivalent to the well-known YBG equation for the liquid state:

$$\left(k_{\mathrm{B}}T\frac{d}{dR}-F(R)\right)g(R)=\int dR'\,F(R')\,M^{(3)}(R,R'), \tag{3.20}$$

where $M^{(3)}(R,R')$ is a spherical average of the three-particle correlation function for CG sites determined from atomistic simulations. The YBG equation [Hansen and McDonald 1986] is an exact equation relating the equilibrium two- and three-particle distribution functions to the pair force field governing the interactions within the system that generated the given distribution functions. For this simple one-component system and according to the assumptions stated above, the MS-CG method assumes that the CG pair and triplet distribution functions determined by atomistic MD simulations were generated by an interaction potential comprised of central pair potentials between CG sites according to Equation 3.16. The MS-CG method then solves for this pair force by inverting the YBG equation [Noid et al. 2007]. Of course the CG distribution functions computed from the atomistic simulations were not generated by a pair force field between CG sites, but rather by an atomically detailed potential energy function governing the interactions between the atoms defining the CG sites. As a consequence of these assumptions, the resulting MS-CG model will not be perfectly consistent with the underlying atomistic model, but an approximation to an exactly consistent MS-CG model free of such assumptions.

Equation 3.20 clearly demonstrates that the MS-CG method explicitly incorporates information regarding three-particle structural correlations in determining the optimized pair force, $F(R)$, between CG sites. If this information were not incorporated into the MS-CG method (i.e., if $M^{(3)}(R,R')=0$), then $F(R)$ would equal $k_{\mathrm{B}}T\,d[\ln g(R)]/dR$, the conventional two-body mean force. The above analysis indicates that by treating both two- and three-particle correlations the MS-CG force field provides the optimal approximation to the exact many-body CG PMF given the assumed decomposition of the CG force field [Noid et al. 2007, 2008a, 2008b]. The interested reader is referred to work by Noid et al. (2007) for further details and discussion of this result.

Although the present example illustrated the importance of many body structural correlations for accurate CG models of simple systems, the preceding analysis generalizes for complex molecular systems with both bonded and nonbonded interactions. If the MS-CG force field includes both bonded and nonbonded interactions, the MS-CG variational principle parameterizes CG interactions by considering many-body structural correlations between sites in distinct bonded interactions, between sites in distinct nonbonded interactions, and even between sites that are in distinct bonded and nonbonded interactions, as has been previously described through a more general analysis [Noid et al. 2008a].

The MS-CG methodology also provides a systematic means to increase the dynamical accuracy of the CG model [Izvekov and Voth 2006a]. In particular, since the MS-CG algorithm determines an approximation to the many-body PMF for the CG sites, one can also examine instantaneous deviations (i.e., fluctuations) of the exact atomically detailed forces from the CG PMF force in the MS-CG model. In turn, correlation functions of these force fluctuations can be calculated and a generalized Langevin (or simpler Langevin) equation for the dynamics of the MS-CG model can be formulated. An approach along these lines has been shown [Izvekov and Voth 2006a] to provide greater dynamical accuracy in the MS-CG model than simply running Newtonian MD on the CG PMF. The latter is correct for statistical sampling of the MS-CG distribution functions and the calculation of thermodynamic properties, but as with all CG models, the dynamics will not be "real" (i.e., "real" in the sense that the dynamical trajectory of the CG system follows that of the corresponding CG coordinates as instantaneously calculated in the all-atom MD simulation). The approach developed in Izvekov and Voth's study (2006a) provides one way to circumvent this problem.

3.3 APPLICATIONS

Subsequent to the introduction of the MS-CG method [Izvekov and Voth 2005a, 2005b], this particular approach to coarse-graining has been applied as of this writing to develop CG models for a wide range of biological and condensed matter processes, including the equilibrium fluctuations of lipid bilayers [Izvekov and Voth 2005a, 2006b], the interactions between lipids and transmembrane proteins [Shi, Izvekov, and Voth 2006], the spatial and dynamic heterogeneity of ionic liquids [Wang et al. 2006; Wang and Voth 2005], the structure and dynamics of small peptides [Liu and Voth 2007; Zhou et al. 2007], the properties of monosaccharides [Liu, Izvekov, and Voth 2007], and the aggregation of soot nanoparticles [Izvekov, Violi, and Voth 2005]. More significantly, though, these preliminary applications have clearly demonstrated that the MS-CG method is a robust and systematic methodology for developing computationally efficient CG models of complex systems that can quantitatively reproduce the structural and thermodynamic properties of more computationally expensive all-atom models. The present section reviews applications of the MS-CG method for studying the equilibrium structure and fluctuations of small peptides [Zhou et al. 2007] and lipid bilayers [Izvekov and Voth 2006b].

3.3.1 MS-CG Models of Peptides

Zhou et al. (2007) have successfully employed the MS-CG method to develop quantitatively accurate CG models for the pentadecamer alanine peptide, Ala_{15}, and the β-hairpin peptide, V_5PGV_5. For both peptides, each amino acid (other than glycine and proline) was represented with one CG site describing the side chain and three amino acids describing the peptide backbone (Zhou et al. (2007) also considered a CG representation of the peptide backbone with one site per amino acid). Figure 3.2 presents CG structures of the Ala_{15} and V_5PGV_5 peptides. Although they are not indicated in Figure 3.2, the CG models included explicit CG water molecules that were represented with one CG site each. Atomistic MD simulations of each peptide were performed and atomically detailed configurations were sampled from these simulations. An ensemble of CG configurations was generated by mapping each sampled atomistic configuration onto a CG configuration according to Equation 3.2. Distribution functions for the CG sites were then calculated from these mapped configurations. Figure 3.3 presents the resulting RDFs for select CG sites in the peptide backbone of Ala_{15} and V_5PGV_5 as the dashed curves in the top and bottom panels of the figure, respectively. In Figure 3.3, NBB, CBB, and OBB indicate CG sites that represent the nitrogen atom, the α-carbon atom, and the carbonyl group of each amino acid, respectively, and CGW indicates the CG sites representing water molecules.

The MS-CG variational principle was applied to parameterize an optimized force field describing the interactions between sites in the CG model according to the theory presented above. It was empirically determined that the bonded interactions between adjacent CG sites could be accurately modeled by considering Boltzmann statistics and treating each bonded interaction as an independent degree of freedom [Zhou et al. 2007]. Recent theoretical developments outlined earlier indicate how these bonded interactions can be systematically incorporated into the general variational MS-CG theory [Noid et al. 2008b]. The nonbonded interactions between CG sites were modeled with central pair potentials that were determined by solving (in a least-squares sense) the overdetermined system of equations in Equation 3.14, with $\mathbf{f}_I$ including only the forces on CG site I resulting from atomistic nonbonded forces.

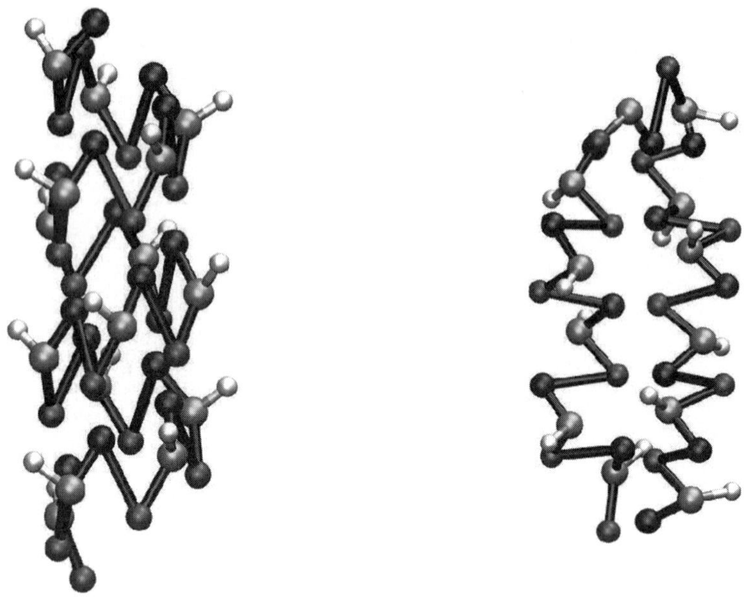

FIGURE 3.2 Coarse-grained (CG) representations of the α-helical alanine pentadecamer, Ala_{15} (left), and the β-hairpin peptide, V_5PGV_5 (right). Except for glycine and proline, one CG site describes the side chain and three CG sites describe the peptide backbone of each amino acid.

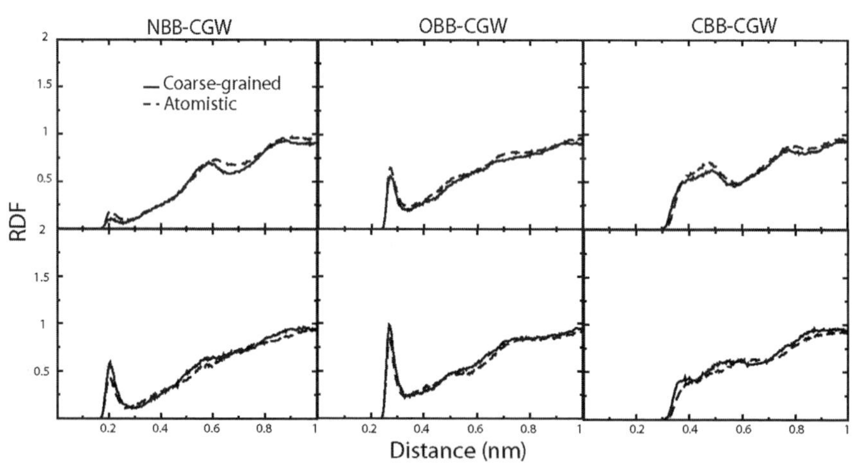

FIGURE 3.3 Comparison of radial distribution functions (RDFs) for select CG sites computed from the simulations of the all-atom models (dashed) and from the simulations of the MS-CG models (solid) for Ala_{15} (top row) and V_5PGV_5 (bottom row).

MD simulations of the CG peptide models were then performed using the resulting MS-CG force fields. The computationally efficient low-resolution MS-CG models for the two peptides accurately reproduced the structures observed in simulations of the high-resolution atomistic models. In particular, the backbone conformation of the CG models remained within ~1 Å rms deviation from the backbone conformation observed in the atomistic MD simulations used in parameterizing the MS-CG models. Figure 3.3 presents the RDFs for the NBB, CBB, OBB, and CGW sites computed from simulations of the MS-CG models for Ala_{15} and V_5PGV_5 as the solid curves in the top and bottom panels, respectively. Figure 3.3 clearly demonstrates that the MS-CG models for both peptides quantitatively reproduced the pair distributions generated by the atomistic models. The reader is directed to the work of Zhou et al. (2007) for further details and analysis.

3.3.2 MS-CG Model of Mixed Lipid–Cholesterol Bilayers

In an early demonstration of the power and flexibility of the MS-CG method to describe complex systems, Izvekov and Voth (2006b) developed a quantitatively accurate MS-CG model for a solvated mixed bilayer system consisting of cholesterol and dimyristoylphosphatidylcholine (DMPC) in a 1:1 ratio. An atomistic MD simulation of this system was performed in the isothermal isobaric ensemble with a united atom force field. The explicit solvent MS-CG bilayer model represented each DMPC lipid with 13 CG sites, each cholesterol molecule with four CG sites, and each water molecule with a single CG site. Figure 3.4 presents atomistic and CG representations of a DMPC lipid (a) and a cholesterol molecule (b). For both molecules in Figure 3.4, the darker larger spheres indicate the coordinates of the CG sites defined by the CG mapping, while the lighter smaller spheres indicate the united atom atomistically detailed representation. The CH, PH, E1 and E2, SM, and ST sites represent the choline group, the phosphate group, ester groups one and two of the glycerol moiety, aliphatic hydrocarbon groups in the lipid tails, and the terminal aliphatic group of the lipid tails, respectively, in DMPC lipid molecules. Cholesterol molecules are represented by CA, CB, CC, and CD sites, in order from top to bottom in Figure 3.4b. In particular, CA sites include the polar hydroxyl group and reside near the surface of the bilayer, while CD sites include the cholesterol hydrophobic tail and reside near the center of the bilayer.

An ensemble of CG configurations was then generated by mapping each atomically detailed configuration sampled from the atomistic MD simulation onto a CG configuration according to Equation 3.2. The MS-CG variational principle was employed to determine an optimized force field describing the interactions between CG sites from these configurations. The MS-CG force field included both bonded and nonbonded interactions that were represented with radially symmetric pair potentials. For example, angle interactions between triples of CG sites were mimicked with radially symmetric pair interactions between next-nearest-neighbor sites in the same molecule. As discussed earlier, recent work has demonstrated that the MS-CG variational principle can be employed to determine many-body interactions between three or more CG sites [Noid et al. 2008b]. These pair potentials were parameterized by solving (in a least-squares sense) the system of equations in Equation 3.14.

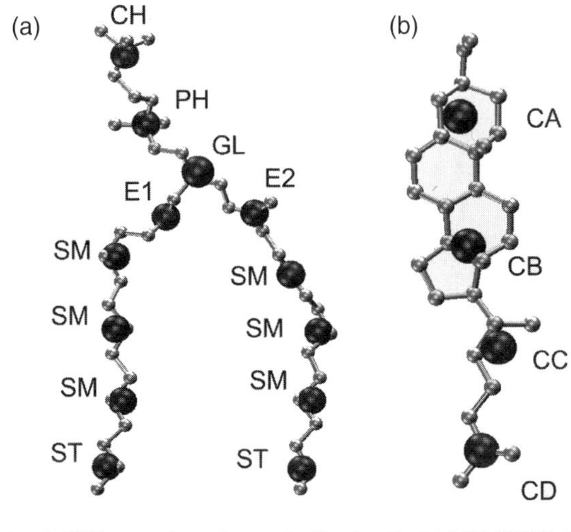

FIGURE 3.4 Coarse-grained (CG) mapping of atomically detailed (a) DMPC lipid (13 sites) and (b) cholesterol molecule (four sites). CG sites are indicated by darker larger spheres, while the united atom representations of the molecules are indicated by the lighter smaller spheres.

This system of equations was supplemented by an additional set of equations for the atomistic virial to ensure that the MS-CG model reproduced the thermodynamic pressure [Izvekov and Voth 2006b]. Figure 3.5 presents as dashed curves the resulting MS-CG nonbonded pair potentials for interactions involving select CG sites in the DMPC lipid tails. The interactions between CG water molecules were determined from a previous study [Izvekov and Voth 2005b].

Simulations of the MS-CG model for the solvated lipid bilayer were performed and distribution functions for the CG sites were computed from these simulations. The MS-CG model quantitatively reproduced the pair structure of both the lipid bilayer and also the water–bilayer interface; these data have been previously published in a study by Izvekov and Voth (2006b). Importantly, the MS-CG model accurately reproduced the pair distribution functions for cholesterol embedded in the bilayer. The two methyl groups projecting from the conjugated carbon rings are an important structural feature of cholesterol that distinguish the two faces of the molecule. For example, in a study by Pandit, Jakobsson, and Scott (2004) it was suggested that differences in the "roughness" of the two faces of cholesterol may play an important role in the formation of interfaces between lipid domains. As is clear from Figure 3.4b, the present CG mapping of cholesterol does not emphasize this asymmetry. Remarkably, the MS-CG model accurately reproduced the interactions between DMPC and CG cholesterol molecules because the underlying atomically detailed interactions are systematically built into the effective interactions at the CG level in the MS-CG method. The reader is directed to word by Izvekov and Voth (2006b) for further details and analysis of this study.

3.3.3 MS-CG Solvent-Free Bilayer Model

As discussed in the introduction to this chapter, CG models include the essential biological and physical features of a particular system while averaging over nonessential details. In the applications described above, the MS-CG models represented each molecule in a particular system with a reduced number of sites while averaging over the details of the atoms that are represented by the sites. The interactions between these atoms are incorporated into the effective interactions between CG sites. As discussed earlier, if no approximations are introduced into the method, the MS-CG variational principle determines the exact many-body CG PMF, which incorporates both energetic and entropic effects. The equilibrium distribution of CG sites generated by simulations of a CG model employing this potential will be consistent with the equilibrium distribution generated by the underlying atomistic model [Noid et al. 2007, 2008a].

The MS-CG theory applies in principle for a wide range of CG mappings. In particular, this theory applies for CG models in which the mapping removes entire molecules from the description of the system. Consequently, the MS-CG variational principle provides a systematic and rigorous

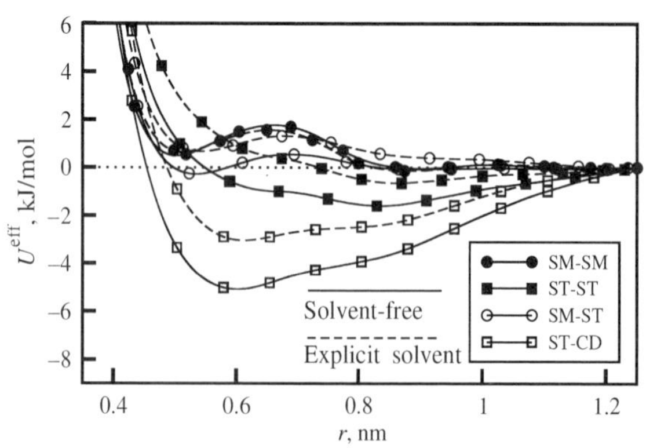

FIGURE 3.5 Comparison of select nonbonded pair potentials describing the interactions of CG sites in the DMPC lipid tails for the explicit CG solvent (dashed) and solvent-free (solid) MS-CG models of the mixed cholesterol–DMPC bilayer.

methodology for completely integrating out explicit solvent molecules and developing solvent-free CG models in which the effects of the solvent are incorporated into effective interactions between the remaining CG sites. Because explicit water solvents often represent a significant fraction of the system (and thus the computational burden) of atomically detailed models, solvent-free CG models are particularly appealing for investigating phenomena that do not explicitly involve the solvent [Brannigan and Brown 2004; Brannigan, Lin, and Brown 2006].

The MS-CG method can therefore be utilized to develop a solvent-free model for the mixed DMPC–cholesterol bilayer considered above. The CG mapping for the DMPC and cholesterol molecules is the same as before (illustrated in Figure 3.4). However, in the solvent-free model the mapping does not determine coordinates for CG water molecules. The total forces on the atoms in the lipid and cholesterol molecules, including the forces from the solvent, are thus employed in the MS-CG variational method to parameterize the effective force field describing the interactions between the CG lipid and cholesterol molecules in the solvent-free model. Despite the absence of explicit CG solvent molecules, if no approximations are introduced into the method, the distribution of CG sites in the solvent-free MS-CG model will be identical to the distribution of CG sites determined from simulations of the atomically detailed explicit solvent model.

In the same manner as described above for the explicit solvent MS-CG model, the many-body CG PMF for the solvent-free model was approximately decomposed into bonded and nonbonded interactions that were described by radially symmetric pair potentials. These pair potentials were parameterized by solving (within a least-squares sense) the overdetermined system of equations in Equation 3.14. As for the earlier explicit solvent model, additional equations for the instantaneous virial were included in this system of equations to reproduce the thermodynamic pressure. Figure 3.5 presents as solid curves the effective CG pair potentials describing nonbonded interactions between CG sites in the solvent-free MS-CG model. The interactions between the same CG sites in the explicit solvent MS-CG bilayer model are presented as the dashed curves in the figure. It is clear that the interactions between CG sites are significantly more attractive in the solvent-free MS-CG model than in the explicit solvent MS-CG model for the same system. The pair potentials between other CG sites are also more attractive in the solvent-free MS-CG model than in the explicit solvent MS-CG model, although this increased attraction is less pronounced for the interactions involving sites in the DMPC head group. In effect, the interactions with water molecules that generated a positive surface tension in the explicit solvent model have been incorporated into the solvent-free MS-CG model (that does not include an explicit water–bilayer interface) as an increased effective attraction between lipid and cholesterol molecules within the bilayer.

Simulations of the solvent-free MS-CG bilayer model were performed and distribution functions for the CG sites were calculated. Figure 3.6 compares select distribution functions computed from simulations of the MS-CG solvent-free model (dashed curves) with the distribution functions computed from simulations of the atomically detailed explicit solvent model (solid curves). As in previous cases, the solvent-free MS-CG bilayer model reproduced with quantitative accuracy the pair distributions of the atomically detailed explicit solvent model. The largest deviations are in the CA–CA cholesterol site–site RDF (cf. Figure 3.4b) because of the relatively primitive four-site CG representation of the cholesterol molecules.

Previous research has demonstrated the difficulty in developing solvent-free bilayer models that are both structurally stable and that also remain fluid. The increased molecular attraction that is necessary to maintain the structural integrity of the bilayer in the absence of an explicit solvent can easily result in gel-like phase behavior [Brannigan, Lin, and Brown 2006]. Importantly, the solvent-free MS-CG bilayer remained fluid despite the increased attraction between lipid tails. Table 3.1 compares the diffusion constants for the cholesterol and lipid molecules in the all-atom model with the diffusion constants in the explicit CG solvent and solvent-free MS-CG models. Although the diffusion of DMPC and cholesterol is too rapid in the explicit solvent MS-CG model, the diffusion in the implicit solvent MS-CG model is comparable to that observed in the atomistic model. Table 3.1 also compares the mechanical properties of the bilayer computed from simulations of the atomistic model

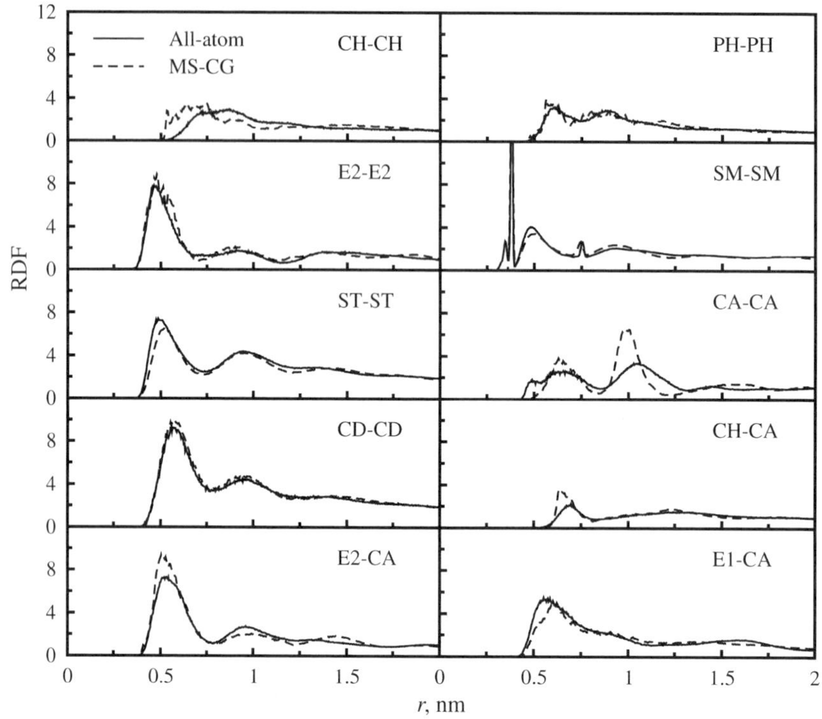

FIGURE 3.6 Comparison of radial distribution functions (RDFs) for select CG sites computed from the simulations of the all-atom model (solid) and the solvent-free MS-CG model (dashed).

TABLE 3.1
Selected Properties of the MS-CG Lipid Bilayer

	Atomistic	Explicit CG Solvent	Solvent-Free
DMPC diffusion constant (m^2/s)	2.6×10^{-12}	1.7×10^{-10}	2.1×10^{-11}
Chol diffusion constant (m^2/s)	2.6×10^{-12}	1.6×10^{-10}	2.2×10^{-11}
Bulk modulus (Pa)	3.5×10^{8}	2.1×10^{8}	N/A
Bending modulus (J)	2.6×10^{-19}	2.5×10^{-19}	2.9×10^{-19}
Area compressibility (J/m^2)	2.6	2.2	5.5

and from simulations of both MS-CG models. Both the explicit and solvent-free MS-CG bilayer models quite reasonably reproduce the mechanical properties of the atomistic bilayer model.

3.4 CONCLUSIONS AND FUTURE DIRECTIONS

The present chapter has reviewed a growing body of work demonstrating that the MS-CG method [Chu, Izvekov, and Voth 2006; Chu et al. 2007; Izvekov and Voth 2005a, 2005b] is a powerful and systematic computational methodology [Iuchi, Izvekov, and Voth 2007; Izvekov and Voth 2005a, 2005b, 2006a, 2006b; Liu, Izvekov, and Voth 2007; Liu and Voth 2007; Shi, Izvekov, and Voth 2006; Wang et al. 2006; Wang and Voth 2005; Zhou et al. 2007] that is founded upon a rigorous statistical mechanical theory [Ayton, Noid, and Voth 2007b; Noid et al. 2007, 2008a, 2008b]. If no approximations are introduced into the method, the MS-CG variational principle provides a novel computational algorithm for determining the exact many-body CG free energy surface (PMF) for a given atomically detailed model. Simulations of a CG model employing this PMF as a potential energy function will generate an equilibrium coordinate distribution of CG sites that is equal to the distribution of CG sites determined by the atomistic model [Noid et al. 2007, 2008a]. In this case, the CG simulation is intended for the purpose of statistical sampling only. Moreover, the MS-CG method provides a systematic algorithm for determining the optimal approximation of a given form to the exact many-body PMF. This is possible because the MS-CG method explicitly incorporates information describing many-body structural correlations determined from atomically detailed simulations [Noid et al. 2007]. The MS-CG theory can also serve as a basis for achieving more correct dynamical behavior (e.g., self-diffusion) in the CG model [Izvekov and Voth 2006a]. The present chapter has illustrated the MS-CG method by presenting quantitatively accurate, computationally efficient CG models for peptides [Zhou et al. 2007] and lipid bilayers [Izvekov and Voth 2005a, 2006b].

The present work also indicates a number of directions that will be important for the future development of the MS-CG method. In particular, following the success of the method for relatively short peptides [Zhou et al. 2007], future work will extend the present methodology for larger and more complex protein systems. Also, because the MS-CG variational principle systematically incorporates both energetic and entropic affects, deriving the principles that define transferability of the MS-CG force field to other thermodynamic conditions is a priority. Future work will generalize the MS-CG variational principle to transfer the MS-CG force field between thermodynamic state points and also to apply the method to other statistical mechanical ensembles. Because the MS-CG method relies upon accurate sampling of the atomistic model, current work also focuses on developing more sophisticated sampling methodologies that may be directly incorporated into the MS-CG theory. Finally, additional work is under way employing the structure of the MS-CG theory to determine CG mappings that are systematically optimized for describing a particular system or process.

ACKNOWLEDGMENTS

This research was supported by grants from the National Science Foundation (Grant No. CHE-0628257) and the National Institutes of Health (GM063796). WGN acknowledges funding from the National Institutes of Heath through a Ruth L. Kirschstein National Research Service Award postdoctoral fellowship (Grant No. 5 F32 GM076839-02). The authors gratefully acknowledge Dr. I. F. Thorpe for kindly providing Figure 3.2 and Figure 3.3. The authors also gratefully acknowledge the contributions of Prof. H. C. Andersen and Drs. J.-W. Chu, V. Krishna, I. F. Thorpe, P. Liu, Y. Wang, and J. Zhou to the work that has been reviewed in the present chapter.

REFERENCES

Allen, M. P., and Tildesley, D. J. 1987. *Computer Simulation of Liquids*. Clarendon: Oxford.

Ayton, G. S., Noid, W. G., and Voth, G. A. 2007a. Multiscale modeling of biomolecular systems: In serial and in parallel. *Curr. Opin. Struct. Biol.* 17, 192–98.

———. 2007b. Systematic coarse-graining of biomolecular and soft–matter systems. *MRS Bull.*, 32, 929–934.

Bahar, I., and Rader, A. J. 2005. Coarse-grained normal mode analysis in structural biology. *Curr. Opin. Struct. Biol.* 15, 586–92.

Blood, P. D., and Voth, G. A. 2006. Direct observation of BAR domain-induced membrane curvature via molecular dynamics simulation. *Proc. Natl. Acad. Sci., U.S.A.* 103, 15068–72.

Brannigan, G., and Brown, F. L. H. 2004. Solvent-free simulations of membrane bilayers. *J. Chem. Phys.* 120, 1059–71.

Brannigan, G., Lin, L. C. L., and Brown, F. L. H. 2006. Implicit solvent simulation models for biomembranes. *Eur. Biophys. J.* 35, 104–24.

Brooks, B. R., Bruccoleri, R. E., Olafson, B. D. et al. 1983. Charmm: A program for macromolecular energy, minimization, and dynamics calculations. *J. Comp. Chem.* 4, 187–217.

Buchete, N. V., Straub, J. E., and Thirumalai, D. 2004. Continuous anisotropic representation of coarse-grained potentials for proteins by spherical harmonics synthesis. *J. Mol. Graph. Model.* 22, 441–50.

Case, D. A., Cheatham, T. E., Darden, T. et al. 2005. The Amber biomolecular simulation programs. *J. Comp. Chem.* 26, 1668–88.

Chacon, P., Tama, F., and Wriggers, W. 2003. Mega-Dalton biomolecular motion captured from electron microscopy reconstructions. *J. Mol. Biol.* 326, 485–92.

Chu, J. W., Ayton, G. S., Izvekov, S., and Voth, G. A. 2007. Emerging methods for the multiscale simulation of biomolecular systems. *Mol. Phys.* 105, 167–75.

Chu, J. W., Izvekov, S., and Voth, G. A. 2006. The multiscale challenge for biomolecular systems: coarse-grained modeling. *Mol. Sim.* 32, 211–18.

Demmel, J. 1997. *Applied Numerical Linear Algebra.* Philadelphia. SIAM.

Freddolino, P., Arkhipov, A., Larson, S., McPherson, A., and Schulten, K. 2006. Molecular dynamics simulations of the complete satellite tobacco mosaic virus. *Structure* 14, 437–49.

Frenkel, D., and Smit, B. 2002. *Understanding Molecular Simulation.* San Diego: Academic Press.

Gallop, J. L., Jao, C. C., Kent, H. M. et al. 2006. Mechanism of endophilin N-BAR domain-mediated membrane curvature. *Eur. Mol. Bio. Org.* 25, 2898–910.

Gnanakaran, S., Nymeyer, H., Portman, J., Sanbonmatsu, K. Y., and Garcia, A. E. 2003. Peptide folding simulations. *Curr. Opin. Struct. Biol.* 13, 168–74.

Groot, R. D., and Warren, P. B. 1997. Dissipative particle dynamics: Bridging the gap between atomistic and mesoscopic simulation. *J. Chem. Phys.* 107, 4423–35.

Hammes-Schiffer, S. 2006. Hydrogen tunneling and protein motion in enzyme reactions. *Acc. Chem. Res.* 39, 93–100.

Hansen, J. P., and McDonald, I. R. 1986. *Theory of Simple Liquids.* San Diego: Academic Press.

Iuchi, S., Izvekov, S., and Voth, G. A. 2007. Are many-body polarization effects important in liquid water? *J. Chem. Phys.* 126, 124505.

Izvekov, S., Parrinello, M., Burnham, C. J., and Voth, G. A. 2004. Effective force fields for condensed phase systems from ab initio molecular dynamics simulation: A new method for force-matching. *J. Chem. Phys.* 120, 10896–913.

Izvekov, S., Violi, A., and Voth, G. A. 2005. Systematic coarse-graining of nanoparticle interactions in molecular dynamics simulation. *J. Phys. Chem. B* 109, 17019–24.

Izvekov, S., and Voth, G. A. 2005a. A multiscale coarse-graining method for biomolecular systems. *J. Phys. Chem. B* 109, 2469–73.

———. 2005b. Multiscale coarse-graining of liquid state systems. *J. Chem. Phys.* 123, 134105.

———. 2006a. Modeling real dynamics in the coarse-grained representations of condensed phase systems. *J. Chem. Phys.* 125, 151101.

———. 2006b. Multiscale coarse-graining of mixed phospholipid/cholesterol bilayers. *J. Chem. Theor. Comp.* 2, 637–48.

Jorgensen, W. L., Maxwell, D. S., and Tirado-Rives, J. 1996. Development and testing of the OPLS all-atom force field on conformational energetics and properties of organic liquids. *J. Am. Chem. Soc.* 118, 11225–36.

Karplus, M., and McCammon, J. A. 2002. Molecular dynamics simulations of biomolecules. *Nat. Struct. Biol.* 9, 646–52.

Kusumi, A., and Suzuki, K. 2005. Toward understanding the dynamics of membrane-raft-based molecular interactions. *Biochim. Biophys. Act.-Mol. Cell Res.* 1746, 234–51.

Levitt, M. 1976. A simplified representation of protein conformations for rapid simulation of protein folding. *J. Mol. Biol.* 104, 59–107.

Levitt, M., and Warshel, A. 1975. Computer simulation of protein folding. *Nature* 253, 694–98.

Lindahl, E., and Edholm, O. 2000. Mesoscopic undulations and thickness fluctuations in lipid bilayers from molecular dynamics simulations. *Biophys. J.* 79, 426–33.

———. 2001. Molecular dynamics simulations of NMR relaxation rates and slow dynamics in lipid bilayers. *J. Chem. Phys.* 115, 4938–50.

Liu, P., Izvekov, S., and Voth, G. A. 2007. Multiscale coarse-graining of monosaccharides. *J. Phys. Chem. B* 111, 11566–75.

Liu, P., and Voth, G. A. 2007. Smart resolution replica exchange: An efficient algorithm for exploring complex energy landscapes. *J. Chem. Phys.* 126, 045106.

Liwo, A., Czaplewski, C., Pillardy, J., and Scheraga, H. A. 2001. Cumulant-based expressions for the multibody terms for the correlation between local and electrostatic interactions in the united-residue force field. *J. Chem. Phys.* 115, 2323–47.

MacKerell, A. D., Bashford, D., Bellott, M. et al. 1998. All-atom empirical potential for molecular modeling and dynamics studies of proteins. *J. Phys. Chem. B* 102, 3586–616.

Marrink, S. J., de Vries, A. H., and Mark, A. E. 2004. Coarse grained model for semiquantitative lipid simulations. *J. Phys. Chem. B* 108, 750–60.

Masuda, M., Takeda, S., Sone, M. et al. 2006. Endophilin BAR domain drives membrane curvature by two newly identified structure-based mechanisms. *EMBO J.*, 25, 2889–97.

Mervis, J. 2001. Advanced computing-NSF launches TeraGrid for academic research. *Science* 293, 1235–37.

Noid, W. G., Chu, J.-W., Ayton, G. et al. 2008a. The multiscale coarse-graining method. I. A rigorous bridge between atomistic and coarse-grained models. *J. Chem. Phys.* (in press).

Noid, W. G., Chu., J. W., Ayton, G. S., and Voth, G. A. 2007. Multiscale coarse-graining and structural correlations: Connections to liquid state theory. *J. Phys. Chem. B* 111, 4116–27.

Noid, W. G., Liu, P., Wang, Y. et al. 2008b. The multiscale coarse-graining method. II. Numerical implementation for coarse-grained molecular models. *J. Chem. Phys.* (in press).

Pandit, S. A., Jakobsson, E., and Scott, H. L. 2004. Simulation of the early stages of nano-domain formation in mixed bilayers of sphingomyelin, cholesterol and dioleylphosphatidylcholine. *Biophys. J.* 87, 3312–22.

Phillips, J. C., Braun, R., Wang, W. et al. 2005. Scalable molecular dynamics with NAMD. *J. Comp. Chem.* 26, 1781–802.

Reed, D. A. 2003. Grids, the TeraGrid, and beyond. *Computer* 36, 62–68.

Sanbonmatsu, K. Y., Joseph, S., and Tung, C. 2005. Simulating movement of tRNA into the ribosome during decoding. *Proc. Natl. Acad. Sci., U.S.A.* 102, 15854–59.

Shi, Q., Izvekov, S., and Voth, G. A. 2006. Mixed atomistic and coarse-grained molecular dynamics: Simulation of a membrane bound ion channel. *J. Phys. Chem. B* 110, 15045–48.

Simons, K., and Toomre, D. 2000. Lipid rafts and signal transduction. *Nat. Rev. Mol. Cell Biol.* 1, 31–39.

Smith, W., and Forester, T. R. 1996. DL_POLY_2.0: A general-purpose parallel molecular dynamics simulation package. *J. Mol. Graph.* 14, 136–41.

Snow, C. D., Nguyen, N., Pande, V. S., and Gruebele, M. 2002. Absolute comparison of simulated and experimental protein-folding dynamics. *Nature* 420, 102–106.

Tama, F., Valle, M., Frank, J., and Brooks, C. L., 3rd. 2003. Dynamic reorganization of the functionally active ribosome explored by normal mode analysis and cryo-electron microscopy. *Proc. Natl. Acad. Sci., U.S.A.* 100, 9319–23.

Tozzini, V. 2005. Coarse-grained models for proteins. *Curr. Opin. Struc. Biol.* 15, 144–50.

Van der Spoel, D., Lindahl, E., Hess, B. et al. 2005. GROMACS: Fast, flexible, and free. *J. Comp. Chem.* 26, 1701–18.

van Gunsteren, W. F., Burgi, R., Peter, C., and Daura, X. 2001. The key to solving the protein-folding problem lies in an accurate description of the denatured state. *Angew. Chem. Int. Ed.* 40, 351–55.

Villa, E., Balaeff, A., Mahadevan, L., and Schulten, K. 2004. Multiscale method for simulating protein–DNA complexes. *Multiscale Model. Simul.* 2, 527–53.

Voth, G. A. 2006. Computer simulation of proton solvation and transport in aqueous and biomolecular systems. *Acc. Chem. Res.* 39, 143–50.

Wang, Y., Izvekov, S., Yan, T., and Voth, G. A. 2006. Multiscale coarse-graining of ionic liquids. *J. Phys. Chem. B* 110, 3564–75.

Wang, Y., and Voth, G. 2005. Unique spatial heterogeneity in ionic liquids. *J. Am. Chem. Soc.* 127, 12192–93.

Warshel, A. 2002. Molecular dynamics simulations of biological reactions. *Acc. Chem. Res.* 35, 385–95.

Wriggers, W. 2004. Spanning the length scales of biomolecular simulation. *Structure* 12, 1–2.

Wu, X., Milne, J. L., Borgnia, M. J. et al. 2003. A core-weighted fitting method for docking atomic structures into low-resolution maps: Application to cryo-electron microscopy. *J. Struct. Biol.* 141, 63–76.

Zhou, J., Thorpe, I. F., Izvekov, S., and Voth, G. A. 2007. Coarse-grained peptide modeling using a systematic multiscale approach. *Biophys. J.* 92, 4289–303.

4 A Model for Lipid Bilayers in Implicit Solvent

Grace Brannigan
Center for Molecular Modeling, University of Pennsylvania

Frank L.H. Brown
Department of Chemistry and Biochemistry and Department of Physics, University of California at Santa Barbara

CONTENTS

4.1 INTRODUCTION

Countless processes in cellular biology depend upon lipid bilayer membranes. In addition to their role in segregating and compartmentalizing subcellular components and organelles, membranes are essential to the production and storage of chemical energy and facilitate biochemical reactions via a vast array of membrane-associated proteins [Lodish et al. 1995]. The structure and functioning of many of these proteins is strongly dependent upon the bilayer environment [Haltia and Freire 1995]. Furthermore, synthetic model membrane systems have found use in varied technological and medical applications [Tien and Ottova-Leitmannova 2003]. It is not surprising that, in addition to many experimental and theoretical biophysical studies [Bloom, Evans, and Mouritsen 1991; Evans and Skalak 1980; Lipowsky and Sackmann 1995; Nagle and Tristram-Nagle 2000], extensive computational resources have been devoted to studying lipid bilayers and related systems.

Atomistic simulations [e.g., Chiu et al. 2002; Feller 2000; Marrink, Berkowitz, and Berendsen 1993; Pastor 1994; Pitman et al. 2005; Smondyrev and Berkowitz 1999; Tobias, Tu, and Klein 1997] can provide insight into the microscopic properties of lipids, cholesterol, transmembrane proteins, etc. within a bilayer environment. Computational limitations, however, severely restrict both the systems that may be successfully studied and questions that may be fully answered by atomic simulations. Mixing of multiple lipid species and/or cholesterol, for example, requires time scales much longer than those currently practical with atomistic simulations. Also, atomistic simulation of a protein in a lipid environment necessarily corresponds to an unnaturally high protein density (due to the periodic boundary conditions and small system sizes). Determining the effective interactions among multiple membrane proteins at biologically reasonable densities is currently out of the question for fully atomic models.

One approach toward overcoming both time scale and length scale issues is the reduction of computational complexity for the molecules to be simulated (coarse-graining). Typically, this means replacing the full set of atoms in the lipids, water, proteins, etc. being studied with a smaller set of effective particles—in essence combining multiple atoms into effective submolecular building blocks. A system with fewer particles can be run for a longer period of time, addressing the time-scale issues. Alternatively, if each lipid is represented by fewer particles than an atomic description, more lipids can be included in simulations, addressing the length-scale issues. For instance, one molecule of the lipid 1-palmitoyl-2-oleoyl-phosphatidylcholine (POPC) contains 134 atoms. Using the guideline that one lipid is fully hydrated by 30 water molecules [Nagle et al. 1996], each atomistic lipid molecule therefore requires simulation of at least 224 atoms. In this chapter we describe a coarse-grained model for lipids that uses five particles or "beads" per lipid, and hydrates the lipids implicitly. We conservatively expect an acceleration of two orders of magnitude with such a simplified representation. (In addition to the naive particle-counting arguments presented above, the time step for coarse-grained models is typically significantly longer than that required for atomistic models, which leads to additional computational savings.)

Approaches to coarse-graining lipids are wide ranging. Some models explicitly simulate solvent (water) while others use implicit solvent. Some models invoke a realistic molecular architecture while others represent a lipid by a single chain of beads. Some models are developed to reproduce general elastic and/or mesoscopic properties of bilayers [Goetz and Lipowsky 1998; Ayton et al. 2001; Rekvig et al. 2003; Drouffe, Maggs, and Leibler 1991; Cooke, Kremer, and Deserno 2005; Farago 2003; Brannigan and Brown 2004], whereas others are calibrated to reproduce more microscopic features, like the area per lipid under vanishing tension [Marrink, de Vries, and Mark 2004; Shelley et al. 2001; Shinoda, Devane, and Klein 2007] or pair distribution functions between various molecular groups [Izvekov and Voth 2005]. While it might be hoped that the reproduction of microscopic behavior will necessarily lead to correct physics at the mesoscale, relatively little work has been done to establish this hypothesis. The delicate interplay between microscopic interactions and the collective behavior they imply in condensed phase systems is poorly understood. Our own approach has been driven by an interest in testing elastic models for bilayer behavior. The model described in this chapter was calibrated to reproduce elastic and mesoscopic bilayer properties with minimal computational effort. Our lipids are easily tuned (by adjusting the few model parameters) to allow for the systematic investigation of phenomena as a function of bilayer elastic properties. One disadvantage of our approach is the rather distant connection between our lipids and the actual molecules that comprise experimental bilayer systems.

Coarse-grained models do not typically provide detailed insight into the hydrophobic effect; they almost always treat water as a single spherical particle without the necessary geometric details to faithfully reproduce water structure. Most coarse-grained models do, however, use explicit solvent as a convenient means to enforce a bilayer stabilizing interfacial tension between solvent and hydrocarbon tails. Controlling the interfacial tension directly, without recourse to explicit solvent, would seem (if possible) a more direct route to the same end. While many implicit solvent models for lipids have been developed [Drouffe, Maggs, and Leibler 1991; Noguchi and Takasu 2001; Wang and Frenkel 2005; Farago 2003; Brannigan and Brown 2004; Cooke, Kremer, and Deserno 2005; Zehl et al. 2006], the model presented in this chapter is unique in that it actually calibrates to this interfacial tension, which for oil and water has been estimated at 20–50 mJ/m^2 [Marsh 1996; Israelachvili 1992].

In this chapter we present a model for lipids that self-assembles and targets continuum elastic properties (the bending modulus and the stretching modulus), along with the stress profile (distribution of stresses throughout the membrane), and the interfacial oil–water tension at the bilayer surface. The model is very efficient (it uses no solvent and only five bead chains per lipid) and is consequently appropriate for simulations requiring large patches of membrane. Because the stress profile for this model approximates fully atomic profiles quite well, it is especially well suited for simulations of molecules (such as coarse-grained proteins) embedded within the membrane.

We present two applications of the model; both are computational tests of analytical theories. The first involves the height and thickness fluctuations of a bilayer; this study functions both as a test of an elasticity theory developed by us and as a means to measure various elastic properties of the model membranes. We find that the present model results in bilayers with reasonable elastic properties; most notably the interfacial tension compares very well with three other atomistic simulations. The second application involves the study of the membrane's response to insertion of transmembrane proteins. We find that the simulated membrane deformation profiles are well described by our analytical model. Implications for membrane-mediated protein aggregation are discussed.

This chapter briefly summarizes the key results obtained in Brannigan, Philips and Brown (2005); Brannigan and Brown (2006) and Brannigan and Brown (2007). Interested readers may consult the original literature for details absent from the discussion below.

4.2 METHODS

A model lipid is a chain comprised of three types of bead: hydrophobic beads, an interface bead, and a hydrophilic bead (Figure 4.1). Hydrophilic beads form the head group region and hydrophobic beads form the tail. Joining hydrophobic regions and hydrophilic regions is an interface bead. The hydrophobic effect (to the extent that we capture it) is concentrated within the interface beads of the system. All beads interact via the following repulsive potential (the repulsive part of a Lennard–Jones interaction)

$$U_{\text{core}}(r) = c_{\text{core}}\left(\frac{\sigma}{r}\right)^{12}, \tag{4.1}$$

where c_{core} is a positive energetic constant, r is the distance between the two bead centers, and $\sigma = 0.75$ nm is the bead diameter. This is the only interaction between hydrophilic–hydrophilic pairs, so that the heads experience a positive pressure when in the bilayer, mimicking the solvation

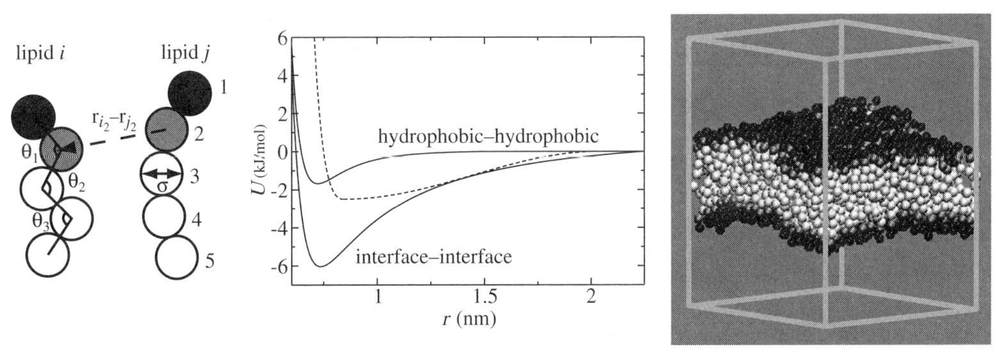

FIGURE 4.1 (Left) Lipids are composed of three bead types: hydrophilic (black), interfacial (gray), and hydrophobic (white). All applications presented in this chapter use the molecular architecture pictured here, but the model is easily extended to multiple hydrophobic chains of any length. (Center) (solid lines) Plot of interaction between two hydrophobic beads, compared to interaction between two interface beads. Hydrophobic-interface is treated as hydrophobic–hydrophobic. (Dashed line) $V_{\text{att}}(r) + V_{\text{rep}}(r)$ from Cooke, Kremer, and Deserno (2005) (assuming $\sigma = 0.75$ nm, $\varepsilon = 2.49$ kJ/mol, $\omega_c = 1.2$ nm). In their model this potential mimics solvent by acting among all tail beads. (Right) Sample conformation of tensionless membrane with 800 lipid molecules ($c_{\text{bend}} = 19.25$ kJ/mol).

pressure of a real bilayer [Israelachvili 1992]. Hydrophobic–hydrophobic and hydrophobic–interface pairs experience the attractive portion of a Lennard–Jones potential as well, to reproduce the van der Waals attractions among carbon groups in a lipid chain [Israelachvili 1992]:

$$U_{\text{tail}}(r) = -c_{\text{tail}}\left(\frac{\sigma}{r}\right)^6, \tag{4.2}$$

where c_{tail} is a positive energetic constant. Finally, the interfacial tension between oil and water is mimicked by a soft attraction among interface beads:

$$U_{\text{int}}(r) = -c_{\text{int}}\left(\frac{\sigma}{r}\right)^2, \tag{4.3}$$

where c_{int} is a positive energetic constant. All three potentials are cut off (and shifted to avoid discontinuities): U_{core} and U_{tail} are cut off at 2σ (15 Å) and U_{int} is cutoff at 3σ (22.5 Å). Truncation of U_{int} is not an approximation, but rather an intentional aspect of the model. The interactions between all possible bead pairs are summarized in Table 4.1.

Similar forms of U_{tail} and U_{core} can be found in most coarse-grained models for lipids. U_{int} is unusual, without precedent in lipid simulations. It is known that Lennard–Jones interactions alone cannot stabilize a fluid membrane in the absence of solvent: the membrane crystallizes at low temperatures and sublimes upon heating [Drouffe, Maggs, and Leibler 1991; Cooke, Kremer, and Deserno 2005]. Existing implicit solvent models all differ in their approach to circumvent this problem. For models based entirely around pair-potentials, it is empirically clear [Farago 2003; Brannigan and Brown 2004; Brannigan, Phillips, and Brown 2005; Cooke, Kremer, and Deserno 2005] that a potential must be introduced that is somewhat softer and longer ranged than the standard 6–12 interactions. Without any fundamental justification for the functional form chosen, we rationalize U_{int} by the fact that it yields stable fluid membranes with the proper continuum properties, particularly the interfacial tension. Furthermore, although we don't know what functional form the "hydrophobic" potential should take in an implicit solvent description, we do know where it should be acting: at the interface between hydrophobic and hydrophilic particles. Our model concentrates all nonstandard interbead potentials (U_{int}) to the interface region.

TABLE 4.1
Interaction Potentials for Various Bead Pairs

	Hydrophilic	Interfacial	Hydrophobic
Hydrophilic	$c_{\text{core}}\left(\frac{\sigma}{r}\right)^{12}$	$c_{\text{core}}\left(\frac{\sigma}{r}\right)^{12}$	$c_{\text{core}}\left(\frac{\sigma}{r}\right)^{12}$
Interfacial		$c_{\text{core}}\left(\frac{\sigma}{r}\right)^{12} - c_{\text{int}}\left(\frac{\sigma}{r}\right)^{2}$	$c_{\text{core}}\left(\frac{\sigma}{r}\right)^{12} - c_{\text{tail}}\left(\frac{\sigma}{r}\right)^{6}$
Hydrophobic			$c_{\text{core}}\left(\frac{\sigma}{r}\right)^{12} - c_{\text{tail}}\left(\frac{\sigma}{r}\right)^{6}$

These potentials do not apply to intramolecular pairs separated by less than three bonds, which interact solely through U_{bend}. Parameters used in the presented applications are: $c_{\text{core}} = 1.1$ kJ/mol, $c_{\text{tail}} = 2.75$ kJ/mol, $c_{\text{int}} = 8.25$ kJ/mol, $\sigma = 0.75$ nm.

All results in this chapter were obtained using the following values for the coefficients: $c_{core} = 1.1$ kJ/mol, $c_{tail} = 2.75$ kJ/mol, and $c_{int} = 8.25$ kJ/mol. The coefficients for the Lennard–Jones terms (c_{core} and c_{tail}) result in a quantitatively similar potential to that introduced by Goetz and Lipowsky (1998) in the simulation of *solvated* lipids with similar architecture to our own. The coefficient chosen for the interfacial potential (c_{int}) results in stable fluid membranes with interfacial tensions that correspond to theoretical predictions. A plot of the hydrophobic–hydrophobic (or hydrophobic–interface) interaction compared to the interface–interface interaction is shown in Figure 4.1. The coefficients are chosen such that the minima of each potential fall within an angstrom of each other, which has implications for bilayer stability (the molecule is shaped like a cylinder as opposed to a cone). The interface–interface interaction is, however, three times as deep as the hydrophobic–hydrophobic interaction and falls off much more slowly. Also shown is the hydrophobic and Lennard–Jones potential used by Cooke, Kremer, and Deserno (2005); in their model these potentials act between all tail beads and there is no interface bead. The two potentials are very similar for medium to long distances between beads, but our potential is much stronger at short range and overall the potential is much wider—concentrating the stabilizing effect of the soft potential into a single bead at the interface between hydrophobic tails and "water."

The chapters of this book provide numerous methods for coarse-graining the internal degrees of freedom of molecules. Our approach is relatively crude, due to the fact that we target continuum properties of the membrane and not microscopic realism. For the studies presented in this chapter, we use three hydrophobic beads, with each bead representing about four to five carbon groups. Bonds are constrained at 1σ (0.75 nm), and there is no cost for dihedral rotations. Chain flexibility is governed purely through a bond angle energy,

$$U_{bend} = c_{bend} \cos\theta, \tag{4.4}$$

where θ is one of three bond angles on the molecule and c_{bend} is a positive energetic constant. Our model is therefore similar to that developed by Goetz and Lipowsky (1998), and slightly more complicated than three-bead models such as those developed by Noguchi and Takasu (2001), Wang and Frenkel (2005), Farago (2003), Cooke, Kremer, and Deserno (2005), and Zehl et al. (2006).

We ran small simulations ($N = 128$ lipids) with $c_{bend} = 19.25$ kJ/mol to verify self-assembly (Figure 4.2). These simulations used Metropolis Monte Carlo in the constant area per molecule ($\Sigma = 0.62\ \text{nm}^2$), constant volume ($V/N = 6.1\ \text{nm}^3$), and constant temperature ($T = 298$ K) ensemble, with periodic boundary conditions in all directions. We started with many different random configurations of molecules, and found that all systems rapidly aggregated. In addition, a few systems aggregated parallel to the *xy* plane, leading to an intact bilayer spanning the periodic boundaries of the simulation box.

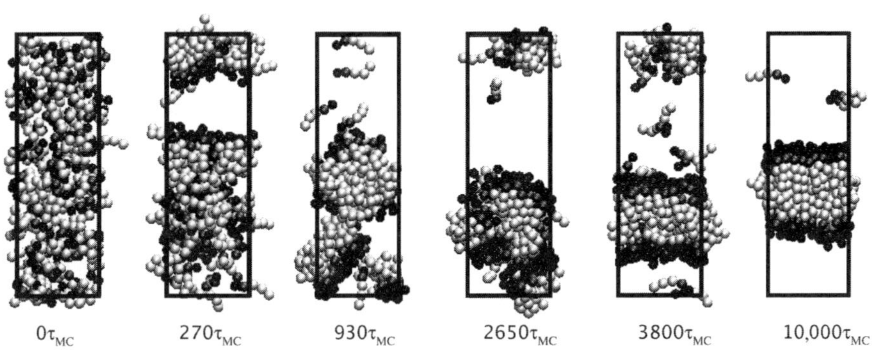

FIGURE 4.2 Self-assembly of a bilayer patch of 128 lipids ($c_{bend} = 19.25$ kJ/mol) in a box with constant dimensions $L_x = L_y = 6.45$ nm, $L_z = 18.75$ nm and periodic boundary conditions. The chosen area corresponds to that assumed by a preassembled bilayer at zero tension. Each Monte Carlo time unit τ_{MC} includes (on average) an attempt to translate each bead in the simulation 100 times.

Using preassembled bilayers with $N = 800$ lipids, we then turned to constant zero tension simulations to measure equilibrium properties. The constant tension, constant volume ensemble was maintained by randomly attempting a move that shrank the box area and extended the box height, or vice versa [Brannigan and Brown 2004; Venturoli and Smit 1999]. Only interface beads were rescaled with the box; all other beads on a given molecule were translated so that intramolecular distances did not change and the bond constraints weren't violated. All of the following simulations in this chapter use the constant zero tension ensemble (although those in the applications section use a lower temperature, 281 K, to speed up equilibration).

This model has a few parameters that, combined with molecular shape, can be used to tune the model. For simplicity we concentrated on varying the molecular bending coefficient, c_{bend}, as a means to tune behavior. It was found that bilayers are stable over a wide range of c_{bend} values and molecular flexibility is strongly correlated to the area per molecule, an important quantity for comparison to experimental systems. Furthermore, cholesterol is significantly stiffer than phospholipids. Increasing c_{bend} can be viewed as an approximate way to mimic the addition of cholesterol within our coarse-grained representation.

The zero tension area per molecule was tuned through the molecular bending rigidity, c_{bend}: stiffer chains lead to smaller areas per molecule. At $T = 298$ K, the area per molecule ranges from 0.57 to 0.68 nm^2 as c_{bend} is tuned from 27.5 to 13.75 kJ/mol. In comparison, zero tension areas for model lipids range from about 0.596 nm^2 for dimyristoylphosphatidylcholine (DMPC) to 0.725 nm^2 for dioleolphosphatidylcholine (DOPC) [Nagle and Tristram-Nagle, 2000]. This agreement in magnitude is unremarkable because we are free to set σ to whatever value we choose. However, the range of lipids that can be represented is a positive attribute of this model.

A linearly elastic sheet can be described by

$$\gamma = \frac{k_A (L^2 - A_0)}{A_0}, \tag{4.5}$$

where γ is the surface tension, k_A is the sheet's area compressibility modulus, L^2 is the total projected area of the sheet, and A_0 is the area of the sheet at zero tension. k_A can be measured by several different methods; one can apply various tensions and measure the resulting areas, or conversely hold the box constant at several areas and measure the resulting tensions [Goetz and Lipowsky 1998]. Additionally, one can use one simulation at zero applied thermodynamic tension, and correlate fluctuations in box area with fluctuations in mechanical tension [Brannigan, Phillips, and Brown 2005]. Alternatively one can measure the area fluctuations in one simulation at zero applied tension [Feller and Pastor 1999] and apply the formula

$$k_A = \frac{k_B T \langle L^2 \rangle}{\langle (L^2)^2 \rangle - \langle (L^2) \rangle^2}; \tag{4.6}$$

this is perhaps the most common method applied to date. Finally, as discussed in the Applications section, k_A can hypothetically be determined from the thickness fluctuation spectrum, but there is only fair agreement between values measured this way and measured via the area fluctuations. We find that all other methods yield comparable values of k_A for the coarse-grained model presented in this chapter. For values of c_{bend} ranging from 13.5 to 27.5 kJ/mol (and $T = 298$ K) we find that $k_A = 40 - 224$ mJ/m^2. This is in good agreement with single-component phospholipid bilayers, for which k_A typically ranges from 60 to 270 mJ/m^2 [Rawicz et al. 2000; Sackmann 1995].

In this chapter we present two methods for estimating the interfacial tension between water and hydrophobic lipid tails, Γ. Since all the interactions between oil and water are incorporated within

U_{int}, we define the effective interfacial tension by the interfacial contribution to the virial tension [Frenkel and Smit 2002]:

$$\Gamma = \frac{1}{2}\sum_{i<j}\left\langle \frac{r_{ij}^2 - 3z_{ij}^2}{2L^2 r_{ij}} \frac{\partial U_{\text{int}}(r_{ij})}{\partial r_{ij}} \right\rangle, \tag{4.7}$$

where the sum is over all distinct pairs of interface beads. The factor of ½ originates from the two interfaces found in a bilayer membrane. Figure 4.3 displays measurements of Γ for systems with 800 lipids. For various values of c_{bend} (and hence various values of the area per molecule), Γ ranges from 35 to 70 mJ/m². Theoretical estimates of Γ [Marsh 1996; Israelachvili 1992] range from 20 to 50 mJ/m², so the chosen form for U_{int} yields reasonable numbers for this important physical constant. A second method for measuring the interfacial tension, through the fluctuation spectrum, is discussed in the Applications section.

The average surface tension for the entire bilayer (not just the interface with water) vanishes in a zero tension simulation, but the microscopic surface tension density varies as you move across the membrane in the direction of the bilayer normal. The surface tension density at a given height z in the bilayer is the difference between normal and tangential pressures:

$$\gamma(z) = P_{\text{n}}(z) - P_{\text{t}}(z), \tag{4.8}$$

where $\gamma(z)$ is the surface tension density, P_{n} is the normal pressure, and P_{t} is the tangential pressure. The integral of $\gamma(z)$ over the height of the simulation box is the surface tension of the bilayer, γ. (The expression $\gamma(z)$ represents the opposite of what is typically called the "lateral pressure profile" in the literature.)

Measurements of stress profiles (the surface tension density as it varies throughout the bilayer normal) have become increasingly commonplace [Goetz and Lipowsky 1998; Lindahl and Edholm 2000b; Gullingsrud and Schulten 2004; Carillo-Tripp and Feller 2005; Patra 2005], since the stress

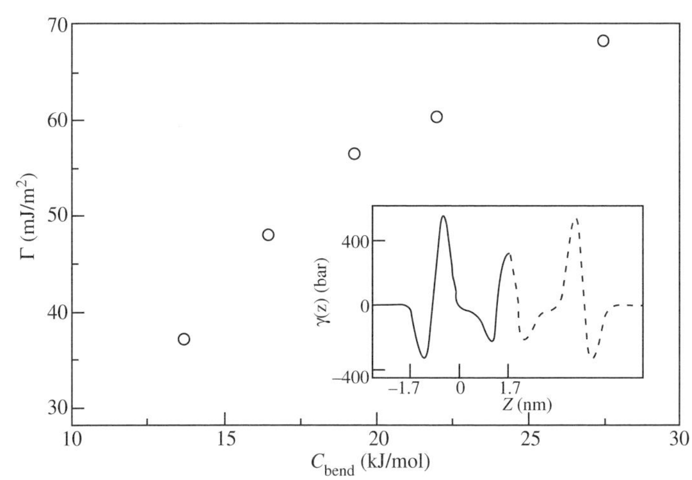

FIGURE 4.3 Interfacial tension as defined in Equation 4.7, as a function of c_{bend}. Inset: Stress profile for systems with 128 molecules showing the same pattern of peaks and valleys as observed in a similar solvated model [Goetz and Lipowsky 1998]. Profiles correspond to systems with $c_{\text{bend}} = 19.25$ kJ/mol. The x axis on the stress profile plot indicates how to take moments of the stress profile: the origin is placed at the neutral surface (point where stress vanishes) and only one half of the bilayer profile is used.

profile provides a convenient continuum description of membrane heterogeneity in the direction normal to the bilayer surface. The stress profile typically contains a valley at the headgroups due to the hydration pressure of the heads (and in some cases electrostatic repulsion), a peak at the oil–water interface due to the oil water interfacial tension, another broad valley over the tail groups due to the positive pressure of the hydrophobic core, and (sometimes) a central peak where the two leaflets meet. The value of Γ essentially amounts to the integral underneath the outer peaks. Changes in stress profile have been proposed as mechanisms of biological action for small molecules such as anesthetics [Cantor 1997] and alcohols [van den Brink-van der Laan et al. 2004] as well as influential lipids such as polyunsaturated fatty acids [Carillo-Tripp and Feller 2005] and cholesterol [Patra 2005]. In our model, the placement of the beads experiencing the pseudohydrophobic force (U_{int}) is strategic: by concentrating this force in the interface beads we reproduce the pattern of peaks and valleys seen in solvated and atomistic models. The inset of Figure 4.3 shows the stress profile for a system with 128 lipids. The stress profile was measured as in Goetz and Lipowsky's (1998) study; it is important to measure it in a small system because of the smoothing effect caused by undulations in larger systems. The stress profile shown in the inset of Figure 4.3 has peaks ranging over a magnitude of about 400 bar, which is equivalent to that measured by atomistic simulations [Lindahl and Edholm 2000b; Gullingsrud and Schulten 2004].

The model can be simply extended to rigid cylindrical inclusions to represent proteins (Figure 4.5). The inclusions are rigid assemblies of hydrophobic, interfacial, and hydrophilic beads, with the same energetics that govern those particles in model lipids. The use of the interface bead is appropriate here because many proteins have aromatic residues that specifically prefer the interface [de Planque and Killian 2003]. Protein radius R is varied through the number of concentric rings of beads that comprise the protein, and protein half-thickness, D, is determined by the number of layers of hydrophobic beads. Proteins were not allowed to tilt in the studies presented here. We stress that this is not a proper coarse-grained model for any particular protein, and that it has more similarity to the models for carbon nanotubes studied by Nielsen et al. (2005). The model is designed for the study of *membrane* behavior in the presence of rigid protein inclusions—the idealized cylindrical geometry is simple to analyze and interpret and was chosen for these reasons.

4.3 APPLICATIONS

The model is especially useful for situations in which large membrane patches are required; the use of implicit solvent and a simple molecular architecture enables the simulation of large numbers of lipid molecules at modest computational expense. As noted above, our picture is appropriate for studies involving insertion of amphipathic bodies (such as proteins) into the membrane, because it reproduces the stress profile of actual bilayers. Here we present two applications: fluctuations of a homogeneous membrane and membrane deformations around a cylindrical protein inclusion. The study involving the homogeneous membrane is compared with similar studies using atomistic simulations, while the heterogeneous study would be prohibitively expensive for atomistic simulations.

4.3.1 Fluctuations of Homogeneous Membranes

At physiological temperatures lipid bilayers exhibit fluctuations in shape (height) and thickness. Such fluctuations are associated with numerous biophysical phenomena including: steric repulsions between proximal bilayers [Israelachvili and Wennerstrom 1992; Lipowsky 1991], shape fluctuations of the red blood cell [Brochard and Lennon 1975], cellular motility [Peskin, Odell, and Oster 1993], and entropically driven interactions between integral membrane proteins [Bruinsma and Pincus 1996; Golestanian, Goulian, and Kardar 1996]. Fluctuations of bilayer systems have been investigated over the past 10 years, using both coarse-grained and atomistic models. Recently, we suggested a theoretical formalism for analyzing these fluctuations [Brannigan and Brown 2006].

This section compares our theory with our coarse-grained simulation data and existing data from three atomistic systems. We find that our coarse-grained model behaves qualitatively identically to the fully atomic systems. Quantitative analysis suggests our model's elastic parameters fall within the range seen in real bilayer systems, indicating our coarse-grained model provides a faithful generic representation of lipid bilayers at the elastic/mesoscopic level.

At the continuum level and assuming small fluctuations from a flat reference state, bilayer shape is described by a height field, $h(\mathbf{r})$, which represents the local deviation of the bilayer midplane from the global average, and a thickness field, $t(\mathbf{r})$, which represents the local deviation of the bilayer half-thickness from its average value t_0. $\mathbf{r}$ represents the xy position on the reference plane.

Defining the Fourier transform pair:

$$g_{\mathbf{q}} = \frac{1}{L}\int d\mathbf{r} g(\mathbf{r}) e^{-i\mathbf{q}\cdot\mathbf{r}}, \tag{4.9}$$

$$g(\mathbf{r}) = \frac{1}{L}\sum_{q} g_{\mathbf{q}} e^{i\mathbf{q}\cdot\mathbf{r}} \tag{4.10}$$

for an arbitrary function g, we expect [Brannigan and Brown 2006], for a membrane under zero tension, the following spectrum of fluctuations in height and thickness

$$\left\langle \left| h_{\mathbf{q}} \right|^2 \right\rangle = \frac{k_B T}{k_c q^4} + \frac{k_B T}{2(k_\lambda + \gamma_\lambda q^2)}, \tag{4.11}$$

$$\left\langle \left| t_{\mathbf{q}} \right|^2 \right\rangle = \frac{k_B T}{k_c q^4 - 4k_c \zeta q^2 / t_0 + k_A / t_0^2} + \frac{k_B T}{2(k_\lambda + \gamma_\lambda q^2)}, \tag{4.12}$$

where k_c is the bending rigidity of the bilayer, c_0 is the spontaneous curvature of each monolayer, $\zeta \equiv c_0 - c_{0_\Sigma}\Sigma_0$, and c_{0_Σ} is the area derivative of the spontaneous curvature, evaluated at the zero tension area, Σ_0. Also, k_λ is a harmonic binding constant and γ_λ is the renormalized interfacial tension. These expressions represent height and thickness fluctuations over all wavelengths (the first term in each expression represents long wavelength fluctuations, while the second term in each expression represents short wavelength "protrusion" fluctuations). The short wavelength portion and the long wavelength portion are separated into two pieces because the total observable ($\langle |h_q|^2 \rangle$ or $\langle |t_q|^2 \rangle$) is composed of two independent contributions, a long wavelength bending field and a short wavelength protrusion field. These two fields are uncoupled in this description [although they do not have to be, see Brannigan and Brown, 2006], and so $\langle |h_q|^2 \rangle$, for instance, is simply composed of the variance of its two independent contributions.

In Figure 4.4, we present height and thickness fluctuation spectra for three fully atomic systems (DPPC, GMO, and SM) and our own coarse-grained model as presented in this chapter. Details on how these fluctuation spectra were obtained are available in the original articles [Lindahl and Edholm 2000a; Marrink and Mark 2001; Chiu et al. 2003], as well as outlined in Brannigan and Brown's study (2006). Equation 4.11 and Equation 4.12 describe all four data sets very well. The parameters used fitting the data in Figure 4.4 are presented in Table 4.2.

The bending modulus k_c multiplies a q^4 term in both the height and thickness fluctuations. This is the standard bending term that originates from Helfrich energetics [Helfrich 1973; Safran 1994]; it is energetically unfavorable to bend a symmetric bilayer away from a flat state or to bend a monolayer away from its state of preferred intrinsic curvature. k_c values span an order of magnitude for the atomistic systems. It is imperative to be able to access bending rigidities in the order of those seen in

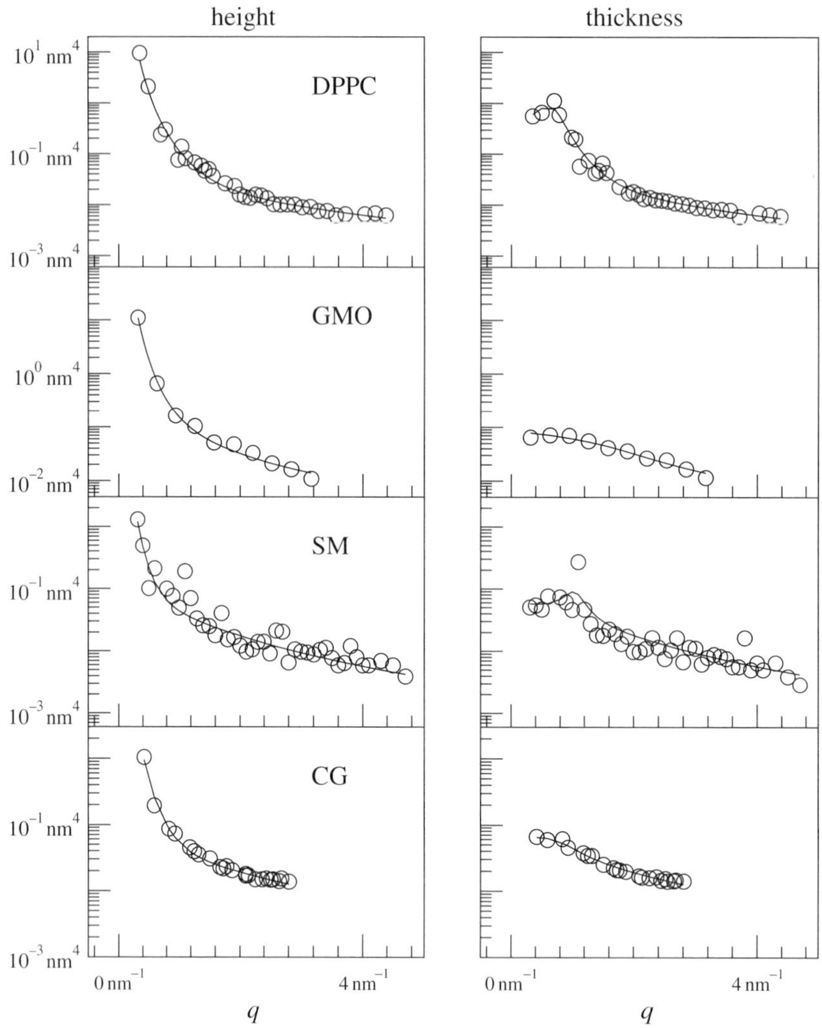

FIGURE 4.4 Height ($\langle|h_q|^2\rangle$) and thickness ($\langle|t_q|^2\rangle$) fluctuation spectra for DPPC [Lindahl and Edholm 2000a], GMO [Marrink and Mark 2001], SM [Chiu et al. 2003], and our coarse-grained model. Simulation data are displayed as circles. The lines represent best fits of the data to Equation 4.11 and Equation 4.12. Parameters used in the fits are displayed in Table 4.2.

TABLE 4.2
Material Properties of DPPC, GMO, SM, and CG Bilayers, as Measured from the Height and Thickness Fluctuation Spectra and Equation 4.11 and Equation 4.12

System	k_λ (10^{-20} J/nm^4)	γ_λ (mJ/m^2)	k_c (10^{-20} J)	k_A (mJ/m^2)	ζ (1/nm)
DPPC	14	15	4.3	36	0.32
GMO	2.9	13	3.8	920	1.1
SM	4.5	21	41	3300	1.3
CG	6.3	13	14	690	0.20

In particular, k_A refers to the area compressibility modulus as measured by the thickness fluctuation spectrum, which does not always agree with k_A measured from the area fluctuations.

phospholipid bilayers, because the bending rigidity centers prominently in all phenomenon related to bilayer fluctuations, as well as those involving mechanical bending deformation of the bilayer. Since our model is generally intended to be applicable to phospholipids ($k_c \sim 4 \times 10^{-20}$ J), we might consider that the value presented for k_c in the CG model ($k_c = 14 \times 10^{-20}$ J) is somewhat high. It is not, however, outside the range of values for phospholipid bilayers measured experimentally (DMPC at higher temperatures, for instance, has a comparable bending rigidity [Seifert and Lipowsky 1995]) and, more importantly, it does not represent the lower limit of what is feasible with this model. As described in Brannigan, Phillips, and Brown's study (2005), increasing the temperature and lowering c_{bend} can lower the bending rigidity to 10^{-20} J. The study described here uses a lower temperature because bilayers with higher bending moduli equilibrate and sample more quickly due to their diminished mesoscale fluctuations.

Surface tensions are expected to accompany q^2 terms [Safran 1994], so the term featuring ζ (which is usually positive) is acting like a negative surface tension and encouraging thickness waves. This term stems from the spontaneous curvature of each leaflet, and the corresponding tendency of the membrane to make "hourglass" shapes. Since each outward bulge (which is favorable for positive spontaneous curvature) is countered by an inward depression (which is unfavorable for positive spontaneous curvature), the overall free energy for an hourglass conformation might be expected to vanish. However, due to lipid volume conservation, molecules in outward bulges have a lower area per molecule, and there are consequently more molecules comprising an outward bulge than an inward depression. This imbalance leads to a net favoring of the hourglass conformation; the conformation is most favorable at wavelengths that correspond to the renormalized spontaneous radius of curvature of the leaflet, resulting in a peak in the thickness fluctuation spectrum. This peak is especially prominent in the DPPC and SM data sets, and was previously attributed to poor sampling [Lindahl and Edholm 2000a]. In the CG data the measured value of ζ corresponds to a peak at a wavelength of 15 nm, or the longest wavelength measured. Consequently there is no nonmonotonic behavior observed for the CG data. This suggests that the CG molecules are slightly more cylindrical than DPPC or SM.

We see that at the longest wavelengths, the thickness fluctuation approaches the constant $k_BT/(k_A/t^2_0)$. This term originates from the harmonic binding of the two leaflets, so that a thickness fluctuation (unlike a height fluctuation) is constrained at the longest wavelengths. Simple solvent-free coarse-grained models for lipids have traditionally been more successful at targeting the area compressibility modulus k_A than the bending rigidity k_c; rigid molecular models such as those of Farago (2003) or Brannigan and Brown (2004) achieve a k_A value comparable to that of phospholipid bilayers without additives, but the bending rigidities of these models correspond to phospholipid bilayers with at least 30% added cholesterol. k_A is central to theories involving area or thickness deformations, and several theories [Evans 1974; Rawicz et al. 2000; Brannigan and Brown 2005] suggest that k_c is derived from k_A (which would imply that targeting k_A is necessary for targeting k_c). It is worth observing that k_A, as measured from the CG model (Table 4.1), corresponds to an average area fluctuation of 10%, so these membranes are not incompressible. k_A ranges over two orders of magnitude for the four systems. We note, however, that k_A measured from the spectrum varies somewhat from k_A measured in a more traditional fashion, from area fluctuations (for the CG model, $k_A = 180$ mJ/m^2). Nonetheless, a similar range is seen among values extracted from area fluctuations, and the differences between the two methods of measurement are not systematic across the four types of lipid.

The short wavelength portions of the fluctuation spectra contain two terms already used in the long wavelength portion of the thickness spectrum: a surface tension term and a constant. In this case the origin of the surface tension term is much more transparent: it stems from the oil–water interfacial tension. γ_λ is the renormalized [Lipowsky and Grotehans 1994; Goetz, Gompper, and Lipowsky 1999] interfacial tension. Because protrusions change the area of the membrane, they experience a modified interfacial tension: $\gamma_\lambda = \pi c_{pr} \Sigma_0 \Gamma^2 / k_B T$, where Lipowsky and Grotehans estimate $c_{pr} = 0.067$. This yields a range of $\Gamma = 11–18$ mJ/m^2 for the four systems examined.

Our model's interfacial tension falls within the range of that experienced by atomistic systems (but below our earlier "microscopic" estimate for Γ using the interfacial contribution to the Virial tension). We also note that, because γ_λ varies so little from system to system, it is an important quantity to consider when developing a generic model for lipids. Finally, the present range in γ_λ over the four models is much narrower than that resulting from earlier methods of interpreting these spectra, and is consequently more consistent with our physical intuition that the interfacial tension between hydrophobic tails and water should vary only minimally from one lipid bilayer to another.

The remaining constant (k_λ) is not measured by experiments, and consequently it is difficult to infer what a "good" value is, and how much range is acceptable. It is even more difficult given that k_λ stems from a somewhat abstract concept (it binds the protrusions to the longer wavelength modes). We can easily say, however, that it is a microscopic quantity, and that therefore some variation among different species of lipid is to be expected. We have obtained a reasonable number for it in our coarse-grained model (as compared to the range of DPPC, GMO, and SM).

4.3.2 Deformations of Membranes around Rigid Proteins

The hydrophobic region of a transmembrane protein need not naturally span the same thickness as the oily core of the membrane. When two components of a membrane naturally span hydrophobic regions of a different size, they are said to be "hydrophobically mismatched." When the hydrophobic region of the reference molecule (usually the protein) is shorter than the average thickness of the surrounding membrane, the molecule is "negatively" mismatched; the reverse scenario is called "positive mismatch." The large interfacial tension between water and hydrophobic residues or hydrocarbon groups causes the membrane to deform, shielding the hydrophobic regions from water [Mouritsen and Bloom 1984; de Planque and Killian 2003]. The resulting shape or "deformation profile" of the bilayer around the perturbation (the membrane thickness as a function of distance r from the protein center) depends upon the shape of the perturbation (i.e., the protein) and the physical properties of the membrane. Interestingly, though interfacial tension drives the membrane to deform to match the protein, it is largely inconsequential for determining the shape of the deformation profile [Huang 1986].

Some interesting questions regarding hydrophobic mismatch that are addressed by this section are:

First, are membrane deformation profiles influenced by bending energetics? If so, we expect possible nonmonotonic behavior in the deformation profile as predicted by Huang (1986). A sufficiently strong aversion to bending can cause the membrane to overshoot its preferred homogeneous equilibrium thickness before returning to normalcy at large distances from the perturbation. Nonmonotonic profiles have been observed in coarse-grained simulations [Venturoli, Smit, and Sperotto 2005; Nielsen et al. 2005; Brannigan and Brown 2006] but are generally not detectable in atomistic simulations [Jensen and Mouritsen 2004], which employ too small a simulation box to observe such phenomena.

Second, do the same parameters implicated in long wavelength fluctuations determine deformation profiles around a mismatched protein? In other words, does k_c measured from the previous section dictate the cost of bending around a protein? Some studies [Harroun et al. 1999] have proposed that microscopic elastic constants, distinct from the macroscopic parameters used in interpreting bilayer fluctuations, are required to describe protein-induced deformations.

Third, what implications do elastic membrane models have for the process of protein aggregation? We show below that the elastic model discussed in the previous section (with minor alterations) is very successful in predicting membrane behavior around a single transmembrane protein. It is natural to ask how this same elastic medium will respond to the presence of multiple proteins and, in particular, how elastic properties of the membrane influence the effective attractions and repulsions between embedded proteins.

To investigate these questions we ran several simulations of proteins embedded in membranes (different protein radii and thickness were considered). Two are shown in Figure 4.5, at both a

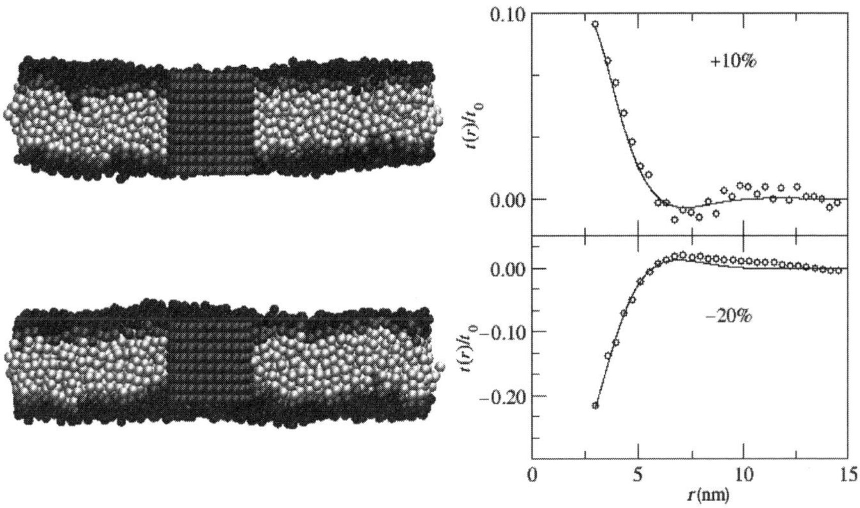

FIGURE 4.5 (Left) Hydrophobic mismatch induced by a protein that is shorter than the membrane is thick (bottom) and by a protein that is longer than the membrane is thick (top). The proteins are composed of double layers of rigidified chains of model lipids; the bottom protein has two hydrophobic beads per chain while the top protein has three hydrophobic beads. Bottom case corresponds to −20% mismatch, while top case corresponds to + 10% mismatch; due to the smaller magnitude of the mismatch the effects of the protein are more subtle in the top case. When the thickness of the membrane is measured quantitatively (see right panel), it is clear that membrane deformation is occurring in both cases. (Right) Thickness deformation profiles for membranes with a rigid protein centered around $r = 0$. Percentage refers to the degree of mismatch between the protein and the membrane. Circles are data points, lines are predictions (not fits) from the theory of Brannigan and Brown (2007). Nonmonotonic behavior has been observed in other studies [Nielsen et al. 2005; Venturoli, Smit, and Sperotto 2005], and is thought to be a direct result of including the bending contribution in the Hamiltonian.

positive ($\sim +10\%$) and a negative ($\sim -20\%$) mismatch. There were approximately 3200 lipids, depending on the size of the protein, so that the average dimensions of the simulation box were about 30×30 nm. Simulations were run in the constant zero tension, constant volume, constant temperature ensemble, with $T = 281$ K. The thickness deformation profiles for $R = 3.0$ nm (both positively and negatively mismatched) are shown in Figure 4.5. The nonmonotonic behavior is clearly visible in both the positively and negatively mismatched profiles, indicating that bending energetics are influencing the profile. The role of bending deformations is significant with regard to protein dimerization, which can in theory reduce the cost associated with hydrophobic mismatch. The sharp transition region that occurs between two proteins that are just beginning to overlap deformation profiles is highly unfavorable in Helfrich energetics, leading to a barrier to dimerization.

The solid lines in both figures correspond to predictions generated using an analytical theory consistent with that presented for the fluctuation spectrum. We consider that the protein causes a deformation in the bending field rather than in the protrusion field. Since these two fields are uncoupled, the protrusion field does not enter into the average deformation profile. This choice is consistent with a viewpoint that protrusions represent nonspecific microscopic noise, rather than particular deformations that can be induced by perturbing the membrane.

Two modifications must be made to the fluctuation theory when predicting the average profile induced by a protein. The first results from the fact that the protein represents a rip in the membrane, and the membrane is no longer a closed surface. Therefore two terms (the spontaneous curvature and Gaussian curvature terms) no longer vanish. The second stems from our observation that lipid volume is not always conserved in the vicinity of a protein, and we introduce a new parameter, $\eta = c_0 - c_{0_\Sigma}\Sigma_0 - c_{0_v}v_0$, where c_{0_v} is the volume derivative of the spontaneous curvature.

For our prediction we therefore require k_c, k_A, k_G, ζ, c_0, and η. The parameters k_c, k_A, and ζ are available from the fluctuation spectrum of homogeneous systems (Table 4.2). c_0 and k_G can both be derived from the stress profile (Figure 4.3, inset) using the equations [Safran 1994]

$$k_c c_0 = -\int \gamma(z) z dz, \tag{4.13}$$

$$k_G = 2\int \gamma(z) z^2 dz, \tag{4.14}$$

where the moments are taken over one leaflet centered on the neutral surface (surface where the tension vanishes), as indicated in Figure 4.3. From these measurements we obtained $c_0 = 0.098\ \text{nm}^{-1}$ and $k_G = -0.47 k_c$. This value for c_0 suggests that the leaflets prefer to have a radius of curvature of about 10 nm, which is both significantly greater than molecular length scales (so the lipids will not be forming micelles) and comparable to mesoscopic length scales (so the bulging effect will be visible in our simulations). The measurement for k_G represents one of the first using simulations, and corresponds well to theories predicting similar scaling for k_G and k_c [Safran 1994].

The value of η (and k_G as well) can be determined from the shape of the membrane right at the boundary between the protein and the membrane. This method involves measuring the first and second derivatives of the membrane as it approaches the protein, for proteins of several different radii. The theoretical foundations are beyond the scope of this chapter, but are detailed in Brannigan and Brown's study (2007). The measurement of η resulted in $\eta = -1.1\ \text{nm}^{-1}$. We found that k_G measured using this method compared very favorably with k_G measured using the stress profile: both methods gave $k_G \sim -k_c/2$.

Using this set of parameters we were able to predict the deformation profiles around 14 proteins, both positively mismatched and negatively mismatched, each over seven protein radii. This was done by minimizing the free energy for each protein using an Euler–Lagrange equation, and solving the resulting fourth-order differential equation. Only the boundary conditions (thickness and volume deformation at the boundary) were allowed to change from protein to protein.

Aranda-Espinoza et al. (1996) propose a phenomenon closely related to the bulging effect discussed in the previous sections. Recall that that effect dealt with thickness deformations in an hourglass shape; the membrane prefers the regions with positive curvature (outward bulge) and has a disinclination for the regions with negative curvature (inward bulge). Once two proteins are in the membrane, however, a region of positive curvature is not necessarily matched by a region of negative curvature: if the proteins are spaced the proper distance apart, the membrane could assume the positive spontaneous curvature in between the two proteins, and be flat everywhere else. According to Aranda-Espinoza et al.'s theory, for membranes where each leaflet has a positive spontaneous curvature, negatively mismatched proteins would prefer a *finite* spacing, such that they could have regions of positive spontaneous curvature between the proteins. This scenario, however, is not consistent with available experimental data. In Brannigan and Brown's study (2007), the parameters extracted from the systems with one protein were used to predict the potential energy between two proteins as a function of distance; this was done by elastic simulation of the Hamiltonian on a lattice. The elastic sheet, with two deformations representing proteins, was allowed to relax under the Hamiltonian. The resulting average energy was recorded for a range of protein separations, giving us an estimate for the membrane-mediated interaction energy between two cylindrical proteins. As predicted, we observed a kinetic barrier to dimerization at about a 4 nm separation, for small proteins with a radius of 0.75 nm. Moreover, we found that the minimum in the interaction energy was at finite spacing (6 nm) if no Gaussian curvature was included, and was at the minimum spacing if Gaussian curvature was included (Aranda-Espinoza et al. neglected Gaussian curvature). This suggests that Gaussian curvature energy may drive dimerization.

Our results provide insights into the three questions outlined at the beginning of this section. We have significant evidence that the nonmonotonic behavior of the membrane is a real effect, and that therefore bending energetics partly determine the shape of the deformation profile. This result has the further ramification that two proteins likely experience a kinetic barrier when dimerizing. We can predict these deformation profiles using the long-wavelength elastic constants extracted from the membrane, providing evidence that the elastic constants determined over mesoscopic length scales can be meaningfully applied to determine bilayer behavior over microscopic distances. Finally, the Gaussian curvature modulus for monolayers was found to be nonvanishing, with the further implication that dimerization is most favorable for two proteins with large mismatches.

4.4 CONCLUDING REMARKS

The interfacial tension between oil and water drives bilayer formation. Furthermore, it strongly influences the interactions between the bilayer and other amphipathic molecules (such as embedded proteins). Remarkably, few coarse-grained models calibrate the interfacial tension to reality (or even explicitly consider its existence).

In this chapter we have presented a model that aims to reproduce some aspects of the interfacial tension between oil and water. Unlike models that consider only two species of particle (hydrophobic and polar), our model acknowledges that hydrophobic groups in the presence of water interact differently than in the presence of oil. Soft medium-range attractions among interface beads mimic the interfacial tension between oil and water, both in magnitude and in the effect being localized to the polar–apolar dividing surface of the bilayer (stress profile). Two estimates for the interfacial tension (Γ and γ_λ) are shown to correspond well with theoretical predictions and atomistic simulations.

Other continuum properties of bilayers formed using this model fall within the range of biological relevance. The bending rigidity and area compressibility modulus are both reasonable; the bending rigidity can be tuned via the molecular bending coefficient. This agreement allows us to make sensible analysis of bilayer deformations induced by embedded proteins. We find that embedded "proteins" or inclusions deform the membrane in agreement with a theory proposed by Aranda-Espinoza et al. (1996) and extended by us [Brannigan and Brown 2007]. This theory is perfectly consistent with that used to measure the elastic constants from the fluctuation spectrum.

Our potential for mimicking the oil–water interfacial tension, U_{int}, has been successfully used for implicitly solvating a model very similar to that of Goetz and Lipowsky (1998). It is not yet clear whether it could be used to implicitly solvate other models (coarse-grained or otherwise), but it satisfies the two known requirements for mimicking the hydrophobic effect: it is soft, and it is localized at the polar/nonpolar interface. Furthermore, the potential is simple and easily implemented.

4.5 ACKNOWLEDGMENTS

We thank Peter Philips for his contributions to his work, which was supported by the Petroleum Research Fund of the American Chemical Society (grant No. 42447-G7) and by the National Science Foundation (grant No. CHE-0321368 and grant No. CHE-0349196). F.B. is an Alfred P. Sloan Research Fellow and a Camille Dreyfus Teacher-Scholar.

REFERENCES

Aranda-Espinoza, H., Berman, A., Dan, N., Pincus, P., and Safran, S. A. 1996. Interaction between inclusions embedded in membranes. *Biophys. J.* 71:648–56.

Ayton, G., Bardenhagen, S. G., McMurty, P., Sulsky D., and Voth G. A. 2001. Interfacing continuum and molecular dynamics: An application to lipid bilayers. *J. Chem. Phys.* 114:6913–24.

Bloom, M., Evans, E., and Mouritsen, O. G. 1991. Physical properties of the fluid lipid–bilayer component of cell-membranes: A perspective. *Q. Rev. Biophys.* 24:293–397.

Brannigan, G., and Brown, F. L. H. 2004. Solvent-free simulations of fluid membrane bilayers. *J. Chem. Phys.* 120:1059–71.

———. 2005. Composition dependence of bilayer elasticity. *J. Chem. Phys.* 122:074905.

———. 2006. A consistent model for thermal fluctuations and protein-induced deformations in lipid bilayers. *Biophys. J.* 90:1501–20.

———. 2007. Contributions of Gaussian curvature and nonconstant lipid volume to protein deformation of lipid bilayers. *Biophys. J.* 92:864–76.

Brannigan, G., Phillips, P. F., and Brown, F. L. H. 2005. Flexible lipid bilayers in implicit solvent. *Phys. Rev. E* 72:011915.

Brochard, F., and Lennon, J. F. 1975. Frequency spectrum of flicker phenomenon in erythrocytes. *J. Phys. (Paris)* 36:1035–47.

Bruinsma, R., and Pincus, P. 1996. Protein aggregation in membranes. *Curr. Opin. Solid State Mater. Sci.* 1:401–406.

Cantor, R. 1997. The lateral pressure profile in membranes: A physical mechanism of general anesthesia. *Biochemistry* 36:2339–44.

Carillo-Tripp, M., and Feller, S. E. 2005. Evidence for a mechanism by which omega-3 polyunsaturated lipids may affect membrane protein function. *Biochemistry* 44:10164–69.

Chiu, S.W., Jakobsson, E., Mashl, R. J., and Scott, H. L. 2002. Cholesterol-induced modifications in lipid bilayers: A simulation study. *Biophys. J.* 83:1842–53.

Chiu, S.W., Vasudevan, S., Jakobsson, E., Mashl, R. J., and Scott, H. L. 2003. Structure of sphingomyelin bilayers: A simulation study. *Biophys. J.* 85:3624–35.

Cooke, I. R., Kremer, K., and Deserno, M. 2005. Tunable generic model for fluid bilayer membranes. *Phys. Rev. E* 72:011506.

de Planque, M. R. R., and Killian, J. A. 2003. Protein–lipid interactions studied with designed transmembrane peptides: Role of hydrophobic matching and interfacial anchoring (Review). *Mol. Membr. Biol.* 20:271–84.

Drouffe, J.-M., Maggs, A. C., and Leibler, S. 1991. Computer simulations of self-assembled membranes. *Science* 254:1353–56.

Evans, E. 1974. Bending resistance and chemically induced moments in membrane bilayers. *Biophys. J.* 14:923–31.

Evans, E., and Skalak, R. 1980. *Mechanisms and thermodynamics of biomembranes.* Boca Raton: CRC Press.

Farago, O. 2003. "Water-free" computer model for fluid bilayer membranes. *J. Chem. Phys.* 119:596–605.

Feller, S. E. 2000. Molecular dynamics simulations of lipid bilayers. *Curr. Opin. Colloid Interface Sci.* 5:217–23.

Feller, S. E., and Pastor, R. W. 1999. Constant surface tension simulations of lipid bilayers: The sensitivity of surface areas and compressibilities. *J. Chem. Phys.* 111:1281–87.

Frenkel, D., and Smit, B. 2002. *Understanding molecular simulation.* 2nd ed. New York: Elsevier.

Goetz, R., Gompper, G., and Lipowsky, R. 1999. Mobility and elasticity of self-assembled membranes. *Phys. Rev. Lett.* 82:221–24.

Goetz, R., and Lipowsky, R. 1998. Computer simulations of bilayer membranes: Self-assembly and interfacial tension. *J. Chem. Phys.* 108:7397–409.

Golestanian, R., Goulian, M., and Kardar, M. 1996. Fluctuation-induced interactions between rods on a membrane. *Phys. Rev. E* 54:6725–34.

Gullingsrud, J., and Schulten, K. 2004. Lipid bilayer pressure profiles and mechanosensitive channel gating. *Biophys. J.* 86:3496–509.

Haltia T., and Freire, E. 1995. Forces and factors that contribute to the structural stability of membrane proteins. *Biochim. Biophys. Acta* 1228:1–27.

Harroun, T. A., Heller, W. T., Weiss, T. M., Yang, L., and Huang, H. W. 1999. Theoretical analysis of hydrophobic matching and membrane-mediated interactions in lipid bilayers containing gramicidin. *Biophys. J.* 76:3176–85.

Helfrich, W. 1973. Elastic properties of lipid bilayers: Theory and possible experiments. *Z. Naturforsch.* 28c:693–703.

Huang, H. W. 1986. Deformation free energy of bilayer membrane and its effect on gramicidin channel lifetime. *Biophys. J.* 50:1061–70.

Israelachvili, J. 1992. *Intermolecular and surface forces.* New York: Wiley.

Israelachvili, J., and Wennerstrom, H. 1992. Entropic forces between amphiphilic surfaces in liquids. *J. Phys. Chem.* 96:520–31.

Izvekov, S., and Voth G. A. 2005. A multiscale coarse-graining method for biomolecular systems. *J. Phys. Chem. B* 109:2469–73.

Jensen, M., and Mouritsen, O. G. 2004. Lipids do influence protein function: The hydrophobic matching hypothesis revisited. *Biochim. Biophys. Acta* 1666:205–26.

Lindahl, E., and Edholm, O. 2000a. Mesoscopic undulations and thickness fluctuations in lipid bilayers from molecular dynamics simulations. *Biophys. J.* 79:426–633.

———. Spatial and energetic-entropic decomposition of surface tension in lipid bilayers from molecular dynamics simulations. *J. Chem. Phys.* 113:3882–93.

Lipowsky, R. 1991. The conformation of membranes. *Nature* 349:475–81.

Lipowsky, R., and Grotehans, S. 1994. Renormalization of hydration forces by collective protrusion modes. *Biophys. Chem.* 49:27–37.

Lipowsky, R., and Sackmann, E. 1995. *Structure and dynamics of membranes.* Amsterdam: Elsevier.

Lodish, H., Baltimore, D., Berk, A., Zipursky, S. L., Matsudaira, P., and Darnell, J. 1995. *Molecular cell biology.* 3rd ed. New York: Scientific American Books.

Marrink, S.-J., Berkowitz, M., and Berendsen, H. J. C. 1993. Molecular dynamics simulation of a membrane water interface: The ordering of water and its relation to the hydration force. *Langmuir* 9:3122–31.

Marrink, S.-J., de Vries, A., and Mark, A.E. 2004. Coarse grained model for semiquantitative lipid simulations. *J. Phys. Chem. B* 108:750–60.

Marrink, S.-J., and Mark, A.E. 2001. Effect of undulations on surface tension in simulated bilayers. *J. Phys. Chem. B* 105:6122–27.

Marsh, D. 1996. Lateral pressure in membranes. *Biochim. Biophys. Acta* 1286:183–223.

Mouritsen, O. G., and Bloom, M. 1984. Mattress model of lipid–protein interactions in membranes. *Biophys. J.* 46:141–53.

Nagle, J. F., and Tristram-Nagle, S. 2000. Structure of lipid bilayers. *Biochim. Biophys. Acta* 1469:159–95.

Nagle, J. F., Zhang, R., Tristram-Nagle, S., Sun, W. J., Petrache, H. I., and Suter, R. M. 1996. X-ray structure determination of fully hydrated L alpha phase dipalmitoylphosphatidylcholine bilayers. *Biophys. J.* 70:1419–31.

Nielsen, S. O., Ensing, B., Ortiz, V., Moore, P. B., and Klein, M. L. 2005. Lipid bilayer perturbations around a transmembrane nanotube: A coarse grain molecular dynamics study. *Biophys. J.* 88:3822–28.

Noguchi, H., and Takasu, M. 2001. Fusion pathways of vesicles: A Brownian dynamics simulation. *J. Chem. Phys.* 115:9547–51.

Pastor, R. W. 1994. Molecular-dynamics and Monte-Carlo simulations of lipid bilayers. *Curr. Opin. Struct. Biol.* 4:486–92.

Patra, M. 2005. Lateral pressure profiles in cholesterol–DPPC bilayers. *Eur. Biophys. J.* 35:79–88.

Peskin, C. S., Odell, G. M., and Oster, G. F. 1993. Cellular motions and thermal fluctuations: the Brownian ratchet. *Biophys. J.* 65:316–24.

Pitman, M. C., Grossfield, A., Suits, F., and Feller, S. E. 2005. Role of cholesterol and polyunsaturated chains in lipid–protein interactions: Molecular dynamics simulation of rhodopsin in a realistic membrane environment. *J. Am. Chem. Soc.* 127:4576–77.

Rawicz, W., Oldbrich, K. C., McIntosh, T., Needham, D., and Evans, E. 2000. Effect of chain length and unsaturation on elasticity of lipid bilayers. *Biophys. J.* 79:328–39.

Rekvig, L., Kranenburg M., Vreede, J., Hafskjold, B., and Smit, B. 2003. Investigation of surfactant efficiency using dissipative particle dynamics. *Langmuir* 19:8195–205.

Sackmann, E. 1995. In *Structure and dynamics of membranes*, ed. R. Lipowsky and E. Sackmann, vol. 1. Amsterdam: Elsevier.

Safran, S.A. 1994. *Statistical thermodynamics of surfaces, interfaces and membranes.* Boulder: Westview Press.

Seifert, U., and Lipowsky, R. 1995. In *Structure and dynamics of membranes*, ed. R. Lipowsky and E. Sackmann, vol. 1. Amsterdam: Elsevier.

Shelley, J. C., Shelley, M. Y., Reeder, R. C., Bandyopadhyay, S., and Klein, M. L. 2001. A coarse grain model for phospholipid simulations. *J. Phys. Chem. B* 105:4464–70.

Shinoda, W., Devane, R., and Klein, M. L., 2007. Multi-property fitting and parameterization of a coarse grained model for aqueous surfactants. *Mol. Sim.* 33:27–36.

Smondyrev, A. M., and Berkowitz, M. L. 1999. Structure of dipalmitoylphosphatidlycholine/cholesterol bilayer at low and high cholesterol concentrations: Molecular dynamics simulation. *Biophys. J.* 77:2075–89.

Tien, H. T., and Ottova-Leitmannova, A. 2003. *Planar lipid bilayers (BLMs) and their applications.* Amsterdam: Elsevier.

Tobias, D. J., Tu, K. C., and Klein, M. L. 1997. Atomic-scale molecular dynamics simulations of lipid membranes. *Curr. Opin. Colloid Interface Sci.* 2:15–26.

van den Brink-van der Laan, E., Chupin, V., Killian, J. A., and de Kruijff, B. 2004. Small alcohols destabilize the KcsA tetramer via their effect on the membrane lateral pressure. *Biochemistry* 43:5937–42.

Venturoli, M., and Smit, B. 1999. Simulating the self-assembly of model membranes. *PhysChemComm* 10:45–49.

Venturoli, M., Smit, B., and Sperotto, M. M. 2005. Simulation studies of protein-induced bilayer deformations, and lipid-induced protein tilting, on a mesoscopic model for lipid bilayers with embedded proteins. *Biophys. J.* 88:1778–98.

Wang, Z. J., and Frenkel, D. 2005. Modeling flexible amphiphilic bilayers: A solvent-free off-lattice Monte Carlo study. *J. Chem. Phys.* 135:234711.

Zehl, T., Wahab, M., Mogel, H.H., and Schiller, P. 2006. Monte Carlo simulations of self-assembled surfactant aggregates. *Langmuir* 22:2523–27.

5 Coarse-Grained Dynamics of Anisotropic Systems

L. Paramonov, M.G. Burke, and S.N. Yaliraki
Institute for Mathematical Sciences and Department of Chemistry, Imperial College London

CONTENTS

5.1 INTRODUCTION

Complex systems exhibit a rich behavior over a range of length and time scales that is not always possible to reach with accurate atomistic simulations. A complementary strategy is to abandon the detailed atomic potentials and replace them instead with simpler models. Such coarse-grained models have a rich history and have had successes in recent years in the modeling of polymer [Glotzer and Paul 2002] and biological systems [Ayton, Noid, and Voth 2007; Brannigan, Lin, and Brown 2006; Tozzini 2005; Nielsen et al. 2004]. These vary from simple to more detailed and have been successful in predicting structural properties as well as thermodynamic phase behavior. The successes of these models often have in common an accurate representation of the architecture and shape of the particles they are modeling. The importance of particle shape in the thermodynamic properties of the system was recognized early on, starting from Onsager's (1949) realization that anisotropic particles will order to maximize their excluded volume entropy when it overcomes their configurational entropy above a critical volume fraction.

Anisotropic descriptions that can capture the effects of shape have been a natural choice for coarse-grained models of colloidal and mesoscopic particles. Examples include the thermodynamics of liquid crystal phases, reviewed in work by Wilson (2005) and Care and Cleaver (2005). Chains of elastically connected anisotropic and isotropic CG sites [Ilnytskyi and Wilson 2001] are also used in the modeling of polymers [Lyulin et al. 1998] as well as dendrimers [Wilson, Ilnytskyi, and Stimson 2003] and more complicated structures [Schoppe and Heermann 1999]. In granular media, single grains are often approximated by ellipsoids [Kakalios 2005; Dziugys and Peters 2001]. Elliptic particles are also in use in modeling of other mesoscale materials [Hafner et al. 2006]. Recent advances in the synthesis of particles at the nanoscale with high aspect ratios such as carbon nanotubes and nanorods [Bauer 2004] and spontaneous molecular organization to self-assembled monolayers [Ulman 1996] motivate renewed interest in this area since control of the shape and architecture of the structure is intimately related to its function in electronic and optical applications.

Anisotropic shapes appear naturally in the applications described above, but they are also encountered when groups of atoms are considered as units that move in unison as part of more complicated structures. For example, anisotropic particles have been used in the modeling of lipid chains [Ayton et al. 2001; Brannigan, Tamboli, and Brown 2004]. The side chains of proteins are modeled as ellipsoids in the coarse-grained United Atom Force Field by Scheraga and co-workers [Liwo et al. 1997], together with regular isotropic Lennard–Jones (LJ) sites for the rest of the protein. This is also the case in other force fields [Fogolari et al. 1996; Kolinski and Skolnick 2004]. Evidence is accumulating that shows that once shape is well represented, many of the properties of protein units and their assemblies can be captured, especially when their mechanical properties play a role [Bahar and Rader 2005; Ming et al. 2002]. Identifying rigid blocks from atomic descriptions can offer a natural way to coarse-grain the physicochemical properties from the bottom-up. We have recently described [Hemberg, Yaliraki, and Barahona 2006; Yaliraki and Barahona 2007] a procedure based on mechanical and graph theoretical concepts [Jacobs 1998; Jacobs and Thorpe 1995; Jacobs et al. 2001], through which simplified structural descriptions can be derived rather than postulated. The induced partitioning brings to fore the importance of "shape" in biomolecules, since the volume enclosed by these units is rarely spherical but, rather, anisotropic, with important consequences for system properties.

In describing such shapes, anisotropic coarse-grained potentials are a natural choice since they properly capture the intrinsic constraints introduced by the interactions and avoid the need for artificial long-range potentials or other schemes. Ellipsoids are appealing as a first approximation to anisotropic shape because they provide enough complexity, can be reduced to spheres, which is important for testing models, and can be represented by quadratic forms, which mean they have very useful algebraic properties.

Anisotropic potentials have a long history [Berne and Pechukas 1972; Gay and Berne 1981; Perram et al. 1996; Allen et al. 1993]. The most widely used elliptic potentials have been extensions [Allen et al. 1993] of the Gay-Berne (GB) potential [Gay and Berne 1981], a generalization of the work of Berne and Pechukas (1972), where the shapes of two identical spheroids are represented by stretched Gaussian distributions. However, some of the drawbacks of this approach have been recognized early on, namely that it is not generally applicable to any mixture and it does not reduce to the regular LJ interactions for spherical particles. Unrealistic features of the phase diagram of the GB fluid have also been found [Wilson 2005] recently, which become more pronounced with growing ellipticity.

The lack of generality prompted another approach based on more geometric considerations suggested by Perram and co-workers [Perram and Wertheim 1985; Perram et al. 1996]. In their approach, called the Elliptic Contact Function (ECF), the shapes of two elliptic particles are represented by quadratic forms. The ECF can serve as an overlap criterion for two ellipsoids of any shape which can be solved efficiently as an optimization problem in an elegant way by exploiting the algebraic representation of ellipsoids [Perram et al. 1996]. Additionally, the ECF gives the exact position of the contact point of the surfaces of the two ellipsoids when they are in tangent contact. As a consequence, this approach has been used mainly in Monte Carlo (MC) simulations of hard ellipsoids [Care and Cleaver 2005]. Although valid for any mixture of particles, the iterative nature of the optimization task, which is required for each pair at each step, is a drawback for molecular dynamics (MD) simulations. Furthermore, the potential that ensued had unrealistic well depths and did not lead to isotropic behavior at large distances.

Although the thermodynamic aspects of anisotropic systems have been well studied, surprisingly little work has been done on the dynamics of elliptic particles. Shape is important not only to structural or equilibrium properties, but also to dynamical processes. For example, ellipsoids have been shown to pack more densely at jammed configurations than spherical particles [Donev, Torquato, and Stillinger, 2005; Donev et al. 2004] and exhibit non-Gaussian diffusion, which is observable in lower dimensions [Han et al. 2006] and predicted to be amplified in the presence of external forces [Grima and Yaliraki forthcoming]. In biological systems, shape is linked to function and the dynamical responses that ensue that are as important to understand.

The lack of the extent of explicit dynamics may be due to the complexities involved in the anisotropic potentials described above coupled with the difficulties in propagating orientational degrees of freedom (DOF). Anisotropic coarse-grained sites representing rigid molecules or rigid groups of atoms can be described in terms of "hard" or "soft" potentials. While in "hard" anisotropic potentials the interaction energy is infinite when the CG sites overlap and is zero otherwise, in "soft" anisotropic potentials the interaction energy is a smooth continuous function of the relative position and orientation. While the hard anisotropic potentials are mainly used in MC simulations, the soft anisotropic potentials can be used in both MD and MC simulations. We have recently shown [Paramonov and Yaliraki 2005] that the GB potential, the most successful soft potential, leads to undesirable deviations of shape and volume of interacting elliptic particles and, as a result, to artificial "ordering forces" under certain orientations, which align elliptic particles into ordered relative configurations. The GB potential might perform well in modeling of liquid crystal phases of almost aligned identical particles, but for modeling of dynamic systems, such as lipid membranes with many different types of interacting CG sites, another anisotropic potential must be used.

When considering the geometric approaches, apart from usually leading to hard potentials, their use has additionally been hampered by computing the distance of closest approach correctly and efficiently, a cumbersome task often solved by brute force. In the spherical case this is straightforward, as the potential remains isotropic after coarse-graining and the interactions are simply functions of the interparticle radial distance. In the general nonspherical case, the interaction can be approximated as a force acting at the points of minimum distance between the particles and in the direction of the minimum distance vector. This force can then be resolved into a radial part acting on the center of mass of the particles and a torque acting about the center of mass. Ideally, one would like the distance of closest approach, which is considered the best measure to be faithful to the shape aspects of the systems (Figure 5.1). Its extremum properties can in principle lead to analytic forces and torques which are essential for MD. In addition we would like the approach to be general for any mixture and any orientation with soft interactions. It would hence be desirable to capture the shape properties correctly and at the same time have a systematic way to check how close the approximation is to the true answer, that is, to have an estimate of the error in comparison with the original problem.

We first show a method to obtain the exact distance of closest approach for all pairs at once of any mixture and orientation [Burke and Yaliraki 2006]. We then discuss a more geometric approach inspired by the ECF. We showed that although the ECF does not give the distance of closest approach, it does give the closest directional distance across the intercenter distance and bounds the exact distance from above and from below, hence giving a good approximation of the exact distance [Paramonov and Yaliraki 2005]. This allows for very efficient and stable dynamics. It was further shown that the suggested potential can be used for modeling of any mixture of any elliptic and spherical particles without deviation of their shape and volume. Simple analytical formulas for forces and torques due to the potential make it especially attractive for MD simulations. The potential now is isotropic at large distances, has realistic widths of the potential wells, and captures the complexity associated with the orientation dependence.

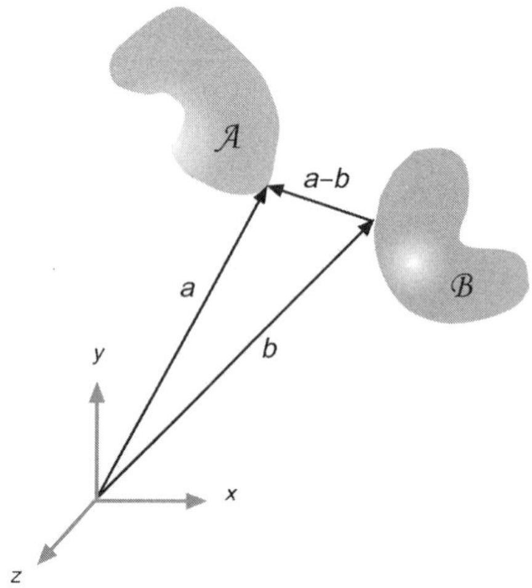

FIGURE 5.1 The interactions of anisotropic particles are determined by their distance of closest approach. This distance unlike those with spherical symmetry, typically does not coincide with the intercenter vector joining the particles.

In the methodology section we give details on how to obtain these distances, which can then be used in an anisotropic potential. We then discuss certain issues that make anisotropic potentials different from isotropic ones for dynamics, namely cutoffs and propagation of DOF where rotational motion is coupled to translational motion. We propose good frameworks for dynamics that avoid such problems. In the application section we show how efficient and stable MD simulations can be performed in the dynamics and assembly of complex systems by giving examples from amphiphilic systems.

5.2 METHODOLOGY

5.2.1 PRELIMINARIES

Ellipsoids can be defined by a Cartesian equation or alternatively by a quadratic form as $A(\vec{x}) = (\vec{x} - \vec{r}) \cdot \hat{A} \cdot (\vec{x} - \vec{r}) = 1$, where $\hat{A}$ is the matrix which gives the size and orientation of the ellipsoid centered at point $\vec{r}$ (Figure 5.2a). Note that all our discussions throughout this work are in three dimensions, but we illustrate only two dimensions in Figure 5.2 for clarity. $\hat{A}$ is defined by $\hat{A} = \sum_{i=1,2,3} a_i^{-2} \vec{u}_i \otimes \vec{u}_i$ where the $\vec{u}_i$ s are the unit vectors along the semiaxes of the ellipsoid. Similarly, for another ellipsoid centered at point $\vec{s}$, $\hat{\mathrm{B}}(\vec{\mathrm{x}}) = (\vec{x} - \vec{s}) \cdot \hat{\mathrm{B}} \cdot (\vec{x} - \vec{s}) = 1$ such that $\hat{B} = \sum_{i=1,2,3} b_i^{-2} \vec{v}_i \otimes \vec{v}_i$. The distance of closest approach, $\vec{d}(A,B)$, between two elliptic particles A and B is given by $\vec{d}(A,B) = \min_{x_a, x_b} \|\vec{x}_a - \vec{x}_b\|$ subject to the constraint that $A(\vec{x}_a) = 1$ and $B(\vec{x}_b) = 1$, namely that $\vec{x}_a$ and $\vec{x}_b$ are points on the surfaces of ellipsoids A and B, respectively.

5.2.2 GEOMETRIC APPROACH TO POTENTIALS BASED ON THE DISTANCE OF CLOSEST APPROACH

5.2.2.1 The Exact Distance of Closest Approach by Sum of Squares Decomposition (SOS) Theory

The problem of finding the distance of closest approach can be solved exactly [Burke and Yaliraki 2006] by a reformulation to search for the optimal sum of squares decomposition. Recasting a problem to finding a description as sum of squares decomposition is an elegant and computationally feasible way to solve a large class of problems that can be posed as nonnegative conditions [Parrilo

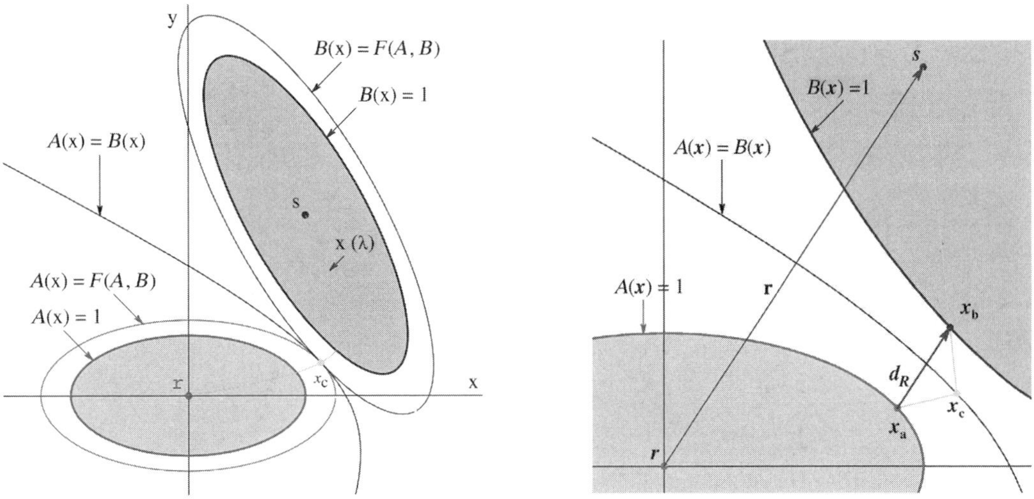

FIGURE 5.2 (a) The ellipsoid particles A and B, given by $A(\mathbf{x}) = 1$ and $B(\mathbf{x}) = 1$, are centered at points r and s, respectively. We consider the local reference frame attached to the center of ellipsoid A aligned along its semi-axis. Although our discussion and calculations in the text are for three dimensions, we illustrate for simplicity in this figure the concepts behind the ECF approach in the two-dimensional case. The original ellipsoids are scaled up or down by $F(A, B)$ until they touch each other tangentially at the contact point $\boldsymbol{x}_c$. The value of the ECF is illustrated by the scaled ellipsoids $A(\boldsymbol{x}_c) = F(\boldsymbol{A}, \boldsymbol{B})$ and $B(\boldsymbol{x}_c) = F(\boldsymbol{A}, \boldsymbol{B})$, used by permission Paramonov and Yaliraki. (b) The directional distance of closest approach, d_R, is obtained directly from the ECF, and is a good approximation to the true distance of closest approach, but remains parallel to the intercenter vector of the two ellipsoids. Using the distance d_R leads to several advantages for the ensuing potential.

2000]. It is important to note that being a sum of squares is a sufficient but not necessary condition for the function to be negative. However, in many cases this is often enough to give the exact answer or an answer reasonably close. The answer is guaranteed to be a lower bound and a proof is obtained when it is exact. The reason this problem is doable is that searching for the coefficients of the SOS is equivalent to a semidefinite program [Vandenberghe and Boyd 1996], a class of modern convex optimization techniques which are highly efficient. The two steps of finding the SOS polynomial and finding its minimum, in this case the distance of closest approach, can be performed simultaneously by solving a semidefinite program. By using the duality, we also obtain the proof that the minimum distance between the two geometrical shapes has been obtained together with the location of the points on the surfaces of the objects. For details on the theory and its implementation for this problem, see Burke and Yaliraki's study (2006).

For ellipsoids, the search for a globally nonnegative representation is now trivial since the norm functions will always be quadratic forms and hence sum of squares. Another advantage of this approach is that we can obtain the minimum distances for all pairs of ellipsoids in any current configuration at once in a single step. In general, the SOS approach is not limited to elliptic particles but provides a useful general mathematical framework for studying interactions between arbitrarily shaped coarse-grained particles, convex as well as nonconvex, such as those depicted in Figure 5.1, as long as there is a polynomial description for their shape.

5.2.3 The Directional Distance of Closest Approach

An alternative approach that gives an estimate of the distance of closest approach by obtaining the closest directional distance is given in this section. The advantage of this approach is that it leads to more efficient dynamics but remains a good approximation to the exact distance (Figure 5.3). It can be considered as a generalization of the ECF approach [Perram and Wertheim 1985; Perram et al.

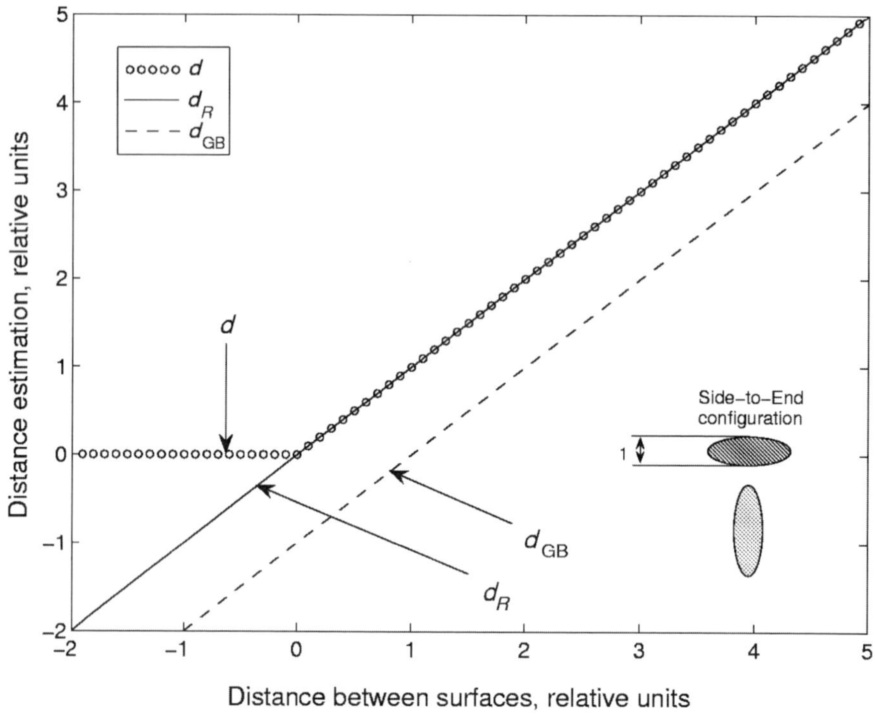

FIGURE 5.3 The directional distance of closest approach, d_R, the exact distance of closest approach, d, and the equivalent distance from the GB potential, between two ellipsoids under a particular orientation. Notice that d_R is a very good approximation to the true distance (the deviation occurs inside the contact point) while the GB gives poor agreement.

1996], discussed in the Introduction. It keeps the positive features of the ECF and removes most of its drawbacks. The key idea is that the directional distance of closest approach of two elliptic surfaces $\vec{d}_R$ (see Figure 5.2b) measured along the vector between the centers of the quadratic forms representing the shapes of two particles can be derived directly from the ECF. We have shown that this value allows for a close estimation of the true distance of closest approach of surfaces of two ellipsoids from above and from below [Paramonov and Yaliraki 2005]. It can be obtained from the ECF value as $\vec{d}_R(A,B) = \vec{R}(1 - F(A,B)^{-1/2})$, where $\vec{R}$ is the intercenter vector joining the centers of the two ellipsoids. The ECF, $F(A,B)$, can be efficiently evaluated as an optimization problem $F(A,B) = \max_{\lambda} \min_{x} S(\vec{x},\lambda)$ where $S(\vec{x},\lambda) = \lambda A(\vec{x}) + (1-\lambda)B(\vec{x})$ and λ is a parameter that can vary between 0 and 1 [for details on the implementation of this see Perram et al. 1996 and Paramonov and Yaliraki 2005]. When the two ellipses are in contact, $F(A,B) = 1$ and $\vec{d}_R$ is zero. When the ellipses overlap, the directional distance becomes negative. A major advantage of using this directional distance to build a potential is that it approaches the intercenter vector $\vec{R}$ at large distances and hence ensures that the potential goes to the correct isotropic limit at large separations.

A potential based on this distance d_R can now be built, which becomes isotropic at large separations of two elliptic particles and has a realistic width of the potential well. To allow for variable and adjustable depth of the potential well, close but different repulsive and attractive shapes of the particles can be used, similar in spirit to the Derjaguin approximation in colloidal science. For LJ interactions, the potential now is given by

$$U(A,B) = 4\varepsilon_0 \left[\left(\frac{\sigma_0}{d_R(A_1,B_1) + \sigma_0} \right)^{12} - \left(\frac{\sigma_0}{d_R(A_2,B_2) + \sigma_0} \right)^{6} \right],$$

where σ_0 has the meaning of the characteristic length and is responsible for the width of the potential well, while ε_0 is the energy parameter that gives the depth of the potential well. Notice that the potential depends on the size and orientation of the ellipsoids. This technique allows for the reproduction of a complex interaction profile of two different particles. We give an example in more detail below.

The convergence speed of the iterative solution of the ECF task is further increased by a close initial estimate of the contact point position, such that only a few simple iterations are required to achieve the desired accuracy. We have also developed a noniterative form of obtaining d_R which leads to even more efficient estimates with a small cost to accuracy.

Another important advantage of the potential above is that it gives analytic expressions for the forces and torques [Paramonov and Yaliraki 2005], a necessity for efficient MD simulations. Additionally, it obeys the corresponding conservation laws.

5.2.4 Obtaining Parameters

There are several ways one can parameterize the potential, including the use of experimental values or radial distribution functions. Here, we discuss a relatively simple fitting procedure to an underlying atomistic LJ potential. The fitting can be done as a least-squares fit of a weighted sum of square distances of the potential depth, position of the root $U_{LJ}=0$, and half-depth width of the potential well of all "i-j" configurations. In an example of the pair potential of anthracene and peropyrene molecules, the aligned configurations were chosen to be represented more accurately than the misaligned ones [see Figure 8 in Paramonov and Yaliraki 2005]. Clearly, this can be chosen based on the application in mind. The target potential was then averaged over rotations of the molecules about their intercenter vector for each i to j configuration. This approach can reproduce, at least qualitatively, a complex interaction profile. Note that this scheme still reduces the number of parameters in comparison, for example, to the GB potential approach. However, more sophisticated approaches could be used to get more accurate parameters and work in this direction is in progress.

5.2.5 Dynamics on Anisotropic Potentials

5.2.5.1 Cutoffs

In any MD simulation, deciding on the cutoff of the potential is an important part of the calculation. The choice of cutoffs when using anisotropic potentials needs to be made with care. Apart from the fact that the anisotropic potentials are longer-ranging, the most important reason is that the equipotential surfaces of anisotropic potentials are not spherical (see for instance Figure 1 in de Miguel et al.'s work (1996)). If a particle "i" is entering a cutoff region of a particle "j" at one point and leaving at another, there is an energy difference between these states due to the integral of the force/torque along the path of the particle inside the cutoff region. As a result, the total energy of the system would not be conserved if one were to follow the usual practice of spherical truncation of the potential at some intercenter distance before the evaluation of the potential itself. In an MC simulation this may not cause numerical problems, but it will have serious implications for MD simulations where the forces are spherically truncated instead. The spherical truncation does not automatically mean truncation of the potential at some equipotential surface anymore as in isotropic potentials and an anisotropic cutoff technique must be used, which, in general, may significantly increase the volume of the cutoff region and the simulation efficiency.

An advantage of using the directional distance $\vec{d}_R$ is that we take advantage of a natural cutoff implied by its properties. The attractive and the repulsive parts of the pair potential are cut off at different distances, which allows the total number of times of resolving the ECF task to be reduced and ensures that the conservation of energy is obeyed.

5.2.5.2 Propagation of Rigid Body Motion

Once anisotropic interaction sites are introduced into the MD simulation, their orientation in space becomes as important as their position and both the representation of the orientational DOFs and the actual propagation algorithm of those DOFs need to be addressed. The propagation of the rotational DOF can be simplified with the use of simplectic integrators and the quaternion representation. We apply a simplectic algorithm developed by Omelyan (1998) to the propagation of the rotational DOF of the elliptic rigid bodies.

We use the angular velocity in the body reference frame and quaternion parameters in the algorithm. The usage of angular velocity in the body frame is the most convenient choice for the numerical propagation of the Euler equations. The usage of quaternions allows avoidance of singularities in the representation of the orientation of a rigid body that are encountered in the usual Euler representation and, additionally, allows for the fast calculation of rotation matrices without evaluation of slow trigonometric functions [Allen and Tildesley 1987]. In turn the algorithm shows very good energy conservation properties. A consequence of the good energy conservation properties of the propagation algorithm is that the integration time step can be increased.

5.3 APPLICATION: SELF-ASSEMBLY OF AMPHIPHILIC MOLECULES

The methodology described above is general and can be applied to a variety of problems. We illustrate here how this approach leads to efficient and stable dynamics that can simulate complex behavior such as self-assembly in amphiphiles, and dynamics of micelles and lipid bilayers. It is possible to simulate behavior at different length scales and observe the emergence of continuum behavior with the formation of undulations (see Figure 5.4). Here, the lipid head is represented by a sphere while the chain is represented by an ellipsoid. Coarse-grained water particles are represented by spherical particles. These simulations are performed on a standard desktop without any code optimization or parallelization. The standard deviation of the fluctuations of the total energy is about 0.01 of a percent compared to the average kinetic energy under normal conditions of 1 bar and 300 K without significant drift of the total energy even for long simulations and/or big systems. A stable simulation time step applied in most simulations is 0.01 in reduced time, which compares favorably with other coarse-grained models.

5.4 CONCLUDING REMARKS

We have presented a methodology for the simulation of coarse-grained anisotropic particles or sites. It is generally applicable to any mixture of shape and size particles, soft or hard interactions, and

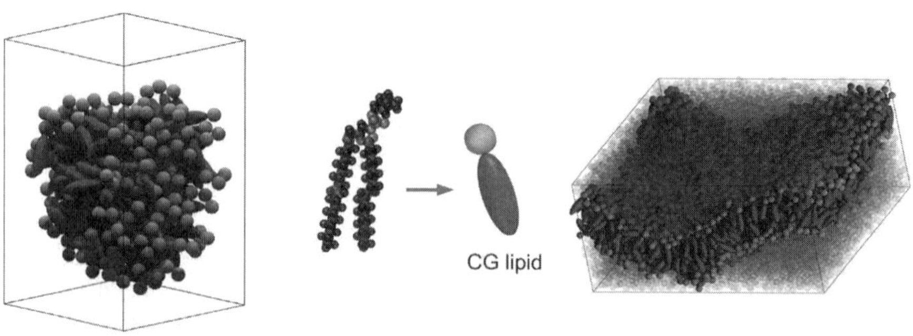

FIGURE 5.4 (See color insert following page 238.) Snapshots from MD simulations of coarse-grained amphiphiles (middle) in coarse-grained water. The potential is versatile to model mixtures of elliptic shapes and the dynamics stable to allow for self-assembly (left) and undulations in model membranes (right), a signature of the emergence of continuum behavior.

can be efficiently applied to large systems. The analytical expression of forces and softer potentials allow for MD simulations with large time steps and system sizes. We have shown a methodology for obtaining the exact distance of closest approach together with a series of controlled approximations which increase the efficiency of the simulation. Ultimately, this interplay of accuracy and efficiency will depend on the particular application. We have shown the emergence of mesoscopic and continuum properties. Coupled with other methodologies for deriving these coarse-grained units from the atomistic detail, rather than postulating them, can increase the power of this approach. It could then provide a bridge from atomistic to continuum behavior. Work in this direction is in progress [Yaliraki and Barahona 2007]. It is also clear that no methodology can outperform all others when considering complex problems. We expect the approach we have described here to be more useful when shape plays an important role in the system dynamics.

ACKNOWLEDGMENT

Funding from GSK and the ONR (U.S.A.) is gratefully acknowledged.

REFERENCES

Allen, M. P., Evans, G. T. et al. 1993. Hard convex body fluids. *Adv. Chem. Phys.* 86:1–166.

Allen, M. P., and Tildesley, D. J. 1987. *Computer Simulation of Liquids.* Oxford: Oxford University Press.

Ayton, G., Bardenhagen, S. G. et al. 2001. Interfacing continuum and molecular dynamics: An application to lipid bilayers. *J. Chem. Phys.* 114 (15):6913–24.

Ayton, G., Noid, W. G., and Voth, G. A. 2007. Multiscale modeling of biomolecular systems: In serial and in parallel. *Curr. Opin. Struct. Biol.* 17:192–98.

Bahar, I., and Rader, A. J. 2005. Coarse-grained normal mode analysis in structural biology. *Curr. Opin. Struct. Biol.* 15 (5):586–92.

Bauer, L. A. 2004. Biological applications of high aspect ratio nanoparticles. *J. Mater. Chem.* 14 (15): 516–26.

Berne, B. J., and Pechukas, P. 1972. Gaussian model potentials for molecular interactions. *J. Chem. Phys.* 56 (8):4213–16.

Brannigan, G., Lin, L.C.L., and Brown, F.L.H. 2006. Implicit solvent simulation models for biomembranes. *Eur. Biophys. J.* 35 (2):104–24.

Brannigan, G., Tamboli, A. C., and Brown, F. L. 2004. The role of molecular shape in bilayer elasticity and phase behavior. *J. Chem. Phys.* 121 (7):3259–71.

Burke, M. G., and Yaliraki, S. N. 2006. Exploring model energy and geometry surfaces using sum of squares decompositions. *J. Chem. Theor. Comp.* 2 (3):575–87.

Care, C. M., and Cleaver, D. J. 2005. Computer simulation of liquid crystals. *Rep. Progr. Phys.* 68 (11): 2665–700.

de Miguel E., Martin del Rio, E. et al. 1996. Effect of the attractive interactions on the phase behavior of the Gay-Berne liquid crystal model. *J. Chem. Phys.* 105 (10):4234–49.

Donev, A., Cisse, I. et al. 2004. Improving the density of jammed disordered packings using ellipsoids. *Science* 303 (5660): 990–93.

Donev, A., Torquato, S., and Stillinger, F. H. 2005. Neighbor list collision-driven molecular dynamics simulation for nonspherical hard particles. I. Algorithmic details. *J. Comp. Phys.* 202 (2):737–64.

Dziugys, A., and Peters, B. 2001. An approach to simulate the motion of spherical and non-spherical fuel particles in combustion chambers. *Granular Matter* 3 (4):231–65.

Fogolari, F., Esposito, G. et al. 1996. Modeling of polypeptide chains as C-alpha chains, C-alpha chains with C-beta, and C-alpha chains with ellipsoidal lateral chains. *Biophys. J.* 70 (3):1183–97.

Gay, J. G., and Berne, B. J. 1981. Modification of the overlap potential to mimic a linear site–site potential. *J. Chem. Phys.* 74 (6):3316–19.

Glotzer, S. C., and Paul, W. 2002. Molecular and mesoscale simulation methods for polymer materials. *Annu. Rev. Mater. Res.* 32:401–36.

Grima, R., and Yaliraki, S. N. 2007. Brownian motion of an asymmetrical particle in potential field. *J. Chem. Phys.* 127:084511.

Hafner, S., Eckardt, S. et al. 2006. Mesoscale modeling of concrete: Geometry and numerics. *Comput. Struct.* 84 (7):450–61.

Han, Y., Alsayed, A. M. et al. 2006. Brownian motion of an ellipsoid. *Science* 314: 626.

Hemberg, M., Yaliraki, S. N., and Barahona, M. 2006. Stochastic kinetics of viral capsid assembly based on detailed protein structures. *Biophys. J.* 90:3029–42.

Ilnytskyi, J., and Wilson, M. R. 2001. A domain decomposition molecular dynamics program for the simulation of flexible molecules with an arbitrary topology of Lennard-Jones and/or Gay-Berne sites. *Comp. Phys. Com.* 134 (1):23–32.

Jacobs, D. J. 1998. Generic rigidity in three-dimensional bond-bending networks. *J. Physics A Math. Gen.* 31:6653–68.

Jacobs, D. J., Rader, A. J. et al. 2001. Protein flexibility predictions using graph theory. *Proteins* 44 (2):150–65.

Jacobs, D. J., and Thorpe, M. F. 1995. Generic rigidity percolation: The pebble game. *Phys. Rev. Lett.* 75 (22): 4051–54.

Kakalios, J. 2005. Resource Letter GP-1: Granular physics or nonlinear dynamics in a sandbox. *Am. J. Phys.* 73 (1): 8–22.

Kolinski, A., and Skolnick, J. 2004. Reduced models of proteins and their applications. *Polymer* 45 (2): 511–24.

Liwo, A., Oldziej, S. et al. 1997. A united-residue force field for off-lattice protein-structure simulations .1. Functional forms and parameters of long-range side-chain interaction potentials from protein crystal data. *J. Comput. Chem.* 18 (7):849–73.

Lyulin, A. V., Al-Barwani, M. S. et al. 1998. Molecular dynamics simulation of main chain liquid crystalline polymers. *Macromolecules* 31 (14):4626–34.

Ming, D., Kong, Y. et al. 2002. Domain movements in human fatty acid synthase by quantized elastic deformational model. *Proc. Natl. Acad. Sci. U.S.A.* 99 (12):7895–99.

Nielsen, S. O., Lopez, C. F. et al. 2004. Coarse grain models and computer simulation of soft materials. *J. Phys.: Cond. Matter* 16:R481–512.

Omelyan, I. P. 1998. Algorithm for numerical integration of the rigid-body equations of motion. *Phys. Rev. E* 58 (1): 1169–72.

Onsager, L. 1949. The effects of shape on the interaction of colloidal particles. *Ann. N. Y. Acad. Sci.* 51:627.

Paramonov, L., and Yaliraki, S. N. 2005. The directional contact distance of two ellipsoids: Coarse-grained potentials for anisotropic interactions. *J. Chem. Phys.* 123 (19):194111.

Parrilo, P. 2000. *Structured Semidefinite Programs and Semialgebraic Geometry Methods in Robustness and Optimisation*. Pasadena: California Institute of Technology.

Perram, J. W., Rasmussen, J. et al. 1996. Ellipsoid contact potential: Theory and relation to overlap potentials. *Phys. Rev. E* 54 (6):6565–72.

Perram, J. W., and Wertheim, M. S. 1985. Statistical-mechanics of hard ellipsoids. 1. Overlap algorithm and the contact function. *J. Chem. Phys.* 58 (3):409–16.

Schoppe, G. and Heermann, D. W. 1999. Alternative off-lattice model with continuous backbone mass for polymers. *Phys. Rev. E* 59 (1):636–41.

Tozzini, V. 2005. Coarse-grained models for proteins. *Curr. Opin. Struct. Biol.* 15 (2):144–50.

Ulman, A. 1996. Formation and structure of self-assembled monolayers. *Chem. Rev.* 96 (15):1533–54.

Vandenberghe, L. and Boyd, S. 1996. Semidefinite programming. *SIAM Rev.* 38:49.

Wilson, M. R. 2005. Progress in computer simulations of liquid crystals. *Int. Rev. Phys. Chem.* 24 (3–4): 421–55.

Wilson, M. R., Ilnytskyi, J. M., and Stimson, L. N. 2003. Computer simulations of a liquid crystalline dendrimer in liquid crystalline solvents. *J. Chem. Phys.* 119 (6):3509–15.

Yaliraki, S. N., and Barahona, M. 2007. Chemistry across scale: from molecules to cells. *Phil. Trens. Roy. Soc. A* 365:2921–34.

6 State-Point Dependence and Transferability of Potentials in Systematic Structural Coarse-Graining

Qi Sun, Jayeeta Ghosh, and Roland Faller
Department of Chemical Engineering & Materials Science, University of California

CONTENTS

6.1 INTRODUCTION

Coarse-graining, the systematic mapping of higher- to lower-resolution models, has been the focus of scientific research in a variety of fields over recent decades [1–7]. Especially the area of polymer simulations has been a hotbed of activity, as the large variety of important length scales in polymers still pose a formidable challenge for simulations [6,7]. Another attractive area for the application of such techniques is the understanding of the glass transition where time-scale problems become truly prohibitive. In these and several other application fields systematic connections between different length and time scales are necessary, in order to keep the identity of the chemical species and extend into large enough scales. To this end a variety of systematic computational coarse-graining techniques have been devised [1,6–10] where simulations on more than one length scale are combined in order to get a better understanding of the system as a whole. It has even been proposed that simulations on both scales can be performed in one single simulation box [11–13], which we do not discuss in this chapter.

There are many reasons for applying coarse-graining schemes to polymer simulations. The overall structure of a chemically specific polymeric system should be reproduced without resorting to the computationally expensive local atomistic detail. We typically also want to improve the speed and memory requirements of simulations, which would enable us to perform simulations of more and longer chains. Especially, simulations of long chain systems are necessary as there are qualitative differences in the behavior of short and long chains [14]. For long chains the relaxation rates increase faster than cubically with the degree of polymerization [15], making it extremely challenging to perform meaningful; that is, equilibrated, detailed atomistic simulations. Even if it were unproblematic to perform such huge simulations, their actual usefulness would be questionable as the amount of data they are producing would be very difficult to analyze.

In order to compare directly to experiments; that is, get not only the scaling with, for example, chain length correct but also the prefactors, one needs mesoscale models that do not represent a generic material like "a polymer", but we need to keep the chemical identity of a specific polymer. So we need a combination of atomistic and mesoscale models that can be mapped uniquely onto one another.

Another area where advanced simulation techniques are indispensable is the area of the glass transition. The transition from a low viscosity liquid to an amorphous solid; that is, a glass, upon cooling in some systems is one of the largely unresolved problems of statistical physics [16,17]. The glass transition expresses itself in a strong increase in many relaxation times. A number of theoretical models have been proposed to explain such behavior: There has been speculation about an underlying thermodynamic glass transition [18] at the so-called Kauzmann temperature T_K where only one minimum energy basin is accessible to the system. Such arguments date back to Gibbs and diMarzio in the 1950s [19,20]. It is also not clear if different behaviors in different systems are due to fundamental differences and if so which fundamental differences are at play. Computer simulations would be an ideal tool to embark on such studies as they allow the behavior of a prescribed model to be solved very exactly. However, low-temperature simulations in glass-forming systems have generally been avoided, and extrapolations are routinely used to infer low temperature behavior [21–24]. This is due to the problem that simulations near or even below a glass transition are exceedingly difficult, and the results have to be considered with caution with respect to their equilibration. The relevant time scales below T_g are way too long to be sampled by conventional molecular dynamics, which may obtain several 100 ns at best [25,26]. In order to address fundamental questions of glasses, we are required to employ advanced molecular simulation techniques and coarse-graining clearly is one of the obvious options.

6.2 ITERATIVE BOLTZMANN INVERSION

6.2.1 General

One of the earliest approaches of systematic structural coarse-graining for a polymeric system was attempted by Tschöp et al. [1]. This technique is based on detailed quantum chemical calculations of a few monomers in vacuum to obtain the relative energies of energetically favorable local conformations. These distributions are used to perform single-chain Monte Carlo simulations in vacuum. The corresponding distributions of "super atom" bond lengths, angles, and torsions are recorded. A super bond (super angle, etc.) in this respect is a hypothetical bond between super atoms (or mapping points). These super atoms are fictitious anchoring points along the chain which serve the role of interaction centers on the mesoscale. The coarse-grained simulation will only contain these super atoms; that is, the super bonds, super angles, etc., will contain the effective interactions on that length scale. In order to correctly obtain a potential from the measured distributions (on the atomistic scale), they have to be weighted by the corresponding Jacobians between coordinate systems. For bond lengths, for example, this Jacobian leads to a factor of r^2, coming from the transformation from spherical to Cartesian coordinates. In general the Jacobian is the derivative of one coordinate set with respect to the other. Then these distributions are "Boltzmann-inverted" to obtain molecular potentials; that is, a potential is derived from the temperature-weighted logarithm of the distribution. This is exact as in vacuum the free energy difference obtained by this procedure equals the potential energy on the coarse-grained scale:

$$V(\varsigma) = -k_B T \ln p(\varsigma). \tag{6.1}$$

Here ς can stand for super bond lengths, bond angles, and torsions after the Jacobian correction, respectively. In order to be able to derive the potential to obtain the corresponding forces, smoothing techniques like local splines or running averages are used. This technique can directly only be used for single chains in vacuum as otherwise the identity between potential and free energy cannot be assumed. The Iterative Boltzmann Inversion (IBI) therefore uses this free-energy-inspired approach and turns it into an iterative technique to separate free energy and potential energy terms [8,27–29]. It additionally can now be used to calculate nonbonded interactions between different chains.

Coarse-graining always is a way of information reduction; that is, we try to describe a system with fewer degrees of freedom while keeping the essential information unchanged. To be specific, let $\{r\}$ be a set of detailed variables and $\{R\}$ a set of coarse-grained variables. These are in our case positions of interaction centers. As we restrict ourselves here to static structural coarse-graining, we can neglect momenta. The corresponding potential energies are then $V(\{r\})$ and $V'(\{R\})$, where $\{R\}$ are functions of $\{r\}$: $\{R(\{r\})\}$. It is clear that this function is not uniquely invertible; that is, no unique function exists that maps $\{R\}$ back onto $\{r\}$, as (many) different sets of $\{r\}$ lead to the same set $\{R\}$. We actually want the $\{R\}$ to be representative of the average atomistic positions $\{r\}$. Therefore the "potential" on the coarse-grained side has to include entropic effects [12]. The IBI is now one way to develop $V'(\{R\})$. The route that we are taking is to analyze detailed atomistic simulations and optimize the potential to reproduce the structure of an atomistic simulation. In concentrated solutions or melts which are the typical applications of the IBI, the structure of the system—especially on small scales—is defined by the interplay of the potential and packing effects stemming from high densities. A melt or solution of polymers or another system has first to be simulated in atomistic detail to obtain a pair distribution function and all the internal functions (super bonds, etc.) described above. We have to take care in performing this simulation as all flaws of the detailed simulation(s) will be "inherited" by the coarse-grained model. The technique is local in interaction distance, which means that for every iteration a one-to-one correspondence between the effects at a distance r_0 and the potential $V(r_0)$ (or force $-d_r V(r)\big|_{r=r_0}$) at the same distance is assumed. It is clear from this approach that the resulting potential is completely numerical, as every distance is optimized independently. It is advantageous to enforce continuity of the potential by using weighted local averages. Cross-dependencies of the different potentials, for example, super bond and super angle which fundamentally have to be expected to be present, can often be eliminated by the proper choice of mapping points [7,30]. In studies of polyisoprene and polystyrene [27,31] it has been shown that if the atomistic units represented by super atoms are rigid, it leads to well defined, almost harmonic, potentials for the bonds. In that case the cross-dependencies can be neglected as the distances between different super atoms essentially do not change. Moreover, torsion potentials on the coarse-grained scales are normally not taken into account as the atomistic torsions, which define the conformation of a macromolecule, are essentially represented by the super angles. So, we can limit ourselves to potentials and distribution functions that only depend on a single property, like the distance or an angle. As the IBI is an iterative procedure an initial guess for the potential is needed. In one-component liquid systems, we start by taking a simple inverse of $g_0(r)$ obtained in the atomistic simulation, the "target" radial distribution function (RDF) that leads to a free energy. Due to the numerical nature of the functions, we have to use a grid. For nonbonded potentials a resolution of 0.01 times the repulsive core of the interaction as measured at its zero passage is a good choice (i.e., 0.01σ in case of a Lennard–Jones potential); angles are normally resolved to within 1 or 2°. For mixed systems, we normally start with the two (or more) one-component systems. Simulating the initial potential will yield a new radial distribution function $g_1(r)$, which is different from $g_0(r)$ as it combines packing and potential effects. Thus, the potential needs to be improved, which is done by a correction term from Boltzmann inverting the difference in the radial distribution functions $-k_BT \ln[g_0(r)/g_1(r)]$. We iterate this procedure until the distributions of the coarse-grained model and the atomistic model coincide within a described tolerance, also the super angles can be iterated this way. Figure 6.1 illustrates this procedure. The super bonds are ideally just simple harmonic potentials by choice of the mapping points [7]. It is noteworthy that the optimizations for the inter- and the intrachain parameters can be either performed together in one combined procedure or done separately, as the mutual effects between the two are weak [29]. It has theoretically been shown that if two pair potentials yield the same radial distribution function, they differ only by a constant. This is called Henderson's theorem [32]. This means that if we can reproduce the radial distribution function we obtain the optimal; that is, the unique potential. Structural coarse-graining has therefore one single solution for a given average structure [33]. Two technical points are worth mentioning. On the scale of nearest neighbors, the local packing of super atoms is the strongest effect in a dense liquid, so the optimization of the potential should focus on local interactions. Experience shows that it is efficient to start the optimization process with this short distance region. Only after the super atom RDF of the direct neighborhood describes the atomistic RDF reasonably well does the optimization of the other regions start [27]. It is efficient to run two to three independent optimizations of different distances in series which focus on increasingly larger distances. Second, during the iteration, we can apply different weighting functions w_i for the correction terms. This is no fundamental problem as in the IBI essentially the physical meaning of the iteration scheme is irrelevant. It is just one way to approach, relatively effectively, a potential representing the system. The magnitude of this weighting function depends on how far the current super atom RDF deviates from the atomistic RDF. The weighting function is set to 1 when the deviation is about 30% or more from the target. After a few iterations a series of parallel runs with values of weighting functions $w_i = 0.125$, 0.25, and 0.5 can be used to find an optimum starting point for the next step. This minimizes the time to the next step, but obviously at the cost of more computer power [27].

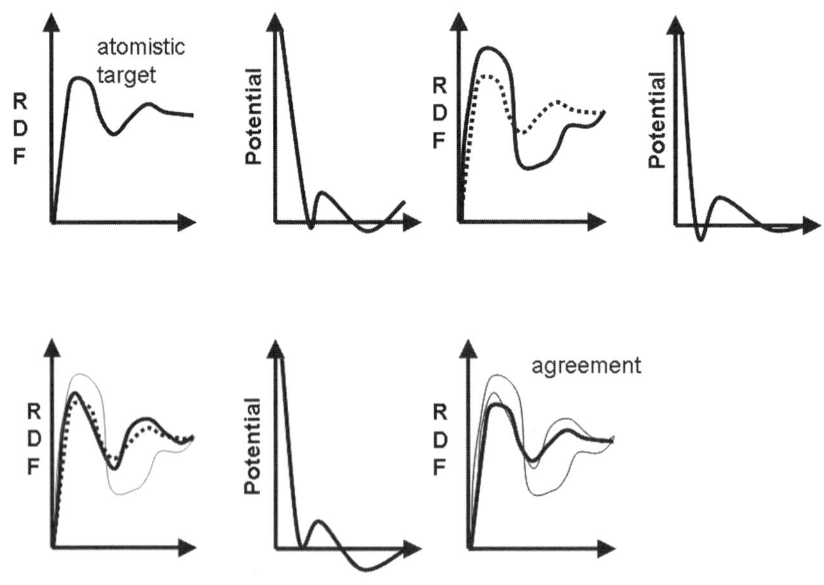

FIGURE 6.1 Illustration of the radial distribution function approaching its target in an Iterative Boltzmann Inversion.

One typically obtains a rapid increase in accuracy over the course of the optimization. The entropy part of the free energy corresponds to the multitude of local conformations subsumed in one mesoscale position of a super atom. Sometimes, the effective size of the monomer—indicated by the point where the RDF starts to deviate from zero—is initially somewhat overestimated [27] and the potentials are too hard; that is, the slope of the potential is too steep. Generally, mesoscale potentials are softer than atomistic potentials which, for example, can also be seen in dissipative particle dynamics [34]. Often the local structure is overly pronounced as both packing effects and the potential lead to local structure [29]. This is where the iteration has to do "most work" in order to separate the packing on the mesoscale from the potential, which has to include the packing on the microscale.

6.2.2 Optimization of Heterogeneous Systems

If we are dealing with a heterogeneous system, like a binary polymer melt of polymers *A* and *B*, which we use here as an illustrative example, we can sort the interaction into self- or homointeractions (*A–A*, *B–B*) and the heterointeraction (*A–B*). Since the self-interaction in the polymer melts is not necessarily the same as in the pure polymer, there are three target RDFs to be optimized in addition to any bonded interaction. This stems from the fact that interactions between like particles (e.g., *A–A*) at a distance r may be mediated by the unlike particles (cf. Figure 6.2). We have correspondingly more target functions, $N+N(N-1)/2$ in an N-component mixture. It is a good idea to optimize the pure systems first, which gives us a reasonable estimate of the self-interactions in the mixed system. Then we start optimizing the heterointeraction where the *A–A* and *B–B* interactions are initially left untouched and only the *A–B* interaction is tuned. Only when this one is satisfactory and we thus have a reasonable estimate of all three interactions do we come back to the optimization of all three target functions at once. It is clearly effective to start the optimization with those potentials that are least affected by changes to all other ones. This is not trivial a priori knowledge, but is typically found during the initial steps of the optimization. There are two reasonable possibilities for starting points for the heterointeraction. We can either, as for the homointeractions, start with a direct inversion of the radial distribution function, or we can start with some kind of a mixing rule. This may be a geometrical or an algebraic average; these averages themselves, however, sometimes may show even qualitatively a wrong system behavior (see below) [35]. We also need two (or more if we do not have a binary mixture) sets of bond length and angle potentials. These can normally be directly transferred from the homosystems as they depend only weakly on the mixing [29]. This

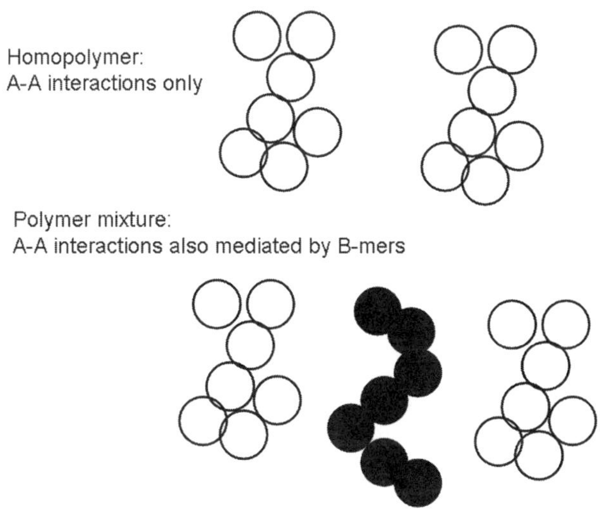

FIGURE 6.2 Explanation of the necessity of independent optimization of mixtures. At intermediate distances even homointeractions may be mediated by polymers of the other type.

should of course be verified for the specific system under study. The whole procedure relies on serially executed iterations; there is no step that can be executed in parallel. In dense systems individual distributions depend on the full set of potentials through higher-order correlation.

6.2.3 Influences of the Environment

From the discussion above and as a direct consequence of Henderson's theorem, it is clear that any coarse-grained potential is dependent on the thermodynamic state at which it was optimized. This applies to concentration, pressure, and especially temperature, even if we have a case, for example, in a glassy system, where the RDF is (almost) temperature independent, and the corresponding potential will be temperature dependent. To illustrate this, we consider two points in the RDF at distance r_1 and r_2, where without loss of generality $g(r_1) > g(r_2)$. For simplicity, we assume that the RDFs are exactly identical. In this case the direct IBI leads to $V_0(r_1) < V_0(r_2)$ due to the monotonicity of the logarithm. If we now assume that we do not need to iterate, as the direct inversion often leads to a reasonable approximation of the final potential, we find that $V(r_2) - V(r_1) = k_BT \ln g(r_2)/g(r_1)$. Thus, the potential differences scale in first approximation with the temperature at which the iteration is performed if the RDF is unchanged. This clearly means that we have to be careful about how large a temperature range a potential can be applied to. This is somewhat in contrast to the case of atomistic simulations where the potentials have—at least limited—physical meaning. This does not apply in the coarse-grained case; these potentials should just be regarded as reproducing a given structure to a certain extent.

This limited range of applicability not only applies to temperature, but can, for polymers in the concentration space, lead to even qualitatively wrong results. Polymers in the melt and in the solution have different scaling behaviors of the molecule dimensions with chain length [14]. The radius of gyration (R_G) in a polymer melt scales with the square root of the degree of polymerization, and in a dilute solution this scaling is closer to 0.6 [14]. As we normally do not want to keep the solvent in the coarse-grained model, it is questionable if we can use the same model over such a concentration range. On the other hand, it has been shown that the model is stable under change of degree of polymerization both for melts and for solutions [31], which is probably the greatest strength of the technique. The scaling of characteristic length scales with degree of polymerization has for melts and solutions been shown to be well in agreement with experiments and theoretical expectations.

6.2.4 Postoptimization: Pressure Correction

So far, we have just been concerned with direct structural optimization. In many systems it turns out that the pressure is too high in a coarse-grained simulation that is often performed under constant number of particle, volume, and temperature (NVT) conditions and the correct (experimental) density [8,29]. It is clear that this effectively is a result of an imperfect representation of the systems, as fundamentally the thermodynamics have to be correct if the structure is perfectly correct [32,33]. There are, however, several error sources, which are difficult to avoid. One of the most severe is an imperfect atomistic simulation as the coarse-grained simulation always inherits all flaws of the atomistic simulation. Another problem is the simulation being run at constant volume and thermodynamic properties not being explicitly present in the merit function of the optimization. In light of this, it is often desirable to postoptimize the system in order to correct the pressure after the structural optimization is complete; of course we have to avoid deteriorating the structural agreement.

The potential of an electrostatically neutral system is always attractive at long ranges due to the dispersion interactions. As we are dealing in this chapter exclusively with short-range nonelectrostatic interactions on the mesoscale, we choose an attractive linear tail function as a weak perturbation to the previously optimized potential [8]:

$$\Delta V_{pc}(r) = A_{pc}(1 - r / r_{cut}). \tag{6.2}$$

A typical initial value is $A_{pc} = -0.1k_BT$. This correction fulfills the following condition: $\Delta V_{pc}(r_{cutoff}) = 0$; clearly the potential as well as the correction has to vanish at the cutoff in order to avoid instabilities. This is the same as in atomistic simulations [36]. The corrected potential is then taken as an initial guess for a second round of structural optimization of the potential again using the IBI. Typically, the shape of the potential, and with it the forces, are very similar to the force field optimized without pressure corrections. The pressure correction should focus on the potential region beyond nearest neighbors ($r > 1$ nm), which is not very crucial for the structural fit. It shifts the whole potentials downward, thus strengthening long-range attraction. If we ended up with too low a pressure after the initial optimization, a positive correction to weaken the intermediate range attraction would help solve the problem.

We will now discuss the fundamentals of the IBI and its state-point dependence, and discuss two illustrative examples in the next two chapters.

6.3 APPLICATION OF THE ITERATIVE BOLTZMANN INVERSION TO THE ORGANIC GLASS FORMER *ORTHO*-TERPHENYL OVER A WIDE RANGE OF TEMPERATURES

6.3.1 ATOMISTIC SIMULATIONS OF *ORTHO*-TERPHENYL

Ortho-terphenyl is a small organic molecule and a very good glass former. It is an ideal model system to study glass transition both experimentally as well as with simulations [37–42]. The molecule consists of three connected benzene rings; the two outer rings are in *ortho* positions to each other, connected to the central ring (cf. Figure 6.3). As described above, structural coarse-graining always starts with atomistically detailed simulations. Here we performed atomistic molecular dynamics simulations using the GROMACS v3.2 molecular simulation suite [43]. The structure of OTP is very simple; it only contains 18 heavy atoms. Nonetheless, devising an appropriate computer model for OTP is challenging. The atomistic model needs to include accurately intermolecular as well as intramolecular interactions, and it should be simple and fast for simulations of the glassy behavior. The atomistic OTP model used here is based on the model developed by Kudchadkar et al. [40], including sites for every carbon, with its corresponding hydrogen if present. We only summarize the main characteristics here, since the model details as well as the exact specifications of the atomistic simulations can be found elsewhere and are not very relevant for the discussion here [39]. The C–C bond lengths are fixed at 0.141 nm; angle potentials keep a 120° angle between three carbons, and improper dihedrals keep the rings planar. The Lennard–Jones potential has been developed for benzene [40]. Torsions are modeled by repulsive Lennard–Jones interactions between the outer rings and not by a regular torsion potential; the parameters have been adapted to reproduce experimental data [40]. The system contains 800 OTP molecules in a cubic box under periodic boundary conditions in the number of particle, pressure, and temperature (NPT) ensemble. It was studied in a wide range of temperatures, from below the experimental glass transition to well above. These simulations suggest a glass transition temperature around 260 K for bulk OTP based on the density as a function of temperature [39] compared to the experimental value of 243 K [41,42]. Considering the short runs of only a few nanoseconds, this agreement should be considered very good and it is completely expected that the computational T_g is higher than the experimental one. The structure of the atomistic simulations depends weakly on temperature between 230 and 418 K. This leads us to the initial assumption that the coarse-grained model should also be only weakly temperature dependent. This will turn out to be incorrect.

6.3.2 IMPLEMENTATION OF A MESOSCALE MODEL OF *ORTHO*-TERPHENYL

In the mesoscale model of *ortho*-terphenyl, each benzene ring is replaced by a single interaction center (a super atom) in the geometric center of the benzene ring [38] (Figure 6.3). This arrangement

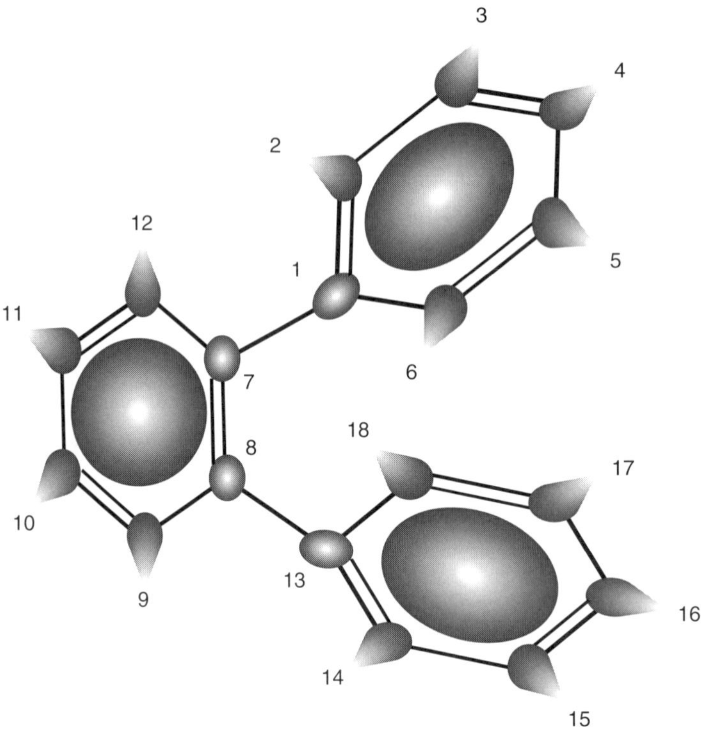

FIGURE 6.3 The coarse-graining sites for *ortho*-terphenyl are represented by the large spheres in the centers of the ring.

satisfies the requirement [7] that distances between bonded super atoms should exhibit single peak distributions. The single peak distribution can be modeled by a single Gaussian curve, which defines a harmonic bond potential. The mesoscale model does not use any angle or dihedral potential. Instead we implement a fictitious bond potential between the two outer rings to represent the coarse-grained angle; this is similar to the conventional fixing of the H–H bond length in SPC water [44]. For the nonbonded potentials, we use the IBI method described above, where OTP is modeled as a trimer of type 1-2-1. This means that the two outer rings are required to have the same interaction leading to three interdependent radial distribution functions to be optimized. Iteration was deemed converged when all coarse-grained RDFs were within 5% of the atomistic target; that is, the maximum difference was not larger than 5% of the target value. The optimization was performed independently at different temperatures. We selected temperatures in the glassy (230 K) and liquid (300 K) region for comparison. In all iterations we ensured that the simulation is equilibrated by monitoring the dependence of the radial distribution function on simulation time [38,45]. We use the DLPOLY [46] molecular dynamics simulation package for the mesoscale simulations. All mesoscale simulations were initialized with the final configuration of an atomistic run in the liquid range at 300 K. The mesoscale simulations have been performed under NVT conditions at the density that is equal to the density of atomistic simulation at 300 K (1084 kg/m^3).

6.3.3 Comparison of Potentials Obtained at Different Temperatures

We performed simulations over a range of temperatures with the mesoscale potential, which was optimized at 300 K. The immediate result was that the system does not form a glass but crystallizes. This clearly shows that using a mesoscale, systematically coarse-grained potential outside its "comfort zone" may lead to qualitatively wrong results. We have to note that the density is slightly too low since all simulations, optimizations, and production runs on the coarse-grained level are

NVT. An additional independent iteration was therefore performed at 230 K in the glassy range. The application of the resulting potential from this optimization in the glassy range now leads to the formation of a glass at lower temperatures, as desired [38]. We can thus not simply define a glass transition temperature in the coarse-grained model. There is ambiguity over which range of temperature we can apply the optimized potentials. The liquid state potentials lead to a crystallization transition; this is not too surprising as the glass transition is essentially a dynamic phenomenon, which we cannot expect to be reproduced without dynamic coarse-graining. A canonical solution to this would be to optimize a coarse-grained potential at every single state point. That, however, somewhat defeats the purpose of coarse-graining as we want to minimize the use of atomistic simulations. Additionally, the potential optimization in the glassy range clearly becomes questionable as the fundamentally required equilibration of the atomistic simulation cannot be achieved in a strict sense any more. So, even if we can define a glass transition using one of the mesoscale models optimized in the glassy range, the transition temperature will vary between the mesoscale model and the atomistic model, as well as between different mesoscale models. For example, for the mesoscale OTP model developed at 300 K we do not find any transition at 250 K or above in contrast to the atomistic transition at 260 K. Structure factors at 230 K show clear evidence of the above-cited crystal formation for the potential developed at 300 K (Figure 6.4). We carried out direct simulations using both potentials (at 230 K and at 300 K) in the temperature region where the atomistic simulation predicts the transition; that is, at 260 K. The potential optimized at the higher temperature does not lead to crystallization at 260 K. It still produces overly pronounced structures, whereas the lower temperature optimization leads to a realistic representation of the 260 K atomistic data. It is noteworthy in this respect that the atomistic RDFs are very similar for OTP over the whole range of temperatures, thus the higher temperature leads to a more pronounced potential, as discussed above.

A similar crystallization tendency has actually been found for systematically coarse-grained polystyrene melt [37]. A mesoscale polystyrene model was optimized against an atomistic simulation of the polymeric glass former atactic polystyrene at 450 K [27]. The dimensionless temperature $T=1$ corresponds to the melt at 450 K, at which the system was optimized. For $T \geq 0.8$

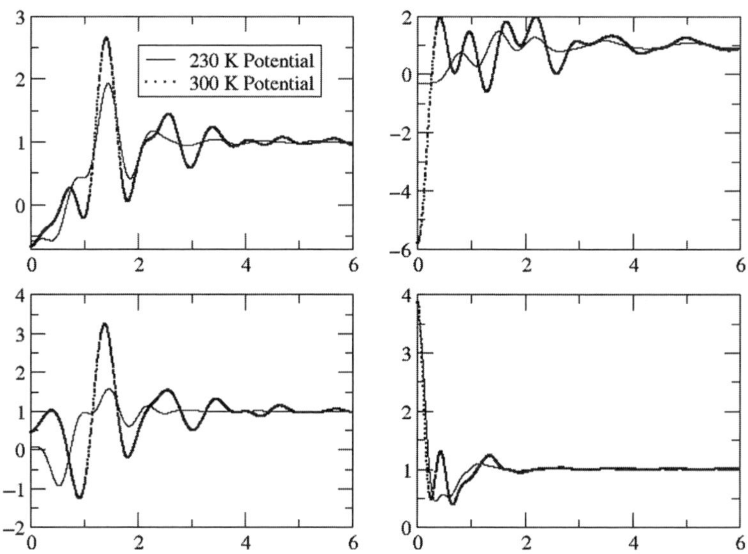

FIGURE 6.4 Temperature dependence of the OTP potentials. OTP structure factors at 230 K using a 230 K and a 300 K potential, top left outer rings, top right inner rings with outer rings, bottom left inner rings, bottom right center of mass.

we observe a clear melt structure and with lowering temperature the structure becomes more pronounced. The temperature effect on the mesoscale model is clearly stronger than expected for a glass-forming system and it has a tendency to crystallize at around $T=0.35$. This again shows clearly that the analysis of a mesoscale model under changed conditions has to be taken with caution.

6.4 OPTIMIZATION OF POLYISOPRENE–POLYSTYRENE BLENDS

6.4.1 Systematically Optimized Potentials

As a second example of a systematic optimization, let us discuss a polymer blend. Here the concentration dependence will be the major issue. As an example, we use a blend of polyisoprene and polystyrene. Starting from an atomistic model of a 50:50 (by weight) blend of polyisoprene and polystyrene oligomers of chain length 15 monomers each, we investigate the local structure, particularly the radial distribution function, bond, and angle distributions. The application of the IBI to binary blends requires three RDFs, in this case, PI–PI, PS–PS, and PI–PS pairs. The iteration process of the binary blend turns out to be more demanding than the melt optimizations as the different RDFs are interrelated to some degree. The mesoscale PI–PS melt mixture systems have been optimized and simulated in the NVT ensemble. The underlying atomistic simulations have been performed at 450 K and ambient pressure [47]. Under these conditions the system should be miscible but close to the phase separation as a slight increase in chain length experimentally leads to a lamellar structure. The temperature in the mesoscale is maintained by a Langevin thermostat with a friction constant $\Gamma=1.0\tau^{-1}$ where $\tau=\sqrt{(\sigma^2 m)/\varepsilon}$ [48], whereas atomistically we use a Berendsen thermostat [49]. All systems use periodic boundary conditions in orthorhombic boxes. The initial conditions for the mesoscale simulations are directly transferred from an atomistic simulation; that is, the final snapshot of the atomistic simulations, which is a one-phase blend, is taken, and the corresponding positions for the mesoscale interaction centers are calculated in the same way as the super atoms for the optimization. This conformation is then directly the initial condition for the mesoscale simulation. The equations of motion are integrated with a time step $\Delta t=0.005\tau$. This was chosen as it is the smaller of the two time steps used in mesoscale simulations of pure polyisoprene and pure polystyrene [27,31]. The bond and angle potentials have been directly transferred from the pure systems. The actual optimization of the nonbonded potentials is based on a system of 36 PI chains and 24 PS chains with 15 monomers each. Atomistic pair distribution functions show a clear tendency of self-aggregation but no full phase separation for this system. The coarse-grained simulations show a phase separation setting in about 15 monomers, which is fully developed at 30 monomers. With increasing chain length of both chains the ensuing interfaces become sharper, indicating a stronger phase separation tendency [50]. If we keep one chain length fixed and increase the other one, we also trigger a phase separation and eventually change the morphology [29]. It turns out that the system shows a lamellar morphology at close to equal weight concentrations. With increasing imbalance in chain length keeping the numbers of chains the same; that is, increasing one chain length and changing the monomer concentration, one observes transitions to cylindrical and spherical morphologies where the minority component constitutes the cylinder or sphere in the background of the continuous majority phase [50]. Two things are noteworthy here: first, as we optimized only at one concentration we cannot be perfectly certain that the other concentrations are correctly represented. We actually verify by comparison that the PS–PS interaction is different in a pure PS melt and in a 50–50 by weight mixture with PI [27,29]. Even the intramolecular angle potentials are weakly different, whereas the bond potentials are identical [29]. The general trends in this model should be very reliable; the actual transition lengths in a morphology diagram are probably only semiquantitatively correct [50]. Second, we can observe the complete phase separation and development of the final morphology starting out from a randomly distributed initial condition. This is a tremendous speed-up in comparison to atomistic simulations, where at best the onset of phase separation can be identified. For lamellar morphologies it turns out that the system often develops a close to cylindrical intermediate first, which then turns into a lamellar system with defects and finally into a lamellar system. As in these systems, the dynamics have not been optimized and it is difficult to assign time scales to the processes. On the other hand, we can find the mechanisms by which the phase transition occurs. This work suggests that the phase separation driven by chain length in PI–PS blends sets in around a chain length of 10–15 monomers. He et al. reported experimentally a miscible PI–PS blend at chain lengths of PI with 13 monomers and PS with 15 monomers, in qualitative agreement with our data [53]. In a dielectric study of *cis*-PI/PS blends at weight concentration of 50:50, clear phase separation at a PI chain length of 32 monomers mixed with PS chain lengths of 24 monomers was observed at temperatures between 380 and 440 K [54]. So the systematically coarse-grained model can actually obtain semi-quantitative agreement with experiment. Lamellar, cylindrical, and spherical morphologies are observed as the chains arrange themselves in minimum free energy configurations to minimize contact surfaces between PI and PS monomers. This has been studied earlier in detail, for copolymer systems [51,52]. As noted above, the interfaces become sharper with increasing chain lengths. By measuring density profiles, a transition length of PI–PS lamellae; that is, a surface roughness, can be calculated as, for example, the distance over which the density of PI varies from 10 to 90% of the maximum PI density. These transition lengths at chain lengths of 30, 60, and 80 monomers of both species are 2.13, 1.37, and 0.87 nm, respectively [50].

6.4.2 Failure of Empirical Mixing Rules

Empirical mixing rules have been very successful in atomistic molecular simulations over the decades and are nowadays so standardized that they are hardly questioned at all [48]. In structural coarse-graining they have normally not been used although it would be very tempting to apply them in order to minimize the optimization effort. A recent study has been performed to test the applicability of empirical mixing rules for systematically optimized polymer blends [35]. The test system is again the PI–PS blend, which at a 50:50 weight concentration should phase separate at a chain length around 10–20 monomers. Based on the independently optimized pure systems, interaction models for the heterointeraction were developed for the blend. Test simulations using both algebraic and geometric averages of the homointeractions as approximations for the heterointeractions have been applied. We tested the following potentials:

$$V_{\mathrm{alg},AB} = \frac{1}{2}(V_{AA} + V_{BB}), \tag{6.3}$$

$$V_{\mathrm{geo},AB} = \sqrt{V_{AA}V_{BB}}\,. \tag{6.4}$$

The first one, the average, is the simplest possible mixing rule and the second one is inspired by the standard Lorentz–Berthelot mixing rule of atomistic simulations [48], which itself goes back to a geometric mixing rule of second virial coefficients. Here, the interaction distances have not been redefined at all as the functions are purely numerical. At large distances where the PS–PS interaction becomes slightly negative we set the mixture potential to zero in both cases. Final results of systems which were set up in morphologies equilibrated by the systematically mapped potential show that the systems are now miscible at long chain lengths (cf. Figure 6.5), which is in disagreement with experiments. The empirical mixing rules clearly underestimate the phase separation tendency and lead to mixed systems in contrast to experimental results reiterating clearly that an independent optimization of the heterointeractions is necessary, as mixture potentials cannot be easily deduced from the homopolymer case.

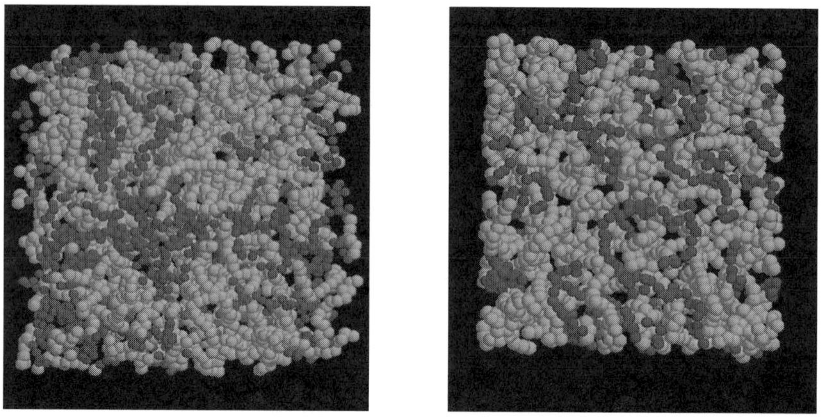

FIGURE 6.5 (See color insert following page 238.) Snapshots of no phase separation in PI–PS mixtures if we use an algebraic (left, system 36 PI and 24 PS 60mers) or geometric (right, 36 PI 45mers and 24 PI 30mers) mixing rule in qualitative disagreement with experiments and systematically optimized models.

6.5 CONCLUSIONS

It is clear that structural coarse-graining in general, and the IBI in particular, is a strong, even essential, tool in the multiscale modeling toolbox. The main strength lies in the polymer field, where the models developed for short chains can be used for longer chains. On the other hand it has become very clear that the transferability of structurally coarse-grained models over different thermodynamic conditions is very limited. Essentially, we always have to perform an atomistic simulation at all conditions to be sure. It is evident that transferring structurally coarse-grained potentials across different phases fails. Although Henderson's theorem guarantees the uniqueness and existence of a potential for any given radial distribution function, we still should take thermodynamic properties, for example, pressure, into account in the optimization as the reliability of radial distribution functions from atomistic simulations may be problematic due to equilibration issues.

From a technical standpoint it has become clear that the intrachain distributions, bonds, angles, etc., are very easy to reproduce and often an actual optimization is not necessary; rather the direct inversion is sufficient. For these distributions also the transferability to other conditions succeeds in many cases when the nonbonded interactions cannot be transferred.

Note, that the overall density is kept fixed in the NVT ensemble used in all mesoscale simulations presented here. In the future, studies under constant pressure will be needed in order to further investigate the state dependence of systematically structurally coarse-grained models.

The outlook for the near future is probably that methods will come online that explicitly take the state point into account and yield potentials which are dependent, for example, on distance, temperature, and concentration.

ACKNOWLEDGMENTS

We thank Petra Träskelin for a critical reading of the manuscript. Some of the work described here was supported by the US Department of Energy, Office of Science, Office of Advanced Scientific Computing Research under grant no. DE-FG02-03ER25568. Computer time at the National Energy Research Supercomputer Center is gratefully acknowledged as well.

REFERENCES

1. Tschöp, W., Kremer, K., Batoulis, J., Bürger, T., and Hahn, O. 1998. Simulation of polymer melts. I. Coarse-graining procedure for polycarbonates. *Acta Polymer.* 49:61–74.

2. Marrink, S. J., de Vries, A. H., and Mark, A. E. 2004. Coarse grained model for semi-quantitative lipid simulation. *J. Phys. Chem. B* 108:750–60.
3. Bai, D., and Brandt, A. 2001. Multiscale computation of polymer models. In *Multiscale Computational Methods in Chemistry and Physics*, eds. A. Brandt, K. Binder, and J. Bernholc. Amsterdam: IOS Press.
4. Akkermans, R. L. C., and Briels, W. J. 2001. A structure-based coarse-grained model for polymer melts. *J. Chem. Phys.* 114:1020–31.
5. Ayton, G., and Voth, G. A. 2002. Bridging microscopic and mesoscopic simulations of lipid bilayers. *Biophys. J.* 83:3357–70.
6. Baschnagel, J., Binder, K., Doruker, P. et al. 2000. Bridging the gap between atomistic and coarse-grained models of polymers: Status and perspectives. *Adv. Polymer Sci.* 152:41–156.
7. Faller, R. 2007. Coarse-grain modeling of polymers. In *Reviews in Computational Chemistry*, vol. 23. eds. K. B. Lipkowitz, and T. R. Cundari. New York: Wiley.
8. Reith, D., Pütz, M., and Müller-Plathe, F. 2003. Deriving effective meso-scale coarse graining potentials from atomistic simulations. *J. Comput. Chem.* 24:1624–36.
9. Milano, G., and Müller-Plathe, F. 2005. Gaussian multicentred potentials for coarse-grained polymer simulations: Linking atomistic and mesoscopic scales. *J. Polym. Sci.* B 43:871–75.
10. Müller-Plathe, F. 2003. Scale-hopping in computer simulations of polymers. *Soft Materials* 1:1–31.
11. McCoy, J. D., and Curro, J. G. 1998. Mapping of explicit atom onto united atom potentials. *Macromolecules* 31:9362–68.
12. Christen, M., and van Gunsteren, W. F. 2006. Multigraining. An algorithm for simultaneous fine-grained and coarse-grained simulation of molecular systems. *J. Chem. Phys.* 124:154106.
13. Praprotnik, M., Kremer K., and Delle Site, L. 2007. Adaptive molecular resolution via a continuous change of the phase space dimensionality. *Phys. Rev.* E 75:017701.
14. Doi, M., and Edwards, S. F. 1986. *The Theory of Polymer Dynamics*. Oxford: Clarendon Press.
15. Strobl, G. 1997. *The Physics of Polymers*. Berlin: Springer.
16. Ediger, M. D., Angell, C. A., and Nagel, S. R. 1996. Supercooled liquids and glasses. *J. Phys. Chem.* 100:13200–12.
17. Debenedetti, P. G., and Stillinger, F. H. 2001. Supercooled liquids and the glass transition. *Nature* 410:259–67.
18. Mezard, M., and Parisi, G. 1999. Thermodynamics of glasses: A first principles computation. *Phys. Rev. Lett.* 82:747–50.
19. Gibbs, J. H. 1956. Nature of the glass transition in polymers. *J. Chem. Phys.* 25:185–86.
20. Gibbs, J. H., and DiMarzio, E. A. 1958. Nature of the glass transition and the glassy state. *J. Chem. Phys.* 28:373–83.
21. Büchner, S., and Heuer, A. 1999. Potential energy landscape of a model glass former: Thermodynamics, anharmonicities, and finite size effects. *Phys. Rev.* E 60:6507–18.
22. Coluzzi, B., Mezard, M., Parisi, G., and Verrocchio, P. 1999. Thermodynamics of a binary mixture glass. *J. Chem. Phys.* 111:9039–52.
23. Coluzzi, B., Parisi, G., and Verocchio, P. 2000. Lennard–Jones binary mixture: A thermodynamical approach to glass transition. *J. Chem. Phys.* 112:2933–44.
24. Coluzzi, B., Parisi, G., and Verrocchio, P. 2000. Thermodynamical liquid–glass transition in a Lennard–Jones binary mixture. *Phys. Rev. Lett.* 84:306–309.
25. Scott, H.L. 2002. Modeling the lipid component of membranes. *Curr. Opin. Struct. Biol.* 12:495–502.
26. Nicolas, G., Stanley, H. E., and Francesco, S. 2005. Relation between the high density phase and the very-high density phase of amorphous solid water. *Phys. Rev. Lett.* 94:107803.
27. Sun, Q., and Faller, R. 2005. Systematic coarse-graining of atomistic models for simulation of polymeric systems. *Comp. Chem. Eng.* 29:2380–85.
28. Sun, Q., and Faller, R. 2006. Crossover from unentangled to entangled dynamics in a systematically coarse-grained polystyrene melt. *Macromolecules* 39:812–20.
29. Sun, Q., and Faller, R. 2006. Systematic coarse-graining of a polymer blend: Polyisoprene and polystyrene. *J. Chem. Theor. Comp.* 2:607–15.
30. Faller, R. 2004. Automatic coarse graining of polymers. *Polymer* 45:3869–76.
31. Faller, R., and Reith, D. 2003. Properties of polyisoprene: Model building in the melt and in solution. *Macromolecules* 36:5406–14.
32. Henderson, R.L. 1974. Uniqueness theorem for fluid pair correlation-functions. *Phys. Lett. A* 49: 197–98.
33. Jain, S., Garde, S., and Kumar, S. K. 2006. Do inverse Monte Carlo algorithms yield thermodynamically consistent interaction potentials? *Ind. Eng. Chem. Res.* 45:5614–18.

34. Groot, R. D., and Warren, P. B. 1997. Dissipative particle dynamics: Bridging the gap between atomistic and mesoscopic simulation. *J. Chem. Phys.* 107:4423–35.
35. Sun, Q., Pon, F.R., and Faller, R. 2007. Multiscale modeling of polystyrene in various environments. *Fluid Phase Equilibria* 261:35–40.
36. Frenkel, D., and Smit, B. 1996. *Understanding Molecular Simulation: From Basic Algorithms to Applications.* San Diego: Academic Press.
37. Ghosh, J., Wong, B. Y., Sun, Q., Pon, F.R., and Faller, R. 2006. Simulations of glasses: Multiscale modeling and density of states Monte Carlo simulations. *Molecular Simulation* 32:175–84.
38. Ghosh, J., and Faller, R. 2006. Modelling of the glass transition of ortho-terphenyl in bulk and thin films. In *Mechanics of Nanoscale Materials and Devices (Proceedings of the 2006 MRS Meeting)*, ed. S. Asif. Warrendale, PA: MRS.
39. Ghosh, J., and Faller, R. 2006. A comparative molecular simulation study of the glass former ortho-terphenyl in bulk and free standing films. *J. Chem. Phys.* 125:044506.
40. Kudchadkar, S. R., and Wiest, J. M. 1995. Molecular-dynamics simulations of the glass former ortho-terphenyl. *J. Chem. Phys.* 103:8566–76.
41. Greet, R. J., and Turnbull, D. 1967. Glass transition in o-terphenyl. *J. Chem. Phys.* 46:1243–51.
42. Naoki, M., and Koeda, S. 1989. Pressure–volume–temperature relations of liquid, crystal, and glass of o-terphenyl. Excess amorphous entropies and factors determining molecular mobility. *J. Phys. Chem.* 93:948–55.
43. Lindahl, E., Hess, B., and van der Spoel, D. 2001. GROMACS 3.0: A package for molecular simulation and trajectory analysis. *J. Mol. Model.* 7:306–17.
44. Berendsen, H. J. C., Postma, J. P. M., van Gunsteren, W. F., and Hermans, J. 1981. In *Intermolecular Forces*, ed. B. Pullman, 331–342. Dordrecht: Reidel
45. Ghosh, J., and Faller, R. 2007. State point dependence of systematically coarse-grained potentials. *Molecular Simulation* 33:759–67.
46. Smith, W., and Forester, T. 1996. DL_POLY_2.0: A general-purpose parallel molecular dynamics simulation package. *J. Molec. Graphics* 14:136–41.
47. Pon, F. R., Sun, Q., and Faller, R. 2007. Static and dynamic heterogeneities in polyisoprene–polystyrene blends by molecular simulation. submitted.
48. Allen, M. P., and Tildesley, D. J. 1987. *Computer Simulation of Liquids.* Oxford: Clarendon Press.
49. Berendsen, H. J. C., Postma, J. P. M., van Gunsteren, W. F., DiNola, A., and Haak, J. R. 1984. Molecular dynamics with coupling to an external heat bath. *J. Chem. Phys.* 81:3684–90.
50. Sun, Q., and Faller, R. 2007. Phase separation in polyisoprene/polystyrene blends by a systematically coarse-grained model. *J. Chem. Phys.* 126:144908.
51. Bates, F. S., and Fredrickson, G. H. 1990. Block copolymer thermodynamics: Theory and experiment. *Annu. Rev. Phys. Chem.* 41:525–57.
52. Frischknecht, A., and Fredrickson, G. H. 1999. Microphase boundaries and chain conformations in multiply branched diblock copolymers. *Macromolecules* 32:6831–36.
53. He, Y., Lutz, T. R., Ediger, M. D., Pitsikalis, M., Hadjichristidis, N., and von Meerwall, E. D. 2005. Miscible polyisoprene/polystyrene blends: Distinct segmental dynamics but homogenous terminal dynamics, *Macromolecules,* 38:6216–26.
54. Se, K., Takayanagi, O., and Adachi, K. 1997. Dielectric study of miscibility in weakly segregated polymer blends. *Macromolecules,* 30:4877–81.

7 Systematic Approach to Coarse-Graining of Molecular Descriptions and Interactions with Applications to Lipid Membranes

Teemu Murtola
Laboratory of Physics and Helsinki Institute of Physics, Helsinki University of Technology

Ilpo Vattulainen
Department of Physics, Tampere University of Technology

Mikko Karttunen
Department of Applied Mathematics, The University of Western Ontario

CONTENTS

7.1 INTRODUCTION

Most physical systems exhibit interesting phenomena over a wide range of length and time scales. Typically a single method is not able to cover the full range, but different approaches are needed for different scales. In computational modeling, the main practical limitation comes from the available computing capacity: detailed models can only be studied over limited time and length scales. For example, electronic structure calculations are only practical up to tens or hundreds of atoms, and classical molecular dynamics (MD) simulations are just beginning to push the limit of a million atoms. The time scales that can be studied are at most hundreds of nanoseconds.

To achieve larger scales, simplified, coarse-grained (CG) models are needed. The central idea in coarse-graining is to find simplified representations for the systems while preserving the essential physics. These simpler models can then be studied on longer time and length scales to complement the picture given by more detailed models. Conceptually, it would be attractive to construct the CG model systematically; that is, by utilizing information from more detailed models to construct the model as rigorously as possible. Examples of methods that can be used in this context include Boltzmann inversion,[1,2] inverse Monte Carlo (IMC),[3,4] the so-called GENERIC formalism for non-equilibrium thermodynamics,[5,6] and matching of forces in the two models.[7–9] Boltzmann inversion and IMC use structural data to match the models to one another.

Coarse-graining of any system consists of two stages. First, one has to define the CG degrees of freedom for representing the system. A very typical choice is to use semiatomistic models where small groups of atoms are described as single interaction sites. So far, the effects of these choices have been studied relatively little. The second stage, after defining the CG representation, is to determine the effective interactions used in the CG model. The latter issue has been studied rather extensively during the last decade (see below). For the sake of focus, the examples here are mainly related to particle-based models of membrane systems.

The choice of the degrees of freedom is intimately related to the ability of the model to describe different phenomena. More specifically, the combination of the degrees of freedom and the form of the interactions is related to this ability, but practical considerations typically limit the interactions to a relatively simple form. For example, to make a particle-based model computationally tractable, the interactions are typically pairwise and isotropic. This may then limit the applicability of the model if the CG particles are not approximately spherical.

Typically, the description is selected using intuitive or heuristic arguments, with the aim of finding the molecular degrees of freedom that are most relevant for the processes at hand. There have been a few attempts at a more systematic approach: Schulten and co-workers have proposed using a neural-network-like algorithm for reproducing the molecular topology based on a single static structure,[10,11] and we have used self-organizing maps (SOMs) for finding the most relevant conformations in an atomistic trajectory.[12] The idea of using clustering methods, and SOMs in particular, for coarse-graining is discussed in more detail in Section 7.2.

For the interaction part, a significant amount of interest has been devoted to the development and applications of semiatomistic models.[13–15] These can be divided into two major classes depending on the treatment of solvent: for implicit solvent methods, there are no solvent particles, and the solvent effects are either just included in the effective interactions or they are included by solving hydrodynamic equations.[16–19] Such models can be used to study much larger systems than models with explicit solvent, but for more complex geometries, hydrodynamic effects are difficult to include. For explicit solvent, several models have been proposed, differing mostly in the way the interactions are constructed (see, for example, Refs. 14 and 15 for reviews). Recently, a versatile model constructed by Marrink et al.[20] has been successfully applied to various studies of lipid systems.[21–25] This model is parameterized using thermodynamic quantities such as partition coefficients between water and oil. The semiatomistic models constructed by Voth and coworkers[26–28] employ the force-matching approach for constructing the interactions, and are discussed separately below. Another typical approach in this context is the use of dissipative particle dynamics (DPD) with soft interaction potentials to look at behavior at long length and time scales.[29–32] In addition to membranes, DPD has been widely applied in the context of polymer simulations.[33–35]

An alternative approach is to use simpler theoretical models that include only the most relevant degrees of freedom.[36–49] Early examples of such models include lattice phospholipid models by Doniach[37] and Pink et al.,[38] and similar theoretical models were later constructed for more complex systems including binary[39] and ternary[40] systems. These models have been able to describe the qualitative features of phase diagrams. To describe the translational degrees of freedom better, Zuckermann and coworkers[41–45] have developed an off-lattice model where the molecules are described by hard-core particles with spin-type internal degrees of freedom. With a particular choice of interaction parameters, they have designed models whose phase diagrams are in agreement with the experimental ones for phospholipid bilayers containing cholesterol or lanosterol.[41–44] In addition, lateral diffusion in the bilayers has been successfully described.[45] In similar spirit, we have constructed simple models for phospholipid/cholesterol bilayers, deriving the interactions systematically from atomistic models.[46,47] Other approaches in the literature include field-theoretical models that have been applied to study phospholipid/cholesterol bilayers[48] and phase separation in mixed phospholipid bilayers.[49]

With most of the above approaches, the main difficulty is the proper construction of the interactions. A conceptually attractive alternative would be to construct the interactions by matching selected properties of the CG model to more detailed simulations of the same system. Several approaches have been proposed based on this idea, such as force matching[7–9] and structure matching.[1–4]

The method of Izvekov and Voth[8,9] uses the configurations and forces from a detailed simulation to perform a least-squares fit to obtain the CG forces. In the context of biological systems, the method has been successfully applied to constructing semiatomistic models of pure phospholipid bilayers[26] and phospholipid/cholesterol bilayers.[27] In addition, it has been applied to model a membrane protein in atomistic detail within a semiatomistically modeled bilayer.[28]

Several other methods have been proposed based on fitting the interactions to the average structural properties of the detailed simulations.[1–4] For dilute systems, it may be possible to obtain good structural properties by using either the potential of mean force between two particles or the Ornstein–Zernike equation together with a closure relation to approximate the pair potential.[50] However, it has been shown that already for relatively simple systems (a mixture of water and ethanol), iterative approaches are needed to accurately reproduce the structure at high densities.[51] Perhaps the simplest iterative method is Boltzmann inversion, used commonly in polymer physics.[1,2] The IMC method improves the convergence of the iteration by taking into account correlations between the effects of interaction modifications, at a cost of computer time needed per iteration. The IMC approach has been used by Lyubartsev et al. to describe ions solvated in water,[3,4,52] to study the binding of different ions to DNA,[53] and to model self-assembly and behavior of phospholipids in water.[54] Elezgaray and Laguerre have also used a structural optimization method to construct a CG model for a pure phospholipid bilayer.[55] Shelley et al.[14,56] employed a similar approach for determining some of the interactions in their model. Our IMC work on models for phospholipid/cholesterol bilayers[46,47] is discussed in more detail in Section 7.3.

7.2 SELECTING COARSE-GRAINED DEGREES OF FREEDOM

As discussed in the Introduction, the choice of the degrees of freedom for a CG model can significantly alter the descriptive power of the model. However, this choice is typically made based on intuitive or heuristic arguments, and the effects have been studied relatively little. Conceptually, it would be attractive to utilize information from more detailed models in the process of choosing the CG description. The conformations seen in the atomistic simulation provide one alternative for doing this. For example, in protein science it has been noted that different types of conformations have very different properties that have implications in stability and evolution.[57–59] One possibility for using the conformational information could be to use clustering methods[60] to classify the observed conformations. Clustering methods describe the data using relatively few model conformations, and hence they produce an abstraction of the data. This feature can be used to extract general characteristics of the data, as well as to find the most relevant conformational states. Such information could then be used to select a minimal description that is able to reproduce the most important conformational features.[12]

In this section, we discuss how SOMs could be used for the purpose outlined above. A SOM is not exactly a clustering method, but can rather be thought of as an advanced visualization tool. A SOM is an unsupervised machine-learning approach that requires very little a priori knowledge of the data that are being analyzed. It is based on ideas similar to neural networks;[61] that is, the map is *trained* with the input data, and in this process it learns to represent the typical features in the data. Discussion on the relationship between SOMs and other clustering methods is postponed until Section 7.2.3, after the presentation of the method. The basics of SOMs are presented in Section 7.2.1, and Section 7.2.2 then gives an overview of the results when SOM is applied to a lipid bilayer system. Finally, a brief discussion of the method is included in Section 7.2.3. The focus here is on the presentation of the method and the results, and the reader is referred to Ref. 12 or the book by Kohonen[62] for technical details.

7.2.1 Self-Organizing Maps

A SOM is a powerful software tool for the visualization of high-dimensional data.[62] Conceptually, it is a mapping from high-dimensional input data vectors into a low-dimensional (usually one- or two-dimensional) grid of so-called neurons. The output space does not have any particular physical interpretation, but rather it is an aid in visualizing the data. In the present case, the input space is the conformational space of the molecules of interest. The conformations must be described by an n-dimensional vector, for example, by using dihedral angles formed by four subsequent atoms (see Figure 7.1a for a definition). Each neuron has an associated model vector, and all input data vectors are mapped to the neuron that has the most similar model vector. The core of the SOM method is the construction of such a mapping. This is done through so-called *training*, and the aim is to construct the mapping such that similar data vectors are mapped to the same or neighboring neurons. Hence, groups of similar neurons on the map should describe a group of similar data vectors in the original

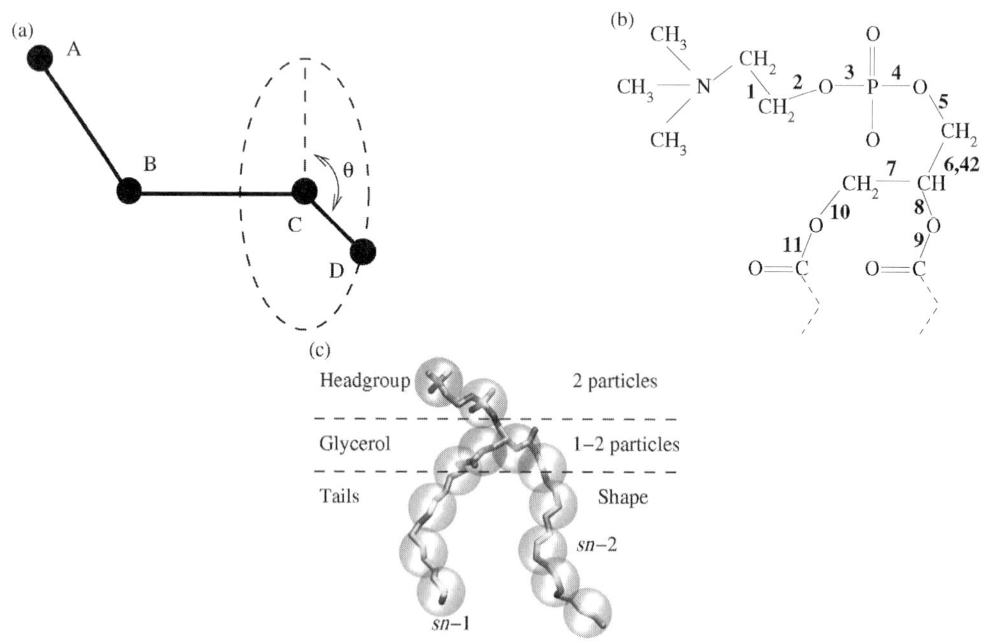

FIGURE 7.1 (a) The dihedral angle θ formed by atoms A–D defines the orientation of the vector C–D as seen by an observer looking along B–C with B–A pointing up. (b) Numbering of PLPC headgroup dihedral angles. (c) Schematic representation of SOM-derived CG model[12] (see text for details). The three regions were considered separately based on the different SOMs.

data; that is, the map should be a low-dimensional representation of the topology of the input data. The low dimensionality of the output space makes the visualization of the results simple and thus makes it possible to quickly find relevant information about the system. These visualization properties are the main advantage of the SOM approach.

The general algorithm for constructing an SOM is given as Algorithm 1. Before the algorithm can be run, some choices have to be made. The selection of map structure and training parameters are specific to the SOM method, but data representation needs to be selected for all clustering methods. These choices are discussed in more detail below, but let us first go through the algorithm.

Algorithm 1. Construction of a self-organizing map.
Require: Data representation, map structure, training parameters ($h_{i,j}(t)$, length of training) selected (see text)
Ensure: Final model vectors in $\boldsymbol{m}_i$

```
N ⇐ length of training
Initialize model vectors m_i (see text)
For t = 1 to N do
    x ⇐ random input data vector
    BMU ⇐ index of the model vector most similar to x
    For each model vector m_i do
        m_i = m_i + h_BMU,i(t)[x − m_i]
    end for
end for
```

First, the model vectors are initialized. The simplest procedures are random and linear initialization.[62] In random initialization, the components of the model vectors are given random values between the minimum and maximum values in the data, while in linear initialization the model vectors are placed on a regular lattice on a two-dimensional plane (one-dimensional if the map is linear) that is oriented such that the variance of the data, projected to the plane, is maximal. This plane is spanned by the eigenvectors of the data covariance matrix corresponding to the largest eigenvalues.[62] In principle, the initialization method should not affect the results, but careful initialization may lead to faster convergence in the training phase. For the example in the next section, linear initialization was used.

The main part of the algorithm is the training loop. It repeatedly presents the data vectors to the map, and after each data vector the model vectors are improved. At each step, the best-matching unit (BMU) of the data vector is moved towards the data vector. In addition to the BMU, the model vectors of neighboring neurons are also updated, although by a smaller amount. This constrains the nearby neurons to be similar. The amount by which the different neurons are moved is controlled by the neighborhood function $h_{i,j}(t)$, which is a decreasing function of the distance between the neurons with indices i and j, as well as the time t that indicates how much of the training has passed. Typically, the neighborhood function is written as

$$h_{i,j}(t) = \alpha(t) g(\|r_i - r_j\|; t),$$

where $\alpha(t) \in [0, 1]$ is a learning rate, and $g(r; t)$ is a shape function, which is typically taken to be Gaussian with a time-varying variance $\sigma^2(t)$, scaled such that $g(0; t) = 1$. The vectors $\boldsymbol{r}_i$ are the positions of the neurons on the low-dimensional grid. Both $\sigma(t)$ (called the neighborhood radius) and $\alpha(t)$ are decreasing functions of time such that in the beginning of the training the map organizes rapidly, while towards the end of the training more and more detailed features of the map are tuned.

There is also a batch version of the training algorithm that represents all data vectors to the map before modifying the model vectors. Each model vector is then updated as the weighted average of the data vectors that have a BMU within the neighborhood radius, using the values of the neighborhood functions $h_{i,j}(t)$ as weights. However, using periodic variables is much more problematic in this approach.

The SOM Toolbox for MatLab[63] can be freely downloaded from http://www.cis.hut.fi/projects/somtoolbox/, and it is straightforward to get started. Minor modifications to the software are needed to take into account the periodic nature of the dihedral angles that are used to represent the conformations, but otherwise this toolbox provides an out-of-the-box implementation of the most commonly needed tasks and visualizations.

Data representation and map structure. To use SOM (or most other clustering methods) for any data, one has to choose how to represent the items as n-dimensional vectors. Typically, Euclidean distance between these vectors is used to measure the similarity of the items, which should be taken into account when selecting the representation. The results of the analysis can only be as good as the underlying representation of the data allows. Hence, selecting the data representation is perhaps the most important step in applying the SOM. For conformational data, using dihedral angles is a common choice,[12,64,65] but other alternatives may be better in certain cases.[60] For example, the dihedral angles are rather sensitive to local conformational changes and similarities, but not very sensitive to the apparent cancellation of changes in two locations. For visualization, it is preferable that the chosen variables be used to reconstruct the conformation adequately.

A two-dimensional hexagonal grid of neurons is a standard choice for the map structure. For the size of the map, an absolute upper limit is given by the number of training samples, because there should be a sufficient number (preferably at least a few hundred) of training samples per neuron. Below this limit, one can then choose a size for the map such that a desirable level of detail is achieved. Generally, a larger map is better (up to the limit imposed by the training data), but it is also computationally heavier for large data sets and thus a compromise is typically needed. The training parameters also influence the level of detail (see below). One dimension is typically chosen significantly larger than the other to allow the map to orient itself properly.[62] For the application in the next section, a map size of 48×72 neurons was used for most of the studies. Training such a map (with 400,000 conformations, each having 12 dihedrals) takes a few days on a standard desktop computer.

Training parameters. The features of the map after training are mainly influenced by three factors: the behavior of the neighborhood function $h_{i,j}(t)$ as a function of time, the length of training, and the training data itself. As discussed above, the behavior of the neighborhood function is parameterized by two quantities: the learning rate $\alpha(t)$ and the neighborhood radius $\sigma(t)$. The effects of these parameters have been studied in detail in Ref. 12 in the context of lipid systems. Qualitative rules were derived for how the different parameters affect the results (see below for a summary). Also, reasonable initial values for these quantities are given. After training a map with these values, one can then see whether the level of detail is proper for the use one needs, and possibly fine-tune the values for the application one has in mind.

For the training parameters, the most important thing is to have the initial neighborhood radius large enough to allow for the initial organization of the map, and also to have a sufficiently large final neighborhood radius for good visualization properties. Too small a radius allows the neighboring neurons to differ from each other too much to see clear regions on the map. Further, long enough training (in practice, slow enough variation of the training parameters) should be used to avoid trapping into a local optimum for the map. Full automation of this process is unfortunately difficult because of the qualitative nature of the SOM and the large amount of human interpretation needed to get full advantage of the map. However, the exact values of the training parameters are not very important, since a rather wide range of values leads to very similar maps.

The choice of training data can also have an effect on the resulting map since the map can be only as good as the training data. Generally, the number of configurations that is used in training is not as crucial as a good sampling of the conformational space.[12] The conformational dynamics; that is, the trajectories formed by the conformations of each molecule over time on the map, can be used to qualitatively judge the sampling: if the trajectories of individual molecules cover only a small part of the map, this indicates that the sampling may not be good enough.

7.2.2 Application to Lipid Bilayers

To illustrate the SOM approach, let us now briefly look at results from a lipid bilayer system. As a model system for the applicability of SOM, we have used a 50 ns MD simulation of a 1-palmitoyl-2-linoleoyl-*sn*-glycero-3-phosphatidylcholine (PLPC) bilayer.[66,67] PLPC was chosen because the double bonds in the *sn*–2 chain can give rise to interesting conformational features, and this choice also allows us to compare our results to those from earlier studies in Ref. 65. A comprehensive presentation of the results can be found in Ref. 12, and here we discuss them mainly in the context of coarse-graining.

In principle, it is possible to include all dihedral angles in a PLPC molecule to the SOM analysis simultaneously, but in practice training a large enough map requires a long time and a large amount of memory.[12] Hence, it is useful to reduce the complexity of the conformational space by studying only a subset of the dihedral angles. We did this for three different groups: the headgroup, the glycerol region, and the diunsaturated *sn*–2 chain. We also confirmed a posteriori that the different groups are not correlated (more than what follows from the overlap between the headgroup and the glycerol regions), justifying independent analysis of the different parts.[12] Here, we present the results for the headgroup region in detail, and briefly discuss the results for the other regions. The conformation of the headgroup region was described using the dihedral angles indicated in Figure 7.1b.

An overview of the trained headgroup SOM is visualized in Figure 7.2. Figure 7.2a shows the so-called U-matrix, and Figure 7.2b shows the so-called component planes. In the U-matrix, neurons are colored based on their average similarity to their neighbors. This gives a clear overview of the general structure of the map. In light regions the model vectors of neighboring neurons are similar, and darker regions mark larger differences. Hence, large light areas correspond to clusters; that is, groups of neurons that are similar to each other and dissimilar to other neurons. The component planes, on the other hand, show the dihedral angles of the model vectors in an easily readable way. Each small figure shows the values of a single dihedral angle for each model vector, and the color range visualizes the angle range that is actually used by the model vectors. Figure 7.2a also shows manually constructed boundaries for several clusters on the map, and reconstructed conformations for selected neurons within these clusters. The neurons were chosen approximately in the middle of the cluster, and the conformations are qualitatively similar for all neurons within a cluster. It should be noted that the conformations are all shown in the same orientation (the glycerol backbone is in identical position for all of them), but in practice they may prefer a different orientation in the bilayer. In fact, the glycerol backbone has clearly different average orientation for different conformations mapped to different regions in the map (T. Murtola, unpublished results).

The different visualizations in Figure 7.2 show that the map is divided into four major regions, and inside these regions there are some smaller clusters. First, there is a prominent boundary in the map, particularly distinct in the U-matrix, extending diagonally from the lower left corner to the upper right corner. The upper half is further divided into two regions, most clearly visible in the component planes. The clusters inside these major regions are distinguished by different orientations of the P–N vector in the headgroup and, in some cases, different orientation of the beginning of the *sn*–2 chain. Finally, in the boundary region, there is a two-part region where dihedral 8 is significantly different from the rest of the map. There are no clear clusters within this region, indicating that the value of dihedral 8 is the main determining factor for this class of conformations.

The lower part of the map has large homogeneous clusters, while the upper part is scattered with small clusters and regions with no clear clusters at all. From the component planes in Figure 7.2b it is evident that the major regions are differentiated by different values of dihedrals 6, 7, and 8. These dihedral angles are located in the region where the headgroup is attached to the glycerol backbone, and thus the boundary reflects the conformational degrees of freedom of the headgroup with respect to the glycerol region. This is also visible in the conformations visualized in Figure 7.2a: the relative orientations of the phosphate group and the atoms in the glycerol backbone show three different major classes, partitioned in accordance with the component planes. There are no

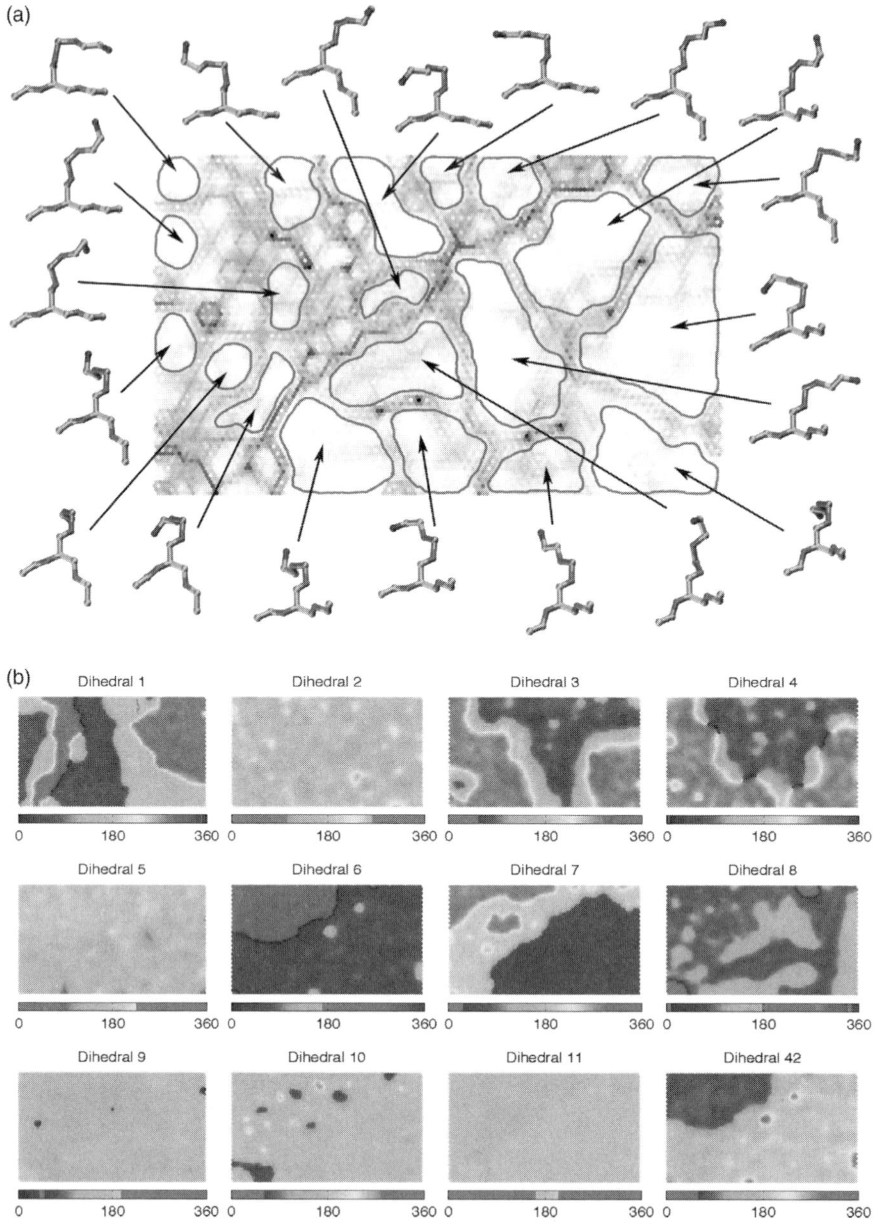

FIGURE 7.2 (See color insert following page 238.) Visualizations of SOM trained with PLPC headgroup data.[12] (a) U-matrix. Neurons are shaded based on similarity to their neighbors; lighter shades indicate higher similarity. Several regions of high similarity (selected manually) are encircled. For each region, the conformation of one model vector is visualized. (b) Component planes. Each small figure shows the value of one dihedral for all model vectors on the map (see fig. 7.1b for definitions of the numbers). The color bars under each plane show the range of values that are present on the map; unused values are in gray.

conformations visualized from the region where dihedral 8 is different, but a careful inspection of these conformations shows that they are characterized as having the first carbon of the *sn*–2 chain and the *sn*–1 carbon atom on the same side of the *sn*–2 C–O bond.

On crossing the major boundaries in the U-matrix, we see a major change in more than one dihedral angle, which explains why the boundaries are so clearly visible. This also suggests that there are high potential barriers associated with the crossings, and therefore we should see fewer conformational transitions across the boundary than inside the regions. This is also confirmed by studying the dynamics of the molecules on the map: there are very few crossings across the diagonal boundary, or from/to the region characterized by dihedral 8.[12]

The glycerol region contains many of the same dihedral angles as the headgroup region. Hence, it is not very surprising that the structure of the glycerol SOM is very similar to the headgroup case (results not shown here, see Ref. 12). The map is divided into four classes similar to those of the headgroup map, but the clusters within these groups are distinguished by the different orientations of the first bonds of the tails.

In the case of the diunsaturated *sn*–2 chain, the structure of the U-matrix is markedly different from the other two cases (see Ref. 12). The map has a large number of small clusters, but the boundaries of these clusters cannot be clearly defined. There are also areas that are difficult to classify as belonging to any cluster.

The differences in the structure of the map, compared with the other two cases, can be explained by the conformational space accessible to the chain. There are a large number of probable conformations of the chain, characterized by a major change in one or more of the dihedral angles. As there is only a limited number of neurons on the map, it is not possible to have a separate cluster for each of these conformations. Other measures also support the conclusion that a larger map may be needed to have clear clusters on the map.[12] However, a careful look at the component planes shows that despite the problems with the size of the map, the SOM is able to form a rather good representation of the conformational space of the chain: each region of the map is characterized by a specific combination of the different possible values for the dihedral angles, and the boundaries for these domains are rather sharp. The relative areas of the different regions are also comparable to the probability of observing a particular value of the angle.

Let us now discuss an example of how the above information from the SOMs could be used in constructing CG models. The ability of the SOM to find the most relevant states with only minimal human intervention could be particularly useful for the selection of the necessary degrees of freedom. Here, we discuss construction of a semiatomistic model. We separately focus on each part of the molecule, and select a minimal CG description that is able to represent the most relevant conformations of the lipid as represented by the maps. Other possible applications are discussed in Section 7.2.3.

The headgroup map shows that there is a fairly small set of specific conformations in this part of the molecule. The most important differences between the conformations are related to the orientation of the P–N vector with respect to the glycerol backbone. To be able to describe these conformations, the CG headgroup should have at least two particles that define the direction of the P–N vector. This finding is in line with atomic-scale MD simulations and experiments that have demonstrated the importance of the P–N vector orientation for electrostatic properties at the membrane–water interface.[68,69]

The glycerol region is very similar to the headgroup region, again showing the importance of the connection between the P–N vector and the glycerol backbone. The direction of the first bonds of the tails is also different in different parts of the map. Hence, the glycerol region itself does not contain any significant internal degrees of freedom, and only acts as a joint between the headgroup and the tails. Hence, it can be described by 1–2 particles. Two particles help to distinguish the tails from each other, although this could also be achieved by a careful choice of bending potentials and other interactions. Two particles could also make it easier to describe the relative orientation of the glycerol and the P–N vector, but again, this could also be achieved by a proper choice of intramolecular interactions.

Finally, the *sn*–2 map is very homogeneous, showing a lack of any specific important conformations. This indicates that the most pronounced effect of the double bonds is to induce generic disorder in the tail region instead of promoting a set of typical conformations. Hence, the general shape of the tails is the most important feature to consider in the CG description, and the double bonds can be included by appropriate intramolecular interactions. Figure 7.1c summarizes the model. To complete the model, we would also need to determine the interactions for the different particles, but such analysis is beyond the scope of this discussion. However, it is interesting to note that the successful CG model developed by Marrink et al.[20] incorporates the features discussed above. That model was constructed based on experiences from the atomistic models, and the similarity demonstrates the possibilities the SOMs have to offer in this context.

7.2.3 Discussion

The SOM results discussed here are very preliminary, in particular those concerning coarse-graining applications. However, the potential of SOM has been scarcely explored in the context of conformational analysis: there has been some work on analysis of 3D structure of amino acid sequences,[64] and prior to our work, Hyvönen et al. also made some preliminary studies of SOM in the conformational analysis of lipids,[65] but we are unaware of any other studies that directly addressed this issue. For example, most of the SOM work in the field of protein science focuses on analysis on sequence level: SOM has been applied to, for example, classification of sequences within a protein family,[70] identification of overrepresented motifs in sequences,[71,72] prediction of HIV protease cleavage sites,[73] and a study of ammonium salts as ligands at the neuronal nicotinic acetylcholine receptor.[74]

The SOM does have its limitations in applications to systematic coarse-graining: the method is intrinsically qualitative in nature, so human interpretation is always needed. Using large maps to analyze complex structures is also impractical from the computational point of view because of the time needed for training the map. Analyzing the structures in parts could at least partially solve the latter problem, but this also introduces more work for the human analyst. Despite these limitations, there are several interesting directions for future research.

It could be possible to use the clusters on a single-molecule SOM (such as those in the previous section) to construct a CG model that includes the conformational information as internal states of the molecule. The SOMs could also be used to visually assess the correlations between the conformations of neighboring molecules, and this information could then be used either as a check for the CG model or as input for construction of the interactions.

The SOM approach could also be used for studies beyond single-molecule level. Possible applications include classification of conformations of phospholipid–cholesterol pairs,[75–77] sphingolipid–cholesterol complexes in lipid raft membranes,[78,79] studies of structures of lipid complexes bridged together by salt,[80,81] characterization of conformations of carbohydrate moieties in glycolipids,[82] and the exploration of conformational degrees of freedom associated with peptides attached to membranes.[83] The main difficulty in this approach is the proper selection of variables for describing the conformations of molecular complexes. However, with a carefully selected set of variables the SOM could give valuable insight into the interaction between the molecules, without the need to make a priori assumptions. This could also provide valuable insight into how to select the CG description so as to reproduce these features as well as possible.

The discussion in this chapter is not completely specific to SOM, and it is possible that other clustering methods could also be applied to a similar end. The main advantage of SOM over other clustering methods is the ease of visualization of the results. Further, this visualization can be made in such a way that it is generally easy for a human analyst to find nontrivial characteristics in the data. This makes it easier to assess the validity of the results than with traditional clustering approaches, where the output typically consists of just the model vectors for the final clusters, and possibly some general quality measure(s). In many ways, SOM is more qualitative than many other methods, but it also gives more freedom for a careful interpreter to get around the limitations of the method. For SOM, some care is also needed in selecting the size of the map and the training parameters, and some experimentation may be required to take full advantage of the approach. Effective interpretation of the results may also require some effort if one has no previous experience of similar methods.

7.3 INTERACTIONS IN COARSE-GRAINING

In addition to the CG description, the effective interactions play a central role in determining the properties of the model. As with the description, obtaining the interactions systematically from a more detailed model is conceptually attractive. Several different approaches for determining the interactions systematically were discussed in the Introduction. Here, we focus on structural inversion; that is, the structural properties of the detailed model are used to determine the interactions for the CG model. The structure is typically represented using radial distribution functions $g(r)$ (RDFs), which are given by

$$g(r) = \frac{V}{N_p A(r)} \langle S(r) \rangle, \tag{7.1}$$

where $S(r)$ is the number of particle pairs with separation r, V is the total volume of the system, N_p is the total number of particle pairs, and $A(r)$ is the surface area of a sphere of radius r. The normalization factor is chosen such that an ideal gas has $g(r) \equiv 1$. The RDFs are taken as the input for the inversion because of a theorem proved by Henderson: for a given thermodynamic state of a system, there is one-to-one correspondence between the RDFs and isotropic pairwise interactions.[84] The RDFs describe the structure of the system at short range, and they contain information about, for example, the occurrence of different types of particles in the neighborhood of others. They can also be used to characterize the phase behavior of the system. Hence, the CG model should at least qualitatively reproduce the structure and the phase behavior of the original detailed model.

In this section, we focus on the IMC method, which is one method for achieving the RDF inversion. First, we present the method itself, and then move on to discuss an application for constructing a very simple model for phospholipid/cholesterol bilayers. For other applications of the method, see the references listed in the Introduction. Finally, we present a short discussion of the method in the context of the application. General discussion follows in Section 7.4.

7.3.1 Inverse Monte Carlo

7.3.1.1 The Method

IMC is one method for iteratively adjusting the effective interactions until they match the target radial distribution functions.[3,4] The potentials are represented by a piecewise constant grid approximation and the RDFs are calculated with the same accuracy. For simplicity, let us consider a single-component system with a pairwise interaction potential $V(r)$. The potential is then approximated as

$$\mathrm{V(r)} = \mathrm{V}_\alpha, \quad (\alpha - 0.5)h < r < (\alpha + 0.5)h, \quad \alpha = 0, 1, \ldots,$$

where h is the size of the potential bins. Also, let S_α be the number of particle pairs whose distance is within thc bin α. The Hamiltonian of the system is then

$$H = \sum_\alpha S_\alpha V_\alpha.$$

The value $<S_\alpha>$ is also tied to the RDF within the same bin (see Equation 7.1).

A simple version of the IMC algorithm is described as Algorithm 2. The algorithm first initializes the interactions with a simple initial guess (see below for more discussion). It then proceeds to improve the approximation: during each iteration, the derivatives of $<S_\alpha>$ with respect to the V_β can be calculated for all pairs (α, β) using statistical mechanics relationships.[3,4] These can then be used to express the changes in $<S_\alpha>$ to the first order in terms of changes in V_β as

Algorithm 2. Simple version of inverse Monte Carlo.
Require: Grid approximation of RDFs in $\boldsymbol{g}^0$
Ensure: Final potentials V reproduce the given RDFs

$\boldsymbol{V}^0 \Leftarrow -k_BT \ln g^0$
$\boldsymbol{S}^0 \Leftarrow$ denormalized g^0
$i \Leftarrow 0$
repeat
  $i \Leftarrow i+1$
  Perform a simulation with the interactions $\boldsymbol{V}^{i-1}$, calculate $<S_\alpha>$, $<S_\alpha S_\beta>$
  Solve $<\boldsymbol{S}>-\boldsymbol{S}^0 = -A\Delta\boldsymbol{V}$ for $\Delta\boldsymbol{V}$, where $A_{\alpha,\beta} = -(<S_\alpha S_\beta> - <S_\alpha><S_\beta>)/k_BT$
  Calculate g by normalizing $<\boldsymbol{S}>$
  $\boldsymbol{V}^i \Leftarrow \boldsymbol{V}^{i-1} + \Delta\boldsymbol{V}$
until $\|\boldsymbol{g} - \boldsymbol{g}^0\| <$ chosen cutoff
$\boldsymbol{V} \Leftarrow \boldsymbol{V}^i$

$$\Delta\langle\mathbf{S}\rangle = A\Delta V, \quad A_{\alpha\beta} = \frac{\partial S_\alpha}{\partial V_\beta} = -\frac{\langle S_\alpha S_\beta\rangle - \langle S_\alpha\rangle\langle S_\beta\rangle}{k_BT}, \tag{7.2}$$

where $\boldsymbol{S}$ and $\boldsymbol{V}$ are the vectors formed from the values S_α and V_β, respectively. When the desired $\Delta<\boldsymbol{S}>$ is known, Equation 7.2 can be inverted to calculate the interactions for the next iteration. The procedure is then repeated until convergence is achieved.

Generalization to multicomponent systems is straightforward and bonded interactions can also be handled. However, in the case of bonded interactions, the matrix A has a zero eigenvalue for each bonded interaction. The physical reason is that the bonded interactions can only be determined up to a constant, and thus the eigenvectors for the zero eigenvalues correspond to situations where a constant is added to one or more bonded potential. Because of these zero eigenvalues, Equation 7.2 does not have a unique solution. A simple remedy is to constrain the sum of the changes of each bonded potential to be zero. This constraint can be easily added to Equation 7.2: if $Av=0$, $\|v\|=1$ (i.e., v is a normalized eigenvector of A with a zero eigenvalue), then the solution to the equation $(A+vv^T)x = b - vv^Tb$ satisfies $Ax=b$ and $v^Tx=0$.

7.3.1.2 Practical Considerations

Initial interactions. The initial convergence of the IMC process depends on the initial guess for the effective interactions. Typically, the potential of mean force is used ($V^{(0)}{}_\alpha = -k_BT \ln g_\alpha$).[3,4] This works for simple systems, but the more complex the model, the worse the performance of this initial guess and the convergence of the iteration become. In these cases, some regularization of the procedure is needed (see below). For complex systems, some other way of constructing the initial potentials may become necessary. This issue is discussed in more detail in Section 7.3.2.3.

Regularization. As noted above, some regularization of the procedure may be needed if the initial guess for the interaction is poor. Further, straightforward inversion of Equation 7.2 requires good sampling of the four-particle correlation function $<S_\alpha S_\beta>$, which is very time consuming with a large number of effective interactions. If this quantity contains too much noise, the noise is transferred to the potentials and may lead to instabilities in the iteration. This is particularly a problem with bins where the RDFs are small but nonvanishing, because their sampling requires extensive simulations. However, the linear form of Equation 7.2 makes it easy to use a wide variety of regularization procedures to solve these problems.

For the first problem, the simplest solution is to multiply the $\Delta\boldsymbol{S}$ in Equation 7.2 by a small factor r between 0 and 1 to make the changes small enough such that the linear approximation is valid.[4]

The value of r can then be increased after a better approximation has been constructed to obtain faster convergence.

To reduce the effect of noise in the matrix A, a more elaborate scheme is needed. First, we note that solving Equation 7.2 is equivalent to solving the minimization problem

$$\min_{\Delta V} f(\Delta V), \quad f(\Delta V) = \left\| A\Delta V - \Delta S \right\|^2. \tag{7.3}$$

In this form, it is easy to add regularization terms to the problem: instead of minimizing $f(\Delta V)$, we minimize

$$\tilde{f}(\Delta V) = f(\Delta V) + a\left\| R\Delta V \right\|^2, \tag{7.4}$$

where a is a regularization parameter and R is a regularization operator. The simplest forms of R are a diagonal matrix and the second derivative operator. In these cases, the regularization either limits the magnitude of changes in the potentials or prefers changes that keep the potentials smooth. The best value for a can be selected by trial and error, and it can also be changed during the iteration. Due to the linear nature of the problem, the computational cost of the new problem is comparable to the original one. Such a regularization enables us to use fewer Monte Carlo steps during the initial phase of the iteration and hence to speed up the calculation considerably. When the iteration has converged within the numerical accuracy of the Monte Carlo simulations, the number of steps can be increased to refine the effective interactions further.

Thermodynamic constraints. To minimize finite-size effects to the effective interactions, the simulations during the IMC procedure should be carried out with a system that is identical in size to the system from which the target RDFs were determined.[3] In some cases, the effective potentials produced in this way do not generalize to larger systems.[47] For example, the effective interactions may become too attractive to maintain a reasonably uniform density in the system. Instead, larger systems form dense clusters separated by empty space, which is typically unphysical.

Situations where thermodynamic properties, particularly the pressure, of the CG model do not match the underlying atomistic model have also been encountered in other coarse-graining approaches.[1,9] Proposed solutions to the problem include iterative adjustment of the pressure followed by reoptimization of the interactions[1] and imposing additional constraints on the instantaneous virial produced by the effective interactions.[9]

Typically, the pressure of the CG model is used to impose additional constraints on the effective interactions. For a two-dimensional system (such as the one discussed in Section 7.3.2), the equivalent quantity is the surface tension γ, defined as

$$\gamma = \frac{1}{V}\left(\langle E_{\text{kin}} \rangle + \frac{1}{2}\left\langle \sum_{i<j} f_{ij} r_{ij} \right\rangle \right), \tag{7.5}$$

where $\langle E_{\text{kin}} \rangle = Nk_BT$ is the average kinetic energy in the system, and the latter term is the virial. Here, N is the number of particles, V is the area of the system, and f_{ij} and r_{ij} are the force and the distance, respectively, between particles i and j. If this quantity is close to zero or negative in the simulations on small systems, then the system may not be stable for larger system sizes.

The IMC procedure seems to favor a certain value of γ, so it is not possible to tune the surface tension first and then reoptimize the interactions as was done in Ref. 1. Instead, the value of γ has to be constrained during the IMC iteration. This can be achieving by solving the minimization problem (Equation 7.3) (or the regularized version) at each iteration with the additional constraint that γ should be equal to the value we specify.[47] For a linear problem, such a constraint is easily implemented.

One should note that the surface tension defined by Equation 7.5 cannot be directly related to the surface tension in the detailed model. This is because the effective potentials are in general volume dependent. Hence, the correct value of γ is not necessarily the same as the surface tension in the atomistic simulations. For membrane systems, there are also some issues with the small size of the atomistically simulated systems.[85] Because of these considerations, the value for γ has to be selected using other quantities. For the results below, we have used the experimental area compressibility of the bilayer to this end.[47]

7.3.2 Results for Phospholipids/Cholesterol Bilayers

Let us now illustrate the possibilities of the IMC method with an application to phospholipid/cholesterol bilayers.[47] The primary goal is to construct a model to study large-scale structural properties of a bilayer consisting of dipalmitoylphosphatidylcholine (DPPC) and cholesterol. The degree of coarse-graining must be high to achieve the interesting length scales. To this end, we describe each cholesterol molecule by a single particle, and each phospholipid molecule is described by three particles: one for the headgroup and the glycerol parts, and two others for the two tails. Each particle in the model describes the center-of-mass (CM) position of the corresponding group of atoms. Hence, there are four types of particles, and a total of seven distinct nonbonded interactions (several similar pairs can be combined to a single interaction[47]). We also include bonded interactions for all pairs in the DPPC molecule, but we do not have any bending potentials. The whole system is projected onto a two-dimensional plane, and the solvent degrees of freedom are integrated out. Some examples of the effective interactions constructed by IMC are shown in Figure 7.3a. The interactions are soft and smooth, and no simple analytic function can be fitted to the curves adequately. It should also be noted that the interactions are dependent on the cholesterol concentration. As our focus is on structural properties, we use Monte Carlo simulations to calculate the equilibrium properties of the system. The simulations are conducted in the canonical ensemble, and the area per molecule is fixed to the average area per molecule in the underlying MD simulations.

In the model, we assume that the two leaflets of the bilayer interact only weakly, and hence we focus on one of them. Further, we neglect the out-of-plane fluctuations of the bilayer and focus on the lateral organization only. These assumptions allow us to construct a two-dimensional model.

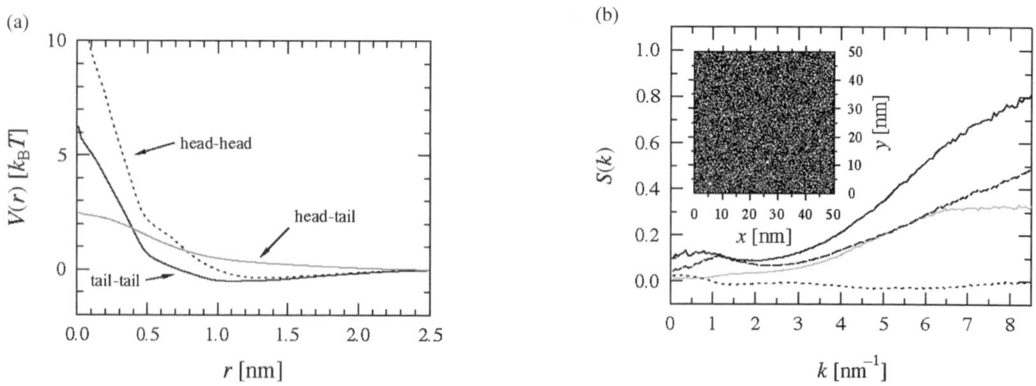

FIGURE 7.3 (a) Effective interactions for selected particle pairs in pure DPPC system. (b) Static structure factors for pure DPPC system. The solid black line shows the total structure factor, and the other lines show the head–head (solid gray line), tail–tail (dashed line), and head–tail (dotted line) structure factors. The inset shows a snapshot of the system. In the snapshot, only the tail particles are shown. The CG simulations were run with 16,384 molecules; that is, with 16 times the linear size of the original 128 lipid system.[47]

We also assume that the interactions between the different kinds of particles can be adequately described by pairwise and radially symmetric effective potentials.

This very simple CG model can be used to study the behavior of the bilayer system at much larger length scales than those accessible by more detailed simulation methods. In comparison to atomistic simulations, the speedup is of the order of seven orders of magnitude.[47] For example, the Monte Carlo simulations for the results below were mostly conducted for systems whose linear size was in the range 80–110 nm. Such systems contained 39,424–49,152 particles, and reasonable statistics (over tens of microseconds) could be obtained within a few days on a standard desktop computer.

The long-range structural properties of such a model are conveniently characterized by the static structure factor defined by

$$S(\mathbf{k}) = \frac{1}{N}\left\langle \sum_{i=1}^{N}\sum_{j=1}^{N} \exp\left\{-i\mathbf{k}\cdot(\mathbf{r}_j - \mathbf{r}_i)\right\} \right\rangle, \tag{7.6}$$

where N is the number of particles, $\mathbf{r}_i$ are the positions of the particles, and $\mathbf{k}$ is a vector in the reciprocal space. The structure factor can also be calculated for a subset of particles by restricting the summations to these particles. The structure factors are related to the RDFs by Fourier transforms, and hence in principle carry the same information. However, the long-range structure is more conveniently characterized using the structure factors.

7.3.2.1 Pure Phospholipid Bilayer

Figure 7.3b shows the circularly averaged static structure factors for the pure DPPC system. In addition to the total structure factor calculated over all pairs of particles, structure factors calculated over the head–head, head–tail, and tail–tail pairs are shown. The inset shows a snapshot of the system, displaying the locations of the tail particles. The head particles are not shown for clarity.

The most interesting feature of the total static structure factor in Figure 7.3b is the small broad peak at around $k \approx 1\ \text{nm}^{-1}$. A more careful look at the different components of the structure factor reveals that the peak is completely due to the tail–tail structure factor. Such a peak indicates that the tail density should have some density variation with a characteristic length scale around $l \approx 6$ nm. An examination of the snapshot in Figure 7.3b indeed shows such small variations. The feature does not change with longer equilibration or larger system size, and a similar peak can be seen in the 5% cholesterol concentration (see below). Further, a similar peak is still present when the cutoff of the effective interactions is changed or the effective interactions are derived from a larger atomistic simulation.[47] These facts strongly indicate that the peak is not an artifact of the simulations but an intrinsic feature of the system.

The presence of such a feature in the CG simulations prompted us to look more carefully at the underlying atomistic simulations to check whether such denser areas are present there as well. The atomistic simulations can also give a more detailed view of the nature of these density variations, assuming that the simulations are long enough to show them. However, the length scale associated with the density variation is similar to the size of the bilayer in the MD simulations, so only limited information can be gained from the atomistic simulations. The atom-scale MD simulations for a small bilayer of 128 lipids were found to show density variation reminiscent of the behavior of the CG model, but the box size limits the usefulness of these simulations.

The above result for transient domain formation in a single-component lipid bilayer was indeed first found by the CG model. In a way, this can be considered as a prediction which can be tested through more extensive atomistic simulations for a considerably larger bilayer system. For this purpose, we employed a fully hydrated single-component DPPC bilayer containing 1152 lipid molecules.

The larger atomistic MD systems indeed do show variations in the area per tail, in agreement with the CG model.[86] The length scales of these fluctuations range up to the size of the system (<20 nm), and the time scales are of the order of 10 ns. The atomistic system also allows us to characterize the denser regions in more detail, and we find that these regions are generally more ordered.

There are also highly ordered transient domains that occur predominantly within the denser regions, but on shorter time scales (<100 ps–10 ns, depending on the size of the domain). There are also some differences between the quantitative results from MD and the CG model: the peak in $S(k)$ is much higher in the MD simulations, and it extends up to the system size. Reasons for the latter can include the relatively small size of the MD simulation box, and in part effects from the cutoff of the effective interactions.[47] For the former, it may be the case that in the smaller MD simulations the density fluctuations do not have enough space to develop fully. Also the lack of description of the ordering of the lipid tails may limit the ability of the CG model to describe the phenomenon.

7.3.2.2 Bilayers with Cholesterol

Figure 7.4 shows the static structure factors for the CG model at different cholesterol concentrations. For clarity, only the total, head–head, tail–tail, and cholesterol–cholesterol structure factors are shown. In addition, a snapshot of cholesterol organization is shown for each concentration. As the cholesterol concentration is increased, the system equilibrates more and more slowly (all the simulations were started from an independent random configuration). This is likely to be the reason for the small peak in the structure factors at smallest k for the 30% concentration: the height of the peak is still slowly decreasing.

Both the snapshots and the structure factors clearly show that the distribution of cholesterol is not uniform at 13 and 20% concentrations. In contrast, the system is homogeneous at 5 and 30%

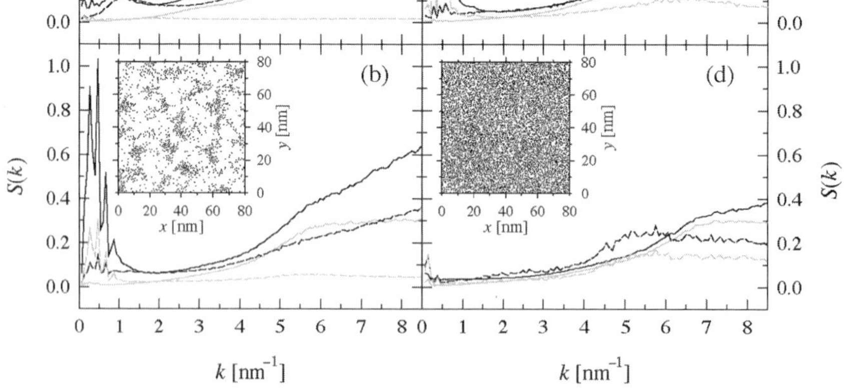

FIGURE 7.4 Static structure factors[47] for different cholesterol concentrations: (a) 5%, (b) 13%, (c) 20%, and (d) 30%. The solid black line shows the total structure factor, and the other lines show the head–head (solid gray line), tail–tail (dashed black line), and cholesterol–cholesterol (dashed gray line) structure factors. The other pairs are not shown for clarity. The insets show snapshots of the systems. In the snapshot, only the cholesterol particles are shown. For the 30% concentration, the peak at the smallest k values is probably due to very slow equilibration of the system.

concentrations. This behavior is in line with the phase diagram of the system:[87] at intermediate concentrations, there should be coexistence of two phases.

For the systems containing cholesterol, it is much more difficult to observe domains in the atomistic systems, compared to the pure DPPC. This is due to the fact that the cholesterol molecules need to diffuse to the domains instead of just altering the local packing. Hence, for the domains to form, very long simulations would be needed, in particular since the addition of cholesterol slows down diffusion.[75] For this reason, it is probably not feasible to confirm the presence of such domains with atomistic simulations.

Even with the CG model, long correlation times of structures at large length scales make it computationally very demanding to determine the precise shape of the $S(k)$ at small k values. Hence, we limit our discussion to the more qualitative aspects. The peaks in the $S(k)$ in the 13 and 20% concentrations indicate a length scale of the order of 10 nm or more, which is in agreement with the snapshots. The domains are most clearly seen for cholesterol, but small peaks are also visible in all the other structure factors, although they are very small for the head particles.

The length scales associated with the domains are comparable to the size of the simulation box, even in the CG simulations. Further, the shape of the $S(k)$ peak can change between independent runs, suggesting that the correlation times for the largest structures are very long. Hence, a significant amount of computer resources (or, alternatively, more efficient Monte Carlo moves) would be needed to determine the shape of the peak accurately. For the same reason, reliable quantitative analysis of the cholesterol distribution would require significantly longer simulations. In the light of the results for the pure DPPC system, we would also need to check whether the effect of the potential cutoff affects the exact results.[47] Since we are focusing on more qualitative aspects, such simulations have not been performed. However, the value of the $S(k)$ in Figure 7.4b and Figure 7.4c seems to be significantly lower for the smallest k values than for the maximum of the peak, indicating that the size of the simulation box may not be the limiting factor for the size of the domains in the present simulations.

We also note that the 5% case shows a peak at $k \approx 1$ nm^{-1} very similar to the pure DPPC case. For the higher concentrations, no evidence of such a peak is found. This indicates that such density variations are a specific property of the phase at low cholesterol concentrations. This is perhaps not surprising because cholesterol increases packing, and thus there is less space to allow for such variation. The lack of the peak for concentrations where coexistence is expected could be explained by the fact that the coexisting domains seem to be of the same order of magnitude in size as the tail density variation. Hence, the formation of denser and more ordered domains in the cholesterol-poor phase could be suppressed by the presence of a nearby region of the other phase.

7.3.2.3 Discussion

The presented results, in particular those for the pure DPPC bilayer, provide a prime example of the benefits that can be gained by coarse-graining. The constructed model allows us to easily study the structure of the system at large scales and to focus on the most important aspects of the system. The knowledge gained can then be used to guide subsequent, more quantitative studies of the system. The fact that the domains seen in the CG DPPC model were later confirmed with an atomistic simulation gives confidence in that other qualitative results given by this approach, such as the formation of cholesterol-rich domains, are also correct. Further, constructing the model based on a different simulation with slightly different interactions gives very similar results.[47] As for the interpretation of the quantitative results, one has to be more careful. For example, the cutoff of the interactions, as well as other factors, can have an effect on the quantitative behavior.[47] Also, the fact that equilibration is very slow for the higher cholesterol concentrations (see above) makes it difficult to make any quantitative conclusions. It is possible that the snapshots shown in Figure 7.4 are only intermediate states in the process of complete phase separation, or that the domains are thermal fluctuations amplified by competition between thermal disorder and the ordering tendency of cholesterol.[47]

We have also studied the limits of the effective interactions in describing the system in a cholesterol concentration that is not the one where the interactions were constructed. This issue of *transferability* is further discussed in the General Discussion, so let us here just summarize the results for the present model. The main result is that qualitative behavior is mainly determined by the interactions and not the concentration. For example, interactions determined for a homogeneous phase (5 and 30% concentrations) give a homogeneous phase also within the coexistence region (13 and 20%). Similarly, potentials from the coexistence region also show formation of domains at the homogeneous concentrations. Nevertheless, in the latter case the heterogeneity decreases significantly, which is qualitatively correct. Transferability is better within a single phase: all changes, for example, in the $S(k)$, are in the right direction, but the magnitudes are generally too small. In conclusion, the effective interactions cannot be used for mapping the phase boundaries, but the general trends are correct when the potentials are transferred to a nearby concentration.

Poor transferability of the potentials between different regions of the phase diagram can be easily understood: the degrees of freedom of the model do not include any internal states for the particles, but the typical conformations change quite remarkably when the cholesterol concentration is increased, due to the ordering effect of cholesterol.[75,88–90] Due to the way the model is constructed, such differences need to be incorporated in the effective interactions. Hence, for example the effective interactions derived in the cholesterol-rich phase are tuned to describe the interactions of ordered chains, and are thus unable to adequately represent systems where the chains are less ordered.

Inclusion of the ordering of the tails in the model could help with the transferability issues. The simplest way would be to do this in the spirit of the phenomenological model of Nielsen et al.;[42] that is, by giving each chain two possible states, describing an ordered and a disordered state. The results of Nielsen et al. suggest that already such a simple description of the conformational states can give the correct form for the phase diagram. Systematic construction of interactions for such a model could give additional insight into the system.

Finally, it is interesting to compare the results to those of a simpler model in which each molecule was described by a single particle.[46] The most striking difference is that the simpler model does not show any density variation in the pure DPPC case. As both the CG and atomistic simulations show, these denser domains have the most pronounced effect in the tail region. Hence, the simpler model, where the tails are not treated separately, may miss crucial structural information that is needed to obtain the effect. For systems with cholesterol, the conclusions are qualitatively identical for both models, except for the presence of the pure DPPC peak in the 5% system for the more complex model. However, with the more complex model the cholesterol-poor regions are virtually depleted of cholesterol, which is a much stronger effect than that seen with the simpler model. This is also reflected in the height of the peaks in the $S(k)$. The difference may be attributed to the ability of the new model to better describe the tail–cholesterol interactions, which leads to a stronger preference for domain formation. Together these results highlight the effect that the degrees of freedom selected for the CG model can have on the properties of the model.

7.4 GENERAL DISCUSSION AND CONCLUDING REMARKS

In this review, we have discussed approaches for systematical coarse-graining. We focused on two separate topics: using clustering methods to select the CG description and constructing the effective interactions by matching the radial distribution functions with IMC. For both cases, we presented results from one application to membrane systems, and then discussed the methods in the context of these applications. Let us now finally compare the presented methods to other approaches in the literature.

SOMs show promise for helping in the selection of the description, but more research is needed to make this a reality. However, there are very few other methods available for this purpose. Schulten and co-workers have also proposed a neural network-based algorithm for constructing

CG models;[10,11] however, their approach is quite different from the clustering-based idea described here. They construct the CG description by trying to reproduce the topology of a single atomistic structure, while our approach uses a full MD trajectory, and gives information on the structure of the available conformational space. Both approaches have their advantages: topology is likely one of the most important features for large macromolecules such as protein complexes, but there can be situations when a single structure cannot adequately describe the degrees of freedom of the system. The clustering-based approach is also the only option if one wants to construct models where the conformations are described by, for example, internal states of the particles.

In general, our approach differs from most other CG lipid models in the level of coarse-graining: instead of semiatomistic models, we try to take a step further and look only at supramolecular organization. The advantage offered by the simpler models is speed: they are several orders of magnitude faster than either atomistic or semiatomistic simulations. However, the high degree of coarse-graining also limits the applicability of the models severely, and the interaction parameters are very specific to the system and the thermodynamic state. Nevertheless, the results speak for themselves: such simple models can be very advantageous for accessing large-scale structural properties, at least qualitatively. The fact that the results of the CG model were later confirmed by atomistic simulations for the pure DPPC case is particularly convincing.

Currently, CG models use one (or a combination of several) of the three main approaches for determining the interactions (see Introduction): the interactions are either determined using structure or forces from atomistic simulations, or they are tuned by a semiempirical approach to match experimental data. Each of these has its advantages and disadvantages.

The differences related to computational complexity are the most apparent. Finding the optimal match for the forces is computationally much easier than matching the structure accurately. For the semiempirical approach, no atomistic simulations are needed, but on the other hand, tuning the interaction parameters can be time-consuming. On the other hand, a general semiempirical model can then be used for a wide range of systems without the need to reparameterize it.

There are also other methods to achieve structural inversion besides IMC, so let us now briefly delve into comparison of these methods. The major drawback of the IMC method is that the four-particle correlation functions should be well sampled during each iteration step. This limits the applicability of the method to systems with relatively few distinct interactions as the required memory and computational time grow very fast with an increasing number of interactions. The Boltzmann inversion scheme, frequently used in polymer physics,[1,2] can provide an alternative to IMC for more complex systems, because it is computationally much lighter than IMC. This is because in the Boltzmann inversion only the RDFs are used in the iteration. This method neglects the correlations between the different RDFs, and can thus converge poorly for more complex systems. However, such an a priori comparison is complicated by the fact that the Boltzmann inversion does not rely on a linearization of the changes in RDFs, in contrast to IMC. Hence, a systematic comparison would be needed to ascertain which of the methods works better in different cases. It could also be possible to make technical improvements to the IMC procedure to improve its speed, but these could have an effect on the convergence, and hence a good compromise would require experimentation with each particular system under study.[47]

One main problem that slows IMC down is the use of the potentials of mean force as the initial interactions. For simple systems, they work relatively well, but for more complex systems they may produce a structure that is too wide off the target to allow for any sensible refinement (e.g., the system may be in the wrong phase). Even if refinement is possible, the linear approximation used in the IMC may not work that well if the desired change in the structure is too large. This then requires very strong regularization and hence many iterations. Hence, it could be worth the effort to construct better initial interactions for more complex systems. One possibility would be to use the Ornstein–Zernike integral equation together with a closure relation[91] to obtain the initial approximations for the potentials, and then fine-tune these with the IMC method. The interactions given by the force matching procedure could also be used for the initial interactions, since several studies have shown that they qualitatively reproduce the correct structure.[8,9,26,27] Hence, only a few iterations should be required to obtain convergence. Such studies could also yield valuable insight into the similarities and differences of the different approaches. An interesting alternative for constructing the initial interactions has also been suggested by Elezgaray and Laguerre: one could first optimize the effective interactions for simpler systems where the interaction centers are solvated in water, and use the resulting interactions as initial guesses for the next model.[55]

Aside from computational complexity, there are also more fundamental differences between the different approaches. As noted in Section 7.3.2.3, the issue of *transferability*; that is, the dependence of the effective interactions on the thermodynamic state of the system, is one of the major concerns. Typically, the interactions derived at a certain temperature, density, and concentration can only be used to describe a limited vicinity of that state point. It is generally difficult to determine the limits for such transferability of the effective interactions a priori; that is, if the desired behavior is not known. In principle, different effective interactions should be constructed for all different state points, but this can be a time-consuming process. This is one reason why the coarse-graining process should be as straightforward and as fast as possible. For these reasons, studies of the limitations of the transferability are crucial for understanding the applicability of CG models.

Finally, there is the question of *representability* of the effective interactions. Namely, for a particular many-body system, there usually is no set of effective pairwise interactions that could reproduce all quantities of interest.[92] For example, if we construct a model to reproduce the structure, one should be very careful to draw any conclusions from quantities not related to the structure without considering the issue of representability. In the case of DPPC/cholesterol bilayers, discussed in Section 7.3, we have found that by tuning the surface tension constraint we can produce a wide range of interactions that give essentially the same RDFs.[47] In general, it is not trivial to say which of them is best for a given purpose. A systematic comparison of the different coarse-graining approaches could be interesting from this point of view.

ACKNOWLEDGMENTS

This work was supported by the Academy of Finland, National Graduate School in Nanoscience (TM), Natural Sciences and Engineering Council of Canada (NSERC), and the Emil Aaltonen Foundation. We also thank the Finnish IT Center for Science, the HorseShoe (DCSC) supercluster computing facility at the University of Southern Denmark, and the Shared Hierarchical Academic Research Computing Network (SHARCNET: www.sharcnet.ca) for computer resources.

REFERENCES

1. Reith, D., M. Pütz, and F. Müller-Plathe. 2003. Deriving effective mesoscale potentials from atomistic simulations. *J. Comp. Chem.* 24:1624–36.
2. Faller, R. 2004. Automatic coarse graining of polymers. *Polymer* 45:3869–76.
3. Lyubartsev, A.P., and A. Laaksonen. 1995. Calculation of effective interaction potentials from radial distribution functions: A reverse Monte Carlo approach. *Phys. Rev. E* 52:3730–37.
4. Lyubartsev, A.P., M. Karttunen, I. Vattulainen, and A. Laaksonen. 2003. On coarse-graining by the Inverse Monte Carlo method: Dissipative particle dynamics simulations made to a precise tool in soft matter modeling. *Soft Materials* 1:121–37.
5. Öttinger, H.C. General projection operator formalism for the dynamics and thermodynamics of complex fluids. 1998. *Phys. Rev. E* 57:1416–20.
6. Español, P., M. Serrano, and H. C. Öttinger. 1999. Thermodynamically admissible form for discrete hydrodynamics. *Phys. Rev. Lett.* 83:4542–45.
7. Ercolessi, F., and J. B. Adams. 1994. Interatomic potentials from 1st-principles calculations: The force-matching method. *Europhys. Lett.* 26:583–88.
8. Izvekov, S., M. Parrinello, C. J. Burnham, and G. A. Voth. 2004. Effective force fields for condesed phase systems from ab initio molecular dynamics simulation: A new method for force-matching. *J. Chem. Phys.* 120:10896–913.

9. Izvekov, S., and G. A. Voth. 2005. Multiscale coarse graining of liquid-state systems. *J. Chem. Phys.* 123:134105.
10. Arkhipov, A., P. L. Freddolino, and K. Schulten. 2006. Stability and dynamics of virus capsids described by coarse-grained modeling. *Structure* 14:1767–77.
11. Arkhipov, A., P. L. Freddolino, K. Imada, K. Namba, and K. Schulten. 2006. Coarse-grained molecular dynamics simulations of a rotating bacterial flagellum. *Biophys. J.* 91:4589–97.
12. Murtola, T., M. Kupiainen, E. Falck, and I. Vattulainen. 2007. Conformational analysis of lipid molecules by self-organizing maps. *J. Chem. Phys.* 126:054707.
13. Venturoli, M., and B. Smit. 1999. Simulating the self-assembly of model membranes. *PhysChemComm* 10:1–5.
14. Nielsen, S.O., C. F. Lopez, G. Srinivas, and M. L. Klein. 2004. Coarse grain models and the computer simulations of soft materials. *J. Phys.: Condens. Matter* 16:R481–512.
15. Venturoli, M., M. M. Sperotto, M. Kranenburg, and B. Smit. 2006. Mesoscopic models of biological membranes. *Phys. Rep.* 437:1–54.
16. Brannigan, G., L. C.-L. Lin, and F. L. H. Brown. 2006. Implicit solvent simulation models for biomembranes. *Eur. Biophys. J.* 35:104–24.
17. Cooke, I.R., K. Kremer, and M. Deserno. 2005. Efficient tunable generic model for fluid bilayer membranes. *Phys. Rev. E* 72:011506.
18. Lazaridis, T. 2005. Implicit solvent simulations of peptide interactions with anionic lipid membranes. *Proteins: Struct. Funct. Genet.* 58:518–27.
19. Lazaridis, T., B. Mallik, and Y. Chen. 2005. Implicit solvent simulations of DPC micelle formation. *J. Phys. Chem. B* 109:15098–106.
20. Marrink, S.J., A. H. de Vries, and A. E. Mark. 2004. Coarse grained model for semi-quantitative lipid simulations. *J. Phys. Chem. B* 108:750–60.
21. Marrink, S.J., and A. E. Mark. 2003. The mechanism of vesicle fusion as revealed by molecular dynamics simulations. *J. Am. Chem. Soc.* 125:11144–45.
22. Marrink, S.J., and A. E. Mark. 2003. Molecular dynamics simulation of the formation, structure, and dynamics of small phospholipid vesicles. *J. Am. Chem. Soc.* 125:15233–42.
23. Marrink, S.J., and A. E. Mark. 2004. Molecular view of hexagonal phase formation in phospholipid membranes. *Biophys. J.* 87:3894–900.
24. Faller, R., and S. J. Marrink. 2004. Simulation of domain formation in DPLC-DSPC mixed bilayers. *Langmuir* 20:7686–93.
25. Marrink, S.J., J. Risselada, and A. E. Mark. 2005. Simulation of gel phase formation and melting in lipid bilayers using a coarse grained model. *Chem. Phys. Lipids* 135:223–44.
26. Izvekov, S., and G. A. Voth. 2005. A multiscale coarse-graining method for biomolecular systems. *J. Phys. Chem. B* 109:2469–73.
27. Izvekov, S., and G. A. Voth. 2006. Multiscale coarse-graining of mixed phospholipid/cholesterol bilayers. *J. Chem. Theory Comput.* 2:637–48.
28. Shi, Q., S. Izvekov, and G. A. Voth. 2006. Mixed atomistic and coarse-grained molecular dynamics: Simulation of a membrane-bound ion channel. *J. Phys. Chem. B* 110:1545–48.
29. Kranenburg, M., M. Venturoli, and B. Smit. 2003. Molecular simulations of mesoscopic bilayer phases. *Phys. Rev. E* 67:060901(R).
30. Ortiz, V., S. O. Nielsen, D. E. Discher, M. L. Klein, R. Lipowsky, and J. C. Shillcock. 2005. Dissipative particle dynamics simulations of polymersomes. *J. Phys. Chem. B* 109:17708–14.
31. Shillcock, J.C., and R. Lipowsky. 2005. Tension-induced fusion of bilayer membranes and vesicles. *Nat. Mater.* 4:225–28.
32. Illya, G., R. Lipowsky, and J. C. Shillcock. 2005. Effect of chain length and asymmetry on material properties of bilayer membranes. *J. Chem. Phys.* 122:244901.
33. Kremer, K., and F. Müller-Plathe. 2001. Multiscale problems in polymer science: Simulation approaches. *Mater. Res. Bull.* 26:205–10.
34. Kremer, K., and F. Müller-Plathe. 2002. Multiscale simulation in polymer science. *Mol. Sim.* 28:729–50.
35. Nikunen, P., I. Vattulainen, and M. Karttunen. 2007. Reptational dynamics in dissipative particle simulations of polymer melts. *Phys. Rev. E* 75:036713.
36. Mouritsen, O.G., B. Dammann, H. C. Fogedby, J. H. Ipsen, C. Jeppesen, K. Jørgensen, J. Risbo, M. C. Sabra, M. M. Sperotto, and M. J. Zuckermann. 1995. The computer as a laboratory for the physical chemistry of membranes. *Biophys. Chem.* 55:55–68.

37. Doniach, S. 1978. Thermodynamic fluctuations in phospholipid bilayers. *J. Chem. Phys.* 68:4912–16.
38. Pink, D.A., T. J. Green, and D. Chapman. 1980. Raman scattering in bilayers of saturated phosphatidylcholines. Experiment and theory. *Biochemistry* 19:349–56.
39. Ipsen, J.H., G. Karlström, O. G. Mouritsen, H. Wennerström, and M. J. Zuckermann. 1987. Phase equilibria in phosphatidylcholine–cholesterol system. *Biochim. Biophys. Acta* 905:162–72.
40. Elliott, R., I. Szleifer, and M. Schick. 2006. Phase diagram of a ternary mixture of cholesterol and saturated and unsaturated lipids calculated from a microscopic model. *Phys. Rev. Lett.* 96:098101.
41. Nielsen, M., L. Miao, J. H. Ipsen, O. G. Mouritsen, and M. J. Zuckermann. 1996. Random-lattice models and simulation algorithms for the phase equilibria in two-dimensional condensed systems of particles with coupled internal and translational degrees of freedom. *Phys. Rev. E* 54:6889–905.
42. Nielsen, M., L. Miao, J. H. Ipsen, M. J. Zuckermann, and O. G. Mouritsen. 1999. Off-lattice model for the phase behavior of lipid–cholesterol bilayers. *Phys. Rev. E* 59:5790–802.
43. Nielsen, M., J. Thewalt, L. Miao, J. H. Ipsen, M. Bloom, M. J. Zuckermann, and O. G. Mouritsen. 2000. Sterol evolution and the physics of membranes. *Europhys. Lett.* 52:368–74.
44. Miao, L., M. Nielsen, J. Thewalt, J. H. Ipsen, M. Bloom, M. J. Zuckermann, and O. G. Mouritsen. 2002. From lanosterol to cholesterol: Structural evolution and differential effects on lipid bilayers. *Biophys. J.* 82:1429–44.
45. Polson, J.M., I. Vattulainen, H. Zhu, and M. J. Zuckermann. 2001. Simulation study of lateral diffusion in lipid–sterol bilayer mixtures. *Eur. Phys. J. E* 5:485–97.
46. Murtola, T., E. Falck, M. Patra, M. Karttunen, and I. Vattulainen. 2004. Coarse-grained model for phospholipid/cholesterol bilayer. *J. Chem. Phys.* 121:9156–65.
47. Murtola, T., E. Falck, M. Karttunen, and I. Vattulainen. 2007. Coarse-grained model for phospholipid/cholesterol bilayer employing inverse Monte Carlo with thermodynamic constraints. *J. Chem. Phys.* 126:075101.
48. Khelashvili, G.A., S. A. Pandit, and H. L. Scott. 2005. Self-consistent mean-field model based on molecular dynamics: Application to lipid–cholesterol bilayers. *J. Chem. Phys.* 123:034910.
49. Shi Q., and G. A. Voth. 2005. Multi-scale modeling of phase separation in mixed lipid bilayers. *Biophys. J.* 89:2385–94.
50. Lagüe, P., M. J. Zuckermann, and B. Roux. 2001. Lipid-mediated interactions between intrinsic membrane proteins: Dependence on protein size and lipid composition. *Biophys. J.* 81:276–84.
51. Silbermann, J. R., S. H. L. Klapp, M. Schoen, N. Chennamsetty, H. Bock, and K. E. Gubbins. 2006. Mesoscale modeling of complex binary fluid mixtures: Towards an atomistic foundation of effective potentials. *J. Chem. Phys.* 124:074105.
52. Lyubartsev, A.P., and A. Laaksonen. 1997. Osmotic and activity coefficients from effective potentials for hydrated ions. *Phys. Rev.* E 55:5689–96.
53. Lyubartsev, A. P., and A. Laaksonen. 1999. Effective potentials for ion–DNA interactions. *J. Chem. Phys.* 111:11207–215.
54. Lyubartsev, A.P. 2005. Multiscale modeling of lipids and lipid bilayers. *Eur. Biophys. J.* 35:53-61.
55. Elezgaray, J., and M. Laguerre. 2006. A systematic method to derive force fields for coarse-grained simulations of phospholipids. *Comp. Phys. Comm.* 175:264-68.
56. Shelley, J. C., M. Y. Shelley, R. C. Reeder, S. Bandyopadhyay, and M. L. Klein. 2001. A coarse grain model for phospholipid simulations. *J. Phys. Chem. B* 105:4464–70.
57. Li, H., R. Helling, C. Tang, and N. Wingreen. 1996. Emergence of preferred structures in a simple model of protein folding. *Science* 273:666–69.
58. Li, H., C. Tang, and N. S. Wingreen. 1997. Nature of driving force for protein folding: A result from analyzing the statistical potential. *Phys. Rev. Lett.* 79:765–68.
59. Emberly, E. G., J. Miller, C. Zeng, N. S. Wingreen, and C. Tang. 2002. Identifying proteins of high designability via surface-exposure patterns. *Proteins: Struct. Funct. Genet.* 47:295–304.
60. Hamprecht, F. A., C. Peter, X. Daura, W. Thiel, and W. F. van Gunsteren. 2001. A strategy for analysis of (molecular) equilibrium simulations: Configuration space density estimation, clustering, and visualization. *J. Chem. Phys.* 114:2079–89.
61. Haykin, S. 1999. *Neural Networks: A Comprehensive Foundation*. New Jersey: Prentice-Hall.
62. Kohonen, T. 2001. *Self-organizing maps,* vol. 30 of Springer series in information science, 3rd ed. Berlin: Springer.
63. Vesanto, J., J. Himberg, E. Alhoniemi, and J. Parhankangas. 1999. Self-organizing maps in Matlab: The SOM toolbox. In *Proceedings of the Matlab DSP Conference 1999* (Espoo, Finland), pp. 35–40.

64. Schuchhardt, J., G. Schneider, J. Reichelt, D. Schomburg, and P. Wrede. 1996. Local structural motifs of protein backbones are classified by serf-organizing neural networks. *Prot. Eng.* 9:833–42.
65. Hyvönen, M. T., Y. Hiltunen, W. El-Deredy, T. Ojala, J. Vaara, P. T. Kovanen, and M. Ala-Korpela. 2001. Application of self-organizing maps in conformational analysis of lipids. *J. Am. Chem. Soc.* 123:810–16.
66. Kupiainen, M., E. Falck, S. Ollila, P. Niemelä, A. A. Gurtovenko, M. T. Hyvönen, M. Patra, M. Karttunen, and I. Vattulainen. 2005. Free volume properties of sphingomyelin, DMPC, DPPC, and PLPC bilayers. *J. Comput. Theor. Nanosci.* 2:401–13.
67. Ollila, S., M. T. Hyvönen, and I. Vattulainen. 2007. Polyunsaturation in lipid membranes: Dynamic properties and lateral pressure profiles. *J. Phys. Chem. B* 111:3139–50.
68. Langner, M., and K. Kubica. 1999. The electrostatics of lipid surfaces. *Chem. Phys. Lipids* 101:3–35.
69. Saiz, L., and M. L. Klein. 2002. Electrostatic interactions in a neutral model phospholipid bilayer by molecular dynamics simulations. *J. Chem. Phys.* 116:3052–57.
70. Andrade, M. A., G. Casari, C. Sander, and A. Valencia. 1997. Classification of protein families and detection of the determinant residues with an improved self-organizing map. *Biochemistry* 76:441–450.
71. Mahony, S., D. Hendrix, T. J. Smith, and A. Golden. 2005. Self-organizing maps of position weight matrices for motif discovery in biological sequences. *Artif. Intell. Rev.* 24:397–413.
72. Mahony, S., D. Hendrix, A. Golden, T. J. Smith, and D. S. Rokhsar. 2005. Transcription factor binding site identification using the self-organizing map. *Bioinformatics* 21: 1807–14.
73. Yang, Z. R., and K. C. Chou. 2003. Mining biological data using self-organizing map. *J. Chem. Inf. Comput. Sci.* 43:1748–53.
74. Ayers, J. T., A. Clauset, J. D. Schmitt, L. P. Dwoskin, and P. A. Crooks. 2005. Molecular modeling of mono- and bis-quaternary ammonium salts as ligands at the alpha4beta2 nicotinic acetylcholine receptor subtype using nonlinear techniques. *AAPS J.* 25:E678–E685.
75. Falck, E., M. Patra, M. Karttunen, M. T. Hyvönen, and I. Vattulainen. 2004. Lessons of slicing membranes: Interplay of packing, free area, and lateral diffusion in phospholipid/cholesterol bilayers. *Biophys. J.* 87:1076–91.
76. Pandit, S. A., D. Bostick, and M. L. Berkowitz. 2004. Complexation of phosphatidylcholine lipids with cholesterol. *Biophys. J.* 86:1345–56.
77. Pandit, S. A., S. Vasudevan, S. W. Chiu, R. J. Mashl, E. Jakobsson, and H. L. Scott. 2004. Sphingomyelin–cholesterol domains in phospholipid membranes: Atomistic simulation. *Biophys. J.* 87:1092–100.
78. Niemelä, P., S. Ollila, M. T. Hyvönen, M. Karttunen, and I. Vattulainen. 2007. Assessing the nature of lipid raft membranes: From atomic-scale structure to lateral pressure profiles. *PLoS Comput. Biol.* 3:304–12.
79. Aittoniemi, J., P. Niemelä, M. T. Hyvönen, M. Karttunen, and I. Vattulainen. 2007. Insight into the putative specific interactions between cholesterol, sphingomyelin and palmitoyl-oleoyl phosphatidylcholine. *Biophys. J.* 92:1125–37.
80. Gurtovenko, A. A., M. Miettinen, M. Karttunen, and I. Vattulainen. 2006. Effect of monovalent salt on cationic lipid membranes as revealed by molecular dynamics simulations. *J. Phys. Chem. B* 109: 21126–34.
81. Böckmann, R. A., and H. Grubmüller. 2004. Multistep binding of divalent cations to phospholipid bilayers: a molecular dynamics study. *Angew. Chem. Int. Ed.* 43:1021.
82. Róg, T., I. Vattulainen, and M. Karttunen. 2005. Modeling glycolipids: Take One. *Cell. Mol. Biol. Lett.* 10:625–30.
83. Jensen, M. Ø., O. G. Mouritsen, and G. H. Peters. 2004. Simulations of a membrane-anchored peptide: Structure, dynamics, and influence on bilayer properties. *Biophys. J.* 86:3556–75.
84. Henderson, R.L. 1974. A uniqueness theorem for fluid pair correlation functions. *Phys. Lett. A* 49: 197–98.
85. Feller, S. E., and R. W. Pastor. 1999. Constant surface tension simulations of lipid bilayers: The sensitivity of surface areas and compressibilities. *J. Chem. Phys.* 111:1281–87.
86. Murtola, T., T. Róg, E. Falck, M. Karttunen, and I. Vattulainen. 2006. Transient ordered domains in single-component phospholipid bilayers. *Phys. Rev. Lett.* 97:238102.
87. Vist, M. R., and J. H. Davis. 1990. Cholesterol/dipalmitoylphosphatidylcholine mixtures: ^{2}H nuclear magnetic resonance and differential scanning calorimetry. *Biochemistry* 29:451–64.
88. Róg, T, and M. Pasenkiewicz-Gierula. 2001. Cholesterol effects on the phosphatidylcholine bilayer nonpolar region: A molecular simulation study. *Biophys. J.* 81: 2190–202.

89. Hofsäß, C., E. Lindahl, and O. Edholm. 2003. Molecular dynamics simulations of phospholipid bilayers with cholesterol. *Biophys. J.* 84:2192–206.
90. Jedlovszky, P., and M. Mezei. 2003. Effect of cholesterol on the properties of phospholipid membranes. 1. Structural features. *J. Phys. Chem. B* 107:5311–21.
91. Hansen, J.-P., and I. R. McDonald. 1986. *Theory of Simple Liquids*, 2nd ed. London: Academic Press.
92. Louis, A.A. 2002. Beware of density dependent pair potentials. *J. Phys.: Condens. Matter* 13:9187–206.

8 Simulation of Protein Structure and Dynamics with the Coarse-Grained UNRES Force Field

Adam Liwo, Cezary Czaplewski, and Stanisław Ołdziej
Baker Laboratory of Chemistry and Chemical Biology, Cornell University
University of Gdańsk

Ana V. Rojas
Baker Laboratory of Chemistry and Chemical Biology, Cornell University
Department of Physics and Astronomy, Louisiana State University
Center for Computation and Technology, Louisiana State University

Rajmund Kaźmierkiewicz and Mariusz Makowski
Baker Laboratory of Chemistry and Chemical Biology, Cornell University
University of Gdańsk

Rajesh K. Murarka and Harold A. Scheraga
Baker Laboratory of Chemistry and Chemical Biology, Cornell University

CONTENTS

8.1 INTRODUCTION

Simulations of protein structure were originally based on a hard-sphere potential coupled with the requirement to avoid steric interactions [Ramachandran, Ramakrishnan, and Sasiekharan 1963; Némethy and Scheraga 1965]. This approach evolved into the use of physically more realistic pairwise interatomic potentials, initially to compute structure [De Santis et al. 1963; Brant and Flory 1965; Levitt and Lifson 1969; Momany et al. 1975] and, subsequently, to compute folding pathways with molecular dynamics (MD) [McCammon, Gelin, and Karplus 1977]. The underlying assumption in these simulations was the thermodynamic hypothesis proposed by Anfinsen (1973), viz., that the native structure of a protein is the one for which the free energy of the protein, and its surroundings, is a minimum. Regardless of whether internal (torsional angle) or external (Cartesian) coordinates are used to describe protein structure, the number of such degrees of freedom and of pairwise interactions for even a moderate-size protein are too large for simulations with available computer software and hardware. To compound the difficulty, MD with an all-atom description of a polypeptide chain cannot achieve the millisecond-to-second time scale of folding of all but the very fast-folding proteins with currently available algorithms and computers. Therefore, coarse-grained models are resorted to, in which the number of degrees of freedom and the number of interaction centers are considerably smaller than those involved in all-atom simulations.

The development of a coarse-grained approach in our laboratory was based on averaging over the less important, fast degrees of freedom, retaining only the most important ones [Liwo et al. 2001]. This was accomplished with the development of a united-residue (UNRES) representation of the polypeptide chain (Figure 8.1) as the initial step in a hierarchical approach in which the resulting coarse-grained structure is subsequently converted to an all-atom one [Kaźmierkiewicz, Liwo and Scheraga 2002, 2003]. The search of the coarse-grained conformational space was initially carried out with a conformational space annealing (CSA) procedure [Lee, Scheraga and Rackovsky

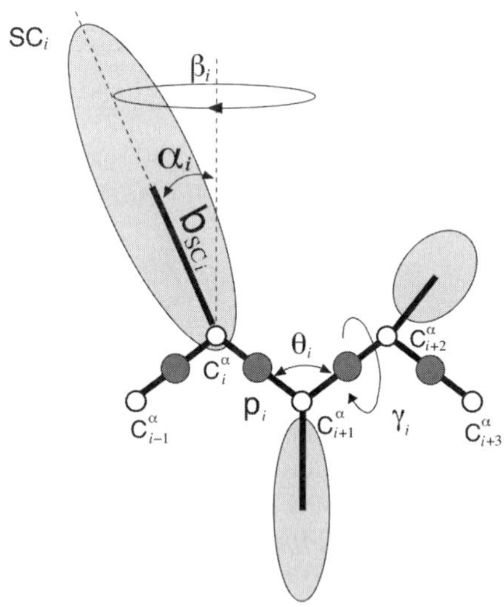

FIGURE 8.1 The UNRES model of polypeptide chains. The interaction sites are side-chain centroids of different sizes (SC) and the peptide-bond centers (p) indicated by shaded circles, whereas the α-carbon atoms (small empty circles) are introduced only to assist in defining the geometry. The virtual $C^\alpha \ldots C^\alpha$ bonds have an "unstrained" length of 3.8 Å, corresponding to a planar trans peptide group; the virtual-bond (θ) and dihedral (γ) angles are variable. Each side chain is attached to the corresponding α-carbon with a virtual bond with the average length b_{SCi}, depending on residue type, variable "bond angle", α_{SCi}, formed by SC_i and the bisector of the angle defined by C^α_{i-1}, C_i^α, and C^α_{i+1}, and with a variable "dihedral angle" β_{SCi} of counterclockwise rotation about the bisector, starting from the right side of the C^α_{i-1}, C_i^α, C^α_{i+1} frame. The virtual-bond lengths were fixed in the applications of the force field involving energy minimization but were made variable for the purpose of MD calculations with UNRES. (From Liwo, A. et al., *J. Chem. Phys.*, 115, 2323–2324, 2001. With permission.)

1997], and subsequently by MD [Khalili et al. 2005a, 2005b; Liwo, Khalili, and Scheraga 2005] and replica exchange (RE) [Sugita and Okamoto 1999; Nanias, Czaplewski, and Scheraga 2006] methods. Initially, our focus was on the computation of the native structure [Liwo et al. 1999], but subsequently our interest expanded to include the computation of folding pathways [Liwo, Khalili, and Scheraga 2005] and kinetics [Khalili, Liwo, and Scheraga 2006]. In this chapter, the UNRES model and its parameterization, together with the basis of the CSA, RE, and MD procedures, and simulations based on these, are described.

8.2 METHODS

8.2.1 Theory of the UNRES Force Field

In the UNRES model, a polypeptide chain is represented as a sequence of α-carbon atoms (C^{α}) with attached united side chains (SC) and united peptide groups (p), each of the latter being positioned in the middle between two consecutive C^{α} atoms, as shown in Figure 8.1.

The effective energy function is a sum of different terms corresponding to interactions between the SC ($U_{SC_iSC_j}$), SC and p ($U_{SC_ip_j}$), and p ($U_{p_ip_j}$) sites, as well as local terms corresponding to bending of virtual-bond angles θ (U_b), side-chain rotamers (U_{rot}), virtual-bond torsional (U_{tor}), and double-torsional (U_{tord}) terms, virtual-bond-stretching (U_{bond}) terms, correlation terms ($U_{corr}^{(m)}$) pertaining to coupling between backbone–local and backbone–electrostatic interactions [Liwo et al. 2001] (where m denotes the order of correlation), and a term accounting for the energetics of disulfide bonds (U_{SS}) [Chinchio et al. 2007]. Each of these terms is multiplied by an appropriate weight, w. The effective energy function is given by Equation 8.1.

$$U = w_{SC}\sum_{i<j} U_{SC_iSC_j} + w_{SCp}\sum_{i\neq j} U_{SC_ip_j} + w_{pp}\sum_{i<j-1} U_{p_ip_j} + w_{tor}\sum_{i} U_{tor}(\gamma_i)$$
$$+ w_{tord}\sum_{i} U_{tord}(\gamma_i,\gamma_{i+1}) + w_b\sum_{i} U_b(\theta_i) + w_{rot}\sum_{i} U_{rot}(\alpha_{SC_i},\beta_{SC_i},\theta_i)$$
$$+ \sum_{m=3}^{6} w_{corr}^{(m)}U_{corr}^{(m)} + w_{bond}\sum_{i} U_{bond}(d_i) + w_{SS}\sum_{i} U_{SS;i}. \tag{8.1}$$

The prototype of the effective energy function given by Equation 8.1 is the free energy of the corresponding all-atom chain, confined to a given coarse-grained geometry. We have termed this effective energy as a restricted free energy (RFE); it should be noted that it can be understood as the potential of mean force (PMF) of the coarse-grained chain. If we partition the variables describing the geometry of the all-atom chain and the surrounding solvent into two orthogonal sets: the coarse-grained variables contained in vector **X** and the secondary variables contained in vector **Y**, which appear in the all-atom but not in the coarse-grained representation, the RFE is expressed by Equation 8.2.

$$F(\mathbf{X}) = -RT\ln\left\{\frac{1}{V_Y}\int_{\Omega_Y}\exp\left[-\frac{E(\mathbf{X};\mathbf{Y})}{RT}\right]dV_{\mathbf{Y}}\right\}, \tag{8.2}$$

where $E(\mathbf{X};\mathbf{Y})$ is the original (all-atom) energy function, R is the gas constant, T is the absolute temperature, $\Omega_{\mathbf{Y}}$ denotes the space spanned by the secondary variables **Y**, and $V_{\mathbf{Y}}$ is the volume of this space.

The secondary degrees of freedom **Y** include solvent degrees of freedom, side-chain torsional angles, and the dihedral angles λ for rotation of the peptide groups about the $C^{\alpha}\ldots C^{\alpha}$ virtual bonds [Nishikawa, Momany, and Scheraga 1974]. The quantity $\exp[-F(\mathbf{X})/RT]$ defines the relative probability of occurrence of a coarse-grained conformation and, consequently, the RFE is the best prototype of the effective coarse-grained energy function for physics-based prediction of protein structure from amino-acid sequence and prediction of thermodynamic quantities. To use the RFE as a potential energy in MD simulations, it must be assumed that the motion of the secondary degrees of freedom is much faster than the motion of the coarse-grained degrees of freedom. An alternative physics-based principle termed the force-matching method, in which the forces acting on coarse-grained sites are matched to those computed by using all-atom MD, has been developed recently [Izvekov and Voth 2005].

The exact RFE can be evaluated only by numerical integration over the space of secondary variables and it is, therefore, necessary to simplify it to obtain a practical effective energy function. Liwo et al. (2001) implemented Kubo's cluster-cumulant approach [Kubo, 1962] to simplify the RFE. The vector $\mathbf{Y}$ of the secondary variables is divided into parts $\mathbf{y}_1, \mathbf{y}_2, \ldots, \mathbf{y}_n$, n being the number of interaction sites in the reduced model, such that $\mathbf{y}_i$ pertains to the secondary degrees of freedom that define the motion within the ith interaction site of the reduced model. The total energy of the all-atom system is then expressed in terms of the energies of interaction of the atoms belonging to the interaction sites in the reduced model, as illustrated in Figure 8.2.

$$E(\mathbf{X};\mathbf{Y}) = \sum_{k=1}^{n}\sum_{i\in I_k} E_{ik}(\mathbf{X};\mathbf{y}_k) + \sum_{k=1}^{n}\sum_{l=1}^{k-1}\sum_{i\in I_k}\sum_{j\in I} E_{ik;jl}(\mathbf{X};\mathbf{y}_k;\mathbf{y}_l) = \sum_{i=1}^{n} \varepsilon_i(\mathbf{X};\mathbf{z}_i), \qquad (8.3)$$

where the sets $\{I_1, I_2, \ldots, I_n\}$ contain the indices of all atoms assigned to interaction site $1, 2, \ldots, n$, respectively, $N = n(n+1)/2$ is the total number of component energies $\varepsilon_1, \varepsilon_2, \ldots, \varepsilon_N$ at the all-atom level of description, and $\mathbf{z}_i = \mathbf{y}_i$ or $\mathbf{z}_i = (\mathbf{y}_k, \mathbf{y}_l)$, depending on whether ε_i is an intra- or intersite energy. Each component energy is either a sum of all interatomic interactions within a given extended site or between two extended sites. By inserting Equation 8.3 into Equation 8.2 and splitting the

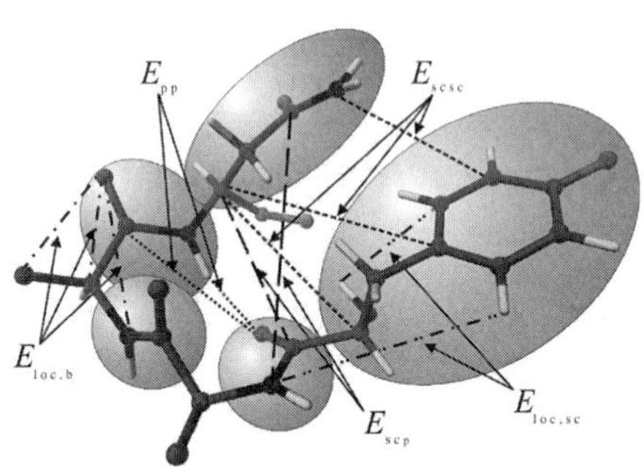

FIGURE 8.2 Splitting of the energy of interaction of the polypeptide chain into component energies for polypeptide chains, applied in the derivation of the UNRES force field. The atoms of two side chains and three peptide groups of the portion of the polypeptide chain shown in the picture are embedded in shaded ellipsoids; the C^{α} atoms are parts of the UNRES SC sites, as shown in the picture. Different line styles mark different interactions; for clarity only some of the interactions are shown. The energies of interaction between side chains (including the water molecules of their solvation spheres, which are not shown for clarity), E_{SCSC}, contribute to $(U_{SC_iSC_j})$ of Equation 8.1. Similarly, the energies of the interactions between the atoms of the peptide groups separated in sequence by more than one residue (E_{pp}), as well as between the peptide groups and side chains (E_{SCp}) contribute to $U_{p_ip_j}$ and $(U_{SC_ip_j})$, respectively; E_{pp} also contributes to $U_{corr}^{(m)}$ that involve third- and higher-order factors comprising E_{pp} and $E_{loc,b}$. The interactions between the atoms of a side chain and those of the adjacent peptide groups ($E_{loc,SC}$) contribute to U_{rot} of Equation 8.1, and the energies of interaction within fragments composed of the C^{α}, H^{α}, C^{β}, and H^{β} atoms of a given residue as well as the atoms of the two peptide groups adjacent to the C^{α} atom ($E_{loc,b}$) contribute to U_b (the first-order factor in the expansion of the RFE that includes only $E_{loc,b}$), U_{tor}, U_{tord} (the second- and third-order factor in the expansion of the RFE, respectively, that include the $E_{loc,b}$ energies of two or three adjacent residues, respectively) and to the $U_{corr}^{(m)}$ correlation terms.

RFE into cluster-cumulant functions, $\langle\langle \varepsilon_{i_1}, \varepsilon_{i_2}, \ldots, \varepsilon_{i_k} \rangle\rangle_f$, containing gradually increasing numbers of component energies, Equation 8.2 becomes Equation 8.4.

$$F(\mathbf{X}) = \sum_i \langle\langle \varepsilon_i \rangle\rangle_f + \sum_{i<j} \langle\langle \varepsilon_i \varepsilon_j \rangle\rangle_f + \sum_{i<j<k} \langle\langle \varepsilon_i \varepsilon_j \varepsilon_k \rangle\rangle_f + \ldots + \langle\langle \varepsilon_1 \varepsilon_2 \ldots \varepsilon_N \rangle\rangle_f, \tag{8.4}$$

where the cluster-cumulant functions are expressed by Equation 8.5.

$$\langle\langle \varepsilon_{i_1} \varepsilon_{i_2} \ldots \varepsilon_{i_k} \rangle\rangle_f = \sum_{l=1}^{k} \sum_{\substack{i_{m_1} < i_{m_2} < \ldots < i_{m_l} \\ m_i \in [1 \ldots k]}} (-1)^{k-l} \langle\langle \varepsilon_{i_{m_1}}, \varepsilon_{i_{m_2}}, \ldots, \varepsilon_{i_{m_l}} \rangle\rangle \tag{8.5}$$

and

$$\langle \varepsilon_{i_1} \varepsilon_{i_2} \ldots \varepsilon_{i_k} \rangle\rangle = -\frac{1}{\beta} \ln \left\{ \frac{1}{V_{\mathbf{y}_I}} \int_{\Omega_I} \exp\left[-\beta \sum_{l=1}^{k} \varepsilon_{i_l}(\mathbf{X}; \mathbf{z}_{i_l}) \right] dV_{yI} \right\} \tag{8.6}$$

(where $\beta = 1/RT$ and $V_{\mathbf{y}_I}$ is the volume of the subspace spanned by the variables $\mathbf{y}_{i_1}, \mathbf{y}_{i_2}, \ldots, \mathbf{y}_{i_k}$) is the RFE containing only a subset of component interactions.

Each of the first-order factors $\langle\langle \varepsilon_i \rangle\rangle_f$ is the RFE corresponding to component interaction i. These factors are the potentials of mean force of the interaction between isolated sites which can, in turn, be identified with $U_{\mathrm{SC}_i\mathrm{SC}_j}$, $U_{\mathrm{SC}_i\mathrm{p}_j}$, $U_{\mathrm{p}_i\mathrm{p}_j}$ in Equation 8.1, as well as the potentials of mean force of the local interactions within isolated sites, which can be identified with U_{b} and U_{rot} of Equation 8.1, respectively. The definition of all the first-order factors matches the definitions of the energy terms and the way that these terms were derived from all-atom potentials in early united-residue force fields [Levitt 1976]. The factors $\langle\langle \varepsilon_i \varepsilon_j \rangle\rangle_f$ contain the RFEs of pairs of component interactions minus the sums of the RFEs of single-component interactions. They can be regarded as a correlation term pertaining to component interactions i and j, reflecting the coupling between the secondary degrees of freedom pertaining to these interactions. Likewise, the third-order factors reflect the coupling between three component interactions, which cannot be expressed in terms of the first- and second-order factors. Factors of order greater than 1 can be identified with multi-body interactions. Now, if the expansion given by Equation 8.4 is truncated at $k < N$, we obtain an approximate expression for the RFE, which contains configuration integrals over parts of the all-atom system and not over the entire system (Equation 8.6).

In the present UNRES [Ołdziej et al. 2004a; Liwo et al. 2007], $k = 4$ appears to be sufficient to capture the basic features of the energetics of a polypeptide chain, even though we have used $k = 6$ in the past [Liwo et al. 2001, 2004]; the higher-order correlations originally seemed to be necessary for chain ordering in β-sheets, but are no longer needed. The correlation terms correspond to the coupling between backbone–local conformational states of two (U_{tor}) or three (U_{tord}) neighboring residues, respectively, and between backbone–local conformational states and electrostatic interactions between backbone peptide groups ($U_{\mathrm{corr}}^{(m)}$); the latter appear necessary to reproduce regular structures, such as α-helices and β-sheets [Liwo et al. 2001].

The functional forms of factors of order 1 and those of simple factors of higher order can usually be deduced based on the analogy with all-atom force fields (e.g., it is reasonable to assume a Lennard–Jones-type expression for $U_{\mathrm{SC}_i\mathrm{SC}_j}$ and a sum of sines and cosines for U_{tor}); these functional forms can be fitted to the factors computed numerically. To derive approximate functional expressions for more complex factors corresponding to the correlation terms ($U_{\mathrm{corr}}^{(m)}$) in UNRES, these factors were expanded [Liwo et al. 2001] into Kubo's generalized cumulant series [Kubo 1962], keeping the lowest-order nonzero cumulant for each factor as a template of the corresponding effective energy term.

The terms corresponding to factors of order greater than 1 (the multibody terms) decrease asymptotically with increasing temperature as $1/T^{m-1}$, where m is the order of a factor [Liwo et al. 2007]; that is, the higher the order of a factor, the quicker it decays with temperature. This is reasonable because higher-order factors act as formers of ordered structure [Liwo et al. 2001, 2004], and the stability of ordered structures should decrease with increasing temperature. The temperature dependence of the force field was (and could be) ignored in earlier applications of UNRES as a potential energy function for global optimization but was introduced in recent applications to compute thermodynamic quantities of protein folding [Liwo et al. 2007].

8.2.2 Parameterization of Individual RFE Terms

The integrals of Equation 8.6 are related to the probability of a given configuration of a part of the system (e.g., the probability density distribution of the side-chain centers of a pair of residues of a given type as a function of distance and orientation [Liwo et al. 1997a] or the probability density distribution of virtual-bond angles θ [Liwo et al. 1997b]) determined from structural databases; that is, to the so-called knowledge-based potentials, as expressed by Equation 8.7.

$$W(\mathbf{X}) = -RT \ln \frac{\rho(\mathbf{X})}{\rho_0(\mathbf{X})}, \tag{8.7}$$

where $\rho(\mathbf{X})$ is the probability density of the fragment at a configuration given by the variables $\mathbf{X}$, and $\rho_0(\mathbf{X})$ is the reference probability density in the absence of interactions. It should be noted that $W(\mathbf{X})$ depends not only on the interactions within a subsystem corresponding to a given factor of the RFE but also on the context of the entire system. It is assumed that the interactions from the other parts of the system are averaged out provided that the probability density distributions are computed using a sufficient number of structural data.

Originally [Liwo et al. 1997a, 1997b], all RFE components except $U_{p_i p_j}$ were determined from protein crystal data using Equation 8.7 to relate the probability distributions to the potentials of mean force. But the populations of the 20 naturally occurring amino acids are not uniformly distributed, giving rise to a different quality of the statistics for different amino acids; also, such knowledge-based information corresponds only to folded proteins and, consequently, does not provide information about the early stages of folding. Therefore, two alternative methods were subsequently used to compute the RFE components.

In the first method, to parameterize the backbone–electrostatic ($U_{p_i p_j}$), torsional (U_{tor}), double-torsional (U_{tord}), and correlation ($U_{corr}^{(m)}$) terms in UNRES, we computed the RFE factors of model systems corresponding to the given terms (Equation 8.6) by direct numerical quadrature of the integrand over their energy surfaces computed at the quantum-mechanical ab initio level [Ołdziej et al. 2003; Liwo et al. 2004]. Kozłowska et al. (2007a, 2007b) recently extended this treatment to the virtual-bond-angle-bending (U_b) and side-chain rotamer (U_{rot}) potentials. The numerical quadrature of the Boltzmann factor (Equation 8.6) was carried out over the angles λ for rotation of the peptide groups about the virtual $C^\alpha \ldots C^\alpha$ bonds [Nishikawa, Momany, and Scheraga 1974] and over the angle χ for rotation about the bonds of the side chains. The analytical expressions were then fitted to the numerically computed RFE factors by the nonlinear least-squares method. The parameters of $U_{corr}^{(m)}$ were subsequently refined together with the determination of the energy-term weights (Equation 8.1) by optimization of the free-energy function as described in Section 8.2.3 to compensate for the approximate character of the analytical expressions.

A second method to obtain physics-based energy terms in coarse-grained potentials is to determine the RFE surfaces from all-atom MD simulations; this approach is desirable for systems with many degrees of freedom, for example, the side chains in water. Recently Makowski et al. (2007) determined physics-based $U_{SC_i SC_j}$ potentials by using umbrella-sampling MD simulations in explicit water to compute the PMFs of homodimers of molecules that model hydrophobic amino acid side chains; these potentials will ultimately replace the old knowledge-based potentials determined from the Protein Data Bank (PDB) [Liwo et al. 1997a]. Analytical expressions for the potentials of mean force, based on a Gaussian model of a hydrophobic hydration shell [Makowski et al. 2007], were subsequently fitted to the PMFs. The resulting analytical potentials reproduce the contact and solvent-separated minima as well as the desolvation maxima in the potentials of mean force for different orientations of the side chains [Makowski et al. 2007]. Similar calculations are in progress for amphiphilic side chains [Makowski et al. in preparation].

8.2.3 Optimization of the Complete Free-Energy Function to Achieve Foldability

To complete the parameterization of the effective energy function (after the determination of the parameters of the individual terms of the cumulant expansion as described in Section 8.2.2), the relative weights (the *w*s of Equation 8.1) of the energy terms must be determined. Initially [Lee et al. 2001; Pillardy et al. 2001b], our procedure for weight determination and also refinement of parameters of the analytical expressions for the individual terms in Equation 8.1 was based on results of the studies of Shakhnovich and coworkers [Sali, Shakhnovich, and Karplus 1994] and Wolynes and coworkers [Goldstein, Luthey-Schulten, and Wolynes 1992], who suggested that it is sufficient to require a large energy gap between the native structure and the lowest-energy non-native structure. Consequently, using a number of training proteins, we optimized the energy gap between the lowest-energy native-like structure and the lowest-energy nonnative structure and, simultaneously, the *Z*-score defined as the difference between the mean energy of the native-like structures and the mean energy of the nonnative structures divided by the standard deviation of the energy of the nonnative structures [Lee et al. 2001; Pillardy et al. 2001b]. Decoy sets were generated using the CSA method for searching the conformational space [Lee, Scheraga and Rackovsky 1997], and the parameters were optimized by using an iterative procedure [Lee et al. 2001; Pillardy et al. 2001b]. Both the set of decoys and the set of native-like structures were updated in each iteration of the procedure. While this algorithm worked well for proteins with simple topologies, we found it difficult to optimize the energy function for proteins with more complicated folds, for which the native-like structures are far from low-energy conformations found with initial-guess parameters. Therefore, we subsequently developed a different method (hierarchical optimization) for optimizing a potential function based not only on energy-ranking of native-like structures with respect to nonnative structures, but also on energy ranking of all structures, depending on their native-likeness [Liwo et al. 2002].

In our hierarchical optimization method [Liwo et al. 2002], the conformational space is discretized into levels, each of which is characterized by a certain degree of similarity between given fragment(s) of the molecule to the native structure. To construct a hierarchy, we define the elementary fragments of the molecule [Ołdziej et al. 2004a, 2004b; Liwo et al. 2007] as secondary-structure elements, and the hierarchy levels of the algorithm in terms of elementary fragments. The construction of a hierarchy is illustrated in Figure 8.3. The order of formation of native-like structure with increasing level number should follow the folding pathway if such information is available about the corresponding training protein. Otherwise, the optimal order of formation of native-like structure can be deduced by trial-and-error, as done in our earlier work [Ołdziej et al. 2004a, 2004b]. The hierarchical-optimization procedure is executed by minimizing a target function consisting of penalties for violation of the preset boundaries of the free-energy gaps between levels at selected temperatures [Ołdziej et al. 2004a, 2004b; Liwo et al. 2007].

Initially [Ołdziej et al. 2004a, 2004b], the decoys used to compute the free energies were generated by using the CSA method with the "local CSA" extension that enabled us to sample the region of the experimental structure [Lee et al. 2001] because the force field was designed to predict native structures of proteins as the lowest-energy conformations [Ołdziej et al. 2005]. However, decoys generated by this search do not form a canonical ensemble and, consequently, the resulting force field can be used only for prediction by global optimization of potential energy, while its

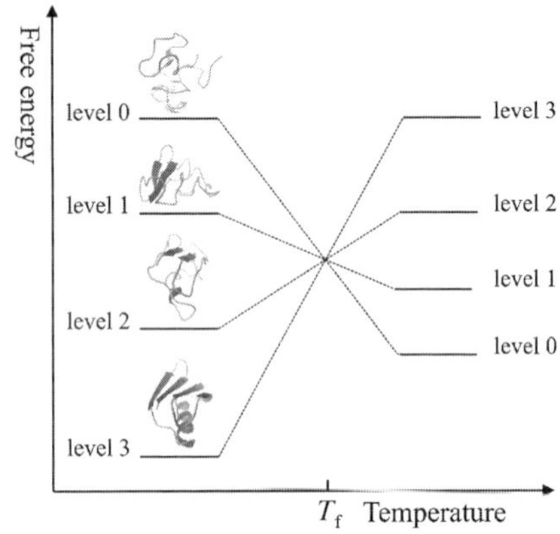

FIGURE 8.3 Schematic illustration of ordering of the energy levels, which is the goal of the algorithm for optimizing the potential function, using the 1EM7 protein of the IGG family, the structure of which consists of two β-hairpins packed to a middle α-helix. The figure illustrates the latest version of the method described in Liwo et al. (2007). Only one representative has been selected to represent each of the structural levels, but it should be stressed that, except for the native level (level 3), the levels contain quite diverse conformations. Below the folding-transition temperature (T_f), the nonnative level (level 0) has the highest free energy, the conformations with only the native C-terminal β-hairpin formed (level 1) have a lower free energy, next are the conformations in which the middle part of the N-terminal β-strand joins the β-hairpin and the middle α-helix starts to form and, finally, the native-like structures with all structural elements formed have the lowest free energy. Above the folding-transition temperature the free-energy relations are reversed and at the folding-transition temperature the free energies should be approximately equal. Such ordering of the free energies of structural levels also guarantees the appearance of a sharp peak in the heat capacity and a jump in native-likeness at T_f [Liwo et al. 2007]. The structural hierarchy was constructed based on the experimental data for folding of proteins of the IGG family [Scalley et al. 1997].

application to compute thermodynamic properties and kinetics of folding is not straightforward. In particular, because of the use of noncanonical decoys, we had no control of the folding-transition temperature. Therefore, recently, Liwo et al. (2007) reoptimized the UNRES force field by using decoys generated by multiplexing replica-exchange MD simulations (Section 8.2.4) at temperatures covering the folding-transition temperature range. Such a procedure also enables us to reproduce thermodynamic properties of folding, in particular the sharp peak in the heat capacity at the folding-transition temperature.

8.2.4 Search of Conformational Space with UNRES

We have developed or adopted a variety of conformational search algorithms with UNRES: global optimization to find the lowest-energy conformation [Liwo et al. 1999; Czaplewski et al. 2004a], MD to compute folding trajectories [Liwo, Khalili, and Scheraga 2005], and replica-exchange MD as a robust tool to generate canonical ensembles at various temperatures [Nanias, Czaplewski, and Scheraga 2006]. These techniques are outlined below.

Among the most effective global optimization algorithms for use with UNRES are stochastic Monte Carlo-based algorithms: Monte Carlo with minimization [Li and Scheraga 1987, 1988], conformation-family Monte Carlo (CFMC) [Pillardy et al. 2000], and two hybrid algorithms: (i) the self-consistent basin-to-deformed-basin mapping [Pillardy et al. 1999], which combines deformation of the potential-energy function with a Monte Carlo search, and (ii) CSA [Lee, Scheraga and Rackovsky 1997; Czaplewski et al. 2004a]. CSA is a hybrid approach combining the global character of a genetic algorithm and the fast convergence of local minimization. The CSA algorithm proved the most effective, and we originally applied it as a search engine. Later [Czaplewski et al. 2004b] we enhanced it to include the search for the optimal topology of disulfide bonds.

In order to study the pathways and kinetics of protein folding and to obtain canonical ensembles (which enable us to study folding thermodynamics), we extended UNRES to mesoscopic MD (UNRES/MD) [Khalili et al. 2005a, 2005b]; recently [Chinchio et al. 2007] we also extended this treatment to handle dynamic formation, breaking, and rearrangement of disulfide bonds by introducing a bimodal potential, with one minimum corresponding to the bonded state of a cystine pair and another one corresponding to the nonbonded state, and a barrier between them. Because the geometry of an UNRES polypeptide chain is not uniquely defined by the Cartesian coordinates of the interacting sites (i.e., the SC and p centers), the Lagrange formalism has been applied to derive the equations of motion for MD calculations with the UNRES model of polypeptide chains using virtual-bond $C^{\alpha} \ldots C^{\alpha}$ and $C^{\alpha} \ldots SC$ vectors as a set of generalized coordinates. Furthermore, the virtual bonds are represented as elastic rods with uniformly distributed masses [Khalili et al. 2005b]. Nonconservative forces (the friction and random forces) coming from the solvent were included in the framework of Langevin dynamics [Khalili et al. 2005a]; we also applied the Berendsen thermostat [Berendsen et al. 1984] to carry out isokinetic runs without explicit friction and random forces. To numerically integrate the nondiagonal equations of motion implied in UNRES dynamics, Khalili et al. (2005a) developed a modified version of the stochastic analogue of the velocity Verlet algorithm (which is symplectic and time-reversible) [Guarnieri and Still 1994] for UNRES Langevin dynamics simulations, and applied the velocity-Verlet algorithm [Swope et al. 1982] with the Berendsen thermostat [Berendsen et al. 1984].

Canonical MD can be used for estimating thermodynamic and kinetic properties as well as for conformational sampling but, in practice, it can easily be trapped in local minima separated by high energy barriers, and thus is not an effective method for studying rough free-energy landscapes of proteins. One of the most effective sampling methods, the RE method (also known as exchange MC [Hukushima and Nemoto 1996] or parallel tempering [Hansmann 1997]), was initially developed to improve sampling in glassy systems in statistical physics. Recently, the RE method has been applied extensively in protein-folding simulations [Sugita and Okamoto 1999; Nanias, Czaplewski, and Scheraga 2006].

The replica exchange MD (REMD) method combines the idea of simulated annealing and Monte Carlo methods, and is one of the generalized-ensemble algorithms that perform a random walk in energy space due to a free random walk in temperature space. In the REMD method, n replica systems, each in the canonical ensemble, and each at a different temperature, are simulated. At given intervals, swaps, or exchanges, of the configurational variables between systems are tried and accepted with the Metropolis criterion. The multiplexing variant of the REMD method (MREMD) [Rhee and Pande 2003; Nanias, Czaplewski, and Scheraga 2006] enhances sampling by multiplexing the replicas with a number of independent MD runs at each temperature.

The replica-exchange technique enables us to evaluate ensemble-based quantities (free energy, energy, heat capacity, Root Mean Square Deviation (RMSD), etc.). To compute the averages from the results of simulations carried out at different temperatures, we developed a version [Liwo et al. 2007] of the weighted histogram analysis method (WHAM) [Kumar et al. 1992].

8.3 APPLICATIONS AND RESULTS

8.3.1 Physics-Based Prediction of Protein Structures

The underlying principle of all physics-based protein-structure prediction methods is the thermodynamic hypothesis formulated by Anfinsen (1973). Given an amino acid sequence and a fast and accurate approximation of the effective potential-energy function, the goal of physics-based protein-structure prediction methods is to find the basin in the energy surface with the lowest free energy at a given (room) temperature; this basin should contain closely overlapping conformations (such as an ensemble of conformations resulting from protein structure determination by NMR) which should correspond to the native structure. However, in most approaches, the physics-based protein structure prediction is formulated in terms of a search for the lowest-energy conformation [Scheraga et al. 2002] and, consequently, protein-structure prediction becomes a problem of global optimization of an effective potential-energy function. A strong limitation of this approach is that it takes care only of the depth (energy) of a basin and, thus, is equivalent to finding the most probable conformations at a temperature of 0 K, while ignoring their volume (conformational entropy), which is often a decisive factor at room temperature. Consequently, conformations from the lowest-energy basin might have a negligible probability to appear in canonical simulations at room temperature, and a force field optimized to find native-like structure as lowest-energy ones could fail in producing any quantity of native-like structures in canonical simulations at room temperature [Liwo, Khalili, and Scheraga 2005].

Our protein structure prediction methodology based on the search of the lowest-potential-energy conformations by CSA was tested on benchmark proteins [Ołdziej et al. 2004a] as well as in critical assessment of techniques for protein structure prediction (CASP) experiments [Liwo et al. 1999; Pillardy et al. 2001a; Ołdziej et al. 2005]. During the CASP3 (1998) and CASP4 (2000) experiments, we were able to predict fragments or, rarely, whole structures of 60- to 100-residue all-α-helical proteins and, in CASP4, only started to predict the structures of $\alpha+\beta$- and β-proteins [Pillardy et al. 2001a], because of the introduction of additional cumulant-based correlation terms to the UNRES force field. In CASP5 (2002), by including double-torsional terms and using our hierarchical approach to optimize the force field, we extended the applicability of our methodology to predict 70-residue fragments of $\alpha+\beta$-proteins; however, there were problems in achieving the correct packing topology of the β-strands [Ołdziej et al. 2005]. During CASP6 (2004), with the introduction of the force field parameterized by using four training proteins simultaneously [Ołdziej et al. 2005], we were able to predict $70-80$ residue fragments or whole structures of $\alpha+\beta$-proteins and up to 140-residue fragments within 6 Å RMSD cutoff of all-α-helical proteins with correct secondary-structure element packing. The length of segments that were correct in terms of topology was up to 208 residues for all-α-helical proteins. Our best predictions from CASP5 and CASP6 are shown in Figure 8.4. However, we found that the CSA global optimization method, which was being used up to this point to search conformational space, reached the limit of its usefulness mainly because of computational cost related to larger proteins. In CASP6, we successfully reduced the computational cost of the global conformational search by using secondary-structure prediction to generate the starting structures. However, this modification meant introducing some elements of knowledge-based information, which we want to avoid in a physics-based approach. Another problem, which arises from the use of potential-energy-based protein-structure prediction, is the inability to correctly predict structures of all-β proteins. All search methods, such as CSA, which are based on potential-energy minimization, will always favor formation of α-helices over that of β-structures because (for a given part of the sequence of a protein) it is an energetically more favorable structure than β-strands or a hairpin. Therefore, the UNRES force field with the CSA search procedure was efficient in finding native-like structures of all-α-helical and most $\alpha+\beta$ proteins but usually failed for all-β proteins. The energy function can be adjusted to produce β-structures as more favorable in energy, but such an operation usually provides a potential which is not able to produce α-helices.

Because of the limitations of global optimization of a potential-energy function as a tool for protein-structure prediction, Liwo et al. (2007) recently developed an approach based on searching for the basin with the lowest free energy at room temperature. To find this basin, MREMD simulations are run at a range of temperatures containing the folding-transition temperature followed by histogram reweighting to calculate the weights of conformations at a temperature of choice. Next, a cluster analysis is carried out to partition the ensemble and compute the probability (P_i) of each subensemble by Boltzmann summation, as given by Equation 8.8.

$$P_i = \frac{\sum_{k\in\{i\}} \exp(\omega_k - \beta U_k)}{\sum_{k=1}^{N} \exp(\omega_k - \beta U_k)}, \tag{8.8}$$

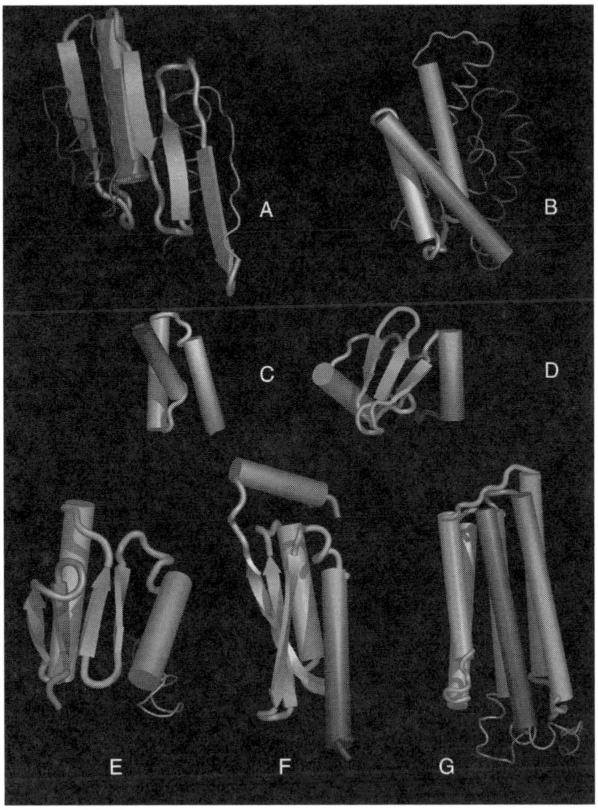

FIGURE 8.4 (See color insert following page 238.) The best predicted structures of CASP5 targets T0149_2 (the C-terminal domain of T0149) (A) and T0129 (B) and CASP6 targets T0215 (C), T0281 (D), T0223_2 (the second domain of T0223) (E), T0230 (F), and T0198 (G). The correctly predicted parts of the structures are marked as cylinders (for α-helices), ribbons (for β-strands), and thick lines (loops and other unstructured regions), and the incorrectly predicted parts are shown as thin lines. The chains are colored from blue to red from the N to the C terminus. (From Ołdziej, S. et al., *Proc. Natl. Acad. Sci. U.S.A.*, 102, 7547–7550, 2005. With permission.)

where N is the total number of conformations, $\{i\}$ is the set of conformations that belong to cluster (basin) i, U_k is the UNRES energy of conformation k, and ω_k is the weighting factor of that conformation obtained by WHAM processing of MREMD results [Liwo et al. 2007]. The clusters are ranked as candidate predictions following the decrease of P_i.

The above approach was tested in CASP7 (2006). Because, at that time, the corresponding force field was only in its early stage of development, it could handle only α-helical proteins. Our best prediction was target T0300, for which we correctly predicted the structure of a 64-residue fragment.

8.3.2 Dynamics and Kinetics of Protein Folding

The most straightforward application of mesoscopic MD is the study of folding pathways. Using the early version of UNRES MD [Liwo, Khalili, and Scheraga 2005], we investigated the folding pathways of a number of α-helical proteins (1BDD, 1CLB, 1LQ7) and an α + β protein (1E0G), starting each simulation from the fully extended chain. We found that UNRES/MD provides about 4000 time speed-up compared to all-atom MD, which enables us to study folding in real computational time. The largest (α-helical) protein folded thus far contains 75 residues. An interesting feature of the simulated folding pathways of 1E0G was that folding started by formation of α-helical structures, and the β-sheet fragment formed later by unfolding some of the α-helices because of favorable non-local interactions. However, the force field used in that study could be too biased towards α-helical structures. A sample trajectory of folding 1E0G is shown in Figure 8.5.

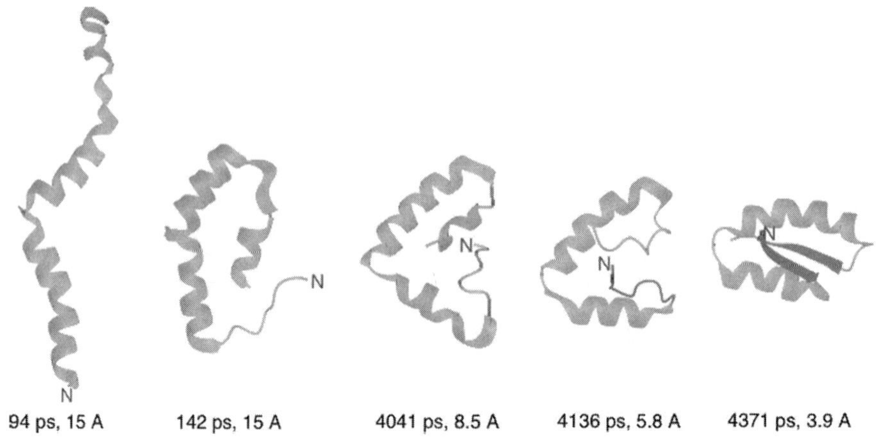

FIGURE 8.5 Example of a folding pathway of 1E0G obtained in Langevin dynamics simulations with UNRES. The N-terminus of the chain is marked for tracing purposes. (From Liwo, A. et al., *Proc. Natl. Acad. Sci. U.S.A.*, 102, 2362–2366, 2005. With permission.)

The comparatively short time necessary to obtain a single trajectory long enough to observe folding (about 5–30 hours with a single AMD processor, depending on protein size) prompted us to simulate the kinetics of folding of the N-terminal fragment of the B-domain of protein A [Khalili, Liwo, and Scheraga 2006]. The structure of this protein is a three-helix bundle. We carried out Langevin dynamics simulations of 400 trajectories, and found that two major routes are present, a fast one leading to the folded state and a slow one with mispacked intermediates on the way. The kinetic equations derived in terms of changes of the RMSD from the native structure as a function of time showed a biexponential behavior and fully conformed to the model encompassing a direct fast route and a slow route through an intermediate.

8.3.3 Multichain Proteins

Since many naturally occurring proteins are oligomeric, consisting of multiple polypeptide chains of the same or differing sequences, the native conformation of a protein monomer often arises as a result of many interchain interactions important in the folding and stabilization of the native structure. Therefore, to successfully predict the three-dimensional structure of an oligomeric protein, it is necessary to take into account not only the interactions within each monomer, but also the interactions between the different chains in the multichain system. For this purpose, we used the UNRES force field to treat multiple-chain complexes [Saunders and Scheraga 2003a, 2003b; Rojas, Liwo, and Scheraga 2007]. Since the intra- and interchain interactions have the same physical basis, the forms, parameters, and weights of the interchain terms are exactly the same as those for single-chain terms in UNRES. The only difference is that, in the case of interchain interactions, the interacting sites are not connected to the same backbone. Therefore, the local interactions and the correlation terms of UNRES involving all adjacent residues do not contribute to the interchain potential energy.

The treatment of multiple-chain complexes with UNRES was initially implemented in combination with the CSA and CFMC search methods [Saunders and Scheraga 2003a, 2003b]. This implementation took advantage of the fact that oligomeric proteins containing multiple chains of identical sequence often adopt symmetrical arrangements. The presence of such symmetries was used to increase the speed of local minimizations by constraining the search to those conformations with certain symmetry. Furthermore, since, in a symmetric structure, the internal conformations are the same for all chains, the contributions to the energy for intrachain interactions have to be calculated for only one chain, thus reducing the time needed for energy evaluations and gradient calculations. The multichain implementation contemplated the possibility of applying different levels and different types of symmetry (no symmetry, one-chain affine, two-chain affine, screw, rotational, and shared-axis symmetries). The implementation was tested on two α-helical proteins and proved to be very efficient in finding their native structures when symmetry constraints were imposed [Saunders and Scheraga 2003a].

Later on, an application combining the multiple-chain UNRES treatment with the MD technique was also developed [Rojas, Liwo, and Scheraga 2007]. In this latter application, no symmetry constraints were imposed. To carry out the MD simulations, it was necessary to mimic the peptide concentration, which was done by confining the system within a soft sphere; that is, an extra term, U_{conf}, was added to the potential, causing each interacting site (either a peptide group or a side chain) to feel an attractive force toward the center of the sphere whenever it moved outside the boundary of the sphere. The radius of the sphere and the number of peptide chains in the solution determines the peptide concentration of the simulated solution. The total UNRES potential energy of an oligomeric protein is then given by Equation 8.9.

$$U = \sum_k U_k^{\text{single chain}} + \sum_k \sum_{l<k} U_{kl}^{\text{inter chain}} + U_{\text{conf}}, \tag{8.9}$$

where $U_k^{\text{single chain}}$ is the contribution to the potential energy due to the internal interactions in the kth chain (as described in Section 8.2.1), $U_{kl}^{\text{inter chain}}$ is the contribution due to the interactions between the kth and lth chain, and U_{conf} is a restraining potential confining the system to a given volume.

The generalization of the equations of motion for single chain proteins to the multichain system is straightforward since the different chains are coupled only through the UNRES potential energy U (the kinetic energy of a specific chain does not contain any dependence on the coordinates from a different chain). As in the case of UNRES MD for single-chain proteins, Berendsen dynamics and a method based on Langevin dynamics were implemented. Both methods were able to fold two α-helical dimers with 48 and 33 amino acids per chain, respectively [Rojas, Liwo, and Scheraga 2007] from the extended conformation.

8.4 CONCLUSIONS AND OUTLOOK

In order to compute the native structures of globular proteins, and the pathways leading to them, a physics-based hierarchical approach, based on a coarse-grained united-residue (UNRES) model of the polypeptide chain, was developed. Initially, it was parameterized primarily on protein structures from the PDB, and the CSA method was used in the conformational search to reach the *region* of the global minimum of the UNRES potential energy. The results of the CSA search were converted to all-atom structures to enable the search for the global minimum to continue with an all-atom potential function. This approach was partially successful in successive CASP exercises in computing the structures of α-helical and $\alpha+\beta$ proteins.

With this partial success, and the recognition that the entropic effect should be included in the search algorithm to enable free energy, rather than potential energy, to control the folding process, Langevin dynamics was adopted with UNRES, with the expectation that the removal of the fast degrees of freedom in the formulation of UNRES would overcome the MD time-step problem to some extent. With this MD approach, a partial success was achieved in computing pathways and native structures of single-chain and multiple-chain α-helical proteins (one as large as 75 residues) and one $\alpha+\beta$ protein, with an advanced penetration into the MD folding-time range.

To capitalize on this success of the Langevin dynamics approach, steps are being taken to improve the force field by replacing the reliance of some remaining UNRES potential-energy terms on the PDB with physics-based quantum-mechanical calculations, by including entropy (presently by introducing temperature dependence of the effective energy function), and by replacing CSA by MD with UNRES. It is expected that this improved physics-based approach will provide reliable structures for whole (and larger) proteins instead of only parts of proteins, and better folding pathways and kinetics than achieved heretofore with earlier versions of UNRES.

ACKNOWLEDGMENTS

This work is supported by grants from the National Institutes of Health (GM-14312), the National Science Foundation (MCB00-03722), the Fogarty Foundation (TW1064), and grants 1 T09A 099 30 and DS 8372-4-0138-7 from the Polish Ministry of Science and Higher Education. Mariusz Makowski was supported by a grant from the "Homing" program of the Foundation for Polish Science (FNP). This research is conducted by using the resources of (a) our 800-processor Beowulf cluster at the Baker Laboratory of Chemistry and Chemical Biology, Cornell University, (b) the National Science Foundation Terascale Computing System at the Pittsburgh Supercomputer Center, (c) the John von Neumann Institute for Computing at the Central Institute for Applied Mathematics, Forschungszentrum Juelich, Germany, (d) the Beowulf cluster at the Department of Computer Science, Cornell University, (e) the resources of the Center for Computation and Technology at Louisiana State University, which is supported by funding from the Louisiana legislature, (f) our 45-processor Beowulf cluster at the Faculty of Chemistry, University of Gdańsk, (g) the Informatics Center of the Metropolitan Academic Network (IC MAN) in Gdańsk, and (h) the Interdisciplinary Center of Mathematical and Computer Modeling (ICM) at the University of Warsaw.

REFERENCES

Anfinsen, C. B. 1973. Principles that govern the folding of protein chains. *Science* 181:223–30.

Berendsen, H. J. C., J. P. M. Postma, W. F. van Gunsteren, A. DiNola, and J. R. Haak. 1984. Molecular dynamics with coupling to an external bath. *J. Chem. Phys.* 81:3684–90.

Brant, D. A., and P. J. Flory. 1965. The configuration of random polypeptide chains. II. Theory *J. Am. Chem. Soc.* 87:2791–800.

Chinchio, M., C. Czaplewski, A. Liwo, S. Ołdziej, and H. A. Scheraga. 2007. Dynamic formation and breaking of disulfide bonds in molecular dynamics simulations with the UNRES force field. *J. Chem. Theory Comput.* 3:1236–48.

Czaplewski, C., A. Liwo, J. Pillardy, S. Ołdziej, and H. A. Scheraga. 2004a. Improved conformational space annealing method to treat β-structure with the UNRES force-field and to enhance scalability of parallel implementation. *Polymer* 45:677–86.

Czaplewski, C., S. Ołdziej, A. Liwo, and H. A. Scheraga. 2004b. Prediction of the structures of proteins with the UNRES force field, including dynamic formation and breaking of disulfide bonds. *Protein Eng. Des. Sel.* 17:29–36.

De Santis, P., E. Giglio, A. M. Liquori, and A. Ripamonti. 1963. Stability of helical conformations of simple linear polymers. *J. Polymer Sci.* A1:1383–404.

Goldstein, R. A., Z. A. Luthey-Schulten, and P. G. Wolynes. 1992. Protein tertiary structure recognition using optimized Hamiltonians with local interactions. *Proc. Natl. Acad. Sci. U.S.A.* 89:9029–33.

Guarnieri, F. and W. C. Still. 1994. A rapidly convergent simulation method: Mixed Monte Carlo/stochastic dynamics. *J. Comput. Chem.* 15:1302–10.

Hansmann, U.H.E. 1997. Parallel tempering algorithm for conformational studies of biological molecules. *Chem. Phys. Lett.* 281:140–50.

Hukushima, K., and K. Nemoto. 1996. Exchange Monte Carlo method and application to spin glass simulations. *J. Phys. Soc. Jpn.* 65:1604–608.

Izvekov, S., and G. A. Voth. 2005. A multiscale coarse-graining method for biomolecular systems. *J. Phys. Chem. B* 109:2469–73.

Kaźmierkiewicz, R., A. Liwo and H. A. Scheraga. 2002. Energy-based reconstruction of a protein backbone from its α-carbon trace by a Monte Carlo method. *J. Comput. Chem.* 23:715–23.

———. 2003. Addition of side chains to a known backbone with defined side-chain centroids. *Biophys. Chem.* 100:261–80. Erratum: *Biophys. Chem.* 106, 91 (2003).

Khalili, M., A. Liwo, A. Jagielska and H. A. Scheraga. 2005a. Molecular dynamics with the united-residue (UNRES) model of polypeptide chains. II. Langevin and Berendsen-bath dynamics and tests on model α-helical systems. *J. Phys. Chem. B* 109:13798–810.

Khalili, M., A. Liwo, F. Rakowski, P. Grochowski, and H. A. Scheraga. 2005b. Molecular dynamics with the united-residue (UNRES) model of polypeptide chains. I. Lagrange equations of motion and tests of numerical stability in the microcanonical mode. *J. Phys. Chem. B* 109:13785–97.

Khalili, M., A. Liwo and H. A. Scheraga. 2006. Kinetic studies of folding of the B-domain of staphylococcal protein A with molecular dynamics and a united-residue (UNRES) model of polypeptide chains. *J. Mol. Biol.* 355:536–47.

Kozłowska, U., A. Liwo, and H. A. Scheraga. 2007a. Determination of virtual-bond-angle potentials of mean force for coarse-grained simulations of protein structure and folding from ab initio energy surfaces of terminally-blocked glycine, alanine, and proline. *J. Phys. Cond. Matter.* 19:285203.

Kozłowska, U., A. Liwo, G. Maisuradze and H. A. Scheraga. 2007b. Determination of side-chain-rotamer potentials of mean force for coarse-grained simulations of protein structure and folding from AM1 energy surfaces of terminally-blocked amino-acid residues. *J. Phys. Chem. B.*, submitted.

Kubo, R. 1962. Generalized cumulant expansion method. *J. Phys. Soc. Japan* 17:1100–20.

Kumar, S., D. Bouzida, R. H. Swendsen, P. A. Kollman, and J. M. Rosenberg. 1992. The weighted histogram analysis method for free-energy calculations on biomolecules. I. The method. *J. Comput. Chem.* 13:1011–21.

Lee, J., D. R. Ripoll, C. Czaplewski, J. Pillardy, W. J. Wedemeyer, and H. A. Scheraga. 2001. Optimization of parameters in macromolecular potential energy functions by conformational space annealing. *J. Phys. Chem. B* 105:7291–98.

Lee, J., H. A. Scheraga, and S. Rackovsky. 1997. New optimization method for conformational energy calculations on polypeptides: Conformational space annealing. *J. Comput. Chem.* 18:1222–32.

Levitt, M. 1976. A simplified representation of protein conformations for rapid simulation of protein folding. *J. Mol. Biol.* 104:59–107.

Levitt, M., and S. Lifson. 1969. Refinement of protein conformations using a macromolecular energy minimization procedure. *J. Mol. Biol.* 46:269–79.

Li, Z., and H. A. Scheraga. 1987. Monte Carlo-minimization approach to the multiple-minima problem in protein folding. *Proc. Natl. Acad. Sci. U.S.A.* 84:6611–15.

———. 1988. Structure and free energy of complex thermodynamic systems. *J. Mol. Struct. (Theochem)* 179:333–52.

Liwo, A., P. Arłukowicz, C. Czaplewski, S. Ołdziej, J. Pillardy, and H. A. Scheraga. 2002. A method for optimizing potential-energy functions by a hierarchical design of the potential-energy landscape. Application to the UNRES force field. *Proc. Natl. Acad. Sci. U.S.A.* 99:1937–42.

Liwo, A., C. Czaplewski, J. Pillardy, and H. A. Scheraga. 2001. Cumulant-based expressions for the multibody terms for the correlation between local and electrostatic interactions in the united-residue force field. *J. Chem. Phys.* 115:2323–47.

Liwo, A., M. Khalili, C. Czaplewski, S. Kalinowski, S. Ołdziej, K. Wachucik, and H. A. Scheraga. 2007. Modification and optimization of the united-residue (UNRES) potential energy function for canonical simulations. I. Temperature dependence of the effective energy function and tests of the optimization method with single training proteins. *J. Phys. Chem. B* 111:260–85.

Liwo, A., M. Khalili, and H. A. Scheraga. 2005. Molecular dynamics with the united-residue (UNRES) model of polypeptide chains; test of the approach on model proteins. *Proc. Natl. Acad. Sci. U.S.A.* 102:2362–67.

Liwo, A., J. Lee, D. R. Ripoll, J. Pillardy, and H. A. Scheraga. 1999. Protein structure prediction by global optimization of a potential energy function, *Proc. Natl. Acad. Sci. U.S.A.* 96:5482–85.

Liwo, A., S. Ołdziej, C. Czaplewski, U. Kozłowska, and H. A. Scheraga. 2004. Parameterization of backbone-electrostatic and multibody contributions to the UNRES force field for protein-structure prediction from ab initio energy surfaces of model systems. *J. Phys. Chem. B* 108:9421–38.

Liwo, A., S. Ołdziej, M. R. Pincus, R. J. Wawak, S. Rackovsky, and H. A. Scheraga. 1997a. A united-residue force field for off-lattice protein-structure simulations. I. Functional forms and parameters of long-range side-chain interaction potentials from protein crystal data. *J. Comput. Chem.* 18:849–73.

Liwo, A., M. R. Pincus, R. J. Wawak, S. Rackovsky, S. Ołdziej, and H. A. Scheraga. 1997b. A united-residue force field for off-lattice protein-structure simulations. II: Parameterization of local interactions and determination of the weights of energy terms by Z-score optimization. *J. Comput. Chem.* 18:874–87.

Makowski, M., E. Sobolewski, C. Czaplewski, A. Liwo, S. Ołdziej, J. No, and H. A. Scheraga. 2007. Simple physics-based analytical formulas for the potentials of mean force for the interaction of amino acid side chains in water. 3. Calculation and parameterization of the potentials of mean force of pairs of identical hydrophobic side chains. *J. Phys. Chem. B* 111:2925–31.

McCammon, J. A., B. R. Gelin, and M. Karplus. 1977. Dynamics of folded proteins. *Nature* 267:585–90.

Momany, F. A., R. F. McGuire, A. W. Burgess, and H. A. Scheraga. 1975. Energy parameters in polypeptides. VII. Geometric parameters, partial atomic charges, nonbonded interactions, hydrogen bond interactions, and intrinsic torsional potentials for the naturally occurring amino acids. *J. Phys. Chem.* 79:2361–81.

Nanias, M., C. Czaplewski, and H. A. Scheraga. 2006. Replica exchange and multicanonical algorithms with the coarse-grained united-residue (UNRES) force field. *J. Chem. Theor. Comput.* 2:513–28.

Némethy, G., and H.A. Scheraga. 1965. Theoretical determination of sterically allowed conformations of a polypeptide chain by a computer method. *Biopolymers* 3:155–84.

Nishikawa, K., F. A. Momany, and H. A. Scheraga. 1974. Low-energy structures of two dipeptides and their relationship to bend conformations. *Macromolecules* 7:797–806.

Ołdziej, S., U. Kozłowska, A. Liwo, and H. A. Scheraga. 2003. Determination of the potentials of mean force for rotation about $C^{\alpha}...C^{\alpha}$ virtual bonds in polypeptides from the ab initio energy surfaces of terminally-blocked glycine, alanine, and proline. *J. Phys. Chem. A* 107:8035–46.

Ołdziej, S., J. Łągiewka, A. Liwo, C. Czaplewski, M. Chinchio, M. Nanias, and H. A. Scheraga. 2004a. Optimization of the UNRES force field by hierarchical design of the potential-energy landscape: III. Use of many proteins in optimization. *J. Phys. Chem. B* 108:16950–59.

Ołdziej, S., A. Liwo, C. Czaplewski, J. Pillardy, and H. A. Scheraga. 2004b. Optimization of the UNRES force field by hierarchical design of the potential-energy landscape: II. Off-lattice tests of the method with single proteins. *J. Phys. Chem. B* 108:16934–49.

Ołdziej, S., C. Czaplewski, A. Liwo, M. Chinchio, M. Nanias, J. A. Vila, M. Khalili, Y. A. Arnautova, A. Jagielska, M. Makowski, H. D. Schafroth, R. Kazmierkiewicz, D. R. Ripoll, J. Pillardy, J. A. Saunders, Y.-K. Kang, K. D. Gibson, and H. A. Scheraga. 2005. Physics-based protein-structure prediction using a hierarchical protocol based on the UNRES force field-test with CASP5 and CASP6 targets. *Proc. Natl. Acad. Sci. U.S.A.* 102:7547–52.

Pillardy, J., C. Czaplewski, A. Liwo, J. Lee, D. R. Ripoll, R. Kaźmierkiewicz, S. Ołdziej, W. J. Wedemeyer, K. D. Gibson, Y. A. Arnautova, J. Saunders, Y.-J. Ye, and H. A. Scheraga 2001a. Recent improvements in prediction of protein structure by global optimization of a potential energy function. *Proc. Natl. Acad. Sci. U.S.A.* 98:2329–33.

Pillardy, J., C. Czaplewski, A. Liwo, W. J. Wedemeyer, J. Lee, D. R. Ripoll, P. Arłukowicz, S. Ołdziej, Y. A. Arnautova, and H. A. Scheraga. 2001b. Development of physics-based energy functions that predict medium-resolution structure for proteins of the α, β, and α/β structural classes. *J. Phys. Chem. B* 105:7299–311.

Pillardy, J., C. Czaplewski, W. J. Wedemeyer, and H. A. Scheraga. 2000. Conformational family Monte Carlo (CFMC): An efficient computational method for identifying the low-energy states of a macromolecule. *Helv. Chim. Acta* 83:2214–30.

Pillardy, J., A. Liwo, M. Groth, and H. A. Scheraga. 1999. An efficient deformation-based global optimization method for off-lattice polymer chains; self-consistent basin-to-deformed-basin mapping (SCBDBM). application to united-residue polypeptide chains. *J. Phys. Chem.* 103:7353–66.

Ramachandran, G. N., C. Ramakrishnan, and V. Sasiekharan. 1963. Stereochemistry of polypeptide chain configurations. *J. Mol. Biol.* 7:95–99.

Rhee, Y. M., and V. S. Pande. 2003. Multiplexed-replica exchange molecular dynamics method for protein folding simulation. *Biophys. J.* 84:775–86.

Rojas, A. V., A. Liwo, and H. A. Scheraga. 2007. Molecular dynamics with the united-residue force field: Ab initio folding simulations of multichain proteins. *J. Phys. Chem. B* 111:293–309.

Sali, A., E. Shakhnovich, and M. Karplus. 1994. How does a protein fold? *Nature* 369:248–51.

Saunders, J. A., and H. A. Scheraga. 2003a. Ab initio structure prediction of two α-helical digomers with a multiple-chain united-residue force field and global search. *Biopolymers* 68:300–17.

———. 2003b. Challenges in structure prediction of oligomeric proteins at the united-residue level: Searching the multiple-chain energy landscape with CSA and CFMC. *Biopolymers* 68:318–32.

Scalley, M. J., Q. Yi, H. Gu, A. McCormack, J. R. Yates III, and D. Baker. 1997. Kinetics of folding of the IGG binding domain of peptostreptoccocal protein L. *Biochemistry* 36:3373–82.

Scheraga, H. A., J. Pillardy, A. Liwo, J. Lee, C. Czaplewski, D. R. Ripoll, W. J. Wedemeyer, and Y. A. Arnautova. 2002. Evolution of physics-based methodology for exploring the conformational energy landscape of proteins. *J. Comput. Chem.* 23:28–34.

Sugita, Y., and Y. Okamoto. 1999. Replica-exchange molecular dynamics method for protein folding. *Chem. Phys. Lett.* 314:141–51.

Swope, W. C., H. C. Andersen, P. H. Berens, and K. R. Wilson. 1982. A computer simulation method for the calculation of equilibrium constants for the formation of physical clusters of molecules: Application to small water clusters. *J. Chem. Phys.* 76:637–49.

9 Coarse-Grained Structure-Based Simulations of Proteins and RNA

Alexander Schug, Changbong Hyeon,* and José N. Onuchic*
Center for Theoretical Biological Physics,
University of California, San Diego

CONTENTS

9.1 PROTEINS

Proteins perform a variety of complex tasks in biological systems. Examples include oxygen transport in blood, catalytic function, cell signaling, and membrane transports. In spite of their broad

* Indicates shared first authorship.

utility, their basic composition is quite simple: linear polymers of amino-acid sequences. In solution, the amino-acid sequences encode all the necessary information to adopt a unique three-dimensional structure, the *native state*. The transition from an unstructured state (unfolded, U) to the native state (folded, F) is a self-organizing process called *protein folding*.

The available phase space of a protein increases exponentially with sequence length. Given the vast phase space of a protein and the surrounding solvent molecules, it appears remarkable that protein folding happens on the milliseconds to seconds timescale.[1] The uniqueness of the native state indicates it occupies the global minimum of the free-energy landscape.[2] The *principle of minimal frustration* introduced the idea of a sufficiently smooth energy landscape with sufficiently small energetic frustration, which offers a protein multiple folding pathways.[3–7] Native interactions dominate the energy landscape generating a funnel-like shape highly biased towards the native state. Evolution selected protein sequences to be reasonably good folders. Otherwise, if the energy landscape underlying the folding process were too rugged, the process would be expected to be far slower and a protein might express glass-like properties and become trapped in nonnative conformations.

Together, the *principle of minimal frustration* and the *funneled energy landscape* are the theoretical foundation for *structure-based simulations*.[8,9] In *structure-based simulations*, inspired by the work of Go,[10] the energetics of the system are determined by the interactions in the native state. By fully removing energetic frustration from these models, the simulations do not quantitatively reproduce energetic properties like folding barriers and protein stability. These are not robust properties of the folding mechanism and can be easily changed by mutation or changes in the environment. On the other hand, these models are very successful in reproducing the geometric properties of the landscape and thereby of the folding mechanism.

Commonly, one coarse-grains the protein structure to a C_α-bead presentation. Coarse-graining simplifies the Hamiltonian to a very concise formulation and reduces computational demands. A concise Hamiltonian in the form of a *minimal model* is desirable, as it condenses the complexity of the protein-folding problem to very few parameters. Additionally the computational demands are highly reduced. One can simulate roughly 1 ns of molecular dynamical (MD) on the all-atom level for a protein of 60 AA including the surrounding solvent within a single CPU-day.[†] Conformational transitions like folding or unfolding happen on the millisecond (or slower) time scale. In contrast, one can easily observe a multitude of folding/unfolding events in a C_α-structure-based model on a single day of CPU time (Figure 9.1c), which illustrates the gain of several orders of magnitude in speed.

Due to their simplicity, structure-based models provide valuable insight into the folding process, validating the underlying concepts. Straightforward MD simulations on structure-based models easily provide sufficient sampling to calculate free-energy profiles for small and mid-sized proteins. For two-state folders, the folding temperature (T_F) is defined as the temperature, at which the unfolded and folded state are equipopulated. The folding barrier in the free-energy profiles between those two states at T_F has been shown to give good qualitative agreement with the experimentally measured folding times for different proteins.[11,12] Φ-values geometrically characterize the protein in the transition-state ensemble (TSE). Φ-values calculated from structure-based simulations have been shown to be in good agreement with experimental measurements. For example, modeling two circular permutants by structure-based simulations[13] could reproduce available experimental data on circular permutants on the two state folders α-spectrin SH3 domain (SH3)[14] and chymotrypsin inhibitor 2 (CI2).[15]

Going beyond folding, structure-based simulations have been used to describe conformational transitions of proteins.[16,17] These transitions are often the molecular basis for their biological function. This interconnection of molecular motion with their biological relevance will become a research topic of much interest in the years to come. Current examples are work on molecular motors,[18] the Rop-dimer,[19,20] or adenylate kinase.[21]

† GROMACS [23] simulates about 1 ns of MD on a 60 AA protein with explicit solvent within 1 cpu-day (2 GHz).

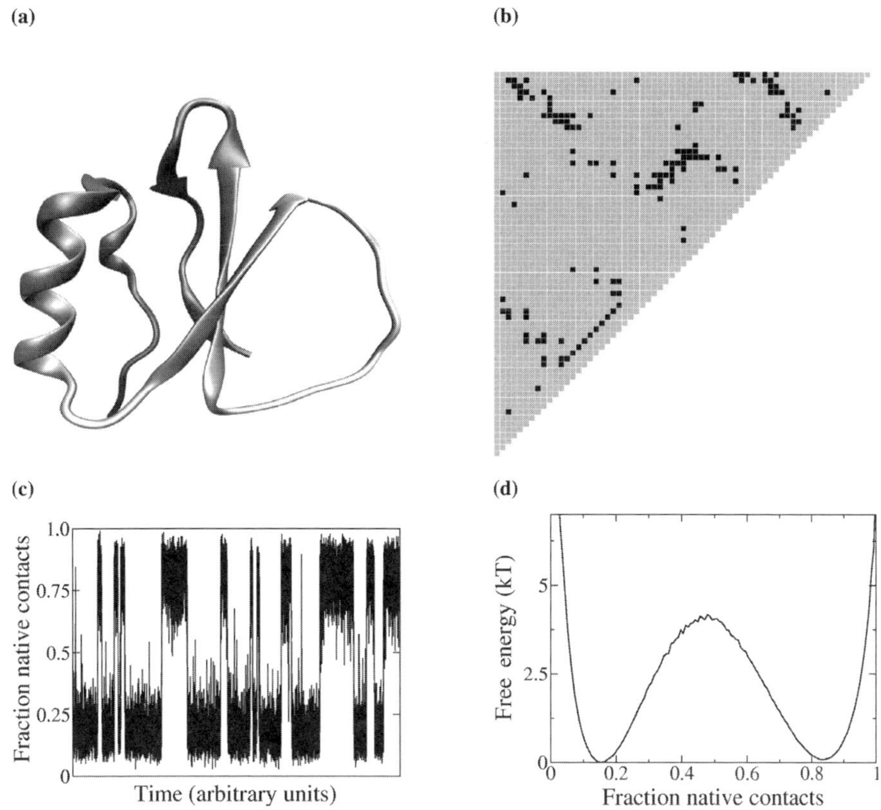

FIGURE 9.1 Typical setup of a simulation on the example of CI2. The tertiary structure of CI2 (a) contains both an alpha-helix and a multistranded beta-sheet. This structure defines the native interactions, plotted as the native contact map (b). In the native contact map, each black dot indicates that the amino acids *i* and *j* (at their respective matrix position) are interacting in the native state. The native contacts and the corresponding native distances are crucial for the setup of structure-based simulations. A part of a typical simulation around T_F is shown in (c). One can observe multiple transitions between the unfolded state, which has a low fraction of native contacts formed, and the folded state, where most of the native contacts are formed. This part took 6 h on a standard CPU (2 GHz). A free-energy profile calculated from one set of simulations at different temperatures is shown in (d). The folding barrier between the folded and unfolded state is around 4 kT.

9.1.1 Methods and Standard Techniques for Structure-Based Simulations

9.1.1.1 Mathematical Description

Structure-based simulations have a very simple functional form:

$$V(\vec{X}) = \sum_{\text{atoms}} K_{\text{m}}(x-\bar{x})^2 + \sum_{\text{bonds}} K_r(r-r_o)^2 + \sum_{\text{angles}} K_\theta(\theta-\theta_0)^2 + \sum_{\text{dihedral}} K_\phi^{(n)}\left[1-\cos(n(\phi-\phi_0))\right]$$

$$+\sum_{i<j-3}\left[\varepsilon_{ij}\left[5\left(\frac{\sigma_{ij}}{r_{ij}}\right)^{12} - 6\left(\frac{\sigma_{ij}}{r_{ij}}\right)^{10}\right] + \bar{\varepsilon}\left(\frac{\sigma_{ij}}{r_{ij}}\right)^{12}\right].$$

Here, $\vec{X}$ describes a conformation of the protein's C_α atoms with the bond lengths of neighboring atoms *r*, the angles θ, the dihedral angles Φ, and the distance r_{ij} between any two atoms. The constants used are $K_r = 100k_BT/\text{Å}^2$, $K_\theta = 20k_BT$, $K_\phi^{(1)} = 1k_BT$ and $K_\phi^{(3)} = 0.5k_BT$ (k_B is the Boltzmann constant and T is the temperature). An attractive interaction $\varepsilon_{ij} = 1k_BT$, $\bar{\varepsilon} = 0$ is assigned to natively interacting residues as defined by the contact map, while $\varepsilon_{ij} = 0$, $\bar{\varepsilon} = 1k_BT$ enforce a hard-sphere potential for nonnatively interacting residues. σ_{ij} gives the native distances between the interacting residues in case of native interaction or is set to 4 Å for noninteracting residues. For multimers, one does not want the individual monomers to move too far apart. To allow the monomers disassociation but prevent them from moving too far apart, a weak center of mass force can be additionally introduced with the force constant $K_m = 25 \cdot 10^{-6}k_BT/\text{Å}^2$ and $\bar{x} = \sum_{\text{atoms}} x / N$ (the xs give the positions of the atoms).[11]

9.1.1.2 Native Contacts

Each interacting pair i, j ($i+3<j$) of residues in the native state is called a *contact*. All contacts together are called the *contact map*. Commonly, the software CSU is used to calculate the contact map.[22] In structure-based simulations, each formed contacts gives a stabilizing energy of ε. During simulations a contact is considered formed when the distance between the interacting atoms is shorter than 1.2 times their native distance.

Native contacts also define the natural reaction coordinate Q for structure-based simulations. Q is the fraction of formed native contacts ($Q=0$ in the fully unfolded state and $Q=1$ in the fully folded state). Folding and unfolding of a protein can be considered a diffusive process of an ensemble of protein conformations over a low-dimensional order parameter. The choice of Q appears plausible, because coarse-grained simulations with a reduced level of energetic frustration have a strongly funneled energy landscape and Q is directly correlated with the stabilizing energy. A more thorough discussion of this choice of reaction coordinate can be found in Ref. 23.

9.1.1.3 Free-Energy Landscape and Folding Barrier

The free energy of a state assigns each macrostate over a reaction coordinate q the probability of occupation $(e^{-\beta G(q)}) / (\sum_q e^{-\beta G(q)})$ ($\beta = k_BT$ with k_B being the Boltzmann constant and T the temperature). In structure-based simulations Q; that is, the fraction of native contacts, is commonly used as the reaction coordinate. It covers transitions between U and F well. To gain a meaningful free-energy landscape with minimal errors, one needs to gain sufficient sampling in the MD simulations. As transitions between F and U are rare events, the relevant TSE region is accordingly only briefly occupied (has an accordingly high free energy).

Optimized sampling is obtained by long simulations at the folding temperature T_F, where U and F are equally populated. To estimate the location of T_F, we run several short simulations over a range of temperatures. For $T < T_F$ the folded state will be populated after equilibration, for $T > T_F$ the unfolded state.‡ After estimating T_F, long simulations with many folding/unfolding events ensure good sampling.§

The simulations at different temperatures can be combined into a free-energy profile using the weighted histogram analysis method (WHAM) algorithm[24,25] (for example see Figure 9.1). The folding barrier is between the F and U state at T_F. Kinetic theory connects this folding barrier to the folding rates $k_F \propto \exp(-\Delta G/kT)$ (k_F is the folding rate, ΔG the free-energy barrier at T_F). These folding barriers are in good agreement with experimentally measured folding rates.[11,12]

9.1.1.4 Φ-Value Analysis

Φ-value analysis is an experimental technique to characterize the TSE, which can be extended to simulations. Φ-values estimate the contribution of each residue to the TSE relative to the folded

‡ Enforcing (un)folding through choice of temperature is called kinetic (un)folding.

§ As a rule of thumb, 20 folding/unfolding transitions can be considered good sampling.

state[26] and are used to characterize changes in the TSE upon mutations.[27] In simulations, Φ_i for the residue i is calculated as

$$\Phi_i = \frac{P^i_{\mathrm{TSE}} - P^i_{\mathrm{U}}}{P^i_{\mathrm{F}} - P^i_{\mathrm{U}}},$$

where P^i_X is the probability of formation of contacts involving residue i in the state X, where X is the TSE, U, or F state.[28] ¶

By calculating Φ-values, one can construct a structural description of the TSE. A value of 1 means that the corresponding residue has all its contacts formed and is very native-like in the TSE, while a value of 0 means that this residues has no contacts formed in the TSE and can be considered unfolded.

9.1.2 Example Applications

9.1.2.1 Simple Example: Investigation on CI2

The CI2 protein is a 64-residue protein with an alpha-helix packed against a beta-sheet forming a hydrophobic core (see Figure 9.1a). It is one of the most thoroughly studied proteins, both experimentally and computationally. Here, we will use CI2 as a basic example for the application of structure-based simulations.

In this example we download the structure with the pdb-cde 2ci2 (http://www.pdb.org).[29] The software CSU[22] is used to obtain a native contact map for this structure (Figure 9.1b). Using the information from the native structure, we can set up our structure-based simulations. The functional form of structure-based simulations is pretty straightforward. Therefore almost any MD software package can be used to run the actual simulations.

We estimate the folding temperature T_{F} by running a simulation at the set of temperatures 0.8, 0.9, 0.95, 1.0, 1.05, 1.1, and 1.2 (in reduced units), starting both from the folded and unfolded states (an unfolded structure can be quickly generated by running at a very high temperature like $T=2.0$). For $T<T_{\mathrm{F}}$ we see, after equilibration, occupation of the folded state. The opposite happens when starting from the unfolded state. These simulations suggest $T_{\mathrm{F}}=1.05$ and we start a long simulation at this temperature. Figure 9.1c displays time vs. Q for six hours of simulation. One can observe several transitions between the folded ($Q\approx 0.8$) and unfolded states ($Q\approx 0.2$).

It is now possible to evaluate the simulation and derive quantitative data about folding. Calculating the Φ-values provides information about the TSE. The free-energy profile (Figure 9.1d) is calculated using the WHAM. This is in general a first step to a more detailed understanding of the folding behavior for the investigated system.

9.1.2.2 Circular Permutants on SH3 and CI2

Φ-values are calculated by measuring the effect single-point mutations have on folding kinetics and the TSE. An inherent problem is the change in charge or sterics of the mutated amino acids as this may disturb the composition of the TSE. An intriguingly different way to ask questions about the nature of the TSE are circular permuted proteins. They are possible, when the C- and N-termini of the investigated protein are spatially close. In this case, both termini can be chemically linked and another peptide bond can be cleaved to generate new termini. This procedure is equivalent to a circular shift of the amino acid sequence. Obviously, the location of the new ends will determine the properties of the resulting protein and its TSE.

¶ U, F, and TSE are defined by the free-energy landscape. Each is located around an extremum with an energetic difference up to 1 kT. U is defined as the region around the minimum at $Q_{\mathrm{U}}\approx 0.2$ and F around the minimum at $Q\approx 0.8$. TSE lies around the highest energy on the minimum free-energy pathway between U and F at $Q_{\mathrm{TSE}}\approx 0.45$. Q always gives the fraction of native contacts.

CI2 has been circular, permuted by linking residues 3 and 63 by a disulfide bond and cleaving residues 40–41 to generate new termini (perm40-41).[30] Both Wild Type (WT) and perm40-41 show experimentally two-state folding behavior with a folding nucleus around Ala_{16}. In C_α structure-based simulations, the circular permutation does not significantly change the TSE (though the linked region shows different Φ-values) and the probability of each contact being formed in the TSE correlates strongly (0.84) between both simulations.[13]

SH3 consists mainly of one large beta-sheet with a polarized TSE.[14] Two different circular permutants of SH3 have been experimentally studied, with the cleaving located between residues 14 and 15 (perm14-15) and 42 and 43 (perm42-53).[31] Both cases have the cut in a sensitive region of the TSE, where chain connectivity is considered important. Not surprisingly, the TSE of the WT and the permutants differ strongly. Structure-based simulations show changed Φ-value distribution[13] in good agreement with the experimental results (see Figure 9.2).

9.1.2.3 Multiple Routes on the β-Trefoil Family

The principle of minimal frustration states that no energetic traps compete with the native state. Consequently, the topology of the native fold dominates the folding mechanism. Members of the β-trefoil family have highly similar topologies yet show different folding mechanisms.[32–36] The β-trefoil fold is composed of three trefoils each made from β-strands (not shown). The loops connecting these strands vary in size and contribution to function.

In structure-based simulations with a C_α model, the folding barrier is too high to observe transitions in conventional MD simulations. Therefore, advanced sampling techniques are used in which the forces occurring during the simulation are modified to facilitate folding/unfolding transitions.[37] Three different folding routes are observed. A direct route proceeds by formation of contacts within the turn regions, which gradually lead to the formation of tertiary structure. In a second route two subsets of contacts compete with each other. This results in unfolding a subset of contacts during the folding process, which are later refolded. A third route observes a folding nucleus around the C- and N-termini. All three routes are in agreement with experimental data.[38] Which folding route an individual member of the β-trefoil family prefers is determined by small changes in energetics and topology and is outside the scope of structure-based simulations. However, although structure-based simulations are energetically unfrustrated, they manage to reproduce complex folding behavior.

9.1.3 Summary on Proteins

The principle of minimal frustrations and the funneled energy landscape are the theoretical foundations for structure-based simulations. These coarse-grained simulations are a concise and

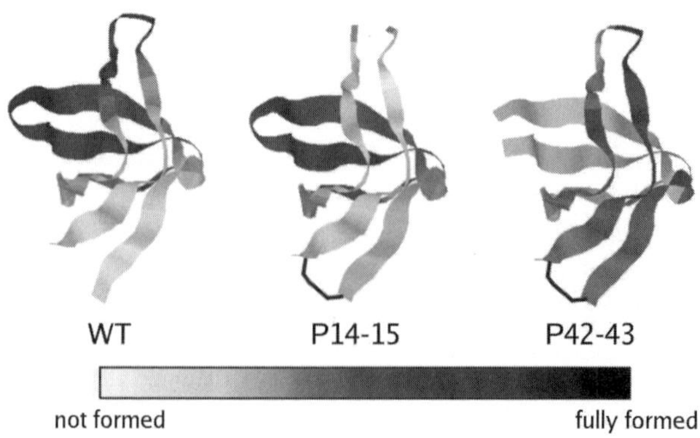

FIGURE 9.2 Circular permutants on WT SH3 and the two circular permutants perm 14-15 and perm 42-43 show, in agreement with experimental data, changed TSE in structure-based simulations. The different shades indicate the different degree of structural formation in the TSE.

inexpensive way to address questions of protein folding. As presented, results from these simulations are in good agreement with experimental data.

It is possible to scale structure-based simulations between the C_α and the all-atom level. These added layers of complexity can describe, for example, side chain packing or mutation in more detail but are computationally more expensive and their higher complexity makes them more difficult to interpret.

Going beyond folding, questions about the interconnection of biological function and the molecular level arise. What is the molecular basis of biological function and how does it work in detail? Structure-based simulations can describe conformational transitions by extending the usual single-funneled energy landscape to contain multiple basins for the different conformations of a protein.

9.2 RNA

The discovery of the self-splicing activity of *Tetrahymena thermophila* ribozyme[39–41] in early 1980s fundamentally changed our perspective on RNA from that suggested by the central dogma of biology. As well as transmitting the information from DNA to proteins as a messenger of genetic codes, RNA plays a number of active enzymatic and regulatory roles during both translation and transcription. To function as successful catalysts, RNAs adopt well-defined three-dimensional structures either spontaneously or in association with other biomolecules. Thus, along with the protein folding, much effort has been devoted to understand the underlying principle of the RNA folding problem.[42–45]

From the theoretical side, RNA structure predictions at secondary structure level have substantially benefited from the relatively straightforward pairing rule among the four nucleotides: adenine pairs with uracil, and guanine pairs with cytosine or uracil. Also, the experimentally compiled thermodynamic data of base stacking free energies[46] with loops, bulges, and internal loops allow us to find the minimum free-energy conformation at the secondary structure level using the dynamic programming algorithm.[47,51] The explicit three-dimensional structure beyond the secondary structure level that describes only the base pairings, however, is extremely difficult to determine because of a lack of quantitative knowledge on tertiary interactions, which requires the full understanding of the nontrivial counterion interaction with the electrostatic environment formed by the highly charged phosphate-sugar backbone as well as the polymeric nature of RNA molecules.[44,45]

Although one can gain insight into the RNA dynamics directly from brute-force all-atom molecular dynamics simulations, the computational cost for RNA is generally higher than for proteins, given the same number of building components, as the average size of a single nucleotide is at least three times larger than amino acids. Additionally, RNA dynamics are slower than protein dynamics, so one needs to simulate longer. RNA folding is also extremely sensitive to the counterion concentration and its valence, thus a careful incorporation of counterions is required to observe meaningful dynamics. Despite all these difficulties, one can study the dynamics of interest at larger time and length scales by reducing the configurational degrees of freedom. Like structure-based models of proteins, the RNA molecules can be coarse-grained. Here, we review the two novel coarse-grained models for RNA, pioneered by Hyeon and Thirumalai,[48,53,54] which were used to study the dynamics of several RNA molecules under varying temperature and mechanical force conditions.

9.2.1 Three-Interaction Site (TIS) Model[48,49]

9.2.1.1 Model

One of the natural ways of coarse-graining the nucleotides is to use the three interaction sites corresponding to the phosphate (*P*), the ribose (*S*), and the base (*B*) group.[48,49] In this representation

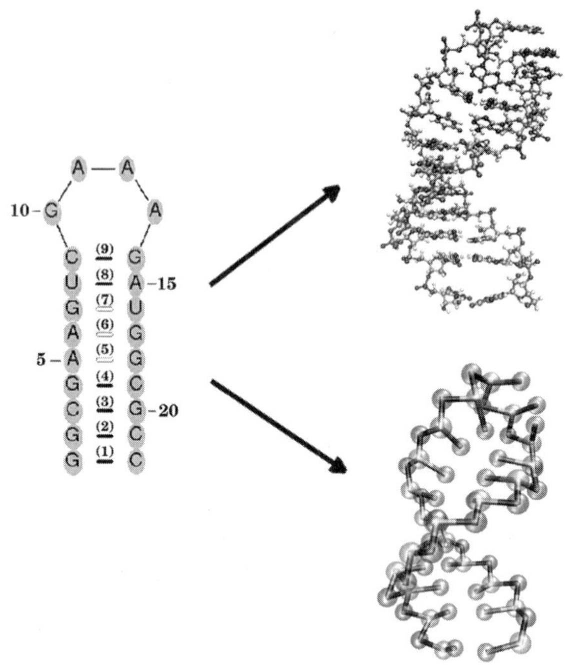

FIGURE 9.3 Coarse-graining of RNA hairpins using the TIS model. The left picture displays a schematic representation of RNA. The upper right picture shows an all-atom representation and the lower right picture is the TIS model.

the RNA backbone is reduced to the polymeric structure [-$(P\text{-}S)_n$-] with the B being covalently bound to the S center. The TIS model represents the RNA molecule with N-nucleotides using the 3N-interaction centers. In the PDB structure of P5GA RNA hairpin,[50] studied here as an example for the TIS model, the first nucleotide does not contain the P group, thus P5GA hairpins in the TIS representation contain in total (3N−1)-interaction sites. The secondary structure and the lowest energy structure using the minimal model are shown in Figure 9.3. The total energy of an RNA conformation is written as

$$
\begin{aligned}
V_{\mathrm{T}} &= V_{\mathrm{BL}} + V_{\mathrm{BA}} + V_{\mathrm{DIH}} + V_{\mathrm{STACK}} + V_{\mathrm{NON}} + V_{\mathrm{ELEC}} \\
&= \left[\sum_{i=1}^{2N-2} \frac{1}{2} k_r \left(\left| \vec{r}_{(SP)_{i+1}} - \vec{r}_{(SP)_i} \right| - \left(R_{SP}^o\right)_i \right)^2 + \sum_{i=1}^{N} \frac{1}{2} k_r \left(\left| \vec{r}_{B_i} - \vec{r}_{S_i} \right| - \left(R_{BS}^o\right)_i \right)^2 \right] \\
&\quad + \left[\sum_{i=1}^{2N-3} \frac{1}{2} k_\theta \left(\theta_i - \theta_i^o\right)^2\right] + \left[\sum_{i=1}^{2N-4} \left(A_{1i}^{\eta} + B_{1i}^{\eta} + C_{1i}^{\eta} + A_{2i}^{\eta} \cos(\phi_i - \phi_i^o + \phi_i^{\eta}) \right.\right. \\
&\quad \left.\left. + B_{2i}^{\eta} \cos 3(\phi_i - \phi_i^o + \phi_i^{\eta}) + C_{2i}^{\eta} \sin(\phi_i - \phi_i^o + \phi_i^{\eta}) \right)\right] \\
&\quad + \left[\sum_{i=1}^{n_{\max}} \left(\Delta G_i(T) \times e^{-\alpha_{st}\left\{\sin^2(\phi_{1i}-\phi_{1i}^o)+\sin^2(\phi_{2i}-\phi_{2i}^o)+\sin^2(\phi_{3i}-\phi_{3i}^o)+\sin^2(\phi_{4i}-\phi_{4i}^o)\right\}} \times e^{-\beta_{st}\left\{(r_{ij}-r_{1i}^o)^2+(r_{i+1,j-1}-r_{2i}^o)^2\right\}} \right.\right. \\
&\quad \left.\left. \times e^{-\gamma_{st}\left\{\sin^2(\psi_{1i}-\psi_{1i}^o)+\sin^2(\psi_{2i}-\psi_{2i}^o)\right\}} \right)\right]
\end{aligned}
$$

$$+\left[\left\{\sum_{i=1}^{N-1}\sum_{j=i+1}^{N} V_{B_iB_j}^{\text{att}}(r) + \sum_{i=1}^{N-1}\sum_{j=1}^{2N-1}{}' V_{B_i(SP)_j}^{\text{att}}(r) + \sum_{i=1}^{2N-4}\sum_{j=i+3}^{2N-1} V_{(SP)_i(SP)_j}^{\text{att}}(r)\right\}\Delta_{ij}\right.$$

$$\left.+\left\{\sum_{i=1}^{N-1}\sum_{j=i+1}^{N} V_{B_iB_j}^{\text{rep}}(r) + \sum_{i=1}^{N}\sum_{j=1}^{2N-1}{}' V_{B_i(SP)_j}^{\text{rep}}(r) + \sum_{i=1}^{2N-4}\sum_{j=i+3}^{2N-1} V_{(SP)_i(SP)_j}^{\text{rep}}(r)\right\}\left(1-\Delta_{ij}\right)\right]$$

$$+\left[\sum_{i,j}^{N} \frac{z_{P_i} z_{P_j} e^2}{4\pi\varepsilon_0\varepsilon_r r} e^{-r/l_D}\right].$$

In the above equation, (i) and (ii) V_{BL} and V_{BA} enforce structural connectivity and back-bone rigidity using the harmonic potentials with $k_r = 20$ kcal/(mol × Å^2) and $k_\theta = 20$ kcal/(mol × rad^2), respectively. The notation $(SP)_i$ denotes either the backbone bead S or P, depending on the value of i ($i = 1, 2, \ldots, 2N-1$). V_{BL} maintains the distance between the covalently bonded beads (P_iS_i, S_iP_{i+1}, and B_iS_i) around the distances $\{R_{SP}^o\}$ and $\{R_{BS}^o\}$ in the PDB structure. The angle formed between three successive beads along the sugar-phosphate backbone is restrained around the value of θ_i^o in the PDB structure. (iii) V_{DIH} is the dihedral angle potential that describes the ease of rotation around the angle formed between four successive beads along the sugar-phosphate backbone ($S_{i-1}P_iS_iP_{i+1}$ or $P_iS_iP_{i+1}S_{i+1}$). The ith dihedral angle, ϕ_i, which is the angle formed between the two planes defined by four successive beads i to $i+3$ is from $\cos\phi_i = (\vec{r}_{i+1,i} \times \vec{r}_{i+1,i+2}) \cdot (\vec{r}_{i+2,i+1} \times \vec{r}_{i+2,i+3})$. In the TIS model, the right-handedness of RNA is realized by appropriate choices of the parameters in the dihedral potential. Based on the angles in the PDB structure, one of the three types of dihedral potentials, *gauche* (+) ($g^+, 0 < \phi_i^o < 2\pi/3$, $\phi_i^{g^+} = \pi/3$), *trans* (t, $2\pi/3 < \phi_i^o < 4\pi/3$, $\phi_i^t = \pi$), and *gauche* (−) ($g^-, 4\pi/3 < \phi_i^o < 2\pi, \phi_i^{g^+} = 5\pi/3$) is assigned to each position along the backbone. The parameters (in kcal/mol) for $\eta = t$, g^+, and g^- are assigned as $A_{1i}^{g^+} = -A_{2i}^{g^+} = 1.0$, $B_{1i}^{g^+} = B_{2i}^{g^+} = 1.6$, $C_{1i}^{g^+} = -C_{2i}^{g^+} = 2.0$, $A_{1i}^{g^-} = -A_{2i}^{g^-} = 1.0$, $B_{1i}^{g^-} = B_{2i}^{g^-} = 1.6$, $C_{1i}^{g^-} = C_{2i}^{g^-} = 2.0$, $A_{1i}^t = A_{2i}^t = 1.2$, $B_{1i}^t = B_{2i}^t = 1.2$, and $C_{1i}^t = C_{2i}^t = 0.0$. To account for the flexibility in the loop region, the dihedral angle barrier is halved for the parameter values for $19 \le i \le 24$. (iv) Simple RNA secondary structures, such as hairpins, are largely stabilized by stacking interactions (V_{STACK}) whose context-dependent values are known.

The folded P5GA RNA hairpin is stabilized by nine hydrogen bonds between the base pairs, including two GA mismatch pairs. Thus, there are in total eight native stacking interactions in P5GA. The strength of the sequence-dependent interaction is experimentally measured as $\Delta G_i(T) = \Delta H_i - T\Delta S_i$ where the values of ΔH_i and ΔS_i are taken from the thermodynamic data compiled by Turner and coworkers.[51] For the stacking interactions including the mismatch of GA pairing, the energy associated with GU is replaced for the GA base pair. The stacking interactions originate from hydrogen bonds and quantum-mechanical interactions between the bases. The orientation and the distance between the neighboring bases and between the base pairs are maintained as a consequence of stacking interactions that are designed using the geometrical parameter sets $\{\phi\}$, $\{r\}$, and $\{\psi\}$ (see Figure 9.4); the bond angles $\{\phi\}$ are $\phi_{1i} \equiv \angle S_i B_i B_j$, $\phi_{2i} \equiv \angle B_i B_j S_j$, $\phi_{3i} \equiv \angle S_{i+1} B_{i+1} B_{j-1}$, and $\phi_{4i} \equiv \angle B_{i+1} B_{j-1} S_{j-1}$ where j refers to the index of complimentary base to the base i in the native state; the distances between two paired bases involving the stacking interaction are $r_{ij} = |B_i - B_j|$ and $r_{i+1j-1} = |B_{i+1} - B_{j-1}|$; ψ_{1i} and ψ_{2i} are the dihedral angles formed by the four beads $B_iS_iS_{i+1}B_{i+1}$ and $B_{j-1}S_{j-1}S_jB_j$, respectively. The superscript o refers to angles and distances in the PDB structure. The values of α_{st}, β_{st}, and γ_{st} are 1.0, 0.3 Å^{-2}, and 1.0, respectively. (v) The fifth term describes the hydrophobic nonbonded interaction between the purine/pyrimidine groups. The prime in the second term in the V_{NON} denotes the condition that excludes the interaction

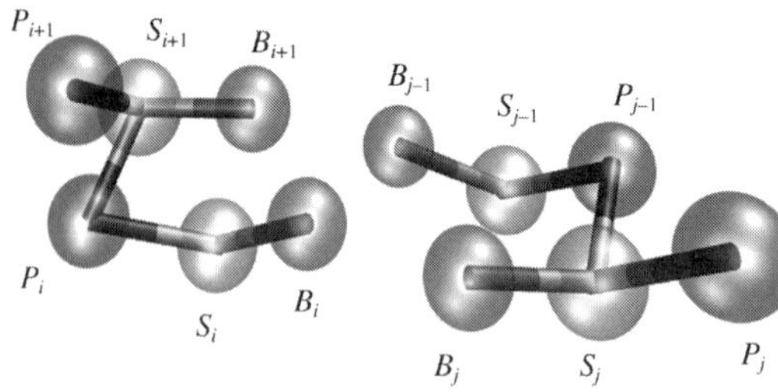

FIGURE 9.4 Geometry of base pairs involved with stacking interactions.

between B_i and S_j in the same nucleotide. In our model, a native contact ($\Delta_{ij} = 1$) exists between two noncovalently bound beads, provided they are within a cutoff distance R_C (=7 Å) at the native state. Two beads whose distance at the native state is beyond R_C are considered to be nonnative ($\Delta_{ij}=0$). For a native contact, $V^{\text{att}}_{\xi_i\eta_j}(r) = C_h^{\xi_i\eta_j}\,[(r^o_{ij}/r)^{12} - 2(r^o_{ij}/r)^6]$ where r^o_{ij} is the distance between beads in the PDB structure and $C_h^{\xi_i\eta_j} = 1.5$ kcal/mol for all native contact pairs, except for the $B_{10}B_{13}$ base pair associated with the formation of the hairpin loop, for which $C_h^{B_{10}B_{13}} = 3.0$ kcal/mol. The additional stability for the base pair associated with loop formation is similar to Turner's thermodynamic rule for the free-energy gain at the tetraloop region. For nonnative pairs, $V^{\text{rep}}_{\xi_i\eta_j}(r) = C_R\,[(a/r)^{12} + (a/r)^6]$, with $a = 3.4$ Å and $C_R = 1$ kcal/mol. The value of $C_h^{\xi_i\eta_j}$ (=1.5 kcal/mol) has been chosen so that the hairpin undergoes a first-order transition from unfolded states with an experimentally reasonable melting temperature. (vi) The charges on the phosphate groups are efficiently screened by the counterions so that in the folded state the destabilizing contribution of the electrostatic potential is relatively small. However, the nature of the RNA conformation (especially tertiary interactions) can be modulated by changing counterions. For simplicity, one can assume that the electrostatic potential between the phosphate groups is pairwise-additive, and use the Debye-Huckel potential, which accounts for the screening by counterions and hydration effects. The charge on the phosphate ion is given by $z_{P_i} = -1$, the dielectric constant is $\varepsilon_r = \varepsilon/\varepsilon_0$, and the Debye length is $l_D = \sqrt{\varepsilon_r k_B T / 8\pi k_{\text{elec}} e^2 I}$ with $k_{\text{elec}} = (1/4\pi\varepsilon_0) = 8.99 \times 10^9$ JmC^{-2} where the ionic strength is calculated as $I = (1/2)\sum_i z_i^2 c_i$. The parameters $c_i = 100$ mM for monovalent salt and $\varepsilon_r = 5$ are used in the simulations. A small dielectric constant ($\varepsilon_r = 5$) is typically used inside the RNA molecule when the nonlinear Poisson-Boltzmann equation is solved to calculate the electrostatic potential.[52] At $T \sim 300$ K, the electrostatic repulsion between the neighboring phosphate groups at $r \sim 6$ Å is ~ 1 kcal/mol.

9.2.1.2 Simulations

The configurations of an RNA molecule can be efficiently sampled by using a Langevin equation at low frictions. The equation of motion for the reaction centers is integrated using a Verlet algorithm.

$$m\frac{d^2\vec{r}_i}{dt^2} = -\zeta_L \frac{d\vec{r}_i}{dt} - \vec{\nabla}_i V(\{\vec{r}\}) + \vec{\Gamma}_i(t).$$

By using a typical value for the mass of a bead in a nucleotide (B, S, or P), $m \sim 100$–160 g/mol, the average distance between the adjacent beads $a = 4.6$ Å, the energy scale $\varepsilon_h = 1$–2 kcal/mol, and the natural time is $\tau_L = (ma^2/\varepsilon_h)^{1/2} = 1.6$–2.8 ps. Stable simulations are performed with $\zeta_L = 0.05 m\tau_L^{-1}$ and $h = 0.0025\tau_L$. Note, however, that the purpose of the *underdamped* dynamics simulations, in which the inertia term dominates over the friction term, is to study the thermodynamic behavior. The under-damped dynamics are in principle quite different from the over-damped dynamics (consider, for example, the solutions of a simple harmonic oscillator at greatly different friction constant). By reducing the friction, the transition dynamics are accelerated, facilitating the samplings of multiple configurations within the available computational time. The dynamics are artificial because of the greatly reduced friction; one cannot convert the integration time step into real times. To study folding/unfolding kinetics in real times, the *overdamped* simulations are performed using a Brownian dynamics algorithm.

$$\zeta_H \frac{d\vec{r}_i}{dt} = -\vec{\nabla}_i V(\{\vec{r}\}) + \vec{\Gamma}_i(t).$$

The natural time for the overdamped simulations is $\tau_H = \zeta_H a^2/k_B T = (\zeta_H \tau_L/m)(\varepsilon_h/k_B T)\tau_L$. The 10^6 simulation time step ($10^6 \times 0.02\ \tau_H$) is converted to 3.5 μs in real time when a single integration time step $h = 0.02\tau_L$, $\zeta_H = 50m\tau_L^{-1}$, $\tau_L = 2$ ps, $\varepsilon_h = 1$ kcal/ mol and $T = 290$ K are used.

9.2.1.3 Exploring the Energy Landscape of RNA Hairpins under Varying Temperature and Force Using TIS Model

As a result of thermodynamic samplings, the diagram of states in the (f, T) plane (Figure 9.5), constructed with the multiple histogram reweighting technique,[24,25] shows that P5GA hairpin behaves as a "two-state" folder.[48] In the absence of force ($f = 0$ pN), the folding/unfolding transition midpoint is at $T_m = 341$ K by using the fraction of native contact (Q) as an order parameter. At $T = 290$ K the equilibrium force required to unfold the P5GA is ~7 pN. As force increases, T_m decreases monotonically, so that the transition midpoints (T_m, f_m) form a phase boundary separating the folding and unfolding states. The phase boundary is sharp at low T_m and large f_m, but becomes fuzzy when the force is weak. When T and f are varied, even the simplest possible RNA hairpin shows an extremely rich dynamical behavior in response to the variations over the external conditions. (i) The hairpins undergo a first-order transition from the native basin of attraction (NBA) to the unfolded basin of attraction (UBA) at critical force. The calculated one-dimensional free-energy profiles as a function of R shows great variations in the height of the barrier separating the two basins of attractions (see Figure 9.6); that is, NBA and UBA, hence the great variation in the hopping times between the NBA and UBA along the (T_m, f_m).[48] The transition becomes strongly first order at low T_m and high f_m. (ii) Unfolding of RNA at constant force occurs in steps; that is, sequentially unzipping from the ends, with long pauses in a number of discrete intermediates, whereas the thermal unfolding occurs in a stochastic manner. (iii) Force-quench refolding occurs much more slowly than the temperature-quench refolding. Detailed analysis of RNA refolding trajectory shows that RNA refolds via multiple steps upon force quench, and that the search process of the native loop-dihedral angles are the critical event to form a loop. Once the loop is correctly formed, the rest of the refolding comes to completion through the zipping process. (iv) In response to the mechanical force, the position of the transition state moves to the small R; that is, the TSE becomes more native-like when the molecule is pulled at a large constant force or equivalently at a large loading rate. Combining the TIS model of RNA hairpins with the worm-like chain model describing the stiff and flexible handle, further study has been done as well on the linker effects on RNA force-extension curves.[49]

9.2.2 Self-Organized Polymer (SOP) Model[53–56]

9.2.2.1 Model

The recent single-molecule force experiments have been used to unravel the RNA from simple hairpins.[57,58] To study the global dynamics of RNAs with a higher complexity beyond simple hairpins, one can use an even more simplified representation and energy function than the TIS model. The model has to be realistic enough to take into account the interactions that stabilize the native fold, yet simple enough that within the computational time one can trace the transition dynamics of the large sized molecule. To this end, a new class of versatile coarse-grained structure-based self-organized

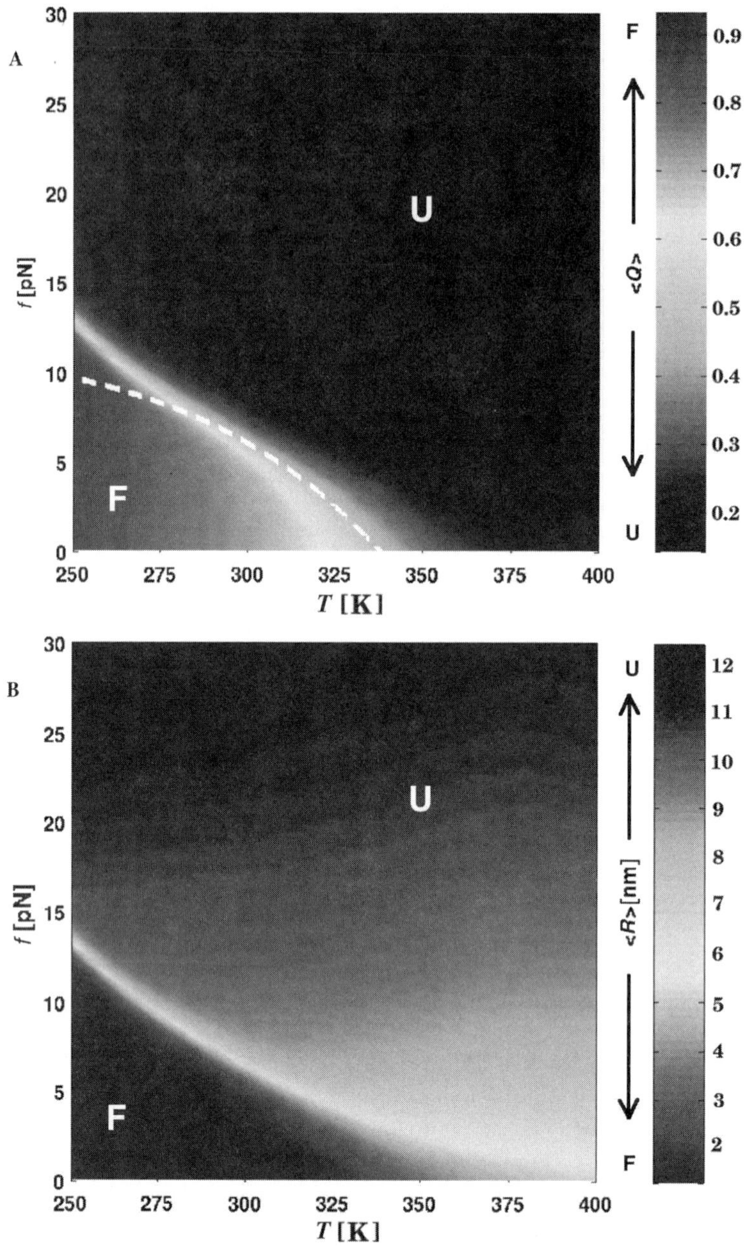

FIGURE 9.5 (f,T)-phase diagram of P5GA RNA hairpins. A uses Q; that is, the fraction of native contacts, as the order parameter; B uses the end-to-end distance $\langle R \rangle$.

polymer (SOP) model is well suited to understand the dynamics at the spatial resolution that the single-molecule force spectroscopy for large ribozymes and proteins can provide. The reasons for using the SOP model in force spectroscopy applications are the following: (i) Forced unfolding and force-quench refolding lead to large conformational changes of $\sim$10–100 nm. Currently, single-molecule experiments (laser optical tweezers or atomic force microscopy) cannot resolve structural changes below a few nanometres. As a result, details of the rupture of hydrogen bonds or local tertiary contacts between specific bases cannot be discerned from force-extension curve (FEC) or the dynamics of end-to-end distance (R) alone. Because only large changes in R are monitored, it is not crucial to

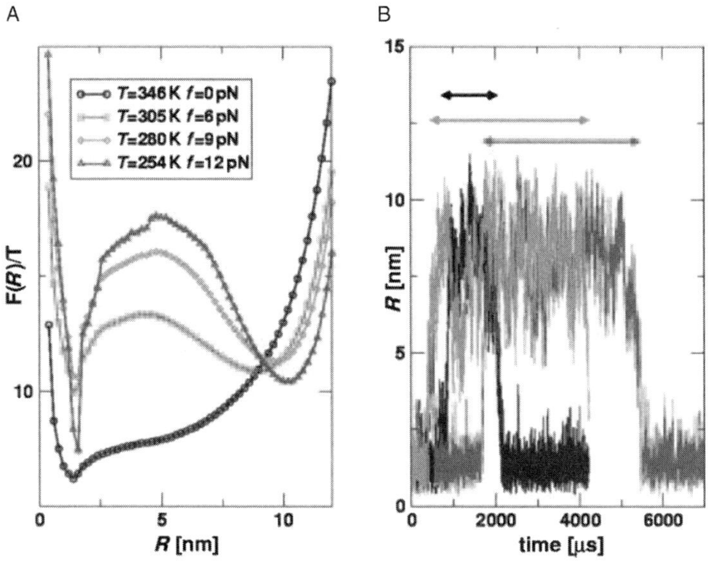

FIGURE 9.6 Hopping transitions along the phase boundary. (A) Free-energy profiles F(R) along the phase boundary (T_m, f_m). The barrier separating the folded and unfolded basins of attraction increases at low T_m and high f_m values. (B) Time traces of T and f are 305 K and 6 pN, respectively. The arrows indicate the residence times in the unfolded basin of attraction for three dynamic trajectories.

include minor perturbative details of the local interactions such as bond angle and various dihedral angle potentials. As suggested in the literature on normal-mode analysis,[59–61] the inclusion of small details only affects the higher frequency modes, and the global dynamics mainly determined by the low frequency normal modes are robust as long as the topological constraints are not changed. (ii) In the context of mechanical unfolding as well as the folding of proteins, many of the details of unfolding and folding pathways can be accurately computed by taking into account only the interactions that stabilize the native fold.[62] Previous studies also suggested that it is crucial to take into account chain connectivity and attractive interactions that faithfully reproduce the contact map of a fold. (iii) Electrostatic interactions are pivotal in RNA. However, under physiological condition, counterion concentration (100–200 mM of monovalent salt) is large enough to effectively screen the electrostatic repulsion between the phosphate groups. Thus, due to effective screening (small Debye length), the repulsive electrostatic potential between P groups is effectively short-ranged.

The SOP representation of RNA only retains chain connectivity and favorable attractive interactions between the sites that stabilize the native fold. Each interaction center represents the center of mass of a nucleotide. In terms of the coordinates of RNA with N nucleotide, the total energy potential is written as follows.

$$V_{\mathrm{SOP}} = V_{\mathrm{FENE}} + V_{\mathrm{NON}}$$

$$= -\sum_{i=1}^{N-1} \frac{k}{2} R_0^2 \log\left(1 - \frac{\left(r_{i,i+1} - r_{i,i+1}^{o}\right)^2}{R_0^2}\right)$$

$$+\sum_{i=1}^{N-3}\sum_{j=i+3}^{N} \varepsilon_h \left[\left(\frac{r_{ij}^0}{r_{ij}}\right)^{12} - 2\left(\frac{r_{ij}^0}{r_{ij}}\right)^{6}\right]\Delta_{ij} + \sum_{i=1}^{N-2} \varepsilon_l \left(\frac{\sigma *}{r_{i,i+2}}\right)^6 + \sum_{i=1}^{N-3}\sum_{j=i+3}^{N-3} \varepsilon_l \left(\frac{\sigma}{r_{ij}}\right)^6 \left(1-\Delta_{ij}\right).$$

The first term is for the chain connectivity. The finite extensible nonlinear elastic (FENE) potential is used with $k = 20$ kcal/(mol $\times$ Å^2), $R_0 = 0.2$ nm, and $r_{i,i+1}$ is the distance between neighboring beads at i and $i+1$. $r^o_{i,i+1}$ is the distance in the native structure. The use of the FENE potential is more advantageous than the standard harmonic potential, especially for the forced stretching, because the fluctuations of $r_{i,i+1}$ are strictly restricted around $r^o_{i,i+1}$ with variations of $\pm R_0$ to produce the worm-like chain behavior. The Lennard-Jones (LJ) potential is used to account for interactions that stabilize the native topology. Native contact is defined for the bead pairs whose distance is $r_{ij} < R_C$ $(= 1.4$ nm$)$ in the native state for $|i-j| > 2$. We use $\varepsilon_h = 0.7$ kcal/mol for the native pairs and $\varepsilon_l = 1$ kcal/mol for nonnative pairs. In the current version, we have neglected the nonnative attractions that will not qualitatively affect the results because, under tension, such interactions are greatly destabilized. To ensure the noncrossing of the chain, we set $\sigma = 7$ Å. Only for i, $i+2$ pairs we set $\sigma^* = 3.5$ Å to prevent the flattening of the helical structures when the overall repulsion is large. There are five parameters in the SOP force field. Of these, the results are sensitive to the precise values of $\varepsilon_h/\varepsilon_l$ and R_C. We have discovered that the quantitative results are insensitive to R_C as long as it is in the physical range that is determined by the RNA contact maps. In principle, the ratio of $\varepsilon_h/\varepsilon_l$ can be adjusted to obtain realistic values of critical forces ($f = 10$–15 pN for small hairpins at $T = 300$ K). For simplicity, we choose a uniform value of ε_h for all RNA constructs, which can apparently be made sequence-dependent and ion-implicit if one wants to improve the simulation results. Surprisingly, the SOP force field, with the same set of parameters, can be used to obtain near-quantitative results for RNA molecules of varying native topology.

The time spent to calculate LJ forces scales as $\sim O(N^2)$. Drastic savings in computational time can be achieved by truncating the forces due to the LJ potential for interaction pairs with $r_{ij} > 3r^0_{ij}$ or 3σ to zero. We refer to the model as the SOP model because it only uses the polymeric nature of the biomolecules with the crucial topological constraints that arise by the specific fold. For probing forced unfolding of RNA (or proteins) it is sufficient to include attractive interactions only between contacts that stabilize the native state. We believe none of the results will change qualitatively if this restriction is relaxed; that is, if nonnative interactions are also taken into account.

9.2.2.2 Simulations

The strategy for simulating RNA dynamics using the SOP model is identical to that in the TIS model except that the coarse-grained interaction center is larger. As a consequence, the range of interaction and the strength of energy scale are readjusted. The SOP model can also be applied to the protein. Table 9.1 summarizes the parameters used in the recent application of the SOP model to the Brownian dynamics simulations of RNA[53,5453,54] and proteins.[53,55,5653,55,56]

9.2.2.3 Probing the Pathways and Kinetic Barriers of RNA from Hairpins to Ribozymes

Similar to proteins, RNA folds into the unique three-dimensional structure to function in the cellular system. During the folding of RNA, many building blocks, such as hairpins, hairpins including bulges, internal loops, and multiloop structures, consisting of ribozymes or ribosome with higher complexity, first independently form stable secondary structures, and the tertiary interaction, for example, kissing loop, ribose zipper, loop-bulges interaction, etc., subsequently patches these secondary structures into the functional three-dimensional structure. The self-organization of the nucleotide sequence to the well-defined native structure is expected to occur on the rugged energy landscape. The recent single-molecule force experiments begin providing glimpses to the biomolecular energy landscape by unraveling the individual molecule by mechanical force. However, it is difficult to analyze the forced-unfolding dynamics by relying on the force–extension curve that only gives the information of mechanical response of the molecule along the end-to-end distance coordinate. For better understanding of the single-molecule force experiments, the SOP model, despite its simplicity, can be useful since it allows one to directly visualize the molecular response of a large molecule to the external force within a computational time.

TABLE 9.1
Parameter Set of the SOP Model for RNA and Proteins

	RNA	Proteins
R_0	2 Å	2 Å
k	20 kcal/(mol Å^2)	20 kcal/(mol Å^2)
R_C	14 Å	8 Å
ε_h	0.7 kcal/mol	1–2 kcal/mol
ε_l	1 kcal/mol	1 kcal/mol
σ	7 Å	3.8 Å
ζ	100 $m\tau_L^{-1}$	50 $m\tau_L^{-1}$
τ_L	4 ps	3 ps

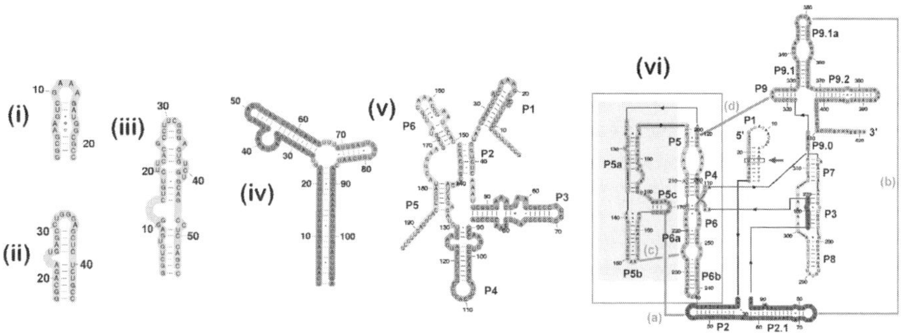

FIGURE 9.7 Secondary structure map of RNA molecules whose mechanical unfolding is studied using the SOP model. Each structure displays (i) P5GA, (ii) TAR-RNA, (iii) 55-nt domain IIa of HCV genome, (iv) prohead RNA of ϕ29 DNA packaging motor, (v) 195-nt *Azoarcus* ribozyme, and (vi) *Tetrahymena thermophila* ribozyme.

In previous works (see also Figure 9.7 for the RNA structures studied),[53,54] the unfolding simulations of RNA using the SOP model, after (i) the parameter tuning with respect to the P5GA hairpins (see Table 9.1), have confirmed as well as predicted the force response of various RNAs. (ii) The TAR-RNA (PDB code: 1uud) containing a single bulge unfolds via a metastable intermediate. (iii) The 55-nt domain IIa of the hepatitis C viral (HCV) genome (PDB code: 1p5m) unfolds via three sequential kinetic intermediates. (iv) Three-way junction prohead RNA (pRNA), which is part of the ϕ 29 DNA packaging motor, exhibits three intermediates under tension. In comparison to the corresponding native contact map of each RNA, all these results (ii)–(iv)) show that the forced-unfolding pathways for simple RNA are encoded in the contact maps. However, when the molecular structure becomes more complicated, the kinetic pathway variation starts emerging. (v) The 195-nt *Azoarcus* ribozyme consists of six subdomains. Depending on the loading rate, the variation in both the number of peaks in FEC and unfolding pathways were predicted. (vi) The 407-nt *T.* ribozyme showed two different unfolding pathways when it was simulated with 3000 times larger loading rate, and one of the unfolding pathways analyzed by the order of subdomain unfolding was consistent with the experimental analysis made by Bustamante and coworkers. The qualitative agreement with the experimentally analyzed unfolding pathway despite the 3000 times larger loading rate, which is inevitably used because of the molecular size, is due to the fact that the force propagation rate at r_f (SIM) $\cong 3000 \times r_f$ (LOT) where r_f (SIM) and r_f (LOT) are the loading rates for the simulation and the experiment, respectively, is greater than the molecular relaxation time of the RNA subdomain.

It is important to note that the SOP model reduces the computational cost, so that one can perform a forced-unfolding simulation as close as possible to the near-experimental condition. It is remarkable that, for the proteins stretched by atomic force microscope (AFM) whose loading rate is typically $\sim 10^3$ greater than the loading rate of laser optical tweezers, one can perform a stretching simulation using the SOP model practically on the same time scales.[53]

Other than the mechanical unfolding of various RNA molecules, the SOP model showed an enormous potential in the various different fields such as the folding dynamics of RNA,[53] the unfolding/refolding dynamics of GFP,[53] the allosteric transition of GroEL,[55] and the structural origin of allosteric regulation in the kinesin dimer.[56]

9.3 CONCLUDING REMARKS

As MD simulations at the all-atom level are both computationally expensive and difficult to interpret, reduced models play an essential role in understanding macromolecule folding and function. Coarse-grained models aim at understanding the basic principles of molecular processes, as they enable access to longer time scales (or larger systems) and the concise formulation of the Hamiltonian represents the basic and robust properties of the investigated system.

The presented examples of coarse-graining both for proteins and RNA address multiple questions. The funneled free-energy landscape is the theoretical framework of structure-based simulations. The results from these simulations, like the folding mechanism or the folding rates, are in good agreement with available experimental data. It is possible to expand these structure-based models by increasing additional atoms or adding multiple basins to accommodate structural transitions. The presented coarse-grained models for RNA can be expanded to various fields such as the folding and unfolding dynamics of both RNA and proteins, allosteric regulation for GroEL, or structural transitions in the kinesin dimer.

This framework of structure-based simulations can decipher the principles of selforganization. The interconnection of molecular motions and biological processes makes them suitable to answer questions of direct biological relevance, such as protein–protein or protein RNA/DNA interactions, allosteric control, signaling, or molecular machines, and unravels the secrets of the involved physical principles.

ACKNOWLEDGMENT

One of us (CH) would like to acknowledge Professor Dave Thirumalai for useful comment on the RNA part. This work was funded by the National Science Foundation-sponsored Center for Theoretical Biological Physics (grants Phy-0216576 and 02255630) and also by grant MCB-0543906.

REFERENCES

1. Kubelka, J., Hofrichter, J., and Eaton, W. A. 2004. The protein folding "speed limit". *Curr. Opin. Struct. Biol.* 14:76–88.
2. Anfinsen, C. B. 1973. Principles that govern the folding of protein chains. *Science* 181:223–30.
3. Bryngelson, J. D., Onuchic, J. N., Socci, N. D., and Wolynes, P. G. 1995. Funnels, pathways, and the energy landscape of protein folding: A synthesis. *Proteins* 21:167–95.
4. Clementi, C., Nymeyer, H., and Onuchic, J. N. 2000. Topological and energetic factors: What determines the structural details of the transition state ensemble and en-route intermediates for protein folding? An investigation for small globular proteins. *J. Mol. Biol.* 298:937–53.
5. Frauenfelder, H., Sligar, S. G., and Wolynes, P. G. 1991. The energy landscapes and motions of proteins. *Science* 254:1598–603.
6. Lyubovitsky, J. G., Gray, H. B., and Winkler, J. R. 2002. Mapping the cytochrome c folding landscape. *J. Am. Chem. Soc.* 124:5481–85.
7. Onuchic, J. N., and Wolynes, P. G. 2004. Theory of protein folding. *Curr. Opin. Struct. Biol.* 14:70–75.

8. Nymeyer, H., Garcia, A. E., and Onuchic, J. N. 1998. Folding funnels and frustration in off-lattice minimalist protein landscapes. *Proc. Natl. Acad. Sci. U.S.A.* 95:5921–28.
9. Onuchic, J. N., Luthey-Schulten, Z., and Wolynes, P. G. 1997. Theory of protein folding: The energy landscape perspective. *Annu. Rev. Phys. Chem.* 48:545–600.
10. Ueda, Y., Taketomi, H., and Go., N. 1978. Studies on protein folding, unfolding, and fluctuations by computer simulation. ii. A three-dimensional lattice model of lysozyme. *Biopolymers* 17:1531–48.
11. Chavez, L.L., Onuchic, J. N., and Clementi, C. 2006. Quantifying the roughness on the free energy landscape: Entropic bottlenecks and protein folding rates. *J. Am. Chem. Soc.* 126:8426–32.
12. Clementi, C., and Plotkin, S. S. 2004. The effects of nonnative interactions on protein folding rates: Theory and simulation. *Protein. Sci.* 13:1750–66.
13. Clementi, C., Jennings, P. A., and Onuchic., J. N. 2001. Prediction of folding mechanism for circular-permuted proteins. *J. Mol. Biol.* 311:879–90.
14. Martinez, J. C., Pisabarrol, M. T., and Serrano, L. 1998. Obligatory steps in protein folding and the conformational diversity of the transition state. *Nat. Struct. Biol.* 5:721–29.
15. McPhalen, C. A., and James, M. N. 1987. Crystal and molecular structure of the serine proteinase inhibitor CI-2 from barley seeds. *Biochemistry* 26:261–69.
16. Best, R., Chen, Y., and Hummer, G. 2005. Slow protein conformational dynamics from multiple experimental structures: The helix/sheet transition of arc repressor. *Structure* 13:1755–63.
17. Okazaki, K., Koga, N., Takada, S., Onuchic, J. N., and Wolynes, P. G. 2006. Multiple-basin energy landscapes for large-amplitude conformational motions of proteins: Structure-based molecular dynamics simulations. *Proc. Natl. Acad. Sci.* 103:11844–49.
18. Koga, N., and Takada, S. 2006. Folding-based molecular simulations reveal mechanisms of the rotary motor F1-ATPase. *Proc. Natl. Acad. Sci. U.S.A.* 103:5367–72.
19. Levy, Y., Cho, S. S., Shen, T., Onuchic, J. N., and Wolynes, P. G. 2005. Symmetry and frustration in protein energy landscape: A near degeneracy resolves the rop dimer-folding mystery. *Proc. Natl. Acad. Sci. U.S.A.* 102:2373–78.
20. Schug, A., Whitford, P. C., Levy, Y., and Onuchic, J. N. 2007. Mutations as trapdoors to two competing native conformations of the Rop-dimer. *Proc. Natl. Acad. Sci. U.S.A.* 104:17674–79.
21. Whitford, P. C., Miyashita, O., Levy, Y., and Onuchic, J. N. 2006. Conformational transitions of adenylate kinase: Switching by cracking. *J. Mol. Biol.* 366:1661–71.
22. Sobolev, V., Sorokine, A., Prilusky, J. Abola, E. E., and Edelman, M. 1999. Automatic analysis of interatomic contacts in proteins. *Bioinformatics* 15:327–32.
23. Cho, S. S., Levy, Y., and Wolynes, P. G. 2006. P versus Q: Structural reaction coordinates capture protein folding on smooth landscapes. *Proc. Natl. Acad. Sci. U.S.A.* 103:586–91.
24. Ferrenberg, A. M., and Swendsen, R. H. 1988. New Monte Carlo technique for studying phase transitions. *Phys. Rev. Lett.* 61:2635–38.
25. Kumar, S., Bouzida, D., Swendsen, R. H., Kollman, P. A., and Rosenberg, J. M. 1992. The weighted histogram analysis method for free-energy calculations on biomolecules. I. The method. *J. Comp. Chem.* 13:1011–21.
26. Fersht, A. R. 1994. Characterizing transition-states in protein-folding-an essential step in the puzzle. *Curr. Opin. Struct. Biol.* 4:79–84.
27. Lindberg, M., Tangrot, J., and Oliveberg, M. 2002. Complete change of the protein folding transition state upon circular permutation. *Nat. Struct. Biol.* 9:818–822.
28. Levy, Y., Wolynes, P. G., and Onuchic, J. N. 2004. Protein topology determines binding mechanism. *Proc. Natl. Acad. Sci. U.S.A.* 101:511–16.
29. McPhalen, C. A., and James, M. N. 1987. Crystal and molecular structure of the serine proteinase inhibitor ci-2 from barley seeds. *Biochemistry* 26:261–69.
30. Otzen, D. E., and Fersht, A. R. 1998. Folding of circular and permuted chymotrypsin inhibitor 2: Retention of the folding nucleus. *Biochemistry* 37:8139–49.
31. Viguera, A. R., Serrano, L., and Wilmanns, M. 1996. Different folding transition states may result in the same native structure. *Nat. Struct. Biol.* 3:874–80.
32. Clementi, C., Jennings, P. A., and Onuchic, J. N. 2000. How native state topology affects the folding of dihydrofolate reductase and interleukin-1beta. *Proc. Natl. Acad. Sci. U.S.A.* 97:5871–76.
33. Finke, J. M., and Jennings, P. A. 2001. Aggregated states in the folding of interleukin-1β. *J. Biol. Phys.* 27:119–31.
34. Heidary, D. K., Gross, L. A., Roy, M., and Jennings, P. A. 1997. Evidence for an obligatory intermediate in the folding of Interleukin-1β. *Nat. Struct. Mol. Biol.* 4:725–31.

35. Murzin, A. G., Lesk, A. M., and Chothia, C. 1992. β-trefoil fold, pattern of structure and sequence in the kunitz inhibitors interleukins-1β and 1α and fibroblast growth factors. *J. Mol. Biol.* 223:531–43.
36. Varley, P., Gronenborn, A. M., Christensen, H., Wingfield, P. T., Pain, R. H., and Clore, G. M. 1993. Kinetics of folding of the all-beta sheet protein interleukin-1 β. *Science* 260:1110–13.
37. Gosavi, S., Chavez, L. L., Jennings, P. A., and Onuchic, J. N., 2006. Topological frustration and the folding of interleukin-1 β. *J. Mol. Biol.* 357:986–96.
38. Chavez, L. L., Gosavi, S., Jennings, P. A., and Onuchic, J. N. 2006. Multiple routes lead to the native state in the energy landscape of the β-trefoil family. *Proc. Natl. Acad. Sci. U.S.A.* 5:10254–58.
39. Kruger, K., Grabowski, P. J., Zaug, A. J., Sands, J., Gottschling, D. E., and Cech, T. R. 1982. Self-splicing RNA: Auto-excision and auto-cyclization of the ribosomal-RNA intervening sequence of *Tetrahymena*. *Cell* 31:147–57.
40. Cech, T. R., Zaug, A. J., and Grabowski, P. J. 1981. In vitro splicing of the ribosomal-RNA precursor of tetrahymena-involvement of a quianosine nucleotide in the excision of the intervening sequence. *Cell* 27:487–96.
41. Guerrier-Takada, C., and Altman, S. 1984. Catalytic activity of an RNA molecule prepared by transcription in vitro. *Science* 223:285–86.
42. Tinoco, Jr., I., and Bustamante, C. 1999. How RNA folds. *J. Mol. Biol.* 293:271–81.
43. Brion, P. and Westhof, E. 1997. Hierarchy and dynamics of RNA folding. *Annu. Rev. Biophys. Biomol. Struct.* 26:113–37.
44. Thirumalai, D., Lee, N., Woodson, S. A., and Klimov, D. K. 2001. Early events in RNA folding. *Annu. Rev. Phys. Chem.* 52:751–62.
45. Thirumalai, D., and Hyeon, C. 2005. RNA and protein folding: Common themes and variations. *Biochemistry* 44:4954–70.
46. Dima, R. I., Hyeon, C., and Thirumalai, D. 2005. Extracting stacking interaction parameter for RNA from the data set of native structures. *J. Mol. Biol.* 347:53–69.
47. Hofacker, I. 2003. Vienna RNA secondary structure server. *Nucleic Acids Res.* 31:3429–31.
48. Hyeon, C., and Thirumalai D. 2005. Mechanical unfolding of RNA hairpins. *Proc. Natl. Acad. Sci. U.S.A.* 100:10249–53.
49. Hyeon, C., and Thirumalai, D. 2006. Forced-unfolding and force-quench refolding of RNA hairpins. *Biophys. J.* 124:3410–27.
50. Rudisser, S., and Tinoco Jr., I. 2000. Solution structure of cobalt(III)hexammine complexed to the GAAA tetraloop, and metal-ion binding to GA mismatches. *J. Mol. Biol.* 295:1211–23.
51. Mathews, D. H., Sabina, J., Zuker, M., and Turner, D. H. 1999. Expanded sequence dependence of thermodynamic parameters improves prediction of RNA secondary structure. *J. Mol. Biol.* 288:911–40.
52. Misra, V. K., and Draper, D. E. 2001. A thermodynamic framework for Mg^{2+} binding to RNA. *Proc. Natl. Acad. Sci. U.S.A.* 98:12456–61.
53. Hyeon, C., Dima, R. I., and Thirumalai, D. 2006. Pathways and kinetic barriers in mechanical unfolding and refolding of RNA and proteins. *Structure* 14:1633–45.
54. Hyeon, C., and Thirumalai, D. 2007. Mechanical unfolding of RNA: From hairpins to structures with internal multiloops. *Biophys. J.* 92:731–43.
55. Hyeon, C., and Thirumalai, D. 2006. Dynamics of allosteric transitions in GroEL. *Proc. Natl. Acad. Sci. U.S.A.* 103:18939–44.
56. Hyeon, C., and Onuchic, J. N. 2007. Internal strain regulates the nucleotide binding site of the kinesin leading head. *Proc. Natl. Acad. Sci. U.S.A.* 104:2175–80.
57. Liphardt, J., Onoa, B., Smith, S. B., Tinoco Jr., I., and Bustamante, C. 2001. Reversible unfolding of single RNA molecules by mechanical force. *Science* 292:733–37.
58. Onoa, B., Dumont, S., Lipardt, J., Smith, S. B., Tinoco Jr., I., and Bustamante, C. 2003. Identifying kinetic barriers to mechanical unfolding of the *T. thermophila* ribozyme. *Science* 299:1892–95.
59. Tirion, M. M. 1996. Large amplitude elastic motions in proteins from a single-parameter, atomic analysis. *Phys. Rev. Lett.* 77:1905–908.
60. Haliloglu, T., Bahar, I., and Erman, B. 1997. Gaussian dynamics of folded proteins. *Phys. Rev. Lett.* 79:3090–93.
61. Bahar, I. and Rader, A. J. 2005. Coarse-grained normal mode analysis in structural biology. *Curr. Opin. Struct. Biol.* 15:586–92.
62. Klimov, D. K., and Thirumalai, D. 2000. Native topology determines force-induced unfolding pathways in globular proteins. *Proc. Natl. Acad. Sci. U.S.A.* 97:7254–59.

10 On the Development of Coarse-Grained Protein Models: Importance of Relative Side-Chain Orientations and Backbone Interactions

N.-V. Buchete
School of Physics, University College Dublin

J.E. Straub
Chemistry Department, Boston University

D. Thirumalai
Biophysics Program, Institute for Physical Science and Technology, University of Maryland

CONTENTS

10.1 INTRODUCTION

Coarse-grained simulations of biomolecular systems are important not only for allowing computational studies of larger systems and for longer times—which is their main advantage over detailed, all-atom methods—but also for understanding the underlying principles that govern the structure and dynamics of biomolecules. The principle of Occam's razor (i.e., adopting the simplest model that explains an observed phenomenon) has long been a useful concept in studies of complex systems such as proteins and nucleic acids. In the light of recent advances in studying more and more complex biomolecular systems, from multiprotein complexes to molecular motors or to the whole ribosome [1–5], the need for simple yet accurate coarse-grained models has become vital to our ability to understand and simulate them.

The principal challenge in coarse-graining of proteins is the development of interaction potentials between the corresponding interaction sites in a low-resolution model. The development of such functions is an active area of research in both bioinformatics and computational molecular biology [6–11]. In this chapter, we present a personal perspective on the relative strengths and weaknesses of several approaches to coarse-graining proteins, from the simplest C_α-based models developed over three decades ago [12,13] to more recent yet complex models [14], focusing in detail on coarse-grained models that can account for aspects of proteins that are crucial to their specific structural and dynamic properties: the relative distances and orientations of side chains [15,16] and interactions involving the backbone [17].

10.2 METHODS

10.2.1 How Simple is "Too Simple"?

To accurately represent the thermodynamics of protein folding, protein–protein interactions, and protein aggregation, coarse-grained models must capture the relative importance of the dominant basins that characterize the fundamental features of the free energy landscape. The simplest coarse-grained models for proteins are typically built at residue level, as a simple string of beads, with single interaction sites per residue located typically at the C_α coordinates. In the simplest case, the interaction potentials between residues are square wells, with constant, nonzero values if the residues are close (i.e., "in contact", typical cutoff distance of 5.6 Å or less) and zero if they are separated by larger distances. These minimalist "contact potentials" are the most widely used class of coarse-grained models of proteins, and many versions have been proposed [7,18,19]. However, a recent analysis [20] has demonstrated that the most popular contact potentials have a relatively high level of uncertainty: a reduction of their detail level from using a 20-letter alphabet to only 7–9 residue types does not significantly alter their specific properties.

At a higher level of detail, the distance-dependence of the residue–residue interaction can be modeled either for several distance ranges [21,22] or by using continuous functions [23]. A new class of more complex interresidue potentials for proteins, accounting for the relative residue–residue orientations, has also been developed based on an extensive analysis of interresidue interaction data extracted from structural databases of proteins [15]. The introduction of orientational dependence represented a significant advance over earlier models that included only amino-acid distances [22,23]. The first 20-letter version of the new potentials [15] has been subsequently improved by including one more interaction site located on the backbone, allowing for the treatment of side chain–backbone and backbone–backbone interactions in the same framework [17].

The new distance- and orientation-dependent coarse-grained potentials were shown to have sufficient continuity properties to make possible their decomposition using a spherical harmonic analysis (SHA) method [25]. The resulting smooth, continuous interactions were represented using separate spherical harmonic expansions of the orientation-dependent potentials for short-, medium-, and long-range interactions [17, 25]. The new potentials significantly improved the discrimination of correct protein structures from incorrect ones, a key step in predicting folded protein structures [14,17]. They have also been used to develop an analog, single-site-per-molecule, orientation-dependent potential for treating intermolecular interactions in neat water molecules [25].

Coarse-grained simulations of polyalanine employing knowledge-based, distance-, and orientation-dependent potentials [26] have been used to study the helix-coil transition in peptides. That work demonstrates the effectiveness of using knowledge-based interaction potentials to determine both the native structure of a given peptide and the thermodynamics associated with the folding transition [27]. An alternative phenomenological "united atom" coarse-grained model using a representation based on four sites per residue has been successfully used in computer simulations of the helix-coil transition for peptides with various sequences [28]. Alternative "united residue" models of residue–residue interactions account for relative orientations by modeling side chains as Gay-Berne ellipsoids [29] and parameterize their interactions either by using a statistical approach [30] or by starting from physics-based considerations [31–34]. Other approaches to coarse-graining proteins, employing elastic networks, and normal mode analysis [34–37] have also been developed and tested.

10.2.2 Coarse-Grained Models for Peptide Chains

Figure 10.1 illustrates models at different levels of complexity that are commonly used in coarse-grained representations of polypeptides and proteins. All models aim at employing the simplest description that can provide a high degree of accuracy and realism in modeling the particular

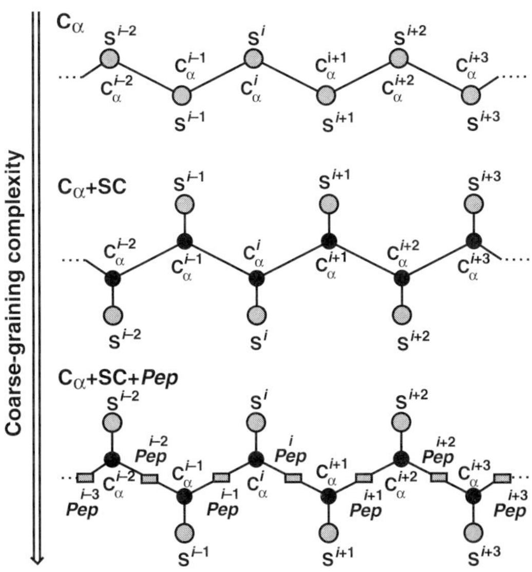

FIGURE 10.1 Models of increasing complexity for coarse-graining peptide and protein interactions. While the goal is attaining the simplest effective model (top), an accurate description of the backbone must include both the C_α positions and information regarding side-chain locations (middle) and even specific sites for backbone–backbone and backbone–side chain interactions (bottom) [25].

properties of systems of interest. The simplest models used for residue–residue interactions consider the peptides as simple chains of interacting beads (as in Figure 10.1, top) [39].

To account for the various sizes and specific packing features of the 20 different amino acid types, more detailed models must be used when estimating the relative magnitudes of residue-residue interactions. In such models (e.g., Figure 10.1, middle) [25] the backbone is described using one type of noninteracting backbone site located at the positions of the C_α atoms, with a second type of interacting side chain site, S^i, which corresponds to each side chain (SC). The S^i interaction centers are typically located at the geometric center (GC) of the heavy atoms in each side chain, with the exception of *Gly*, where it coincides with the position of the C_α atom. The backbone sites $C_\alpha{}^i$ are used to describe the backbone structure, but only the S^i interaction centers are considered to interact with each other. Similar models have been successfully used to obtain contact-based side chain–side chain (SC–SC) interaction potentials, distance-dependent potentials [22,23], and, more recently, other potentials dependent on both relative SC–SC distances and orientations [15,16,29]. While relatively effective in recognizing native-like conformations, these models do not allow for the explicit treatment of side chain–backbone and backbone–backbone interactions. Recent estimates [17] suggest that the number of backbone–backbone contacts can range from 12% to as much as 35% depending on the protein class (e.g., α, β, mixed α/β) and the topological interaction level along the sequence that is considered (e.g., $i-j \geq 3$, $i-j \geq 4$). The importance of including the backbone interactions is also supported by the results of previous statistical derivations of backbone potentials that used virtual bond and torsion angles [40], as well as secondary structure information [41].

Our work has employed a more complex model (Figure 10.1, bottom) that includes an additional interaction center located on the backbone [12] at the GC of each peptide bond (*Pep*). In this description, the local conformation of a certain residue "*i*" is sufficiently well described by the corresponding $C_\alpha{}^i$, S^i, and Pep^i interaction centers. Results from tests using various standard decoy sets [42] suggest that this level of complexity offers important improvements over the simpler $C_\alpha{}^i + \text{SC}$ model in the ability to recognize native-like states of proteins.

10.2.3 Local Reference Frames for Side Chains

The relative orientation of side chains plays a critical role in determining the magnitude of their interactions [15,40]. To extract quantitative parameters for the relative orientational dependence of coarse-grained potentials from structural databases of proteins (e.g., such as the Protein Data Bank, PDB [43]) we define local reference frames (LRFs) for each amino acid type by using the approach summarized below [15].

Figure 10.2 illustrates an example for the interaction between an *Arg* and an *Ile* side chain. To fully describe their relative orientations, a right-handed LRF is attached to each side chain by using the reference points P_1, P_2, and P_3, that uniquely define the orientation of the LRF, and a fourth point (usually denoted by S^i for the *i*th side chain) that specifies the location of the LRF_i origin. For side chains, the positions of the three reference points P_1, P_2, and P_3 are identified with the positions of the C_α, C_β, and C_γ atoms [15]. The position of the interaction centers S_i are identified with the GC of the heavy atoms in the side chain. Exceptions to this rule are made for the special cases of *Gly*, *Ala*, *Cys*, *Ser*, *Ile*, and *Val* [15,17]. These definitions have the advantage that, while being side-chain dependent, the positive O_z axis is always oriented away from the local backbone while the positive O_y axis points toward more "remote" C_γ atoms in the SC. For small side chains, O_y will point towards the next SC on the backbone sequence.

Significant improvements can be achieved by adding a virtual backbone interaction center in the middle of the peptide bond (termed "*Pep*", see Figure 10.1, bottom). The inclusion of this additional 21st interaction center was motivated by the observation that folded structures are stabilized by a substantial number of side chain–backbone contacts [15,20]. For *Pep*, the positions of the three reference points P_1, P_2, and P_3 are identified with the positions of the carbonyl C atom, its O atom and the peptide bond N atom. The interaction center S_i for *Pep* is located in the middle of its C–N peptide

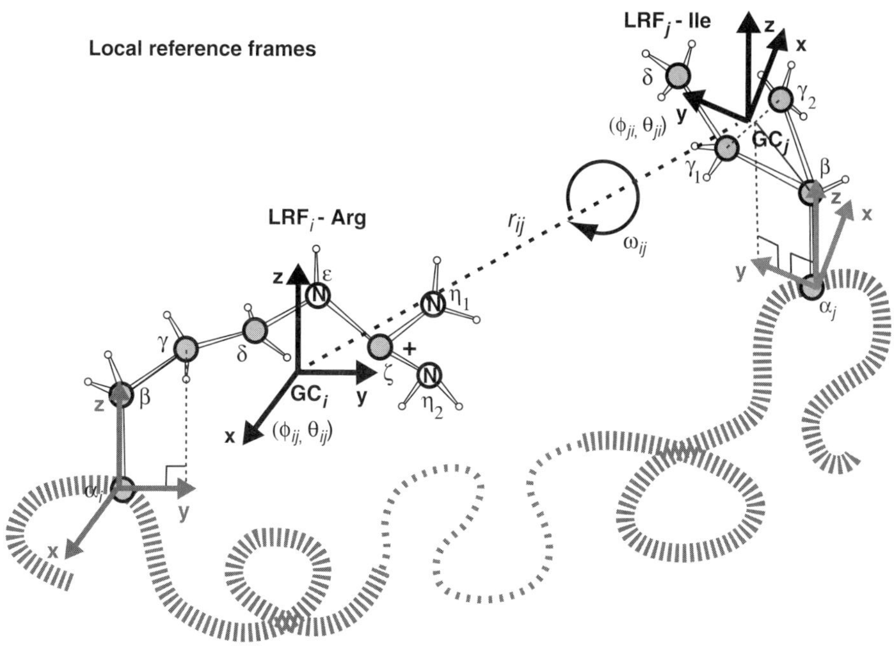

FIGURE 10.2 Quantifying the relative distances and orientations of protein side chains. The orientations of side-chain-dependent local reference frames (LRFs) [15] are first obtained (depicted as gray) using the coordinates of three heavy atoms for each side chain (C_α, C_β, and C_γ atoms, when available), being backbone independent (with the exception of *Gly* and *Ala* residues). Finally, the LRFs are translated such that their origin matches the GC of the heavy atoms in the side chains (depicted as black LRFs). The backbone trace is illustrated schematically as a gray, interrupted strip. Note that six parameters are sufficient to fully describe the relative positions of LRF$_i$ with respect to LRF$_j$: r_{ij}, θ_{ij}, ϕ_{ij}, θ_{ji}, ϕ_{ji}, and ω_{ij}. Due mainly to the limited available statistics on *i-j* contacts, the dependence on the torsion angle ω_{ij} around the r_{ij} distance is usually averaged over [17].

link. These definitions of the LRFs permit the quantitative detailed analysis of relative coordination probabilities (e.g., for hydrogen bonding) as well as of hydropathic effects in side-chain packing [14]. Note that, with the exception of *Gly* and *Ala* residues, our LRF definitions are backbone independent, which is a significant distinction compared to other recently developed orientation-dependent potentials [16].

As shown in Figure 10.2, six parameters are sufficient to fully describe the relative positions of an LRF$_i$ with respect to an LRF$_j$: r_{ij}, θ_{ij}, ϕ_{ij}, θ_{ji}, ϕ_{ji}, and ω_{ij}. However, due mainly to the limited available statistics on *i-j* contacts, the dependence on the torsion angle ω_{ij} around the r_{ij} distance is averaged over in our potentials [17]. Other studies, using backbone-dependent definitions and complete sets of Euler angles for describing the relative orientations of LRFs for all the amino acid pairs, have shown that it is possible to extract statistical potentials that depend on all the orientational degrees of freedom between residues [16]. This represents an important direction for the future development of orientation-dependent knowledge-based potentials.

10.2.4 Extracting Distance- and Orientation-Dependent Probability Densities

Using the LRFs defined and illustrated in Figure 10.2, one can analyze the wealth of experimentally solved structures deposited in protein structural databases such as the PDB [43] and extract probability densities for SC–SC interactions, $P^{ij}_{\mathrm{DO}}(r_{ij},\theta_{ij},\phi_{ij})$, which describe both their relative distance and orientation dependence (i.e., notation "DO" as in P^{ij}_{DO}).

10.2.5 The "Boltzmann Device": From Probabilities to Interaction Potentials

To extract the statistical potentials, one assumes that the known protein structures from protein databases such as PDB [43] correspond to classical equilibrium states. The SC-SC potentials can be related to probability densities for relative distances and orientations by the relation

$$U_{\mathrm{DO}}^{ij}(r_{ij},\theta_{ij},\phi_{ij}) = -kT \ln\left[\frac{P^{ij}(r_{ij},\theta_{ij},\phi_{ij})}{P_{ref}(r,\theta,\phi)}\right]. \tag{10.1}$$

Equation 10.1 is a generalization of the similar relation, known as the "Boltzmann device" [23,24], between pair distributions and distance-dependent statistical potentials [15]. An important further simplification comes from the ansatz that the orientational probability distributions for amino acids of type "j" around the amino acids "i" are independent of the distributions of amino acids of type "i" around the ones of type "j" [15].

$$P_{\mathrm{DO}}^{ij}(r_{ij},\theta_{ij},\phi_{ij},\theta_{ji},\phi_{ji}) = P_{\mathrm{DO}}^{ij}(r_{ij},\theta_{ij},\phi_{ij})\cdot P_{\mathrm{DO}}^{ji}(r_{ji},\theta_{ji},\phi_{ji}). \tag{10.2}$$

This relationship (Equation 10.2) between orientational probabilities, used in conjunction with the Boltzmann device, leads immediately to the pairwise additivity property of the total statistical potentials

$$\mathrm{U}_{\mathrm{DO}}^{ij}(\mathrm{r}_{ij},\theta_{ij},\phi_{ij},\theta_{ji},\phi_{ji}) = \mathrm{U}_{\mathrm{DO}}^{ij}(\mathrm{r}_{ij},\theta_{ij},\phi_{ij}) + \mathrm{U}_{\mathrm{DO}}^{ji}(\mathrm{r}_{ji},\theta_{ji},\phi_{ji}). \tag{10.3}$$

A pairwise additive description of the side chain–side chain potentials offers a conceptual simplification, allowing one to account easily for effects of mutations. It also allows for the further quantitative analysis of potentials corresponding to each amino acid pair by using a general decomposition in spherical harmonic functions [25].

10.2.6 "Smoothed" Coarse-Grained Interactions: Spherical Harmonic Analysis and Synthesis

An important factor that affects the accuracy of statistical potentials is the sparsity of the available data, known as "the problem of small data sets." Sippl [23] noted that dividing the SC–SC pair frequencies by both side-chain type and distance intervals results in situations when the available data is too small for conventional statistical procedures. Statistical methods have been derived to introduce "sparse data correction" methods that construct the corrected probability densities as linear combinations between the measured data and the reference probabilities [15,23]. When relative orientations are being considered, the discontinuous properties of the statistical potentials can be alleviated by using the method of spherical harmonic analysis (SHA) and synthesis (SHS) of the discrete potentials [25].

For each interaction range, the angular dependent U_{DO} potentials are functions of the θ and ϕ polar angles defined in the corresponding LRFs of the amino acids (Figure 10.2). These potential functions can be decomposed as

$$U(\theta,\phi) = \sum_{m,n} c_{mn} Y_{nm}(\theta,\phi), \tag{10.4}$$

where Y_{nm} are complex spherical harmonics and c_{mn} are the expansion coefficients [25]. This formula is valid for functions $U(\theta, \phi)$ that have well-behaved continuity properties over the entire angular range. In practice, it is convenient to use a series with real even and odd eigenfunctions, namely,

$$U(\theta,\phi)=\sum_{m,n}\left[a_{mn}Y_{nm}^{\text{odd}}(\theta,\phi)+b_{mn}Y_{nm}^{\text{even}}(\theta,\phi)\right]. \tag{10.5}$$

To analyze the data extracted from protein structures, if N is the number of grid points corresponding to sampling the data along the θ angle, we use $2(N\text{–}1)$ grid points for ϕ. These sampling points are placed on the following equiangular grid defined by:

$$\begin{aligned}\theta_i &= i\cdot\pi/(N-1)-\pi/2, \qquad i=0,1,\ldots,N-1,\\ \phi_j &= j\cdot\pi/(N-1), \qquad j=0,1,\ldots,2N-1.\end{aligned} \tag{10.6}$$

Assuming that the angular-dependent potential function is sufficiently smooth, one can perform its SHA and find the corresponding coefficients [25]:

$$a_{mn}=\alpha_{mn}\int_0^{2\pi}\int_{-\pi/2}^{\pi/2}U(\theta,\phi)P_n^m(\cos\theta)\cos(m\phi)\cos\theta d\theta d\phi \tag{10.7a}$$

and

$$b_{mn}=\alpha_{mn}\int_0^{2\pi}\int_{-\pi/2}^{\pi/2}U(\theta,\phi)P_n^m(\cos\theta)\sin(m\phi)\cos\theta d\theta d\phi, \tag{10.7b}$$

where

$$\alpha_{mn}=\frac{2n+1}{2\pi}\cdot\frac{(n-m)!}{(n+m)!}$$

and P_n^m are the associated Legendre functions [44,45].

Alternatively, if the coefficients a_{nm} and b_{nm} are known, one can reconstruct the corresponding smooth potential function $U(\theta,\phi)$ using the SHS formula:

$$U(\theta,\phi)=\sum_{n=0}^{N}{\sum_{m=0}^{n}}'P_n^m(\cos\theta)\left[a_{mn}\cos(m\phi)+b_{mn}\sin(m\phi)\right]. \tag{10.8}$$

The prime notation on the sum indicates [44] that the first term corresponding to $m=0$ must be multiplied by 0.5. The effect of this orientational potential smoothing can be visualized in Figure 10.6 as discussed below.

10.2.7 Reference Probabilities

Another consequence of applying the "Boltzmann device" [23,24,46,47] is the need to choose an appropriate reference probability for obtaining the distance- and orientation-dependent potentials using Equation 10.1. The negative logarithm of the result gives the statistical potential for the relative SC–SC orientations in units of kT. The orientation-dependent potentials were typically derived for short- (i.e., $r_{ij}\in[2$ Å, 5.6 Å$]$), medium- ($r_{ij}\in[5.6$ Å, 9.2 Å$]$), and long-range interaction shells ($r_{ij}\in[9.2$ Å, 12.8 Å$]$) [17]. One method is to consider as reference $P_{\text{ref}}(r,\theta,\phi)$ probabilities that are averaged over all the (i,j) interaction types [15,17]. With this choice, the reference average probabilities are not side-chain-specific, which is consistent with the random mixing approximation of side chains. Alternatively, a uniform orientational distribution of side chains was considered as a reference and proven to be an effective choice for constructing similar statistical potentials [16]. That study employed a backbone-related definition of LRFs and thus the relative performance improvement resulting from this choice alone cannot be directly estimated.

10.3 APPLICATIONS

10.3.1 DECOY TESTS

To assess the efficacy of the reconstructed orientation- and distance-dependent potentials, tests were performed to discriminate the native state from multiple decoy sets [17,42]. Decoys are artificially generated structures resembling native protein folds yet having different properties (e.g., different side-chain orientations and/or backbone conformations) [42] to the native folds. In cases where the native structure is experimentally resolved at high resolution, databases of decoys provide useful test cases for assessing the ability of new interaction potentials or even simple scoring functions to correctly discriminate the native state from the decoys. A standard quantitative measure used in these test cases is the Z-score. In general, the Z-score of a statistical quantity x (e.g., in our case the energy, U, or the root mean square distance, RMSD) is defined as

$$Z_x = \frac{x - \langle x \rangle}{\sigma_x}, \tag{10.9}$$

where σ_x is the standard deviation and $\langle x \rangle$ is the mean of the distribution of x values. Figure 10.3a, c, d illustrates the calculation of Z-scores for the set of 654 decoys of calbindin (PDB code *3icb*) [42]. In all cases, a good scoring function will lead to a negative Z-score. In Figure 10.3a, we show the radial energy term computed using our method. The strong similarity between this plot and the energies depicted in Figure 1 of Samudrala and Levitt [42] confirms that our radial potentials are essentially the same as those used by other groups. In Figure 10.3c, the circle shows the position of the native state and the diamond shows the position of the decoy structure that has the lowest energy. For a potential or scoring function that is efficient in discriminating the native state of a protein from a set of decoys, it is expected that (i) the native state (circle) corresponds to the lowest interaction energy and (ii) the decoy with the lowest energy (diamond) should have as small an RMSD as possible. Both criteria are important and we find that both the Z_U and Z_{RMSD} scores defined here are useful quantitative measures of the performance of the potential energy function. These Z-scores are proportional to the distance of the corresponding parameter of interest (depicted in Figure 10.3a, c, d as interrupted lines) from the mean values of their distributions (solid lines), in units of standard deviations σ. As shown in Figure 10.3b for a typical test case, there is no significant correlation between Z_{RMSD} scores and Z_U scores calculated for either the distance-dependent statistical potential (U_D) or the distance- and orientation-dependent type (U_{DO}). To compare the performance with and without the *Pep* interaction center of the interaction potentials on sets of decoy structures, we used both Z_U and Z_{RMSD} scores [15,17].

10.3.2 DEPENDENCE ON SECONDARY STRUCTURE

Figure 10.4a shows the relation between the fraction of local contacts and the contact order (CO) [48] calculated for the main types of protein secondary structures (i.e., α, β, and α/β). The CO definition employed here is

$$\mathrm{CO} = \frac{1}{N_c} \sum_{\langle i,j \rangle}^{N_c} \Delta S_{i,j}, \tag{10.10}$$

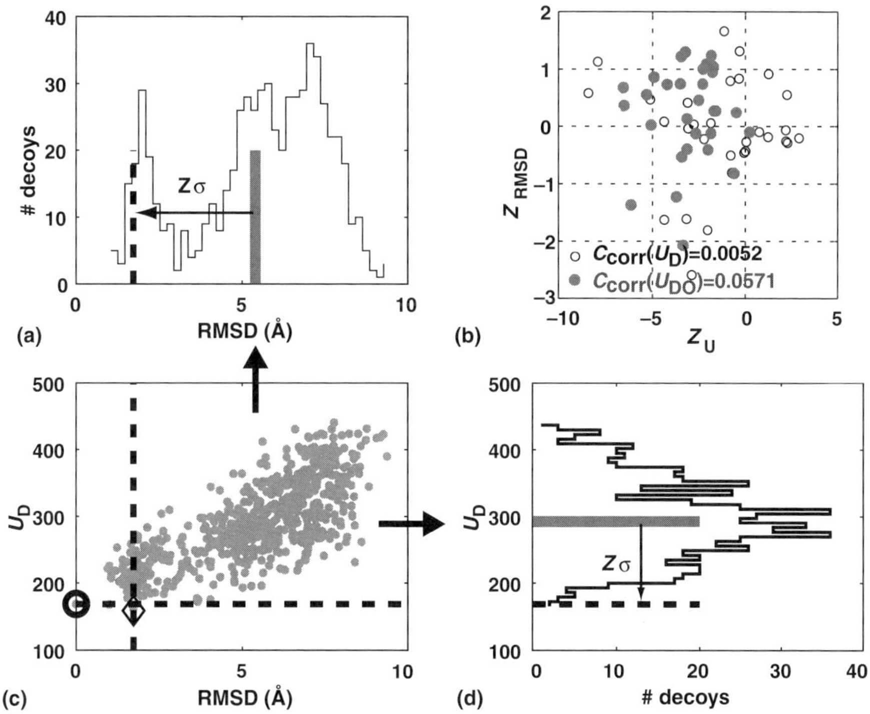

FIGURE 10.3 *Z*-scores for decoy sets of protein structures. The distance-dependent energy (U_D) for the set of 654 decoys of Calbindin (PDB code *3icb*) is plotted (c) as a function of the C_α RMSD. The RMSD (a) and energy (d) *Z*-scores are proportional to the distance from the mean values of their distributions (solid lines) of the corresponding value for the native fold (depicted as interrupted lines, (a), (c) and (d)) in units of standard deviations σ. There is generally no significant correlation between RMSD and energy Z-scores (b), so both must be considered in the data analysis [15,42].

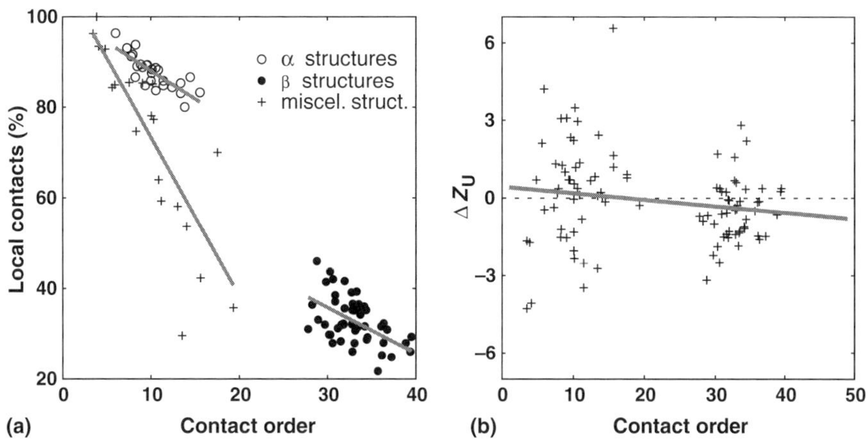

FIGURE 10.4 (a) Relationship between the fraction of local contacts and the contact order (CO) [48] calculated for the main types of protein secondary structures (i.e., α, β, and α/β). (b) The dependence of the difference $\Delta Z_U = Z_U(U_{DO}) - Z_U(U_D)$ for the energy *Z*-scores on the contact order (CO) values. Negative ΔZ_U values correspond to better performing scores for the distance- and orientation-dependent potentials (U_{DO}) [15]. Linear fits of ΔZ_U are shown (solid lines) to emphasize the trends.

where N_c is the total number of contacts in a given structure and $\Delta S_{i,j}$ is the sequence separation, in residues, between contacting residues i and j. Two residues are considered to be in contact when $|i-j|>1$ and any of the heavy atoms of residue i is within 3.75 Å of any of the heavy atoms of residue j [48]. When the CO is small the protein structure presents mostly local contacts (i.e., $|i-j|<6$), and when CO is larger, the contacts are nonlocal [15, 48]. As shown in Figure 10.4a, the CO values are a very good measure for discriminating between α-helical- and β-sheet-containing structures in our decoy tests. In Figure 10.4b the dependence of the difference $\Delta Z_E = Z_E(U_{DO}) - Z_E(U_D)$ for the energy Z-scores on the CO values is shown. Negative ΔZ_E values correspond to better performing scores for the distance- and orientation-dependent potentials (U_{DO}). The lines represent linear fits. While the trends are not very strong, it is observed that for the energy $Z_E(U_{DO})$ scores the novel U_{DO} statistical potentials perform better than the U_D potentials for longer proteins, presenting large COs, such as the β-sheet structures from the "ig-structal" and "ig-structal-hires" decoy sets [15,42]. These results suggest that details of side chain–backbone interactions should be included in statistical potentials for short or α-helical proteins with a high content of local contacts.

10.3.3 Importance of Backbone Interactions

The importance of explicit inclusion of interactions involving the backbone is supported by recent estimates of the fraction of side chain–backbone and backbone–backbones interactions found in various protein structural classes. Consistent with previous estimates [15], the data in Table 10.1 show that when considering the fraction of side chain–backbone (SC-*Pep*) contacts, using an extra interaction side *Pep* located on the backbone (see Figure 10.1), there is a high SC-*Pep* fraction of contacts in all protein structural classes [20] as defined for the CATH database [49] (v.3.0.0, May 2006). However, mainly-α structures have a 20% higher fraction of SC–SC contacts as compared to mainly-β structures, at the expense of *Pep-Pep* interactions [20]. To assess the performance of statistical potentials that include the *Pep* centers, we compared the results obtained using U_{DO-20} and U_{DO-21} interaction parameters [17]. The values of Z_U computed using U_{DO-21} are found to be more negative than for U_{DO-20} for all decoy sets except one ("4state"). For a large majority of decoy sets (84% when considering the energy score Z_U), the performance was improved by including the backbone interaction centers [42].

10.3.4 Smoothing Effect of the SHA/SHS Method

The continuous representation of the U_{DO} potentials obtained with the SHA/SHS method [25] described above, while more convenient than the original discrete version, is nevertheless different. Figure 10.5 shows a direct comparison of Z_E scores calculated using the U_{DO-21} potentials, including the 21st *Pep* interaction center, and the U_{DO-21s} potentials reconstructed from expansion coefficients with the SHS method. The effects of the SHA/SHS method on the energy Z_U score

TABLE 10.1
Fractions of Side Chain–Backbone (SC-*Pep*) Contacts in Typical Protein Structural Classes (as defined for the CATH Database, v.3.0.0)

	Class 1 α [1877]	Class 2 β [1839]	Class 3 α+β [3956]	Class 4 *misc.* [162]
SC-SC	53.14% (0.26)	32.10% (0.07)	41.05% (0.07)	37.44% (1.09)
SC-*Pep*	36.73% (0.12)	39.83% (0.07)	39.17% (0.08)	42.95% (0.68)
Pep-Pep	10.13% (0.18)	28.07% (0.09)	19.78% (0.07)	19.61% (0.66)

Note: The numbers of representative protein structural domains used are given in square brackets. The standard errors estimated for each fraction of contacts are shown in regular brackets.

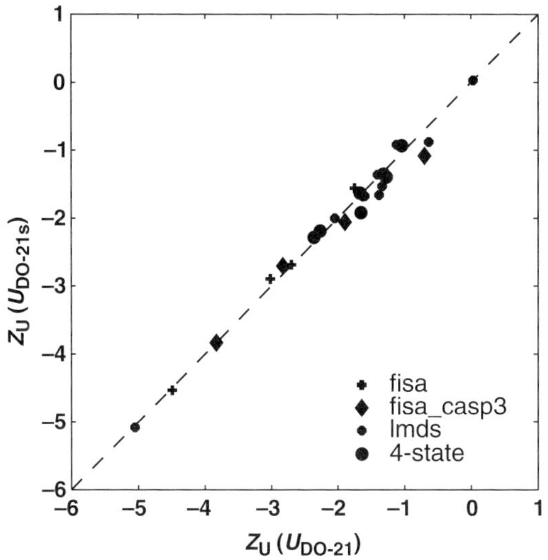

FIGURE 10.5 Effect of smoothing on the dependence on relative side-chain orientations. For all the decoy sets [42] analyzed [17], a very good correlation is observed between the energy Z_U scores obtained using a 21×21 interaction scheme (i.e., U_{DO-21}, including the *Pep* interaction sites) and the Z_U scores calculated after applying the spherical harmonic analysis/synthesis (SHA/SHS) method (U_{DO-21s}). Interestingly, the Z_U scores obtained using U_{DO-21s} are in fact slightly better (more negative) [17].

are relatively small, and a very good correlation is observed between Z_U scores obtained using the 21×21 interaction scheme (U_{DO-21}) and the Z_U scores calculated using the smooth U_{DO-21s} potentials. Noticeably, the Z_U scores obtained using U_{DO-21s} are marginally better (more negative). Although there is an intrinsic information loss introduced by the SHS/SHA procedure [44,50], the resulting potential smoothing improves the performance of the orientation-dependent potentials. These results show that the coarse-graining of the orientational potential using the SHA does not lead to a loss of accuracy and could enhance the native fold recognition ability [25].

From a computational point of view, there are potential benefits both for free energy calculations and for coarse-grained dynamical simulations that employ continuous and smoother statistical potentials. The memory requirements for storing the spherical harmonic coefficients, as opposed to the raw orientational data, are smaller, and the values of the potentials can be readily computed for any values of the θ and ϕ orientational parameters specified over the entire spherical domain. The continuous distance- and orientation-dependent statistical potentials could be instrumental in developing more efficient computational methods for protein structure prediction as well as for Monte Carlo or molecular dynamics simulations of coarse-grained models of peptides and proteins [17,25].

To illustrate the smoothing effect, Figure 10.6 depicts 3D representations of several distance- and orientation-dependent potentials (U_{DO-21s}): (a) a 12×24 equiangular grid and (b)–(i) a 96×192 equiangular grid, for the SHS procedure. The orientation-dependent potentials were derived for short- (i.e., $r_{ij} \in [2$ Å, 5.6 Å$]$), medium- ($r_{ij} \in [5.6$ Å, 9.2 Å$]$), and long-range interaction shells ($r_{ij} \in [9.2$ Å, 12.8 Å$]$). In Figure 10.6d–f the corresponding orientational potential values extracted for *Pep-Pep* contacts in each of these three interaction ranges are illustrated. Short-range interactions extracted for *Asp-Lys* (Figure 10.6a–c), *Gly-Gly* (Figure 10.6g), *Ala-Gly* (Figure 10.6h), and *Ala-Ala* (Figure 10.6i) are also illustrated. Note that for the graphical representation in Figure 10.6c the magnitude of the interaction potentials is proportional to both the radius from the center of each LRF and to the color (red for repulsive and blue for attractive regions) [25]. These types of 3D representations, with valleys corresponding to attractive regions and elevated regions for repulsive areas, illustrate better the roughness of the statistical energy surfaces introduced by the inclusion

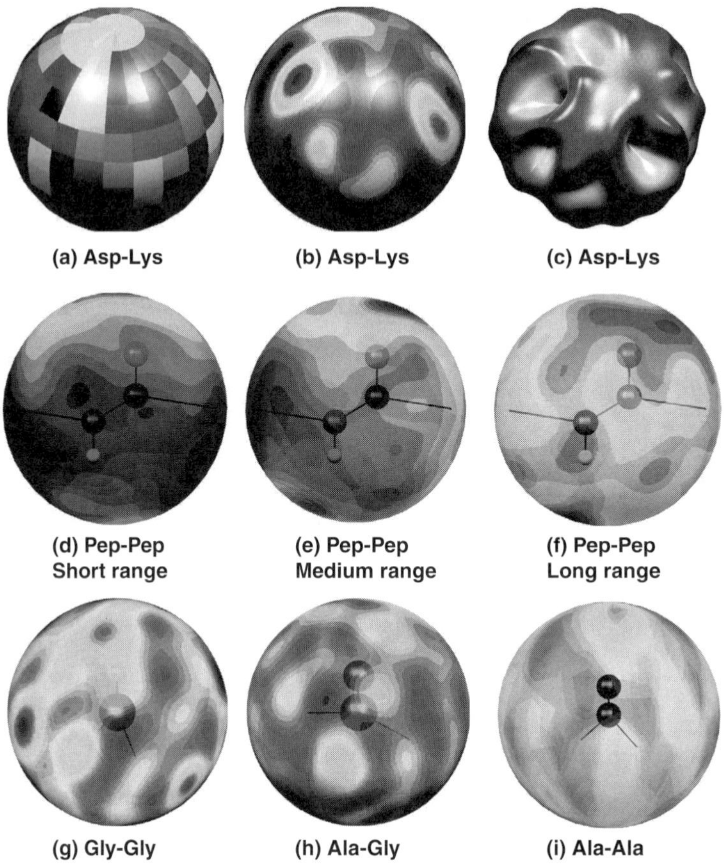

FIGURE 10.6 (See color insert following page 238.) Three-dimensional representations of the distance- and orientation-dependent statistical potential fields for side chain–side chain and side chain–backbone interactions [17]. Short-range orientation-dependent potentials $U_{\text{DO-21s}}$ are shown using: (a) a 12×24 equiangular grid and (b)–(i) a 96×192 equiangular grid, for the SHS procedure. In the graphical representation (c) for *Asp-Lys* short-range contacts [25], the magnitude of the interaction potentials is proportional to both the radius from the center of each local reference frame and to the color (red for repulsive and blue for attractive regions). These types of 3D representations, with valleys corresponding to attractive regions, illustrate better the roughness of the statistical energy surfaces. The backbone–backbone potentials (*Pep-Pep*) are shown for the smooth short-range (d), middle-range (e), and long-range (f) interaction ranges. *Asp-Lys* (a)–(c), *Gly-Gly* (g), *Ala-Gly* (h), and *Ala-Ala* (i) potentials are also shown for short-range interactions. The relative orientation of the *Pep*, *Gly*, and *Ala* atoms with respect to the orientation-dependent potential values are also shown.

of orientational dependence in each interaction type. Note also that, unlike contact potentials [20], the SC–SC orientational potentials are not generally symmetric (i.e., the angular distribution of *Gly* around *Ala* is not the same as the distribution of *Ala* around *Gly* residues).

10.4 CONCLUDING REMARKS

For successful simulations of coarse-grained models of biomolecules such as proteins, lipids, and nucleic acids, the greatest challenge remains the construction of *accurate* and *transferable* coarse-grained potential functions. Demand for a high level of accuracy requires that in certain tests, including recognition of native state from alternative "decoy" structures, the coarse-grained potentials should have an excellent performance. Some of the most successful potential functions for native state recognition in proteins are "knowledge based" [9,10,51], with statistical parameters that take extensive advantage of the experimentally solved X-ray and NMR structures from databases such as PDB/RCSB [43]. In this chapter, we presented methods that improve and extend the contact- and distance-dependent statistical potentials to include detailed information about relative side-chain orientations [15] and backbone interactions [17,20] in proteins [14]. Other simplified descriptions of residue–residue interactions accounted for relative orientations by modeling side chains as ellipsoids and attained promising results in decoy tests [30] and even in coarse-grained molecular simulations of specific systems [31,32,34,52]. Recent studies of statistical potentials dependent on relative side-chain orientations, in a manner similar to the methods presented here, raise important questions regarding the optimal choices of LRFs and reference orientational probabilities that must be addressed in future investigations [16]. The challenge remains to refine and improve the coarse-grained statistical potentials, without unnecessarily increasing their complexity, so they can be used in a transferable manner in coarse-grained simulations that can go beyond strictly structural aspects such as native state recognition, and address realistically more global thermodynamic properties of protein conformation transitions and protein–protein interactions. Significant advances are being made both in the field of coarse-grained potential development for proteins, using statistical [10,16,30,51] and physics-based parameterization approaches [32–34,54] and in molecular simulation methodologies [53–62]. That work promises to bridge the current gap between atomic and coarse-grained levels of description and enable transferable, reliable, and efficient methods for multiscale simulations [27,32,63–72].

ACKNOWLEDGMENTS

This work was supported by a grant from the National Science Foundation through grant numbers NSF-CHE-05-14056 (DT) and NSF-CHE-03-16551 (JES) and partially by the Intramural Research Program of the NIH, NIDDK (NVB). NVB is thankful to Dr. Gerhard Hummer and Dr. Attila Szabo for helpful discussions and support during the preparation of this manuscript, and to the SFI/HEA Irish Centre for High-End Computing (ICHEC) for computational facilities.

REFERENCES

1. Fisher, M. E., and Y. C. Kim. 2005. Kinesin crouches to sprint but resists pushing. *Proc. Natl. Acad. Sci. U.S.A.* 102:16209–14.
2. Miyashita, O., J. N. Onuchic, and P. G. Wolynes. 2003. Nonlinear elasticity, proteinquakes, and the energy landscapes of functional transitions in proteins. *Proc. Natl. Acad. Sci. U.S.A.* 100:12570–75.
3. Okazaki, K., N. Koga, S. Takada, et al. 2006. Multiple-basin energy landscapes for large-amplitude conformational motions of proteins: Structure-based molecular dynamics simulations. *Proc. Natl. Acad. Sci. U.S.A.* 103:11844–49.
4. Prag, G., H. Watson, Y. C. Kim, et al. 2007. The Vps27/Hse1 complex is a GAT domain-based scaffold for ubiquitin-dependent sorting. *Dev. Cell* 12:973–86.
5. Trylska, J., V. Tozzini, and J. A. McCammon. 2005. Exploring global motions and correlations in the ribosome. *Biophys. J.* 89:1455–63.
6. Donald, J. E., W. W. Chen, and E. I. Shakhnovich. 2007. Energetics of protein–DNA interactions. *Nucleic Acids Res.* 35:1039–47.
7. Murphy, J., D. W. Gatchell, C. Prasad, et al. 2003. Combination of scoring functions improves discrimination in protein–protein docking. *Proteins* 53:840–54.
8. Oliveberg, M., and E. I. Shakhnovich. 2006. Folding and binding: The conformational repertoire of proteins: folding, aggregation and structural recognition. *Curr. Opin. Struct. Biol.* 16:68–70.
9. Skolnick, J. 2006. In quest of an empirical potential for protein structure prediction. *Curr. Opin. Struct. Biol.* 16:166–71.
10. Zhang, Y., A. K. Arakaki, and J. Skolnick. 2005. TASSER: An automated method for the prediction of protein tertiary structures in CASP6. *Proteins* 61 (Suppl. 7): 91–98.
11. Zhou, Y., H. Y. Zhou, C. Zhang, et al. 2006. What is a desirable statistical energy function for proteins and how can it be obtained? *Cell Biochem. Biophys.* 46:165–74.

12. Levitt, M., and A. Warshel. 1975. Computer simulation of protein folding. *Nature* 253:694–98.
13. Tanaka, S., and H. A. Scheraga. 1976. Medium- and long-range interaction parameters between amino acids for predicting three-dimensional structures of proteins. *Macromolecules* 9:945–50.
14. Buchete, N. V., J. E. Straub, and D. Thirumalai. 2004. Development of novel statistical potentials for protein fold recognition. *Curr. Opin. Struct. Biol.* 14:225–32.
15. Buchete, N. V., J. E. Straub, and D. Thirumalai. 2003. Anisotropic coarse-grained statistical potentials improve the ability to identify native-like protein structures. *J. Chem. Phys.* 118:7658–71.
16. Miyazawa, S., and R. L. Jernigan. 2005. How effective for fold recognition is a potential of mean force that includes relative orientations between contacting residues in proteins? *J. Chem. Phys.* 122:024901.
17. Buchete, N. V., J. E. Straub, and D. Thirumalai. 2004. Orientational potentials extracted from protein structures improve native fold recognition. *Protein Sci.* 13:862–74.
18. Hinds, D. A., and M. Levitt. 1992. A lattice model for protein structure prediction at low resolution. *Proc. Natl. Acad. Sci. U.S.A.* 89:2536–40.
19. Tobi, D., G. Shafran, N. Linial, et al. 2000. On the design and analysis of protein folding potentials. *Proteins* 40:71–85.
20. Buchete, N. V., J. E. Straub, and D. Thirumalai. 2008. Dissecting contact potentials for proteins: Relative contributions of individual amino acids. *Proteins* 70:119–30.
21. Miyazawa, S., and R. L. Jernigan. 1996. Residue-residue potentials with a favorable contact pair term and an unfavorable high packing density term, for simulation and threading. *J. Mol. Biol.* 256:623–44.
22. Tobi, D., and R. Elber. 2000. Distance-dependent, pair potential for protein folding: Results from linear optimization. *Proteins* 41:40–46.
23. Sippl, M. J. 1990. Calculation of conformational ensembles from potentials of mean force. An approach to the knowledge-based prediction of local structures in globular proteins. *J. Mol. Biol.* 213:859–83.
24. Sippl, M. J. 1995. Knowledge-based potentials for proteins. *Curr. Opin. Struct. Biol.* 5:229–35.
25. Buchete, N. V., J. E. Straub, and D. Thirumalai. 2004. Continuous anisotropic representation of coarse-grained potentials for proteins by spherical harmonics synthesis. *J. Mol. Graph. Model.* 22:441–50.
26. Buchete, N. V., J. E. Straub, and D. Thirumalai. 2004. Orientation-dependent coarse-grained potentials derived by statistical analysis of molecular structural databases. *Polymer* 45:597–608.
27. van Giessen, A. E., and J. E. Straub. 2005. Monte Carlo simulations of polyalanine using a reduced model and statistics-based interaction potentials. *J. Chem. Phys.* 122:024904.
28. van Giessen, A. E., and J. E. Straub. 2006. Coarse-grained model of coil-to-helix kinetics demonstrates the importance of multiple nucleation sites in helix folding. *J. Chem. Theory Comput.* 2:674–84.
29. Gay, J. G., and B. J. Berne. 1981. Modification of the overlap potential to mimic a linear site-site potential. *J. Chem. Phys.* 74:3316–19.
30. Mukherjee, A., P. Bhimalapuram, and B. Bagchi. 2005. Orientation-dependent potential of mean force for protein folding. *J. Chem. Phys.* 123:014901.
31. Das, P., S. Matysiak, and C. Clementi. 2005. Balancing energy and entropy: A minimalist model for the characterization of protein folding landscapes. *Proc. Natl. Acad. Sci. U.S.A.* 102:10141–46.
32. Liwo, A., M. Khalili, and H. A. Scheraga. 2005. Ab initio simulations of protein-folding pathways by molecular dynamics with the united-residue model of polypeptide chains. *Proc. Natl. Acad. Sci. U.S.A.* 102:2362–67.
33. Liwo, A., S. Oldziej, M. R. Pincus, et al. 1997. A united-residue force field for off-lattice protein-structure simulations. 1. Functional forms and parameters of long-range sidechain interaction potentials from protein crystal data. *J. Comput. Chem.* 18:849–73.
34. Rojas, A. V., A. Liwo, and H. A. Scheraga. 2007. Molecular dynamics with the United residue force field: ab initio folding simulations of multichain proteins. *J. Phys. Chem. B* 111:293–309.
35. Bahar, I., and A. J. Rader. 2005. Coarse-grained normal mode analysis in structural biology. *Curr. Opin. Struct. Biol.* 15:586–92.
36. Lu, M. Y., B. Poon, and J. P. Ma. 2006. A new method for coarse-grained elastic normal mode analysis. *J. Chem. Theory Comput.* 2:464–71.
37. Tama, F., and C. L. Brooks. 2006. Symmetry, form, and shape: Guiding principles for robustness in macromolecular machines. *Annu. Rev. Biophys. Biomolec. Struct.* 35:115–33.
38. Zheng, W. J., and B. R. Brooks. 2005. Normal-modes-based prediction of protein conformational changes guided by distance constraints. *Biophys. J.* 88:3109–17.
39. Honeycutt, J. D., and D. Thirumalai. 1990. Metastability of the folded states of globular proteins. *Proc. Natl. Acad. Sci. U.S.A.* 87:3526–29.

40. Bahar, I., and R. L. Jernigan. 1996. Coordination geometry of nonbonded residues in globular proteins. *Folding and Design* 1:357–70.
41. Miyazawa, S., and R. L. Jernigan. 1999. Evaluation of short-range interactions as secondary structure energies for protein fold and sequence recognition. *Proteins* 36:347–56.
42. Samudrala, R., and M. Levitt. 2000. Decoys 'R' Us: A database of incorrect conformations to improve protein structure prediction. *Protein Sci.* 9:1399–401.
43. Berman, H. M., J. Westbrook, Z. Feng, et al. 2000. The Protein Data Bank. *Nucleic Acids Res.* 28:235–42.
44. Adams, J. C., and P. N. Swarztrauber. 1997. Spherepack 2.0: A model development facility. *NCAR Tech. Note, NCAR/TN-436-STR.*
45. Arfken, G. B., and H.-J. Weber. 2005. *Mathematical methods for physicists*. Boston: Elsevier.
46. Ben-Naim, A. 1997. Statistical potentials extracted from protein structures: Are these meaningful potentials? *J. Chem. Phys.* 107:3698–706.
47. Thomas, P. D., and K. A. Dill. 1996. Statistical potentials extracted from protein structures: How accurate are they? *J. Mol. Biol.* 257:457–69.
48. Dima, R. I., and D. Thirumalai. 2002. Exploring the propensities of helices in PrP(C) to form beta sheet using NMR structures and sequence alignments. *Biophys. J.* 83:1268–80.
49. Pearl, F., A. Todd, I. Sillitoe, et al. 2005. The CATH domain structure database and related resources Gene3D and DHS provide comprehensive domain family information for genome analysis. *Nucleic Acids Res.* 33:D247–51.
50. Adams, J. C., and P. N. Swarztrauber. 1999. Spherepack 3.0: A model development facility. *Monthly Weather Rev.* 127:1872–78.
51. Bradley, P., L. Malmstrom, B. Qian, et al. 2005. Free modeling with Rosetta in CASP6. *Proteins* 61 (Suppl. 7): 128–34.
52. Hyeon, C., and D. Thirumalai. 2005. Mechanical unfolding of RNA hairpins. *Proc. Natl. Acad. Sci. U.S.A.* 102:6789–94.
53. Miyashita, O., P. G. Wolynes, and J. N. Onuchic. 2005. Simple energy landscape model for the kinetics of functional transitions in proteins. *J. Phys. Chem. B* 109:1959–69.
54. Makowski, M., A. Liwo, K. Maksimiak, et al. 2007. Simple physics-based analytical formulas for the potentials of mean force for the interaction of amino acid side chains in water. 2. Tests with simple spherical systems. *J. Phys. Chem. B* 111:2917–24.
55. Best, R. B., Y.-G. Chen, and G. Hummer. 2005. Slow protein conformational dynamics from multiple experimental structures: The helix/sheet transition of Arc repressor. *Structure* 13:1755–63.
56. Buchete, N. V., and G. Hummer. 2007. Coarse master equations for peptide folding dynamics. *J. Phys. Chem. B.* (in press, http://dx.doi.org/10.1021/jp0761665).
57. Buchete, N. V., and G. Hummer. 2007. Peptide folding kinetics from replica exchange molecular dynamics. *Phys. Rev.* E 77:030902
58. Hummer, G., and I. G. Kevrekidis. 2003. Coarse molecular dynamics of a peptide fragment: Free energy, kinetics, and long-time dynamics computations. *J. Chem. Phys.* 118:10762–73.
59. Karanicolas, J., and C. L. Brooks. 2003. Improved Go-like models demonstrate the robustness of protein folding mechanisms towards non-native interactions. *J. Mol. Biol.* 334:309–25.
60. Kim, J., J. E. Straub, and T. Keyes. 2006. Statistical-temperature Monte Carlo and molecular dynamics algorithms. *Phys. Rev. Lett.* 97:050601.
61. Sriraman, S., L. G. Kevrekidis, and G. Hummer. 2005. Coarse master equation from Bayesian analysis of replica molecular dynamics simulations. *J. Phys. Chem. B* 109:6479–84.
62. Tan, R. K. Z., A. S. Petrov, and S. C. Harvey. 2006. YUP: A molecular simulation program for coarse-grained and multiscaled models. *J. Chem. Theory Comput.* 2:529–40.
63. Arkhipov, A., P. L. Freddolino, and K. Schulten. 2006. Stability and dynamics of virus capsids described by coarse-grained modeling. *Structure* 14:1767–77.
64. Ayton, G. S., W. G. Noid, and G. A. Voth. 2007. Multiscale modeling of biomolecular systems: In serial and in parallel. *Curr. Opin. Struct. Biol.* 17:192–98.
65. Chen, Y. G., and G. Hummer. 2007. Slow conformational dynamics and unfolding of the calmodulin C-terminal domain. *J. Am. Chem. Soc.* 129:2414–15.
66. Mukherjee, A., and B. Bagchi. 2006. Anomalous orientation-dependent effective pair interaction among histidine and other amino acid residues in metalloproteins: Breakdown of the hydropathy scale index. *Biochemistry* 45:5129–39.

67. Nguyen, H. D., and C. K. Hall. 2004. Molecular dynamics simulations of spontaneous fibril formation by random-coil peptides. *Proc. Natl. Acad. Sci. U.S.A.* 101:16180–85.
68. Noid, W. G., J. W. Chu, G. S. Ayton, et al. 2007. Multiscale coarse-graining and structural correlations: connections to liquid-state theory. *J. Phys. Chem. B* 111:4116–27.
69. Shi, Q., S. Izvekov, and G. A. Voth. 2006. Mixed atomistic and coarse-grained molecular dynamics: Simulation of a membrane-bound ion channel. *J. Phys. Chem. B* 110:15045–48.
70. Shih, A. Y., A. Arkhipov, P. L. Freddolino, et al. 2006. Coarse grained protein–lipid model with application to lipoprotein particles. *J. Phys. Chem. B* 110:3674–84.
71. Tozzini, V., J. Trylska, C. E. Chang, et al. 2007. Flap opening dynamics in HIV-1 protease explored with a coarse-grained model. *J. Struct. Biol.* 157:606–15.
72. Whitford, P. C., O. Miyashita, Y. Levy, et al. 2007. Conformational transitions of adenylate kinase: Switching by cracking. *J. Mol. Biol.* 366:1661–71.

11 Characterization of Protein-Folding Landscapes by Coarse-Grained Models Incorporating Experimental Data

Silvina Matysiak
Institute for Computational Engineering and Science, The University of Texas at Austin

Cecilia Clementi
Department of Chemistry, Rice University

CONTENTS

11.1 INTRODUCTION

The detailed characterization of protein folding poses outstanding challenges to both theory and experiment. Relevant time scales in the folding process span several orders of magnitude from fast vibrations to global structural transitions [1]. The heavy computational requirements of simulating protein dynamics with atomistic details do not allow the direct investigation of time scales significantly larger than nanoseconds for biologically relevant problems [2–6]. Theoretical studies aimed to explore long time scales and system sizes inaccessible to atomistic simulations are mostly based on coarse-grained models that "renormalize" groups of atoms (typically on the scale of one amino acid) into effective degrees of freedom [4–15]. By sacrificing the atomistic details, coarse-grained models can explore larger time and length scales, providing a good starting point to identify relevant regions on the vast configurational landscape of protein systems. However, the reduction in complexity comes with a cost: most coarse-grained models are built on approximations and assumptions that may affect the accuracy, and the results need to be carefully evaluated through comparison with experimental data. In particular, the coarse-graining of proteins for folding studies is based on the assumption that, because of its highly cooperative nature, the essential features of the folding process do not strongly depend on each single atomic detail. Generally, coarse-graining approaches for protein folding are formulated within the "free energy landscape perspective" [16], which advocates the use of statistical mechanics to organize the multitude of the protein's microstates in terms of a minimal number of collective parameters.

Existing problems in the coarse-grained modeling scenario for protein folding can be roughly divided in two classes: either the models artificially remove most of the interactions between pairs of residues that are not "in contact" in the native state (i.e., the so-called Gō-like models) [8–11], or they compromise with the minimal frustration principle [16] and/or realistic protein geometry (e.g., simple on-lattice models). We discuss here an alternative approach that we have developed in the last few years, and that can, at least approximately, address both these problems. In particular, we have recently proposed a new realistic coarse-grained model (henceforth called DMC, from the initials of the authors [17]) that "naturally" incorporates sequence details and energetic frustration into an overall minimally frustrated protein-folding landscape. The model is coupled with an optimization procedure to design the parameters of the effective protein Hamiltonian to fold into a desired native structure (within 1 Å root-mean-squared deviation (rmsd) from the crystal structure). The ability of the DMC model, despite its simplicity, to smoothly fold a protein sequence into its correct native structure relies on a "mean-field-like" definition of effective interactions. Previous models have been proposed to correct for the physicochemical information lost in the complete removal of energetic frustration associated with a Gō-like protein model [8,18–20]. These models correctly reproduce the secondary structure and global fold, although oftentimes they produce a folded structure largely deviating from the correct native state. The main improvements of the DMC model with respect to previous ones are: (i) it allows a protein to be folded smoothly into a desired structure without removing nonnative interactions and (ii) it takes into account, at least approximately, the chemical and geometrical diversity of amino acids. Recent applications of this modeling technique have shown that it contains the main physicochemical ingredients that are responsible for shaping the folding landscape, at least for the problems that have been investigated [17,21,22].

Significant advancement in the characterization of protein dynamics and function can be achieved by combining the complementary strength and applicability of experimental and theoretical studies. Because of the high dimensionality of the parameter space, several sets of effective parameters able to fold a protein model into a desired protein structure can be generally obtained from the application of the DMC coarse-graining technique [17]. In order to define an *optimal* set of effective parameters that enable the model to best mimic the behavior of the real protein under consideration, we have developed an inverse statistical mechanic procedure to "anchor" a coarse-grained model to experimental data [21,23]. Unlike previous studies, the goal of this procedure is not just the a posteriori comparison of theory and experiment, but rather their a priori integration. This method has allowed us to reconstruct in simulation the overall free-energy landscape for protein S6, using a Gō-like model as an initial reference, which is in perfect agreement with all experimentally measured free-energy differences [23]. Moreover, we have recently shown that, when dressed with this inverse procedure, the DMC coarse-grained model can correctly predict the mutations triggering misfolding and aggregation in protein S6 [21].

In the following we first describe the basic ideas and the practical implementation of the DMC model and the inverse statistical mechanics procedure to incorporate experimental data into it. The performance of this coarse-graining approach is then illustrated in a couple of recent applications.

11.2 METHODS

11.2.1 Das, Matysiak, Clementi (DMC) Coarse-Grained Model

The main idea at the base of the model is that effective model parameters can be designed in a simplified protein geometry by enforcing minimum-energy requirements for the native state of a protein [24–27], and by choosing the functional form of the interaction potential function to maximize the native-state packing [24]. The latter requirement stems from the fact that native protein structures are extremely packed [28–31], and mutations introducing cavities generally produce destabilization [32,33]. Upon coarse-graining, the precious information on the specific shape and size of the amino acids is lost. Previous models have been mostly based on isotropic interactions in the coarse-grained coordinates, implicitly assuming a spherical geometry for every coarse-grained residue. This assumption introduces an unphysical source of frustration, as many different maximally compact configurations can be obtained for a chain of spherical beads. It has been shown that geometric considerations on the native state packing play an important role in determining the native structure [34]. In order to take geometric effects into account, at least approximately, a protein is represented as a chain of interacting beads of different shapes, centered in the C_α coordinates as illustrated in Figure 11.1. The nontrivial amino acid geometry is reintroduced in the DMC model by using a distribution of equilibrium distances for each amino acid pair, rather than a constant value.

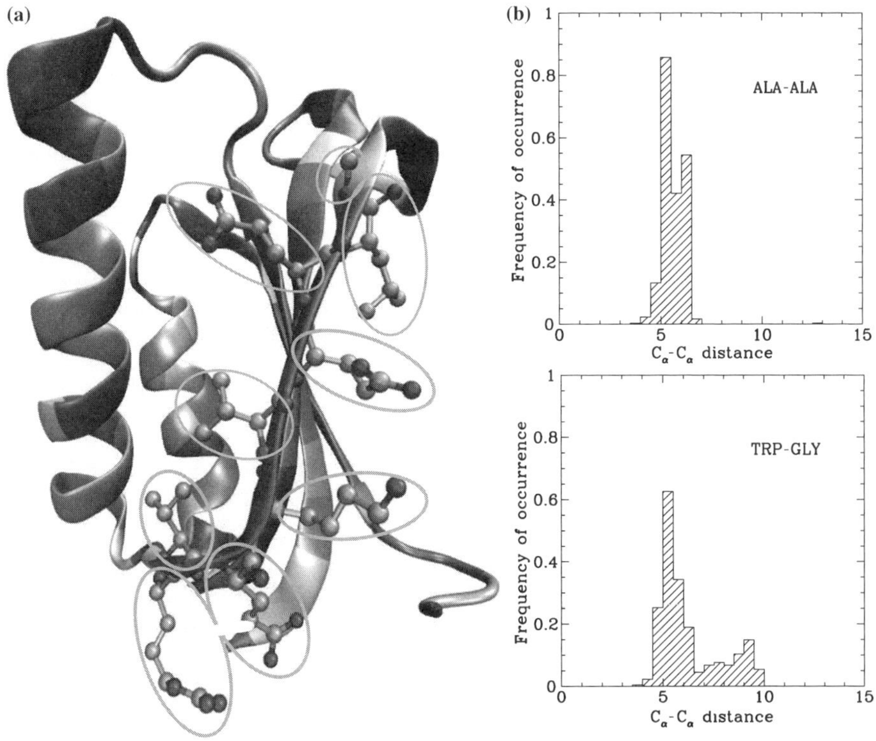

FIGURE 11.1 (a) Structure of protein S6. Each residue is colored with a different shade of gray depending on the amino-acid type. As an illustration of the coarse-graining procedure, residues of β-strand 2 are represented in atomistic detail to highlight the nontrivial amino-acid geometry. (b) Distribution of C_α–C_α distances between different pairs of amino acids (ALA-ALA in the top figure, TRP-GLY in the bottom) for a distance greater than five residues along the protein sequence, as obtained by considering all the native contacts present in a set of nonredundant protein structures.

The potential energy function comprises a local and nonlocal term, $V = V_{\text{local}} + V_{\text{nonlocal}}$, where V_{local} encompasses bond, angle, and torsional energy terms and is designed to have its absolute minimum in the native state. The complete removal of frustration from the local energy term is not critical for folding and can be alternatively designed as a statistical potential. Following Das et al. [17], V_{nonlocal} is chosen in the form:

$$V_{\text{nonlocal}} = \sum_{\substack{i,j=1 \\ i<j-3}}^{N} \varepsilon(a_i, a_j)\left[5\left(\frac{\sigma_{ij}}{r_{ij}}\right)^{12} - \delta(a_i, a_j)6\left(\frac{\sigma_{ij}}{r_{ij}}\right)^{10}\right] \tag{11.1}$$

for a protein of N residues. The amino acid sequence is encoded in the parameter a_i that associates an amino acid's type to each residue i. As mentioned above, in order to capture some information about the geometric complementarity of interacting residues, the value of the parameter σ_{ij} for a pair of residues (i, j) is extracted from a distribution $P(\sigma; a_i, a_j, |i-j|)$. This distribution is obtained from a statistical analysis of a large database of nonredundant protein structures and is illustrated in Figure 11.1b for the instances of residue pairs ALA-ALA and TRP-GLY.

The shape of the distribution changes depending on the distance between the interacting residues, due to local steric effects. The distribution $P(\sigma; a_i, a_j, |i-j|)$ is obtained by considering the distance between C_α–C_α over all the native contacts formed by residues of type a_i and a_j with a specific distance $|i-j|$ along the protein sequence. More specifically, three different types of distribution are considered for different relative distances between residues i and j along the protein: $|i-j|=4$, $|i-j|=5$ and $|i-j|>5$.

The parameter $\varepsilon(a_i, a_j)$ in Equation 11.1 is positively defined, while the parameter $\delta(a_i, a_j)$ can take one of two values: $\delta(a_i, a_j) = 1$ if the interaction between the pair residues is attractive, or $\delta(a_i, a_j) = 0$ if it is repulsive. The set of parameters $(\{\varepsilon\}, \{\delta\})$ is generated through an iterative procedure as follows.

(i) A set of non-native-like structures (decoys) $\{\Gamma_i\}, i = 1, \ldots, N_{\text{decoys}}$ is generated. Decoys are defined as any protein configuration with rmsd > 3 Å from the native crystal structure. The initial library of decoys is obtained by performing a short MD simulation with a random choice of the effective parameters $\varepsilon(a_i, a_j)$ and $\delta(a_i, a_j)$. Additional decoys are added to the decoy library after every iteration (see point (iv)).

(ii) A set of effective parameters $(\{\varepsilon\}, \{\sigma\})$ is defined by enforcing the maximum energy gap criterion [24,26,27] on the library of decoys:

$$\Delta E(\{\varepsilon\}, \{\delta\}) = \max_{\{\varepsilon\},\{\delta\}}[(\min_{i\in(1,\ldots,N_{\text{decoys}})} E_{\Gamma_i}) - E_{\Gamma_{\text{nat}}}], \tag{11.2}$$

where Γ_{nat} is the native structure, with energy $E_{\Gamma_{\text{nat}}}$, and E_{Γ_i} is the energy corresponding to the decoy structure Γ_i. Equation 11.2 is solved by means of a combination of Monte-Carlo and simulated annealing [35] and perceptron learning [36] algorithms.

(iii) Several "heat and quench" unfolding/refolding MD simulations are performed to sample the robustness of the solution.

(iv) The procedure converges if the native state of the protein under consideration is recovered from the unfolding/refolding process. Otherwise, the collapsed nonnative structures visited during the MD simulations are collected and included in the decoy library to perform another iteration.

Essentially, by performing this iterative procedure the energy landscape of the protein is gradually modified from a random heteropolymer to a minimally frustrated polymer, as illustrated in Figure 11.2.

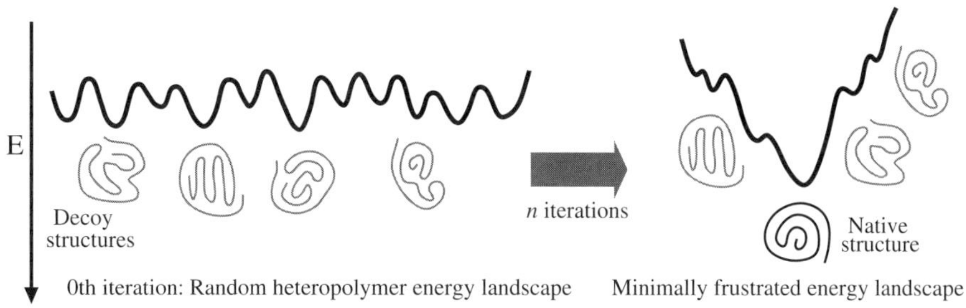

FIGURE 11.2 Energy landscape associated to the protein model is progressively modified during the selection of the effective parameters. The right figure describes the initial energy landscape, when a random set of parameters $(\{\varepsilon\},\{\delta\})$ is used. After n iterations of the optimization procedure (that maximizes the energy gap between the native structure and all the decoy structures, as described in the text), the energy landscape of the protein model becomes minimally frustrated, with the absolute minimum corresponding to the native structure.

11.2.2 Inverse Statistical Mechanics Procedure

Given the high dimensionality of the parameter space, several sets of parameters $(\{\varepsilon\},\{\sigma\})$ that allow the DMC model to fold correctly into its crystal structure (within 1 Å of the X-ray structure) can be generally obtained for a given protein. This leaves us room for an additional refinement, by using experimental data to guide the definition of an optimal set of parameters that can exactly reproduce a set of physical observables.

Following Matysiak et al. [21,23], we assume that there exists a set of optimal parameters $(\{\varepsilon^*\},\{\sigma^*\})$ such that experimentally measured differences in free energy upon mutation can be reproduced exactly (within the error bars). The part of the Hamiltonian that is significantly affected upon mutation is the nonlocal term (see Equation 11.1), which can be expressed as follows:

$$H_0 = \sum_{i \neq j} \varepsilon(a_i, a_j) Q_{ij}(a_i, a_j), \tag{11.3}$$

where $Q_{ij}(a_i, a_j)$ is a function of the C_α–C_α distance between i and j. In order to simplify the notation, in the following we indicate Q_{ij} to represent $Q_{ij}(a_i, a_j)$. Considering mutations that reduce the size of a side chain (so-called *conservative mutations*) we can model a mutation of residue k as a small perturbation of the Hamiltonian. The effect of a mutation at position k is introduced in the DMC model by changing the "color type" of residue k from a_k into a_k'; the mutation can then be expressed as a perturbation δH_k to the wild-type Hamiltonian H_0 as follows:

$$\delta H_k = H_{(a_k \to a_k')} - H_0 = \sum_i [-\varepsilon(a_i, a_k) Q_{ik} + \varepsilon(a_i, a_k') Q'_{ik}]. \tag{11.4}$$

New equilibrium distances $\sigma_{ik}(a_k')$ are extracted from the distribution $P(\sigma; a_i, a_k', |i-k|)$ for each mutated residue k with the constraint $\sigma_{ik}(a_k) \geq \sigma_{ik}(a_k')$. This constraint reflects the fact that only conservative mutations are considered. The modification of the equilibrium distances produces a change in the function Q_{ik}, which becomes Q_{ik}'.

The change of free energy due to the mutation $a_k \to a_k'$ in a selected region of the folding landscape can be expressed as:

$$\frac{\Delta G^X_{a_k \to a_k'}}{RT} = -\ln \left\langle e^{-\frac{\delta H_k}{RT}} \right\rangle^X_{wt}, \tag{11.5}$$

where $\langle \bullet \rangle$ represents a canonical average over the configurational space of the unperturbed system. In order to simplify the notation, in the following we indicate as ΔG_k^X the free energy difference $\Delta G^X_{a_k \to a_k'}$ associated with the mutation $a_k \to a_k'$ at residue k. The canonical averages can be evaluated over the structural ensembles associated with the native state N, the unfolded state U, and the transition state TS. Differences between these free-energy values allow us to estimate changes in the protein stability, $\Delta\Delta G_k^0$, and in the free-energy barrier height, $\Delta\Delta G_k^\dagger$, as follows:

$$\Delta G_k^{\mathrm{N}} - \Delta G_k^{\mathrm{U}} = \Delta\Delta G^0_{\mathrm{sim},k}, \tag{11.6}$$

$$\Delta G_k^{\mathrm{TS}} - \Delta G_k^{\mathrm{U}} = \Delta\Delta G^\dagger_{\mathrm{sim},k}. \tag{11.7}$$

As thermodynamic quantities are generally robust against small perturbations in a protein Hamiltonian, we can use a first-order Taylor expansion around the optimal set of parameters to estimate the change in free energy as a function of the values of the parameters used in the model. This expansion provides a way to estimate the correction to the set of parameters $(\{\varepsilon\},\{\sigma\})$ that enables the protein model to better mimic a set of experimental data:

$$\Delta\Delta G^0_{\mathrm{exp},k} - \Delta\Delta G^0_{\mathrm{sim},k}(\{\varepsilon_0\}) = \sum_{a_m, a_l} \left.\frac{\partial \Delta\Delta G^0_{\mathrm{sim},k}}{\partial \varepsilon(a_m,a_l)}\right|_{\{\varepsilon_0\}} \cdot (\varepsilon^*(a_m,a_l) - \varepsilon^0(a_m,a_l)), \tag{11.8}$$

$$\Delta\Delta G^\dagger_{\mathrm{exp},k} - \Delta\Delta G^\dagger_{\mathrm{sim},k}(\{\varepsilon_0\}) = \sum_{a_m, a_l} \left.\frac{\partial \Delta\Delta G^\dagger_{\mathrm{sim},k}}{\partial \varepsilon(a_m,a_l)}\right|_{\{\varepsilon_0\}} \cdot (\varepsilon^*(a_m,a_l) - \varepsilon^0(a_m,a_l)), \tag{11.9}$$

where $\Delta\Delta G^0_{\mathrm{exp}}$ and $\Delta\Delta G^\dagger_{\mathrm{exp}}$ indicate the experimentally measured free-energy differences upon mutation. By using basic statistical mechanics, the derivatives in Equation 11.8 and Equation 11.9 can be expressed as follows:

$$\frac{\partial \Delta G^X_{\mathrm{sim},k}}{\partial \varepsilon_{a_m,a_l}} = \frac{\left\langle \sum_i \left[Q_{ik}'\delta(a_k',a_k) + \sum_j Q_{ij}\delta(a_j,a_l)(1-\delta_{ik})(1-\delta_{jk}) \right] \delta(a_i,a_m) e^{-\frac{\delta H_k}{RT}} \right\rangle_X}{\left\langle e^{-\frac{\delta H_k}{RT}} \right\rangle_X} - \left\langle \sum_j Q_{ij}\delta(a_i,a_m)\delta(a_j,a_l) \right\rangle_X, \tag{11.10}$$

where $X \in (\mathrm{U, TS, N})$. Equation 11.6 through Equation 11.10 combined give a system of linear equations, where the unknowns are the desired corrections to the model parameters:

$$\delta\varepsilon(a_i,a_j) = \varepsilon^*(a_i,a_j) - \varepsilon^0(a_i,a_j). \tag{11.11}$$

If after the correction it is found that, for a pair of residue types (a_i,a_j) the parameter $\varepsilon(a_i,a_j) < 0$, then the interaction is modified from attractive to repulsive, or vice-versa, by changing the parameter $\delta_0(a_i,a_j)$ to $\delta^*(a_i,a_j) = 1 - \delta_0(a_i,a_j)$.

It is common to have a larger number of unknowns (*L*) than available experimental data (*n*); thus the optimal set of parameters is not uniquely defined by the solution of Equation 11.8 and Equation 11.9. However, other constraints apply: in order for the model to fold the protein to its desired native structure, the parameters $(\{\varepsilon\},\{\sigma\})$ must obey the minimum frustration principle, that is:

$$E_{\Gamma_i}(\{\varepsilon\},\{\delta\}) - E_{\Gamma_{\text{nat}}}(\{\varepsilon\},\{\delta\}) > 0 \tag{11.12}$$

for each decoy structure Γ_i. A solution exists only if it is possible to satisfy the system of Equation 11.8 and Equation 11.9 inside the region of the parameter space delimited by the set of inequalities given by Equation 11.12. The solution of Equation 11.8 and Equation 11.9 can be expressed in the general form:

$$\delta\vec{\varepsilon} = \vec{x} + \sum_{j=1}^{M} \lambda_j \vec{y}_j, \tag{11.13}$$

where $\vec{x}$ is the minimum norm solution of the system of linear Equation 11.8 and Equation 11.9 obtained from a singular value decomposition algorithm, $\{\vec{y}\}$ are the eigenvectors corresponding to the null eigenvalue of the matrix associated with the linear system, $\{\lambda\}$ is any set of real values, and $M = L - n$ is the dimensionality of the invariant space associated with the solutions of Equation 11.8 and Equation 11.9. An optimal solution $(\{\varepsilon^*\},\{\sigma^*\})$ can be found if the intersection of the invariant space and the solution of Equation 11.12 is not null. If this is the case, an optimal solution is defined by the values of $\{\lambda^*\}$ that satisfy the maximum energy gap criterion:

$$\Delta E(\{\lambda\}) = \max_{\{\lambda\}}[(\min_{i\in(1,\ldots,N_{\text{decoys}})} E_{\Gamma_i}(\varepsilon(\lambda))) - E_{\Gamma_{\text{nat}}}(\varepsilon(\lambda))]. \tag{11.14}$$

The parameter selection is implemented by solving Equation 11.8, Equation 11.9, and Equation 11.14 by means of a Monte-Carlo and simulated annealing algorithm [35].

11.3 APPLICATIONS

11.3.1 Folding of SH3 Protein

Protein SH3 is an ideal candidate to investigate the interplay between native and nonnative interactions in the protein-folding landscape. SH3 is a small protein domain consisting of two β-sheets orthogonally packed around a hydrophobic core. The β-sheets are joined by the RT, n-src, and distal loops [37]. Experimental studies [37–40] have established that SH3 folding and unfolding can be modeled by a two-state kinetics. Mutational studies on SH3 wild type [41] have shown that the formation of nonspecific hydrophobic interactions in the mutated protein results in the stabilization of the transition state, speeding up both folding and unfolding rates.

The DMC model (as presented in the Methods section), without including any experimental data to refine the search in the parameter space, was used to obtain several sets of parameters $(\{\varepsilon\},\{\sigma\})$ that allow the model to fold the src-SH3 sequence into its native structure. A first set of parameters, $(\{\varepsilon\},\{\sigma\})$, was obtained after five iterations. In each iteration several heat and quench simulations were performed to check if the parameter set would allow the protein model to fold correctly. In total 16 parameter sets were found that allowed the model to fold back to the native state from unfolded conformations. Not surprisingly, there is a high correlation between the different sets of parameters, with a maximum correlation coefficient of .77. The small deviation on the set of parameters $(\{\varepsilon\},\{\sigma\})$ is reflected in different agreement with experimental data. In the following we present the results obtained with the set of parameters that best reproduce the experimental Φ-values [42].

The free-energy surface obtained from simulation of the DMC model of SH3 has a single free-energy barrier separating the folded and unfolded states, in agreement with experimental findings. The amount of nonnative interactions is not negligible; nevertheless it does not introduce significant roughness in the landscape (see Figure 1 in Ref. [10]). The correlation between simulated and experimental [37,40] Φ-value is $r \sim .85$ (see Figure 3 in Ref. [17]) when the comparison is restricted to mutations that destabilize the protein more than $\Delta\Delta G \sim 4$ kJ/mol [43]. It has been shown that unfrustrated (Gō-like) models can qualitatively reproduce the folding landscape of a large set of proteins. However, it is worth mentioning that when using a plain Gō-like model the correlation between the simulated and experimental Φ-value drops to $R^2 \sim .45$. An intermediate agreement can be found when using a heterogeneous Gō-like model, defined by replacing all nonnative interactions by a hard-core repulsive interaction, and maintaining native interactions as defined in the DMC model. This result suggests that, at least for this protein, nonnative interactions play a role in shaping the folding landscape around the transition state. The inclusion of nonnative interactions affects the Φ-values in two ways: (i) it changes the probability of forming contacts not present in Gō-like models, at different stages of the folding process, and (ii) it changes the energetic balance among the interactions that involves a mutated residue. In agreement with theoretical predictions [44–46], both the energetic heterogeneity and the presence of nonnative energy are found to contribute to increasing the folding rate.

The role of nonnative interactions in the folding of SH3 is apparent from Figure 11.3, where the probability of contact formation of native and nonnative interaction is compared. A main cluster of nonnative contacts is formed at the transition state and involves the residues of the hydrophobic core. It is worth mentioning that the same cluster has been identified experimentally to be important for the folding of SH3 [41].

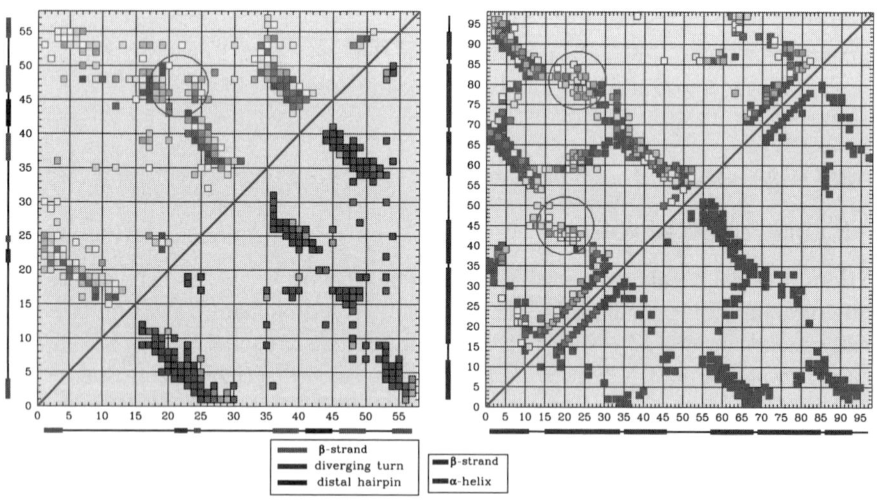

FIGURE 11.3 (See color insert following page 238.) The left figure shows the contact probabilities in the transition state of the folding of SH3 (upper contact map), and in the SH3 folded state (lower contact map). The right figure shows the distribution of contact probabilities in the folded state of S6^{Alz} (upper contact map). Different shades of blue are used for the native interactions, from white to deep blue; the contacts with higher probabilities are in deep blue, lower probabilities are in white. Different shades of red are used for the nonnative interactions, from white to deep red. The main clusters of nonnative interactions are circled in red. (Adapted from Das, P., Matysiak, S., and Clementi, C., *Proc. Natl. Acad. Sci. U.S.A.*, 102, 10141–46, 2005; Matysiak, S., and Clementi, C., *J. Mol. Biol.*, 363, 297–308, 2006.)

11.3.2 Folding of S6 Protein

Ribosomal protein S6 is a 97-residue protein, consisting of four β-sheets and two α-helices surrounding the hydrophobic core. $S6^{wt}$ and its two circular permutants ($S6^{p13\text{-}14}$ and $S6^{p68\text{-}69}$) are two-state folders. Recent experiments have shown that by circular permutation the folding landscape is changed from a diffuse transition state for $S6^{wt}$ to a structurally polarized one for $S6^{p13\text{-}14}$ [47]. Previous work [48] has shown that unfrustrated models can qualitatively reproduce the general features of this protein, although fine-tuning of the model parameters is needed for a quantitative comparison with experiments [23]. However, in order to study the detours on the landscape of protein S6 that are experimentally observed upon mutation of some key residues, or by changing experimental conditions [49–51], a coarse-grained protein model needs to capture the physicochemical diversity of the amino acids and the effect on nonnative interactions. In particular, by mutating all the charged residues into hydrophobic residues (EA41, EI42, RM46, and RV47), β-strand 2 of protein S6 becomes homologous to the Alzheimer's peptide (β-AP). When this set of mutations is performed in $S6^{wt}$ the mutated protein is referred to as $S6^{Alz}$, as it displays a complex reversible aggregation and forms soluble aggregates in its folded state [49]. Experimental findings suggest that oligomerization of $S6^{wt}$ is prevented by the electrostatic repulsion between residues in β-strand 2 from different copies of the same protein, while oligomerization is promoted in the mutated $S6^{Alz}$ protein by favorable interactions between hydrophobic residues [49,51].

The DMC model (described in the Methods section) was used to obtain a set of parameters $(\{\varepsilon\},\{\sigma\})$ able to fold protein $S6^{wt}$ and its circular permutants $S6^{p13\text{-}14}$ and $S6^{p68\text{-}69}$. Experimental data were then incorporated into the model to define a set of optimal parameters $(\{\varepsilon^*\},\{\sigma^*\})$ in the allowed region of the parameter space. The inverse statistical mechanical procedure illustrated in the Methods section was performed until the differences between a set of free-energy differences as obtained in simulations and in experiment were reduced within the corresponding error bars. Experimentally determined free-energy differences for $S6^{wt}$ and $S6^{p13\text{-}14}$ [52] were used in the training phase (i.e., to solve Equation 11.8 and Equation 11.9). The set of optimal parameters thus obtained yields a Pearson correlation coefficient between simulation and experimental data of $R^2 = .79$ (p-value $< 10^{-4}$) and a reduced chi-squared of $\chi^2 = .7$. As experimental data on $S6^{p68\text{-}69}$ were not incorporated into the model, the predictive power of the model equipped with the optimal set of parameters $(\{\varepsilon^*\},\{\sigma^*\})$ to realistically reproduce the folding landscape of S6 can be tested directly by performing a blind test on $S6^{p68\text{-}69}$ [52]. The free-energy differences obtained for $S6^{p68\text{-}69}$ correlate well with experimental data, with a Pearson correlation coefficient between simulation and experimental data of $R^2 = .88$ (p-value $< 10^{-4}$) and reduced chi-squared of $\chi^2 = .5$, comparable to those obtained for $S6^{wt}$ and $S6^{p13\text{-}14}$.

Figure 11.4a shows the projection of the folding free-energy landscape of S6 on the two-dimensional subspace spanned by the reaction coordinates Q and A, defined as the fraction of native and nonnative interactions, respectively. Native contacts are defined as the contacts between any pair of residues formed in the folded state with a probability higher than in the transition state. Contacts occurring with probability $\leq .01$ in the folded state are considered to be nonnative [44]. A single barrier separates the unfolded and folded state for $S6^{wt}$, $S6^{p13\text{-}14}$, and $S6^{p68\text{-}69}$. The number of nonnative interactions formed throughout the folding landscape is not negligible, although it does not introduce significant roughness into the landscape.

11.3.3 Modeling Perturbations on S6 Folding Landscape

As described above, in order to perform a mutation on a residue at position k in the protein sequence, the amino acid color a_k is changed into a different color a_k'. As the size of the side chain is reduced in a conservative mutation, new values for the equilibrium distances $\{\sigma\}$ for each interaction involving the mutated residue k are extracted from the distribution $P(\sigma; a_i, a_k, |i-k|)$. Using this definition, the four mutations EA41, EI42, RM46, and RV47 can be modeled within the DMC model to

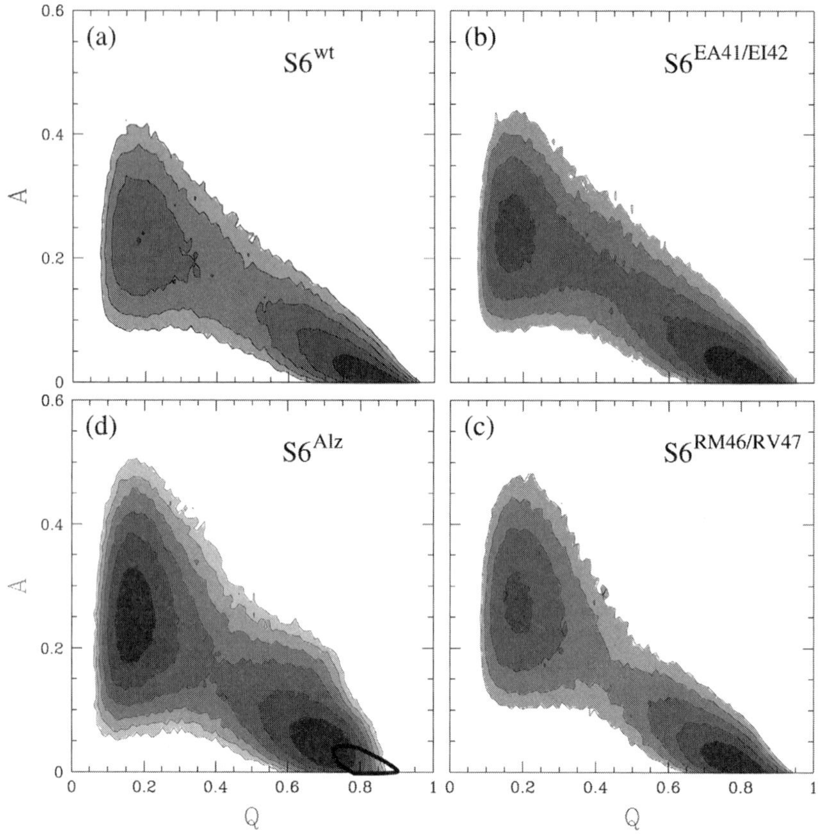

FIGURE 11.4 Free energy plots obtained used the optimal set of parameters (defined by incorporating experimental data) for S6. (a) corresponds to S6^{wt}, (b) to S6$^{EA41/42}$, (c) to S6$^{RM46/47}$, and (d) to S6^{Alz}.

characterize the folding of S6^{Alz}. Simulations at the folding temperature are performed to obtain the folding landscape of S6^{Alz} as shown in Figure 11.4d. The folding landscape associated with S6$^{EA41/EI42}$ and S6$^{RM46/RV47}$ is essentially the same as that of the wild-type protein (see Figure 11.4a–c). On the contrary, the location of the free-energy minimum corresponding to the S6^{Alz} native state is shifted with respect to the native state of S6^{wt}, indicating a difference in the equilibrium structure of the two proteins. This difference is mainly due to the increased flexibility of some regions of the S6^{Alz} native state, as shown in Figure 11.5. Enhanced flexibility is observed in the C-terminal region and around residue 50, locally disrupting the formation of secondary structure. In the native state of S6^{Alz} β-strands 1 and 3, and helix 2 remain structured, whereas the rest of the protein becomes very flexible. Mispacking of β-strand 2 and helix 1 is also observed. This prediction on the changes occurring in the native state of S6^{Alz} is strongly supported by the fact that the per residue deviation between representative conformations of the native state of S6^{wt} and S6^{Alz} obtained from simulations compares remarkably well with the per residue deviation between the corresponding crystal structures (correlation coefficient $R^2 = .86$, p-value $< 10^{-4}$), as shown in Figure 11.5. The sharp "twist and bend" region near residues 46–47 of S6^{wt} is completely destroyed in the native state of S6^{Alz}, in agreement with what is observed in the crystal structure. The small difference in rmsd ($\sim$ 1.5 Å, comparable to the experimental error $\sim$2.2 Å) between the native state representative structure of S6^{Alz} obtained from our simulations and the crystal structure of S6^{Alz} [49] further supports these results. As the changes in the native state of S6^{Alz} detected with the DMC model are fully consistent with all available experimental results, the model allows us to interpret the data in a broader perspective. In particular, the enhanced flexibility in S6^{Alz} may favor the formation of intermolecular contacts, which

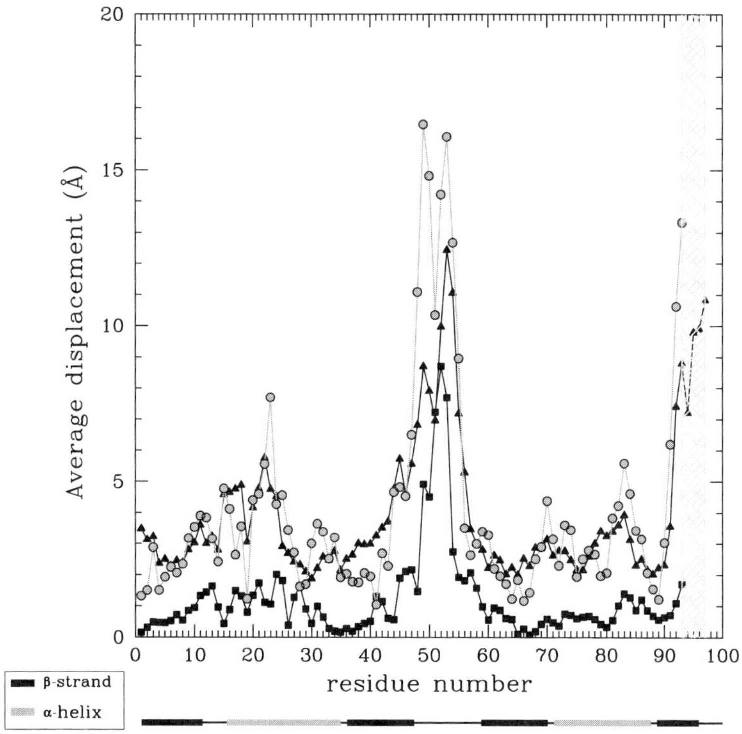

FIGURE 11.5 Average displacement per residue in S6^{Alz} native state as obtained from our study. The filled circles correspond to the per residue difference between the representative (the center of the most populated cluster) native structures of S6^{Alz} and S6^{wt}. Filled squares correspond to the per residue difference between the crystal structure of S6^{Alz} and S6^{wt}. The absence of coordinates in the regions 92–97 in the crystal structure of S6^{Alz} indicates a very large flexibility in this region in agreement with our predictions (shaded area). Triangles correspond to the average displacement per residue over all the conformations belonging to the S6^{Alz} native state. Our model correctly mimics the main conformational changes in S6 upon mutation (the correlation between the displacement between crystal structures and the one observed in our study is $R = 0.86$, $p\text{-value} < 10^{-4}$). (Adapted from Matysiak, S., and Clementi, C., *J. Mol. Biol.*, 363, 297–308, 2006.)

could in turn lead to aggregation. The sharp "twist and bend" region around residue 46–47 in S6^{wt} becomes the most flexible part of the protein in S6^{Alz}, and opens the hydrophobic core of the protein. Our results support the previously proposed hypothesis that a nonoptimal twist between β-strands may be a result of natural evolution to avoid edge-to-edge aggregation between several copies of the same protein [53]. Several misfolded traps are populated in S6^{Alz} in folding simulations at temperatures below the folding temperature [21], suggesting the "gatekeeping" role of the mutated residues, by steering the protein away from "dangerous" conformations.

Nonnative interactions play a major role in changing the folding landscape in S6^{Alz} upon mutation, as shown in Figure 11.3. Two new clusters of nonnative contacts are formed in the transition and native state of S6^{Alz}, indicating a high probability of forming β-sheet conformations different from those observed in S6^{wt}. Interestingly, the stretch of residues 38–53, which participates in the formation of a competing nonnative β-sheet conformation, is also the part of the protein involved in the formation of interprotein interactions in the tetrameric crystal structure of S6^{Alz} [49].

11.4 CONCLUSIONS

Significant progress has been made over the last decade in the definition of coarse-grained models for the characterization of protein-folding dynamics. In this context, we have developed the DMC model, which can be further enhanced when coupled with an inverse procedure to incorporate experimental data into the definition of the model parameters. The DMC model takes into account the amino acid modulation of a protein sequence, and does not a priori remove the contribution of nonnative interactions. The fact that the DMC model effectively takes into account the nonspherical shape of amino acids represents a crucial improvement over previous models. In order to define the optimal set of effective parameters that can best reproduce a set of physical observables, we have developed a procedure to incorporate experimental data into the model, to guide the search in the parameter space. Recent applications show that when the DMC model is dressed with this inverse procedure it can be used as a predictive tool to characterize protein misfolding and aggregation.

In this chapter we have described the essential ideas and technical implementations of the model and the results obtained for a few selected protein systems. The results obtained in these studies are in excellent agreement with experimental data and offer a clue to their interpretation.

The main drawback of the proposed coarse-grained methodology consists in the fact that the model is protein dependent. As the role of the solvent in the protein dynamics is taken into account only implicitly, by "renormalizing" it in the definition of intermolecular interactions, the model's effective parameters reflect the local environment of a given protein, and cannot be reliably used to study the folding of a completely different protein. As water mediates interactions between amino acids, and stabilizes protein structures locally and globally, in order to construct a more "general" protein model the inclusion of explicit water is paramount (at least at the coarse-grained level). Work along these lines is underway.

REFERENCES

1. Creighton, T. E. (1996) *Proteins: Structure and molecular properties.* New York: W. H. Freeman and Company.
2. Duan, Y., and Kollman, P. (1998) Pathways to a protein folding intermediate observed in a 1-microsecond simulation in aqueous solutions. *Science* 282, 740–43.
3. Fersht, A., and Daggett, V. (2002) Protein folding and unfolding at atomic resolution. *Cell* 108, 573–82.
4. Nymeyer, H., and García, A. (2003) Simulation of the folding equilibrium of alpha-helical peptides: A comparison of the Generalized-Born approximation with explicit solvent. *Proc. Natl. Acad. Sci. U.S.A.* 100, 13934–39.
5. Shea, J. E., Onuchic, J. N., and Brooks, C. L. (2002) Probing the folding free energy landscape of the src-SH3 protein domain. *Proc. Natl. Acad. Sci. U.S.A.* 99, 16064–68.
6. Snow, C., Nguyen, H., Pande, V., and Gruebele, M. (2002) Absolute comparison of simulated and experimental protein folding dynamics. *Nature* 420, 102–106.
7. Brown, S., Fawzi, N. J., and Head-Gordon, T. (2003) Coarse-grained sequences for protein folding and design. *Proc. Natl. Acad. Sci. U.S.A.* 100, 10712–17.
8. Clementi, C., García, A., and Onuchic, J. N. (2003) Interplay among tertiary contacts, secondary structure formation and side-chain packing in the protein folding mechanism: All-atom representation study of protein L. *J. Mol. Biol.* 326, 933–54.
9. Clementi, C., Jennings, P., and Onuchic, J. N. (2001) Prediction of folding mechanism for circular-permuted proteins. *J. Mol. Biol.* 311, 879–90.
10. Clementi, C., Jennings, P., and Onuchic, J. N. (2000) How native-state topology affects the folding of dihydrofolate reductase and interleukin-1b. *Proc. Natl. Acad. Sci. U.S.A.* 97, 5871–76.
11. Clementi, C., Nymeyer, H., and Onuchic, J. N. (2000) Topological and energetic factors: What determines the structural details of the transition state ensemble and en-route intermediates for protein folding? An investigation for small globular proteins. *J. Mol. Biol.* 298, 937–53.
12. Honeycutt, J. D., and Thirumalai, D. (1990) Metastability of the folded states of globular proteins. *Proc. Natl. Acad. Sci. U.S.A.* 87, 3526–29.
13. Koga, M., and Takada, S. (2001) Roles of native topology and chain-length scaling in protein folding: A simulation study using Gō-like model. *J. Mol. Biol.* 313, 171–80.

14. Levy, Y., Wolynes, P., and Onuchic, J. N. (2004) Protein topology determines binding mechanism. *Proc. Natl. Acad. Sci. U.S.A.* 101, 511–16.
15. Takada, S. (1999) Gō-ing for the prediction of protein folding mechanism. *Proc. Natl. Acad. Sci. U.S.A.* 96, 11698–700.
16. Onuchic, J. N., Luthey-Schulten, Z., and Wolynes, P. G. (1997) Theory of protein folding: The energy landscape perspective. *Annu. Rev. Phys. Chem.* 48, 545–600.
17. Das, P., Matysiak, S., and Clementi, C. (2005) Balancing energy and entropy: A minimalist model for the characterization of protein folding landscapes. *Proc. Natl. Acad. Sci. U.S.A.* 102, 10141–46.
18. Hubner, I., Oliveberg, M., and Shakhnovich, E. (2004) Simulation, experiment, and evolution: Understanding nucleation in protein S6 folding. *Proc. Natl. Acad. Sci. U.S.A.* 101, 8354–59.
19. Karanicolas, J., and Brooks III, C. (2003) Improved Gō-like models demonstrate the robustness of protein folding mechanisms towards non-native interactions. *J. Mol. Biol.* 334, 309–25.
20. Karanicolas, J., and Brooks III, C. (2002) The origins of asymmetry in the folding transition states of protein L and protein G. *Proteins* 11, 2351–61.
21. Matysiak, S., and Clementi, C. (2006) Minimalist protein model as a diagnostic tool for misfolding and aggregation. *J. Mol. Biol.* 363, 297–308.
22. Heath, A. P., Kavraki, L. E., and Clementi, C. (2007) From coarse-grain to all-atom: Toward multiscale analysis of protein landscapes. *Proteins* 68, 646–61.
23. Matysiak, S., and Clementi, C. (2004) Optimal combination of theory and experiment for the characterization of the folding landscape of S6: How far a minimalist model go? *J. Mol. Biol.* 343, 235–48.
24. Clementi, C., Maritan, A., and Banavar, J. R. (1998) Folding, design, and determination of interaction potentials using off-lattice dynamics of model heteropolymers. *Phys. Rev. Lett.* 81, 3287–90.
25. Minry, L., and Shakhnovich, E. I. (1996) How to derive a protein folding potential? A new approach to an old problem. *J. Mol. Biol.* 264, 1164–79.
26. Van Mourik, J., Clementi, C., Maritan, A., Seno, F., and Banavar, J. (1998) Determination of interaction potentials of amino acids from native protein structures: Tests on simple lattice models. *J. Chem. Phys.* 110, 10123–33.
27. Vendruscolo, M., Minry, L. A., Shakhnovich, E. I., and Domany, E. (2000) Comparison of two optimization methods to derive energy parameters for protein folding: Perceptron and Z score. *Proteins* 41, 192–201.
28. Harpaz, Y., Gerstein, M., and Chothia, C. (1994) Volume changes on protein folding. *Structure* 2, 641–49.
29. Liang, J., and Dill, K. A. (2001) Are proteins well-packed? *Biophys. J.* 81, 751–66.
30. Richards, F. (1977) Areas, volumes, packing, and protein structures. *Annu. Rev. Biophys. Bioeng.* 6, 151–76.
31. Tsai, J., Taylor, R., Chothia, C., and Gerstein, M. (1999) The packing density in proteins: Standard radii and volumes. *J. Mol. Biol.* 2, 253–66.
32. Eriksson, A., Baase, W., Zhang, X., Heinz, D., Blader, M., Baldwin, E., and Matthews, B. (1992) Response of a protein structure to cavity-creating mutations and its relation to the hydrophobic effect. *Science* 255, 178–83.
33. Lim, W. A., Farruggio, D., and Sauer, R. T. (1992) Structural and energetic consequences of disruptive mutations in a protein core. *Biochemistry* 31, 4324–33.
34. Maritan, A., Micheletti, C., Trovato, A., and Banavar, J. R. (2000) Optimal shapes of compact strings. *Nature* 406, 287–90.
35. Kirkpatrick, S., Gelatt, C. D., and Vecchi, M. P. (1983) Optimization by simulated annealing. *Science* 220, 671–80.
36. Krauth, W., and Mezard, M. (1987) Learning algorithms with optimal stability in neural networks. *J. Phys. A* 20, L745–52.
37. Grantcharova, V., Riddle, D., Santiago, J., and Baker, D. (1998) Important role of hydrogen bonds in the structurally polarized transition state for the folding of the src-SH3 domain. *Nat. Struct. Biol.* 5, 714–20.
38. Grantcharova, V., and Baker, D. (1997) Folding dynamics of the src-SH3 domain. *Biochemistry* 36, 15685–92.
39. Martinez, J. C., Pisabarro, T., and Serrano, L. (1998) Obligatory steps in protein folding and the conformational diversity of the transition state. *Nat. Struct. Biol.* 5, 721–29.
40. Riddle, D. S., Grantcharova, V. P., Santiago, J. V., Anl, E., Ruczinski, I., and Baker, D. (1999) Experiments and theory highlight role of native state topology in SH3 folding. *Nat. Struct. Biol.* 6, 1016–24.

41. Cobos, E. S., Filimonov, V. V., Vega, M. C., Mateo, P. L., Serrano, L., and Martinez, J. C. (2003) A thermodynamic and kinetic analysis of the folding pathway of an SH3 domain entropically stabilised by a redesigned hydrophobic core. *J. Mol. Biol.* 328, 221–33.
42. Fersht, A. R., Leatherbarrow, R. J., and Wells, T. N. C. (1986) Quantitative analysis of structural activity relationship in engineered proteins by linear free energy relationships. *Nature* 322, 284–86.
43. De Los Rios, M. A., Muralidhara, B. K., Wildes, D., Sosnick, T. R., Marqusee, S., Wittung-Stafshede, P., Plaxco, K. W., and Ruczinski, I. (2006) On the precision of experimentally determined protein folding rates and phi-values. *Protein Sci.* 15, 553–63.
44. Clementi, C., and Plotkin, S. S. (2004) The effects of nonnative interactions on protein folding rates: Theory and simulation. *Protein Sci.* 13, 1750–66.
45. Plotkin, S. S., and Onuchic, J. N. (2000) Investigation of routes and funnels in protein folding by free energy functional methods. *Proc. Natl. Acad. Sci. U.S.A.* 97, 6509–14.
46. Plotkin, S. S., and Onuchic, J. N. (2002) Structural and energetic heterogeneity in protein folding. I. Theory. *J. Chem. Phys.* 116, 5263–83.
47. Lindberg, M., Tangrot, J., and Oliveberg, M. (2002) Complete change of the protein folding transition state upon circular permutation. *Nat. Struct. Biol.* 9, 818–22.
48. Stoycheva, A. D., Brooks III, C. L., and Onuchic, J. N. (2004) Gatekeepers in the ribosomal protein S6: Thermodynamics, kinetics and folding pathways revealed by a minimalist protein model. *J. Mol. Biol.* 340, 571–85.
49. Otzen, D., Kristensen, O., and Oliveberg, M. (2000) Designed protein tetramer zipped together with a hydrophobic Alzheimer topology: A structural clue to amyloid assembly. *Proc. Natl. Acad. Sci. U.S.A.* 97, 9907–12.
50. Otzen, D., and Oliveberg, M. (1999) Salt induced detour through compact regions of the protein folding landscape. *Proc. Natl. Acad. Sci. U.S.A.* 96, 11746–51.
51. Pedersen, J. S., Christensen, G., and Otzen, D. E. (2004) Modulation of S6 fibrillation by unfolding rates and gatekeeper residues. *J. Mol. Biol.* 341, 575–88.
52. The experimental data for this test have been kindly supplied by Mikael Oliveberg and Magnus Lindberg, Stockholm University.
53. Richardson, J. S., and Richardson, D. C. (2002) Natural b-sheet proteins use negative design to avoid edge-to-edge aggregation. *Proc. Natl. Acad. Sci. U.S.A.* 99, 2754–59.

12 Principles and Practicalities of Canonical Mixed-Resolution Sampling of Biomolecules

Daniel M. Zuckerman
Department of Computational Biology, University of Pittsburgh School of Medicine

CONTENTS

12.1 INTRODUCTION

Although the idea of "refining" coarse-grained (CG) models of proteins into atomically detailed structures is over 30 years old [1], only recently have mixed-resolution algorithms been developed which can perform the task rigorously—i.e., which in principle can lead to equilibrium ensembles of proteins distributed according to the Boltzmann factor of an atomistic force field [2, 3]. Here we will review the basic ideas underlying such mixed-resolution sampling of atomistic force fields, by presenting a series of algorithms including "Resolution Exchange" (ResEx). In addition to the underlying statistical principles, we will also discuss the limitations and potential of various algorithmic variants. The chapter will focus on our own group's work, but mixed-resolution ideas from other groups are certainly of interest (e.g., [4–9]).

Mixed-resolution sampling is intended to address the famous observation by Frauenfelder et al. [10] that proteins exhibit (presumably functional) behaviors on a wide range of length and time scales. If fast-running CG models can sample conformational fluctuations that normally occur on slow time scales, then the addition of atomic detail can be used to sample faster motions. A key point in mixed-resolution sampling is that any CG model will be biased with respect to the desired atomic force field (i.e., it will exhibit differing state populations), and hence rigorous algorithms must correct for this bias.

Although the ResEx algorithms indeed correct for bias introduced by (in principle, arbitrary) CG models, key practical issues remain to be addressed: Does the ensemble of the CG model possess sufficient overlap with the atomistic ensemble? What are the intrinsic and limiting time scales associated with sampling the most coarse-grained model? And what are the intrinsic time scales (if any) associated with the addition of atomic detail? Below, we will find "discouraging" answers to some of these questions for the original ResEx protocol, and therefore alternatives will be proposed. The new approaches appear to be promising for true equilibrium sampling of implicitly solvated proteins.

It is worth stressing that the algorithms discussed here in no way restrict the user's ability to employ elevated temperature, if desired. That is, models that are both CG and simulated at high T are fully accommodated. In fact, such a mixed approach has already been demonstrated [3]; see Figure 12.6.

12.2 ELEMENTS OF THE PROBLEM

12.2.1 Notation and Statement of the Problem

All the algorithms described here rely on a division of the full set of configurational coordinates, denoted $\mathbf{r}$, into two or more subsets. The primary division is between a subset Φ, which constitutes the coordinates of the coarse model, and the remainder, denoted $\mathbf{x}$. Thus we write

$$\mathbf{r} = \{\Phi, \mathbf{x}\}. \tag{12.1}$$

For concreteness, Φ may be viewed as representing the backbone degrees of freedom and $\mathbf{x}$ as the side-chain coordinates. Although these choices are not required for the algorithms in a mathematical sense, we will usually employ them to keep the discussion concrete. In published work, however, we have studied united-atom and all-atom models, in which case the $\mathbf{x}$ coordinates have represented nonpolar hydrogen atoms, while Φ gave united-atom coordinates [2,3].

We will also consider further subdivisions of the $\mathbf{x}$ coordinates—in our case, into sets describing individual side-chains. Thus, for a protein with M residues, we will sometimes write

$$\mathbf{x} = \{\mathbf{x}_1, \mathbf{x}_2, \ldots, \mathbf{x}_M\}, \tag{12.2}$$

and we will also employ coordinate sets intermediate between the most detailed and the most coarse by writing

$$\mathbf{r}_j = \{\Phi, \mathbf{x}_1, \mathbf{x}_2, \ldots, \mathbf{x}_j\}. \tag{12.3}$$

Thus, for example, $\mathbf{r}_4$ could represent the coordinates of the backbone plus the first four side chains.

The goal of all the algorithms considered here is to sample the equilibrium distribution of the full set of coordinates at a specified temperature T—i.e., to sample all-atom configurations $\mathbf{r}$ distributed according to

$$\pi(\mathbf{r}) \propto \exp[-\beta U(\mathbf{r})], \tag{12.4}$$

where $\beta = 1/k_B T$ and U is the potential energy for an atomically detailed biomolecule—presumably a standard force field. (Note that normalization by the partition function will not be required in any of the proposed algorithms, but only ratios of probabilities.)

Auxiliary force fields will also be employed to sample intermediate coordinate sets $\mathbf{r}_j$ with the notation

$$\pi_j(\mathbf{r}_j) \propto \exp[-\beta U_j(\mathbf{r}_j)] \quad \text{and} \quad \pi_0(\mathbf{r}_0) \propto \exp[-\beta U_0(\Phi)]. \tag{12.5}$$

Note that the intermediate force fields U_j (for $j < M$) do not include all atoms, and hence are *not* available in standard software packages. Here we will comment briefly on some design considerations for developing such models—a process which turns out to be less difficult than might be anticipated.

Why are such intermediate models needed? In essence, the outline of many of the algorithms described below is to sample distributions of increasingly larger coordinate sets until the full set is reached: $\pi_0, \pi_1, \pi_2, \ldots, \pi_M = \pi$. This type of strategy is reminiscent of approaches described in Ref. [11]. In our case, the procedure can be viewed as starting with just the backbone (Φ), then adding one side chain ($\mathbf{x}_1$), and so on. Our initial studies did, however, explore the varying "traditional" resolution—i.e., switching between united-atom and all-atom models [2,3].

Finally, one of the algorithms will also require ensembles of individual side-chains $\mathbf{x}_i$, and the corresponding distributions will be denoted by p_i. We will write

$$p_i(\mathbf{x}_i) \propto \exp[-\beta U(\mathbf{x}_i)], \tag{12.6}$$

where, as in Equation 12.4—but unlike Equation 12.5—U refers to a *standard* all-atom force field. The only difference from Equation 12.4 is that only a small set of coordinates are considered, but their full interactions are preserved.

12.2.2 Intermediate "Resolution" Models and Model-Design Considerations

What are the specific models employed? In our most current approach, the "coarsest" model π_0 is an *atomistic* backbone model (an Ala-Gly-Pro heteropolymer), which includes appropriate Ramachandran propensities and backbone hydrogen-bonding. Additional interactions (hydrophobic and Go-like, based on beta carbons when present) are included to maintain stability in the folded state. We have built highly optimized software employing pregenerated libraries of atomistic peptide-plane configurations in order to ensure that full sampling can be obtained in this critical coarsest level of modeling, as will be reported shortly [12]. If this coarsest level cannot be fully sampled, after all, none of the more detailed levels will be converged.

The subsequent models π_1, π_2, ... include full atomistic descriptions for a limited number of side chains (up to the subscripted index), and $\pi_M = \pi$ includes all side chains and hence all atoms of the full peptide or protein. The key question becomes: How do atomistic side chains interact with coarsely modeled side chains possessing only a beta-carbon (C_β) interaction center? The simplest approach, which we are implementing, is to employ the coarse (C_β–C_β) interactions in such cases since beta carbons are present in both species. Once a given pair of side chains is converted to an atomistic model, then the atomistic interactions will apply.

Our strategy may seem surprisingly atomistic on the whole. The reason lies with the algorithms described below, which will be most successful when "overlap" between neighboring model levels is maximal. That is, for the coordinates held in common, it is optimal for the distributions between neighboring levels (π_j and π_{j-1}) to be as similar as possible. Although, in principle, the algorithms will work with arbitrary models, we expect their efficiency will be substantially degraded if the overlap between neighboring levels is poor. Put another way, additional intermediate levels might be required.

12.3 ALGORITHMS AND CRITIQUES

We will now critically review a series of mixed-resolution algorithms we have developed. The "critical" aspect will focus on the likelihood of a given algorithm achieving true equilibrium sampling for proteins. By examining the time scales involved, as well as correlation issues, one can make educated guesses as to the limitations of proposed methods.

A key caveat should be mentioned at the outset: all the approaches are described and critiqued in the context of *implicit solvent* simulation. We emphasize that the ability to perform fully sampled implicit solvent simulations of proteins would be a significant advance. Furthermore, while such approximations are far from perfect, there is evidence that implicit solvent models are improving (e.g., [13–16]). In Section 12.6, we discuss one possible avenue for extending our work to explicit solvent simulation.

12.3.1 Simple Resolution Switching

The simplest rigorous algorithm employing mixed resolutions considers two resolutions levels, high ("M") and low ("0"), governed by $\pi_M(\Phi, \mathbf{x})$ and $\pi_0(\Phi)$. (The subscripts here correspond, respectively, to "H" and "L" in Ref. [2].) In essence, the idea of the algorithm is to use the coarse distribution to generate trial moves solely for the coarse subset of variables, Φ. Note that the two distributions can employ unrelated temperatures. To our knowledge, the algorithm has not been published.

The algorithm can quickly be derived by reviewing the general Metropolis-Hastings algorithm (e.g., [11]); see also related algorithms [17, 18]. As usual, one starts from the equation for detailed balance, $\pi(i)T_{ij} = \pi(j)T_{ji}$, written in shorthand where "i" and "j" represent configurations in the appropriate space. Here π and T are the equilibrium and transition probabilities, respectively. Typically, the transition probability is divided into two parts, the generating probability g and acceptance probability α, so that $T_{ij} = g_{ij}\alpha_{ij}$. One can immediately derive two key results. First, one has the standard acceptance criterion for a trial move from i to j generated by g, namely, $\min[1, R]$, where

$$R \equiv \alpha_{ij} / \alpha_{ji} = \pi(j) g_{ji} / \pi(i) g_{ij}. \tag{12.7}$$

One can also derive the simpler ratio of the overall transition probabilities

$$T_{ij} / T_{ji} = \pi(j) / \pi(i). \tag{12.8}$$

Based on these two equations, we can readily derive an acceptance criterion for the case when a trial move for the full set of coordinates $\mathbf{r} = (\Phi, \mathbf{x})$ is generated by "ordinary" simulation (e.g., Monte Carlo) on the subset Φ based on the potential corresponding to $\pi_0(\Phi)$. That is, the full procedure is to alternate canonical simulation on the full set $\mathbf{r}$ with trial moves consisting of canonical simulation for the subset Φ based on π_0, as sketched in Figure 12.1.

Since the simulation segments employing the full set of coordinates can use any canonical algorithm, we need only specify the acceptance criterion for the Φ moves. In detail, such moves correspond to a trial move from the "old" configuration $\mathbf{r}_{\text{old}} = (\Phi_{\text{old}}, \mathbf{x}_{\text{old}})$ to the trial state $\mathbf{r}_{\text{try}} = (\Phi_{\text{new}}, \mathbf{x}_{\text{old}})$, where it should be noted that the $\mathbf{x}$ coordinates do not change. The key ratio R in the acceptance criterion is itself the product of two ratios. The first is simply the ratio of equilibrium probabilities, $\pi_M(\mathbf{r}_{\text{try}})/\pi_M(\mathbf{r}_{\text{old}})$. The second is the ratio of generating probabilities. The key point here is to recognize that the generating probability for the trial move $\mathbf{r}_{\text{old}} \to \mathbf{r}_{\text{try}}$ is the *full transition probability for the CG simulation trajectory segment* $\Phi_{\text{old}} \to \Phi_{\text{try}}$. This reflects that the CG simulation is simply an ordinary simulation under the potential corresponding to π_0—the whole sequence of trajectory steps has *already been accepted.* Thus, when considered as a trial move in the full

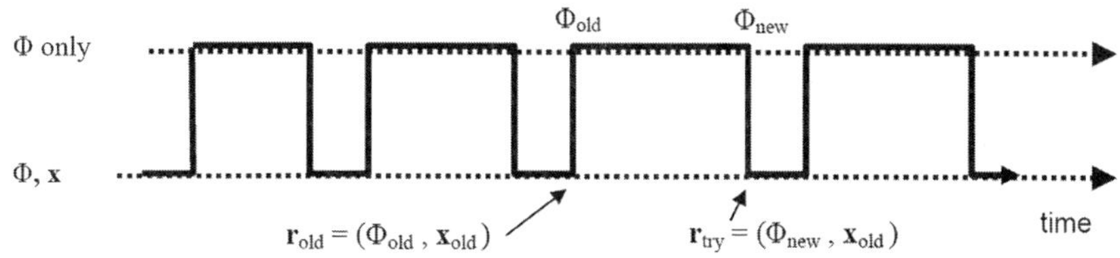

FIGURE 12.1 The resolution switching algorithm, depicted schematically. The full set of coordinates (Φ, **x**) is simulated only for brief "relaxational" segments, while the coarse trial moves (Φ only) are the primary "engine" for exploring conformational space. Coordinates are shown for a trial move governed by Equation 12.9.

space, the ratio of generating probabilities is $g_{ji}/g_{ij} = T_{ji}(\Phi_j \rightarrow \Phi_i)/T_{ij}(\Phi_i \rightarrow \Phi_j) = \pi_0(\Phi i)/\pi_0(\Phi_j)$, where we have used Equation 12.8 above for the ratio of T values, and set "old" $= i$ and "try" $= j$ for the sake of compactness.

Putting together all these ingredients, the final acceptance criterion for a trial move generated by canonical simulation of a subset of coordinates (from $\Phi_{old} \rightarrow \Phi_{try}$ governed by π_0) is min[1, R*], with

$$R^* = [\pi_M(\mathbf{r}_{try})/\pi_M(\mathbf{r}_{old})]/[\pi_M(\Phi_{try})/\pi_M(\Phi_{old})]. \tag{12.9}$$

In words, this is the standard ratio of equilibrium probabilities for the full distribution π_M, corrected by any bias in the CG model for configuration Φ_{try} in preference to Φ_{old}. Note that temperature is included implicitly in the distributions {π}, which are proportional to Boltzmann factors. The procedure just described can employ two different temperatures for levels "*M*" and "0".

Practicality analysis. In practice, this simplest of resolution-switching procedures can easily suffer from the well-known "overlap" problem: if the "M" and "0" models are too different, Φ configurations generated by π_0 simulation will rarely be important for the full "M" system. Quantitatively, this means that high-energy full configurations will tend to be generated and these will be rejected. The next algorithm allows for this problem to be corrected via a ladder of models intermediate between the all-atom and most coarse levels.

12.3.2 Parallel Resolution Exchange

The ResEx algorithm [2] was apparently the first rigorous algorithm exploiting a CG model to enhance sampling in an atomistic model. The algorithm was constructed in analogy to the replica exchange algorithm [19,20] wherein a ladder of simulations performed at different temperatures are coupled via occasional exchange attempts. In standard replica exchange, an exchange of the complete set of coordinates between two ladder levels is accepted or rejected via a Metropolis criterion (see Figure 12.2).

The ResEx approach is readily understood following a straightforward discussion of the general exchange idea. Consider a pair of simulations ("a" and "b") running in parallel (i.e., independently) generating trajectories $\mathbf{r}_a(t)$ and $\mathbf{r}_b(t)$ distributed according to π_a and π_b. If the two simulations are considered to constitute an extended system with configurations $\mathbf{R} = (\mathbf{r}_a, \mathbf{r}_b)$, then the independence implies the distribution takes a simple product form: $\pi_{tot}(\mathbf{R}) = \pi_a(\mathbf{r}_a) \cdot \pi_b(\mathbf{r}_b)$.

We can now employ the Metropolis logic to construct an exchange simulation which will generate this product-form distribution—i.e., to construct a trial move and acceptance criterion which will generate $\pi_{tot}(\mathbf{R})$. Thus, in addition to performing ordinary canonical simulation for each system, we can also employ swapping trial moves $(\mathbf{r}_a, \mathbf{r}_b) \leftrightarrow (\mathbf{r}_b, \mathbf{r}_a)$. Since the generating probability is symmetric for such a move (see Ref. [3]), the g factors will cancel in the acceptance criterion

(Equation 12.7). Hence, only the equilibrium distributions enter the exchange criterion, and one finds that such a swap should be accepted according to min[1, R_{ex}], with

$$R_{ex} = \pi_a(\mathbf{r}_b) \cdot \pi_a(\mathbf{r}_b) / \pi_a(\mathbf{r}_a) \cdot \pi_b(\mathbf{r}_b). \tag{12.10}$$

This notation (following Ref. [20]) makes clear that the approach could readily be generalized to systems with different Hamiltonians, as has already been implemented [21].

To construct a procedure for "resolution exchange," note that the preceding derivation made no assumptions about the total numbers of degrees of freedom in the two distributions—but only assumed implicitly that the swapped coordinates were of commensurate dimensionality. We can therefore immediately generalize the above to cover swaps $\Phi_a \leftrightarrow \Phi_b$ to determine an acceptance criterion with

$$R_{res\text{-}ex} = \pi_M(\Phi_b, \mathbf{x}_a) \cdot \pi_0(\Phi_a) / \pi_M(\Phi_b, \mathbf{x}_a) \cdot \pi_0(\Phi_b), \tag{12.11}$$

which is equivalent to the "unidirectional" trial move $\{(\Phi_a, \mathbf{x}_a), \Phi_b\} \rightarrow \{(\Phi_b, \mathbf{x}_a), \Phi_a\}$ for the extended system. Here $\mathbf{x}_a$ and $\mathbf{x}_b$ represent specific values of the coordinate set $\mathbf{x}$, which is arbitrary and does not refer to a particular side chain.

In exact analogy to replica exchange, a ResEx simulation can be performed with a series of ladder levels, as shown in Figure 12.2 constructed to ensure sufficient overlaps for successful exchange. Therefore, the criterion for exchange of a pair of simulations—Equation 12.11—can be used to construct a full ladder.

Practicality analysis. How would overlapping models $\{0, 1, 2, \ldots, M\}$ actually be constructed? Our first paper on the subject, which employed exchange between united-atom and all-atom models of a peptide, envisioned a series of models embodying a level of resolution uniform throughout the protein (e.g., backbone, backbone + beta carbon, ...). However, such an approach would appear difficult to implement. Therefore, we suggested and tested an "incremental (de)coarsening" strategy [3], as described in Section 12.2.2 for a protein, level "0" could embody solely backbone degrees of freedom with subsequent models reflecting the number of side chains included. The intermediate models would therefore include regular force field interactions between all groups included atomistically—i.e., the

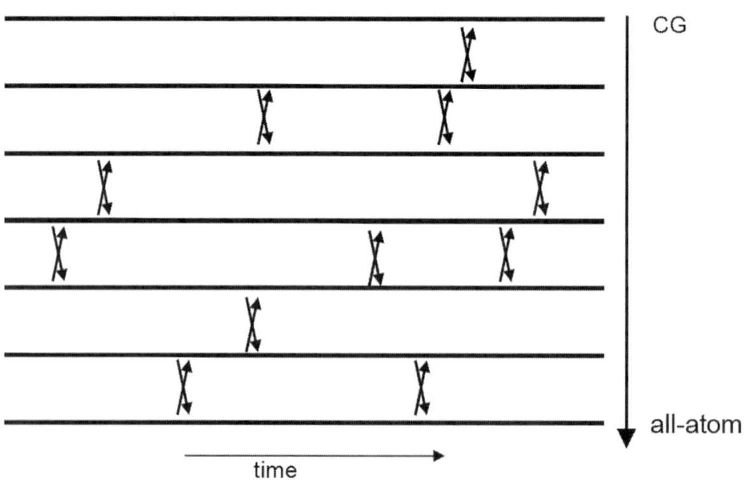

FIGURE 12.2 Schematic representation of the resolution exchange algorithm. Exchanges among different levels of resolution are governed by a Metropolis criterion which preserves canonical distributions at all levels. The vertical arrow indicates increasing resolution, embodied in the increasing number of coordinates. The top level could include just the backbone of a protein, with subsequent lower levels representing the addition of a single side-chain until all atoms have been added.

backbone, modeled as an Ala-Gly-Pro heteropolymer, along with side chains—in addition to auxiliary interactions used to model the missing side chains such as hydrophobicity or Go-like attractions.

Even with a fully exchangeable ladder, the ResEx approach suffers from at least three weaknesses: the "lingering" time scales throughout the ladder, a suboptimal expenditure of computing resources among ladder levels, and the possibility of confounding correlations between coarse and fine coordinates.

(i) *Lingering time scales.* Imagine that the simulation is long enough so that the top level (coarsest model) can be fully sampled. This in itself does not guarantee that full sampling has occurred throughout the ladder. For example, a very coarse model of a protein's backbone could fully sample large-scale fluctuations, but certain rotameric transitions may be very difficult to sample and may "linger" unresolved.

(ii) *Non-optimal expenditure of computing resources.* This is related to the previous concern, and in fact, is common to any parallel exchange protocol: our protocol does not automatically allot resources optimally among ladder levels. A typical parallel exchange simulation allots equal CPU time to each level, but some levels may be faster to converge than others. Thus, obtaining full sampling may require considerable waste of resources.

(iii) *Confounding correlations.* This worry is particular to ResEx (and not temperature-based replica exchange). Certain correlations may be so "confounding" as to prevent sampling of states other than the initial one: for instance, backbone Φ and side-chain $\mathbf{x}$ coordinates could be so correlated that no trial Φ coordinates will be found which are compatible with (i.e., yield low energy when paired with) existing detailed coordinates $\mathbf{x}$. Thus trial moves to other states will be rejected even though such states may be important.

Apparent solutions to these problems are presented in the algorithms below. We believe the question of optimal expenditure of computing resources is amply addressed by the top-down approaches described next. Interestingly, both lingering time scales and confounding correlations are overcome in principle by the "combinatorial decorating" procedure (Section 12.3.4).

12.3.3 Serial/Top-Down Resolution Exchange

Why is a top-down procedure, such as that depicted in Figure 12.3, expected to be important? Recall that the motivation for a ladder of simulations (whether based on temperature, Hamiltonian, or resolution) is the ability of higher levels (e.g., low resolution) to fully and rapidly sample large and slow fluctuations. After all, the lowest level—the target force field and temperature embodied in

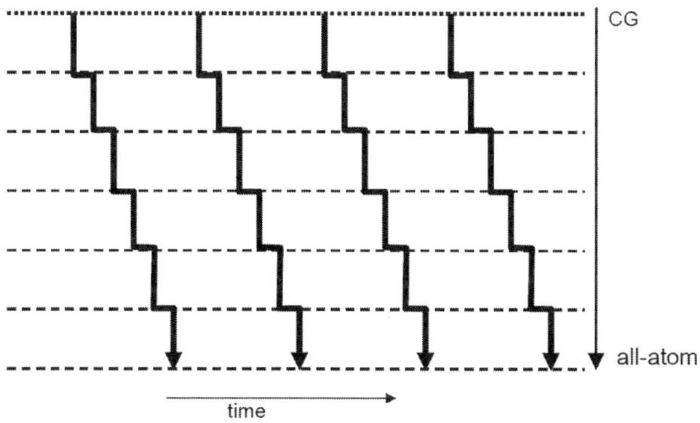

FIGURE 12.3 Schematic depiction of top-down resolution exchange, which avoids excessive simulation at the expensive lower ladder levels. In this protocol, the most coarse-grained model (top level) is fully sampled first. Subsequently, configurations are picked from the top level and atomic detail is added incrementally using a "pseudoexchange" process based on the "J-walking" idea [22]. Canonical sampling is maintained throughout all levels.

Equation 12.4—cannot run nearly fast enough for global sampling. The higher levels are used for sampling large-scale configurational motions, while the low levels can be expected to perform high-quality local sampling.

This "division of labor" between large-scale and local sampling suggests a reexamination of the parallel approach, which in typical cases enforces equal CPU usage at every level. However, it is far from obvious that global and local sampling will take equal amounts of wall-clock time. In many cases, local sampling may be much faster. Ideally, only as much CPU time would be spent at each level as is necessary.

A general strategy for overcoming this hurdle is provided by a top-down approach, which was first advocated by Doll and coworkers in their "J-walking" protocol [22]. For readers familiar with replica exchange, the procedure can be cast in terms of the "pseudo-exchanges" we suggested for ResEx [2,3]. The procedure is as follows: first perform a long simulation fully sampling the top level (e.g., lowest resolution or highest temperature). Then a simulation is begun at the next level down, with pseudoexchange attempts made with *randomly chosen* configurations from the preceding (higher) level. This is repeated until the target level is reached (see Figure 12.3). The use of randomly selected configurations is critical: it provides immediate access to the whole space sampled in the previous trajectory regardless of correlation time scales in the previous level.

Pseudoexchanges were statistically justified in our previous work [2,3], but a simple argument can be given. First, note that the derivations of the exchange acceptance criteria—Equation 12.10 and Equation 12.11—did not assume equivalent numbers of elapsed simulation steps between the two levels attempting a swap. We are therefore free to imagine that the CPU running the higher level is much faster than the lower one. In fact, we can imagine it is so fast that configurations considered in subsequent exchange attempts are completely decorrelated—i.e., it is as if the higher-level configurations (e.g., Φ) are generated independently from the governing distribution (e.g., π_0). This corresponds exactly to choosing configurations randomly from a well-sampled trajectory, and employing the identical acceptance criterion.

The pseudoexchange process would appear to permit a natural allotment of CPU time among levels, if one requires a fixed number of successful pseudoexchanges at each level. Since presumably it is the exchanges which generate statistically independent configurations (in a top-down protocol), such a protocol naturally allots a suitable amount of CPU time to each level.

Practicality analysis. What about parallelization? Our point of view is that every algorithm is intrinsically embarrassingly parallelizable: if a simulation is truly sampling the desired distribution, then multiple independent runs can be set up. Since no exchange-based protocol preserves dynamics, there is no concern that the trajectories are not continuous. Of course, even in the top-down procedure, the top level must be fully sampled, but there is no reason why this initial simulation could not be performed in parallel using multiple CPUs.

Of the three intrinsic problems facing parallel ResEx, as discussed above, two remain even with a top-down procedure: lingering time scales and the possibility of confounding correlations. These are addressed in the decorating procedure.

12.3.4 Library-Based Combinatorial Decorating

The final algorithm we describe appears to overcome all the limitations of the previous approaches, at least in the context of implicit-solvent study of biomolecules. Except for the top-level initial simulation of the $\pi_0(\Phi)$ system (which may be performed in parallel), the approach is completely nondynamical and hence the possibility for confounding correlations (Section 12.3.2) is minimal. Similarly, the use of pregenerated libraries removes lingering time scales (e.g., in side-chain motions; Section 12.3.2) by construction. The "decorating" approach was only sketched briefly in Ref. [3] and is therefore described here in more detail.

Our proposed decorating procedure is combinatorial in nature, based on older build-up ideas in polymer sampling [23–26]. Put simply, after sampling the backbone of a protein, side chains will be added one at a time by a combinatorial and reweighting procedure [3]. That is, the polymer (protein) will be built up in a branching process (side-chain addition) (see Figure 12.4). A new aspect described here is the pregeneration of truly statistical rotamer libraries based on Equation 12.6, which can be used for reweighting. Because there are only 18 amino-acid side chains (glycine and proline are part of the backbone in our protocol), such libraries will be relatively inexpensive to construct. The use of pregenerated libraries simultaneously eliminates both "lingering" dynamical time scales for rotameric transitions and correlations with the backbone. *These libraries need only be generated once and can then be used for all proteins.* One-residue-at-a-time combinatorial addition of side chains, followed by pruning and resampling [11, 24] permits statistical sampling without a combinatorial explosion.

Three ingredients are necessary for "combinatorial decorating," all of which were presented in Section 12.2. First, assume that one possesses a well-sampled ensemble of the coarsest (top level) model $\pi_0(\Phi)$ generated by ordinary, possibly parallel simulation. Additionally, one requires separate libraries for each side chain, characterized by the distributions (probability densities) $p_i(\mathbf{x}_i)$ where $\mathbf{x}_i$ denotes the coordinates for side chain i. These libraries should be distributed according to the force field used for the full protein. As described in Section 12.2, we also have a set of intermediate models characterized by distributions $\pi_j(\mathbf{r}_j) = \pi_j(\Phi, \mathbf{x}_1, \mathbf{x}_2, \ldots, \mathbf{x}_j)$.

The algorithm, sketched in Figure 12.4, is fairly simple to describe: it is the "decoration" of the backbone by side chains. One first generates an ensemble in the π_1 model by combining configurations drawn independently from both the backbone model (π_0) and the statistical rotamer library (p_1). The generating or sampling density in this case is a simple product, namely $\pi_0(\Phi) \cdot p_1(\mathbf{x}_1)$, implying that the resulting configurations must be reweighted according to the weights

$$w(\Phi, \mathbf{x}_1) = \pi_1(\Phi, \mathbf{x}_1) / \pi_0(\Phi) \cdot p_1(\mathbf{x}_1). \quad (12.12)$$

A more practical way to reweight in a combinatorial procedure is to perform "resampling" [11]—still following Equation 12.12—so that an ensemble of configurations results, which is distributed according to π_1 but *without weights*. That is, low-weight configurations are discarded in a statistical manner while maintaining the distribution; see Ref. [11]. Future stages of the decorating process therefore avoid low-weight configurations, and a constant ensemble-size can be maintained. For instance, 10^3 side-chain configurations can be used to decorate, combinatorially, 10^4 backbone configurations—yielding 10^7 configurations each with a weight given by Equation 12.12. These can be resampled to generate 10^4 properly distributed configurations to start the next stage of decorating [via $\pi_1(\Phi, \mathbf{x}_1)$ and $p_2(\mathbf{x}_2)$], and so on.

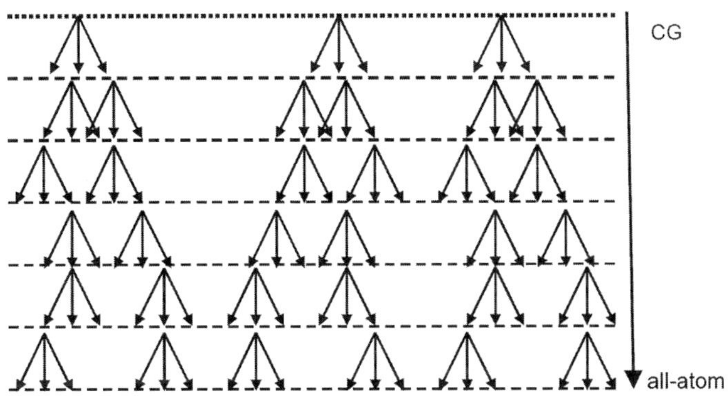

FIGURE 12.4 Library-based combinatorial "decorating." Each level below the backbone "CG" level has one side chain added. In this algorithm, atomic detail is added incrementally, *without dynamical simulation* beyond that required to sample the top level. Rather, assuming the top level is the protein backbone and each level represents one additional side chain, presampled statistical rotamer libraries can be employed to generate side-chain configurations.

Several additional technical points can be made.

1. Side-chain sampling can be optimized on the fly by favoring rotamers which have been preferentially adopted.
2. When adding a side chain to the backbone, additional coordinates need to be specified—i.e., those embodying the six relative degrees of freedom between the backbone and side chain. These additional coordinates can be drawn independently based on knowledge of the force field.
3. It may be advisable to perform canonical "relaxation" simulation after each side-chain addition.
4. If the libraries have been generated according to the force field for the full protein $U(\mathbf{r})$, then in principle the calculation of energy terms internal to the side chains (and, separately, the backbone) can be avoided during the combinatorial process. This improves the way the algorithm scales with system size.

Practicality analysis. The combinatorial decorating procedure corrects the two key defects of the exchange-based approaches. It eliminates lingering time scales through the use of precalculated libraries, and avoids most correlations by construction. The only "danger" is that an important part of configuration space could be eliminated early in the combinatorial process. However, if a sufficiently large ensemble is maintained (10^4 or more configurations), this should not be a worry: multiple rotamers can be maintained for all the side chains. Further, backbone models will be employed which sample small and large fluctuations about a known experimental structure—see Section 12.2.2.

12.5 RESULTS

All of our published results, and those presented below, have employed the top-down ResEx approach described above. Preliminary results from the combinatorial decorating procedure became available only after the preparation of this chapter, and hence are not included. However, the preliminary data indeed fulfilled the expectations described above [27].

12.5.1 Model Systems via Two-Level ResEx: Butane, Di-Leucine Peptide

In our first ResEx results [2], we sought to show that the algorithm worked correctly and that it could be applied to model molecular systems. We therefore studied butane (14 atoms) and di-leucine peptide (two side-chains, 50 atoms). Only two ladder levels were employed in these first studies—an all-atom and a CG model. Full details can be found in our paper [2].

The results were encouraging. As shown in Figure 12.5, our studies with butane showed that the algorithm could produce the correct all-atom distribution even if a poor-quality CG model was used. That is, both CG models used were quite different from the corresponding potential of mean force for the variable used. The results also validated our top-down strategy.

Top-down ResEx was also applied to the more challenging system of 50-atom di-leucine peptide, modeled in all-atom (AA) detail with implicit solvent. In this case, our CG model was simply a united-atom (UA) representation (rather than the backbone model described above). Again, the ResEx algorithm corrected for bias in the CG model since the calculated free-energy difference is actually of the opposite sign in UA vs. AA [2]. Figure 12.5 shows that the ResEx dramatically reduces statistical error. In fact, the efficiency gain of a factor of 15 (including the cost of UA simulation) is the highest we know of in a peptide—even though only two levels were employed.

12.5.2 Peptides: Met-Enkephalin and a 20-Residue Designer Beta Protein

To extend ResEx to larger systems, it was necessary to use additional ladder levels—i.e., to proceed *incrementally* from the top-level CG model to the bottom AA level. Our second paper [3] explained

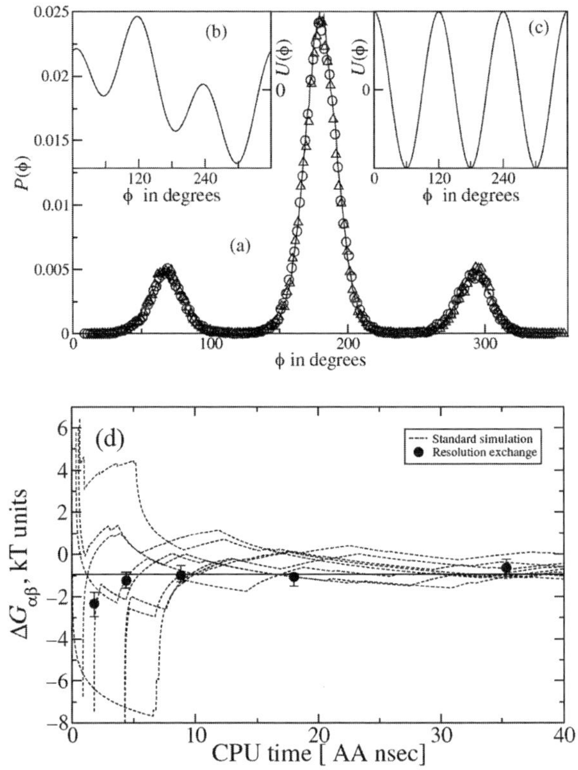

FIGURE 12.5 Top-down resolution exchange results for two model systems. On the **top**, all-atom *butane* is sampled based on two fairly poor coarse potentials, (b) and (c), which describe the central dihedral only. The main plot (a) shows that the algorithm correctly reproduces the true distribution (solid line), where the symbols are ResEx results employing the CG potentials (b) and (c). On the **bottom**, in panel (d), ResEx for *di-leucine peptide* correctly reproduces the all-atom results with much smaller statistical uncertainty. The horizontal axis gives total CPU cost and the error bars show the range of eight independent ResEx runs, which are contrasted with running averages from standard simulation. The quantity being estimated in (d) is the free energy difference between two states of the peptide. (Reprinted from Lyman, E., F. M. Ytreberg, and D. M. Zuckerman. *Phys. Rev. Lett.*, 96, 028105, 2006. With permission.)

a systematic approach for doing so, namely, by adding atomic detail to a peptide one residue at a time. In essence such an approach guarantees sufficient overlap between neighboring ladder levels so that ResEx simulation can be performed effectively.

Our first implementation of this incremental approach, for five-residue (75-atom) met-enkephalin, employed UA as the top-level model. In subsequent lower levels, one residue at a time was converted to an all-atom model. This dramatically improved exchange as shown in Figure 12.6, by comparison to direct conversion from UA to AA. Further, we were also able to use elevated temperature in the higher levels (see Figure 12.6), which presumably can improve efficiency further. Our data showed good agreement with independently produced Ramachandran plots [3].

In the final system studied with ResEx, a 20-residue "designer" beta protein previously considered in Ref. [28], we have implemented a backbone/side-chain division for the first time. The whole study was performed within the UA framework to enable more rapid generation of initial results. Specifically, our top-level CG model was a UA model of the backbone and beta carbons only: we used an Ala-Gly-Pro model, where glycines and prolines were represented explicitly and all other residues were truncated to alanine. Such a model permits backbone hydrogen bonding and good reproduction of Ramachandran propensities. Subsequent lower levels of the ladder were generated by adding one side-chain at a time, until the full protein was represented at the UA level. Good exchange was obtained between all neighboring pairs of levels (unpublished).

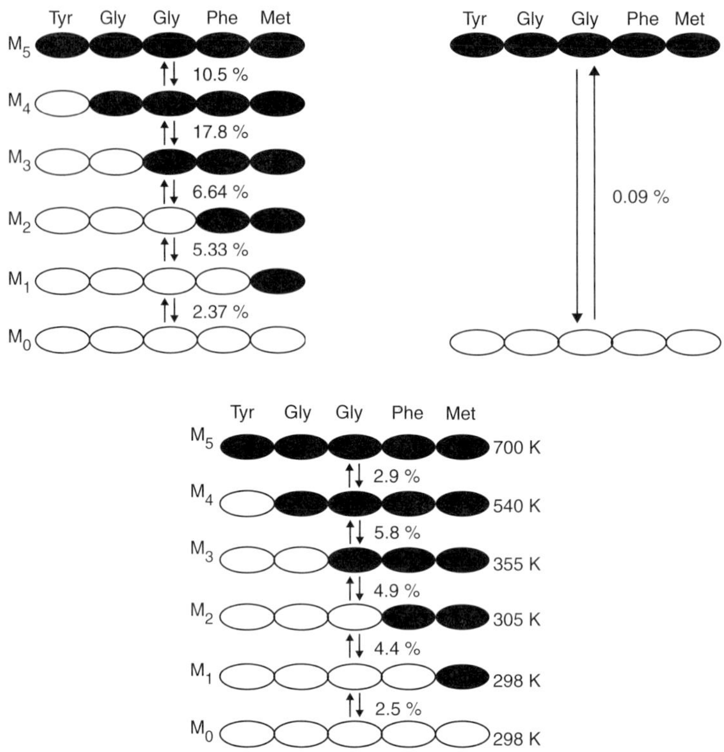

FIGURE 12.6 The incremental (residue-at-a-time) approach to ladder-building for resolution exchange in met-enkephalin. The **upper right** ladder shows the extremely low exchange acceptance rate from the top level (CG) to the bottom (all-atom, AA) at $T = 298$ K. By contrast, on the **upper left**, when a single residue at a time is "de-coarsened," quite reasonable acceptance is obtained at $T = 298$ K. On the **bottom**, changes in resolution are combined with temperature changes. For all the met-enkephalin ladders shown, the top level is a united-atom model of the whole peptide and the bottom level is all-atom, with GBSA implicit solvent [3]. (Reprinted from Lyman, E., and D. M. Zuckerman. *J. Chem. Theory Comput.*, 2, 656–66, 2006. With permission.)

12.6 LOOKING AHEAD

This chapter has focused on our group's efforts to establish methods for statistical-mechanics-based sampling of atomistic models based on auxiliary simulation of CG models. We have made the generation of true canonical atomistic ensembles a priority in our work—i.e., we have not proposed methods that only approximately satisfy criteria for sampling configurations in proportion to a suitable Boltzmann factor. Rather, we have developed a series of algorithms, including "Resolution Exchange," which can produce canonical sampling according to standard atomistic force fields, albeit in implicit solvent. The chapter has attempted to be self-critical, describing practical limitations of the algorithms.

Our main hope for true sampling of full-size proteins (say, 100–200 residues) in implicit solvent lies with the "Combinatorial Decorating" approach. By employing presampled statistical libraries of side chains, this protocol is largely nondynamical and appears to remove any "lingering time scales" (Section 12.3.2). It also seems to overcome potentially serious correlation problems associated with ResEx (Section 12.3.2).

While we believe that true sampling of implicitly solvated proteins is a critically important and extremely difficult challenge, it is worth speculating on the ultimate goal of explicit-solvent sampling. In such a context, it is possible that water molecules can be used to "decorate" an implicit-solvent ensemble of structures in a statistical way. Another possibility is to first generate implicit-solvent structures, followed by relaxation in explicit solvent, leading to a biased (non-Boltzmann) ensemble which may be amenable to reweighting by novel techniques [29].

ACKNOWLEDGMENTS

I am particularly indebted to Dr. Edward Lyman for his contributions to the development of the resolution-exchange approach. Prof. Marty Ytreberg contributed a key technical idea enabling the project to get off the ground in the early days. Other members of the group have provided valuable feedback: Dr. Artem Mamonov and Mr. Bin Zhang. Funding for this work was provided by the NIH (Grant GH070987) and the NSF (Grant MCB-0643456).

REFERENCES

1. Levitt, M., and A. Warshel. 1975. Computer simulation of protein folding. *Nature* 253:694–98.
2. Lyman, E., F. M. Ytreberg, and D. M. Zuckerman. 2006. Resolution exchange simulation. *Phys. Rev. Lett.* 96:028105.
3. Lyman, E., and D. M. Zuckerman. 2006. Resolution exchange simulation with incremental coarsening. *J. Chem. Theory Comput.* 2:656–66.
4. Izvekov, S., and G. A. Voth. 2006. Modeling real dynamics in the coarse-grained representation of condensed phase systems. *J. Chem. Phys.* 125:151101–104.
5. Liu, P., and G. A. Voth. 2007. Smart resolution replica exchange: An efficient algorithm for exploring complex energy landscapes. *J. Chem. Phys.* 126:045106.
6. Mori, G. M. S. D., C. Micheletti, and G. Colombo. 2004. All-atom folding simulations of the villin headpiece from stochastically selected coarse-grained structures. *J. Phys. Chem. B* 108:12267–70.
7. Praprotnik, M., L. D. Site, and K. Kremer. 2005. Adaptive resolution molecular-dynamics simulation: Changing the degrees of freedom on the fly. *J. Chem. Phys.* 123:224106–14.
8. Praprotnik, M., L. D. Site, and K. Kremer. 2007. A macromolecule in a solvent: Adaptive resolution molecular dynamics simulation. *J. Chem. Phys.* 126:134902–908.
9. Lwin, T. Z., and R. Luo. 2005. Overcoming entropic barrier with coupled sampling at dual resolutions. *J. Chem. Phys.* 123:194904–10.
10. Frauenfelder, H., S. G. Sligar, and P. G. Wolynes. 1991. The energy landscapes and motions of proteins. *Science* 254:1598–603.
11. Liu, J. S. 2002. *Monte Carlo Strategies in Scientific Computing.* New York: Springer.
12. Mamonov, A., and D. M. Zuckerman, in preparation; Ytreberg, F. M., S. Kh., Aroutiounion, and D. M. Zuckerman. 2007. Demonstrated convergence of the equilibrium ensemble for a fast united–residue protein model. *J. Chem. Theory Comput.* 3:1860–66.
13. Still, W. C., A. Tempczyk, R. C. Hawley, and T. Hendrickson. 1990. Semianalytical treatment of solvation for molecular mechanics and dynamics. *J. Am. Chem. Soc.* 112:6127–29.
14. Mongan, J., C. Simmerling, J. A. McCammon, D. A. Case, and A. Onufriev. 2007. Generalized born model with a simple, robust molecular volume correction. *J. Chem. Theory Comput.* 3:156–69.
15. Emilio Gallicchio, R. M. L. 2004. AGBNP: An analytic implicit solvent model suitable for molecular dynamics simulations and high-resolution modeling. *J. Comput. Chem.* 25:479–99.
16. Zhu, J., E. Alexov, and B. Honig. 2005. Comparative study of generalized born models: Born radii and peptide folding. *J. Phys. Chem. B* 109:3008–22.
17. Hetenyi, B., K. Bernacki, and B. J. Berne. 2002. Multiple "time step" Monte Carlo. *J. Chem. Phys.* 117:8203–207.
18. Gelb, L. D. 2003. Monte Carlo simulations using sampling from an approximate potential. *J. Chem. Phys.* 118:7747–50.
19. Swendsen, R. H., and J. S. Wang. 1986. Replica Monte Carlo simulation of spin-glasses. *Phys. Rev. Lett* 57:2607–609.
20. Geyer, C., J. Markov. 1991. Chain Monte Carlo maximum likelihood. In *Proceedings of the 23rd Symposium on the Interface.*
21. Sugita, Y., A. Kitao, and Y. Okamoto. 2000. Multidimensional replica-exchange method for free-energy calculations. *J. Chem. Phys.* 113:6042–51.
22. Frantz, D. D., D. L. Freeman, and J. D. Doll. 1990. Reducing quasi-ergodic behavior in Monte Carlo simulation by J-walking: Applications to atomic clusters. *J. Chem. Phys.* 93:2769–84.
23. Rosenbluth, M. and A. Rosenbluth. 1955. Monte Carlo calculation of the average extension of molecular chains. *J. Chem. Phys.* 23:356–59.
24. Wall, F. T. and J. J. Erpenbeck. 1959. New method for the statistical computation of polymer dimensions. *J. Chem. Phys.* 30:634–37.

25. Frauenkron, H., U. Bastolla, E. Gerstner, P. Grassberger, and W. Nadler. 1998. New Monte Carlo algorithm for protein folding. *Phys. Rev. Lett.* 80:3149–52.
26. Grassberger, P. 1997. Pruned-enriched Rosenbluth method: Simulations of theta polymers of chain length up to 1 000 000. *Phys. Rev. E* 56:3682–93.
27. Mamonov, A., X. Zhang, and D. M. Zuckerman. In preparation.
28. Roe, D. R., V. Hornak, and C. Simmerling. 2005. Folding cooperativity in a three-stranded beta-sheet model. *J. Mol. Biol.* 352:370–81.
29. Ytreberg, F. M., and D. M. Zuckerman. 2008. A black-box re-weighting analysis can correct flawed simulation data, *Proc. Nat. Acad. Sci. USA*. 105:7982–87.

13 Pathways of Conformational Transitions in Proteins

Peter Májek and Ron Elber
Department of Computer Science, Cornell University

Harel Weinstein
Department of Physiology and Biophysics and the HRH Prince Alwaleed Bin Talal Bin Abdulaziz Alsaud Institute for Computational Biomedicine, Weill Medical College of Cornell University

CONTENTS

13.1 INTRODUCTION

This chapter is about coarse-graining of pathways and trajectories of proteins in action. In particular we focus on protein switches that flip between different structures; flips that modify their activity. These switches offer another layer of control and are of considerable interest in current fields of study such as systems biology [Alon 2006]. It is the collective behavior and interactions of many protein molecules that is the topic of biological networks, and such switches are essential components of the functional mechanisms represented by the networks.

The way we compute and think about trajectories is quite different from the widely used molecular dynamics (MD) approach in which differential equations of motions are solved by an initial value method. Instead we have chosen, for reasons that will become clearer below, to use a boundary value formulation. Our choice of coarse-graining and of boundary value calculations to study biological switches requires some discussion that we start with an analogy.

A useful comparison to the way we compute and analyze molecular paths is a web search for driving directions. In such a search one specifies the starting and end locations of the drive and seeks a path that requires minimal time. A web engine analyzes the request and outputs a written description of the driving directions and a two-dimensional map of the roads, highlighting the chosen path. Obviously the web instructions are only guidelines that do not take into account extra traffic due to special events, road works that intervene and require a bypass, and other specific circumstances. In short, the paths proposed are coarse-grained and averaged over many actual paths that differ in many details but agree in their overall shape. The coarse-grained (CG) paths miss many details but are nevertheless very useful when making travel plans. Part of their effectiveness is because we understand their limitations and operate accordingly.

Similarly, in molecular biophysics we are frequently provided with starting and end configurations of a protein, captured with experimental techniques. These end points are stable for significant periods of time and can be measured with static techniques. We call the initial conformation $\mathbf{x}_i$ and the final structure $\mathbf{x}_f$. For example, X-ray crystallography or NMR can determine structures of activated and deactivated forms of a protein molecule. However, experimentally elucidating the "driving" pathway that links the two forms and determines the time scale of the transition is considerably more difficult. Structures "in-between" exist for significantly shorter times than the stable end points, and are harder to measure. Computer simulations suggest a useful alternative that, in conjunction with partial experimental data (such as the change of a few distances as a function of time) can provide a comprehensive view of the process. The present chapter is about the computations of such transitions. Let us consider some approaches to molecular simulations and how they can be used to study biomolecular switches.

As mentioned earlier, the most widely used technique for atomically detailed simulations is MD. In the MD approach we consider Newton's equations of motions, $\mathbf{M}\ddot{\mathbf{x}} = \mathbf{F}$, where $\mathbf{M}$ is the mass matrix, and $\mathbf{x}$ the coordinate vector of all the particles that we used to model the system. For proteins, the length of $\mathbf{x}$ (the number of degrees of freedom) is typically of the order of a few thousand. The double dot on top of the vector $\mathbf{x}$ denotes second derivative with respect to time, and $\mathbf{F}$ is the force vector. In MD initial values are used for the integration ($\mathbf{x}$, the coordinate vector and $\mathbf{v} \equiv \dot{\mathbf{x}}$, are given at an initial time, say $t=0$). The numerical solvers of the initial value problem employ numerical integrators such as Verlet (1967):

$$\begin{aligned} \mathbf{x}(t+\Delta t) &= \mathbf{x}(t) + \mathbf{v}(t)\Delta t + \frac{\Delta t^2}{2}\mathbf{M}^{-1}\mathbf{F}(t), \\ \mathbf{v}(t+\Delta t) &= \mathbf{v}(t) + \frac{\Delta t}{2}\mathbf{M}^{-1}[\mathbf{F}(t) + \mathbf{F}(t+\Delta t)] \end{aligned} \tag{13.1}$$

The time step is Δt. Hence, given coordinates and velocities at a particular time we can generate the coordinates and the velocities at slightly later times. By repeating this process N times we can generate a trajectory (a path) of length $N \cdot \Delta t$. The final structure $\mathbf{x}(N \cdot \Delta t)$ is determined by the initial conditions since the algorithm described in Equation 13.1 is deterministic. However, it is hard to predict or to tune with velocity variations the location of $\mathbf{x}(N \cdot \Delta t)$ before doing the actual calculation since the results are very sensitive to the initial value. The hard-to-predict final point is a significant difficulty with the application of MD to biomolecular switches. Unless "all roads lead to Rome," it is not obvious that $\mathbf{x}(N \cdot \Delta t)$ is the final desired configuration of the switch $\mathbf{x}_f$. Hence we may be wasting many cycles and computing unsuccessful trajectories while adjusting the initial conditions until $\mathbf{x}(N \cdot \Delta t)$ is the desired final coordinate vector. Notably, MD does not effectively use all the information at hand (i.e., the structures at the two ends) that could facilitate the study of switching mechanisms. It is therefore not surprising that the use of MD is not optimal for this problem.

Even if the two end points are strong attractors (i.e., "all roads indeed lead to Rome"), the transition may require a large number of integration steps (the basic integration step Δt must be small) making the calculation (again) inefficient. Some techniques [Dellago, Bolhuis, and Geissler 2002] use initial value formulation, starting somewhere in between, to compute rare *short-time* trajectories between strong attractors. This allows for the sampling of trajectories between states separated by large and narrow barriers (activated processes) [Bolhuis et al. 2002]. However these conditions are not satisfied for conformational transitions of the type discussed in this manuscript. An example of a broad barrier that leads to long transitional trajectories is the conversion in the myosin molecule [West and Elber, work in progress]. The time scale for myosin postrecovery (a part of the transition that controls muscle motions) is in milliseconds (10^{-3} s), while the typical size of an integration step in MD is a femtosecond (10^{-15} s). The number of integration steps with MD required to simulate the transition in myosin is hopelessly large (10^{12} s). Why is the step so small? The reason is that many molecular motions are very rapid (e.g., bond vibrations, collisions) so that in order to maintain the numerical stability of an algorithm like Verlet the step has to be significantly smaller than the time scale of the fastest motions in the system. Significant effort was therefore invested [Schlick et al. 1999] into algorithms that increase the time steps, and into factoring out rapid motions. Perhaps the most widely used algorithm that eliminates certain categories of rapid motions is SHAKE [Ryckaert, Ciccotti, and Berendsen 1977], in which the fast bond vibrations are constrained to their ideal values. Typically used for bonds of heavier atoms with hydrogen (x-H), SHAKE allows the increase of the time step by about a factor of two, but probably not more. The problem is that other rapid motions (nonbonded collisions between atoms) remain after the removal of bond vibrations and also require small time steps. The latter are much harder to factor out rigorously since their internal coordinate representation (the identity and distance between a pair of colliding atoms) is changing during the progress of the trajectory. Some atoms that are close to each other at a particular time (colliding) may be separated at a later time of the process in which other atoms may collide. While a special treatment for a collision can be worked out [Ulitsky and Elber 1993, 1994], the change of colliding partners requires complex bookkeeping, which is expensive computationally. Nevertheless, the overall success and wide use of the SHAKE algorithm suggests that other reductions in the details of the spatial description of the system may be useful. Indeed a significant part of this book is about approaches to perform spatial coarse-graining to simplify force calculations, reduce the number of degrees of freedom, and enable longer time and more comprehensive sampling with the simplified spatial description. This reduction is established using a number of physical assumptions and approximations, and it does not solve the problem of double-ended trajectories. Furthermore, atomically detailed description may be necessary to understand many biophysical and biochemical processes. Giving it up with spatial coarse-graining may lose critical elements of it.

The approach we discuss in the next section, which we have used for more than 10 years now, provides a CG description of the path (like driving directions at different resolution), while keeping a complete description of the atomic coordinates of the system. We are able to do this since the boundary value formulation is numerically more stable than initial value solvers. The boundary value representation makes it possible to use much larger path steps than is possible with initial value solvers. One must keep in mind however that the trajectories so produced are approximate.

13.2 THEORY OF BOUNDARY VALUE FORMULATION OF PATHWAYS AND TRAJECTORIES

Newton's equations of motion are usually derived in an analytical mechanics course [Landau and Lifshitz 1976] from the classical action. The classical action, S, is defined as

$$S[\mathbf{x}(t)] = \int_{0,\mathbf{x}_i}^{t,\mathbf{x}_f} L[\mathbf{x}(\tau), \dot{\mathbf{x}}(\tau)]\, d\tau,$$

$$L = \frac{1}{2}\dot{\mathbf{x}}^{\mathrm{T}}\mathbf{M}\dot{\mathbf{x}} - U(\mathbf{x}). \tag{13.2}$$

Then, Newton's equations of motion are derived from the condition that the action, which is a functional of the trajectory, is stationary with respect to variations in the path [Landau and Lifshitz 1976]. The variations do not change the end configurations, illustrating that a solution of Equation 13.2

is indeed a result of a boundary value problem. However, such a derivation will not help in the switching problem. A more straightforward approach to solving Equation 13.2 is to use a finite difference approximation to the integral. For example: $\sum_{j=1}^{N-1}[1/(2\Delta t)](\mathbf{x}_{j+1}-\mathbf{x}_j)^{\mathrm{T}}\mathbf{M}(\mathbf{x}_{j+1}-\mathbf{x}_j)-U(\mathbf{x}_j)\Delta t$. The action is now a function (not a functional!) of all the intermediate structures $\mathbf{x}_j$'s. When the action is stationary with respect to all these coordinates ($\partial S/\partial \mathbf{x}_j = \mathbf{0}\ \ j=2,\ldots,N-1$) then the sequence of coordinates provides a finite difference approximation to a classical trajectory. The approximation is better for smaller Δt but it is stable for any time step. This stability distinguishes the approach from initial value solvers that "explode" at time steps that are too large. There are however good reasons why finding a stationary solution of the classical action is not a popular way of computing trajectories. First the action is not necessarily a minimum but a stationary point (e.g., a classical trajectory can be a saddle point of the action). Searches for stationary points are numerically more difficult than for a minimum. Second, the optimization is of a function of a very large number of variables. If the biomolecular switch is described with K atoms then the number of degrees of freedom for the action optimization is $3\times K\times N$ (K and N are in the thousands for a typical calculation). Third, time is not a good variable for parameterizing or indexing the path. The last surprising statement is in the sense that the parameterization determines *the density of points along the path*, and the time density can be very different from the *spatial* density. Let us return one more time to the driving analogy and parameterize the path according to the time of the drive. A realization of this parameterization is to draw a dot on the map at constant time intervals (say every minute) of the drive. The path may include a segment in which the car is likely to move slowly or even stop and a segment on a free road or the highway in which the car moves very quickly. If we distribute the dots that describe the path equally in time we will have high density at a traffic jam and dilute the description of the path on the highway. Tracking the path visually under these circumstances is not ideal. A better path representation (at least from a visualization perspective) is to have the dots equally spaced, say every 200 m. More precisely, the above suggestions for alternative parameterization means to replace the parameterization of the path with respect to time $\mathbf{x}(t)$ by the parameterization $\mathbf{x}(l)$, where l is the arc-length of the path in mass weighted coordinates, $dl=\left|\mathbf{M}^{1/2}\dot{\mathbf{x}}\right|dt$. The classical action as a function of length is [Landau and Lifshitz 1976]

$$\bar{S}=\int_{\mathbf{x}_i}^{\mathbf{x}_f}\sqrt{2(E-U)}\,dl, \tag{13.3}$$

and in a discrete representation

$$\begin{aligned}\bar{S}\left[\{\mathbf{x}_j\}_{j=1}^{N}\right]&=\sum_{j=1}^{N-1}\frac{1}{2}\left(\sqrt{2(E-U(\mathbf{x}_j))}+\sqrt{2(E-U(\mathbf{x}_{j+1}))}\right)\Delta l_{j,j+1},\\ \Delta l_{j,j+1}&=\left|\sqrt{\mathbf{M}}\mathbf{x}_j-\sqrt{\mathbf{M}}\mathbf{x}_{j+1}\right|.\end{aligned} \tag{13.4}$$

A classical trajectory will be obtained when for all intermediate configurations we have $\partial\bar{S}/\partial\mathbf{x}_j=0$. The parameterization with respect to the arc-length is indeed more convenient computationally than with respect to time. However, we are still faced with the need to compute a stationary action instead of a minimum. Furthermore, Equation 13.3 is always non-negative; as such it has an undesired minimum at $E=U$ in which the first-order variation is discontinuous. This is the classical turning point (zero kinetic energy) in which the trajectory may get stuck. The derivative is not continuous or zero for that path so it is not a true classical trajectory. However, attempts at direct minimization of $\bar{S}$ may pick it up. It is therefore better to work with the Gauss action in length [Elber 2006].

The Gauss action (written originally for the time-dependent formulation as $S_{\text{Gauss}} = \int_0^t [\delta S / \delta \mathbf{x}(t')]^2 dt' = \int_0^t (\mathbf{M}\ddot{\mathbf{x}} + dU/d\mathbf{x})^2 dt$ [Lanczos 1970]) is trivially extended to the length formulation and a finite difference formula as $S^l_{\text{Gauss}} \approx \sum_{j=2}^{N-1} (\partial \bar{S}/\partial \mathbf{x}_j)^{\text{T}} (\partial \bar{S}/\partial \mathbf{x}_j)$ (for explicit formulas of the derivatives of $\bar{S}$ see Appendix II). A direct minimization of the function S^l_{Gauss} will provide an approximate classical trajectory as a function of length. We typically perform this minimization with simulated annealing (SA). Let $\mathbf{y}$ be the vector of the joint set of all coordinates $\mathbf{y} \equiv (\mathbf{x}_1, \mathbf{x}_2, \ldots, \mathbf{x}_N)$, then the SA procedure uses S^l_{Gauss} as the potential energy and $\mathbf{y}$ as variables as follows. We integrate the following stochastic differential equation for the *whole trajectory*: $\ddot{\mathbf{y}} + \gamma\dot{\mathbf{y}} + \partial \mathbf{S}^l_{\text{Gauss}} / \partial \mathbf{y} = \mathbf{R}$, where γ is the friction constant and $\mathbf{R}$ is a Gaussian random force sampled according to the conditions $\langle \mathbf{R} \rangle = \mathbf{0}$ and $\langle \mathbf{R}^2 \rangle = 2\gamma T \delta(t)$. The temperature, T, is reduced monotonically to zero at which point a path with the minimal action is recovered. At finite temperatures it is possible to sample plausible paths to form a collection of trajectories depending on allowed variance in the value of the action [Elber, Meller, and Olender 1999].

A critical advantage of the boundary value calculations compared to MD is that the step can be very large while still providing correct qualitative behavior of a classical trajectory. The numerical stability of the optimization process is poorer with initial value solvers (with the exception of the Backward-Euler algorithm of Peskin and Schlick (1989), which is stable for large steps but still cannot be aimed to a desired product). For example, in the simulation of the folding of cytochrome c [Cardenas and Elber 2003] we have used 1000 length slices to provide a CG description of the folding pathway; the resulting CG path was consistent with numerous experimental observations, but of course approximate. Since the time scale of folding of cytochrome c is in milliseconds, about 10^{12} steps would be required with an initial value solver (straightforward MD). This number of steps is nine orders of magnitude larger than the number of steps we have used in the boundary value formulation.

To start the SA algorithm we need to specify an initial trajectory $\mathbf{y}^0$. The simplest initial guess for a trajectory is a Cartesian linear interpolation between the two end points. However, the energies of structures along linear paths are usually very high since they distort the covalent structure of the protein chain and have significant steric overlaps. These initial guesses require substantial optimization that is not always successful. The generation of the initial paths can benefit from *spatial* coarse-graining which we discuss in Section 13.4.

In addition to an initial guess for the path, the total energy of the system E is needed for the calculation of the action. An obvious try would be to use the average thermal energy, which is a sum of the average potential, $\langle U \rangle$, and average kinetic energy, $\langle K \rangle$ (if we had computed a large number of trajectories then it would make sense to sample from the distribution of these energies instead of using the averages). However, our approximate procedure to compute trajectories introduces a subtle complication. The trajectories with large steps in time or in the arc-length do not include motions with high frequencies [Olender and Elber 1996] (i.e., bond oscillations) and these degrees of freedom do not contribute to the thermal energy. The number of fast degrees of freedom is uncertain since the number of the (transient) collisions is not known. Since the collision between pairs of particles takes only a small fraction of time, the amount of filtering is also uncertain. If we use a lower bound for the filtering and consider only the bonds and the angles of the protein molecule (the number of bonds or angles is of order K—the number of atoms), the thermal energy of the nonfiltered motions is approximately $E \approx (3K - 2K)(k_{\text{B}}T/2) + \langle U \rangle = K(k_{\text{B}}T/2) + \langle U \rangle$. This is the energy that we use for the functional 13.4.

13.3 PATH CONSTRAINTS

In this section, we describe a number of constraints that are imposed during the calculations of the path to ensure correctness, given the typical computing environments used for molecular modeling and simulations. During simulations of large molecules we use Cartesian coordinates for which the equations of motion are simple to write and manipulate. However, the Cartesian representation requires a reference frame. Changes in the reference frame of one structure along the path affect the distance between sequential coordinate sets $\Delta l_{j,j+1}$. Therefore, the same reference frame must be used for all the coordinate sets along the path. Fixing the reference system is achieved by applying the Eckart conditions [Elber 1990]: six constraints on overall translations and rotations on each of the structures along the path,

$$\sum_k m_k \mathbf{r}_k = \mathbf{0} \quad \sum_k m_k \mathbf{r}_k^i \times \mathbf{r}_k = \mathbf{0} \quad \text{or} \quad \sigma_{i=1,\ldots,6} = 0, \tag{13.5}$$

where m_k, $\mathbf{r}_k$, and $\mathbf{r}_k^i$ are the mass, the position, and the initial position of the kth atom, respectively. We assume without loss of generality that the initial position of the center of mass is zero.

To ensure that the protein structures in the set are equally spaced along the trajectory we also use a harmonic restraint (13.6) to keep the distances between all the structures the same. The variation principle of the action as a function of length provides a condition on the motion in the direction perpendicular to the path, but it does not explicitly constrain the motion along the path. Therefore the constraint has no impact on the equations of motion.

$$\eta_1 \sum_j (\Delta l_{j,j+1} - \langle \Delta l \rangle)^2 \quad \langle \Delta l \rangle = \frac{1}{N-1} \sum_{j=1}^{N-1} \Delta l_{j,j+1}. \tag{13.6}$$

The strength of the restraint is controlled by η_1, which should be chosen as high as possible. It should be kept in mind though that η_1 values that are too high would make the equations of motion for the annealing stiff and would require a much smaller and less efficient integration step. The same type of constraint was used in the calculation of approximate minimum energy paths and in the calculations of the steepest descent path [Elber and Karplus 1987; Czerminski and Elber 1990; Jonsson, Mills, and Jacobsen 1998]. Another way of implementing the equidistance constraints is via the formulation of Lagrange's multipliers [E, Ren, and Vanden-Eijnden 2002]. The Lagrange's multiplier approach for dealing with the equidistance constraints was used also in the calculations of minimum energy and minimum free-energy paths [E, Ren, and Vanden-Eijnden 2002; Weinan, Ren, and Vanden-Eijnden 2005].

At the beginning of the calculation, we also use the penalty function $\eta_2 \sum_{j=2}^{N-1} 1/[E - U(\mathbf{x}_j)]$, which makes the algorithm more stable by forcing the potential energy, U, along the whole trajectory to be smaller than the total energy E. Without this additional term, the terms under the square roots in Equation 13.4 may be negative and make the optimization ill-defined when the trajectory is far from the optimal one and the structures are highly distorted. After some annealing, the trajectory converges to a neighborhood of a true classical trajectory and the value of η_2 is gradually reduced to zero. The optimized trajectory is not sensitive to the initial value of η_2.

The final target function that is used in the algorithm is

$$T = \sum_{j=2}^{N-1} \left[\left(\frac{\partial \bar{S}}{\partial \mathbf{x}_j} \right)^{\mathrm{T}} \left(\frac{\partial \bar{S}}{\partial \mathbf{x}_j} \right) \right] + \eta_1 \sum_{j=1}^{N-1} (\Delta l_{j,j+1} - \langle \Delta l \rangle)^2 + \eta_2 \sum_{j=2}^{N-1} \frac{1}{E - U(\mathbf{x}_j)}. \tag{13.7}$$

Stochastic optimization of T is performed similarly to the procedure described for optimization of S^l_{Gauss}, except that the constraints on translations and rotations of the system (13.5) are solved explicitly (the constraints are linear—see appendix of West, Elber, and Shalloway (2007)). We call the optimization of T an SDEL calculation (Stochastic Difference Equation in Length [Elber, Ghosh, and Cardenas 2002]).

The formulation in Equation 13.7 is applicable to any type of dynamics between two fixed end points that can be described by an action. Another choice of action implemented in our MD simulation software MOIL [Elber et al. 1995] is an action that provides approximate most probable Brownian trajectories and the intrinsic reaction coordinate (Steepest Descent Path [Olender and Elber 1997; Elber and Shalloway 2000]). Calculating reaction coordinates with boundary value formulation and action minimization is intriguing, since local calculations of reaction coordinates suffer from similar problems as the calculations of trajectories. For example, they are difficult to direct to desired product states. Other global algorithms for path optimization are available [Ulitsky and Elber 1990; Jonsson, Mills, and Jacobsen 1998; E, Ren, and Vanden-Eijnden 2002], however, they are not based on a global optimization of an action, which makes their calculation less robust. In essence they are similar to the direct optimization of the classical action, which is a saddle point, and equivalent to solving a large number of differential equations simultaneously. The advantage of having a target function to optimize is the global quality control it provides. As long as the function value is reduced, a large step can be accepted. In contrast, solving differential equations requires small steps and locally controlled accuracy. The action that we used for the approximate Brownian trajectories is

$$S = \int_{\mathbf{x}_i}^{\mathbf{x}_f} \sqrt{H_S + \left(\frac{dU}{dx}\right)^T \left(\frac{dU}{dx}\right)}\, dl, \tag{13.8}$$

where the constant H_S is zero for the calculation of minimum energy path [Elber and Shalloway 2000].

Another interesting feature of the boundary value trajectories is the ability to parallelize the code efficiently. This is in contrast to initial value solvers in which only the calculations of the forces can be made parallel at considerable communication cost. In the boundary value formulation every time (or arc-length) slice can be optimized on a different CPU [Zaloj and Elber 2000]. For a detailed description of the parallelization of the algorithm and recent improvements of the SDEL implementation, see Appendix I.

13.4 SPATIAL COARSE-GRAINING

For proteins with several hundreds of amino acids, computing an atomically detailed trajectory starting from an initial guess (e.g., linear interpolation) far from the optimal trajectory can be a formidable task. With the resources available to us we are able to perform simulated annealing runs that optimize the initial trajectory locally but do not perform global search for alternative pathways. The algorithm puts considerable effort into adjusting local positions of all atoms, but it is less effective in relaxing collective variables of the transition that extend over significant length scales. Similar in spirit to multigrid methods [Briggs 1987], it is worth separating the optimization of path to global and local length relaxations. Otherwise the relaxation of global variables will be slowed down by the "noise" of the local variables. In our car-driving analogy this would correspond to an algorithm that tries to calculate the best driving directions by considering all car types and their conditions, experience of the driver and his level of knowledge of the neighborhood, and so on, before having a general appreciation of the driving route. The additional factors can slow down the speed of the calculations considerably, while their benefit is not obvious at the beginning of the calculations. Average properties of cars and drivers are simpler to use and are providing useful pathways. It makes sense to consider first pathways of average properties that will be refined later (if necessary) according to additional information available at the time of evaluation.

We can use a related idea for conformational transitions. First we determine a trajectory for a system of reduced dimensionality (CG) that we believe captures the global characteristics and relaxation of the path. The CG trajectory provides the backbone on which an atomically detailed trajectory is constructed and refined by the SDEL methodology.

Obviously, there are numerous choices of how to coarse-grain atomically detailed systems, and the choice is far from obvious or unique. Spatially CG models have been successfully used for several years to model behavior of complex biomolecular systems. In these CG models a molecule is represented by a reduced set of representative points, where a point corresponds to at least several atoms. A typical reduction of a protein that we use is to keep only the position of the C_α atom of each amino acid. One model potential energy of this reduced representation is [Tirion 1996; Haliloglu, Bahar, and Erman 1997; Xu, Tobi, and Bahar 2003; Lu, Poon, and Ma 2006]

$$U = \frac{\kappa}{2} \sum_{r_{ij} < C} (r_{ij} - r_{ij}^0)^2, \tag{13.9}$$

where κ is a force constant in kcal mol^{-1}/m^2, C is a distance cutoff, and r_{ij}^0 and r_{ij} are the distances between the C_α atoms of residues i and j in the native and the current conformation, respectively. This is the Anisotropic Network Model (ANM) [Xu, Tobi, and Bahar 2003], an extension of the simpler Gaussian Network Model (GNM) [Haliloglu, Bahar, and Erman 1997]. It has been shown that even these simple potentials provide a very good agreement with X-ray experimental B-factors (see for example Yang et al. 2007), and therefore may give adequate descriptions of the system dynamics in the neighborhood of the native conformation.

The quadratic functional form of Equation 13.9 cannot describe multiple minima and the barriers separating them. Hence it is not an adequate model to represent transitions between stable states. To allow the study of transitions with simplified network models Maragakis and Karplus (2005) used the Empirical Valence Bond (EVB) theory of Warshel [Aqvist and Warshel 1993] and generalized the simple ANM. Two ANM models, U_i and U_f, are defined for the reactants and the products, respectively. The EVB computes a new potential U that interpolates between the two models

$$U(\mathbf{x}) \equiv \frac{1}{2}\left(U_i(\mathbf{x}) + [U_f(\mathbf{x}) - \alpha] - \sqrt{\{U_i(\mathbf{x}) - [U_f(\mathbf{x}) - \alpha]\}^2 + 4\beta^2} \right). \tag{13.10}$$

The scalar α is the energy gap between the two minima and β is a coupling constant that helps tune the barrier height and smoothness.

13.5 STOCHASTIC DYNAMICS

For the CG model we consider Brownian dynamics

$$\gamma \dot{\mathbf{x}} = -\nabla U(\mathbf{x}) + \mathbf{R}(t), \tag{13.11}$$

where γ is a friction constant and $\mathbf{R}(t)$ is a random force with normal distribution ($\langle \mathbf{R}(t) \rangle = \mathbf{0}$ and $\langle \mathbf{R}(t_i)\mathbf{R}^T(t_j) \rangle = 2\gamma k_B T \delta(t_i - t_j)\mathbf{I}$). The boundary value formulation in Brownian dynamics settings was discussed by Elber and Shalloway (2000). They showed that the *most probable* trajectory in approximate and discrete variants of Brownian dynamics minimizes the following action (see also Equation 13.8)

$$S_{\mathrm{BD}}(\mathbf{x}_2, \ldots, \mathbf{x}_{N-1} \,|\, \mathbf{x}_1 = \mathbf{x}_i, \mathbf{x}_N = \mathbf{x}_f, H_s) = \sum_{j=1}^{N-1} \sqrt{H_S + \left(\frac{\partial U}{\partial \mathbf{x}_j} \right)^2} \left| \mathbf{x}_{j+1} - \mathbf{x}_j \right|. \tag{13.12}$$

The constant H_S mimics the energy in classical mechanics and can be chosen empirically. If $H_S \to 0$, the optimal trajectory that minimizes S_{BD} is the steepest descent path. On the other hand, if $H_S \to \infty$, then the linear interpolation between $\mathbf{x}_i$ and $\mathbf{x}_f$ (the shortest trajectory) is the optimal path. Varying the parameter H_S provides a set of optimal CG paths with different thermal energies.

The same simulated annealing algorithm as applied for all-atom $S^{l}{}_{\mathrm{Gauss}}$ minimization is used for optimization of S_{BD}. Since the CG model is much simpler than an all-atom model, path searches can be performed comprehensively.

13.6 REFINEMENT OF COARSE-GRAINED TRAJECTORIES TO ATOMICALLY DETAILED PATHS

Once a set of most probable CG trajectories is obtained by minimization of S_{BD} for different values of H_S, we can return to the initial task of finding approximate long time (arc-length) trajectories for the all-atom representation. First, atomically detailed structures are built based on the CG shapes along the trajectories. For a given length slice k and residue j, the following reconstruction is applied: Let C_j^k be a position of the kth C_α atom in the jth frame. For each residue, in each frame, rigid body transformations $\mathrm{T}_{1\to j}^k$, of $C_1^{k-l},\ldots,C_1^{k+l}$ to $C_j^{k-l},\ldots,C_j^{k+l}$, and $\mathrm{T}_{n\to j}^k$, of $C_n^{k-l},\ldots,C_n^{k+l}$ to $C_j^{k-l},\ldots,C_j^{k+l}$, are calculated. The parameter l defines the size of the local neighborhood. If $l=0$, a single C_α atom is considered at a time and only translational transformation can be determined, for $l=1$ the local neighborhood is defined by a triplet of consecutive C_α atoms (which are not linear in proteins) and both the translation and the rotation can be determined uniquely. In the actual implementation, we have used $l=2$, which is more stable in capturing the local neighborhood. The position of a non-C_α atom A in the length slice k, belonging to the residue j, is reconstructed as a linear interpolation of its transformed position from the initial and the final frame:

$$A_j^k = \frac{n-j}{n-1}\mathrm{T}_{1\to j}^k A_1^k + \frac{j-1}{n-1}\mathrm{T}_{n\to j}^k A_n^k . \tag{13.13}$$

After this interpolation of non-C_α atoms the intermediate structures along the trajectory have unreasonably high potential energies, which must be reduced by minimizations before the all-atom SDEL computation can be used for the refinement of the path. The minimizations of the structures find local minima in the neighborhood of the initial structures and therefore do not change the path significantly. There are three processes of minimization: (i) minimization with soft (core) potential to eliminate truly bad van der Waals contacts, (ii) minimization with regular Lennard–Jones potential, and (iii) short MD simulation at 10 K with harmonic restraints on the positions of the C_α to escape undesired local minima. Typical numbers of minimization steps for the glutamate receptor problem described in Section 13.7 are 100, 200, and 1000 for each of the three minimization processes, respectively.

13.7 THE ALLOSTERIC TRANSITION OF THE MGLUR1 RECEPTOR

With a set of plausible trajectories in atomically detailed representation, the SDEL algorithm is executed to obtain physically and energetically sound pathways. In this section we describe the calculation of the conformational transition in the extracellular (ligand-binding) region of the metabotropic glutamate receptor (mGluR). The mGluRs are membrane proteins that mediate the transmission of a signal into the cell after binding a glutamate molecule in the extracellular domain. These receptors belong to class C of G protein-coupled receptors (GPCR); there are three different subgroups of mGluRs, termed I, II, and III, which do not differ much in their overall molecular architecture. Thus, the mGluR receptor is divided structurally into three regions: the extracellular region, the transmembrane region composed of a seven-helix bundle, and the cytoplasmatic region. The extracellular region consists of the ligand-binding region (LBR) and the cysteine-rich domain (CRD) [Muto et al. 2007; Pin et al. 2004, 2005; Pin and Acher 2002].

Experimental structures for the two states of the LBR of the extracellular part of the mGluR1 (belonging to subgroup I of mGluR) are available (PDB entries 1ewk, 1ewt) [Kunishima et al.

2000]. The receptor functions as a homodimer; consisting of 980 residues (490 amino acids per monomer). However, not all residues of the LBR were resolved by X-ray crystallography and in each monomer a loop of approximately 30 residues is missing. We found, however, that MD simulations of the PDB construct maintain a stable structure in nanosecond-length simulations, indicating that the missing loop segments can be ignored for the present study. The protocol of generating the path from the PDB structures is as follows.

The energy of the two PDB structures is minimized. The C_α representation of the molecules is kept and CG reactive trajectories between the two end points are computed with simulated annealing as described in Section 13.2 for different values of H_S (from Equation 13.12). We have used 100 structures distributed along the path to represent the transition. The action S_{BD} is minimized with 100 K steps of simulated annealing. This requires approximately 24 hours on a single modern CPU for this system. The steepest descent path (a minimum of (1.12) with $H_S = 0$) deviates approximately by a 1.25 Å/frame root mean squared distance (RMSD) from the linearly interpolated (LI) trajectory. Figure 13.1 shows optimal CG trajectories for different values of H_S. The trajectories are projected onto a two-dimensional space with a multidimensional scaling technique [Cox and Cox 1994]. As predicted by theory, the higher the value of H_S, the closer the LI path to the optimal trajectory. The potential energy profiles of the optimal trajectories as a function of their arc-lengths are shown in Figure 13.2. Figure 13.3 shows that the length of the optimal trajectory varies from 15.5 Å to 13 Å for different amounts of thermal energy (H_S) of the system.

Once the optimal CG paths for different values of H_S are found, they are refined to atomically detailed trajectories and are locally minimized as discussed in Section 13.5. A three-dimensional projection (not shown) indicates that the atomically detailed refinement moves the trajectories in a direction perpendicular to the manifold defined by H_S (physically it suggests that the refinement focuses on the side-chain positions, while the locations of the C_α atoms are not affected appreciably). To minimize S^l_{Gauss}, at least 10 K steps of SDEL optimization are required.

This SDEL calculation takes approximately 100 hours of parallel computation on 100 CPUs, thus the SDEL part of the overall calculation is approximately 4000 times more expensive than the

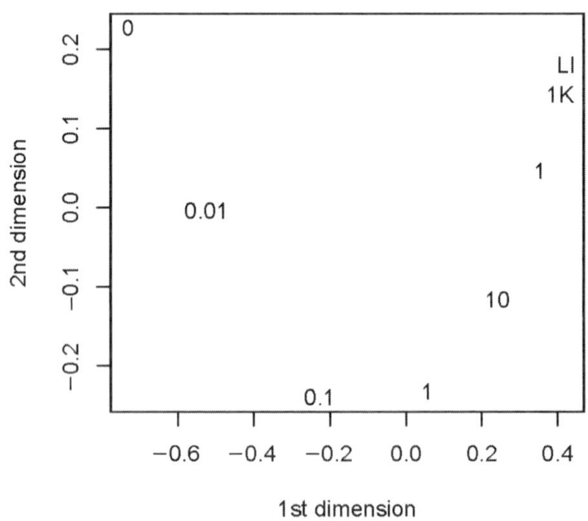

FIGURE 13.1 The distances between optimal coarse-grained trajectories for the transition of the extracellular component of mGluR1. The distances are projected onto a two-dimensional space for better visualization. Each number in the plot corresponds to the trajectory with the given value of H_S in multiples of $K\kappa$ (K is the number of residues and κ is the force constant from the formula (Equation 13.9)). LI represents the linearly interpolated path between the two known conformations. The distance metric, upon which the projection is defined, is a sum of pairwise C_α-RMSD distances between corresponding path structures.

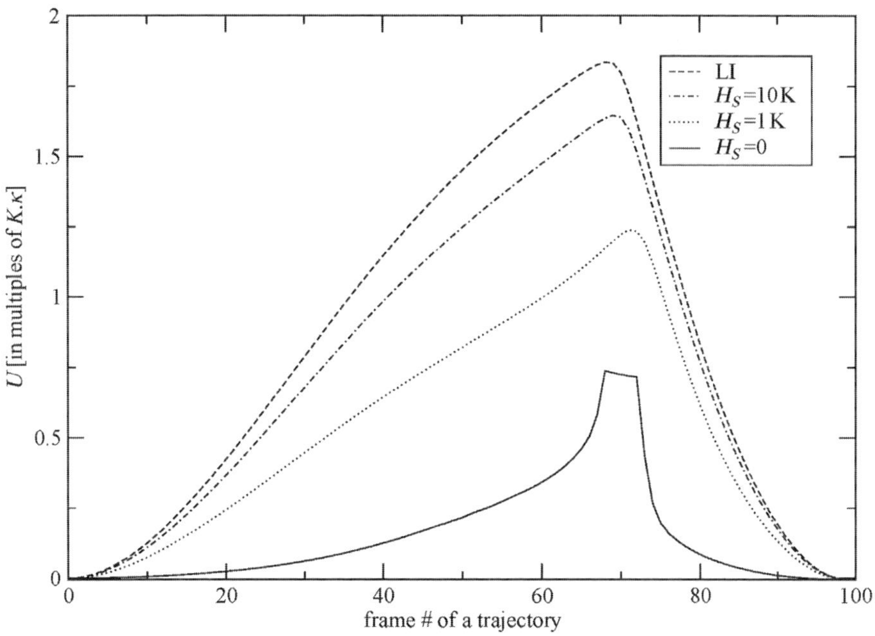

FIGURE 13.2 The potential energy profile of optimal Brownian trajectories of a coarse-grained model for different values of H_S. The transition is of mGluR1. The energy-increasing curves correspond to the optimal trajectories with H_S equal to 0, 1 K, 10 K, and linear interpolation, respectively. The potential and the values of H_S are in multiples of $K\kappa$ (K is the number of residues and κ is the force constant from the Equation 13.9).

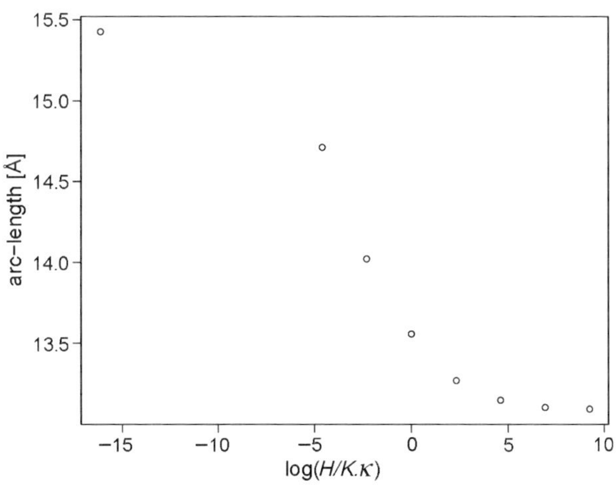

FIGURE 13.3 Arc-length of the optimal trajectory as a function of H_S. The RMSD between the initial and the final structure is approximately 13 Å.

CG preprocessing part (nevertheless, we do believe that the resulting atomically detailed description of the system is important and worth the investment). The most expensive part of the SDEL calculations for systems of this size in our code MOIL [Elber et al. 1995] is the Generalized Born implicit solvation energy. It takes approximately 95% of the SDEL's computation time. The complexity of this calculation is likely to be reduced in the future since there is significant room for improving the GB implementation of MOIL—a project that we intend to pursue. The SDEL protocol tries to find a minimum in $3 \times K \times N$ (for this system $\approx 3 \times 10^6$) dimensional space, which also adds to the complexity of the calculations. Note, however, that the CG dimensionality is smaller by only a factor

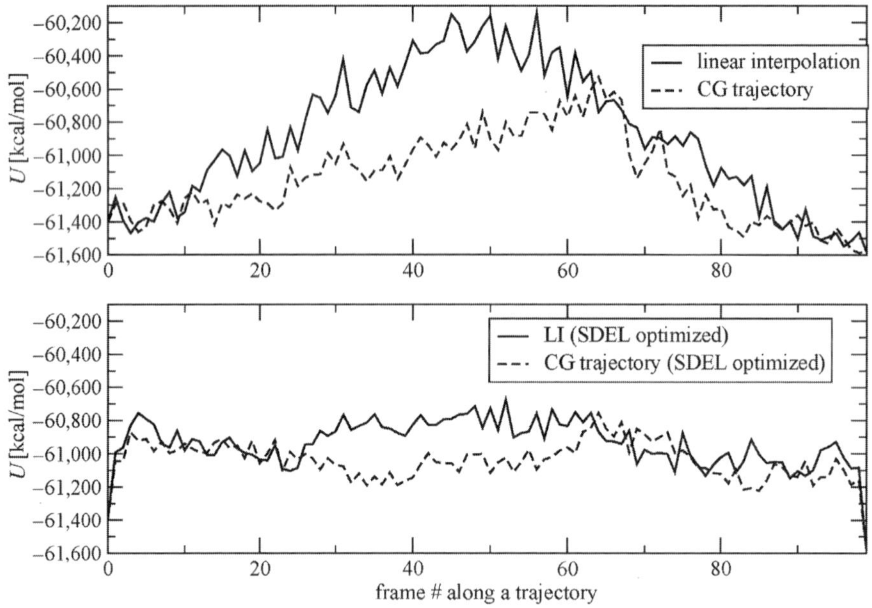

FIGURE 13.4 The energy profile of optimized SDEL trajectories for atomically detailed model of mGluR1. The horizontal axis shows frame index along the trajectory. The vertical axis is the potential energy. The top part shows potential energy profiles for trajectories from which SDEL optimization was started. In the bottom part the potential energy profiles after the SDEL optimization are shown. Solid lines correspond to trajectories optimized from the linear interpolation and dashed lines correspond to trajectories starting from the optimal CG trajectory for $H_S = 0$.

of 10–100, significantly smaller than the factor of 4000 mentioned above, between the calculations of the atomically detailed and CG models. The dominant factor in the latter is the much simpler (smoother) ANM potential compared to atomically detailed potentials. Even when employing simulated annealing for SDEL search, the path search stays in a neighborhood of the initial guess trajectory. The 10 K SDEL steps modify the trajectory of mGluR transition by no more than 0.1 Å/frame,* compared to 1.25 Å/frame obtained by the CG preprocessing, which is clearly more significant.

Figure 13.4 shows the potential energy profiles of optimal paths selected by SDEL. Only profiles for trajectories that were optimized from the LI path and SDP ($H_S = 0$) are shown. The energy profiles of other trajectories refined by SDEL runs have comparable values. The SDEL adds considerable thermal kinetic energy to the SDP path, making the SDEL potential energies higher than the SDP potential energies, and the SDEL path more appropriate for describing the thermal processes. The energy barrier for a trajectory starting from the LI path is somewhat higher than the barrier obtained from a path derived from the SDP, suggesting an improvement in the SDEL trajectory produced from the Steepest Descent Path.

Figure 13.5 shows the simulated annealing history of an SDEL run. The target function, T, rapidly decreases in the first thousand steps; however, further reduction with more steps is considerably slower. The RMSD changes of the trajectory in the last thousands of steps of the minimization are small (order of 10^{-2} Å/frame) and thus it might not be so important to locate the exact global minimum of T if the structural changes are of prime interest. Any substantial differences in the inferences obtained from the SDP- and LI-based paths can be revealed by examining the contacts between the two monomers that substantially change during the transition.

In Figure 13.6, we examine the distance between GLU A 60 (atom OE2) and ARG B 448 (atom NH2). These two atoms are not in contact in either of the two end conformations, but are brought

* The 0.1 Å RMSD per frame is based on C_α atoms only. The all-atom RMSD is approximately twice as large.

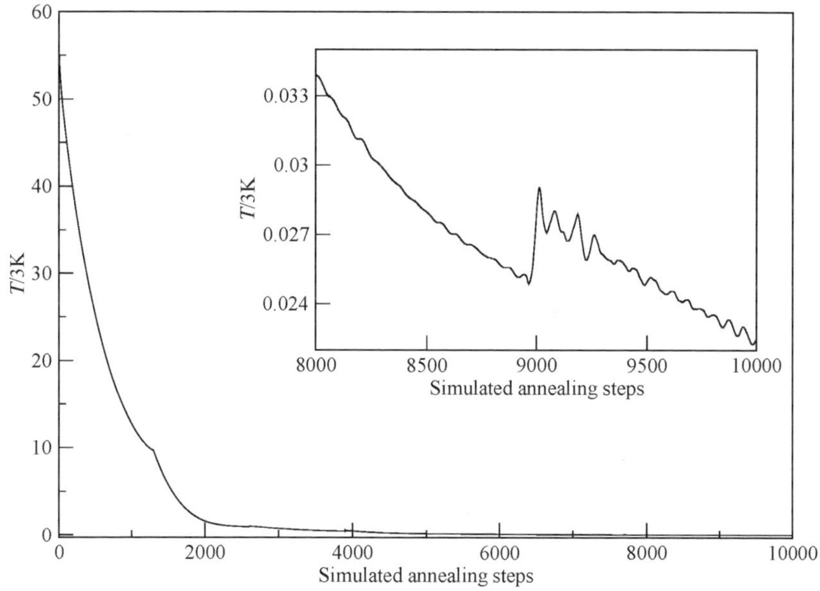

FIGURE 13.5 The simulated annealing profile in an SDEL minimization of the target function T as a function of the number of minimization steps. The example is for a CG trajectory with $H_S = 0$. In the right insert we expand the view of the last 2000 steps. The decrease in the target function is rapid at the beginning, but in the last 2000 steps the target function is decreased by only 33%.

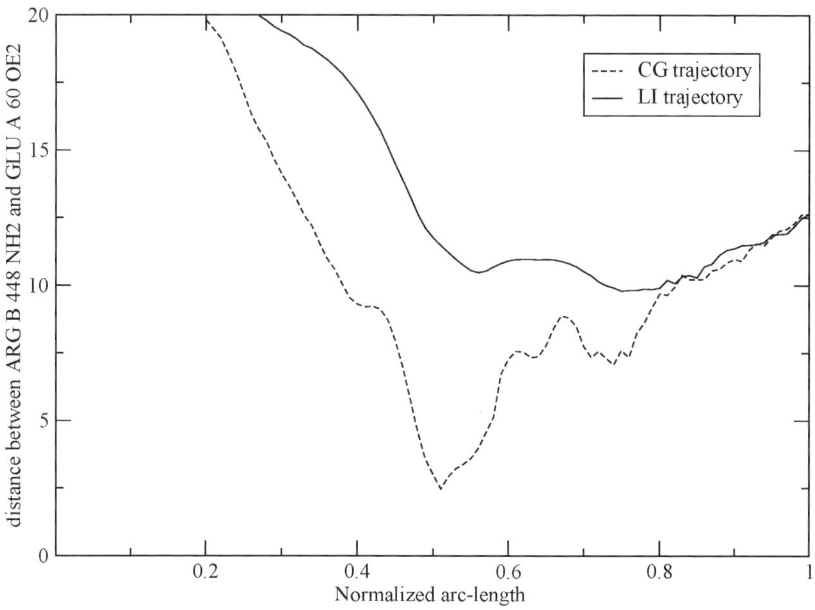

FIGURE 13.6 The distance between GLU A 60 (atom OE2) and ARG B 448 (atom NH2) during the transition between the inactive and active conformation of mGluR1.

together during the transition in the optimal path based on SDP, but not in the path based on LI. Notably, the formation of an intermediate salt bridge may reduce the barrier height for the transition. This evolution is represented in the sequence of structures shown in Figure 13.7 that are taken from points along the path. This sequence illustrates the formation and the breaking of the salt bridge along the path.

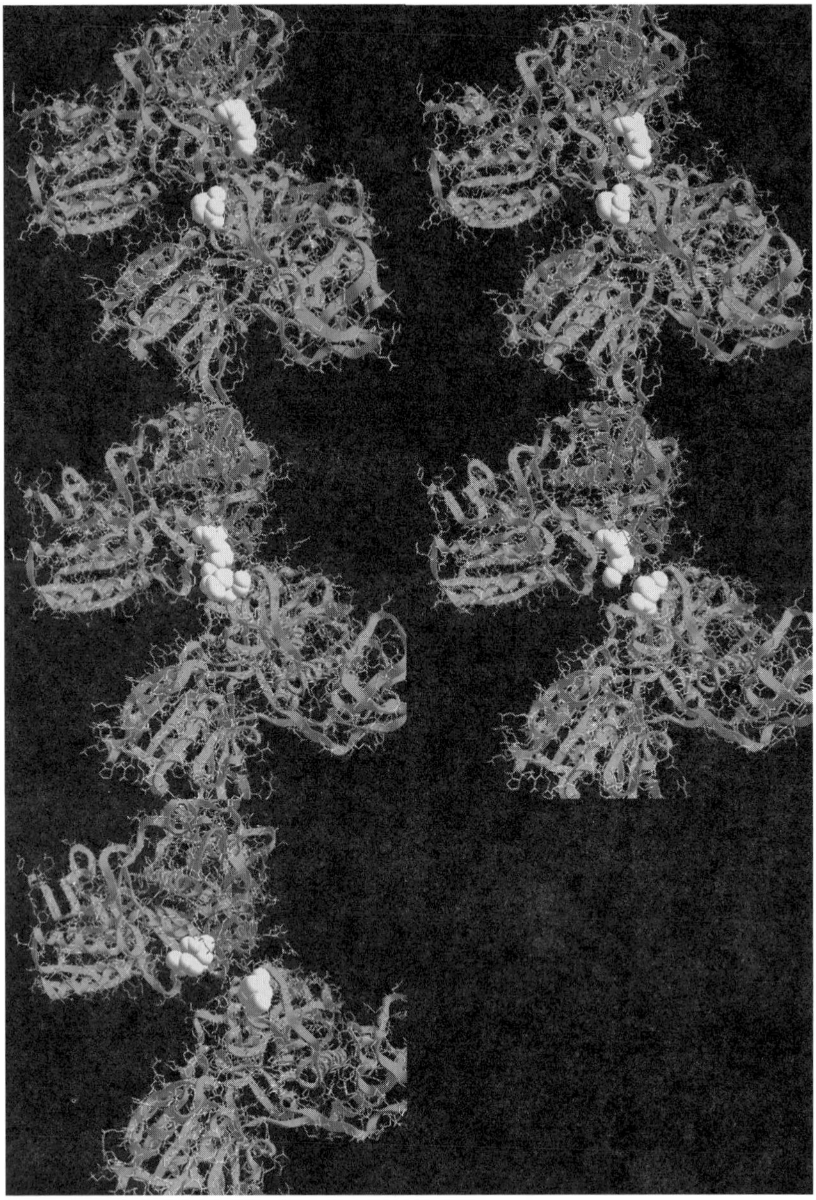

FIGURE 13.7 (See color insert following page 238.) An illustration of the strong coupling between atomically detailed motion and large-scale domain opening. A sequence of events along the transitional pathway is shown starting from the upper left corner (structure 1) continuing to the right (structure 20) and then down. The length slices are shown from an atomically detailed path of 100 slices that was constructed from a coarse-grained model. Only slices 1, 20, 50, 60, and 100 are shown. The atomically detailed event is the *transient formation of a salt bridge* between a glutamic acid (Glu60 in chain A of the dimer) and an arginine (Arg448 in chain B) (yellow space filling model). There is also a large-scale motion that causes a visible separation between the two lobes. The salt bridge is not present at the end points. It assists in lowering the transition barrier. Notably, it is not present in the linearly interpolated path (see text for more details).

13.8 APPENDIX I: PARALLEL CALCULATIONS OF BOUNDARY VALUE PATHWAYS

The calculation of trajectories with the SDEL formulation requires the determination of the paths that minimize the action S^l_{Gauss}, or more precisely minimize the target function T as described in Equation 13.7. The simulated annealing procedure requires the values of $\partial T/\partial \mathbf{x}_j$ $j = 2,\ldots,N-1$ for a gradient-based move. Let us examine the communication and computation required to calculate the $\partial T/\partial \mathbf{x}_j$ values. The most complex part of the function T is S^l_{Gauss} itself. S^l_{Gauss} is a sum of squared norms of $\partial \bar{S}/\partial \mathbf{x}_j$'s, which are functions of $\mathbf{x}_j$, $\mathbf{x}_{j\pm1}$, $\partial U/\partial \mathbf{x}_j$, $U_{j\pm1}$, and U_j. See the exact formula in Appendix II. Here we use an abbreviation $\partial \bar{S}/\partial \mathbf{x}_j = \mathbf{F}(\mathbf{x}_{j-1},\mathbf{x}_j,\mathbf{x}_{j+1},\partial U/\partial \mathbf{x}_j,U_{j-1},U_j,U_{j+1})$, and $\partial \bar{S}/\partial \mathbf{x}_{jk} = \mathrm{F_k}(\ldots)$. After substituting $\partial \bar{S}/\partial \mathbf{x}_j$ into $\partial S^l_{\mathrm{Gauss}}/\partial x_{jm}$ we get

$$\frac{\partial S^l_{\mathrm{Gauss}}}{\partial x_{jm}} = 2\sum_k\left(\frac{\partial \bar{S}}{\partial x_{j-1k}}\frac{\partial^2 \bar{S}}{\partial x_{j-1k}\partial x_{jm}} + \frac{\partial \bar{S}}{\partial x_{jk}}\frac{\partial^2 \bar{S}}{\partial x_{jk}\partial x_{jm}} + \frac{\partial \bar{S}}{\partial x_{j+1k}}\frac{\partial^2 \bar{S}}{\partial x_{j+1k}\partial x_{jm}}\right)$$

$$= 2\sum_k\left(\begin{array}{l}\mathrm{F_k}(\mathbf{x}_{j-2},\mathbf{x}_{j-1},\mathbf{x}_j,\partial U/\partial \mathbf{x}_{j-1},U_{j-2},U_{j-1},U_j)\dfrac{\partial}{\partial x_{j-1k}}\mathrm{F_m}(\mathbf{x}_{j-1},\mathbf{x}_j,\mathbf{x}_{j+1},\partial U/\partial \mathbf{x}_j,U_{j-1},U_j,U_{j+1}) + \\ \mathrm{F_k}(\mathbf{x}_{j-1},\mathbf{x}_j,\mathbf{x}_{j+1},\partial U/\partial \mathbf{x}_j,U_{j-1},U_j,U_{j+1})\dfrac{\partial}{\partial x_{jk}}\mathrm{F_m}(\mathbf{x}_{j-1},\mathbf{x}_j,\mathbf{x}_{j+1},\partial U/\partial \mathbf{x}_j,U_{j-1},U_j,U_{j+1}) + \\ \mathrm{F_k}(\mathbf{x}_j,\mathbf{x}_{j+1},\mathbf{x}_{j+2},\partial U/\partial \mathbf{x}_{j+1},U_j,U_{j+1},U_{j+2})\dfrac{\partial}{\partial x_{j+1k}}\mathrm{F_m}(\mathbf{x}_{j-1},\mathbf{x}_j,\mathbf{x}_{j+1},\partial U/\partial \mathbf{x}_j,U_{j-1},U_j,U_{j+1})\end{array}\right)$$

$$= \mathrm{G}_m(\mathbf{x}_{j\pm2},\mathbf{x}_{j\pm1},\mathbf{x}_j,\partial U/\partial \mathbf{x}_{j\pm1},\partial U/\partial \mathbf{x}_j,\partial^2 U/\partial \mathbf{x}_j^2,U_{j\pm2},U_{j\pm1},U_j), \tag{13.14}$$

where $\mathbf{G}$ is a function that takes its listed inputs and returns a vector $\partial S^l_{\mathrm{Gauss}}/\partial \mathbf{x}_j$. In order to calculate the derivative of T with respect to $\mathbf{x}_j$ we need the position and potential information of five different structures, need to compute forces for three different structures, and compute a Hessian matrix for one structure. In previous studies the following protocol was used: A trajectory represented by N conformations is distributed among P processors, each processor being responsible for updating N/P successive conformations. Suppose that a processor p is responsible for conformations $\mathbf{x}_m,\ldots,\mathbf{x}_{m+N/P-1}$. According to Equation 13.14, the processor p requires positions $\mathbf{x}_{m-2},\ldots,\mathbf{x}_{m+N/P+1}$ to update its conformations. However, the positions of $\mathbf{x}_{m-2},\mathbf{x}_{m-1},\mathbf{x}_{m+N/P},\mathbf{x}_{m+N/P+1}$ are modified on different processors, and these conformations have to be communicated from the processors responsible for their updates. The send and receive communications sum up to $4.3n_{pt}$ floating-point numbers per each step of simulated annealing (where n_{pt} is the number of particles in the system). This amount of communication may contribute significantly to the computation clock-time. Therefore, the forces $\partial U/\partial \mathbf{x}_{m-1}$ and $\partial U/\partial \mathbf{x}_{m+N/P}$ (together with $U_{m-1},U_{m+N/P}$) that are required as an input for the function $\mathbf{G}$ in Equation 13.14 are recomputed on processor p, after the positions of $\mathbf{x}_{m-1},\mathbf{x}_{m+N/P}$ are received. It is recommended to communicate the values of U_{m-2} and $U_{m+N/P+1}$ since on most platforms their computation is more expensive than their communication. The proposed scheme requires $(N/P+2)$ force computations, N/P Hessian matrix computations, and $\approx 4\cdot 3n_{pt}$ floating-point numbers received and sent in each step for each processor. The scheme provides reasonable scaling, unless the number of processors P approaches the number of conformations N. The last limit is approached for a large system (like mGluR1) for which we wish to exploit the benefit of parallelization to the maximum. In these cases we assign a single structure to each processor ($P=N$), then we require $(N/P+2)=3$ force calculations per algorithm step. The number of the force calculations could be reduced to one per step, if we allow for additional communication of $\partial U/\partial \mathbf{x}_{j\pm1}$ from neighboring processors. This reduces the number of forces calculations per step per processor from $(N/P+2)$ to N/P and increases the amount of communication to $6.3n_{pt}$. The actual algorithm as implemented in MOIL uses a slightly different reduction of Equation 13.14, which can be rewritten as

$$\frac{\partial S^{l}_{Gauss}}{\partial x_{jm}} = \mathbf{G}'_m(\mathbf{x}_{j\pm1}, \mathbf{x}_j, \partial \bar{S}/\partial \mathbf{x}_{j\pm1}, \partial U/\partial \mathbf{x}_{j\pm1}, \partial U/\partial \mathbf{x}_j, \partial^2 U/\partial \mathbf{x}_j^2, U_{j\pm1}, U_j), \tag{13.15}$$

where instead of $\mathbf{x}_{j\pm2}, U_{j\pm2}$, derivatives $\partial \bar{S}/\partial \mathbf{x}_{j\pm1}$ are used. This solution is equivalent (up to negligible constants) to the former one in terms of computation, memory, and communication; the advantage is that it can be implemented without requiring extra special cases in the code for the first and last processors and the code is more easily generalizable for the case $\mathrm{mod}(P,N) \neq 0$.

Additional reduction in the computation time can be obtained by transforming the problem of Hessian matrix computation to an additional force computation. This can be done because the Hessian $\partial^2 U/\partial \mathbf{x}_j^2$ is used in calculation of $\partial T/\partial \mathbf{x}_j$ only for a multiplication with some other vector $\mathbf{v}$. The following first-order approximate reduction can be used to compute the product $\partial^2 U/\partial \mathbf{x}_j^2 \cdot \mathbf{v}$ [Eric Vanden-Eijnden, private communication]:

$$\left.\frac{\partial^2 U}{\partial \mathbf{x}^2}\right|_{\mathbf{x}_j} \cdot \mathbf{v} \approx \frac{1}{\alpha}\left(\left.\frac{\partial U}{\partial \mathbf{x}}\right|_{\mathbf{x}_j+\alpha\mathbf{v}} - \left.\frac{\partial U}{\partial \mathbf{x}}\right|_{\mathbf{x}_j}\right), \tag{13.16}$$

The expression in Equation 13.16 becomes accurate for a sufficiently small scalar α. For the MOIL potential energy function with implicit solvent modeling, the calculation of Hessian matrix is approximately 50% more expensive than the calculation of the forces. However, the benefit of introducing this approximate reduction is not only in those 50% of run time, it also makes the code simpler and more understandable, since the formulas for the Hessian calculation are significantly more complex than those for the forces calculation. Derivatives of other terms in function T (from Equation 13.7) can be computed from local information kept on each processor.

An additional piece of global information required on each processor is $\langle \Delta l \rangle$, which is a slowly varying function of the number of optimization steps. Therefore, it suffices to recompute $\langle \Delta l \rangle$ every 10–20 steps of simulated annealing. The parallel computation of $\langle \Delta l \rangle$ can be done classically in 2 log K communication rounds with a total number of $2P$ (single floating-point number) messages passed.

13.9 APPENDIX II: EXPLICIT EXPRESSION FOR THE SDEL ACTION

The exact formulas for SDEL derivatives: i, j are index frames in a trajectory and k, m are indices over degrees of freedom (atoms' positions) in particular frames.

$$\frac{\partial \bar{S}}{\partial x_{im}} = \frac{1}{2}\left(\frac{-\partial U/\partial x_{im}}{p_i}(\Delta l_{i,i+1} + \Delta l_{i-1,i}) + (p_i + p_{i+1})\frac{x_{im} - x_{i+1m}}{\Delta l_{i,i+1}} + (p_{i-1} + p_i)\frac{x_{im} - x_{i-1m}}{\Delta l_{i-1,i}}\right.,$$

where $p_i = \sqrt{2(E - U(\mathbf{x}_i))}$ and $\Delta l_{i,i+1} = |\mathbf{x}_{i+1} - \mathbf{x}_i|$

$$\frac{\partial^2 \bar{S}}{\partial x_{im}\partial x_{ik}} = \frac{1}{2}\left(\begin{array}{l} -\left(\frac{\partial^2 U/\partial x_{im}\partial x_{ik}}{p_i} + \frac{\partial U/\partial x_{im}\,\partial U/\partial x_{ik}}{p_i^3}\right)(\Delta l_{i,i+1} + \Delta l_{i-1,i}) \\ -\frac{\partial U/\partial x_{im}}{p_i}\left(\frac{x_{ik}-x_{i+1k}}{\Delta l_{i,i+1}} + \frac{x_{ik}-x_{i-1k}}{\Delta l_{i-1,i}}\right) - \frac{\partial U/\partial x_{ik}}{p_i}\left(\frac{x_{im}-x_{i+1m}}{\Delta l_{i,i+1}} + \frac{x_{im}-x_{i-1m}}{\Delta l_{i-1,i}}\right) \\ +\delta_{km}\left(\frac{(p_i + p_{i+1})}{\Delta l_{i,i+1}} + \frac{(p_{i-1} + p_i)}{\Delta l_{i-1,i}}\right) \\ -(p_i + p_{i+1})\frac{(x_{im}-x_{i+1m})(x_{ik}-x_{i+1k})}{(\Delta l_{i,i+1})^3} - (p_{i-1} + p_i)\frac{(x_{im}-x_{i-1m})(x_{ik}-x_{i-1k})}{(\Delta l_{i-1,i})^3} \end{array}\right)$$

$$\frac{\partial^2 \bar{S}}{\partial x_{i-1k}\partial x_{im}} = \frac{1}{2}\left(\begin{array}{l} -\frac{\partial U/\partial x_{im}}{p_i}\frac{x_{i-1k}-x_{ik}}{\Delta l_{i,i+1}\Delta l_{i-1,i}} - \frac{\partial U/\partial x_{i-1k}}{p_{i-1}}\frac{x_{im}-x_{i-1m}}{\Delta l_{i,i+1}\Delta l_{i-1,i}} \\ -\frac{p_{i-1}+p_i}{\Delta l_{i,i+1}\Delta l_{i-1,i}}\delta_{km} + (p_{i-1}+p_i)\frac{(x_{im}-x_{i-1m})(x_{ik}-x_{i-1k})}{(\Delta l_{i-1,i})^3} \end{array}\right)$$

$$\frac{\partial^2 \bar{S}}{\partial x_{i+1k}\partial x_{im}} = \frac{1}{2}\left(\begin{array}{l} -\frac{\partial U/\partial x_{im}}{p_i}\frac{x_{i+1k}-x_{ik}}{\Delta l_{i,i+1}\Delta l_{i-1,i}} - \frac{\partial U/\partial x_{i+1k}}{p_{i+1}}\frac{x_{im}-x_{i-1m}}{\Delta l_{i,i+1}\Delta l_{i-1,i}} \\ -\frac{p_{i+1}+p_i}{\Delta l_{i,i+1}\Delta l_{i-1,i}}\delta_{km} + (p_{i-1}+p_i)\frac{(x_{im}-x_{i-1m})(x_{ik}-x_{i+1k})}{(\Delta l_{i,i+1})^3} \end{array}\right),$$

where δ_{km} is Kronecker's delta $\left(\delta_{km} = \begin{cases} 1 & \text{if } k = m \\ 0 & \text{otherwise} \end{cases}\right)$.

$$\frac{\partial S^l_{Gauss}}{\partial x_{jk}} \approx \sum_{i,m} \frac{\partial \left(\frac{\partial \bar{S}}{\partial x_{im}}\right)^2}{\partial x_{jk}} = 2\sum_{i,m}\left(\frac{\partial \bar{S}}{\partial x_{im}}\right)\left(\frac{\partial^2 \bar{S}}{\partial x_{im}\partial x_{jk}}\right)$$

REFERENCES

Alon, U. 2006. *An introduction to systems biology: Design principles of biological circuits.* Vol. 10. Chapman Hall. CRC Mathematical and Computational Biology Series: CRC.

Aqvist, J., and A. Warshel. 1993. Simulation of enzyme reactions using valence bond force fields and other hybrid quantum/classical approaches. *Chemical Reviews* 93: 2523–44.

Bolhuis, P. G., D. Chandler, C. Dellago, and P. L. Geissler. 2002. Transition path sampling: Throwing ropes over rough mountain passes, in the dark. *Annual Review of Physical Chemistry* 53: 291–318.

Briggs, W.L. 1987. *A multigrid tutorial.* Philadelphia: SIAM.

Cardenas, A. E., and R. Elber. 2003. Kinetics of cytochrome C folding: Atomically detailed simulations. *Proteins: Structure Function and Genetics* 51 (2): 245–57.

Cox, T.F., and M.A.A. Cox. 1994. *Multidimensional scaling.* London: Chapman and Hall, London.

Czerminski, R., and R. Elber. 1990. Self avoiding walk between 2 fixed end points as a tool to calculate reaction paths in large molecular systems. *International Journal of Quantum Chemistry* 38 (S24): 167–86.

Dellago, C., P. G. Bolhuis, and P. L. Geissler. 2002. Transition path sampling. In Advances in chemical physics, Vol. 123.

E, W. N., W. Q. Ren, and E. Vanden-Eijnden. 2002. String method for the study of rare events. *Physical Review B* 66 (5): 052301.

Elber, R. 1990. Calculation of the potential of mean force using molecular dynamics with linear constraints: An application to a conformational transition in a solvated dipeptide. *Journal of Chemical Physics* 93 (6): 4312–21.

Elber, R. 2006. Calculations of classical trajectories with boundary value formulation. In *Computer simulations in condensed matter: From materials to chemical biology*, ed. M. Ferrario, G. Ciccotti, and K. Binder. Berlin: Springer.

Elber, R., A. Ghosh, and A. Cardenas. 2002. Long time dynamics of complex systems. *Accounts of Chemical Research* 35 (6): 396–403.

Elber, R., and M. Karplus. 1987. A method for determining reaction paths in large molecules: Application to myoglobin. *Chemical Physics Letters* 139 (5): 375–80.

Elber, R., J. Meller, and R. Olender. 1999. Stochastic path approach to compute atomically detailed trajectories: Application to the folding of C peptide. *Journal of Physical Chemistry B* 103 (6): 899–911.

Elber, R., A. Roitberg, C. Simmerling, R. Goldstein, H. Y. Li, G. Verkhivker, C. Keasar, J. Zhang, and A. Ulitsky. 1995. Moil: A program for simulations of macromolecules. *Computer Physics Communications* 91 (1–3): 159–89.

Elber, R., and D. Shalloway. 2000. Temperature dependent reaction coordinates. *Journal of Chemical Physics* 112 (13): 5539–45.

Haliloglu, T., I. Bahar, and B. Erman. 1997. Gaussian dynamics of folded proteins. *Physical Review Letters* 79:3090–93.

Jonsson, H., G. Mills, and K.W. Jacobsen. 1998. Nudged elastic band method for finding minimum energy paths of transitions. In *Classical and quantum dynamics in condensed phase simulations*, ed. B. J. Berne, G. Ciccotti, and D. F. Coker. Singapore: World Scientific.

Kunishima, N., Y. Shimada, Y. Tsuji, T. Sato, M. Yamamoto, T. Kumasaka, S. Nakanishi, H. Jingami, and K. Morikawa. 2000. Structural basis of glutamate recognition by a dimeric metabotropic glutamate receptor. *Nature* 407 (6807): 971–77.

Lanczos, C. 1970. *The variational principles of mechanics*. University of Toronto Press.

Landau, L. D., and E. M. Lifshitz. 1976. *Mechanics. Vol. 1, Course of theoretical physics*. Oxford: Pergamon.

Lu, M. Y., B. Poon, and J. P. Ma. 2006. A new method for coarse-grained elastic normal-mode analysis. *Journal of Chemical Theory and Computation* 2 (3): 464–71.

Maragakis, P., and M. Karplus. 2005. Large amplitude conformational change in proteins explored with a plastic network model: Adenylate kinase. *Journal of Molecular Biology* 352 (4): 807–22.

Muto, T., D. Tsuchiya, K. Morikawa, and H. Jingami. 2007. Structures of the extracellular regions of the group II/III metabotropic glutamate receptors. *Proceedings of the National Academy of Sciences of the United States of America* 104 (10): 3759–64.

Olender, R., and R. Elber. 1996. Calculation of classical trajectories with a very large time step: Formalism and numerical examples. *Journal of Chemical Physics* 105 (20): 9299–315.

Olender, R., and R. Elber. 1997. Yet another look at the steepest descent path. *Theochem: Journal of Molecular Structure* 398: 63–71.

Peskin, C.S., and T. Schlick. 1989. Molecular dynamics by the backward-Euler method. *Communications on Pure and Applied Mathematics* 42: 1001–31.

Pin, J. P., and F. Acher. 2002. The metabotropic glutamate receptors: Structure, activation mechanism and pharmacology. *Current Drug Targets CNS and Neurological Disorders* 1 (3): 297–317.

Pin, J. P., J. Kniazeff, C. Goudet, A. S. Bessis, J. Liu, T. Galvez, F. Acher, P. Rondard, and L. Prezeau. 2004. The activation mechanism of class-C G-protein coupled receptors. *Biology of the Cell* 96 (5): 335–42.

Pin, J. P., J. Kniazeff, J. Liu, V. Binet, C. Goudet, P. Rondard, and L. Prezeau. 2005. Allosteric functioning of dimeric class C G-protein-coupled receptors. *The FEBS Journal* 272 (12): 2947–55.

Ryckaert, J. P., G. Ciccotti, and H. J. C. Berendsen. 1977. Numerical integration of cartesian equations of motion of a system with constraints: Molecular dynamics of N-alkanes. *Journal of Computational Physics* 23 (3): 327–41.

Schlick, T., R. D. Skeel, A. T. Brunger, L. V. Kale, J. A. Board, J. Hermans, and K. Schulten. 1999. Algorithmic challenges in computational molecular biophysics. *Journal of Computational Physics* 151 (1): 9–48.

Tirion, M. M. 1996. Large amplitude elastic motions in proteins from a single-parameter, atomic analysis. *Physical Review Letters* 77 (9): 1905–908.

Ulitsky, A., and R. Elber. 1990. A new technique to calculate steepest descent paths in flexible polyatomic systems. *Journal of Chemical Physics* 92 (2): 1510–11.

Ulitsky, A., and R. Elber. 1993. The thermal equilibrium aspects of the time-dependent hartree and the locally enhanced sampling approximations: Formal properties, a correction, and computational examples for rare gas clusters. *Journal of Chemical Physics* 98 (4): 3380–88.

Ulitsky, A., and R. Elber. 1994. Application of the locally enhanced sampling (LES) and a mean field with a binary collision correction (cLES) to the simulation of Ar diffusion and NO recombination in myoglobin. *Journal of Physical Chemistry* 98 (3): 1034–43.

Verlet, L. 1967. Computer "experiments" on classical fluids. I. Thermodynamical properties of Lennard-Jones molecules. *Physical Review* 159 (1): 98–103.

Weinan, E., W. Q. Ren, and E. Vanden-Eijnden. 2005. Finite temperature string method for the study of rare events. *Journal of Physical Chemistry B* 109 (14): 6688–93.

West, A. M. A., R. Elber, and D. Shalloway. 2007. Extending molecular dynamics time scales with milestoning: Example of complex kinetics in a solvated peptide. *Journal of Chemical Physics* 126 (14): 145104.

Xu, C. Y., D. Tobi, and I. Bahar. 2003. Allosteric changes in protein structure computed by a simple mechanical model: Hemoglobin T <-> R2 transition. *Journal of Molecular Biology* 333 (1): 153–68.

Yang, L. W., E. Eyal, C. Chennubhotla, J. Jee, A. M. Gronenborn, and I. Bahar. 2007. Insights into equilibrium dynamics of proteins from comparison of NMR and X-ray data with computational predictions. *Structure* 15 (6): 741–49.

Zaloj, V., and R. Elber. 2000. Parallel computations of molecular dynamics trajectories using the stochastic path approach. *Computer Physics Communications* 128 (1–2): 118–27.

14 Insights into the Sequence-Dependent Macromolecular Properties of DNA from Base-Pair Level Modeling

Wilma K. Olson, Andrew V. Colasanti, Luke Czapla, and Guohui Zheng
Department of Chemistry & Chemical Biology, BioMaPS Institute for Quantitative Biology, Rutgers, The State University of New Jersey

CONTENTS

14.1 INTRODUCTION

Some of the most interesting features of double-helical DNA occur at the mesoscopic level; that is, at chain lengths of a few hundred base pairs [Laughlin et al. 2000]. For example, DNA loops of this size bring proteins bound at separate, widely spaced sites on the double helix into close contact, influencing not only the three-dimensional organization of the genome but also the activity of proteins that control, shuffle, and express the genetic message [Semsey, Virnik, and Adhya 2005]. The nucleosome, the protein–DNA assembly that constitutes the basic packaging unit in the nucleus, binds similarly sized pieces of DNA: a 147-bp fragment winds $\sim$1.6 superhelical turns around the core of eight histone proteins in the best-resolved nucleosome core-particle structure [Richmond and Davey 2003].

Deformations of DNA structure in the absence of protein are so limited that chains of a hundred base pairs resemble rigid rods [Hagerman 1988]. The global flexibility of free DNA becomes significant only when the molecule is much longer than the so-called persistence length [Kratky and Porod 1949]; that is, the $\sim$ 150-bp distance over which the direction of the chain is maintained [Hagerman 1988; Olson et al. 1993]. At such chain lengths the sequence-dependent features of the constituent base pairs become magnified, revealing themselves in the higher-order folding of the double helix [Olson 1996]. On the other hand, some regularly repeating sequences of a few hundred base pairs exhibit strong natural curvature [Crothers et al. 1992; Hagerman 1986; Trifonov 1991], adopting circular and superhelical shapes. Such folding is an important facet of biological processes that involve long-range recognition and/or interactions. Indeed, naturally curved DNA can mimic the biological activity of proteins that bend DNA [Bracco et al. 1989].

The sequence-dependent structural properties of DNA play important roles at the mesoscopic level. For example, the judicious placement of curved inserts or sequences that bind a specific protein guides the overall folding of the DNA loops held in place by bacterial repressor proteins [Geanacopoulos et al. 2001; Mehta and Kahn 1999]. Other DNA sequences bias the positioning of nucleosomes [Cao et al. 1998; Lowary and Widom 1998; Thåström et al. 1999; Widlund et al. 1997], binding much more tightly than random DNA of the same base composition to the histone octamer [Tolstorukov et al. 2007]. Some of these so-called "positioning" sequences, with periodically repeating chemical features, also close much more easily than mixed-sequence DNA into tight minicircles [Cloutier and Widom 2004, 2005].

Treatment of the sequence-dependent properties of polymeric DNA lies beyond the scope of conventional all-atom calculations. Although there are a growing number of examples of detailed atomic simulations of the fluctuations of DNA in tight minicircles [Lankas, Lavery, and Maddocks 2006] and in nucleosomes [Bishop 2005; Roccatano, Barthel, and Zacharias 2007; Ruscio and Onufriev 2006], limitations on computer resources preclude routine study of mesoscopic problems of biological interest, such as the effects of nucleotide sequence and chain length on loop formation and nucleosome positioning. Furthermore, most atomic-level DNA simulations reported to date fail to account for a number of critical features of B-DNA structure, such as the sequence-dependent base-pair twist [Olson and Zhurkin 2000], the deformations of the chemical backbone that direct the movement of base pairs [Rohs, Sklenar, and Shakked 2005], and the associated propensities of DNA to undergo large-scale helical transformations [Feig and Pettitt 1998]. Confidence in the computed structures depends upon the extent to which the predictions match these and other critical benchmarks [Olson et al. 2006].

On the other hand, the classic phenomenological representation of DNA as an ideal, inextensible, naturally straight elastic rod with a persistence length of $\sim$ 150 bp is too simple to account for the effects of sequence on mesoscopic chain properties. Such models forfeit the connection between the observed behavior and the actions of the underlying nucleotide components. For example, interpretation of the in vivo looping properties of DNA in terms of the classic model ignores changes in local structure and deformation that are imparted by the binding of proteins and other molecules in the cellular environment, pointing instead to large, physically unrealistic changes in the flexibility and helical repeat of cellular DNA compared to "pure" DNA in solution [Becker, Kahn, and Maher 2005; Bellomy, Mossing, and Record, 1988; Law et al. 1993; Zhang et al. 2006]. An appropriate model must conform closely and in an identifiable manner with the actual characteristic of the real chain; that is, its structural geometry and the potentials impeding configurational deformations, for meaningful interpretation of observed polymeric properties [Flory 1969].

Our goal is to understand the mechanisms by which primary DNA base sequence and tightly bound proteins control the biophysical properties of the long, threadlike molecule under both in vitro and in vivo conditions. The central hypothesis of our work is that primary chemical sequence dictates how DNA folds and responds to stresses arising from its interactions with other molecules. Toward this end, we have developed a new coarse-grained model of DNA [Coleman, Olson, and Swigon 2003;

Olson et al. 1998], in which the base pairs are treated as rigid bodies subject to realistic, knowledge-based energy constraints [Olson et al. 1998] and new computational techniques to determine the minimum-energy configurations [Coleman, Olson, and Swigon 2003], intrinsic dynamics [Matsumoto and Olson 2002], and looping/cyclization propensities [Czapla, Swigon, and Olson 2006] of these molecules. We have learned how the configurational properties of adjacent base pairs depend upon sequence [Gorin, Zhurkin, and Olson 1995; Olson et al. 1998] and thereby contribute to the overall configuration of polymeric DNA [Matsumoto and Olson 2006; Olson, Swigon, and Coleman 2004].

This chapter starts with an analysis of the spatial arrangements of neighboring DNA base pairs, covering mathematical approaches developed to quantify the orientation and displacement of the base pairs as well as structural trends and sequence-dependent features revealed in the database of high-resolution protein–DNA crystal structures. A description of the effects of base sequence on the persistence length and the ring-closure properties of DNA molecules then follows. These findings rest upon a novel set of knowledge-based potentials of dimeric deformability, a matrix treatment of global chain configuration, and Monte Carlo sampling of DNA configuration space. The potentials also provide useful estimates of the nucleosome-positioning propensities of different sequences. The review concludes with a brief commentary on the general utility of a "realistic" base-pair-level treatment of polymeric DNA and some likely new developments in the coming years.

14.2 METHODS

14.2.1 Base-Pair Parameters

One of the simplest ways to monitor large-scale DNA configurational changes is in terms of the local bending, twisting, and displacement of neighboring base pairs. These deformations are analogous to the variation of chemical-bond lengths and angles in all-atom molecular models. Unlike simple valence-angle perturbations, the bending of base pairs entails two components—a so-called tilting motion θ_1 about the short axes of associated side groups and a rolling component θ_2 about the long axes (Figure 14.1). The twisting angle θ_3 relates corresponding vectors in successive base-pair planes and normally differs from the virtual "backbone" torsion defined with respect to the vectors between the centers of the base pairs; only in the absence of bending and translation are the two parameters equivalent. Compared to the simple one-dimensional stretching and compression of chemical bonds, the translational motions of the base pairs involve three components—two in-plane displacements termed shift θ_4 and slide θ_5 and an out-of-plane component called rise θ_6 (see Figure 14.1). The values of these parameters are defined by a matrix-based scheme (described below) that allows for the characterization of base-pair arrangements in known structures and the precise reconstruction of models from these values.

The atoms of each base pair n are expressed in a local, orthogonal coordinate frame $(\mathbf{d}_1^n, \mathbf{d}_2^n, \mathbf{d}_3^n)$ defined in accordance with recently established guidelines [Olson et al. 2001]. The $\mathbf{d}_1^n$ and $\mathbf{d}_2^n$ axes lie in the plane of the base pair, with $\mathbf{d}_1^n$ pointing in the direction of the major groove along the pseudodyad axis of the base pair and $\mathbf{d}_2^n$ running along the long axis of the base pair in the direction of the leading (sequence) strand (strand I in Figure 14.1), parallel to the C1'···C1' vector between complementary residues, and displaced so as to pass through the intersection of $\mathbf{d}_1^n$ with the vector connecting the pyrimidine C6 and purine C8 atoms. The $\mathbf{d}_3^n$ axis is perpendicular to the base-pair plane, pointing in the 5'-3' direction of the leading strand. The coordinate system of the base pair is further characterized by the position of its origin $\mathbf{o}_n$ and the rotation matrix $\mathbf{T}_n$, which relates the local frame on base pair n to that of its predecessor $n+1$.

The matrix $\mathbf{T}_n$ is expressed, following Zhurkin, Lysov, and Ivanov (1979) and El Hassan and Calladine (1995), in terms of a sequence of Euler rotations:

$$\mathbf{T}_n = \mathbf{R}_z\left(\frac{\theta_3}{2} - \phi\right)\mathbf{R}_y(\Gamma)\,\mathbf{R}_z\left(\frac{\theta_3}{2} + \phi\right). \tag{14.1}$$

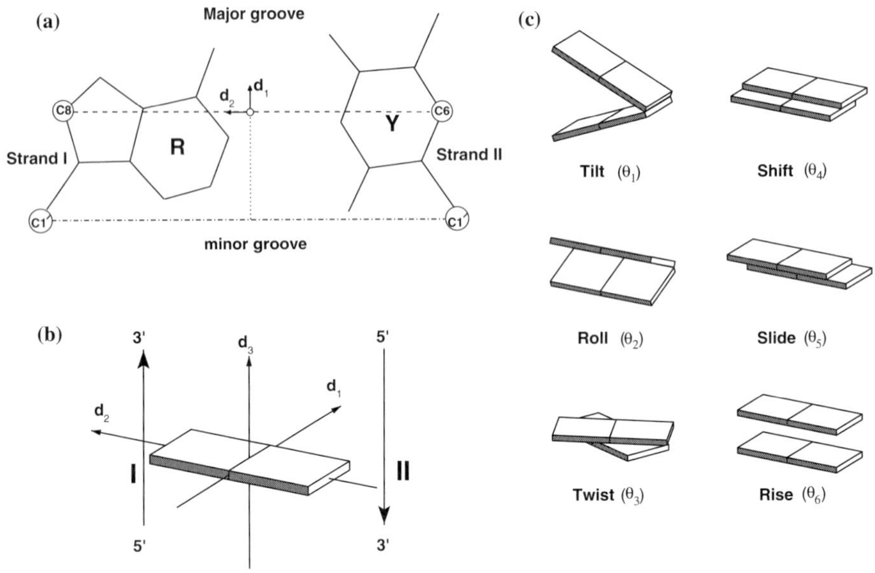

FIGURE 14.1 Illustration relating the chemical structure of the purine (R) and pyrimidine (Y) partners of a Watson-Crick base pair to the coarse-grained representation used to model the sequence-dependent properties of DNA: (a) position of the standard base-pair reference frame ($\mathbf{d}_1$, $\mathbf{d}_2$, $\mathbf{d}_3$) in terms of the identity of the base on the sequence strand I, the locations of the major groove edge and the structural pseudodyad axis (finely dotted vertical line), the direction of the vector joining the C1' atoms on complementary strands (dash-dot line), and the minimal displacement of that vector from the C8 atom on R and the C6 atom on Y (dashed line); (b) block diagram of the base pair showing the embedded reference frame, the directions of complementary strands I and II, and the (shaded) minor-groove edge of the base pair; (c) pictorial definitions depicting positive values of the six rigid-body parameters θ_i ($i = 1$–6) used to describe the orientation and displacement of sequential base pairs.

Here Γ is the net bend angle between base-pair planes; that is, $\Gamma = (\theta_1{}^2 + \theta_2{}^2)^{1/2}$, ϕ is a phase angle that "symmetrizes" the coordinate transformation; that is, $\phi = \tan^{-1}(\theta_1/\theta_2)$, and $\mathbf{R}_\mathbf{u}(\varphi)$ is a matrix describing a rotation by an amount φ about the axis $\mathbf{u}$. The elements of $\mathbf{R}$ are defined, following Jeffreys and Jeffreys (1946), in terms of the value of φ and the components (u_x, u_y, u_z) of the unit vector $\mathbf{u}$:

$$r_{\nu\mu} = (1 - \cos\varphi)\, u_\nu u_\mu - \sin\varphi \sum_\kappa \varepsilon_{\nu\mu\kappa} u_\kappa + \delta_{\nu\mu} \cos\varphi\,, \tag{14.2}$$

where $\delta_{\nu\mu}$ is the Kronecker delta and $\varepsilon_{\nu\mu\kappa} = \pm 1$ when ν, μ, κ are even or odd permutations of 1, 2, 3, respectively, and vanishes otherwise.

Translational parameters are expressed in terms of the projection of the vector $\mathbf{r}_n = (\mathbf{o}_{n+1} - \mathbf{o}_n)$ that links the origins of base pairs n and $n+1$ onto the coordinate axes of the "middle" dimeric frame, an intermediate coordinate frame defined by the halfway rotational operation $\mathbf{T}_n^{1/2}$. That is, $\mathbf{r}_n = \mathbf{T}_n^{1/2}\hat{\rho}_n$, where $\hat{\rho}_n$ is a translational vector with components $(\theta_4^n, \theta_5^n, \theta_6^n)$ in the "middle" frame of the nth base-pair step and $\mathbf{T}_n^{1/2}$ is the matrix that expresses $\hat{\rho}_n$ in the coordinate frame of base pair n; that is,

$$\mathbf{T}_n^{1/2} = \mathbf{R}_\mathbf{z}\left(\frac{\theta_3}{2} - \phi_3^n\right)\mathbf{R}_\mathbf{y}\left(\frac{\Gamma}{2}\right)\mathbf{R}_\mathbf{z}(\phi)\,. \tag{14.3}$$

The subscripts **y** and **z** in this expression as well as in Equation 14.1 refer, respectively, to the long axis and the normal of the "middle" frame.

14.2.2 Chain Model and Dimensions

Base-pair level models of DNA are constructed from the serial products of generator matrices $\mathbf{A}_n$ that incorporate the displacement vectors $\mathbf{r}_n$ and the rotation matrices $\mathbf{T}_n$, which relate coordinate frames on successive base pairs: $\mathbf{A}_{1:N} = \mathbf{A}_1\mathbf{A}_2...\mathbf{A}_{N-1}\mathbf{A}_N$ [Marky and Olson 1994], where

$$\mathbf{A}_n = \begin{bmatrix} \mathbf{T}_n & \mathbf{r}_n \\ \mathbf{0} & 1 \end{bmatrix} \text{ and } \mathbf{A}_{1:N} = \begin{bmatrix} \mathbf{T}_{1:N} & \mathbf{r}_{1:N} \\ \mathbf{0} & 1 \end{bmatrix}. \tag{14.4}$$

Values used to evaluate chain configuration—(i) the end-to-end vector $\mathbf{r}_{1:N+1}$, (ii) the cosine of the angle γ between the normals of terminal base pairs, and (iii) the twisting τ of terminal base pairs—are embedded in $\mathbf{A}_{1:N+1}$ [Czapla, Swigon, and Olson 2006]:

$$\mathbf{r}_{1:N+1} = \begin{bmatrix} \mathbf{I}_3 & \mathbf{0} \end{bmatrix} \mathbf{A}_{1:N+1} \begin{bmatrix} \mathbf{0} \\ 1 \end{bmatrix}, \tag{14.5a}$$

$$\cos\gamma = \begin{bmatrix} 0 & 0 & 1 & 0 \end{bmatrix} \mathbf{A}_{1:N+1} \begin{bmatrix} 0 \\ 0 \\ 1 \\ 0 \end{bmatrix}, \tag{14.5b}$$

$$\mathrm{Tr}\,(\mathbf{T}_{1:N+1}) = \cos\tau\,(1+\cos\gamma) + \cos\gamma. \tag{14.5c}$$

Here $\mathbf{I}_3$ is the identity matrix of order three and the **0**s are null matrices of orders necessary to fill the 3×4 premultiplication and 4×1 postmultiplication vectors. A joining step $N+1$, which is included in these expressions to test for terminal base-pair overlap, is subsequently removed and circles are closed by a step c that connects the Nth to the first base pair; that is, $\mathbf{A}_{1:N}\mathbf{A}_c = \mathbf{I}$, where $\mathbf{I}$ is the 4×4 identity matrix.

14.2.3 Deformational Energy

The deformational energy U of a configuration of DNA is the sum, over n, of the energy of interaction ψ_n of the nth and $(n+1)$th base pairs, $U = \sum_{n=1}^{N} \psi_n$. Here ψ_n is a function of the relative orientation, the displacement, and the chemical composition of base pairs n and $n+1$, and N is the number of base-pair steps that make up the DNA. The known complementarity of Watson–Crick base pairs; that is, the specific association of adenine with thymine (A·T) and guanine with cytosine (G·C), and the antiparallel directions of the sugar-phosphate chains place restrictions on the ψ_n. That is, step parameters are defined such that tilt and shift (θ_1, θ_4) change signs in complementary strands [Dickerson et al. 1989], and the potential $\psi_n(\mathrm{XZ})$ of dimer step XZ determines that of its complement X'Z' [Coleman, Olson, and Swigon 2003; Olson et al. 1998].

The deformability of DNA is based on the range of configurational states found in a nonredundant set of 239 protein–DNA crystal complexes of 2.5 Å or better resolution, taken from the Nucleic Acid Database [Berman et al. 1992]. The dataset includes 101 structures of double-helical DNA bound to enzymes, 121 duplexes in the presence of regulatory proteins, 16 complexes with structural proteins, and one DNA associated with a multifunctional protein [Olson et al. 2006]. The structures have been filtered to exclude over-represented complexes in order to obtain a balanced sample of spatial and functional forms. The dinucleotide samples exclude chemically modified bases, terminal and penultimate base pairs, and side groups attached to nicked backbones. The working dataset also omits base pairs on nucleotides that are attached to modified or mispaired residues. The preferred arrangements and likely fluctuations of base-pair steps are derived from the average properties of the dimeric units in these structures [Olson et al. 1998].

The cost of deformation $\psi_n(\mathrm{XZ})$ of a given base-pair step is expressed by a double summation of elastic terms over the six base-pair step parameters:

$$\psi_n\left(\mathrm{XZ}\right)=\frac{1}{2}\sum_{i=1}^{6}\sum_{j=1}^{6} f_{ij}\left(\mathrm{XZ}\right)\Delta\theta_i^n\Delta\theta_j^n . \tag{14.6}$$

Here $\Delta\theta_i^n=\theta_i^n-\theta_i^0(\mathrm{XZ})$ is the imposed deviation of the ith step parameter θ_i^n at the nth dinucleotide step from the equilibrium rest-state value $\theta_i^0(\mathrm{XZ})$ of the XZ dimer step, and the $f_{ij}(\mathrm{XZ})$ are stiffness constants determined by the XZ sequence. The rest-state values of the dinucleotide steps are equated to the average step parameters of the XZ dimers in the protein–DNA sample; that is, $\theta_i^0(\mathrm{XZ})=\langle\theta_i(\mathrm{XZ})\rangle$ ($i=1$–6), and the stiffness constants are extracted from the pairwise covariance of these variables; that is, the covariance matrix with elements given by the differences between the mean squares and the squares of the means of all pairs of step parameters, $\langle\theta_i(\mathrm{XZ})\theta_j(\mathrm{XZ})\rangle-\langle\theta_i(\mathrm{XZ})\rangle\langle\theta_j(\mathrm{XZ})\rangle$, and equal to the inverse of the 6×6 force-constant matrix $\mathbf{F}(\mathrm{XZ})$ that contains the $f_{ij}(\mathrm{XZ})$ [Olson et al. 1998]. Such an approach accounts for both the sequence-dependent structure of DNA and the correlations of dinucleotide step parameters, which are especially important for "realistic" models of DNA. The model, however, omits consideration of (i) the sequence context of the given dimer; that is, the spatial configuration of a given dimer is assumed to be independent of that of adjacent base-pair steps, (ii) the precise arrangement of complementary purine and pyrimidine bases, such as the propeller and buckle angles that effect base-pair nonplanarity, (iii) the detailed arrangement of the sugar-phosphate backbone, and (iv) the "structure" of the surrounding chemical environment. Backbone and solvent atoms are implicitly treated in the energy terms so that their omission introduces no serious error when duplex deformations are limited to energies of the order of k_BT, where k_B is the Boltzmann constant and T the temperature, and the DNA remains within the limits of the B-DNA family. If distortions are large, such as in a "melted" helix, these atoms should be incorporated, along with complementary base-pair parameters; that is, the rigid-body parameters that describe the orientation and displacement of paired purine and pyrimidine bases, in the DNA model. The energies derived from the protein-DNA crystal set must also be scaled to account for known configuration-dependent properties of DNA in solution (see below).

14.2.4. Configurational Sampling

By denoting the configuration of base-pair step n by the vector Θ_n, with components θ_i^n ($i=1$–6) corresponding to the instantaneous values of the angular and translational parameters at the given step and defining $\Theta^0(\mathrm{XZ})$ as the vector that contains the intrinsic step parameters of dimer XZ, the potential of the step can be expressed in matrix form as $\psi_n(\mathrm{XZ})=(1/2)\Delta\Theta^{\mathrm{T}}\mathbf{F}(\mathrm{XZ})\Delta\Theta$, where $\Delta\Theta=\Theta_n-\Theta^0(\mathrm{XZ})$. To facilitate the sampling of representative chain configurations, each dimeric energy contribution $\psi_n(\mathrm{XZ})$ is reexpressed in terms of a diagonal matrix $\mathbf{D}=\mathbf{QF}(\mathrm{XZ})\mathbf{Q}^{\mathrm{T}}$ and a basis variable set $\Omega_n=\mathbf{Q}\Theta_n$, with elements ω_i^n ($i=1$–6) given by linear combinations of the base-pair step parameters [Czapla, Swigon, and Olson 2006]. Here $\mathbf{Q}$ is the eigenvector matrix specifying the directions of the principal axes of deformation, and the superscript T is used to denote the transpose. Elimination of the cross terms in the energy expression makes it possible to write the probability density function for a single base-pair step, including normalization, as a product of Gaussians. This function can be sampled with a standard Gaussian random-number generator [Press et al. 1986] and a Boltzmann distribution of states can be collected without the necessity of using the Metropolis method [Metropolis et al. 1953]. Such an approach is superior to the Metropolis method in that it is computationally more efficient and does not suffer from correlations between sample points or incomplete coverage of phase space. Gaussian sampling cannot be used, however, if the potential function includes long-range electrostatic terms.

14.2.5 Persistence Length

The persistence length a is computed from the projection of the mean end-to-end vector $\langle \mathbf{r} \rangle$, the so-called persistence vector [Flory 1973], at infinite chain length along the initial direction of the chain; that is, $a = \langle \mathbf{r}_\infty \rangle \cdot \mathbf{r}_1 / |\mathbf{r}_1|$. If the dimeric chain units are independent, $\langle \mathbf{r} \rangle$ can be determined from the product $\mathbf{P}_N = \langle \mathbf{A}_1 \rangle \langle \mathbf{A}_2 \rangle \ldots \langle \mathbf{A}_{N-1} \rangle \langle \mathbf{A}_N \rangle$ of average generator matrices $\langle \mathbf{A}_n \rangle$ [Marky and Olson 1994]. The components of $\langle \mathbf{r} \rangle$, which accumulate in the far right column of $\mathbf{P}_N$, approach limiting values with increasing N, owing to the nonorthogonality of each $\langle \mathbf{T}_n \rangle$ matrix of the flexible duplex [Olson et al. 1993]. Thus the persistence length of DNA can be obtained by calculating the limiting value of the [3,4] matrix element of $\mathbf{P}_N$:

$$a = \lim_{N \to \infty} \begin{bmatrix} 0 & 0 & 1 & 0 \end{bmatrix} \mathbf{P}_N \begin{bmatrix} 0 \\ 0 \\ 0 \\ 1 \end{bmatrix}. \tag{14.7}$$

14.2.6. J-Factor

The J-factor depends on the fraction M_c / M of configurations that meet the criteria of chain closure, namely that (i) the end-to-end vector $\mathbf{r}$ is null, $W(\mathbf{r} = \mathbf{0})$, (ii) the terminal normals are aligned; that is, the cosine of the angle between the normals of the first and last base pairs is unity, given that the vector $\mathbf{r}$ is null, $\Gamma_r(\cos\gamma = 1)$, and (iii) the end-to-end twist is zero, given that the normals are aligned and the vector $\mathbf{r}$ is null, $\Phi_{r,\cos\gamma}(\tau = 0)$ [Flory, Suter, and Mutter 1976]. The product of these probability densities is approximated by choosing three corresponding bounds: (i) the magnitude of $\mathbf{r}$ being less than r_0; (ii) the cosine of the angle γ between the normals of terminal base pairs being greater than $1-\Gamma_0$; and (iii) the magnitude of the end-to-end twist being less than τ_0. Thus, the J factor is given by

$$J = \frac{4\pi}{N_A} W(|\mathbf{r}| \le r_0)\, \Gamma_r(\cos\gamma \ge 1 - \Gamma_0)\, \Phi_{r,\cos\gamma}(\tau \le \tau_0) = \frac{1}{K}\frac{M_c}{M}, \tag{14.8}$$

where $K = 4\pi N_A r_0^3 \Gamma_0 \tau_0 / 3$, N_A is Avogadro's number, M_c is the number of configurations that satisfy the three closure constraints, and M is the total sample size. The bounds used here—$r_0 = 10$ Å, $\Gamma_0 = 0.02$, $\tau_0 = 11.5°$ $(\cos\tau_0 = 1 - 0.02 = 0.98)$—are very restrictive, constraining the trace of $\mathbf{A}_{1:N}$ to values very close to 3 and the radial bound to distances no more than 5% of the contour length of the sampled DNA chains. Previous work [Czapla, Swigon, and Olson 2006] has shown that such bounds yield the most accurate results for $M_c \ge 1000$.

14.2.7 Half-Chain Sampling

Since DNA segments shorter than a persistence length are stiff and the probability that a randomly generated configuration satisfies all three ring-closure criteria is very small, we use a modified half-chain, pairwise-combination technique [Czapla, Swigon, and Olson 2006] in order to enhance the sampling. Rather than generate the configurations of a DNA with N base pairs, we divide the chain into two equal (or nearly equal) pieces and sample L configurations of each half-chain segment separately. By taking all pairwise combinations of both halves we can theoretically achieve L^2 configurations of the full-length chain. Even though configurations of half-chain segments are uncorrelated, the multiplicative combination of a finite number of states introduces some bias in the full-chain ensemble. In order to reduce the number of unnecessary half-chain combinations, we join only those pairs of segments that have a good chance of satisfying the end-to-end ring-closure criteria [Czapla, Swigon, and Olson 2006]. This essential component of the method reduces the number of combinations performed to the order of millions, which is in the neighborhood of the number of half-chains generated. The algorithm keeps track of transformed values of the end-to-end vectors of the first and second half-segments, and combines only those for which the radial overlap condition, $W(|\mathbf{r}| \leq r_0)$ in Equation 14.8, is roughly satisfied. The calculation time runs approximately linearly with L with such an approach even though the possibility of ring closure in all L^2 samples is evaluated.

14.2.8 DNA Threading

The nucleosome-binding affinity of a given DNA sequence is estimated by "threading" the constituent base pairs on the three-dimensional pathway found in the currently best-resolved nucleosome core-particle structure [Richmond and Davey 2003] and calculating a knowledge-based deformation score in terms of the deviations of the base-pair step parameters that make up the structure from their preferred equilibrium values. The total "energy" U of the threaded sequence is expressed as a sum of quadratic terms $U = \sum_{n=1}^{N} \psi_n$, where ψ_n is given by Equation 14.6 and N is the number of base-pair steps that comprise the nucleosome template. Here θ_i^n is the value imposed on the ith step parameter at the nth dinucleotide step of the assumed structure. This approach assumes that the core of histone proteins imposes exactly the same configurational constraints on DNA regardless of base-pair sequence and ignores the occurrence of gaps, for example, small "bubbles" of unbound duplex that may loop away from the surface of the nucleosome [Kulic and Schiessel 2003].

14.3 APPLICATIONS

14.3.1 Knowledge-Based Potentials

The equipotential surfaces in Figure 14.2 illustrate the sequence-dependent deformability and structural interdependence of DNA dimer steps in the selected set [Li 2006] of protein–DNA structures. The contour plots reveal the distinctive equilibrium (average) rest states of the 10 unique dimers [Gorin, Zhurkin, and Olson 1995; Olson et al. 1998] and the strong coupling of rigid-body parameters found in most base-pair steps. By contrast, there is no sequence dependence, bending and twisting are uncoupled, and there are no translational deformations in the classical representation of DNA as an inextensible elastic rod. The ellipses in the figure are projections of the multidimensional potential surface of each dimer on the roll-twist plane obtained from the covariance of two different sets of (θ_2, θ_3) values: (i) a "refined" set of dimer steps (dots), found by iteratively removing outlying states (open circles) of extreme bending, twisting, and stretching; that is, states with one or more step parameters that deviate from their respective mean values by more than three times their root-mean-square deviations before culling [Olson et al. 1998]; and (ii) a "complete" set of structural examples (dots and open circles). The contours correspond to deviations of parameters equal to two times the combined root-mean-square deviations of the θ_i in the selected sets of data and thus encompass $\sim$95% of the reference points. The mean values of roll and twist in the respective datasets are highlighted by thin (solid and dashed) lines, and the contour surfaces by curves of the same style.

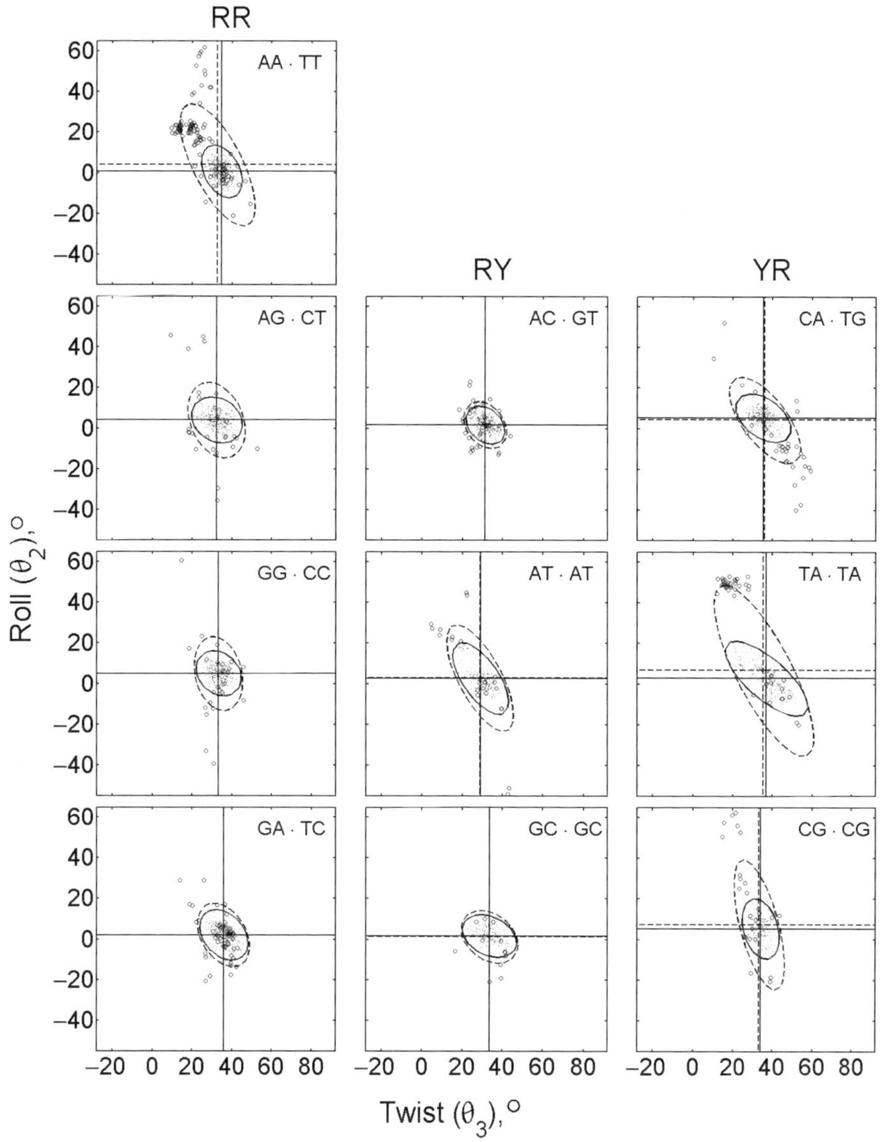

FIGURE 14.2 Collective scatter plots in the roll-twist (θ_2, θ_3) plane of base-pair step parameters found in high-resolution protein–DNA crystal complexes and derived sequence-dependent potentials of the 10 unique dimer steps. Dots correspond to the points used to derive the "refined" potentials (solid contours) and open circles to the states of extreme bending, twisting, and stretching that are included with the preceding points in the "complete" functions (dashed contours). Ellipses are projections of the multidimensional potentials on the (θ_2, θ_3) plane obtained from the 2×2 covariance matrix of observed roll-twist values. Contours correspond to deviations of parameters equal to two times the combined root-mean-square deviations of θ_2 and θ_3. Average values of roll and tilt are highlighted by thin (solid and dashed) lines. The three columns show the respective deformational patterns of individual purine-purine (RR), purine-pyrimidine (RY), and pyrimidine-purine (YR) steps.

As is clear from Figure 14.2, the coupling of roll and twist depends upon sequence and dataset. The positive values of f_{23} associated with the dimer steps produce energy pathways that involve a decrease in one angle and an increase in the other, mimicking the observed variation of step parameters. The extent and direction of parametric coupling reflect the sequence and choice of reference points. For example, whereas all other "complete" dimers tend to deform more easily via roll than twist, the GC·GC steps in the dataset tend to deform via twist slightly more easily than roll. The extreme distortions of DNA found in the crystal complexes similarly reflect sequence: for example, the CA·TG steps show a propensity to take up the large negative roll values associated with the kinking of DNA into the minor groove, while the AA·TT, TA·TA, and CG·CG dimers tend to kink more easily via large positive roll into the major groove. The coupling of TA·TA parameters changes direction when outlying states of extreme roll and twist are included. Consideration of the outlying states also softens the knowledge-based potentials, with some of the most pronounced changes in deformability occurring at AA·TT, AT·AT, and TA·TA steps (note the larger areas spanned by the dashed contours of the "complete" potentials compared to the solid contours of the "refined" potentials for these steps). The relative deformability of CG·CG dimers compared to other base-pair steps also changes substantially if outlying states are considered.

In addition to the roll-twist correlations noted above, roll and twist are frequently coupled to slide, the local displacement of neighboring base pairs along their long axes. Roll-slide coupling is very sensitive to sequence: whereas the roll and slide of CA·TG, TA·TA, and GC·GC dimers show negative correlations in both the "refined" and the "complete" datasets, the parameters are positively correlated at most other base-pair steps [Olson et al. 1998]. By contrast, the f_{35} twist-slide constants are predominantly negative and the correlations of twist and slide are positive (data not shown).

14.3.2 Intrinsic Motions

The molecular images in Figure 14.3 illustrate the pathways of preferred DNA deformation deduced from the known structures. The sets of low-energy librations, which lie along the longest principal axes of the knowledge-based potentials; that is, in the direction of most probable configurational change, are reminiscent of the normal modes of vibration of small molecules. The illustrated motions involve combinations of roll and twist plus varying degrees of translation, dictated by the set of "complete" potentials. As is clear from the images, correlations between roll and twist dominate the preferred movements of the 10 unique base-pair steps. The CA·TG step, however, incorporates significant translational changes along this lowest energy pathway, whereas the GA·TC and AT·AT steps involve essentially no base-pair displacement. The GG·CC step also includes variation in tilt, although the observed changes are substantially lower than those of roll. The illustrated moves correspond to one of the six directions of configurational sampling; that is, linear combinations of base-pair step parameters, used in the Monte Carlo simulation of polymeric structures.

14.3.3 Persistence Length

The values of the persistence length in Table 14.1 show how the dimeric deformability of a given base-pair step influences the global properties of DNA. Each numerical value in the table gives the computed mean extension along the initial direction of a hypothetical, naturally straight homopolymer with a helical repeat of 10.5 bp per turn and local elastic properties corresponding to those deduced for the designated step in the specified structural sample. The force constants are scaled by a factor ζ so that the persistence length of a mixed-sequence DNA homopolymer is ~ 500 Å, or ~ 150 bp; that is, $f_{ij}^{\dagger}(\mathrm{XX}) = \sum_{\mathrm{XZ}} f_{ij}^{\dagger}(\mathrm{XZ})$, where $f_{ij}^{\dagger}(\mathrm{XX})$ is the force constant of the mixed-sequence repeating unit, $f_{ij}^{\dagger}(\mathrm{XZ}) = \zeta f_{ij}(\mathrm{XZ})$ is the scaled force constant of the XZ base-pair step, and the summation is carried out over all 16 possible steps. The values of $f_{ij}^{\dagger}(\mathrm{XZ})$ determine the range of step parameters sampled for the specified dimer and thus the average components of the generator matrices $\langle \mathbf{A}_n \rangle$ used in Equation 14.7 to determine the limiting values of a.

The values of ζ reveal the extent to which the sampled points mimic the average properties of DNA in solution. Interestingly, mixed-sequence homopolymers guided by the potentials of the "complete" dataset have chain extension properties more closely resembling those known to characterize polymeric DNA than chains that are subject to the deformations associated with the

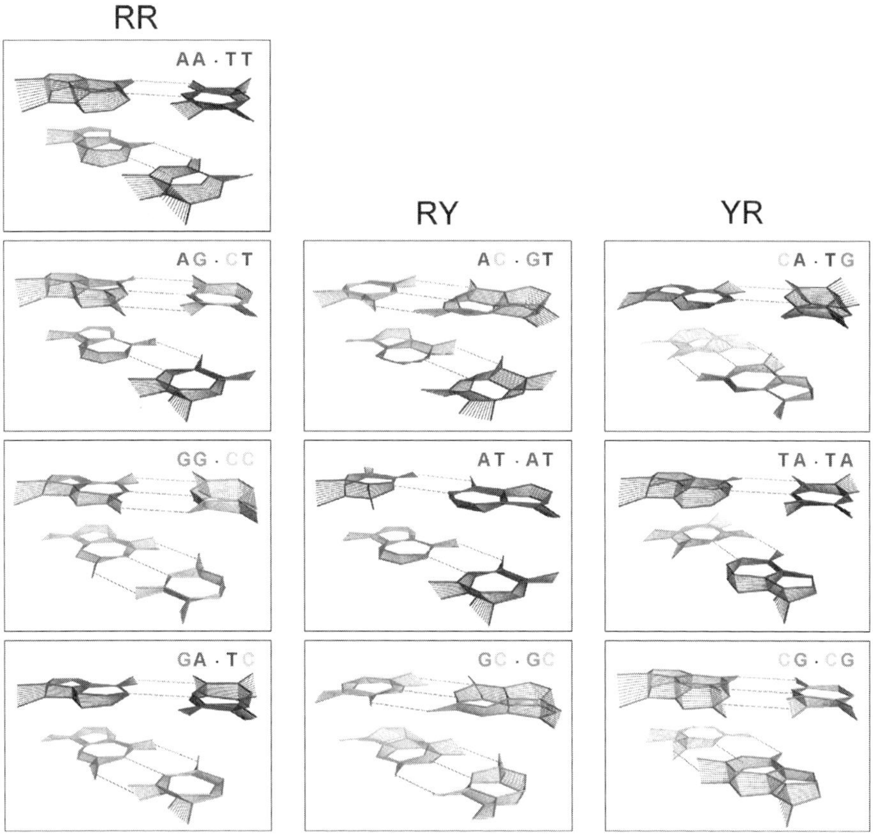

FIGURE 14.3 (See color insert following page 238.) Sequence-dependent motions along the longest principal axes of the 10 unique DNA base-pair steps. Nonequilibrium forms are superimposed on the intrinsic (average) dimer structures. Perturbed states correspond to deformations, at increments of $3\langle\lambda_1^2\rangle^{1/2}$, along the longest principal axes of the "complete" knowledge-based potentials, where λ_1 is the largest eigenvalue of the covariance matrix, and "energies" range from zero to $4.5m^2k_BT$ for displacements of $\pm 3m\langle\lambda_1^2\rangle^{1/2}$. Here m is set to 5 to enhance visualization of structural deformations. Base pairs are represented as ideal Watson–Crick pairs [Olson et al. 2001], with the hydrogen bonds of rest structures denoted by dashed lines. Bases are color-coded according to chemical identity: adenine (red); thymine (blue); guanine (green); cytosine (yellow). Motions are illustrated with respect to the "middle" frame of each step and viewed into the minor groove of the upper 3'-base pair of each miniduplex. Note the correspondence of observed structural variability with the corresponding contour surfaces in Figure 14.2.

"refined" more B-like dataset. That is, ζ is closer to unity for the "complete" homopolymer than the "refined" homopolymer, with the range of accessible configuration space increased by a factor of $1.18 = 0.85^{-1}$ in the former case and $2.0 = 0.5^{-1}$ in the latter case to yield a persistence length of $\sim$500 Å. The persistence lengths of mixed-sequence homopolymers that conform to the unscaled potentials are greater than 500 Å; that is, 592 Å for the "complete" potential and 995 Å for the "refined" potential. Thus, the occasional adoption of extreme configurational states like those induced by the binding of proteins appears, from this perspective, to be necessary to account for the observed persistence length of mixed-sequence DNA in solution.

The data in Table 14.1 further show that AC·GT and GC·GC steps have more pronounced stiffening effects at the polymeric level than other base-pair steps, with longer computed persistence lengths. As is clear from the contours of the scaled potentials in the roll-tilt (θ_2, θ_1) plane (Figure 14.4), these steps bend to a much lesser extent than the other dimers. Furthermore, the AC·GT step is even stiffer than the mixed-sequence homopolymeric repeating unit that yields a persistence length

TABLE 14.1
Persistence lengths of hypothetical, naturally straight DNA homopolymers with knowledge-based elastic properties for individual base-pair steps

Dimeric repeat	Persistence length[a] (Å)	
	"Complete" sample	"Refined" sample
ζ[b]	0.85	0.50
AA·TT	150	395
AG·CT	359	461
GG·CC	298	405
GA·TC	423	395
AC·GT	735	625
AT·AT	193	245
GC·GC	562	454
CA·TG	276	391
TA·TA	87	217
CG·CG	140	269
Mixed sequence	500.5	500.3
Ideal DNA	500.2	

[a] Persistence lengths obtained using Equation 14.7 with average generator matrices $\langle \mathbf{A}_n \rangle$ based on Monte Carlo samples of 10^6 states of the designated dimeric repeating unit subject to the specified knowledge-based potential. All steps assigned an unsheared, naturally straight, B-like rest state with 10.5 bp per helical turn; that is, $\theta_1^0(XZ) = \theta_2^0(XZ) = 0°$, $\theta_3^0 = 34.3°$, $\theta_4^0(XZ) = \theta_5^0(XZ) = 0$ Å, $\theta_6^0 = 3.4$ Å.

[b] Factor used to scale the force constants of each set of knowledge-based potentials so that the persistence length of a mixed-sequence homopolymer is $\sim$500 Å (see text).

of $\sim$500 Å. Similarly, TA·TA steps stand out as being highly bendable at both the global and local levels, although the local bending deformability, as measured by the area within the corresponding energy contours, is somewhat greater for CG·CG compared to TA·TA steps that obey the "complete" potential. The degree of dimeric bending and the values of a based on the "complete" potentials are much more sensitive to sequence than the corresponding values associated with the "refined" functions. Figure 14.4 also includes the contour surface of a dimer subject to the classic elastic-rod model of DNA. Notably, none of the "real" dimers exhibits the bending isotropy assumed in the ideal model. The well-known anisotropy of DNA bending; that is, the preferential bending of base-pair steps via roll rather than tilt [Zhurkin, Lysov, and Ivanov 1979], is clear from the elliptical (as opposed to circular) shapes of the derived contour surfaces.

14.3.4 Distribution Functions and Ring-Closure Propensities

The radial density functions in Figure 14.5 show how the placement of individual dimers affects the range of accessible configurations of a series of 94-bp DNA molecules compared to that of a mixed-sequence DNA of the same chain length. The molecules, which are detailed in Table 14.2, include two fragments, TA-94 and 5S-94, taken from well characterized nucleosome-positioning sequences [Lowary and Widom 1998; Simpson and Stafford 1983] and found to form small minicircles [Cloutier and Widom 2004, 2005], and four sequences—E6-94, E8-94, E13-94, CA-94—used as experimental controls in the determination of the J factor [Cloutier and Widom 2004, 2005].

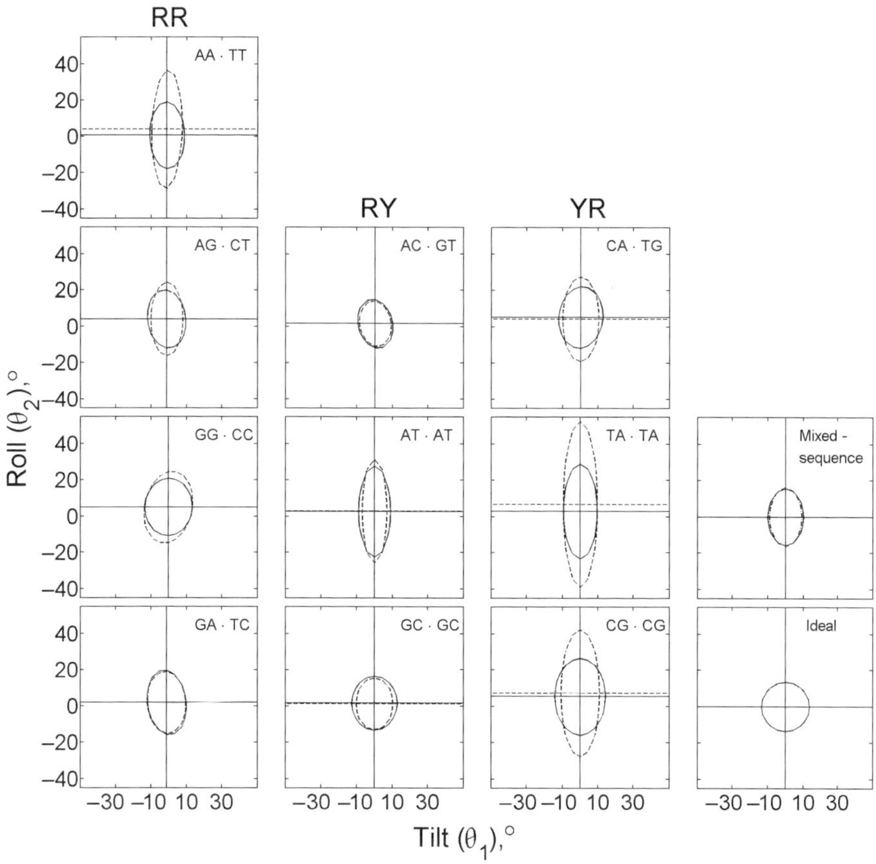

FIGURE 14.4 Contour surfaces in the roll-tilt (θ_2, θ_1) plane of scaled, knowledge-based potentials of the 10 unique base-pair steps (columns 1–3), the dimeric repeating unit of a naturally straight, mixed-sequence DNA homopolymer with force constants averaged over all 16 dimeric potentials and weighted to yield a persistence length a of ~500 Å (column 4, top), and the dimeric repeat of an ideal DNA elastic rod with the same value of a (column 4, bottom). See text and legend to Figure 14.2.

All of the sequences shift the distribution of the end-to-end distance r toward smaller values than those determined for the mixed-sequence chain. Moreover, the ends of the nucleosome-positioning sequences are more likely to come into close contact than the ends of the control sequences. That is, the tails of the distributions formed from the most compact arrangements of the positioning sequences lie closer to zero than the tails of the control sequences. The boundary $r_{0.1}$ delimiting the 10% shortest configurations is smaller for the positioning sequences than the control sequences, and both limits are substantially smaller than the $r_{0.1}$ boundaries for mixed-sequence DNA and an ideal 94-bp DNA model (Table 14.2). Furthermore, the values of $r_{0.1}$ computed with the "complete" potentials are roughly proportional to the negative logarithm of the reported J factors. The likelihood of ring closure is lower for chains that obey the "refined" potentials; that is, the value of $r_{0.1}$ associated with a given sequence is larger compared to that obtained with the "complete" sequence. The values of $r_{0.1}$, however, do not take account of the orientational constraints (see Equation 14.8) that must be met for successful ring closure and considered in the calculations reported below.

The TA-94 and 5S-94 fragments stand out from the other sequences in Table 14.2 in containing flexible dimers with strong bending propensities and appreciable coupling of roll and twist, such as the TA and CG steps of the "complete" potential (highlighted in boldface), that recur approximately in phase with the ~10.5-bp double-helical repeat. The fragments with larger values of $r_{0.1}$ and smaller J factors; that is, greater values of $-\log J$, contain few such steps.

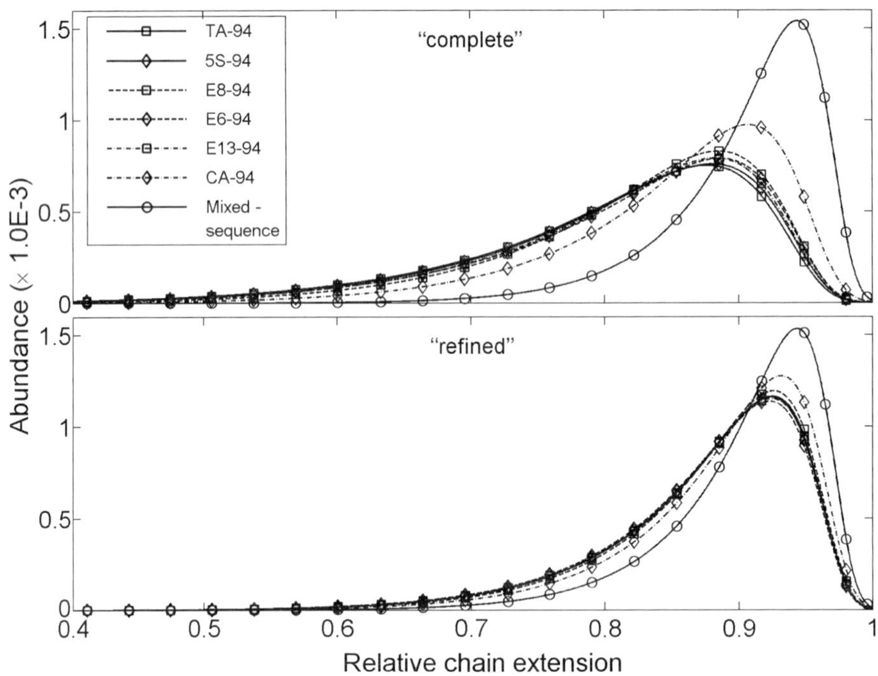

FIGURE 14.5 Distributions of the end-to-end distances r for a series of 94-bp DNA molecules compared to that of a mixed-sequence DNA of the same chain length. The double helix is assumed to be naturally straight in its equilibrium rest state with a 10.5 bp double-helical repeat. Fluctuations of local structure in polymeric sequences are based on Monte Carlo sampling of the scaled, knowledge-based potentials of the 10 unique dimers and the mixed-sequence potential that yields a persistence length of ~500 Å. Radial distributions of 2.5×10^8 sampled configurations are expressed in terms of relative chain extension, r/L_0, where L_0 is the contour length of the fully extended polymer (93 bp steps × 3.4 Å/step = 316.2 Å). Chain sequences are listed in Table 14.2.

Comparison of the computed J factors; that is, ring-closure propensities, of three representative 94-bp sequences—TA-94, E6-94, and CA-94—with experimental data suggests that the elastic properties of "real" DNA lie somewhere between the broad range of states sampled by the "complete" dimeric potentials and the narrower set of step parameters incorporated in the "refined" potentials. For example, the values of log J extracted from 2.5×10^{15} configurations of a naturally straight TA-94 duplex with a 10.5-bp helical repeat, −7.7 when subject to the "complete" potentials and −10.7 when governed by the "refined" functions, bracket the observed value of −9.0. The J factors of a CA-94 chain with the same rest state and simulated on the basis of the two energy functions also bracket the experimental value (log J values of −8.3 "complete" and −10.8 "refined" vs. −10.1 observed). The computed cyclization propensities of E6-94, however, exceed the observed J factors (log J values of −7.7 "complete" and −9.9 "refined" vs. −10.1 observed). The "refined" potential, nevertheless, yields a closer match to experiment in all three cases. Any natural curvature associated with the sequence-dependent rest states and ignored in the present calculations will enhance ring closure [Czapla, Swigon, and Olson 2006]. Notably, the sequence-dependent flexibility of DNA that is treated here increases the J factors of all sequences compared to that of an ideal DNA elastic rod with the same chain length and double-helical repeat, where log J = −11.6.

14.3.5 Nucleosome Positioning

The "cost" of threading the same three sequences on the central 60 base-pair steps of the best-resolved nucleosome core-particle structure [Richmond and Davey 2003] is reported in Figure 14.6.

TABLE 14.2
Sequences and ring-closure properties of representative 94-bp DNA molecules

DNA[a]	Base-pair sequence	$r_{0.1}$ (Å) "Complete" sample	$r_{0.1}$ (Å) "Refined" sample	log J
TA-94	ggccgggtcg TAgcaagctc TAgcaccgct TAaacgcacg TAcgcgctgt cTAccgcgtt tTAaccgcca aTAggatTAc tTAcTAgtct cTAc	206	245	−9.0
5S-94	ggccgacatc cctgaccctt TAaaTAgctT Aactttcatc aagcaagagc cTAcgaccaT Accatgctga aTATAccggt tctcgtccga tcac	208	245	−9.3
E6-94	ggccgtgcgc acgaaatgcT Atgccgaaga ttggatggac atgctTATAa aaggaatccc cagaggTAat ccttgatctg atgatgatcc gccc	211	243	−10.1
E8-94	ggccgtgcgT AgaacTActt tTAttTAtcg cctccacggt gctgatcccc tgtgctgttg gccgtgtTAt ctcgagtTAg TAcgacgtcc gccc	218	247	−10.3
E13-94	ggccgtgcgt tcggTAaggt gcgatggcct catcaaggcg ccaTATAaga tcactcgTAg tgaaaaccTA cccttcattT Aatgttgatc gccc	211	244	−10.2
CA-94	ggccgtccca gcaagctcca ggtgcgccca aacggctgca gacgccctgc acggcagccc aagcgcaccc agagccccct ctccggaatt cacc	231	251	−10.2
Mixed-sequence	xxxxxxxxxx xxxxxxxxxx xxxxxxxxxx xxxxxxxxxx xxxxxxxxxx xxxxxxxxxx xxxxxxxxxx xxxxxxxxxx xxxxxxxxxx xxxx	262	262	
Ideal-94	—	259		−11.6

[a] Sequences and measured J factors taken from Cloutier and Widom (2004, 2005); predicted J factor of ideal DNA taken from Czapla et al. (2006).

The imposed distortions of DNA reflect the close contact with the $(H3{\cdot}H4)_2$ tetramer that is believed to be critical to nucleosome positioning [Hayes, Tullius, and Wolffe 1990; Thåström, Bingham, and Widom 2004]. The 34 settings of each sequence on the crystalline template are described in terms of the displacement, with respect to the central base-pair step, of the nucleotide that is placed on the twofold structural dyad. Here, since the sequences contain an even number of base pairs, the settings are numbered from − 17 to + 17, without a zero entry. In order to extract the contribution of dimeric deformability to positioning, the sequences are assigned an unsheared, naturally straight, B-like rest state with 10.5 bp per helical turn.

Although the crystallographic template accommodates the regularly repeating TA-94 sequence in several relatively low-cost settings (denoted by triangles in Figure 14.6), none of these corresponds to the + 1 setting that aligns most closely with the observed positioning of the 601 sequence, from which TA-94 is derived. That is, none of the local TA-94 minima in the computed scoring profile is in register with the observed setting of nucleosomes on 601, regardless of the choice of scoring function.

The TA-94 fragment shares 83% sequence identity with base pairs 88–181 of the 232-bp 601 sequence, falling in the middle of the stretch found to position nucleosomes. That is, base pair 47 of TA-94 coincides in this alignment with the observed location of the dyad on 601 at base pair 134 (J. Widom, personal communication). By contrast, the predicted sites of nucleosome binding

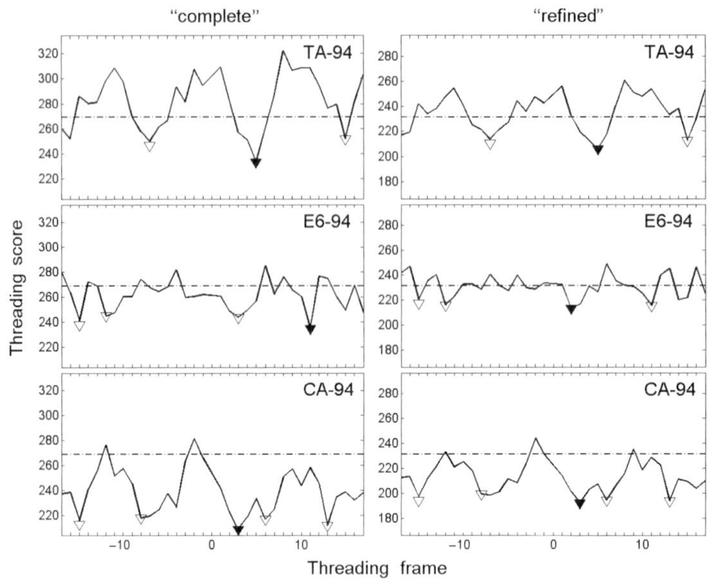

FIGURE 14.6 Deformation profiles of representative DNA sequences "threaded" on the central 60 base-pair steps of the currently best-resolved nucleosome core-particle structure (Richmond and Davey 2003): the TA-94 sequence derived from the 232-bp synthetic high-affinity "601" sequence (top); the "random" E6-94 sequence used as a control in ring-closure experiments (middle); and the CA-94 sequence with CA·TG dimer steps repeated at 10-11-bp intervals (bottom). The major minima in the "energy" profiles, which are denoted by triangles (the filled triangle corresponding to the deepest minimum), are taken as "predicted" nucleosomal dyad positions. The settings are numbered with respect to the center of each sequence; note that there is no zero position. The threading scores of the sequences (black lines) are compared at each test position with the score of a mixed-sequence homopolymer (dashed line). Data are reported for chains subject to both "complete" and "refined" dimeric potentials.

on TA-94 recur at 10–11-bp increments in settings where the naturally flexible TA·TA steps of the sequence easily take up the "kink-and-slide" states of nucleosomal DNA [Tolstorukov et al. 2007], in which roll is negative ($\theta_2 < -10°$) and slide is highly positive ($\theta_5 > 1.5$ Å). The 4-bp discrepancy in the predicted positioning of TA-94 vs. the observed positioning of 601 may reflect limitations of the model in dealing with the sequence-dependent features of the 601 sequence [Tolstorukov et al. 2007] and/or subtle differences in the sequence of TA-94, including the replacement of two GG·CC steps in 601 by phased TA·TA steps in TA-94, that bias the positioning. Indeed, the same predicted nucleosome positions occur with the "complete" and "refined" potentials.

By contrast, there are no deep minima in the scoring profiles of the E6-94 control sequence, although the overall cost of nucleosomal deformation is lower for E6-94 than TA-94 (note the relative displacement of the "energy" profiles with respect to the fixed cost of deforming a mixed-sequence homopolymer (dashed line) on the nucleosome). Thus, there are no intrinsic features in the E6-94 sequence that accommodate the known distortions of nucleosomal DNA in a particular setting.

The cost of deforming the CA-94 sequence on the nucleosome is much lower than that for the other sequences. The CA·TG steps, which repeat at 10–11 bp along CA-94, in phase with the double-helical repeat, accommodate the positive slide found at distorted nucleosomal steps much more easily than any other dimer. The lower cost of sliding contributes, in turn, to the low positioning scores despite the higher cost of bending CA·TG compared to TA·TA steps. In fact, deformations in slide make a contribution to the total positioning score that is comparable to, if not greater than, that from roll [Tolstorukov et al. 2007]. The slight displacement of the CA·TG steps on CA-94 relative to the positions of the TA·TA steps on TA-94 accounts for the 1–2-bp shift in the predicted settings of nucleosomes on CA-94 compared to TA-94.

14.4 CONCLUDING REMARKS

The mathematics used to relate local base-pair structure to global chain configuration underlies successful "realistic" treatment of polymeric DNA. The distribution of accessible chain configurations governs the overall behavior of long DNA fragments, including the likelihood of loop formation and the ease of wrapping on the surface of the nucleosome. The naturally discrete representation of DNA described herein [Coleman, Olson, and Swigon 2003] is general in the sense that any functional description of DNA dimeric geometry can be employed; that is, not just the harmonic form of the knowledge-based potentials that have been extracted from the three-dimensional arrangements of DNA base-pair steps in high-resolution crystal structures [Olson et al. 1998]. The latter functions incorporate the intrinsic structure, the sequence-dependent fluctuations, the anisotropy of DNA deformations, and the known correlations of base-pair step parameters.

These local dimeric features translate into measurable effects at the macromolecular level, giving useful new insights into the contribution of base sequence to the mesoscopic properties of DNA. The examples presented here show how judicious placement of flexible base-pair steps enhances the likelihood of ring closure and lowers the cost of deforming a DNA sequence on the surface of a nucleosome. Thus, the regular repetition of TA·TA steps in phase with the helical repeat of the TA-94 sequence promotes spontaneous ring closure and preferential positioning of nucleosomes on DNA. The bending flexibility of the TA·TA steps gives rise to a relatively high proportion of compact polymer configurations with chain ends close enough to effect cyclization. The relative ease of TA·TA bending, in combination with its coupled propensity to slide, lowers the cost of deforming particular settings of the sequence on the nucleosome [Tolstorukov et al. 2007]. By contrast, the regularly repeated CA·TG steps in the CA-94 sequence inhibit chain cyclization but enhance the wrapping of the sequence on the surface of the nucleosome. These steps, although not as easily deformed via roll as TA·TA dimers, readily take up the costly sliding deformations found in nucleosomal DNA. Other dimers, such as CG·CG steps, which easily bend but resist sliding in the positive sense observed on the nucleosome, could be used in the design of DNA molecules that would preferentially loop rather than form nucleosomes.

Finally, the base-sequence-dependent treatment of DNA offers promise for deciphering, at the molecular level, properties of DNA detected in living systems. For example, the in vivo expression levels of reporter genes in bacteria [Becker, Kahn, and Maher 2005; Bellomy, Mossing, and Record, 1988; Law et al. 1993] reflect the ease of DNA looping in the presence of repressor proteins and architectural factors, such as the bacterial nucleoid protein HU, found in the cell. A base-pair-level representation of DNA can provide "real" physical insights into the control of transcription, going beyond the ideal elastic-rod model used until now to interpret the data and accounting for changes in expression levels in terms of realistic features of DNA and/or bound protein. Preliminary studies of the cyclization properties of DNA in the presence of HU [Czapla, Swigon, and Olson 2008] show that one can account for observed gene expression levels and for the apparent independence of Lac-repressor-mediated expression on chain length without the need to introduce bizarre features in DNA. Consideration of the unique properties of sequence, the placement of proteins, and the various protein-DNA loop types [Swigon, Coleman, and Olson 2006] appear to account for these data.

ACKNOWLEDGMENTS

This research has been generously supported by the U.S. Public Health Service under research grants GM20861 and GM34809. LC gratefully acknowledges GAANN fellowship support from the U.S. Department of Education, AVC traineeship support from the U.S. Public Health Service (training grant GM08319), and fellowship support from the Program in Mathematics and Molecular Biology (Burroughs-Wellcome Fund Interfaces Program).

REFERENCES

Becker, N. A., Kahn, J. D., and Maher III, L. J. 2005. Bacterial repression loops require enhanced DNA flexibility. *J. Mol. Biol.* 349:716–30.

Bellomy, G. R., Mossing, M. C., and Record, Jr., M. T. 1988. Physical properties of DNA *in vivo* as probed by the length dependence of the *lac* operator looping process. *Biochemistry* 27:3900–3906.

Berman, H. M., Olson, W. K., Beveridge, D. L. et al. 1992. The nucleic acid database: A comprehensive relational database of three-dimensional structures of nucleic acids. *Biophys. J.* 63:751–59.

Bishop, T. C. 2005. Molecular dynamics simulations of a nucleosome and free DNA. *J. Biomol. Struct. Dyn.* 22:673–86.

Bracco, L., Kotlarz, D., Kolb, A., Diekmann, S., and Buc, H. 1989. Synthetic curved DNA sequences can act as transcriptional activators in *Escherichia coli*. *EMBO J.* 8:4289–96.

Cao, H., Widlund, H. R., Simonsson, T., and Kubista, M. 1998. TGGA repeats impair nucleosome formation. *J. Mol. Biol.* 281:253–60.

Cloutier, T. E., and Widom, J. 2004. Spontaneous sharp bending of double-stranded DNA. *Mol. Cell.* 14:355–62.

———. 2005. DNA twisting flexibility and the formation of sharply looped protein–DNA complexes. *Proc. Natl. Acad. Sci. U.S.A.* 102:3645–50.

Coleman, B. D., Olson, W. K., and Swigon, D. 2003. Theory of sequence-dependent DNA elasticity. *J. Chem. Phys.* 118:7127–40.

Crothers, D. M., Drak, J., Kahn, J. D., and Levene, S. D. 1992. DNA bending, flexibility, and helical repeat by cyclization kinetics. *Methods Enzymol.* 212:3–29.

Czapla, L., Swigon, D., and Olson, W. K. 2006. Sequence-dependent effects in the cyclization of short DNA. *J. Chem. Theor. Comp.* 2:685–95.

Czapla, L., Swigon, D., and Olson, W. K. 2008. Effects of the nucleoid protein HU on the structure, flexibility, and ring-closure properties of DNA deduced from Monte-Carlo simulations. *J. Mol. Biol.* (in press).

Dickerson, R. E., Bansal, M., Calladine, C. R. et al. 1989. Definitions and nomenclature of nucleic acid structure parameters. *J. Mol. Biol.* 208:787–91.

El Hassan, M. A., and Calladine, C. R. 1995. The assessment of the geometry of dinucleotide steps in double-helical DNA: A new local calculation scheme. *J. Mol. Biol.* 251:648–64.

Feig, M., and Pettitt, B. M. 1998. Structural equilibrium of DNA represented with different force fields. *Biophys. J.* 75:134–49.

Flory, P. J. 1969. *Statistical Mechanics of Chain Molecules*. New York: Interscience Publishers.

———. 1973. Moments of the end-to-end vector of a chain molecule, its persistence and distribution. *Proc. Natl. Acad. Sci. U.S.A.* 70:1819–23.

Flory, P. J., Suter, U. W., and Mutter, M. 1976. Macrocyclization equilibria. 1. Theory. *J. Am. Chem. Soc.* 98:5733–39.

Geanacopoulos, M., Vasmatzis, G., Zhurkin, V. B., and Adhya, S. 2001. Gal repressosome contains an antiparallel DNA loop. *Nat. Struct. Biol.* 8:432–36.

Gorin, A. A., Zhurkin, V. B., and Olson, W. K. 1995. B-DNA twisting correlates with base pair morphology. *J. Mol. Biol.* 247:34–48.

Hagerman, P. J. 1986. Sequence-directed curvature of DNA. *Nature* 321:449–50.

———. 1988. Flexibility of DNA. *Annu. Rev. Biophys. Biophys. Chem.* 17:265–86.

Hayes, J. J., Tullius, T. D., and Wolffe, A. P. 1990. The structure of DNA in a nucleosome. *Proc. Natl. Acad. Sci. U.S.A.* 87:7405–7409.

Jeffreys, H., and Jeffreys, B. S. 1946. *Methods of Mathematical Physics*. Cambridge, UK: Cambridge University Press.

Kratky, O., and Porod, G. 1949. Röntgenuntersuchung Gelöster Fadenmoleküle. *Rec. Trav. Chim. Pays-Bas* 68:1106–22.

Kulic, I. M., and Schiessel, H. 2003. Nucleosome repositioning via loop formation. *Biophys. J.* 84:3197–3211.

Lankas, F., Lavery, R., and Maddocks, J. H. 2006. Kinking occurs during molecular dynamics simulations of small DNA minicircles. *Structure* 14:1527–34.

Laughlin, R. B., Pines, D., Schmalian, J., Stojkovic, B. P., and Wolynes, P. 2000. The middle way. *Proc. Natl. Acad. Sci. U.S.A.* 97:32–37.

Law, S. M., Bellomy, G. R., Schlax, P. J., and Record, Jr., M. T. 1993. *In vivo* thermodynamics analysis of repression with and without looping in *lac* constructs. *J. Mol. Biol.* 230:161–73.

Li, Y. 2006. Understanding DNA–Protein Interactions from the Nucleic-Acid Perspective. PhD thesis, Rutgers, the State University of New Jersey, New Brunswick, NJ.

Lowary, P. T., and Widom, J. 1998. New DNA sequence rules for high affinity binding to histone octamer and sequence-directed nucleosome positioning. *J. Mol. Biol.* 276:19–42.

Marky, N. L., and Olson, W. K. 1994. Configurational statistics of the DNA duplex: Extended generator matrices to treat the rotations and translations of adjacent residues. *Biopolymers* 34:109–20.

Matsumoto, A., and Olson, W. K. 2002. Sequence-dependent motions of DNA: A normal mode analysis at the base-pair level. *Biophys. J.* 83:22–41.

______. 2006. Effects of sequence, cyclization, and superhelical stress on the internal motions of DNA. In *Normal Mode Analysis: Theory and Applications to Biological and Chemical Systems*, ed. Q. Cui, and I. Bahar, 188–211. Boca Raton, FL: Chapman & Hall/CRC Press.

Mehta, R. A., and Kahn, J. D. 1999. Designed hyperstable Lac repressor·DNA loop topologies suggest alternative loop geometries. *J. Mol. Biol.* 294:67–77.

Metropolis, N., Rosenbluth, A. W., Rosenbluth, M. N., Teller, A., and Teller, E. 1953. Equation of state calculations by fast computing machines. *J. Chem. Phys.* 21:1087–92.

Olson, W. K. 1996. Simulating DNA at low resolution. *Curr. Opin. Struct. Biol.* 6:242–56.

Olson, W. K., Bansal, M., Burley, S. K. et al. 2001. A standard reference frame for the description of nucleic acid base-pair geometry. *J. Mol. Biol.* 313:229–37.

Olson, W. K., Colasanti, A. V., Li, Y., Ge, W., Zheng, G., and Zhurkin, V. B. 2006. DNA simulation benchmarks as revealed by X-ray structures. In *Computational Studies of RNA and DNA*, ed. J. Sponer, and F. Lankas, 235–57. Dordrecht, The Netherlands: Springer.

Olson, W. K., Gorin, A. A., Lu, X.-J., Hock, L. M., and Zhurkin, V. B. 1998. DNA sequence-dependent deformability deduced from protein–DNA crystal complexes. *Proc. Natl. Acad. Sci. U.S.A.* 95:11163–68.

Olson, W. K., Marky, N. L., Jernigan, R. L., and Zhurkin, V. B. 1993. Influence of fluctuations on DNA curvature. A comparison of flexible and static wedge models of intrinsically bent DNA. *J. Mol. Biol.* 232:530–54.

Olson, W. K., Swigon, D., and Coleman, B. D. 2004. Implications of the dependence of the elastic properties of DNA on nucleotide sequence. *Philos. Trans. R. Soc. A* 362:1403–22.

Olson, W. K., and Zhurkin, V. B. 2000. Modeling DNA deformations. *Curr. Opin. Struct. Biol.* 10:286–97.

Press, W. H., Flannery, B. P., Teukolsky, S. A., and Vetterling, W. T. 1986. *Numerical Recipes in C*. New York: Cambridge University Press.

Richmond, T. J., and Davey, C. A. 2003. The structure of DNA in the nucleosome core. *Nature* 423:145–50.

Roccatano, D., Barthel, A., and Zacharias, M. 2007. Structu: ¹ flexibility of the nucleosome core particle at atomic resolution studied by molecular dynamics simulatio.. *Biopolymers* 85:407–21.

Rohs, R., Sklenar, H., and Shakked, Z. 2005. Structural and energetic origins of sequence-specific DNA bending: Monte Carlo simulations of papillomavirus E2-DNA binding sites. *Structure* 13:1499–1509.

Ruscio, J. Z., and Onufriev, A. 2006. A computational study of nucleosomal DNA flexibility. *Biophys. J.* 91:4121–32.

Semsey, S., Virnik, K., and Adhya, S. 2005. A gamut of loops: Meandering DNA. *Trends Biochem. Sci.* 30:334–41.

Simpson, R. T., and Stafford, D. W. 1983. Structural features of a phased nucleosome core particle. *Proc. Natl. Acad. Sci. U.S.A.* 80:51–55.

Swigon, D., Coleman, B. D., and Olson, W. K. 2006. Modeling the Lac repressor-operator assembly: The influence of DNA looping on Lac repressor conformation. *Proc. Natl. Acad. Sci. U.S.A.* 103:9879–84.

Thåström, A., Bingham, L. M., and Widom, J. 2004. Nucleosomal locations of dominant DNA sequence motifs for histone-DNA interactions and nucleosome positioning. *J. Mol. Biol.* 338:695–709.

Thåström, A., Lowary, P. T., Widlund, H. R., Cao, H., Kubista, M., and Widom, J. 1999. Sequence motifs and free energies of selected natural and non-natural nucleosome positioning DNA sequences. *J. Mol. Biol.* 288:213–29.

Tolstorukov, M. Y., Colasanti, A. V., McCandlish, D., Olson, W. K., and Zhurkin, V. B. 2007. A novel 'roll-and-slide' mechanism of DNA folding in chromatin. Implications for nucleosome positioning. *J. Mol. Biol.* 371:725–38.

Trifonov, E. N. 1991. DNA in profile. *Trends Biochem. Sci.* 16:467–70.

Widlund, H. R., Cao, H., Simonsson, S. et al. 1997. Identification and characterization of genomic nucleosome-positioning sequences. *J. Mol. Biol.* 267:807–17.

Zhang, Y., McEwen, A. E., Crothers, D. M., and Levene, S. D. 2006. Analysis of *in-vivo* LacR-mediated gene repression based on the mechanics of DNA looping. *PLoS ONE* 1:e136.

Zhurkin, V. B., Lysov, Y. P., and Ivanov, V. I. 1979. Anisotropic flexibility of DNA and the nucleosomal structure. *Nucleic Acids Res.* 6:1081–96.

15 Coarse-Grained Models for Nucleic Acids and Large Nucleoprotein Assemblies

Robert K.-Z. Tan, Anton S. Petrov, Batsal Devkota, and Stephen C. Harvey
School of Biology, Georgia Institute of Technology

CONTENTS

15.1 INTRODUCTION

There are a variety of reasons for using coarse-grained (CG) models. There is, of course, the obvious advantage of computational efficiency: if a CG model can capture the most important features of the structure, then CG simulations can treat larger systems and longer time scales than are accessible for all-atom simulations. In most such cases, one would prefer to use an all-atom description, but the computational demand precludes doing so. Other authors in this volume consider such cases.

Here we deal with the application of CG models to problems where working at low resolution is either required (because of the limited quality and/or quantity of available experimental data), or where a low-resolution description is actually more suitable to the problem than an all-atom treatment would be. In these cases, a CG representation—or a multiscale representation with different levels of detail—is actually more appropriate than a model with atomic resolution. We will consider two cases.

The first is the development of three-dimensional models in those cases where the available experimental data are insufficient to justify the development of all-atom models. As a concrete example, there are a number of macromolecular assemblies that have not yet yielded to studies by X-ray crystallography, but for which there are low-resolution structural data available from cryo-electron microscopy (cryo-EM). In this case, an all-atom model would necessarily contain a fair amount of speculation, and a low-resolution CG treatment is a more honest representation of what is actually known about the structure. Such models can sometimes continue to be useful if crystal structures become available, and we will describe such a case here: the ribosome. Prior to the crystallization of complete ribosomal subunits in 2000,[1,2] there was a substantial body of low-resolution data on the ribosome, including from cryo-EM. We reported the first quantitative description of the structure of the ribosome at low resolution in 1994.[3] There is now a crystal structure for the complete *Escherichia coli* ribosome,[4,5] but it lacks tRNAs and other protein cofactors, so it is not in a biologically relevant state. Fortunately, several functional states of the *E. coli* ribosome have been captured by cryo-EM,[6–8] and the problem then becomes the refinement of all-atom models in the cryo-EM density maps. We will presently show how a combination of CG and all-atom models for RNAs and RNA–protein complexes can be used for refining the structures of the biologically relevant states of the ribosome.

Another place where CG models are more appropriate than all-atom models is when one is examining the mesoscale structure and properties of a biological system. The example that we will consider here involves the elastic properties of double-helical DNA, which determine the range and energetics of conformations of supercoiled DNA, and of DNA packaged into viral capsids. Several groups developed models for DNA supercoiling in the 1980s using either coarse-graining[9–14] or mathematical approaches.[15,16] Here we review newer CG models that have been developed in the past few years to examine the structural and thermodynamic aspects of the packaging of double-helical DNA in bacteriophage.

The two classes of models mentioned above address very different problems, and the coarse-graining is done with very different philosophies. The CG DNA models are parameterized to mimic the known elastic properties of DNA, and they are then used to examine the behavior of DNA in various situations, in an effort to rationalize existing experimental data, and to make predictions that will motivate new experiments. In contrast, the RNA models are used for structure refinement, so they are parameterized empirically, with the aim of optimizing the performance of the refinement algorithms. These differences are seen in the organization of the Methods and Applications sections of this article, which are both subdivided into RNA models and DNA models.

All of the models described here have been developed within the framework of *yammp*,[17] our inhouse molecular modeling package, or using the more recent version *yammp2*, also called YUP (Yammp Under Python).[18]

15.2 METHODS

15.2.1 Coarse-Grained Model for RNA

During the 1980s, a large body of very-low-resolution data was obtained on the structure of the small (30S) subunit of the *E. coli* ribosome. Three critical advances set the stage for everything else that followed. First was the demonstration that ribosomes could be disassembled into their various RNA and protein components, and then reassembled.[19,20] Second was the determination of the 16S RNA secondary structure by Carl Woese and Harry Noller in 1981, based on the comparative analysis of sequences of ribosomal 16S RNAs from different organisms.[21] Third, neutron diffraction experiments initiated by Don Engelman and Peter Moore[22] eventually led to the determination of the relative positions of all 21 proteins of the 30S subunit.[23] The question then became, how is the RNA threaded throughout the protein scaffold? A variety of groups attached photoactivatable chemical cross-linking reagents to different ribosomal proteins, reassembled the ribosomes in the dark, activated the cross-linking with ultraviolet light, then isolated and digested the RNA to determine which regions were cross-linked to which proteins. A number of RNA–RNA contacts were also determined by cross-linking.

Other RNA–protein contacts were determined by footprinting. We recognized that the data described above could be used as restraints in an automated procedure for building three-dimensional models. In particular, intra- and intermolecular contacts determined by cross-linking and footprinting define distance restraints just as interatomic contacts determined by the nuclear Overhauser effect (NOE contacts) can be used to determine structures in nuclear magnetic resonance (NMR) experiments. Our first model contained only the information on the protein positions, the RNA secondary structure, and the available cross-linking data.[24] We subsequently added a unique energy penalty term that restrains the structure to the cryo-EM density map using spherical harmonics[25] and produced a second-generation model using all of the data then available.[3] The details of the energy function and the algorithm are described elsewhere.[18,26,27] Here we provide a brief overview.

Each nucleotide is represented as a single pseudoatom, called a "P-atom" and centered on the phosphorus atom (Figure 15.1). RNA chains of any sequence are modeled as a string of identical P pseudoatoms that are held together by appropriately parameterized pseudobonds and pseudoangles, each of which has the standard quadratic dependence of energy on deformation. The ideal P–P bond length along the backbone is determined by the average interphosphate distance in the crystallographic database, while the variance in the distance distribution σ^2 is used to define the bond-stretching force constant k, using a relationship derivable from the Boltzmann probability distribution function,

$$k = RT/\sigma^2, \tag{15.1}$$

where R and T are the universal gas constant and temperature, respectively. Bond angles are parameterized in the same way. There is also a set of P–P bonds that join pairs of P atoms representing the phosphate groups for pairs of nucleotides that define basepairs in the secondary structure. The ideal bond lengths and force constants for these P–P pairs are different from those for P–P pairs along the backbone. As a consequence, one cannot define bond parameters based on atom types, as is done in standard all-atom programs. This CG model is therefore not easily implemented in a conventional molecular mechanics program. The ability to define energy functions without using atom types is one of the major reasons that we developed *yammp* and YUP; another major advantage of these packages is that they contain a variety of potential energy functions that are not found in conventional programs.

The P atoms have a non-bonded exclusion volume, defined by a semi-harmonic (soft sphere) repulsion. The exclusion diameter of the P-atom is small enough that double helices defined only by P atoms would be hollow, which would allow helices to interpenetrate. Space-filling atoms ("X-atoms") are introduced along the helix axis at the midpoints of basepairs to prevent this (Figure 15.1).

We note that, with this representation, it is not possible to define a one-to-one mapping of the all-atom representation onto the CG representation. The P atoms are located at the positions of phosphorus atoms in the all-atom representation, but the X atoms are located at the midpoint of

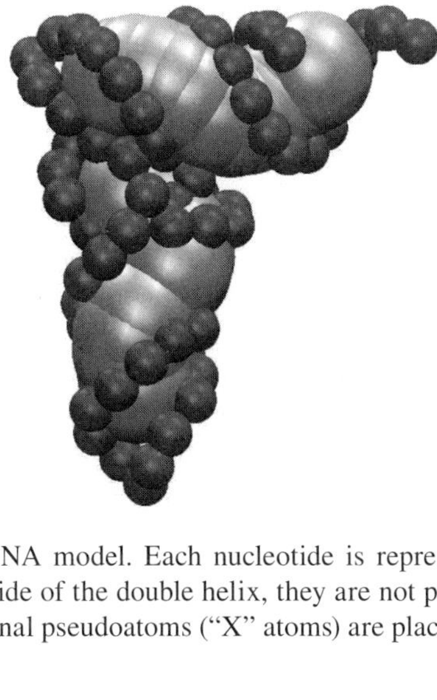

FIGURE 15.1 Coarse-grained RNA model. Each nucleotide is represented by a single "P" pseudoatom (black). Since these lie on the outside of the double helix, they are not properly positioned to provide appropriate volume exclusion, so additional pseudoatoms ("X" atoms) are placed at the center of each base pair for this purpose (gray).

the line connecting two P atoms in a basepair. This is not simply related to the center of mass of the riboses and the bases, so one cannot identify the X atom with any particular component of the nucleotides comprising the base pair. Here again, our CG model is developed in a spirit somewhat different from those where the coarse-graining is based on the joining of groups of atoms into united atoms or pseudoatoms with a defined mapping. This is another reason that *yammp* and YUP were developed.

Double-helical regions require a series of pseudotorsions to guarantee the correct chirality of the double helix. The pseudotorsion is defined by P-atoms *ijkl*, where atoms *i* and *j* are sequential in the 5' to 3' direction along one strand, while atoms *k* and *l* are sequential in the opposite direction (3' to 5') along the other, and atoms *j* and *k* form a Watson–Crick or Hoogsteen base pair that is defined by the secondary structure. The pseudotorsion is given a harmonic energy function, with the ideal value corresponding to the average P–P–P–P pseudotorsion angle found in A-form RNA double helices in the crystallographic database, and the force constant is derived from the variance, as described above for bonds.

When we developed our first models for the 30S subunit, there were only a couple of crystal structures available for the proteins of that subunit. The location of each protein's center of mass was known from the neutron diffraction studies described above, but there were no data that would have enabled us to rotationally orient any of the proteins. Consequently, the proteins were modeled as spheres, with radii defined by their volumes, calculated from their molecular weights and a standard density for hydrated proteins. Each protein was anchored (studded) to the point in space specified by the protein map from the neutron diffraction studies, and the computer then had to thread the RNA through this protein scaffold in such a way that all the distance restraints and other low-resolution data were satisfied. Protein–RNA and RNA–RNA contacts determined by cross-linking and footprinting were expressed as harmonic restraints, with force constants derived from the estimated uncertainties in the measurements. Semiharmonic restraints were used for the spherical harmonic representation of the cryo-EM density map for the complete subunit, with the energy being zero for any pseudoatom inside the surface of the density map, and rising quadratically with distance for pseudoatoms outside that surface.

This approach had the advantage that every term in the energy function was either zero or positive. That is, if every single restraint was satisfied by the model, the energy would be zero, and any deviations would give a positive energy. This facilitated an evaluation of alternative refinement algorithms and protocols. It also enabled us to identify some sets of experimental restraints that were in conflict with one another. These presumably were the result of differences in the conditions of different experiments, or, in some cases, of experimental error. The final product of those early efforts was a set of models, from which we identified a consensus model, along with quantitative statements about the uncertainties in the positions of each part of the model.[3]

15.2.2 Coarse-Grained and All-Atom Modeling of RNA–Protein Complexes

There was an exponential growth in the quantity and quality of experimental data on the ribosome in the late 1990s, making it necessary to update our approach. The most important of these was the result of impressive advances in cryo-EM, permitting, under favorable circumstances, the separation of RNA density from protein density, and the visualization of some of the grooves of the RNA double-helical regions. Many ribosomal proteins also yielded their structures to the efforts of the crystallographers and, in some cases, to NMR spectroscopy. There was also a barrage of new cross-linking and footprinting data, some of which was in clear conflict with earlier data.

We decided not to focus our efforts on modeling the complete ribosome, or even a complete subunit. Instead, we began focusing on areas of particular biological significance, trying to work out the details of key interactions by combining data from crystallography, NMR, cryo-EM, and the lower-resolution methods into all-atom models.[28–33] This required a set of hybrid methods that used a mixture of CG and atomistic models.

The modern era of structural ribosomology began with the appearance of the crystal structures of the *T. thermophilus* 30S subunit[2] and the *H. marismortui* 50S subunit,[1] and matured with the availability of the structure of the complete *E. coli* ribosome.[4,5] By combining data from these structures with the cryo-EM maps of different functional states,[6–8] it is possible to describe subtle conformational changes in near-atomic detail, leading to significant insights into ribosomal structure–function relationships.[8,34–36]

In these studies, we use CG models very much like those described above for initial fitting. The principal modification is the addition of a CG model for the protein components, with one pseudoatom per amino acid, placed at the positions of the alpha carbons. The deformations accompanying the conformational changes between states are generally fairly small, so we often use a Gaussian network model (GNM)[37] to refine the model with respect to the density map. This can be applied to either the aforementioned CG model, or to an all-atom representation, but the final refinement includes all atoms in the ribosome.

Finally, we note that the detail available from the current cryo-EM maps renders the representation of electron density with spherical harmonics described above computationally unfeasible. We have developed a new scoring function that represents every voxel in the cryo-EM density map by a three-dimensional pseudoenergy function, allowing us to refine all-atom GNM models within that detailed density map [Tan et al., submitted]. Some aspects of the hybrid CG/all-atom methodology used in this work are described in a recent review,[27] and further details will soon be published elsewhere [Devkota et al., submitted].

15.2.3 Coarse-Grained Model for DNA

The treatment of double-helical DNA as a chain polymer has a very long history.[38,39] This statistical mechanical approach is a powerful method for describing the bulk properties of DNA in solution, but it lacks sufficient detail to permit examination of the relationships between DNA sequence, flexibility, conformation, and biological function. These issues began to be examined theoretically and computationally in the 1980s, when a number of authors began modeling DNA bending and looping under supercoiling pressure and/or bound to proteins.

A very coarse-grained model was developed by Alex Vologodskii, Maxim Frank-Kaminetskii and their co-workers in the 1980s,[11–13] taking advantage of the observation that the statistical mechanical properties of a very long polymer with a specified persistence length are equivalent to those of a freely jointed chain if each element of the chain has the length of a Kuhn segment, which is twice the persistence length. For DNA, the persistence length is about 500 Å, or 150 basepairs, so each chain segment in these models represents 300 basepairs. At about the same time, Wilma Olson and her collaborators used a variety of mathematical approaches to describe supercoiled DNA and DNA bending by proteins.[15,16,40–42] These are basically continuum approaches, so they are also very-low-resolution models.

We developed the first CG model for DNA that was able to represent the stretching, bending, and twisting deformations of the double helix with single-base-pair resolution.[9,10] Each base pair is represented by three pseudoatoms (Figure 15.2). This number was chosen because three points specify a plane (the base pair plane in the present case), allowing the modeling of variations in rise, roll, tilt, and twist between successive basepairs. The model also allows the inclusion of sequence-dependent ideal values for rise, roll, tilt, and twist, and for sequence-dependent differences in the elastic moduli for each of these deformations. As described in the caption to Figure 15.2, the stretching, bending, and twisting stiffnesses of DNA are represented by the stiffness parameters for bond stretching, angle bending, and pseudotorsional rotation, respectively. The potential function includes a soft sphere volume exclusion term, to prevent chain crossings, and the *yammp* code for modeling supercoiled closed circular DNA allows the user to verify that there have been no chain crossings (no change in linking number) by periodic calculation of the writhe and twist, taking advantage of the fundamental topological conservation law:[43]

$$\text{Link} = \text{Twist} + \text{Writhe}. \tag{15.2}$$

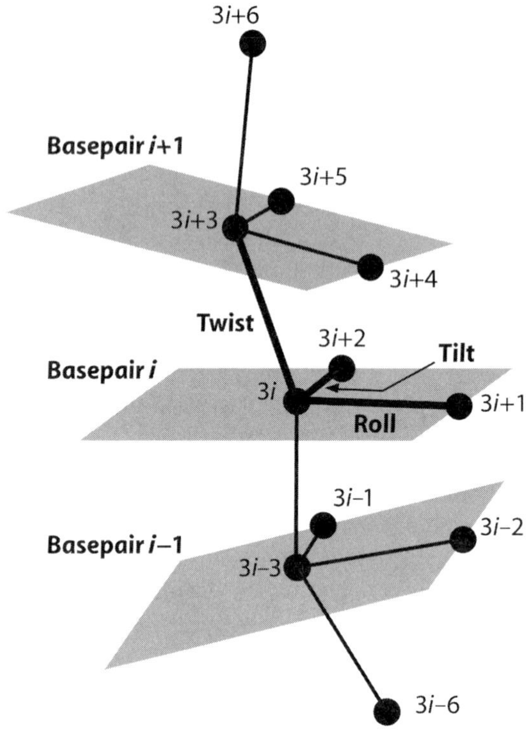

FIGURE 15.2 The 3DNA CG model has three pseudoatoms per base pair. The central pseudoatoms (e.g., atom $3i$) are treated as soft spheres with an effective diameter of 25 Å, to provide for volume exclusion. The backbone pseudoatoms (e.g., atom $3i+1$) and the major groove pseudoatoms (e.g., atom $3i+2$) do not participate in nonbonded interactions, so they have effective radii of zero. For clarity, pseudoatomic radii are not shown graphically in this figure. The DNA elastic modulus for stretching is used to parameterize bonds between successive central pseudoatoms (e.g., $3i$–$3i+3$). Elastic moduli for roll and tilt about the axes indicated in the figure provide the angle force constants (e.g., $3i-2$–$3i$–$3i+3$), while pseudotorsional rotations about the central axis (e.g., $3i+1$–$3i$–$3i+3$–$3i+4$) are parameterized to match the DNA torsional elastic modulus.

A few years ago, we began applying a lower-resolution version of this model to the investigation of packaging of double-helical DNA into bacteriophage capsids, using one pseudoatom to represent 6–10 basepairs (Figure 15.3). Bacteriophage capsids are preformed protein shells with either icosahedral symmetry, or with geometries derived from elongated icosahedra, and DNA is driven into the capsid through one vertex of the icosahedron (the portal) by a molecular motor that consumes about one molecule of ATP for each two basepairs that are packaged. Our work was motivated in part by advances in cryo-EM on mature bacteriophage,[44] which suggested that the average DNA density has the conformation of a coaxial spool, supporting the standard model proposed many years ago.[45,46] We wondered whether individual conformations are well spooled, or if they might be quite varied, with the coaxial motif being an idealization that arises because of image averaging. Furthermore, Carlos Bustamante and his collaborators had measured the force vs. distance curve as the portal motor drives the DNA into the capsid.[47] If we developed a model for packaging, the measurement of force vs. distance would provide an independent check on the quality of the model, and it would permit a decomposition of the work done by the motor into its various components.

At the scale of viral packaging, one does not need a description of DNA down to the level of individual base-pair steps. Our CG model is shown in Figure 15.3. The most important parameter for this model is the bending stiffness, derived from the persistence length. The soft sphere

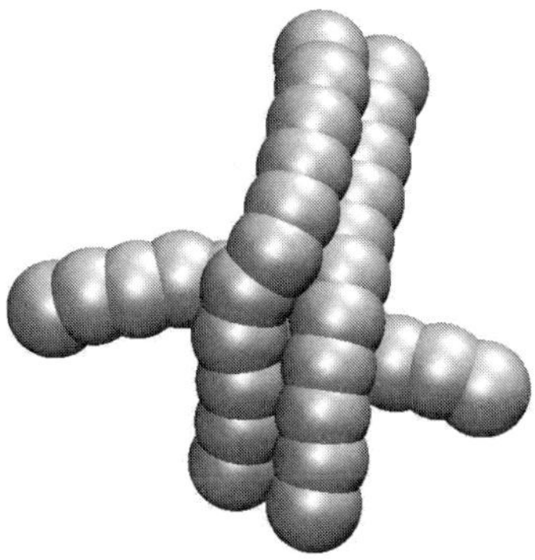

FIGURE 15.3 Coarse-grained model for viral packaging, with one pseudoatom representing six basepairs. Three double-helical fragments of a DNA conformation inside a bacteriophage are shown to illustrate volume exclusion effects, which are represented by a soft nonbonded sphere.

volume exclusion term is based on the observation that, when DNA is tightly packed, the distance between neighboring double helices is about 25–30 Å. With spherical pseudoatoms of this diameter, one can gather 6–10 base pairs into a single pseudoatom without introducing the risk of unwanted strand crossings. This simple elastic model captures many of the essential features of DNA packaging in bacteriophage,[48–50] and we have recently added an interstrand electrostatic interaction term[51] based on experimentally determined potentials of mean force. This nonstandard electrostatic term is not available in conventional molecular mechanics packages, but it was easily implemented in YUP.

15.3 RECENT APPLICATIONS

15.3.1 The Ribosome

As described above, the availability of ribosomal crystal structures and improvements in the quality of reconstructed density maps from cryo-EM have enabled us to tackle a number of interesting problems with a combination of CG and all-atom modeling. Here we briefly review some of the most recent results. Coordinates for the various structures described here are available from the Research Collaboratory for Structural Biology (RCSB), or by request.

We have examined the details of the conformational changes in tRNA that take place during the translational cycle. Aminoacylated tRNA first binds to the ribosome as part of a ternary complex between elongation factor EF-Tu, GTP, and tRNA. Formation of the codon–anticodon complex requires a substantial kinking of the stacked tRNA D-stem and anticodon stem, and we believe that unkinking helps to drive tRNA accommodation, which is a critical step in translational fidelity.[8]

Comparative sequence analysis of ribosomal RNAs provides important information on the evolution of structure–function relationships within the ribosome. We examined the smallest ribosomal RNAs from various organisms, along with those from mitochondria and chloroplasts, proposing a three-dimensional structure for the "minimal ribosome."[52] Later, when Rajendra Agrawal's group determined the cryo-EM density map for the ribosome from bovine mitochondria,[53] we collaborated with them and with Robin Gutell to establish the three-dimensional structure of the large subunit rRNA.[36]

Most recently, we have taken advantage of the availability of the crystal structure of the *E. coli* ribosome[4,5] to build the first all-atom models of the ribosome in two biologically functional states captured in cryo-EM. These structures, immediately before and after accommodation, suggest an active energetic role for tRNA in translational fidelity [Devkota et al., submitted].

15.3.2 Packaging of Double-Helical DNA into Bacteriophage

We have used a stepwise approach to develop a series of successively more detailed and more accurate models for studying DNA packaging in bacteriophage. We first built a coaxially spooled model for a spherical virus, consistent with the standard assumption about the DNA conformation.[48] In a careful search for the minimum energy structure we later found that concentric spooling is energetically favored at $T=0$ K over the classic coaxially spooled model.[49] This was confirmed in subsequent packaging simulations carried out at $T=300$ K.[50,54] The latter two papers also provided the first estimates of the packaging force along the packaging pathway for models derived without fitting any parameters to the experimental force curves, with the finding that there is a substantial entropic penalty upon confining the DNA to the restricted capsid space. There was no electrostatic term in those simulations, but they led to the prediction that the total work done by the packaging motor is divided about equally between the entropic and electrostatic work.

With the introduction of a pseudoatomic representation for the capsid and core proteins, and with the addition of a mean field treatment of DNA–DNA electrostatic interactions, we were finally in a position to rigorously test the model's ability to reproduce the experimental force–distance curves.[51] A test on ϕ29, which was used in the experiments,[47] gave an excellent match to the experimental data,[51] and it also confirmed the predictions described at the end of the previous paragraph. We have also carried out simulations on ε15, resulting in the prediction that the packaging forces are even higher than those in ϕ29.[55]

We have recently finished a series of simulations examining the effects of capsid size and shape and the presence or absence of the cylindrical cores that are believed to assist in organizing the DNA in some viruses.[56] We examined model bacteriophage with spherical capsids, icosahedral capsids, and extended icosahedral capsids; some models have no cylindrical core, while others have a small core, and others have large cores. The simulations represent models for bacteriophages P4, ϕ29, ε15, and T7. In the absence of core proteins, DNA in spherical and icosahedral viruses tends to favor the concentric spool conformation, while slightly elongated capsids produce a folded toroidal DNA conformation, and more elongated capsids lead to twisted toroidal conformations. Small cores do not significantly affect these results, but larger cores lead to coaxially spooled DNA conformations, as would be expected. Most importantly, density reconstructions done from the models match the most recent high-resolution cryo-EM reconstructions,[57,58] demonstrating how the average densities obtained by experimental reconstructions represent idealized average conformations and fail to reveal the virus-to-virus variations in DNA conformation.

Finally, we have addressed the issue of DNA's torsional stiffness, which was omitted in the earlier models because of computational burden, both because of the number of pseudoatoms, and because of substantial increases in the time required for the DNA to reach equilibrium at each step of the packaging. We have found that the torsional stiffness does not have a significant effect on either the favored conformations or on the energetic cost of packaging.[59] This is probably because the end of the DNA that is first introduced into the capsid is unrestrained, and torsional strains are fairly easily relaxed by passage of the twisting deformations along the double helix, in contrast with results obtained when that end of the DNA is constrained to a point on the inside of the capsid.[60]

ACKNOWLEDGMENT

Supported by a grant from the National Institutes of Health (P42 RR12255; C.E. Brooks, III, PI).

REFERENCES

1. Ban, N., Nissen, P., Hansen, J., Moore, P. B., and Steitz, T. A. 2000. The complete atomic structure of the large ribosomal subunit at 2.4 A resolution. *Science* 289:905–20.

2. Wimberly, B. T., Brodersen, D. E., Clemons, Jr., W. M., Morgan-Warren, R. J., Carter, A. P., Vonrhein, C., Hartsch, T., and Ramakrishnan, V. 2000. Structure of the 30S ribosomal subunit. *Nature* 407:327–39.
3. Malhotra, A., and Harvey, S. C. 1994. A quantitative model of the *Escherichia coli* 16 S RNA in the 30 S ribosomal subunit. *J. Mol. Biol.* 240:308–40.
4. Berk, V., Zhang, W., Pai, R. D., and Cate, J. H. 2006. Structural basis for mRNA and tRNA positioning on the ribosome. *Proc. Natl. Acad. Sci. U.S.A.* 103:15830–34.
5. Schuwirth, B. S., Borovinskaya, M. A., Hau, C. W., Zhang, W., Vila-Sanjurjo, A., Holton, J. M., and Cate, J. H. 2005. Structures of the bacterial ribosome at 3.5 A resolution. *Science* 310:827–34.
6. Allen, G. S., Zavialov, A., Gursky, R., Ehrenberg, M., and Frank, J. 2005. The cryo-EM structure of a translation initiation complex from *Escherichia coli. Cell* 121:703–12.
7. Agrawal, R. K., Sharma, M. R., Kiel, M. C., Hirokawa, G., Booth, T. M., Spahn, C. M., Grassucci, R. A., Kaji, A., and Frank, J. 2004. Visualization of ribosome-recycling factor on the Escherichia coli 70S ribosome: Functional implications. *Proc. Natl. Acad. Sci. U.S.A.* 101:8900–8905.
8. Valle, M., Zavialov, A., Li, W., Stagg, S. M., Sengupta, J., Nielsen, R. C., Nissen, P., Harvey, S. C., Ehrenberg, M., and Frank, J. 2003. Incorporation of aminoacyltRNA into the ribosome as seen by cryoelectron microscopy. *Nat. Struct. Biol.* 10:899–906.
9. Tan, R. K.-Z., and Harvey, S. C. 1987. A model for DNA supercoiling. In *Unusual DNA Structures*, ed. R. D. Wells and S. C. New York: Springer.
10. Tan, R. K., and Harvey, S. C. 1989. Molecular mechanics model of supercoiled DNA. *J. Mol. Biol.* 205:573–91.
11. Anshelevich, V. V., and Vologodskii, A. V. 1988. A theoretical study of formation of DNA noncanonical structures under negative superhelical stress. *J. Biomol. Struct. Dyn.* 6:247–59.
12. Klenin, K. V., Vologodskii, A. V., Anshelevich, V. V., and Dykhne, A. M. 1988. Effect of excluded volume on topological properties of circular DNA. *J. Biomol. Struct. Dyn.* 5:1173–85.
13. Klenin, K. V., Vologodskii, A. V., and Anshelevich, V. V. 1989. Variance of writhe for wormlike DNA rings with excluded volume. *J. Biomol. Struct. Dyn.* 6:707–14.
14. Zhurkin, V. B., Ulyanov, N. B., and Gorin, A. A. 1991. Static and statistical bending of DNA evaluated by Monte Carlo simulations. *Proc. Natl. Acad. Sci. U.S.A.* 88:7046–50.
15. Hao, M. H., and Olson, W. K. 1989. Modeling DNA supercoils and knots with BSpline functions. *Biopolymers* 28:873–900.
16. Zhang, P., and Olson, W. K. 1991. Accelerated record keeping Fourier series Monte Carlo Simulations of an isotropic elastic rod model of DNA. *Comput. Polymer Sci.* 1:3–17.
17. Tan, R. K.-Z., and Harvey, S. C. 1993. Yammp: Development of a molecular mechanics program using the modular programming method. *J. Comput. Chem.* 14:455–70.
18. Tan, R. K.-Z., Petrov, A. S., and Harvey, S. C. 2006. YUP: A molecular simulation program for coarse-grained and multiscale models. *J. Chem. Theory Comput.* 2:529–40.
19. Traub, P., and Nomura, M. 1968. Structure and function of E. coli ribosomes. V. Reconstitution of functionally active 30S ribosome particles from RNA and proteins. *Proc. Natl. Acad. Sci. U.S.A.* 59:777–84.
20. Held, W. A., Ballou, B., Mizushima, S., and Nomura, M. 1974. Assembly mapping of 30S ribosomal proteins from *Escherichia coli. J. Biol. Chem.* 249:3103–11.
21. Noller, H. F., and Woese, C. R. 1981. Secondary structure of 16S ribosomal RNA. *Science* 212:403–11.
22. Ramakrishnan, V. R., Yabuki, S., Sillers, I. Y., Schindler, D. G., Engelman, D. M., and Moore, P. B. 1981. Positions of proteins S6, S11 and S15 in the 30 S ribosomal subunit of *Escherichia coli. J. Mol. Biol.* 153:739–60.
23. Capel, M. S., Kjeldgaard, M., Engelman, D. M., and Moore, P. B. 1988. Positions of S2, S13, S16, S17, S19 and S21 in the 30 S ribosomal subunit of *Escherichia coli. J. Mol. Biol.* 200:65–87.
24. Malhotra, A., Tan, R. K., and Harvey, S. C. 1990. Prediction of the three dimensional structure of *Escherichia coli* 30S ribosomal subunit: A molecular mechanics approach. *Proc. Natl. Acad. Sci. U.S.A.* 87:1950–54.
25. Malhotra, A., Tan, R. K.-Z., and Harvey, S. C. 1994. Utilization of shape data in molecular mechanics using a potential based on spherical harmonic surfaces. *J. Comput. Chem.* 15:191–99.
26. Malhotra, A., Tan, R. K., and Harvey, S. C. 1994. Modeling large RNAs and ribonucleoprotein particles using molecular mechanics techniques. *Biophys. J.* 66:1777–95.
27. Mears, J. A., Stagg, S. M., and Harvey, S. C. 2004. Modeling very large RNA assemblies using reduced representations. In *Handbook of RNA Biochemistry*, ed. R. K. Hartmann, A. Bindereif, A. Schön, and E. Westhof, 546–59. Weinheim: Wiley-VCH.

28. Easterwood, T. R., Major, F., Malhotra, A., and Harvey, S. C. 1994. Orientations of transfer RNA in the ribosomal A and P sites. *Nucleic Acids Res.* 22:3779–86.
29. VanLoock, M. S., Easterwood, T. R., and Harvey, S. C. 1997. tRNA/mRNA/rRNA interactions in the Escherichia coli ribosomal decoding site with and without bound aminoglycoside. *Nucleic Acids Symp. Ser.* 36:68.
30. VanLoock, M. S., Easterwood, T. R., and Harvey, S. C. 1999. Major groove binding of the tRNA/mRNA complex to the 16 S ribosomal RNA decoding site. *J. Mol. Biol.* 285:2069–78.
31. Wang, R., Alexander, R. W., VanLoock, M., Vladimirov, S., Bukhtiyarov, Y., Harvey, S. C., and Cooperman, B. S. 1999. Three-dimensional placement of the conserved 530 loop of 16 S rRNA and of its neighboring components in the 30 S subunit. *J. Mol. Biol.* 286:521–40.
32. VanLoock, M. S., Agrawal, R. K., Gabashvili, I. S., Qi, L., Frank, J., and Harvey, S. C. 2000. Movement of the decoding region of the 16 S ribosomal RNA accompanies tRNA translocation. *J. Mol. Biol.* 304:507–15.
33. VanLoock, M. S., Malhotra, A., Case, D. A., Agrawal, R., Penczek, P., Easterwood, T. R., Frank, J., and Harvey, S. C. 2000. A functional interpretation of the cryo-electron microscopy map of the 30S ribosomal subunit from Escherichia coli. In *The Ribosome: Structure, Function, Antibiotics and Cellular Interactions*, ed. R. A. Garrett, S. R. Douthwaite, A. Liljas, A. T. Matheson, P. B. Moore, and H. F. Noller, H. F., 165–71. Washington, DC: ASM Press.
34. Stagg, S. M., Valle, M., Agrawal, R. K., Frank, J., and Harvey, S. C. 2002. Problems with the transorientation hypothesis. *RNA* 8:1093–94.
35. Gao, H., Sengupta, J., Valle, M., Korostelev, A., Eswar, N., Stagg, S. M., Van Roey, P., Agrawal, R. K., Harvey, S. C., Sali, A., Chapman, M. S., and Frank, J. 2003. Study of the structural dynamics of the E coli 70S ribosome using realspace refinement. *Cell* 113:789–801.
36. Mears, J. A., Sharma, M. R., Gutell, R. R., McCook, A. S., Richardson, P. E., Caulfield, T. R., Agrawal, R. K., and Harvey, S. C. 2006. A structural model for the large subunit of the mammalian mitochondrial ribosome. *J. Mol. Biol.* 358:193–212.
37. Haliloglu, T., and Bahar, I. 1998. Coarse-grain simulations of conformational dynamics of proteins: Application to apomyoglobin. *Proteins* 31:271–81.
38. Schellman, J. A. 1974. Flexibility of DNA. *Biopolymers* 13:217–26.
39. Flory, P. J. 1969. *Statistical Mechanics of Chain Molecules*. New York: Wiley.
40. Schlick, T., and Olson, W. K. 1992. Supercoiled DNA energetics and dynamics by computer simulation. *J. Mol. Biol.* 223:1089–1119.
41. Schlick, T., and Olson, W. K. 1992. Trefoil knotting revealed by molecular dynamics simulations of supercoiled DNA. *Science* 257:1110–15.
42. Yang, Y., Tobias, I., and Olson, W. K. 1993. Finite element analysis of DNA supercoiling. *J. Chem. Phys.* 98:1673–86.
43. White, J. H. 1986. Calculation of the twist and the writhe for representative models of DNA. *J. Mol. Biol.* 189:329–41.
44. Cerritelli, M. E., Cheng, N., Rosenberg, A. H., McPherson, C. E., Booy, F. P., and Steven, A. C. 1997. Encapsidated conformation of bacteriophage T7 DNA. *Cell* 91:271–80.
45. Richards, K. E., Williams, R. C., and Calendar, R. 1973. Mode of DNA packing within bacteriophage heads. *J. Mol. Biol.* 78:255–59.
46. Earnshaw, W. C., and Harrison, S. C. 1977. DNA arrangement in isometric phage heads. *Nature* 268:598–602.
47. Smith, D. E., Tans, S. J., Smith, S. B., Grimes, S., Anderson, D. L., and Bustamante, C. 2001. The bacteriophage straight phi29 portal motor can package DNA against a large internal force. *Nature* 413:748–52.
48. Arsuaga, J., Tan, R. K.-Z., Vazquez, M., Sumners, D. W., and Harvey, S. C. 2002. Investigation of viral DNA packaging using molecular mechanics models. *Biophys. Chem.* 101–102:475–84.
49. LaMarque, J. C., Le, T. V., and Harvey, S. C. 2004. Packaging double-helical DNA into viral capsids. *Biopolymers* 73:348–55.
50. Locker, C. R., and Harvey, S. C. 2006. A model for viral genome packing. *Multiscale Modeling Simul.* 5:1264–79.
51. Petrov, A. S., and Harvey, S. C. 2007. Structural and thermodynamic principles of viral packaging. *Structure* 15:21–27.
52. Mears, J. A., Cannone, J. J., Stagg, S. M., Gutell, R. R., Agrawal, R. K., and Harvey, S. C. 2002. Modeling a minimal ribosome based on comparative sequence analysis. *J. Mol. Biol.* 321:215–34.

53. Sharma, M. R., Koc, E. C., Datta, P. P., Booth, T. M., Spremulli, L. L., and Agrawal, R. K. 2003. Structure of the mammalian mitochondrial ribosome reveals an expanded functional role for its component proteins. *Cell* 115:97–108.
54. Locker, C. R., Fuller, S. D., and Harvey, S. C. 2007. DNA organization and thermodynamics during viral packaging. *Biophys. J.* 93:2861–69.
55. Petrov, A. S., Lim-Hing, K., and Harvey, S. C. 2007. Packaging of DNA by bacteriophage epsilon15: Structure, forces, and thermodynamics. *Structure* 15:807–12.
56. Petrov, A. S., Boz, M. B., and Harvey, S. C. 2007. The conformation of double-stranded DNA inside bacteriophages depends on capsid size and shape. *J. Struct. Biol.* 160:241–48.
57. Jiang, W., Chang, J., Jakana, J., Weigele, P., King, J., and Chiu, W. 2006. Structure of epsilon15 bacteriophage reveals genome organization and DNA packaging/injection apparatus. *Nature* 439:612–16.
58. Lander, G. C., Tang, L., Casjens, S. R., Gilcrease, E. B., Prevelige, P., Poliakov, A., Potter, C. S., Carragher, B., and Johnson, J. E. 2006. The structure of an infectious P22 virion shows the signal for headful DNA packaging. *Science* 312:1791–95.
59. Rollins, G. C., Petrov, A. S., and Harvey, S. C. 2008. The role of DNA twist in the packaging of viral genomes. *Biophys. J.* 94:L38–L40.
60. Spakowitz, A. J., and Wang, Z. G. 2005. DNA packaging in bacteriophage: Is twist important? *Biophys. J.* 88:3912–23.

16 Elastic Network Models of Coarse-Grained Proteins Are Effective for Studying the Structural Control Exerted over Their Dynamics

Robert L. Jernigan and Lei Yang
LH Baker Center for Bioinformatics and Biological Statistics
Department of Biochemistry, Biophysics, and
Molecular Biology, Iowa State University

Guang Song
LH Baker Center for Bioinformatics and Biological Statistics
Department of Computer Science, Iowa State University

Ozge Kurkcuoglu and Pemra Doruker
Department of Chemical Engineering and Polymer
Research Center, Bogazici University

CONTENTS

16.1 INTRODUCTION

The functions of biological macromolecules are some of the more intriguing phenomena in nature, which can connect between the atomic and the biological. While more and more structures are being determined experimentally by X-ray crystallography, NMR, electron microscopy, and other techniques Protein Data Bank (PDB) [Berman et al. 2000] with an increasing speed, the structures themselves are not fully informative, and there is a pressing need to understand their dynamics in order to comprehend the functions and mechanisms of these molecular systems. One of the most intuitive and successful approaches to study molecule motions has been molecular dynamics (MD) [Rahman 1964; Stillinger and Rahman 1974; McCammon, Gelin, and Karplus 1977]. It is an important tool and has been used extensively in protein structure determinations and refinements, and simulations of (un)folding pathways and protein dynamics. There are two major challenges in applying MD to study the motions of large macromolecules: (1) the limits of computational power and (2) limits to the information provided by the analyses of the results. In general, there is a huge gap between the time scales of feasible simulations and the durations of biological events. To remedy this, some efforts have been spent on large-scale distributed computing, with the best known example being the folding@home project [Shirts and Pande 2000; Larson, Snow, and Pande 2003], which utilizes tens (or even hundreds) of thousands of processors worldwide. Coarse-grained (CG) models of proteins (lattice models and off-lattice models) and simplified potentials have also been developed in attempts to reduce the complexity, including other approaches such as the United Residue (UNRES) model of Scheraga [Liwo et al. 1997], the C alpha – C beta – Side group (CABS) model of Kolinski (2004), and many others. Since it takes a long time for MD to sample the conformational space sufficiently and to fully explore the collective nature of the dynamics of large macromolecular structures, another approach that is better suited for studying the collective motions of macromolecules is provided by the elastic network models. Also, results from these methods have been validated in direct comparisons with atomic MD simulations [Doruker, Atilgan, and Bahar 2000; Kim et al. 2003] and with multiple structures of the same protein [Yang et al. 2008]. These models originated in the study of rubbery polymeric materials [Flory 1977], and were first applied to proteins by Tirion (1996). Normal mode analysis (NMA) has been a particularly useful tool for the analyses of both the MD trajectories and the motions of the elastic networks; it expresses the motions in terms of some collective variables, known as the normal modes.

One of the most surprising aspects of the applications of elastic network models to proteins is the finding that proteins can be treated as completely uniform materials with excellent results. This results in the view that protein structures are really uniform rubbery bodies with a highly limited repertoire of motions, determined mostly by their shapes. The extent of control of these motions is directly reflected in a very small number (~10–20) of the normal modes as the dominant motions, and this fact is truly remarkable; there is extensive evidence for such control of motions from experiments, for example, see Lee et al. (2006). The most important motions of the normal modes are those between the large domains—most typically hinges and shearing motions—a general characteristic agreeing with the transitions known to occur between the multiple structures of the same protein. The control of these motions is so strong that it immediately becomes appropriate

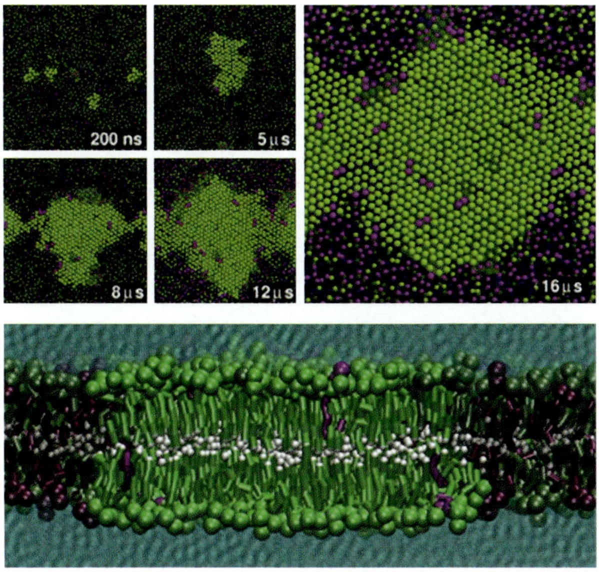

COLOR FIGURE 2.3 Coarse-grained simulation of domain formation in an unsaturated/saturated (diC16:2/diC16 PC 1:3) phospholipid mixture. The saturated lipid is colored green, the unsaturated lipid purple. Upper snapshots: top view of the membrane showing only the second tail beads of the lipids. Large/small beads are used for tails in the gel/fluid phase. A darker shade is used for the underlying monolayer. At $t = 0$ the system was quenched from the fluid phase at $T = 295$ K into the gel–liquid coexistence region at $T = 273$ K. During the first microseconds the system remains in a metastable fluid phase. Small gel-clusters composed of mainly the saturated component are seen to form and disappear. Only when a critical threshold size is exceeded, a gel-cluster remains stable (snapshot at 5 μs) and starts growing, until the system boundaries are reached ($t = 8$ μs). The gel cluster slowly optimizes its interface with the remaining fluid phase, which is enriched in the unsaturated component. Lower figure: final snapshot of the coexisting gel and fluid domain. Light/dark shades distinguish gel/fluid phase lipids. The terminal tail beads are depicted as white/gray.

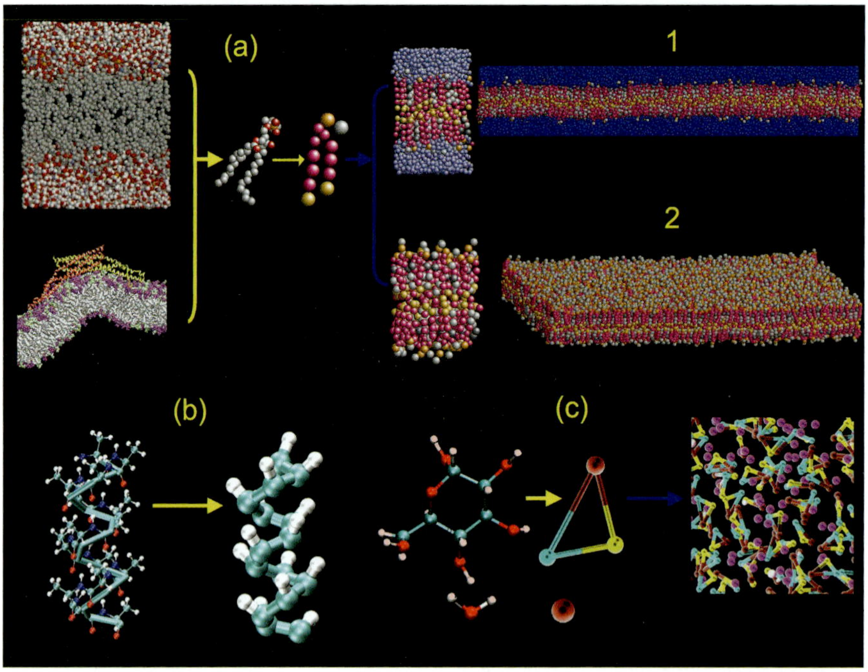

COLOR FIGURE 3.1 A schematic depicting various aspects of the MS-CG methodology for biomolecular systems. (a) The MS-CG method being applied for modeling interactions between lipids, cholesterol, and membrane-bound proteins, such as the N-BAR domain. The MS-CG method has been implemented with both explicit solvent (1) and solvent-free (2) CG models. The MS-CG method also applied for modeling peptides (b) and monosaccharide solutions (c).

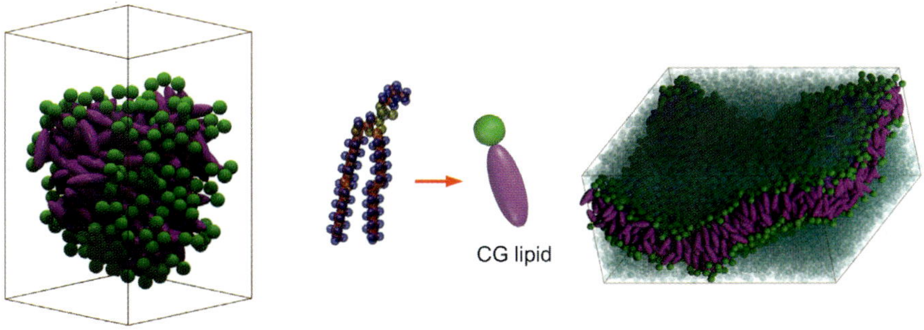

COLOR FIGURE 5.4 Snapshots from MD simulations of coarse-grained amphiphiles (middle) in coarse-grained water. The potential is versatile to model mixtures of elliptic shapes and the dynamics stable to allow for self-assembly (left) and undulations in model membranes (right), a signature of the emergence of continuum behavior.

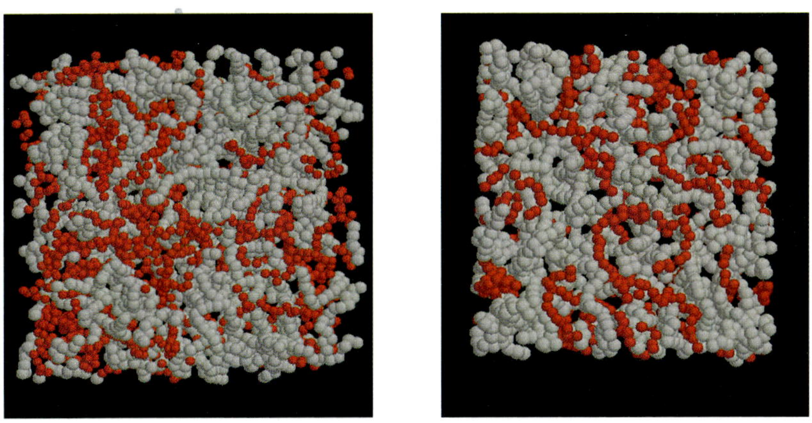

COLOR FIGURE 6.5 Snapshots of no phase separation in PI–PS mixtures if we use an algebraic (left, system 36 PI and 24 PS 60mers) or geometric (right, 36 PI 45mers and 24 PI 30mers) mixing rule in qualitative disagreement with experiments and systematically optimized models.

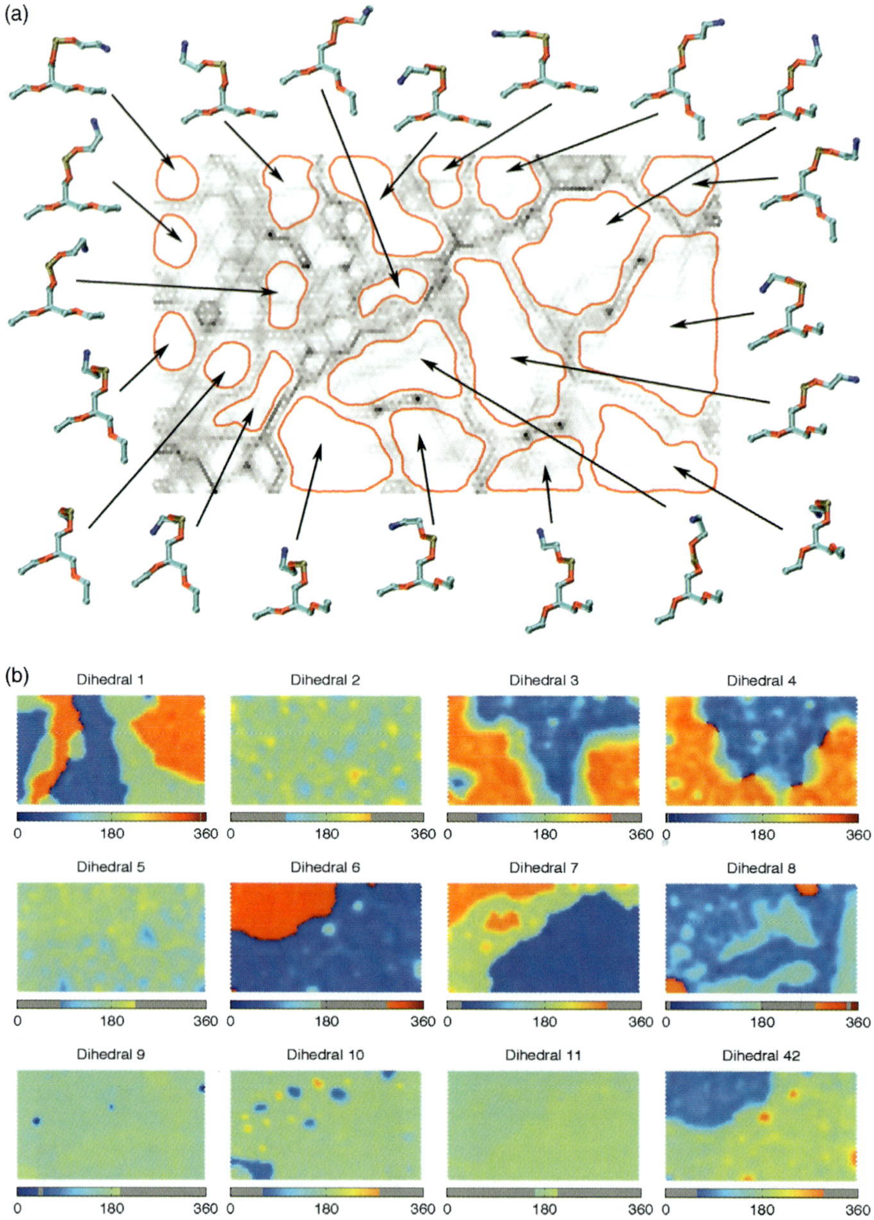

COLOR FIGURE 7.2 Visualizations of SOM trained with PLPC headgroup data.[12] (a) U-matrix. Neurons are shaded based on similarity to their neighbors; lighter shades indicate higher similarity. Several regions of high similarity (selected manually) are encircled. For each region, the conformation of one model vector is visualized. (b) Component planes. Each small figure shows the value of one dihedral for all model vectors on the map (see Figure 7.1b for definitions of the numbers). The color bars under each plane show the range of values that are present on the map; unused values are in gray.

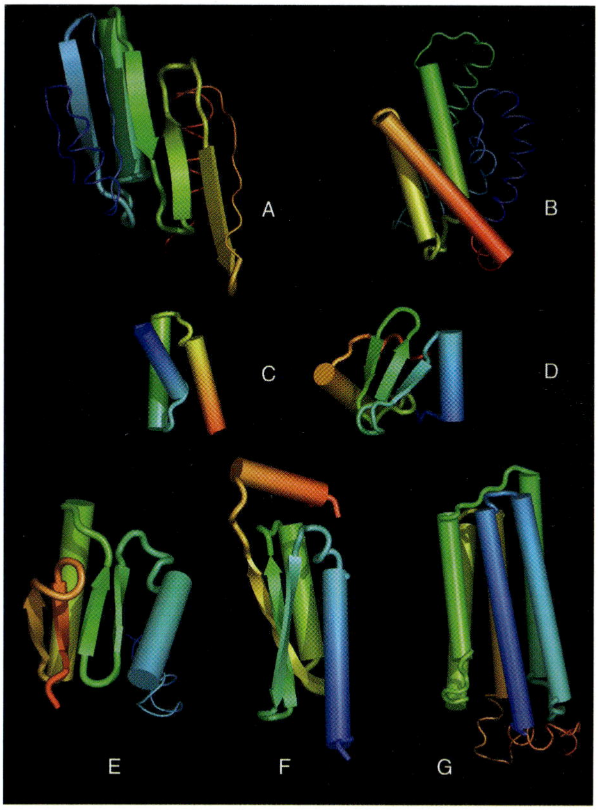

COLOR FIGURE 8.4 The best predicted structures of CASP5 targets T0149_2 (the C-terminal domain of T0149) (A) and T0129 (B) and CASP6 targets T0215 (C), T0281 (D), T0223_2 (the second domain of T0223) (E), T0230 (F), and T0198 (G). The correctly predicted parts of the structures are marked as cylinders (for α-helices), ribbons (for β-strands), and thick lines (loops and other unstructured regions), and the incorrectly predicted parts are shown as thin lines. The chains are colored from blue to red from the N to the C terminus. (From Ołdziej, S. et al., *Proc. Natl. Acad. Sci. U.S.A.*, 102, 7547–7550, 2005. With permission.)

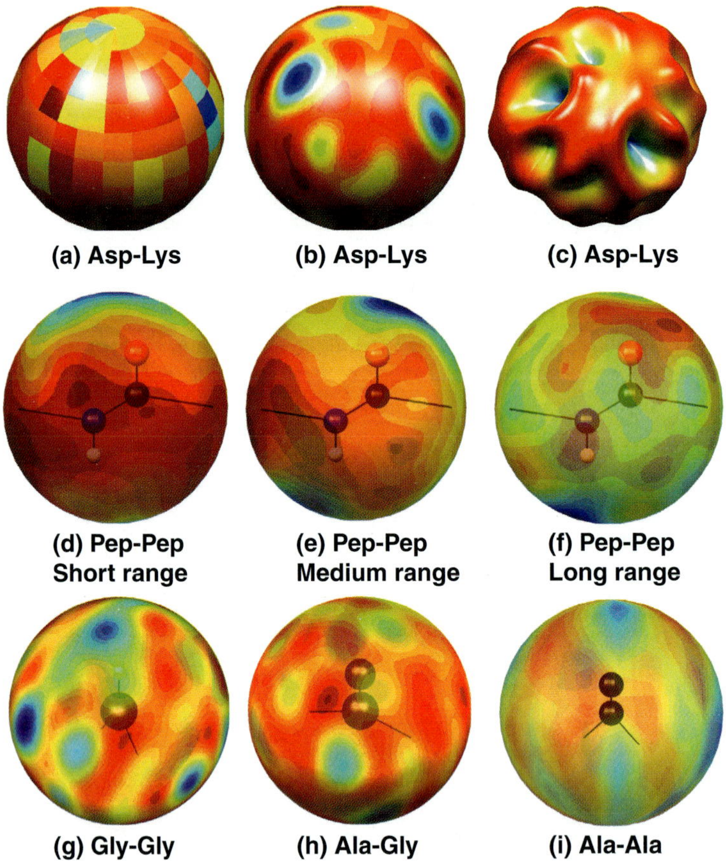

COLOR FIGURE 10.6 Three-dimensional representations of the distance- and orientation-dependent statistical potential fields for side chain–side chain and side chain–backbone interactions [17]. Short-range orientation-dependent potentials $U_{\text{DO-21s}}$ are shown using: (a) a 12×24 equiangular grid and (b)–(i) a 96×192 equiangular grid, for the SHS procedure. In the graphical representation (c) for *Asp-Lys* short-range contacts [25], the magnitude of the interaction potentials is proportional to both the radius from the center of each local reference frame and to the color (red for repulsive and blue for attractive regions). These types of 3D representations, with valleys corresponding to attractive regions, illustrate better the roughness of the statistical energy surfaces. The backbone–backbone potentials (*Pep-Pep*) are shown for the smooth short-range (d), middle-range (e), and long-range (f) interaction ranges. *Asp-Lys* (a)–(c), *Gly-Gly* (g), *Ala-Gly* (h), and *Ala-Ala* (i) potentials are also shown for short-range interactions. The relative orientation of the *Pep*, *Gly*, and *Ala* atoms with respect to the orientation-dependent potential values are also shown.

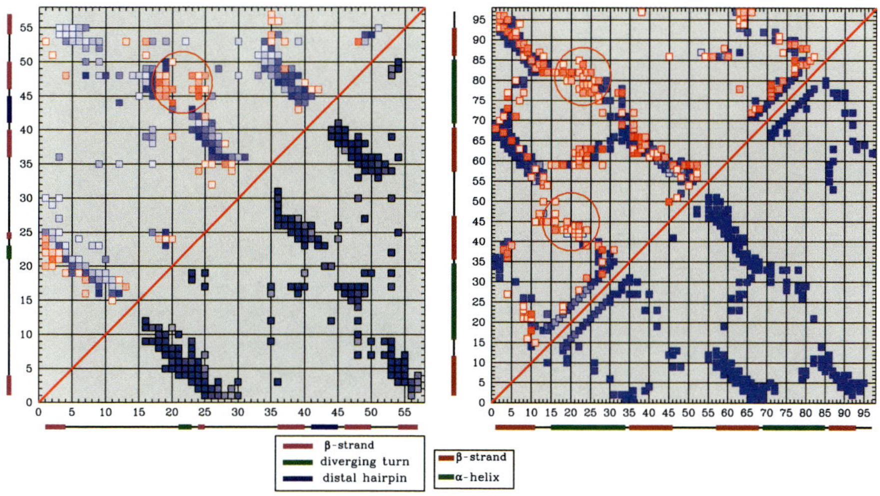

COLOR FIGURE 11.3 The left figure shows the contact probabilities in the transition state of the folding of SH3 (upper contact map), and in the SH3 folded state (lower contact map). The right figure shows the distribution of contact probabilities in the folded state of $S6^{Alz}$ (upper contact map). Different shades of blue are used for the native interactions, from white to deep blue; the contacts with higher probabilities are in deep blue, lower probabilities are in white. Different shades of red are used for the nonnative interactions, from white to deep red. The main clusters of nonnative interactions are circled in red. (Adapted from Das, P., Matysiak, S., and Clementi, C., *Proc. Natl. Acad. Sci. U.S.A.*, 102, 10141–46, 2005; Matysiak, S., and Clementi, C., *J. Mol. Biol.*, 363, 297–308, 2006.)

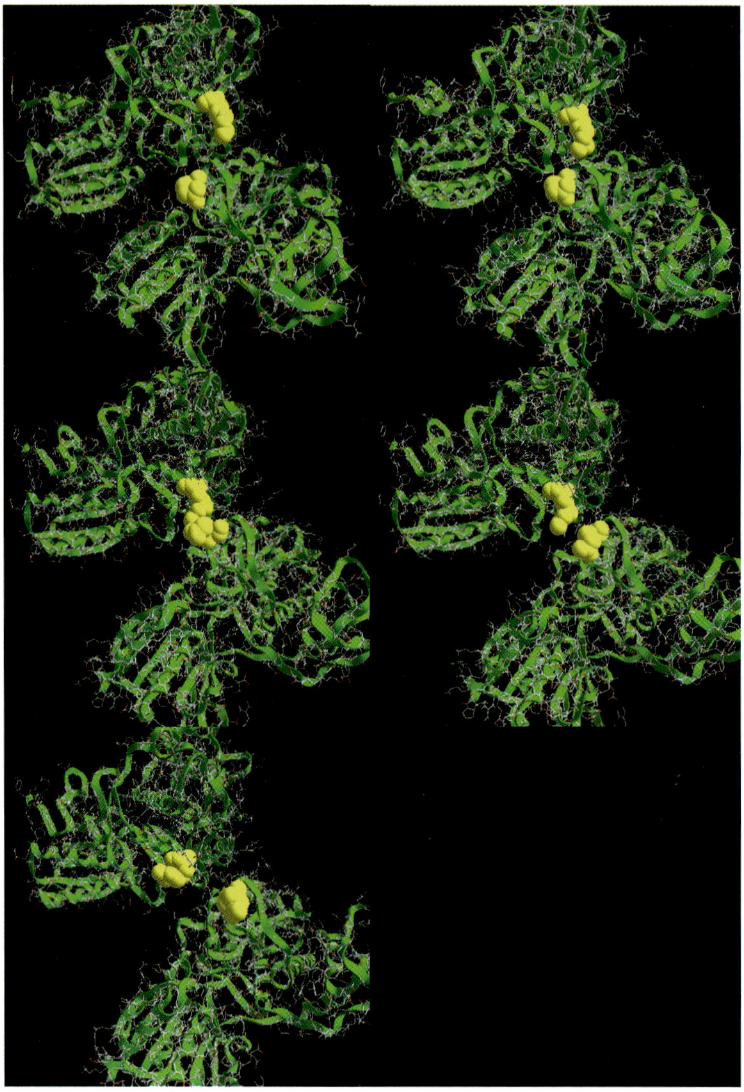

COLOR FIGURE 13.7 An illustration of the strong coupling between atomically detailed motion and large-scale domain opening. A sequence of events along the transitional pathway is shown starting from the upper left corner (structure 1) continuing to the right (structure 20) and then down. The length slices are shown from an atomically detailed path of 100 slices that was constructed from a coarse-grained model. Only slices 1, 20, 50, 60, and 100 are shown. The atomically detailed event is the *transient formation of a salt bridge* between a glutamic acid (Glu60 in chain A of the dimer) and an arginine (Arg448 in chain B) (yellow space filling model). There is also a large-scale motion that causes a visible separation between the two lobes. The salt bridge is not present at the end points. It assists in lowering the transition barrier. Notably, it is not present in the linearly interpolated path (see text for more details).

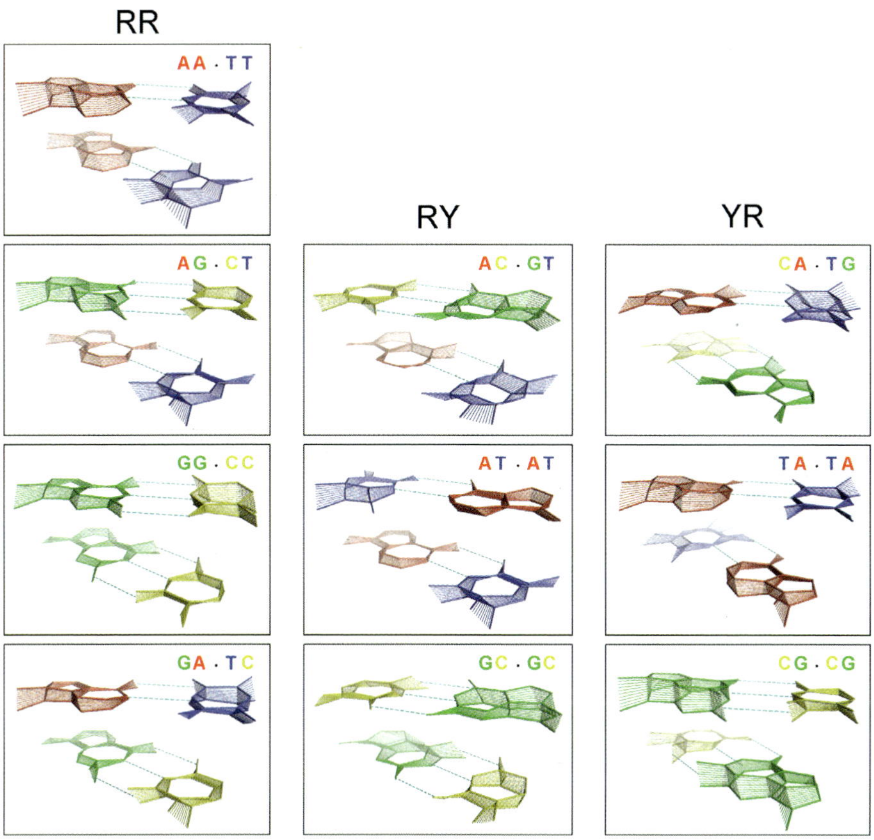

COLOR FIGURE 14.3 Sequence-dependent motions along the longest principal axes of the 10 unique DNA base-pair steps. Nonequilibrium forms are superimposed on the intrinsic (average) dimer structures. Perturbed states correspond to deformations, at increments of $3\langle\lambda_1^2\rangle^{1/2}$, along the longest principal axes of the "complete" knowledge-based potentials, where λ_1 is the largest eigenvalue of the covariance matrix, and "energies" range from zero to $4.5m^2k_BT$ for displacements of $\pm 3m\langle\lambda_1^2\rangle^{1/2}$. Here m is set to 5 to enhance visualization of structural deformations. Base pairs are represented as ideal Watson–Crick pairs [Olson et al. 2001], with the hydrogen bonds of rest structures denoted by dashed lines. Bases are color-coded according to chemical identity: adenine (red); thymine (blue); guanine (green); cytosine (yellow). Motions are illustrated with respect to the "middle" frame of each step and viewed into the minor groove of the upper 3'-base pair of each miniduplex. Note the correspondence of observed structural variability with the corresponding contour surfaces in Figure 14.2.

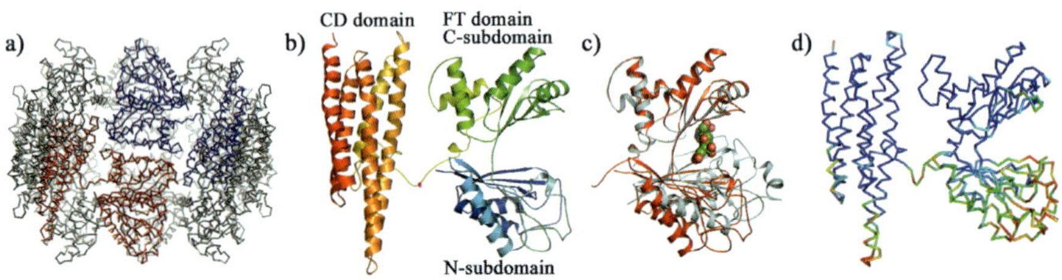

COLOR FIGURE 17.3 Structure and thermal ellipsoids of FTCD. (a) The square doughnut structure of an FTCD octamer. Two subunits are shown in red and blue, respectively. (b) The subunit structure of ligand-free FTCD. Backbone trace color ramped from the N-terminus to the C-terminus. (c) Superposition of the FT domain of human ligand-free FTCD (red) with the structure of the same domain in isolation (cyan) with the product analog, folinic acid (CPK mode), bound in the groove. (d) Rainbow-colored isotropic *B*-factor in the original model. The hotter the color, the larger the *B*-factors. The high flexibility of the N-subdomain, the linker region, and the lower half of the CD domain are evident. (This figure is adopted from figure 1 on page 7870 in Poon, B. K., Chen, X., Lu, M., Vyas, N. K., Quiocho, F. A., Wang, Q., and Ma, *J. Proc. Natl. Acad. Sci. U.S.A.*, 104, 7869, 2007.)

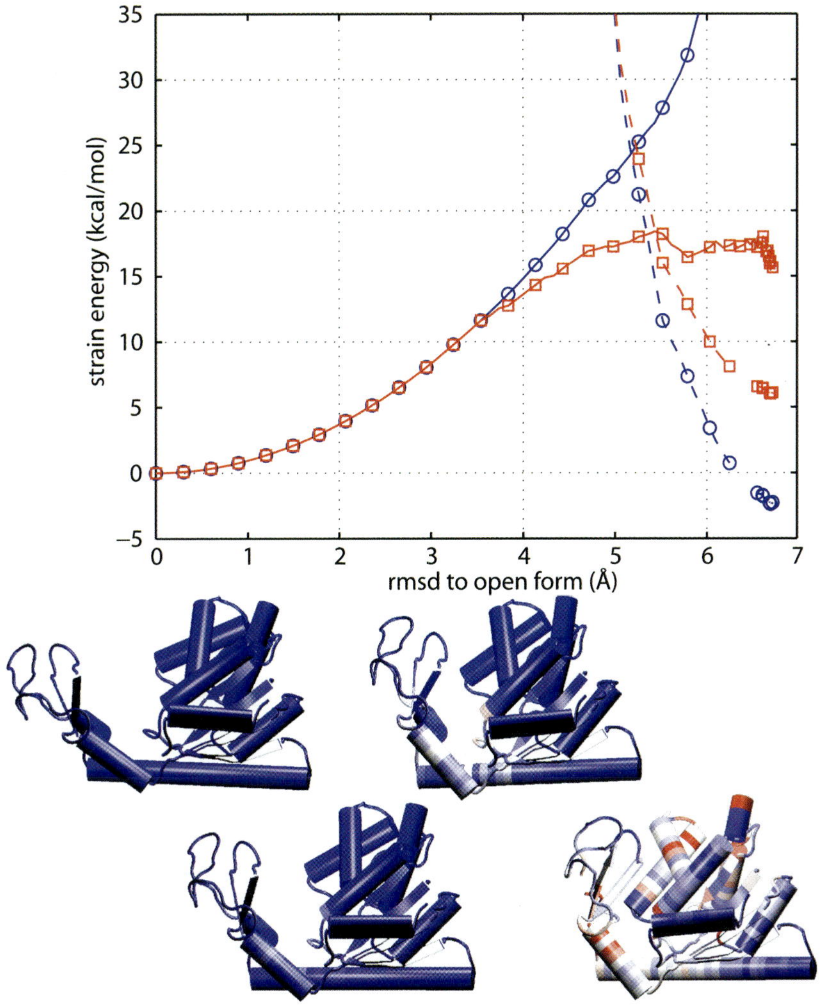

COLOR FIGURE 18.2 Energy surface of adenylate kinase conformational transition calculated from the nonlinear elastic model. Energies of the ligand-unbound state are shown by solid lines, and the ligand bound state are shown by broken lines. Energy profiles without cracking are marked ○ and with cracking are marked □. The cracking effect lowers the energy barrier. The energy surface of the ligand bound states with low surface tension has higher energy than without cracking, since some residues are partially unfolded. Actual conformational change would involve a refolding process; that is, shift from the energy profile with cracking to the energy profile without cracking. The structures along the conformational change pathway are also shown (open form, transition state and closed form from left side). The residues are colored according to the strain energy calculated for the ligand-unbound state; that is, the open form has no strain and the closed form has the largest strain; blue corresponding to no strain and red residues high strain. At the transition state, the strain energy is very localized and it could cause cracking or partial unfolding.

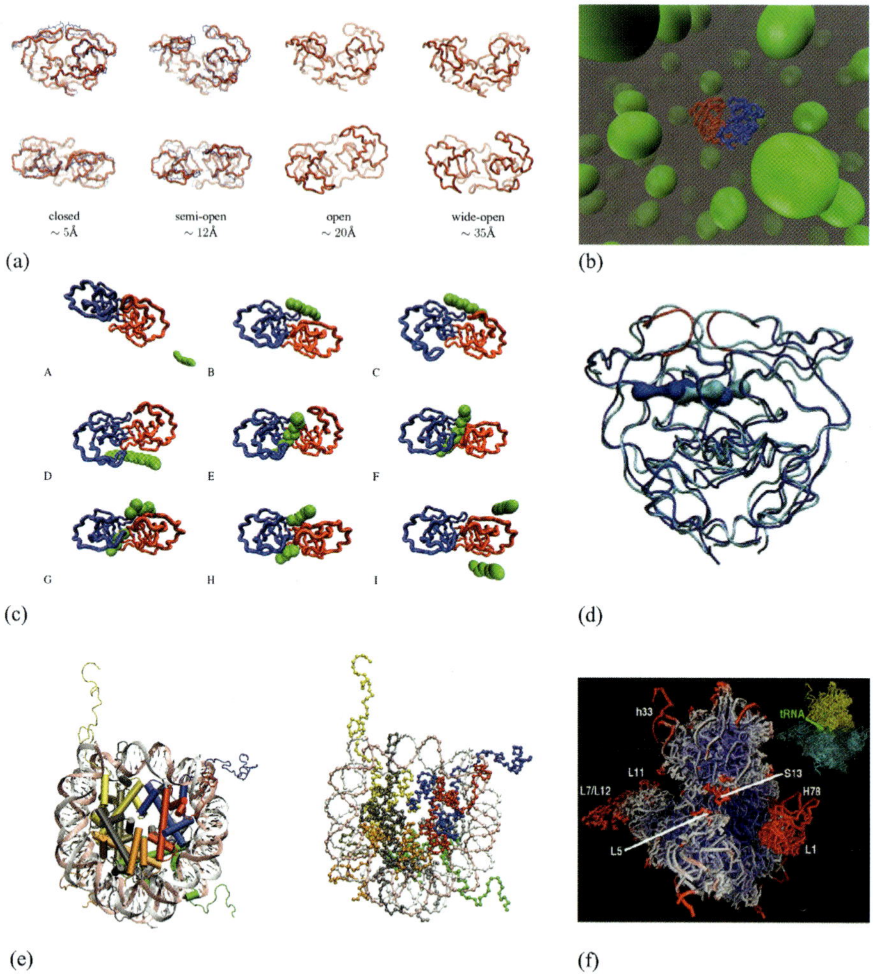

COLOR FIGURE 19.5 (a) Snapshots from the free protease simulation, showing the steps of the flap opening. For the first two steps the experimental structures are available (in blue in the color figure) that superimpose very well on the simulated structures (in red). (b) A snapshot of the simulation in the presence of crowders. (c) Substrate approach (A, B), interaction with the flaps (C, D), substrate adjustment and flap closing (E, F), cleavage and release (G–I). (d) The ligand binding with closed flaps (for small ligands). (e) Coarse-graining of the nucleosome: the all-atom cartoon representation (left) and the one-bead model (right). (f) The one-bead model of the S70 bacterial ribosome.

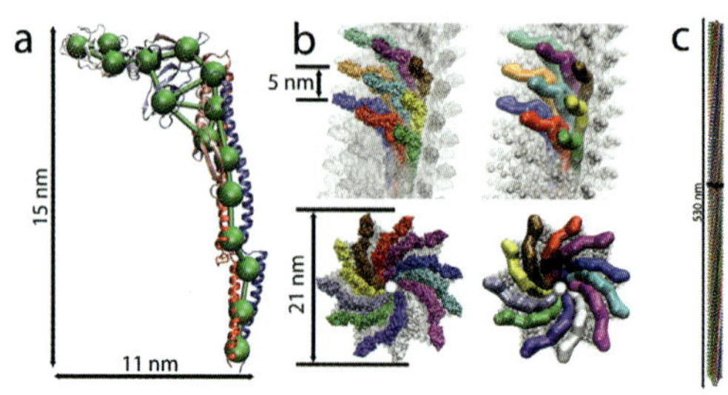

COLOR FIGURE 20.5 CG of the flagellar filament. Unit proteins are represented by 15 CG beads (a). In (b), the flagellar filament viewed from the side and from the top is shown in all-atom (left) and CG (right) representations. A filament segment (1100 monomers) is shown in CG representation in (c). A single helix turn of 11 unit proteins is highlighted in black.

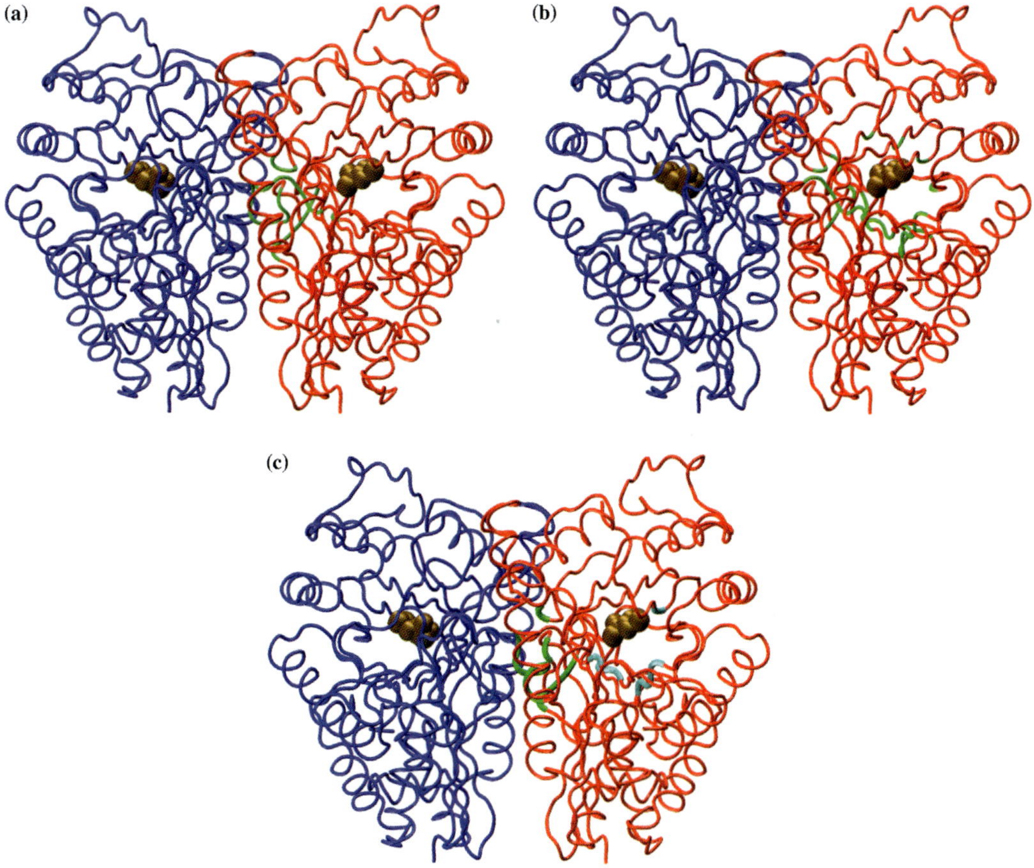

COLOR FIGURE 21.4 (a) Backbone diagram of enolase. Residues with high force constants within the right-hand monomer are shown in black. (b) Following domain separation, residues with high force constants within the right-hand domain are shown in black. (c) Mechanical changes in passing from the monomeric to the dimeric form of enolase. Residues with significantly increased force constants are shown in black and those with significantly decreased force constants in gray (changes are only shown for the right-hand domain).

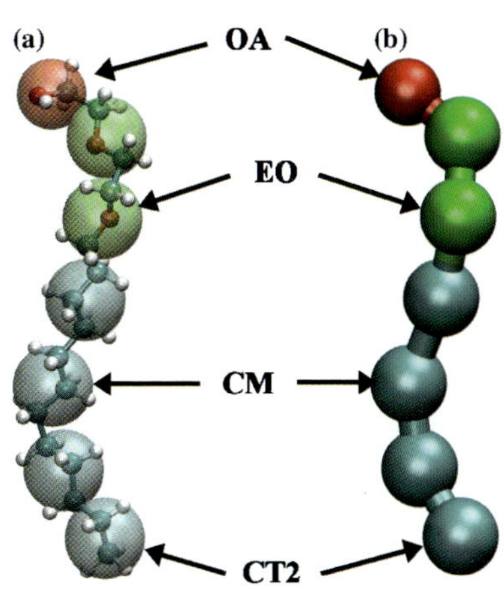

COLOR FIGURE 22.1 Atomistic (a) and coarse-grained (b) representation of $C_{12}E_2$ molecule. The atomic groups, ($HO–CH_2–$), ($–CH_2–O–CH_2–$), ($–CH_2–CH_2–CH_2–$), and ($CH_3–CH_2–$), are referred to as OA, EO, CM, and CT2 segments, respectively.

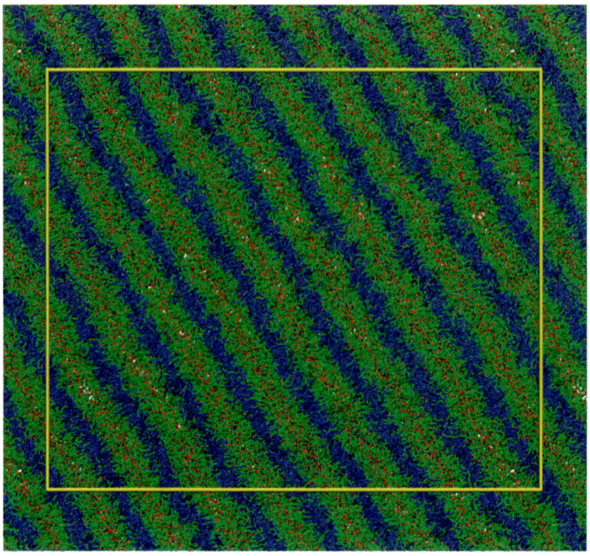

COLOR FIGURE 22.7 Shown here is a snapshot of a simulation box containing 807,360 CG particles composed of $C_{12}E_6$ and water in the lamellar phase after 470 ns of MD simulation started from the hexagonal phase. The hydrophobic tails are shown in blue, the headgroups are shown in green and red, and the water is shown in white. The unit cell is shown in yellow.

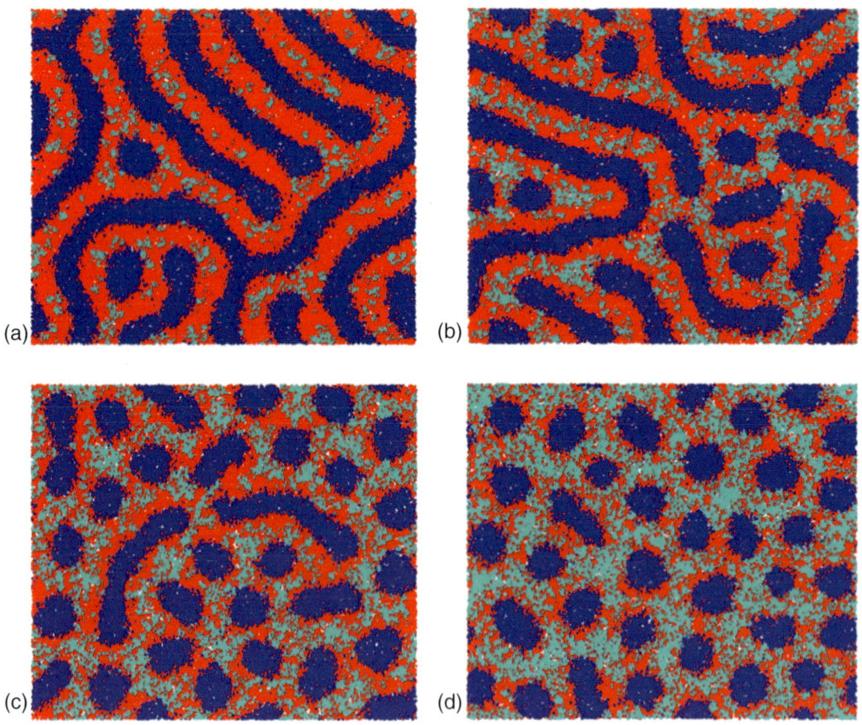

COLOR FIGURE 24.6 Morphology of nanoparticle/copolymer mixtures. In these cross-sections, the A beads are shown in red, the B beads in blue, and the nanoparticle beads in green. As the nanoparticle volume fraction increases, the morphology changes from lamellar to cylindrical. The diblock copolymer is symmetric; the nanoparticles are A-selective and have a radius $R_p = 0.16R_e$ and the volume fraction is $\phi_p = 0.1, 0.2, 0.3$, and 0.4 in (a), (b), (c), and (d) respectively. Parameters: $\chi N = 40$, $\kappa N = 50$, $\sqrt{\bar{N}} = 128$ ($N = 32$, $\Delta L = 0.19R_e$).

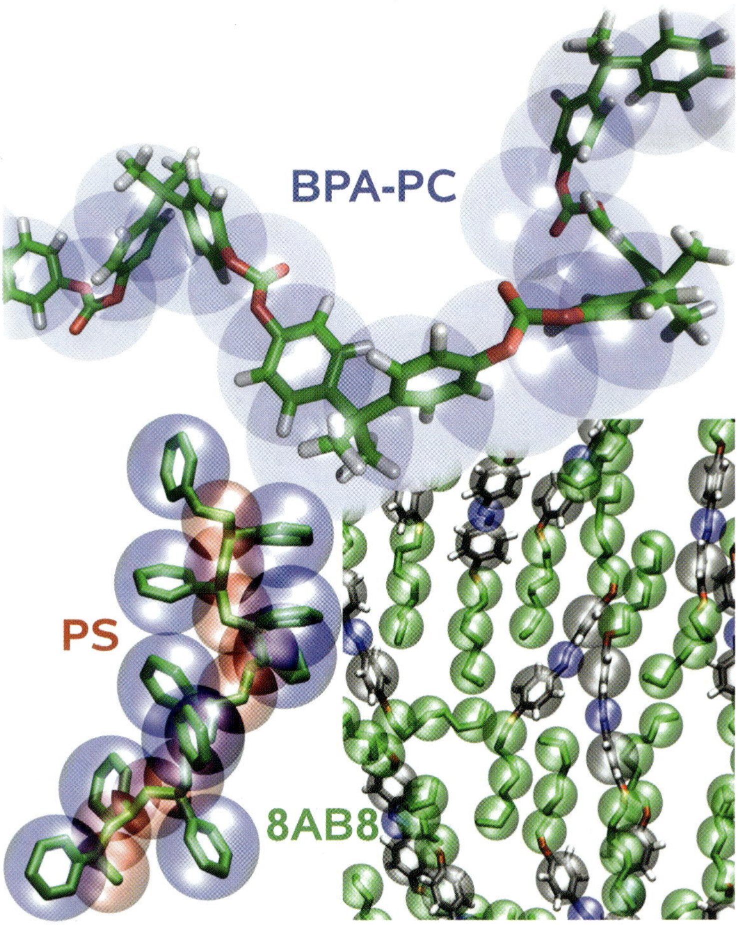

COLOR FIGURE 25.6 Snapshots of selected molecules from CG simulations of BPA-PC, PS, and 8AB8 indicating both CG centers and atomistic coordinates obtained through inverse mapping.

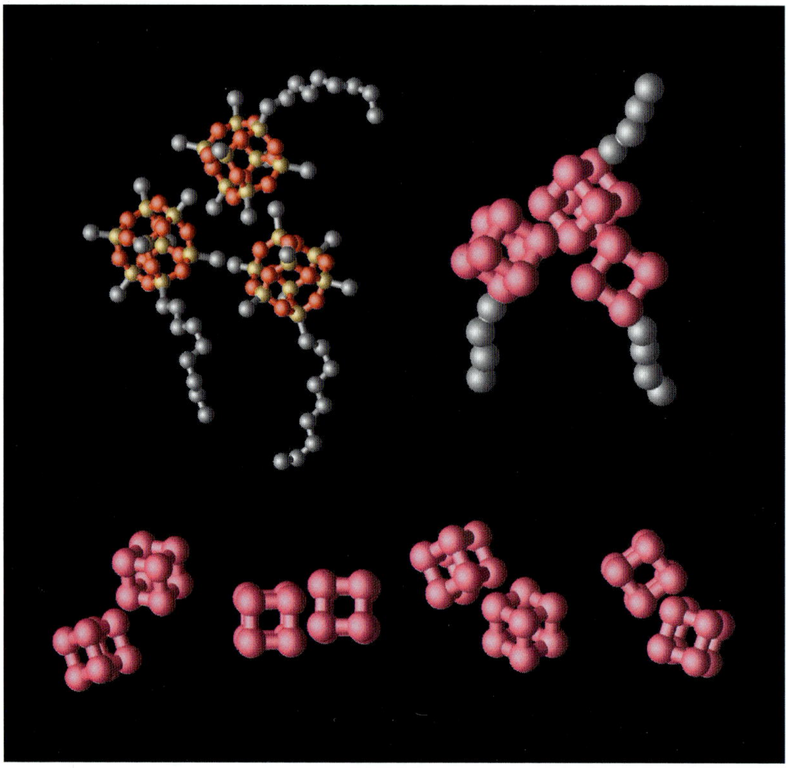

COLOR FIGURE 27.6 (Top) Self-assembled aggregates observed in all-atom and coarse-grained simulations of polyhedral oligomeric silsesquioxane (POSS) nanoparticles functionalized with a single nonyl tether and dissolved in hexane solution. Solvent molecules have been omitted from the atomistic simulation snapshot for clarity. (Bottom) Simulated local packings of coarse-grained POSS nanoparticles functionalized solely with methyl groups. Left to right: corner-corner, face-face, edge-edge, and corner-edge orientations.

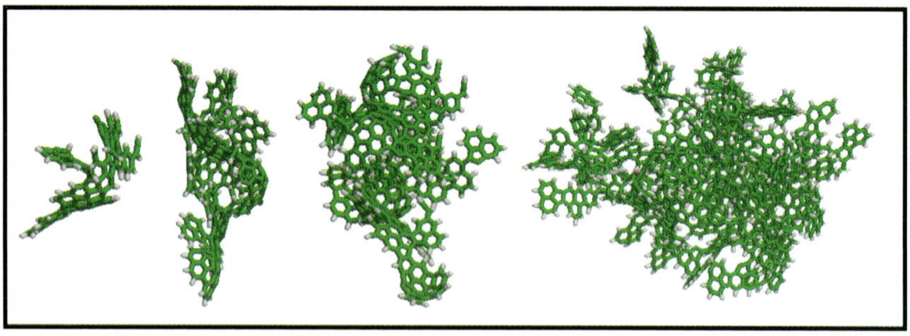

COLOR FIGURE 28.3 Nanoparticles formed with the AMPI code.

to consider these computed dominant motions of a protein to be related to biological mechanisms, for example the processing steps, enzyme reactions, and walking motions along fibrils. The functional atomic parts of the structures thus can even be viewed as decorations on the surfaces of the domains that follow or respond in their motions. This leads to two extremely important conclusions: that motions of the surface components are controlled by the domain motions, whether they are external loops or the atoms of enzyme sites. This point of view remains to be fully validated, but it establishes an important hypothesis for developing models for the further investigation of biological processes in their molecular details. The elastic models also have the further important advantage of not requiring all details of structure to be present in order to derive a given motion. *One of the most important lessons we have learned from these models is that it is not so important to know all the details of a structure in order to calculate its motions, but rather that the structure must be complete, even if the levels of detail provided by a particular representation differ among the parts of a structure* [Ming et al. 2002; Kurkcuoglu, Jernigan, and Doruker 2004].

Prior to atomic calculations, an energy minimization based on some semiempirical potential must be applied to a system. One major drawback of doing this is that it can result in a conformation that can differ significantly from the crystallographic structure. However, for the elastic network models, the problem is avoided entirely since we assume that the crystal structure is the minimum energy form, and certainly the details of the structure are less important for CG structures.

The low-frequency collective motions of these elastic network models are insensitive to the atomic details of the structure and its interactions, as was first demonstrated by Tirion (1996). Therefore, to determine the low-frequency modes of motion of a structure, it is acceptable to use highly coarse-grained models, such as representing each residue with a single point mass and atomic interactions with elastic springs or even several tens of residues as a single point [Doruker, Atilgan, and Bahar 2000]. This affords a major computational advantage, especially for large structures. Either coarse-graining by representing a number of sequential residues by a single point [Kurkcuoglu, Jernigan, and Doruker 2004, 2005] or by treating a domain as a rigid block [Song and Jernigan 2006; Yang, Song, and Jernigan 2007] has been shown to be valid.

Specifically, the Elastic Network Model (ENM) for isotropic fluctuations is usually named the Gaussian Network Model (GNM) [Bahar, Atilgan, and Erman 1997], where only the magnitudes of the fluctuations are considered. Its anisotropic counterpart, where both the magnitude and *direction* of the collective motions are examined, is called the Anisotropic Network Model (ANM) [Atilgan et al. 2001]. The details of these models are described in the Methods section.

Mixed CG elastic network models have been recently introduced to investigate the collective motions of especially large biological systems with high- (atomistic details) and low-resolution (CG) regions [Kurkcuoglu, Jernigan, and Doruker 2005, 2006]. This method has proved to be a computationally efficient tool for the exploration of the dynamic motions of supramolecular assemblages at even atomic detail, difficult to attain with conventional full-atom/empirical potential simulation techniques. In this method, the nodes of the elastic network, either as atoms or coarse-grained groups of residues, that fall within a cutoff distance are linked with harmonic springs varying according to the node size.

16.2 METHODS

16.2.1 GNM (for Magnitudes of Motions)

GNM is the isotropic ENM that only considers the magnitude, but not the direction of the fluctuations of residues [Bahar, Atilgan, and Erman 1997]. For GNM, each residue of the protein is represented by its corresponding alpha carbon, and interacts only with others within a cutoff distance (usually 7–8 Å). For each pair of interacting residues, their connection is simplified as a harmonic force with an equal spring constant. The potential energy V is computed by

$$V = \frac{\gamma}{2} \Delta R^T \Gamma \Delta R, \tag{16.1}$$

where ΔR represents the residue fluctuations vector. The Kirchhoff matrix Γ is given by

$$\Gamma_{ij} = \begin{cases} -1 & \text{if } i \neq j \text{ and } r_{ij} \leq r_c \\ 0 & \text{if } i \neq j \text{ and } r_{ij} > r_c \\ -\sum_{i, i \neq j} \Gamma_{ii} & \text{if } i = j \end{cases}, \tag{16.2}$$

for a pair of residues i and j, where r_c is the cutoff distance. The mean-square fluctuations of residue i are given by

$$\langle u_i^2 \rangle = 3(k_B T/\gamma)\left[\Gamma^{-1}\right]. \tag{16.3}$$

Since the inverse of the Kirchhoff matrix can be expressed in terms of the eigenvalues λ_k and eigenvectors u_k of Γ, the mean-square fluctuations of residue i can also be written as

$$\langle u_i^2 \rangle = 3(k_B T/\gamma)\sum_{k=2}^{n} \lambda_k^{-1} u_{ki}^2. \tag{16.4}$$

There are in total $n-1$ modes generated by GNM (the first eigenvalue, representing rigid body motion being zero) with the eigenvectors representing the magnitudes of fluctuations of the modes, without information regarding the directions of fluctuations.

16.2.2 vGNM (Variable GNM, to Allow Variable Model Contributions)

vGNM is an empirical version of the GNM that takes into account the contribution of rigid-body translation and rotation and the effect of crystal packing by allowing the coefficient of each mode to be variable [Song and Jernigan 2007]. It is postulated that the effect of crystal packing can amplify some modes while suppressing others, although the effect may be difficult to compute directly. This can be realized in B-factor calculations by allowing the weights of the mean-square fluctuations to be variables. In vGNM, the mean-square fluctuations can be expressed as

$$B^{\text{calc}} = w_{\text{trans}} + w_{\text{rot}} \left\| r_i - r_{\text{centroid}} \right\|^2 + \sum_{k=2}^{n_{\text{low}}} w_k u_{ki}^2, \tag{16.5}$$

where u_{ki} is proportional to the magnitude of the motion of atom i in mode k. In this equation, the contribution of translation is simply represented by a uniform weight w_{trans}, the contribution of rotation is proportional to the square of the distance between each alpha carbon and the protein's centroid ; that is, $w_{\text{rot}} \left\| r_i - r_{\text{centroid}} \right\|^2$, and for internal motions, it is postulated that the contribution of each mode is not necessarily proportional to the inverse of its corresponding eigenvalue. Thus, the total fluctuations can be expressed as a sum of all these terms weighted by the parameters w_{trans}, w_{rot}, and w_k for each internal mode. To determine these parameters, least-squares fitting between the calculated B-factors B_i^{calc} and the experimental B-factors B_i^{exp} is required ; that is, to minimize the following function

$$\sum_{i=1}^{n_{\text{low}}} \left(B_i^{\text{calc}} - B_i^{\text{exp}}\right)^2, \tag{16.6}$$

while requiring all weights to be nonnegative. Note that the summation in this function does not run up to the total number of modes, but instead to n_{low}, since the low-frequency modes contribute most to the internal fluctuations. Such a restriction will improve the computational efficiency and also ensure avoiding overfitting of the experimental B-factors with too many parameters. In practice, it has been shown that 20 is about the smallest n_{low} for which the contributions from both translation and rotation seem to have stabilized.

16.2.3 ANM (for Vector Directions of Motions)

ANM is the anisotropic ENM that considers both the magnitudes and the directions of fluctuations of residues [Atilgan et al. 2001]. In ANM, the potential energy V is a function of the displacement vector D

$$V = \frac{\gamma}{2} DHD^T, \tag{16.7}$$

where γ is the spring constant for all identical interactions of all residues falling within a cutoff distance (usually 13–15 Å), and H is the Hessian matrix containing the second derivatives of the energy function. Therefore, the counterpart of the Kirchhoff matrix Γ of the GNM is simply H/γ in the ANM. The Hessian matrix H can be decomposed as

$$H = M\Lambda M^T, \tag{16.8}$$

where Λ is a diagonal matrix with elements equal to the eigenvalues and the eigenvectors from the columns of the matrix M. This decomposition generates $3n-6$ normal modes (the first six modes account for the rigid body translations and rotations of the system) reflecting the vibration fluctuations. The eigenvalues are usually sorted in descending order. Each eigenvalue represents the frequency of the corresponding mode, while its corresponding eigenvector represents the directions and relative magnitudes of the fluctuations of each structural point.

16.2.4 Mixed CG (to Include the Effect of the Whole Structure, while Permitting a Focus on One Particular Part)

The mixed CG of the ENM enables the modeling of a protein structure with different regions having lower and higher resolution [Kurkcuoglu, Jernigan, and Doruker 2004, 2005, 2006]. In such modeling, the dynamics of the interesting functional parts can be analyzed at high resolution, while the remainder is represented at poorer resolution so that the total number of nodes in the system remains sufficiently low for computational efficiency. The mixed CG procedure necessitates the assignment of different cutoff distances and force constants for the different sized nodes (node 1: high-resolution; node 2: low-resolution) to maintain the appropriate flexibility of the protein. This method has been applied using two different types of cg, namely Case 1 with node 1: one-node-per-residue, node 2: multiple residues per node, and Case 2 with node 1: one-node-per-heavy atom, node 2: one-node-per-residue.

For Case 1, uniform CG of the protein structure is performed at a series of levels by retaining n, $n2$, $n/5$, $n/10$, and $n/40$ residues (n is the total number of residues). Such coarse-grained structure can be viewed as a collection of s coarse-grained segments, each containing m residues. Thus the total number of residues in the original structure is $n = sm$. The relationship between the radius of gyration R_G and the segment length m is given by

$$R_G = am^b, \tag{16.9}$$

where a and b are constants that can be obtained by fitting or from the literature [Doruker, Jernigan, and Bahar 2002; Doruker et al. 2002]. The cutoff radius is given as a function of segment length as

$$r_c = 2R_G + R_0, \tag{16.10}$$

where R_0 is typically taken to be 13 Å for $n = 1$.

By neglecting the cross-correlation terms of residue fluctuations, the B-factors for segment s can be approximated by

$$B_s = \frac{\sum_{i=1}^{n} m_i^2 B_i}{(\sum_{i=1}^{n} m_i)^2}, \tag{16.11}$$

where m_i represents the mass of node i.

From a comparison of experimental and theoretical fluctuations, the force constant can be determined for different segment lengths; that is, different levels of coarse-graining, which gives a relationship between cutoff distance and the force constant γ.

Finally, the interesting parts of the structure can be coarse-grained at a higher resolution and the rest at a lower resolution. For the interfacial region, the cutoff distance ($r_{c,12} = \sqrt[3]{(r_{c,1}^3 + r_{c,2}^3)/2}$) and the corresponding force constant ($\gamma_{1,2}$) is determined between node types 1 and 2.

For Case 2, a cutoff value for the high-resolution nodes (atomistic node 1) is determined in the range of 6–9 Å from a comparison of the frequency distributions obtained from atomistic and residue-based ANM for several proteins. And the corresponding values are determined from B-factor data, as for Case 1 [Kurkcuoglu, Jernigan, and Doruker 2005, 2006].

16.2.5 Domain-ENM (for Fixing the Most Cohesive Parts as Rigid Elements of Structure)

All the above approaches use a single spring constant except for the mixed CG where the spring constant varies with the node size. In domain-ENM [Song and Jernigan 2006; Yang, Song, and Jernigan 2007], a larger spring constant is assigned for the intradomain contacts. This conveniently and effectively encodes domain rigidity with a single parameter. It also enables rigid body domain motions to be separated from the low amplitude fluctuations of each rigid domain, thereby permitting the dominant rigid body domain motions to be more easily captured than with the uniform ENMs. This is a structure-based coarse-graining over a region of a structure geometrically rather than over sequential segments, as was done for the mixed CG approach. In order to use this approach, rigid clusters or "domains" have to be determined first. For the case where the end structures are known, such as the "open" and "closed" forms of a protein, this can be done relatively easily by comparing the two end structures to determine the rigid clusters or "domains" (see Song and Jernigan (2006) and Yang, Song, and Jernigan (2007) for details).

16.3 AGREEMENT WITH EXPERIMENTS

16.3.1 *B*-Factors from X-Ray Structures

In GNM, the B-factors can be computed by

$$B_i = 8\pi^2 \langle u_i^2 \rangle / 3, \tag{16.12}$$

which can be compared with the experimental B-factors reported for X-ray structures. The correlation between them is given as

$$\mathrm{corr} = \frac{\sum_{i=1}^{n}\left(B_i - \langle B\rangle\right)\left(B_i^{\exp} - \langle B^{\exp}\rangle\right)}{\left\|B - \langle B\rangle\right\|\left\|B^{\exp} - \langle B^{\exp}\rangle\right\|}. \tag{16.13}$$

The B-factors predicted by GNM are in good agreement with X-ray crystallographic data. For a set of 113 X-ray structures, the average correlation for the agreement between experimental B-factors and those calculated by GNM was about 0.60 [Kundu et al. 2002]. An example is shown in Figure 16.1a. The B-factors can also be computed by ANM from the decomposition of the Hessian matrix, and the correlation with experimental values is generally slightly worse than for the GNM.

16.3.2 Pseudo-B-Factors from NMR Structure Ensembles

The dynamics of NMR structures can be obtained from the ensemble of reported structures by defining pseudo-B-factors that are derived from atomic distances between the individual NMR models reported and the average of the ensemble [Wilmanns and Nilges 1996]. Such pseudo-B-factors are defined as

$$B_i^{\mathrm{pseudo}} = k\left\langle \left\|x_i^j - \langle x_i\rangle\right\|^2 \right\rangle, \tag{16.14}$$

where x_i^j contains the coordinates of the i th atom in the j th NMR model, and k is a scaling constant. The averages $\langle\ \rangle$ are computed over all reported models. For the NMR ensemble, such pseudo-B-factors can be viewed as experimental values, while the predicted values can be obtained by ENM from the individual or averaged model. Correlations between them can be computed using the same correlation equation as for the X-ray B-factors. Our preliminary results show that the pseudo-B-factors of NMR ensembles can be predicted well by GNM. For a large dataset with thousands of

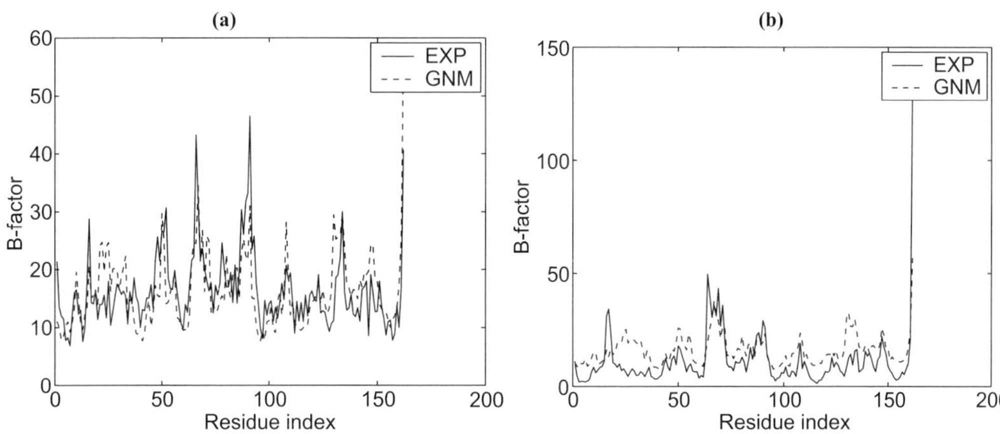

FIGURE 16.1 An example of comparisons between fluctuations represented in X-ray B-factors and NMR ensembles with those predicted by GNM for dihydrofolate reductase. (a) X-ray structure (PDB code: 3dfr). The correlation between experimental B-factors and GNM predicted ones is 0.62. (b) NMR ensemble (PDB code: 1ao8). The correlation between pseudo-B-factors from the NMR ensemble and GNM is 0.73. The values of GNM predicted B-factors and NMR pseudo B-factors have been normalized with the experimental B-factors of the X-ray structure.

NMR structures, the average correlation between pseudo-B-factors and GNM reproduced ones is about 0.70 [Yang, L., Song, G., and Jernigan, R. L., unpublished results], slightly better than correlations observed for the X-ray B-factors. An example is shown in Figure 16.1b. Yang, L.W. et al. [2007] also reported better agreement with NMR than with X-ray data.

16.3.3 Principal Motions (PCs) Matching with Modes

Principal Component Analysis (PCA) is performed on multiple X-ray structures of the same protein, or an NMR ensemble, to capture the principal motions. The input is an n by p coordinate matrix X where n is the number of structures and p is three times the number of residues [Teodoro, Phillips, and Kavraki 2002, 2003]. Each row in X represents the coordinates of the alpha carbons of each structure. For the coordinates X the elements of the covariance matrix C are calculated as

$$c_{ij} = \left\langle \left(x_i - \langle x_i \rangle\right)\left(x_j - \langle x_j \rangle\right) \right\rangle, \tag{16.15}$$

where averages $\langle\rangle$ are computed over all structures. The covariance matrix C can be decomposed as

$$C = P\Delta P^T, \tag{16.16}$$

where the eigenvectors p represent the principal components (PCs) and the eigenvalues are the elements of the diagonal matrix Δ. The eigenvalues are sorted in descending order. Each eigenvalue is directly proportional to the variance it captures through its corresponding PC.

The alignment of a given PC and a given ANM mode is measured by their overlap, which is defined by [Tama and Sanejouand 2001]

$$O_{ij} = \frac{\left|P_i \cdot M_j\right|}{\left\|P_i\right\|\left\|M_j\right\|}, \tag{16.17}$$

where P_i is the ith PC and M_j is the jth mode. A perfect match gives an overlap value of 1. We define the cumulative overlap (CO) between the first k modes and a given PCI as

$$CO(k) = \left(\sum_{j=1}^{k} O_{ij}^2\right)^{1/2}, \tag{16.18}$$

which measures how well the first k modes together can capture the motion of a single PC.

The PC and mode spaces are related by the root mean-square inner product (RMSIP) [Amadei et al. 1999] as

$$\mathrm{RMSIP}(I,J) = \left(\frac{1}{I}\sum_{i=1}^{I}\sum_{j=1}^{J}\left(P_i \cdot M_j\right)^2\right)^{1/2}, \tag{16.19}$$

where P_i is the i th PC and M_j is the j th mode. This RMSIP indicates how well the motion space spanned by the first I PCs can be explained by the first J modes.

Our study shows that the essential protein motions of HIV-1 protease can be identified by PCA from its multiple X-ray or NMR structures and that such key motions are in good agreement with elastic network modes [Yang, L., Song, G., and Jernigan, R. L., 2008]. In summary, we can extract important information about the dynamics by analyzing experimental data. Such data, when combined with the dynamics predicted by ENM, may enable us to develop a deeper comprehension of essential protein motions, and thus may provide valuable insights for future drug design.

Now, we have several significant ENM methods to apply to treat the motions of coarse-grained protein structures, even including the relatively large-scale conformational transitions.

16.4 APPLICATIONS

16.4.1 Protein Transitions, Evidence for Their Controlled Directions

Most protein functions are associated with some conformational changes. A conformational change may, for example, provide the needed extra space to accommodate a ligand; it may inactivate or activate a protein; it may also act as an on/off signal of some biological processes. Therefore, to understand what dictates the direction and magnitude of such conformational changes is clearly important. Although the underlying driving forces for such transitions are external forces such as those arising from new atomic interactions and thermal effects, a simpler and clearer way is needed to explore the intrinsic preferences for transitions that are intrinsic to a structure. Elastic network models do provide some valuable insights. These coarse-grained, structure-based models provide a simple explanation of how the structure of a protein influences its directions of motion and facilitates the realization of its functions. One simple way that this has been demonstrated is by showing that there is usually a significant overlap between the direction of the conformation transition of a protein and its intrinsic low-frequency vibration modes. We have found that the direction of transition required for the realization of a protein's function is usually significantly congruent to its intrinsically favored directions of motions.

16.4.2 Investigation of the Database of Macromolecular Movements

We have applied ENM (as well as domain-ENM) to study the transitions of many of the proteins in Gerstein's Database of Macromolecular Movements [Gerstein and Krebs 1998; Flores et al. 2006] (http://www.molmovdb.org/). There are about 200 pairs of structures in Gerstein's database, classified by their motion scales and types. A few structures were excluded from our dataset since their PDB entries are not specified. The remaining 170 pairs of structures have been used in our analyses [Yang, Song, and Jernigan 2007].

It can be seen from Table 16.1 that ENM is able to produce modes of motion that have significantly large overlaps with the experimentally observed conformational changes. It is noteworthy in Table 16.1 that some classes of motion are more amenable to representation of their conformational transitions by the coarse-grained ENM/s than others, with the most common motion types—hinges and shears—being the types best represented by the domain-ENM.

TABLE 16.1
Analyses of the Conformational Transitions by Motion Types

Motion Type	I. Domain—Fragments	II.A. Domain—Shear	II.B. Domain—Hinge	II. Domain—Other	III. Subunit—Larger Motions
Number of pairs (170 total)	48	27	59	18	18
Concertedness	23.9	37.4	99.7	51.8	46.0
Reduced DOF	81	107	68	79	113
Maximum overlap	0.50	0.58	0.67	0.46	0.50
CSO(20)	0.56	0.70	0.79	0.61	0.60

The numbers shown are the mean values over all the structure pairs for each motion type [Yang et al. 2007].

16.4.3 The Applicability of ENMs for Treating Even Large Conformational Changes

Since the ENM modes reflect the patterns of local fluctuations around an equilibrium state, it is often thought that the applicability of ENMs should be limited only to conformation changes small in magnitude. However, the results in Table 16.1 demonstrate that the applicability of ENMs is not actually so limited by the scale of the conformation changes, or by the size of the proteins. ENMs are able to predict quite accurately how residues may be subdivided into nearly rigid clusters and move collectively, as well as their directions of motions. For many proteins, these "rigid" clusters tend to stay together and move along the initial directions, which explains why the ENMs have been so successful even in explaining large-scale conformation changes. For some other proteins, however, the collective motions may break down quickly after a small conformation change is made. The conformational changes in these cases therefore can hardly be considered to be collective. ENM does not perform so well in such cases.

16.4.4 Finding Transition Pathways Using ENMs

Because ENMs are effective at capturing the collective motions of proteins, it is natural to extend them to find transition pathways. A medium-sized protein may contain some hundreds of residues and huge numbers of degrees of freedoms. The collective motions uncovered by ENMs can greatly reduce the dimensionality of the motion space and make the search for transition pathways between the "open" and "closed" forms of a protein entirely feasible. In the following, we will use diphtheria toxin, a domain-swapping protein, as an example to illustrate how the transition pathways may be found between the monomer form ("closed") and the domain-swapped form ("open").

Domain swapping is a process wherein a monomeric protein can undergo a transition in which two (or more) protein monomers form a dimer (or higher oligomer) by exchanging an identical domain. It was first observed in diphtheria toxin by Bennett, Choe, and Eisenberg (1994). It is one of the mechanisms that proteins use to build oligomeric structures. For domain-swapping proteins, since the domains are rigid and the hinge loops are highly flexible, domain-ENM is an excellent model to study the domain-swapping process [Song and Jernigan 2006; Yang, Song, and Jernigan 2007; Kundu and Jernigan 2004]. A simulation to generate a transition pathway from the monomeric form ("closed") of diphtheria toxin to the domain-swapped form ("open") is as follows [Song and Jernigan 2006]:

1. Begin with the "closed" form and solve for modes 7–12, which represent the important relative rigid body motions of the two individual domains (1–6 are the rigid body motions) of the entire structure.
2. Pick the lowest mode from these that can decrease the number of contacts between the domains. Note each mode has two directions: forward and backward.
3. If such a mode exists, move along that mode in the separation direction for a small step.
4. If no such modes are found, then pick a random mode and move a small step.
5. Use the newly obtained structure as the new starting point; repeat steps 1–5 until the two domains are separated or the iteration reaches its limit.

Note that we limit our search to only six modes (i.e., modes 7–12). We can do so because we know that the domain-swapping process is dominated by the rigid body motions of the two domains, and their relative motions are captured by these six modes. If we did not have this advantage, we would require many more modes, which in turn would make the transition pathways much more difficult to capture.

Figure 16.2 shows the characteristics of one transition pathway from the "closed" monomer to the "open" domain-swapped dimer of diphtheria toxin. In four separate parts we display how several parameters change as the simulation proceeds. These metrics are: (1) the distance between

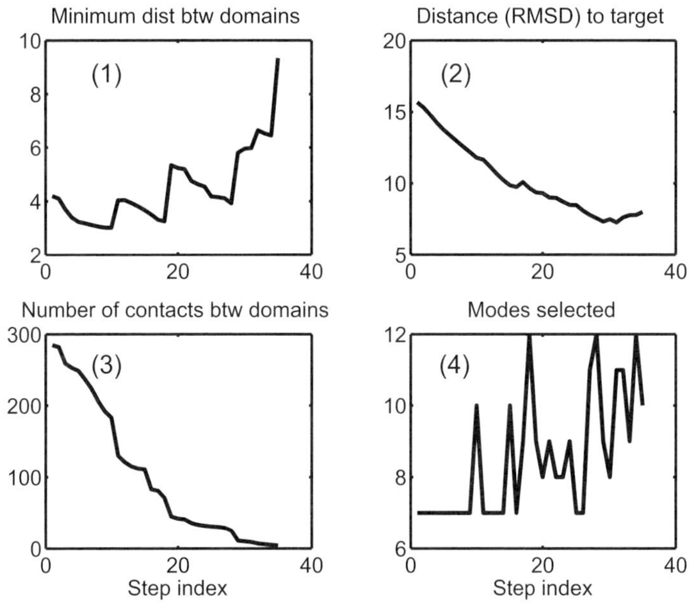

FIGURE 16.2 Statistics of a transition pathway of diphtheria toxin generated by following the algorithm described in the Methods section. The four parts show (1) the closest distance between the two domains, (2) the RMS distance to the final state, (3) the number of interdomain contacts, and (4) the mode selected at each step.

the two domains, (2) the RMS distance to the final state, (3) the number of interdomain contacts, and (4) the mode selected at each step. The distance between the two domains is measured by the distance between the closest pairs of residues, one from each domain. The number of interdomain contacts counts how many pairs of residues between the two domains are closer than 10 Å. From part (3) of Figure 16.2 we see that the number of interdomain contacts indeed decreases over the simulation, as is assumed in our approach. We also require that no interdomain contact distances should be below 3 Å, which acts as a hard-core repulsion. It helps prevent two domains from passing through one another. What is most intriguing about the pathway is that, even though we put in no information about the "open" form (domain-swapped dimer) before the simulation, the structure moves directly towards the "open" form (domain-swapped dimer) as the simulation proceeds: the RMS distance decreases from more than 15 Å to nearly 7 Å. Once it reaches that point, the two domains separate and they may begin to interact with domains of other similar molecules to form a domain-swapped dimer.

Finally, part (4) of Figure 16.2 identifies the mode selected at each step during the simulation. The mode most frequently selected is the 7th mode, the lowest mode (slowest) that corresponds to the separate rigid body motions of the two domains—17 times out of 35 the mode selected is the 7th mode. This agrees with our general expectation that lower index modes are more important.

16.4.5 Enzyme Mechanism for Triosephosphate Isomerase

Triosephosphate isomerase (TIM) is an important enzyme in the glycolytic pathway, catalyzing reversible isomerization of dihydroxyacetone phosphate (DHAP) to D-glyceraldehyde 3-phosphate (GAP). TIM is functional as a homodimer with each monomer comprised of 245 residues and an active site. No cooperativity or allostery between the subunits has been reported. The protonation/deprotonation of the ligand DHAP occurs due to the active site residues on each monomer, namely K13, H95, and E165. Figure 16.3a shows two crystal structures aligned, namely apo TIM (PDB: 8tim) and TIM bound to the inhibitor phosphoglycolohydroxamate (PGH) (PDB: 1tph) [Zhang et al. 1994]. The functional loop 6 (P166-A176) closes over the active site to protect the ligand from solvent exposure during catalysis. However, loop closure is not ligand-gated [Williams and McDermott 1995]; that is, loop closure/opening takes place in apo-TIM as well. When the crystal structures of TIM available in the PDB are aligned, variations mainly in the loop 6 conformation indicative of open/closed or intermediate states are observed. Accordingly, most simulation studies have so far concentrated on the region containing the active site and/or the loop 6 [Joseph, Petsko, and Karplus 1990; Derreumaux and Schlick 1998; Massi, Wang, and Palmer 2006].

For the first time, ANM results for the dimeric TIM reveal large-scale collective deformations of the enzyme that may drive the loop 6 motion [Kurkcuoglu, Jernigan, and Doruker 2005, 2006]. Figure 16.3b,c exhibits alternative conformations in the slowest mode for apo TIM with uniform

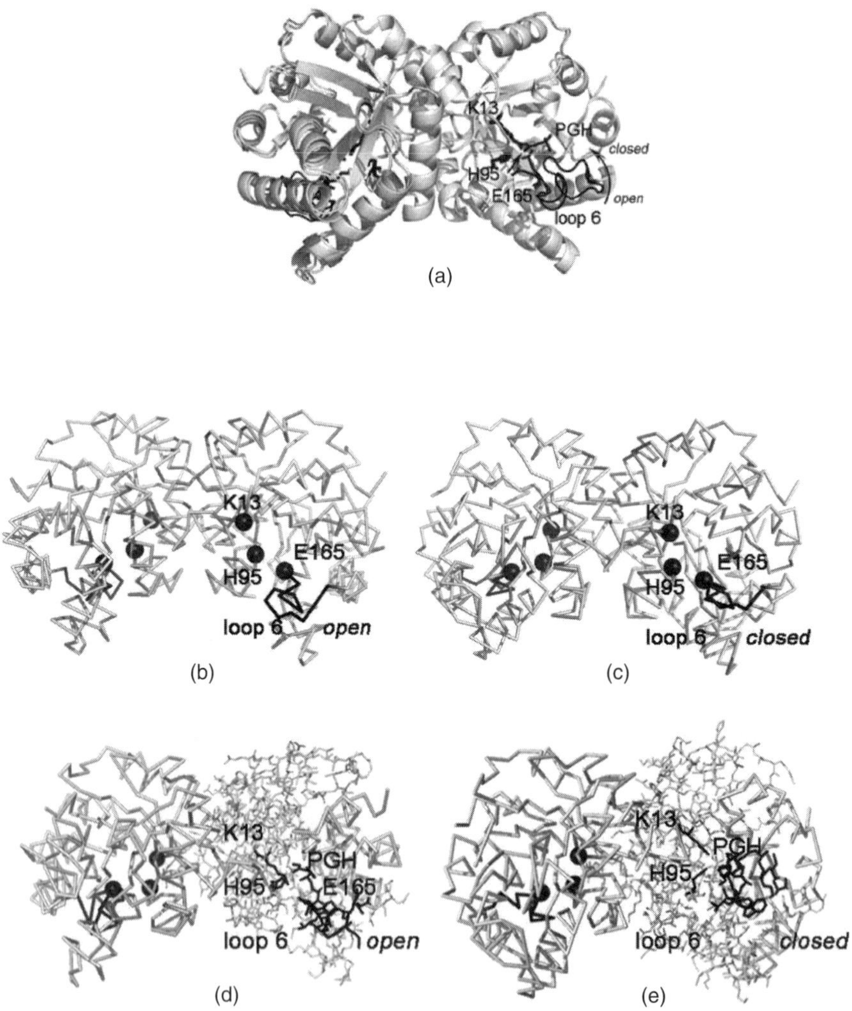

FIGURE 16.3 (a) Aligned crystal structures of apo (light) and PGH-bound (dark) TIM with loop 6, active site residues K13, H95, E165, and inhibitor PGH indicated, (b,c) seventh-mode alternative deformations from ANM using residue-based uniform CG, (d,e) seventh-mode alternative deformations from mixed CG ANM with the high-resolution region including active site residues and loop 6.

CG of the dimer at a single node per residue. Harmonic deformations are exaggerated here for better visualization. Loop 6 closure is clearly observed, which is coupled to the large domain motions. Other collective motions including loop 6 are detected for some other slow modes [not shown here, see Kurkcuoglu, Jernigan, and Doruker 2006].

ANM is applied to the apo- and inhibitor-bound structures of TIM for models with all-atom (all heavy atoms included), residue-based uniform CG, and mixed CG approaches. The *RMSIP*(10,10) values between the all-atom model and the residue-based uniform coarse-graining are quite high, namely 0.87 and 0.97 for atomistic cutoffs of 6 and 9 Å, respectively. Recent MD simulations of 60 ns duration [Cansu, S. and Doruker, P., unpublished results] support the existence of similar global deformations of the dimer structure.

As a test of the new methodology, several mixed CG models (all based on Case 2 type CG) have been applied to TIM; one of them is shown in Figure 16.3d,e. Here the atomistic/high-resolution region encompasses the active site residues and loop 6 of one subunit, whereas the rest of the protein is modeled at low-resolution/one-node-per-residue CG. In general, *RMSIP*(10,10) values between uniform CG and mixed CG models are quite satisfactory, within the range of 0.80–0.89 [Kurkcuoglu, Jernigan, and Doruker 2006]. Thus, NMA on the mixed-resolution network reveals the biologically important low-frequency motions of the active site and its surroundings with atomistic detail. Experimental observations such as the opening/closing of loop 6, displacement of the catalytic base E165 and the 50° rotation of the indole ring W168 upon loop 6's closure [Desamero et al. 2003] are successfully obtained. Figure 16.4a and Figure 16.4b display the seventh and tenth mode shapes of the active site and loop region, respectively, with modes 1–6 being the rigid body translations and rotations. The overlap between the loop closure direction (extracted from the apo and bound X-ray structures) and the seventh/tenth eigenvector directions for loop residues comprising 169–173 is 0.55/0.81. The N and C termini of loop 6 reported as the hinge regions in the opening/closing motion of loop 6 [Joseph, Petsko, and Karplus 1990] are immobile in the tenth mode, whereas the hinge regions are displaced to inner residues in other slow modes including the seventh mode (the slowest and first internal mode).

By looking at changes in the interatomic distances between the active site residues and the ligand DHAP (PDB code: 1ney, jogl, et al., 2003), some key harmonic vibrations in the collective modes can be identified in relation to the catalytic pathway of TIM. Specifically, a proton is transferred from C1 of DHAP to the catalytic base E165 in the first step of TIM catalysis [Guallar et al. 2004]. Figure 16.4c shows the significant changes in interatomic distances in the slowest internal harmonic mode. The alternative distances on the figure are based on an overall RMSD of 1.7 Å between the alternative TIM conformations generated by deformation along the $+/-$ directions of the seventh mode. Here the change in the distance between H1 and OE2 atoms is indicative of the fact that proton transfer from DHAP to E165 is directly enhanced by collective deformations in the direction required for the reaction. Another three distance pairs on the figure relate to hydrogen bonding [Uyar, A., Kurkcuoglu, O., Jernigan, R. L., and Doruker P., unpublished results], indicating other changes that could be monitored experimentally.

In summary, the collective motions extracted for TIM are quite similar at all levels of detail provided that the dimer structure of TIM is included in the calculations. (Our previous monomeric calculations were not satisfactory.) In contrast, when ANM is applied to only the high resolution region of the mixed CG model (Figure 16.3d,e), the resulting slow modes are not satisfactory [Kurkcuoglu, Jernigan, and Doruker 2006]. Thus keeping the whole protein structure in the computational model, regardless of the level or CG, is crucial for obtaining the correct global dynamics. In this respect, the elastic network model provides strong evidence for why TIM functions as a dimer rather than in monomer form. Even though TIM is a relatively small enzyme that could be studied with an all-atom representation of the whole structure, having validated the mixed CG methodology for TIM opens the way for its application to even supramolecular assemblages such as the ribosome.

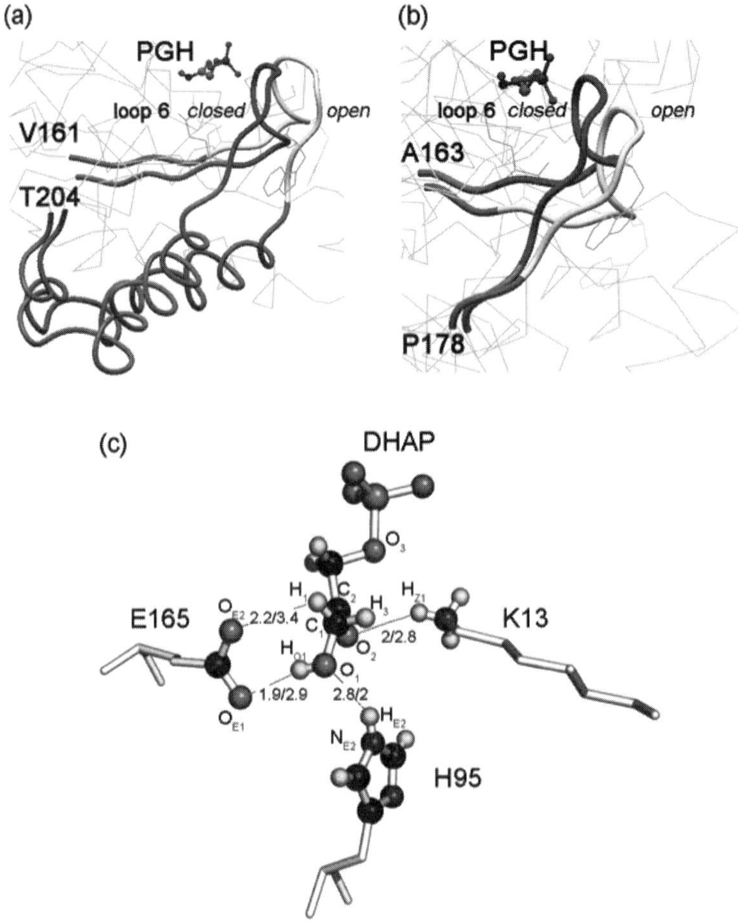

FIGURE 16.4 (a) Deformations of loop 6 from (a) mode 7 and (b) mode 10 for PGH-bound TIM using mixed CG. (c) Changes in interatomic distances at the catalytic site in the first mode.

16.4.6 Processing Mechanisms

16.4.6.1 Reverse Transcriptase

One of the first proteins that we studied [Bahar et al. 1999] with the ENM gave evidence of motions that relate to its processing motions. The slowest internal mode of motion was a hinge at a site between the two enzyme sites, and because it can push the two enzyme sites further apart, it can be utilized to pull the nucleic acid strand ahead by one position. Another slow mode corresponds to a hinge that opens/closes the polymerase site. A coordinated motion alternating between these two forms open/pulling and closed/pushing could account directly for the enzyme's processive motions where one base is copied at the polymerase site and one base is cut off at the RnaseH site.

16.4.6.2 Ribosome

The ribosome is presently the largest known molecular structures and engages in the critical protein synthesis function, carefully copying the information from the messenger RNA through the intermediation of the tRNAs. In our studies of the ribosome with the ENMs a number of interesting aspects of the processing have emerged [Jernigan and Kloczkowski 2007; Wang and Jernigan 2005; Wang et al. 2004]. Three particularly noteworthy features have been observed:

TABLE 16.2
Rigid Body Fraction of Total Motion for All Components of the Ribosome

Ribosome Parts	Rigid Body Fraction of Total Motion
30S	0.40
50S	0.13
A-tRNA	0.73
P-tRNA	0.75
E-tRNA	0.73
mRNA	0.96

1. The intrinsic dominant motions of the ribosome are the ratchet-like rotational motion between the 30S and 50S subunits, which are known from Joachim Frank's EM pictures [Gao et al. 2003; Valle et al. 2003], and these functionally control the translational motions of the mRNA, and tRNA in the interior.
2. The intrinsic computed correlations of the motions of the components are appropriate for the functional motions—the 30S and 50S subunits are completely anticorrelated (rotating in opposite directions); whereas the mRNA and three tRNAs are all positively correlated with one another in their motions since they must move ahead, nearly lock-step in the same direction.
3. Decomposition of the motions of each component in the ribosome structure into its own internal motions and rigid body motions (within the context of the motions of the entire ribosome) is particularly interesting (Tablc 16.2). The mRNA, which is intrinsically the most flexible of all the components of the ribosome, is held nearly fully rigid and moves translationally as a rigid body. This is essential to prevent errors in reading the genetic code from the mRNA script.

16.5 CONCLUSIONS AND FUTURE WORK

There are several different versions of the elastic network models that have been developed—GNM, ANM, vGNM, and domain-ENM. There has been significant experimental confirmation of the validity of these approaches using these various coarse-grained models. ANM performed at mixed levels of CG is a computationally efficient tool useful for obtaining the biologically important collective motions of proteins and their complexes. It can even be used to successfully explore the dynamics in enzyme catalysis whereas a full-atom empirical potential approach with structural constraints might fail, if the motion control is exerted through the domain motions.

We are currently developing a revised version of mixed CG where a single cutoff value for all atomic interactions is employed and the force constants between any pair of nodes (high–high, low–low, and high–low resolution pairs) are determined according to the total number of interatomic interactions surviving between the node pairs. Thus it will not require the prior adjustment of cutoffs and force constants, which may depend on the system size as described above for our previous methodology. Application of the mixed CG technique to supramolecular assemblages, specifically to the ribosome, is ongoing. Preliminary results on the ribosome modeled together with its A-, P-, E-tRNA, and mRNA components reveal the experimentally observed ratchet-like rotation of ribosome's subunits, and strongly attest to the reliability of the model. The mixed CG method may also prove to be an efficient tool for flexible protein–ligand docking if realistic conformations can be generated along the slow modes.

There are of course some uncertainties about the proper tuning of the parameters to reflect the extent of rigidity and cooperativity of elements of structure, which must rely on experimental data, but the elastic network models are remarkably robust and insensitive to their parameter values. The viewpoint taken here is that the functional atoms and loops move almost as decorations, except that, as we have seen here for TIM, and also previously for tubulin [Keskin et al. 2002], the loops can be controlled to move in different directions than the domains, but nonetheless under the control of the domain motions. The elastic network models have proven themselves successful in representing cooperativity as it manifests itself in the protein domain motions. Here we have shown applications ranging from enzyme reactions to the processing motions of reverse transcriptase and the ribosome, and how the normal mode motions of the coarse-grained elastic network models can have direct mechanistic interpretations. These results provide frameworks upon which detailed molecular mechanisms can be constructed. Additional aspects of the ribosome processing steps that are being investigated include the progression of the mRNA and the growing peptide chains through their respective channels in the ribosome interior; likewise we plan to investigate the enzymatic reaction for amino acid addition, and how the atomic motions in this active site are controlled by the domain motions of the ribosome.

Generally the ENMs for coarse-grained models, hand in hand with their normal modes, provide a particularly useful way to sample protein conformations. This approach is also being used for guided MD, which could overcome its intrinsic limitation of not properly representing the high-frequency local excursions along a trajectory.

The future of biology will require simulations on much larger structures—the CG models appear to be the most appropriate way to satisfy these requirements. All of the results suggest that a high-resolution model may not be required to compute motions from a structure, but, of course, how well this can be done depends on the application—enzyme investigations would require atomic knowledge in the vicinity of the active site. Other applications could readily include representation of disordered regions of structure to investigate their influence on the overall motions of a structure. Also the investigation of the influence of specific type of microscopic forces will be important in order to establish biological mechanisms, which usually occur in response to external forces.

ACKNOWLEDGMENTS

We are grateful for support for these projects from NIH grants R01 GM072014 and R01 GM073095. PD and OK acknowledge partial support from TUBITAK Project 104M247 and EU-FP6-ACC-2004-SSA-2 Contract No. 517991.

REFERENCES

Amadei, A., Ceruso, M. A., and Di Nola A. 1999. On the convergence of the conformational basis set obtained by the essential dynamics analysis of proteins' molecular dynamics simulations. *Proteins* 36:419–24.

Atilgan, A. R., Durell, S. R., Jernigan, R. L., Demirel, M. C., Keskin, O., and Bahar, I. 2001. Anisotropy of fluctuation dynamics of proteins with an elastic network model. *Biophys. J.* 80:505–15.

Bahar, I., Atilgan, A. R., and Erman, B. 1997. Direct evaluation of thermal fluctuations in proteins using a single-parameter harmonic potential. *Fold Des.* 2:173–81.

Bahar, I., Erman, B., Jernigan, R. L., Atilgan, A. R., and Covell, D. G. 1999. Collective motions in HIV-1 reverse transcriptase: Examination of flexibility and enzyme function. *J. Mol. Biol.* 285:1023–37.

Bennett, M. J., Choe, S., and Eisenberg, D. 1994. Refined structure of dimeric diphtheria toxin at 2.0 Å resolution. *Proteins* 3:1444–63.

Berman, H. M., Westbrook, J., Feng, Z., Gilliland, G., Bhatand, T. N., Weissig, H., Shindyalov, I. N., and Bourne, P. E. 2000. The protein data bank. *Nucleic Acids Res.* 28:235–42.

Derreumaux, P., and Schlick, T. 1998. The loop opening/closing motion of the enzyme triosephosphate isomerase. *Biophys. J.* 74:72–81.

Desamero, R., Rozovsky, S., Zhadin, N., McDermott, A., and Callender, R. 2003. Active site loop motion in triosephosphate isomerase: T-jump relaxation spectroscopy of thermal activation. *Biochemistry* 42:2941–51.

Doruker, P., Atilgan, A. R., and Bahar, I. 2000. Dynamics of proteins predicted by molecular dynamics simulations and analytical approaches: Application to α-amylase inhibitor. *Proteins* 40:512–24.

Doruker, P., Jernigan, R. L., and Bahar, I. 2002. Dynamics of large protein systems through hierarchical levels of coarse-grained structures. *J. Comp. Chem.* 23:119–27.

Doruker, P., Jernigan, R. L., Navizet, I., and Hernandez, R. 2002. Important fluctuation dynamics of large protein structures are preserved upon coarse-grained renormalization. *Int. J. Quantum Chem.* 90:822–37.

Flores, S., Echols, N., Milburn, D., Hespenheide, B., Keating, K., Lu, J., Wells, S., Yu, E. Z., Thorpe, M., and Gerstein, M. 2006. The database of macromolecular motions: New features added at the decade mark. *Nucleic Acids Res.* 34:D296–301.

Flory, P. J. 1977. Theory of elasticity of polymer networks: The effects of local constraints on junctions. *J. Chem. Phys.* 66:5720.

Gao, H., Sengupta, J., Valle, M., Korostelev, A., Eswar, N., Stagg, S. M., Van Roey, P., Agrawal, R. K., Harvey, S. C., Sali, A., Chapman, M. S., and Frank, J. 2003. Study of the structural dynamics of the *E. coli* 70S ribosome using real-space refinement. *Cell* 113:789–801.

Gerstein, M., and Krebs, W. 1998. A database of macromolecular motions. *Nucleic Acids Res.* 26:4280–90.

Guallar, V., Jacobson, M., McDermott, A., and Friesner, R. A. 2004. Computational modeling of the catalytic reaction in triosephosphate isomerase. *J. Mol. Biol.* 337:227–39.

Jernigan, R. L., and Kloczkowski, A. 2007. Packing regularities in biological structures relate to their dynamics. *Methods Mol. Biol.* 350:251–76.

Joseph, D., Petsko, G. A., and Karplus, M. 1990. Anatomy of a conformational change: Hinged lid motion of the triosephosphate isomerase loop. *Science* 249:1425–28.

Keskin, O., Durell, S. R., Bahar, I., Jernigan, R. L., and Covell, D. G. 2002. Relating molecular flexibility to function: A case study of tubulin. *Biophys. J.* 83:663–80.

Kim, M. K., Li, W., Shapiro, B. A., and Chirikjian, G. S. 2003. A comparison between elastic network interpolation and MD simulation of 16S ribosomal RNA. *J. Biomol. Struct. Dyn.* 21:395–405.

Kolinski, A. 2004. Protein modeling and structure prediction with a reduced representation. *Acta Biochim. Pol.* 51:349–71.

Kundu, S., and Jernigan, R. L. 2004. Molecular mechanism of domain swapping in proteins: An analysis of slower motions. *Biophys. J.* 86:3846–54.

Kundu, S., Melton, J. S., Sorensen, D. C., and Phillips, Jr., G. N. 2002. Dynamics of proteins in crystals: Comparison of experiment with simple models. *Biophys. J.* 83:723–32.

Kurkcuoglu, O., Jernigan, R. L., and Doruker, P. 2004. Mixed levels of coarse-graining of large proteins using elastic network model succeeds in extracting the slowest motions. *Polymer* 45:649–57.

———. 2005. Collective dynamics of large proteins from mixed coarse-grained elastic network model. *QSAR Comb. Sci.* 24:443–48.

———. 2006. Loop motions of triosephosphate isomerase observed with elastic networks. *Biochemistry* 45:1173–82.

Larson, M., Snow, C., and Pande, V. S. 2003. Folding@Home and Genome@Home: Using distributed computing to tackle previously intractable problems in computational biology. In *Modern Methods in Computational Biology*, ed. R. Grant. Hethersett, Norwich: Horizon Press.

Lee, B. M., Xu, J., Clarkson, B. K., Martinez-Yamout, M. A., Dyson, H. J., Case, D. A., Gottesfeld, J. M., and Wright, P. E. 2006. Induced fit and "lock and key" recognition of 5S RNA by zinc fingers of transcription factor IIIA. *J. Mol. Biol.* 357:275–91.

Liwo, A., Oldziej, S., Pincus, M. R., Wawak, R. J., Rackovsky, S., and Scheraga, H. A. 1997. A united-residue force field for off-lattice protein-structure simulations. I. Functional forms and parameters of long-range side-chain interaction potentials from protein crystal data. *J. Comput. Chem.* 18:849–73.

Massi, F., Wang, C., and Palmer, III, A. G. 2006. Solution NMR and computer simulation studies of active site loop motion in triosephosphate isomerase. *Biochemistry* 45:10787–94.

McCammon, J. A., Gelin, B. R., and Karplus, M. 1977. Dynamics of folded proteins. *Nature* 267:585–90.

Ming, D., Kong, Y., Lambert, M. A., Huang, Z., and Ma, J. 2002. How to describe protein motion without amino-acid sequence and atomic coordinates. *Proc. Natl. Acad. Sci. U.S.A.* 99:8620–25.

Rahman, A. 1964. Correlations in the motion of atoms in liquid argon. *Phys. Rev. A* 136:405–11.

Shirts, M. R., and Pande, V. S. 2000. Screen savers of the world, unite! *Science* 290:1903–1904.

Song, G., and Jernigan, R. L. 2006. An enhanced elastic network model to represent the motions of domain swapped proteins. *Proteins* 63:197–209.

———. 2007. vGNM: A better model for understanding the dynamics of proteins in crystals. *J. Mol. Biol.* 369:880–93.

Stillinger, F. H., and Rahman, A. 1974. Improved simulation of liquid water by molecular dynamics. *J. Chem. Phys.* 60:1545–57.

Tama, F., and Sanejouand, Y. H. 2001. Conformational change of proteins arising from normal mode calculations. *Protein Eng.* 14:1–6.

Teodoro, M. L., Phillips, Jr., G. N., and Kavraki, L. E. 2002. A dimensionality reduction approach to modeling protein flexibility. In *Proceedings of the Sixth Annual International Conference on Computational Biology*, 299–308.

———. 2003. Understanding protein flexibility through dimensionality reduction. *J. Comput. Biol.* 10:617–34.

Tirion, M. M. 1996. Large amplitude elastic motions in proteins from a single-parameter, atomic analysis. *Phys. Rev. Lett.* 77:1905–8.

Valle, M., Zavialov, A., Sengupta, J., Rawat, U., Ehrenberg, M., and Frank, J. 2003. Locking and unlocking of ribosomal motions. *Cell* 114:123–134.

Wang, Y., and Jernigan, R. L. 2005. Comparison of tRNA motions in the free and ribosomal bound structures. *Biophys. J.* 89:3399–3409.

Wang, Y., Rader, A. J., Bahar, I., and Jernigan, R. L. 2004. Global ribosome motions revealed with elastic network model. *J. Struct. Biol.* 147:302–14.

Williams, J. C., and McDermott, A. E. 1995. Dynamics of the flexible loop of triosephosphate isomerase: The loop motion is not ligand gated. *Biochemistry* 34:8309–19.

Wilmanns, M., and Nilges, M. 1996. Molecular replacement with NMR models using distance-derived pseudo B factors. *Acta Crystallogr. D* 52:973–82.

Yang, L., Song, G., and Jernigan, R. L. 2007. How well can we understand large-scale protein motions using normal modes of elastic network models? *Biophys. J.* 93:1–10.

Yang. L., Song, G., Carriquiry, A., and Jernigan, R. L. 2008. Close correspondence between the motions from principal component multiple HIV-1 protease structures and elastic network modes. *Structure* 16:321–30.

Yang, L. W., Eyal, E., Chennubhotia, C., Jee, J., Gronenborn, A. M., and Bahar I. 2007. Insight into equlibrium dynamics of proteins from comparison of NMR and X-ray data with computational predictions. *Structure* 15:741–9.

Zhang, Z., Sugio, S., Komives, E. A., Liu, K. D., Knowles, J. R., Petsko, G. A., and Ringe, D. 1994. Crystal structure of recombinant chicken triosephosphate isomerase phosphoglycolohydroxamate complex at 1.8 Å resolution. *Biochemistry* 33:2830–37.

17 Coarse-Grained Elastic Normal Mode Analysis and Its Applications in X-Ray Crystallographic Refinement at Moderate Resolutions

Jianpeng Ma
Baylor College of Medicine

CONTENTS

17.1 INTRODUCTION

Normal mode analvsis (NMA) is a powerful method that has been used to model the harmonic vibrations of protein structures [1–4], especially large-scale deformational motions of supramolecular complexes [5,6]. However, due to the construction and diagonalization of the Hessian matrix, the normal mode calculation can become computationally intractable as the size of the structure increases. One effective way to address this problem is with the recent development of elastic normal mode analysis (eNMA) [7,8], which makes a coarse-grained approximation of the complex atomic structure with a much simpler elastic network and reduces the complex all-atom potential to a simpler Hookean potential. The coarse-grained approximation can be based on either the C_α atoms [8], a subset of C_α atoms [9], a lattice representation of the protein [10], or even nodes placed within low-resolution electron cryomicroscopy maps (cryo-EM) [11,12] that have no correlation to the positions of any real atoms. Even with such a simplification of the structure and potential, eNMA preserves the low-frequency eigenvectors predicted by more accurate molecular mechanics force fields and is as effective as and sometimes more powerful than the standard NMA procedure [5]. An additional benefit to these approximations is that the initial structure does not need to be minimized according to the potential function before the normal modes are calculated.

Regarding the validity of eNMA with drastically simplified potential function, the low-frequency deformational modes are usually assumed to be insensitive to local structural connectivity and more dependent on the overall shape [5]. This was confirmed quantitatively in a paper [13] where we randomized the elements of the Hessian matrix and compared the resulting eigenvectors with those calculated from eNMA and from a molecular mechanics potential. If we kept the information concerning the shape of the molecule by randomizing the values of just the nonzero elements in the Hessian matrix, we were able to retain the subspace spanned by the low-frequency modes. However, if we altered the connectivity information in the Hessian matrix by changing the positions of the zero and nonzero elements, the resulting low-frequency normal modes changed substantially.

While expanding the usefulness of NMA to large systems, the coarse-graining procedure in eNMA [8] introduces an artifact to the eigenvectors known as the "tip effect." Structural components that project from the surface of the protein, like an isolated surface loop, will have unrealistic motions due to the "tip effect." Since these tip regions are less densely packed, there may be an imbalance in the elastic forces among neighboring harmonic oscillators. The result is large displacements in the tip regions with respect to the rest of the protein. Compounding the problem is that because the normal modes are normalized, the motion for the rest of the system will be suppressed. Currently, there is no systematic method to predict which modes will have the "tip effects." In many cases, even the lowest frequency modes may be tainted. If a continuous set of eigenvectors is needed, such as for refinement in X-ray crystallography, the "tip effect" can become a significant problem.

Given that the complex all-atom potentials do not exhibit the "tip effect" and a Hessian matrix with randomized nonzero elements reproduces the subspace spanned by the low-frequency eigenvectors, there must be a potential function between the two extremes that can reproduce the low frequency subspace, not suffer from the "tip effect," and not require an initial minimization. Recently, we designed a potential function for that purpose [14]. To avoid the "tip effect," we changed the eNMA procedure to work in internal coordinates (IC) [15]. By doing so, we prevent the C_α–C_α pseudobond from stretching, which constrains the possible motions for the tip regions.

One important application of NMA is in X-ray crystallographic refinement because large supramolecular complexes often have highly flexible domains that yield poor crystals that do not diffract well. Having fewer data points in the diffraction pattern and a large number of parameters due to the size of the system—each atom requires three coordinate parameters and one thermal parameter for isotropic refinement, or six thermal parameters for anisotropic refinement—the possibility for overfitting is very big. During the early 1990s, methods using low-frequency normal modes in X-ray refinement were proposed [16–21]. Furthermore, normal modes have been applied to other areas of X-ray crystallography, such as molecular replacement [22,23], positional refinement [24], and protein dynamics [25–27].

Even though normal-mode-based refinement in X-ray crystallography seems advantageous in theory, the methods have not gained much traction in practice. The main reason is that traditional NMA is based on an all-atom molecular mechanics force field. As a result, an initial energy minimization must be performed, which shifts the atomic coordinates, especially the residues on the protein surface [3,4], away from the positions determined from the diffraction data. The conventional eNMA does not require initial minimization, but it introduces the "tip effect." The issues with initial minimization and tip effect were only resolved recently [14]. Another reason is that normal-mode refinement is most effective for structures that are intrinsically flexible and exhibit large domain motions. Until recently, such large proteins have been rare. Lastly, the resolution of the diffraction data is important in determining how effective the use of normal modes is to refinement. We believe that normal-mode-based refinement will be most applicable at moderate resolutions (3–4 Å).

This chapter is written primarily based on three seminal papers [13,14,28].

17.2 METHODS

17.2.1 Elastic Normal Mode Analysis with No Tip Effect

First, we describe the procedure for a single chain before generalizing for multiple chains. The method is implemented in internal coordinate (IC) space (see Figure 17.1). For a system with N C_α atoms ($i=1, 2,\ldots, N$), the degrees of freedom in IC are $N-2$ bond angles $\{\theta_i\}$ and $N-3$ pseudodihedral angles $\{\varphi_i\}$, where the angles are indexed starting from the second C_α atom and the dihedrals are indexed starting from the third C_α atom. The conversion from IC to Cartesian coordinates (CC) is given by,

$$
\begin{aligned}
&\mathbf{r}_1=(0,0,0)^T;\mathbf{e}_2=(1,0,0)^T;\mathbf{e}_3=(\cos\theta_2,\sin\theta_2,0)^T;\\
&\mathbf{e}_{i+1}=\cos\theta_i\mathbf{e}_i+\sin\theta_i\cos\varphi_i\frac{(\mathbf{e}_{i-1}\times\mathbf{e}_i)\times\mathbf{e}_i}{\left\|(\mathbf{e}_{i-1}\times\mathbf{e}_i)\times\mathbf{e}_i\right\|}+\sin\theta_i\sin\varphi_i\frac{\mathbf{e}_{i-1}\times\mathbf{e}_i}{\left\|\mathbf{e}_{i-1}\times\mathbf{e}_i\right\|};\\
&\mathbf{r}_{i+1}=\mathbf{r}_i+l_{i+1}\mathbf{e}_{i+1}
\end{aligned}
\tag{17.1}
$$

where l_i is the ith C_α–C_α pseudobond length and $\mathbf{e}_i$ is the directional vector for that bond. The first C_α atom, $\mathbf{r}_1$, is at the origin, the first bond, $\mathbf{e}_2$, is along the x axis, and the first bond angle θ_2 is in the X-Y plane. Conversely, the conversion from CC to IC is given by,

$$
\begin{aligned}
&l_i=\left\|\mathbf{r}_i-\mathbf{r}_{i-1}\right\|;\mathbf{e}_i=\frac{\mathbf{r}_i-\mathbf{r}_{i-1}}{l_i};\theta_i=\cos^{-1}(\mathbf{e}_i\cdot\mathbf{e}_{i+1})\\
&\varphi_i=\begin{cases}\cos^{-1}\left(\left(\dfrac{\mathbf{e}_{i-1}\times\mathbf{e}_i}{\left\|\mathbf{e}_{i-1}\times\mathbf{e}_i\right\|}\right)\cdot\left(\dfrac{\mathbf{e}_i\times\mathbf{e}_{i+1}}{\left\|\mathbf{e}_i\times\mathbf{e}_{i+1}\right\|}\right)\right), & \text{if } \mathbf{e}_{i+1}\cdot(\mathbf{e}_{i-1}\times\mathbf{e}_i)\geq 0\\[2ex]
-\cos^{-1}\left(\left(\dfrac{\mathbf{e}_{i-1}\times\mathbf{e}_i}{\left\|\mathbf{e}_{i-1}\times\mathbf{e}_i\right\|}\right)\cdot\left(\dfrac{\mathbf{e}_i\times\mathbf{e}_{i+1}}{\left\|\mathbf{e}_i\times\mathbf{e}_{i+1}\right\|}\right)\right), & \text{if } \mathbf{e}_{i+1}\cdot(\mathbf{e}_{i-1}\times\mathbf{e}_i)<0\end{cases}
\end{aligned}
\tag{17.2}
$$

The bond angles, υ_i, range from $[0,\pi]$ where $i=2, 3,\ldots,N-1$, and the dihedrals, φ_i, range from $[-\pi,\pi]$ where $\theta_i=3, 4,\ldots,N-1$. To simplify the notation, we combine the bond angles and dihedrals into a single set, $\phi_\alpha=\{\theta_2, \varphi_3, \theta_3, \varphi_4, \theta_4,\ldots, \varphi_{N-1}, \theta_{N-1}\}$, where $\alpha=1, 2,\ldots, 2N-5$ and the relationships between i and α are

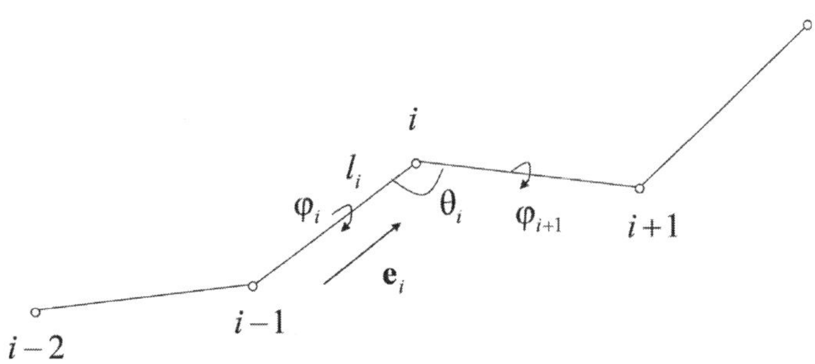

FIGURE 17.1 Schematic illustration of the internal coordinate system used in the new eNMA. (This figure is adopted from Figure 1 on page 465 in Lu, M., Poon, B., and Ma, J. *J. Chem. Theor. Comp.*, 2, 464, 2006.)

$$i = [\alpha/2] + 2 = \kappa(\alpha)$$
$$\alpha^{\varphi}(i) = 2i - 4 \quad (i \geq 3) \tag{17.3}$$
$$\alpha^{\theta}(i) = 2i - 3 \quad (i \geq 2)$$

Using this IC notation, we define a new potential function that is similar to that from eNMA, but with an extra term. This potential is

$$V = \frac{\gamma}{2}\sum_i \sum_j \sigma_{ij}\left(\left|\mathbf{r}_{ij}\right| - \left|\mathbf{r}_{ij}^0\right|\right)^2 + \frac{\omega}{2}\sum_\alpha \left(\phi_\alpha - \phi_\alpha^0\right)^2$$
$$\sigma_{ij} = \begin{cases} 1 & \left|\mathbf{r}_{ij}^0\right| \leq r_c \\ 0 & \left|\mathbf{r}_{ij}^0\right| > r_c \end{cases}; \quad \omega = \lambda \min\left(H_{\alpha\alpha}^0\right) \tag{17.4}$$

where γ is a spring constant normally set to 1.0, $H_{\alpha\alpha}^0$ represents the diagonal elements of the Hessian matrix from the standard eNMA potential in IC, and λ is a scaling factor that applies to only the diagonal term. The scaling factor stiffens the more flexible regions, which removes the "tip effect." Depending on the system, this scaling factor can range from 3.0 to 30.0, where larger values cause the system to be more stiff. Lastly, similar to eNMA, this potential is at its minimum at the current structure so no initial minimization is required. With this new potential, we can calculate the Hessian matrix by

$$H_{\alpha\beta} = \frac{\partial^2 V}{\partial\phi_\alpha \partial\phi_\beta} = \sum_i \sum_j \frac{\partial \mathbf{r}_i}{\partial\phi_\alpha} \cdot \mathbf{h}^{ij} \cdot \frac{\partial \mathbf{r}_j}{\partial\phi_\beta}$$
$$\frac{\partial \mathbf{r}_i}{\partial\phi_\alpha} = \frac{\partial \mathbf{r}_i}{\partial\theta_{\kappa(\alpha)}} = \begin{cases} \left(\mathbf{e}_{\kappa(\alpha)} \times \mathbf{e}_{\kappa(\alpha+1)}\right) \times \mathbf{r}_{\alpha i}, & i > \kappa(\alpha) \\ 0, & i \leq \kappa(\alpha) \end{cases} \quad \phi_\alpha \in \theta, \tag{17.5}$$
$$\frac{\partial \mathbf{r}_i}{\partial\phi_\alpha} = \frac{\partial \mathbf{r}_i}{\partial\varphi_{\kappa(\alpha)}} = \begin{cases} \mathbf{e}_{\kappa(\alpha)} \times \mathbf{r}_{\alpha i}, & i > \kappa(\alpha) \\ 0, & i \leq \kappa(\alpha) \end{cases} \quad \phi_\alpha \in \varphi,$$

where $\mathbf{r}_{\alpha\iota} = \mathbf{r}_\iota - \mathbf{r}_{\kappa(\alpha)}$, and $\mathbf{h}^{ij}$ is the submatrix for the pair i, j in the Hessian matrix in CC. Once the generalized eigenvalue problem is solved for the Hessian matrix, we can convert the eigenvectors in IC to orthonormal vectors in CC by

$$\Delta\mathbf{r}_i^{(k)} = \sum_\alpha \frac{\partial \mathbf{r}_i}{\partial\phi_\alpha} \Delta\phi_\alpha^{(k)},$$
$$\sum_i \Delta\mathbf{r}_i^{(k)} \cdot \Delta\mathbf{r}_i^{(k')} = \delta_{k,k'}, \tag{17.6}$$

where $\Delta\mathbf{r}_i^{(k)}$ is the eigenvector components of the kth mode for the ith C_α atom in CC, $\Delta\phi_\alpha^{(k)}$ is the eigenvector components of the kth mode in IC, the summation over all α, and δ is the Dirac delta function. Finally, to measure the "tip effect," we define a quantitative localization factor, T,

$$T = \sum_i \left(\frac{\left\| \Delta\mathbf{r}_{i+1} - \Delta\mathbf{r}_i \right\|}{\left\| \mathbf{r}_{i+1} - \mathbf{r}_i \right\|} \right)^3, \tag{17.7}$$

where the larger the T, the more prominent the "tip effect."

To generalize this method to proteins with multiple chains, we create a virtual bond connecting the last C_α atom of the preceding chain to the first C_α atom of the following chain. This introduces six more degrees of freedom for each additional chain. Five of these degrees of freedom are internal and the sixth one is the virtual bond length, l, which is the only bond length that is flexible. We redefine ϕ_α to contain these new degrees of freedom as $\phi_\alpha = \left\{\theta_2, \varphi_3, \theta_3, \varphi_4, \theta_4, \ldots, \varphi_{N_1}, \theta_{N_1}, l_{N_1+1}, \varphi_{N_1+1}, \theta_{N_1+1}, \varphi_{N_1+2}, \ldots\right\}$ where N_1 is the number of C_α atoms in the first chain. For the Hessian matrix construction, the virtual bond is handled by

$$\frac{\partial \mathbf{r}_i}{\partial l_{\kappa(\alpha)}} = \mathbf{e}_{\kappa(\alpha)}, \quad i > \kappa(\alpha). \tag{17.8}$$

Additionally, the index order has to be changed accordingly to account for the extra degrees of freedom.

17.2.2 X-Ray Crystallographic Refinement of Anisotropic Thermal Parameters Using Normal Modes

In X-ray crystallography, the diffraction pattern of a structure can be calculated by

$$F_{\mathrm{cal}}(\mathbf{q}) = \sum_j f_j(\mathbf{q}) \exp\left(i\mathbf{q}^{\mathrm{T}} \langle \mathbf{r}_j \rangle\right) \exp\left(-\frac{1}{2}\left\langle \left(\mathbf{q}^{\mathrm{T}} \Delta\mathbf{r}_j\right)^2 \right\rangle\right),$$
$$\mathbf{q} = 2\pi\Theta^{\mathrm{T}}\mathbf{h}, \tag{17.9}$$

where $F_{\mathrm{cal}}(\mathbf{q})$ is the calculated structure factor, $\Theta = (\mathbf{a}^*, \mathbf{b}^*, \mathbf{c}^*)^{\mathrm{T}}$ is a 3×3 matrix that converts CC into fractional coordinates with $\mathbf{a}^*$, $\mathbf{b}^*$, and $\mathbf{c}^*$ being the reciprocal unit cell vectors of the crystal, $\mathbf{h}$ is the Miller index for a lattice point in reciprocal space, f_j is the scattering factor for atom j, and $\mathbf{r}_j$ is the position for atom j. The second exponential is referred to as the Debye–Waller factor, $D(\mathbf{q})$, and represents the thermal fluctuations in the position of the atom. This term can be rewritten and simplified as

$$D(\mathbf{q}) = \exp\left(-\frac{1}{2}\left\langle \left(\mathbf{q}^{\mathrm{T}} \Delta\mathbf{r}_j\right)^2 \right\rangle\right) = \exp\left(-\frac{1}{2}\mathbf{q}^{\mathrm{T}}\mathbf{U}_j\mathbf{q}\right), \tag{17.10}$$

where $\mathbf{U}_j$, the temperature factor, is a 3×3 symmetric matrix representing the mean square displacements for atom j. For full anisotropic refinement, the six independent parameters of $\mathbf{U}_j$ are the thermal parameters. This matrix is positive-definite and can be visualized as an ellipsoid in real space. In the isotropic limit, $\mathbf{U}_j$ is a diagonal matrix where the three diagonal terms are identical, which reduces the number of thermal parameters to one. This special case for $\mathbf{U}_j$ can be visualized as a sphere.

Since a set of normal modes is an equivalent basis set for the system, we can write the displacement of atom j from its equilibrium position as a function of M modes by

$$\Delta \mathbf{r}_j = \mathbf{E}_j \sigma, \tag{17.11}$$

where $\mathbf{E}_j$ is a $3 \times M$ matrix containing the components of the eigenvectors for atom j, and σ is a vector containing the weights that define the contributions of each eigenvector. Combining Equation 17.10 and Equation 17.11, we can express the Debye–Waller factor as a function of normal modes by

$$D(\mathbf{q}) = \exp\left(-\frac{1}{2}\mathbf{q}^{\mathrm{T}}\mathbf{U}_{nm}\mathbf{q}\right) = \exp\left(-\frac{1}{2}\mathbf{q}^{\mathrm{T}}\mathbf{E}_j\langle\sigma\sigma^{\mathrm{T}}\rangle\mathbf{E}_j^{\mathrm{T}}\mathbf{q}\right),$$
$$\Pi \equiv \langle\sigma\sigma^{\mathrm{T}}\rangle. \tag{17.12}$$

In conventional crystallographic refinement, each atom in the structure has independent thermal parameters. However, as shown in Equation 17.12, the thermal parameters are common across the entire structure and reduces to the variances and covariances of the $M \times M$ matrix, Π. To ensure that $\mathbf{U}_j$ remains positive and definite, Π is expressed as a lower triangular matrix, Ω, such that

$$\Pi = \Omega\Omega^{\mathrm{T}}. \tag{17.13}$$

Therefore, the number of thermal parameters for normal-mode-based refinement is $M(M+1)/2$, which is the number of nonzero terms in Ω. These thermal parameters from normal modes are optimized according to a least-squares method by minimizing the function

$$\sum_{\mathbf{h}} w(\mathbf{h})\left(\left|F_{\mathrm{obs}}(\mathbf{h})\right| - \left|F_{\mathrm{cal}}(\mathbf{h})\right|\right)^2, \tag{17.14}$$

where $|F_{\mathrm{obs}}(\mathbf{h})|$ is the diffraction data measured from experiment. Since only the magnitudes are measured, the phases cannot be used during the minimization process. Lastly, because the modified eNMA method only calculates the eigenvectors for the C_α atoms, we extrapolate the normal modes to all the atoms by assuming that all the atoms in a residue move in the same direction as its C_α atom.

While NMA is a powerful method that can describe the intrinsic motions of a structure, the external motions must be characterized as well for crystallographic refinement to be successful. Fortunately, the rigid body motion of the entire structure can be described by the Translation, Libration, and Screw (TLS) method [29]. Implemented in REFMAC5 [30] of the CCP4 suite of crystallographic software [31], the TLS method can model the motion of a rigid body with three 3×3 tensors, each describing the translation, libration, and screw motions, respectively. One final source of anisotropy comes from the crystal and not the atomic positions. However, we can account for this by adding an additional overall anisotropic temperature factor. If we assume that the sources of fluctuations are independent of each other, we can construct the final $\mathbf{U}_j$ for each atom as

$$\mathbf{U}_j = \mathbf{U}_{nm} + s_{\mathrm{tls}}\mathbf{U}_{\mathrm{tls}} + \mathbf{U}_{\mathrm{overall}}, \tag{17.15}$$

where $\mathbf{U}_{nm}$ is from Equation 17.12, $\mathbf{U}_{\mathrm{tls}}$ is from REFMAC5, s_{tls} is a scaling factor, and $\mathbf{U}_{\mathrm{overall}}$ is the overall anisotropic temperature factor. The scaling factor is included because the TLS parameters are determined by an external program and are independent of the minimization of the other parameters in $\mathbf{U}_{nm}$ and $\mathbf{U}_{\mathrm{overall}}$.

With the theory in place to use normal modes to replace all the temperature factors of the protein atoms, we follow the standard procedure for model building where the temperature factors and atomic positions are updated iteratively. To track the progress of the refinement, the R factor

$$R = \frac{\sum_{\mathbf{h}} \left\| F_{\text{obs}}(\mathbf{h}) \right| - k \left| F_{\text{cal}}(\mathbf{h}) \right\|}{\sum_{\mathbf{h}} \left| F_{\text{obs}}(\mathbf{h}) \right|},$$

$$k = \frac{\sum_{\mathbf{h}} \left| F_{\text{obs}}(\mathbf{h}) \right| \left| F_{\text{cal}}(\mathbf{h}) \right|}{\sum_{\mathbf{h}} \left| F_{\text{cal}}(\mathbf{h}) \right|^2} \tag{17.16}$$

is used. For validation purposes, a small percentage of the diffraction data, usually 5–10%, is set aside as the test set, while the rest of the data, the working set, is used for refinement. The *R* factor calculated from the working set is R_{cryst} and the *R* factor calculated from the test set is R_{free}.

17.3 RESULTS

17.3.1 Modified eNMA in Internal Coordinates without Tip Effect

To verify that the modified eNMA method reproduces the subspace of the low-frequency eigenvectors with no contamination of the tip effect, we compared the eigenvectors from the new method to those from conventional eNMA and from CHARMM for a variety of systems [14], one of which was a multichain supramolecular complex, the molecular chaperonin GroEL [32]. The structure is composed of 14 monomers, each with 525 residues, organized into two stacked heptameric rings. The chaperonin utilizes ATP to help other proteins fold correctly. Its structure has been studied extensively and is known to undergo large conformational changes to open and close the chamber in which the folding occurs [33–36].

Without coarse-graining, the Hessian matrix for a system the size of GroEL would not be possible to calculate due to the shear number of atoms. With the conventional eNMA [8], the "tip effect" is quite severe as the *T* values for the first 500 modes show in Figure 17.2b (solid squares). But by calculating the eigenvectors by the modified eNMA [14], the tip effect is dramatically reduced (empty circles). The motional patterns of the modes were also verified by comparing the low-frequency modes with those previously observed [36]. Figure 17.2a shows that the collective motion of the second mode is a stretching motion along the diagonals of the complex.

17.3.2 Refinement of X-ray Anisotropic Thermal Parameters Using Normal Modes

For the refinement of X-ray crystallography, we showed that the normal-mode-based refinement protocol is successful in improving an isotropically-refined model in a previous study [28]. The target system was a 3.42 Å structure of mammalian formiminotransferase cyclodeaminase (FTCD) [37]. Biologically, this protein is involved in linking histidine catabolism and folate metabolism [38], integrating the Golgi complex with the vimentin intermediate filament cytoskeleton [39–41], and causing autoimmune hepatitis [42] and glutamate formiminotransferase deficiency [43]. The protein's structure is similar to GroEL in that there are two stacked rings, but FTCD has eight monomers in two tetrameric rings. Each monomer is composed of two domains, the FT domain and the CD domain. The FT domain is further divided into the N subdomain and the C subdomain. Figure 17.3 shows the structures of FTCD in full complex and in various components. This 0.5 million Dalton (over 16,000 atoms) enzyme also is sufficiently large that coarse-graining is required in order for the normal modes to be calculated on contemporary computers. The normal mode calculation was performed on the biologically relevant molecule, the full octamer, and only the portions of the eigenvectors corresponding to the structure in the asymmetric unit of the crystal, two subunits from two octamers, were kept.

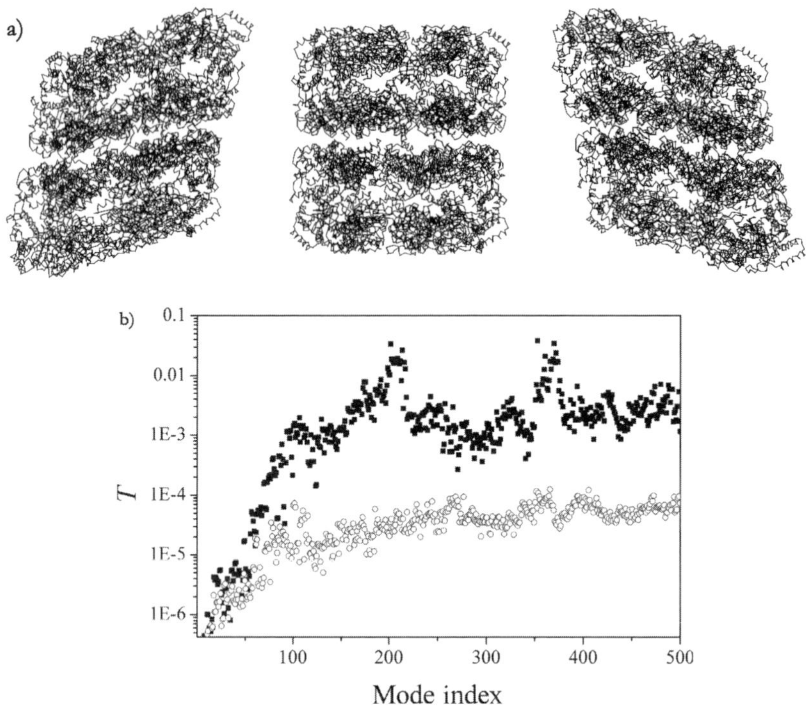

FIGURE 17.2 Results on multisubunit supramolecular complex, the molecular chaperonin GroEL. (a) Motional pattern of the second vibrational mode, which is a stretching mode along the diagonal line of the molecule. (b) Tip effect; the solid squares are for conventional eNMA, and the empty circles are for new eNMA. Note the vertical axis is made in logarithmic scale. (This figure is adopted from Figure 6 on page 469 in Lu, M., Poon, B., and Ma, J. *J. Chem. Theor. Comp.*, 2, 464, 2006.)

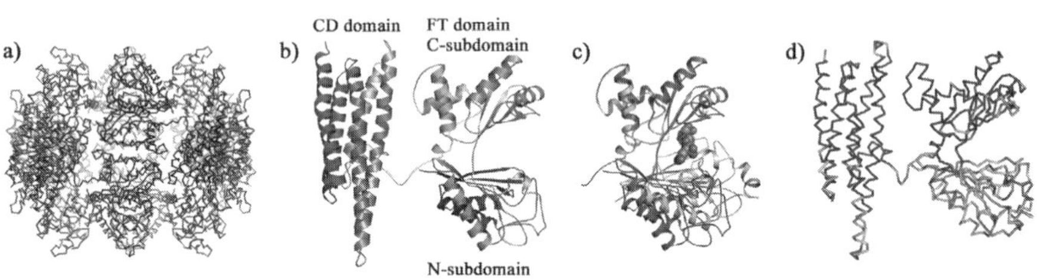

FIGURE 17.3 (See color insert following page 238.) Structure and thermal ellipsoids of FTCD. (a) The square doughnut structure of an FTCD octamer. Two subunits are shown in red and blue, respectively. (b) The subunit structure of ligand-free FTCD. Backbone trace color ramped from the N-terminus to the C-terminus. (c) Superposition of the FT domain of human ligand-free FTCD (red) with the structure of the same domain in isolation (cyan) with the product analog, folinic acid (CPK mode), bound in the groove. (d) Rainbow-colored isotropic *B*-factor in the original model. The hotter the color, the larger the *B*-factors. The high flexibility of the N-subdomain, the linker region, and the lower half of the CD domain are evident. (This figure is adopted from Figure 1 on page 7870 in Poon, B. K., Chen, X., Lu, M., Vyas, N. K., Quiocho, F. A., Wang, Q., and Ma, *J. Proc. Natl. Acad. Sci. U.S.A.*, 104, 7869, 2007.)

Due to the poorly diffracting crystal and size of the structure, it was very difficult to even build the original isotropic model. Only the C_α trace was deposited into the Protein Data Bank (PDB code, 1TT9). However, we were able to obtain the final all-atom, isotropic structure and apply our refinement method. At the start, the R_{cryst} and R_{free} of the original structure refined in

CNS [44] were 24.6 and 28.8%, respectively [37]. After recalculating the initial values using REFMAC5, the R_{cryst} and R_{free} became 23.5 and 28.7%, respectively. After several rounds of iteratively updating the normal-mode-based temperature factors and atomic coordinates, R_{cryst} and R_{free} converged to 24.0 and 24.9%, respectively. This is a significant improvement because the R_{free} is a more accurate measure of the quality of the model. For refinement, the first 50 modes were used, resulting in 1275 normal mode parameters. Compared with over 16,000 thermal parameters for the original isotropic model, there was an order of magnitude decrease in the number of thermal parameters while improving the model and providing an anisotropic description of the thermal fluctuations.

In addition to quantitatively improving the model through the R factors, our method also improves the electron density map, which gives crystallographers a more accurate picture of the structure. Figure 17.4a shows plots of the root mean square deviation (rmsd) of the main-chain atoms of one subunit between the original isotropic model and the final anisotropic model. The other three subunits in the asymmetric unit show the same trend. The peaks signify regions where the biggest changes were made to the original structure.

The first spike occurs around residue 14. As shown in Figure 17.5a, this spike corresponds to a major shift in the main chain coordinates. The 2Fo-Fc omit map for the isotropic model is fragmented, which can make the correct tracing of the backbone unclear. However, after performing normal-mode-based refinement on the structure, the same type of map disambiguates the placement of the main chain and the side chains.

The second spike also corresponds to a shift in the main chain atoms, but is less severe. In both the isotropic and anisotropic models, Figure 17.5b shows that the electron density is clear enough for atoms to be placed with confidence. The spike is a result of centering the atoms within the electron density.

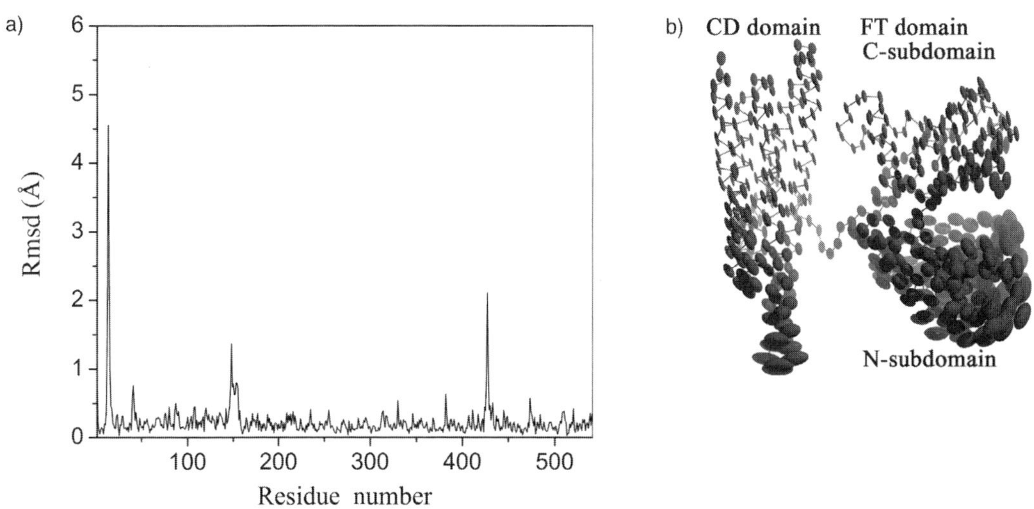

FIGURE 17.4 (a) Structural shifts of the normal-mode-refined new model with respect to the original model. The rmsd (Å) along the chain of a single subunit is shown for the main chains. Three large spikes are evident in both graphs. (This figure is adopted from Figure 3a on page 7871 in Poon, B. K., Chen, X., Lu, M., Vyas, N. K., Quiocho, F. A., Wang, Q., and Ma, *J. Proc. Natl. Acad. Sci. U.S.A.*, 104, 7869, 2007.) (b) Anisotropically refined thermal ellipsoids for a single subunit of FTCD, in the same view as in Figure 17.3d. It is evident that the N-terminal subdomain of FT domain and the lower half of the CD domain are highly flexible. The results for other subunits are very similar due to symmetry constraint. (This figure is adopted from Figure 5a on page 7873 in Poon, B. K., Chen, X., Lu, M., Vyas, N. K., Quiocho, F. A., Wang, Q., and Ma, *J. Proc. Natl. Acad. Sci. U.S.A.*, 104, 7869, 2007.)

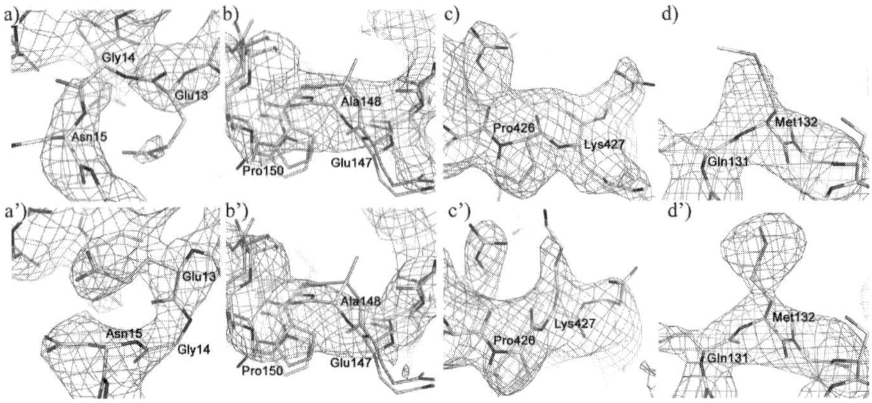

FIGURE 17.5 Examples of large structural adjustments in normal-mode refinement. The top panels are for the original model while the bottom panels are for the new normal-mode model. (a) and (a') Region Glu13-Asn15 superimposed with omit $2F_o-F_c$ map contoured at 1.5σ. (b) and (b') Region Glu147-Pro150 superimposed with omit $2F_o-F_c$ map contoured at 1.0σ. In both panels, the original model (uniform in color) and the new model (grayscaled for chemical groups) are superimposed to highlight the structural shifts. (c) and (c') Region Pro426-Lys427 superimposed with omit $2F_o-F_c$ map contoured at 1.0σ. (d) and (d') Residue Se-Met132 superimposed with omit $2F_o-F_c$ map contoured at 1.5σ. (This figure is adopted from Figure 4 on page 7872 in Poon, B. K., Chen, X., Lu, M., Vyas, N. K., Quiocho, F. A., Wang, Q., and Ma, *J. Proc. Natl. Acad. Sci. U.S.A.*, 104, 7869, 2007.)

Lastly, the third spike represents a rotation of the side chain for residue 427, as shown in Figure 17.5c. Again, in both the isotropic and anisotropic models, the electron density allowed for the placement of atoms. However, after normal-mode-based refinement, the electron density map was changed, which allowed for adjustments to be made that lowered the *R* factors.

While the three spikes showed large changes to the model, many of the improvements were smaller, but the sum total of these improvements allowed us to reach our final model. An example of a smaller but more common improvement is shown in Figure 17.5d. In this case, the electron density map of the original model did not show the positions of the atoms at the tip of the side chain, but after normal-mode-based refinement, the density became visible and allowed for correct placement. Overall, there were about 55 residues for each subunit where the improved electron density map allowed for more confident placement of the atoms.

In addition to the structure shift, Figure 17.4b shows the C_α trace and thermal ellipsoids of one subunit of the final model. It is clear that the distribution of the magnitudes of the ellipsoids is comparable with the original isotropic model (Figure 17.3d). Furthermore, the direction of motion shown by the thermal ellipsoids nicely correlates with the ligand-induced cleft-closing motion (Figure 17.3c). This figure is an example of how powerful normal-mode-based refinement can be. Traditionally, X-ray crystallography is often viewed as providing a snapshot, frozen in time, of the molecule of interest. However, as we have shown, diffraction data contains information about the dynamics of the protein and only with anisotropic models can this information be elucidated.

ACKNOWLEDGMENTS

The author acknowledges support from an NIH grant (GM067801) and a grant from the Welch Foundation.

REFERENCES

1. Brooks, III, C. L., Karplus, M., and Pettitt, B. M. 1988. Proteins: A theoretical perspective of dynamics, structure, and thermodynamics. *Adv. Chem. Phys.* 71:1.
2. McCammon, J. A., and Harvey, S. 1987. *Dynamics of Proteins and Nucleic Acids.* Cambridge: Cambridge University Press.
3. Brooks, B. R., Janezic, D., and Karplus, M. 1995. Harmonic analysis of large systems. I. Methodology. *J. Comput. Chem.* 16:1522.
4. Levitt, M., Sander, C., and Stern, P. S. 1985. Protein normal-mode dynamics: Trypsin inhibitor, crambin, ribonuclease and lysozyme. *J. Mol. Biol.* 181:423.
5. Ma, J. 2004. New advances in normal mode analysis of supermolecular complexes and applications to structural refinement. *Curr. Protein Pept.* Sci. 5:119.
6. Ma, J. 2005. Usefulness and limitations of normal mode analysis in modeling dynamics of biomolecular complexes. *Structure* 13:373.
7. Tirion, M. M. 1996. Large amplitude elastic motions in proteins from a single-parameter, atomic analysis. *Phys. Rev. Lett.* 77:1905.
8. Atilgan, A. R., Durell, S. R., Jernigan, R. L., Demirel, M. C., Keskin, O., and Bahar, I. 2001. Anisotropy of fluctuation dynamics of proteins with an elastic network model. *Biophys. J.* 80:505.
9. Doruker, P., Jernigan, R. L., and Bahar, I. 2002. Dynamic of large proteins through hierarchical levels of coarse-grained structures. *J. Comput. Chem.* 23:119.
10. Doruker, P., and Jernigan, R. L. 2003. Functional motions can be extracted from on-lattice construction of protein structures. *Proteins* 53:174.
11. Ming, D., Kong, Y., Lambert, M., Huang, Z., and Ma, J. 2002. How to describe protein motion without amino-acid sequence and atomic coordinates. *Proc. Natl. Acad. Sci. USA* 99:8620.
12. Tama, F., Wriggers, W., and Brooks, C. L. 2002. Exploring global distortions of biological macromolecules and assemblies from low-resolution structural information and elastic network theory. *J. Mol. Biol.* 321:297.
13. Lu, M., and Ma, J. 2005. The role of shape in determining molecular motions. *Biophys. J.* 89:2395.
14. Lu, M., Poon, B., and Ma, J. 2006. A new method for coarse-grained elastic normal-mode analysis. *J. Chem. Theor. Comp.* 2:464.
15. Kamiya, K., Sugawara, Y., and Umeyama, H. 2003. Algorithm for normal mode analysis with general internal coordinates. *J. Comput. Chem.* 24:826.
16. Diamond, R. 1990. On the use of normal modes in thermal parameters refinement: Theory and application to the bovine pancreatic trypsin inhibitor. *Acta Crystallogr.* A 46:425.
17. Kidera, A., and Go, N. 1990. Refinement of protein dynamic structure: normal mode refinement. *Proc. Natl. Acad. Sci. U.S.A.* 87:3718.
18. Kidera, A., and Go, N. 1992. Normal mode refinement: Crystallographic refinement of protein dynamic structure. I. Theory and test by simulated diffraction data. *J. Mol. Biol.* 225:457.
19. Kidera, A., Inaka, K., Matsushima, M., and Go, N. 1992. Normal mode refinement: Crystallographic refinement of protein dynamic structure. II. Application to human lysozyme. *J. Mol. Biol.* 225:477.
20. Kidera, A., Inaka, K., Matsushima, M., and Go, N. 1992. Normal mode refinement: Crystallographic refinement of protein dynamic structure applied to human lysozyme. *Biopolymers* 32:315.
21. Kidera, A., Matsushima, M., and Go, N. 1994. Dynamic structure of human lysozyme derived from X-ray crystallography: Normal mode refinement. *Biophys. Chem.* 50:25.
22. Suhre, K., and Sanejouand, Y. H. 2004. On the potential of normal-mode analysis for solving difficult molecular-replacement problems. *Acta Crystallogr. D Biol. Crystallogr.* 60:796.
23. Lindahl, E., Azuara, C., Koehl, P., and Delarue, M. 2006. NOMAD-Ref: Visualization, deformation and refinement of macromolecular structures based on all-atom normal mode analysis. *Nucleic Acids Res.* 34:W52.
24. Delarue, M., and Dumas, P. 2004. On the use of low-frequency normal modes to enforce collective movements in refining macromolecular structural models. *Proc. Natl. Acad. Sci. U.S.A.* 101:6957.
25. Kundu, S., Melton, J. S., Sorensen, D. C., and Phillips, Jr., G. N. 2002. Dynamics of proteins in crystals: Comparison of experiment with simple models. *Biophys.* J. 83:723.
26. Kondrashov, D. A., Cui, Q., and Phillips, Jr., G. N. 2006. Optimization and evaluation of a coarse-grained model of protein motion using x-ray crystal data. *Biophys.* J. 91:2760.
27. Kondrashov, D. A., Van Wynsberghe, A. W., Bannen, R. M., Cui, Q., and Phillips, Jr., G. N. 2007. Protein structural variation in computational models and crystallographic data. *Structure* 15:169.

28. Poon, B. K., Chen, X., Lu, M., Vyas, N. K., Quiocho, F. A., Wang, Q., and Ma, J. 2007. Normal mode refinement of anisotropic thermal parameters for a supramolecular complex at 3.42-A crystallographic resolution. *Proc. Natl. Acad. Sci. U.S.A.* 104:7869.
29. Schomaker, V., and Trueblood, K. N. 1968. On the rigid-body motion of molecules in crystals. *Acta Crystallogr.* B 24:63.
30. Murshudov, G. N., Vagin, A. A., and Dodson, E. J. 1997. Refinement of macromolecular structures by the maximum-likelihood method. *Acta Crystallogr. D Biol. Crystallogr.* 53:240.
31. Collaborative Computational Project, Number 4. 1994. The CCP4 suite: Programs for protein crystallography. *Acta Crystallogr. D Biol. Crystallogr.* 50:760.
32. Xu, Z., and Sigler, P. B. 1998. GroEL/GroES: Structure and function of a two-stroke folding machine. *J. Struct. Biol.* 124:129.
33. Sigler, P. B., Xu, Z., Rye, H. S., Burston, S. G., Fenton, W. A., and Horwich, A. L. 1998. Structure and function in GroEL-mediated protein folding. *Annu. Rev. Biochem.* 67:581.
34. Ma, J., and Karplus, M. 1998. The allosteric mechanism of the chaperonin GroEL: A dynamic analysis. *Proc. Natl. Acad. Sci. U.S.A.* 95:8502.
35. Ma, J., Sigler, P. B., Xu, Z., and Karplus, M. 2000. A dynamic model for the allosteric mechanism of GroEL. *J. Mol. Biol.* 302:303.
36. Keskin, O., Bahar, I., Flatow, D., Covell, D. G., and Jernigan, R. L. 2002. Molecular mechanisms of chaperonin GroEL-GroES function. *Biochemistry* 41:491.
37. Mao, Y., Vyas, N. K., Vyas, M. N., Chen, D. H., Ludtke, S. J., Chiu, W., and Quiocho, F. A. 2004. Structure of the bifunctional and Golgi-associated formiminotransferase cyclodeaminase octamer. *EMBO J.* 23:2963.
38. Shane, B., and Stokstad, E. L. R. 1984. Folates in the synthesis and catabolism of histidine. In *Folates and Pterins*, vol. 1, ed. R. L. Blakley and S. J. Benkovic, 433–55. New York: Wiley.
39. Bashour, A. M., and Bloom, G. S. 1998. 58K, a microtubule-binding Golgi protein, is a formiminotransferase cyclodeaminase. *J. Biol. Chem.* 273:19612.
40. Gao, Y. S., Alvarez, C., Nelson, D. S., and Sztul, E. 1998. Molecular cloning, characterization, and dynamics of rat formiminotransferase cyclodeaminase, a Golgi-associated 58-kDa protein. *J. Biol. Chem.* 273:33825.
41. Gao, Y. S., Vrielink, A., MacKenzie, R., and Sztul, E. 2002. A novel type of regulation of the vimentin intermediate filament cytoskeleton by a Golgi protein. *Eur. J. Cell Biol.* 81:391.
42. Lapierre, P., Hajoui, O., Homberg, J. C., and Alvarez, F. 1999. Formiminotransferase cyclodeaminase is an organ-specific autoantigen recognized by sera of patients with autoimmune hepatitis. *Gastroenterology* 116:643.
43. Rosenblatt, D. 1995. Inherited disorders of folate transport and metabolism. In *The Metabolic and Molecular Bases of Inherited Diseases*, ed. C. Scriver, A. Beaudet, W. Sly, and D. Valle, Vol. 2, pp. 3111–28. New York: McGraw-Hill.
44. Brünger, A. T., Adams, P. D., Clore, G. M., DeLano, W. L., Gros, P., Grosse-Kunstleve, R. W., Jiang, J. S., Kuszewski, J., Nilges, M., Pannu, N. S., Read, R. J., Rice, L. M., Simonson, T., and Warren, G. L. 1998. Crystallography & NMR system: A new software suite for macromolecular structure determination. *Acta Crystallogr. D Biol. Crystallogr.* 54:905.

18 Coarse-Grained Normal Mode Analysis to Explore Large-Scale Dynamics of Biological Molecules

Osamu Miyashita and Florence Tama
Department of Biochemistry and Molecular Biophysics, The University of Arizona

CONTENTS

18.1 INTRODUCTION

Biomolecular machines made of proteins and RNAs perform and sustain most functions in our bodies. To elucidate their functional mechanisms, there has been a tremendous effort to obtain structural information for these biological molecules. While structure provides important insights, a deeper understanding could be obtained through examination of their dynamical properties and physical interactions within the system. Thus it is beneficial to complement experimental work by theoretical and computational techniques that can directly examine physical interactions, explore dynamics of the biological molecules, and bring useful atomic-level insights into protein functions.

To computationally study dynamical properties of biological molecules, several approaches can be considered. The most common is the use of molecular dynamics simulations in which the system evolves as a function of time [Karplus and McCammon 2002]. Exploration of molecular motions of biological molecules and their assemblies by this approach has provided significant insights into structure-function relationships. This method can give very detailed information on the dynamics near the native state. However, even though computational techniques and processing power have been improving significantly, the application for large-scale macromolecular assemblies is limited due to the computational complexity of all-atom simulation methods and reaching time scales corresponding to functional motions still remains impractical. An example of such work is the 10 ns simulation of the satellite mosaic virus, which required 10 days of computer time using 256 processors. It would take years to reach longer time scales (ms), which are relevant for large-scale rearrangements of proteins [Freddolino et al. 2006].

An alternative approach to extend the time scale of molecular dynamics simulations is to use coarse-grained models, which enable microsecond time scales to be reached for small proteins [Tozzini 2005]. The simulations typically consider the C_α and P atoms, strung as beads, which considerably reduce the number of atoms necessary for simulation. Details for such models can be found in other chapters of this book. However, these calculations are still computationally expensive to observe large functional motions for large macromolecular assemblies such as the ribosome. Also, use of advanced sampling methods to explore long time scale and large-amplitude conformational changes (e.g., protein folding) are still far from routine.

In order to simulate large and slow conformational rearrangements of large biological molecules, we need to employ alternative techniques. One of these techniques is normal mode analysis (NMA), which is commonly used in physics, and was introduced to structural biology about 20 years ago [Go, Noguti, and Nishikawa 1983; Brooks and Karplus 1983]. In NMA the energy surface is approximated, in other words coarse-grained, as harmonic. Exploration of the normal modes of a molecular system can yield insights, at the atomic level, on the mechanism of large-scale rearrangements of protein/protein complexes, which often occur upon ligand/protein binding. Biological studies employing NMA have generally focused on a few large-amplitude/low-frequency normal modes, which are expected to be relevant to function.

Until recently, NMA applications were limited to small proteins (up to 300 residues). There were two reasons for this limitation. The first one is related to the size of the biological system. The standard protein model used in the calculation consists of classical points of mass with typically one point per atom. Interactions between these atoms are defined by semiempirical force fields. Using these force fields requires an all-atom description to represent the macromolecule, which becomes computationally difficult with increasing system size (see Methods). The second problem is related to the minimization process (see also Methods), which is required before NMA when semiempirical force fields are used. It is particularly detrimental due to the distortion in protein conformation during minimization. Moreover, this process is time consuming.

The applicability of NMA has been advancing by the development of new coarse-grained models. Those coarse-grained models do not require all-atom description to represent the mechanical properties of a system. Thus a subset of atoms could be used to perform NMA and virtually any system size could be studied (of course at coarse-grained level).

NMA and coarse-grained models are approximations. Ideally, it would be best if we simulate biological molecules at full scale and full detail, however in order to study conformational changes of large macromolecules with the computational power available today, alternative approaches are necessary and coarse-grained NMA is quite successful in this aspect. Coarse-grained methods at both molecule and algorithmic levels provide us with tools to extend our work to larger systems. However, one has to be aware that using such coarse-grained models is an approximation, which means that there are limitations to the approach; therefore one needs to be careful in the interpretation of the data. Our philosophy is to take full advantage of the computational power available today and to adjust the level of coarse-grain accordingly in order to represent the system as precisely as possible.

18.2 METHODS

18.2.1 Normal Mode Theory and Analysis

NMA is a relatively mature technique [Goldstein 1950], which has in recent years piqued the interest of researchers due to new algorithmic developments that enable applications to larger systems. In NMA, one approximately represents the dynamics of a molecule as a set of harmonic oscillators. This is beneficial because the motion of a harmonic oscillator can be analytically described. For a harmonic oscillator of a mass m with coordinate x connected to a spring with the spring constant k, the Hamiltonian is:

$$H = \frac{1}{2} m \left(\frac{dx}{dt} \right)^2 + \frac{1}{2} kx^2. \tag{18.1}$$

The dynamics of the particle can then be derived by solving Equation 18.1, as $x = C\cos(\omega t + \phi)$ where C and ϕ are the amplitude and the phase at time $t=0$ and $\omega = (k/m)^{1/2}$ is the angular frequency associate with the vibrational mode.

Unlike a simple harmonic oscillator, the potential energy of biological molecules is complex, and thus the equation of motion cannot be solved analytically. However, if one focuses on the motions in the vicinity of stable conformation, the potential function can be approximated into a simple form. We consider a molecule with N atoms and describe the coordinates of the atoms as $\mathbf{r} = (x_1, y_1, z_1, \ldots z_N)$, where (x_i, y_i, z_i) is the coordinate of atom i. Assuming that we analyze the motion around a stable conformation $\mathbf{r}^0$, where superscript 0 indicates the energy minimum, a Taylor expansion of the potential energy function $U(\mathbf{r})$ around a minimum on the energy surface, $\mathbf{r}^0$, gives:

$$\begin{aligned} U(\mathbf{r}) = U(\mathbf{r}^0) + \sum_i \left. \frac{\partial U}{\partial r_i} \right|_{\mathbf{r}=\mathbf{r}^0} (r_i - r_i^0) + \frac{1}{2!} \sum_{ij} \left. \frac{\partial^2 U}{\partial r_i \partial r_j} \right|_{\mathbf{r}=\mathbf{r}^0} (r_i - r_i^0)(r_j - r_j^0) \\ + \frac{1}{3!} \sum_{ijk} \left. \frac{\partial^3 U}{\partial r_i \partial r_j \partial r_k} \right|_{\mathbf{r}=\mathbf{r}^0} (r_i - r_i^0)(r_j - r_j^0)(r_k - r_k^0) + \ldots. \end{aligned} \tag{18.2}$$

Since the reference structure $\mathbf{r}^0$ is a minimum of the energy function, $\partial U / \partial r_i(\mathbf{r}^0) = 0$. In addition, the potential energy can be defined relative to this reference structure as $U(\mathbf{r}^0) = 0$. Finally, if one considers sufficiently small displacements, terms beyond the second order may be neglected (i.e., harmonic approximation). The approximate potential energy function is given as:

$$U(r) \cong \frac{1}{2} \sum_{ij} \left. \frac{\partial^2 U}{\partial r_i \partial r_j} \right|_{\mathbf{r}=\mathbf{r}^0} (r_i - r_i^0)(r_j - r_j^0). \tag{18.3}$$

Thus the Hamiltonian of the system is then given by:

$$H(\mathbf{r}) \cong K(\mathbf{r}) + U(\mathbf{r}) = \frac{1}{2} \sum_i m_i \frac{dr_i^2}{d^2 t} + \frac{1}{2} \sum_{ij} \left. \frac{\partial^2 U}{\partial r_i \partial r_j} \right|_{\mathbf{r}=\mathbf{r}^0} (r_i - r_i^0)(r_j - r_j^0), \tag{18.4}$$

where K represents the kinetic energy, and m_i represents the mass of the coordinate r_i. For convenience, we rewrite the equation using a mass weighted coordinate, $X_i = m_i^{1/2}(r_i - r_i^0)$:

$$H(\mathbf{X}) \cong \frac{1}{2} \sum_i \frac{dX_i^2}{d^2 t} + \frac{1}{2} \sum_{ij} \left. \frac{\partial^2 U}{\partial X_i \partial X_j} \right|_{\mathbf{X}=\mathbf{X}^0} X_i X_j. \tag{18.5}$$

As we already discussed, in the normal mode theory we represent the dynamics of a biological molecule as a collection of harmonic oscillators. The dynamics is not directly expressed in Cartesian coordinates but in normal-mode coordinates **q**. The two coordinates are related by the transformation matrix **A** as follows: $\mathbf{X}=\mathbf{Aq}$. This relation might be more intuitive in a vector form:

$$\begin{pmatrix} \sqrt{m_1}(x_1-x_1^0) \\ \sqrt{m_1}(y_1-y_1^0) \\ \sqrt{m_1}(z_1-z_1^0) \\ \sqrt{m_2}(x_2-x_2^0) \\ \sqrt{m_2}(y_2-y_2^0) \\ \sqrt{m_2}(z_2-z_2^0) \\ \vdots \\ \sqrt{m_N}(x_N-x_N^0) \\ \sqrt{m_N}(y_N-y_N^0) \\ \sqrt{m_N}(z_N-z_N^0) \end{pmatrix} = \begin{pmatrix} a_{x11} \\ a_{y11} \\ a_{z11} \\ a_{x21} \\ a_{y21} \\ a_{z21} \\ \vdots \\ a_{xN1} \\ a_{yN1} \\ a_{zN1} \end{pmatrix} q_1 + \begin{pmatrix} a_{x12} \\ a_{y12} \\ a_{z12} \\ a_{x22} \\ a_{y22} \\ a_{z22} \\ \vdots \\ a_{xN2} \\ a_{yN2} \\ a_{zN2} \end{pmatrix} q_2 + \cdots \begin{pmatrix} a_{x1n} \\ a_{y1n} \\ a_{z1n} \\ a_{x2n} \\ a_{y2n} \\ a_{z2n} \\ \vdots \\ a_{xNn} \\ a_{yNn} \\ a_{zNn} \end{pmatrix} q_n + \cdots. \tag{18.6}$$

Each column vector of **A** represents a normal mode. The matrix, **A**, needs to be defined in such a way that the new coordinates $\{q_n\}$ are independent from each other in the Hamiltonian. The second term of the original Hamiltonian can be converted by finding a matrix **A** that satisfies $\mathbf{A}^{\mathrm{T}}\mathbf{HA}=\mathbf{L}$, where $H_{ij}=\partial^2 U/\partial X_i \partial X_j$ (Hessian matrix) and **L** is a diagonal matrix also to be determined. In addition, if $\mathbf{A}^{\mathrm{T}}\mathbf{A}=\mathbf{I}$, the first term in Equation 18.2 remains to be the same kinetic energy form. Using these expressions above, the Hamiltonian can be converted into the following form:

$$H(q) \cong \frac{1}{2}\sum_{n=1}^{3N-6} \frac{dq_n^2}{d^2t} + \frac{1}{2}\sum_{n=1}^{3N-6} \omega_n^2 q_n^2, \tag{18.7}$$

where ω_n^2 is the diagonal element nn of the matrix **L**. This Hamiltonian can be solved as a set of independent harmonic oscillators $\{q_n\}$ with corresponding frequencies $\{\omega_n\}$.

In practice, the transformation matrix **A** and the diagonal matrix **L** can be determined by solving the eigenvalue problem, which is to find a vector **a** and a value λ that satisfy $\mathbf{Ha}=\lambda\mathbf{a}$. For **H**, which is a $3N \times 3N$ matrix, we find $3N$ sets of solutions $(\mathbf{a}_1, \lambda_1)$, $(\mathbf{a}_2, \lambda_2)$ … $(\mathbf{a}_{3N}, \lambda_{3N})$. Among them, $3N-6$ normal modes are meaningful—the six normal modes have an eigenvalue equal to 0 and correspond to rigid body translational and rotational motions of the whole system. The solutions are normally sorted in ascending order of the eigenvalue, providing the eigenvalue matrix $\mathbf{L}=\mathrm{diag}(\lambda_1,\lambda_2 \ldots \lambda_{3N-6})$ and associated eigenvector matrix $\mathbf{A}=(\mathbf{a}_1,\mathbf{a}_2,\ldots \mathbf{a}_{3N-6})$.

In summary, as a result of NMA a set of normal mode vectors $\{q_n\}$ and corresponding frequencies $\{\omega_n\}$ are obtained. The nth normal mode variable, q_n, oscillates with the frequency ω_n; that is, the nth eigenvector $(a_{1n}\ a_{2n} \ldots a_{3Nn})^{\mathrm{T}}$ gives the direction and relative amplitudes of atomic displacements in Cartesian space and all those oscillational displacements occur at the same frequency, ω_n. Motions within the system are described as a superposition of those modes. In the case of a simple system such as a water molecule, the resulting normal mode vectors reveal three well-known motions of the water molecule; that is, bending mode, symmetric mode, and asymmetric stretching mode. We should note that the frequency obtained from computation with a detailed force field can be directly related to infrared experiments for which bond stretching can be observed.

From NMA, several dynamical properties can be calculated. As an example, B-factors or temperature factors can be calculated as follows. If the system is in thermal equilibrium, the average of the potential of each mode is equal to $k_B T/2$, where T is the absolute temperature and k_B is the Boltzmann constant; thus

$$\langle q_n^2 \rangle = \frac{k_B T}{\omega_n^2}. \tag{18.8}$$

Using those relations, the B-factor of each atom is given as:

$$B_i = \frac{8\pi^2}{3} \langle (r_i - r_i^0)^2 \rangle = \frac{8\pi^2}{3} \frac{1}{m_i} \sum_{n=1} a_{in}^2 \langle q_n^2 \rangle = \frac{8\pi^2}{3} \frac{k_B T}{m_i} \sum_{n=1} \frac{a_{in}^2}{\omega_n^2}. \tag{18.9}$$

From this equation, it is evident that the largest contribution to the atomic displacement comes from the lowest frequency normal modes (small ω). For the same reason, the lowest frequency modes are expected to be relevant to biological functions because large conformational changes can be induced by perturbations to the system such as ligand binding. In addition, the lowest-frequency eigenvectors represent the most globally distributed or collective motions; that is, a large number of atoms have significant components $(a_{xi}, a_{yi}, a_{zi})^T$, while for high-frequency eigenvectors only a few atoms are involved in the motions. Studies employing NMA generally focus on a few large-amplitude/low-frequency normal modes as they can be used to unveil large conformational changes of biological molecules.

18.2.2 Rotation-Translation-Block (RTB) Method

The application of NMA critically depends on diagonalization of the Hessian and this can be a limiting factor in applying NMA to interesting large molecular systems such as the ribosome, myosin, chaperones, or viruses, among others. The RTB method was introduced to reduce the size of the Hessian by introduction of a simple physical idea: a protein or nucleic chain may be viewed as being comprised of rigid components linked together, such as residues/bases, groups of residues/bases, or more extensive segments of secondary structural elements (see Figure 18.1a) [Durand, Trinquier, and Sanejouand 1994; Tama et al. 2000]. The combination of rotation and translation of these rigid components should provide a good representation of the low-frequency normal modes of the biological system. Thus, in the RTB method, the molecular system is first divided into n_b blocks, each consisting of one or a few consecutive residues/base pairs, etc. Then, the lowest-frequency normal modes of the biological system are obtained as a linear combination of the rotations and translations of these blocks.

In standard approaches, the normal modes of the system are calculated through the diagonalization of the mass-weighted Hessian matrix **H**. In the RTB approach, **H**, the Hessian being diagonalized, is first expressed in a basis set defined by the rotational and translational degrees of freedom of n_b blocks. $\mathbf{H}_b$, the projected Hessian, is given by:

$$\mathbf{H}_b = \mathbf{P}^T \mathbf{H} \mathbf{P}, \tag{18.10}$$

where **P** is the orthogonal $3N \times 6n_b$ matrix built with the vectors associated with the local rotations and translations of each block. By diagonalizing $\mathbf{H}_b$, which is a $6n_b \times 6n_b$ matrix, the normal modes, $\mathbf{A}_P$, are obtained. The corresponding ($3N$) atomic displacements are recovered by

$$\mathbf{A}_p = \mathbf{P}\mathbf{A}_b.$$

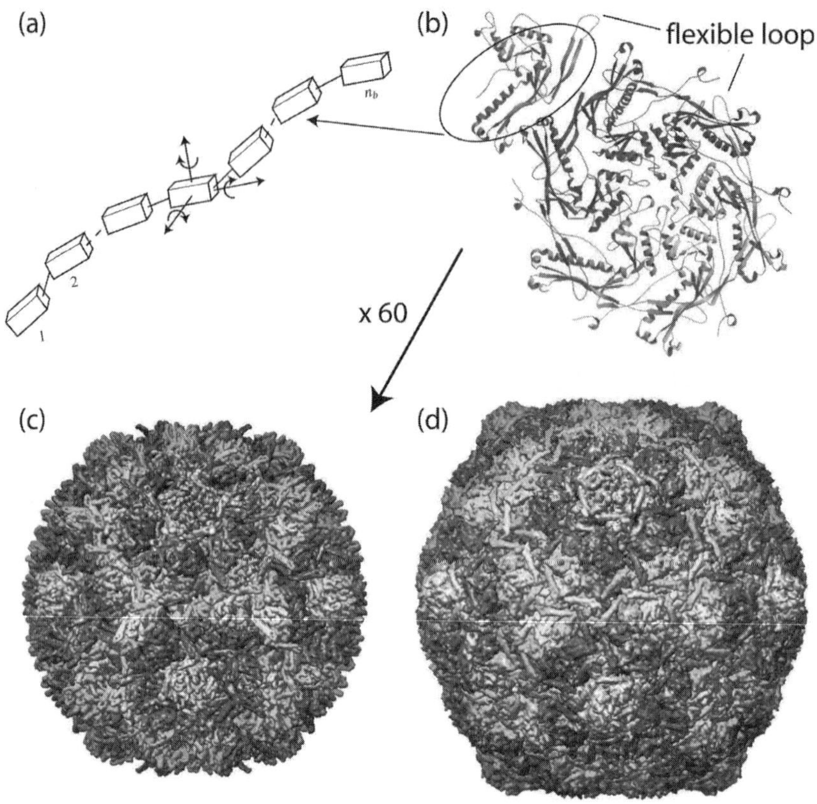

FIGURE 18.1 (a) In the RTB approach, the polypeptidic chain is treated as a collection of rigid blocks, the blocks being made of one residue or more, and only the rotational and translational degrees of freedom of those blocks are considered. (b) Asymmetric unit of HK97, which contains seven copies of the same protein. HK97 is a T = 7 virus; that is, a total of 420 proteins. In the normal mode calculation, each protein is assigned to a block or divided into two blocks (840 total) to take into account the flexibility of the loop. (c, d) Difference in the shape of HK97 (T = 7) between its two known conformations. (Adapted from the Viperdb web site: Shepherd, C. M., I. A. Borelli, G. Lander, P. Natarajan, V. Siddavanahalli, C. Bajaj, J. E. Johnson, C. L. Brooks, and V. S. Reddy, *Nucleic Acids Res.*, 34:D386, 2006.)

Following the above formalism, the actual computational procedure consists of three steps. In the first step, blocks of residues are defined and for each block, α, the corresponding component of matrix $\mathbf{P}$, $\mathbf{U}_\alpha$, is determined and stored. These $6n_b$ vectors form a new basis of small dimension that corresponds to the projector $\mathbf{P}$. In the second step, the Hessian matrix is expressed in this RTB basis, separately for each coupling or diagonal block, $\mathbf{H}_{\alpha\beta}$:

$$\mathbf{H}^b_{\alpha\beta} = \mathbf{U}^T_\alpha \mathbf{H}_{\alpha\beta} \mathbf{U}_\beta. \tag{18.11}$$

The set of $n_b^2 \mathbf{H}^b_{\alpha\beta}$ block-matrices forms the matrix $\mathbf{H}_b$. The construction of $\mathbf{H}_b$ has minimal memory requirements, since the Hessian corresponding to each block; that is, $\mathbf{H}_{\alpha\beta}$, is calculated and projected into the rotation-translation matrix. Therefore, during this step, the largest matrix kept in memory corresponds to the size of one block in the 3D coordinates. The RTB method requires only the small dimension vectors $\mathbf{U}_\alpha$ and the small $6n_b \times 6n_b$ $\mathbf{H}_b$ matrix to be stored. In the last step, $\mathbf{H}_b$ is diagonalized, as in standard methods.

It has been demonstrated that this approach yields very accurate approximations of the low-frequency normal modes of proteins. Studies have also shown that the manner the protein is partitioned into blocks has minimal qualitative consequence on the description of the low-frequency normal modes of the system [Tama et al. 2000].

18.2.3 Conformational Change Pathway

Identifying the pathways for conformational changes in macromolecular systems can be useful to understand their functional mechanism. In particular atomic-level descriptions of the conformational transition process could help to elucidate the molecular basis of the motions. To generate these pathways, tentative models of intermediate structures between the two known conformations need to be built which can be approached in several ways from the stone point of modeling [Schlitter et al. 1993; Guilbert, Perahia, and Mouawad 1995].

NMA has also been applied to describe conformational change pathways [Mouawad and Perahia 1996; Xu, Tobi, and Bahar 2003]. These methods take advantage of the low-energy normal-mode directions of a system between the two end-point states. Here we present an alternative iterative technique.

Before introducing this iterative technique, we should mention how a simple linear approach can be used to describe conformational changes of biological molecules. The displacement vector between two end-point conformations, $\Delta\mathbf{r}$, can be expressed as the superposition of displacements along the normal-mode direction of the system:

$$\Delta\mathbf{r} = \sum_n \mathbf{a}_n q_n, \tag{18.12}$$

since the normal-mode eigenvectors should span the conformational space. The normal-mode amplitudes $\{q_n\}$ are given as:

$$q_n = \mathbf{a}_n \cdot \Delta\mathbf{r}. \tag{18.13}$$

By using some fraction of normal-mode coordinate, q_n, of Equation 18.13 for the deformation, the intermediate structures can be generated (Equation 18.12). However, we should note that using all modes corresponds to simple Cartesian interpolation of the two end-point structures, which often generates physically unrealistic structures. Generally, one finds that a smaller subset of modes $\{\mathbf{a}_n\}$ account for the majority of the conformational deformation between two end-point conformations and this serves as a basis for expressing the, possibly functional, dynamics of the conformational change. These modes are coincident to a few modes with the lowest frequencies [Tama and Sanejouand 2001].

Although the linear interpolation approach just described is often adequate in some instances, to describe conformational changes between two conformationally distinct states requires a nonlinear description due to the anharmonic character of the energy landscape. Conformational change pathways are nonlinear, however normal modes provide only linear motions and therefore such modes cannot provide pathways from one structure to another. Another critical aspect is that displacing too far along the direction given by the lowest-frequency modes, which is however the globally preferential direction of the conformational change, can induce large distortion in the local structure such as bond distances.

The problems arising from the harmonic approximation employed in the NMA can be ameliorated by performing the normal-mode analyses and conformational deformations in an iterative manner [Miyashita, Onuchic, and Wolynes 2003; Tama, Miyashita, and Brooks 2004ab; Miyashita, Wolynes, and Onuchic 2005]. Instead of moving the structure from the initial to the final form directly, the deformation is limited to a small amount, and normal modes are recalculated for the deformed structure. The procedure is as follows: the initial conformation is defined as $C_I = C_0$ and the final state is C_F. At step k, NMA is performed on C_k (k initially taken at $k = I$). The vector difference $\Delta\mathbf{r}^k$ between C_k and C_F is (re)evaluated. The structure C_k is displaced along a linear combination of a few normal modes $\{\mathbf{a}_n^k\}$ toward the final state leading to the next structure C_{k+1}. The amplitude, q_n^k, of the displacements along normal mode n is given by:

$$q_n^k = \mathbf{a}_n^k \cdot \Delta\mathbf{r}^k Q \tag{18.14}$$

where Q is a parameter that determines how far the structure is displaced, 0 equals the current coordinates, and 1 equals the full projection of the current normal mode coordinates onto C_F. A small value of Q such as 0.01 may be used to generate pathways with small distortions. This procedure is repeated until RMSD between the kth iterate and the final conformation cannot be decreased.

18.2.4 The Protein Elastic Model: Tirion Potential

To perform NMA, a potential energy function needs to be defined. Models used in standard calculation to represent biological molecules consist of classical points of mass with typically one point per atom. The energy terms for interactions between these atoms are defined by semiempirical force fields. Using these force fields requires an all-atom description to represent the macromolecule, which becomes computationally difficult with increasing system size. Using such models also requires a minimization of the potential energy before NMA to ensure that the system is at an energy minimum. This process is particularly detrimental due to the change of protein conformation occurring during the minimization. Moreover, it is time consuming and structures with missing residues are difficult to study.

Instead a simplified representation of the potential energy can be introduced for NMA of biological systems. In this representation, the elastic network model, the biological system is described as a three-dimensional elastic network based on the equilibrium distribution of atoms [Tirion 1996]. Amino acids or base pairs may be represented in full atomic detail, or at a more coarse-grained level. For example one mass point per residue [Hinsen 1998], only C_α atoms [Bahar, Atilgan, and Erman 1997; Tama and Sanejouand 2001], or more coarse-grained particle-based models [Doruker, Jernigan, and Bahar 2002] may be used to identify the junctions of the network. These junctions are representative of the mass distribution of the system and are connected together via a simple harmonic restoring force:

$$E(\vec{r}_a, \vec{r}_b) = \begin{cases} \frac{k}{2}(|\vec{r}_a - \vec{r}_b| - |\vec{r}_a^{\,0} - \vec{r}_b^{\,0}|)^2 & \text{for } |\vec{r}_a^{\,0} - \vec{r}_b^{\,0}| \le R_C \\ 0 \text{ for } |\vec{r}_a^{\,0} - \vec{r}_b^{\,0}| > R_C \end{cases}, \tag{18.15}$$

where $\vec{r}_a - \vec{r}_b$ denotes the vector connecting pseudoatoms a and b, the zero superscript indicates the initial configuration of the pseudoatoms, and R_C is a spatial cutoff for interconnections between the particles. The strength of the potential k is a phenomenological constant assumed to be the same for all interacting pairs.

The total potential energy of the molecule is expressed as the sum of elastic strain energies:

$$E_{\text{System}} = \sum_{a,b} E(\vec{r}_a, \vec{r}_b). \tag{18.16}$$

Note that this energy function, E_{System}, is a minimum for any chosen configuration of any system, thus eliminating the need for minimization prior to NMA. Consequently, NMA can be performed directly on crystallographic or NMR structures [Tirion 1996].

Several studies have shown that this Hookean potential is sufficient to reproduce the low-frequency normal modes of proteins as produced by more complete potential energy functions

[Tama and Sanejouand 2001]. The high degree of accord between the modes constructed from these methods suggests that low-frequency normal modes are predominantly a property of the shape of the molecular system [Tama et al. 2003; Tama, Wriggers, and Brooks 2002; Ming et al. 2002]. While this agreement tends to breakdown at high frequencies, there have been many cases showing that collective motions found in the low-frequency modes characterize biologically relevant conformational changes well [Tama and Sanejouand 2001].

18.2.5 Strain Energy Analysis

Originally, the Tirion potential was proposed for NMA. This potential is crude but adequate for the low-frequency motion. Thus, it can be used to analyze mechanical energy of structures along those low-frequency motions. In many respects it is the elastic counterpart of the Go model used in protein-folding simulations.

In the strain energy analysis, we examine how a protein would be strained when it is deformed from its stable structure. Normally, the stable conformation is the one determined by X-ray crystallography. The network definition of the Tirion potential is defined based on this structure, which is then the most stable structure from the definition of the potential (Equation 18.15 and Equation 18.16). Any deformation to the original structure causes increase in the energy; that is, strain.

To examine strain energy quantitatively, the spring constant of the Tirion potential, k, has to be chosen appropriately (note that the normal mode vector does not depend on this parameter but the frequency does). One of the simplest approaches is to adjust it so that the average atomic B-factors from X-ray crystallography and NMA coincide [Tirion 1996; Bahar, Atilgan, and Erman 1997]. It could also be determined from a systematic study of the X-ray crystallography structure database [Kundu et al. 2002]. The B-factor includes not only atomic fluctuation from protein dynamics but also crystal disorder. On the other hand, crystal contact could also affect the B-factor. Thus estimation of the spring constant is not straightforward. There is also an approach to consider a protein as a plastic object [Maragakis and Karplus 2005].

Strain energy analysis can be used to estimate the energetic cost of deforming a protein structure from a stable one to others. In addition, examination of local distribution of the strain energy provides information of effects of conformational fluctuation on the local environment around each of the residues. High strain energy indicates that the local environment of the residue is correlated to the global dynamics of the protein. The strain energy of an atom, i, is defined in Equation 18.17:

$$E_i = \frac{k}{4} \sum_{j}^{|\mathbf{r}_{i,j}|<R_C} \left(|\mathbf{r}_{i,j}| - |\mathbf{r}_{i,j}^0| \right)^2 . \tag{18.17}$$

Strain energy analysis has been performed for large-scale conformational transitions for proteins with known multiple conformational states. Conformational transition from one state to another can be simulated by iterative NMA. We can examine how strain energy distribution changes following the transition, which provides insight into how large-scale conformational transitions induce changes in the local environment [Miyashita, Onuchic, and Wolynes 2003; Tama et al. 2005]. Strain energy analysis has also been applied to small-scale conformational fluctuations of a protein, which are predicted by NMA from an X-ray structure. Such analysis provides information on, when a protein is at the equilibrium, the coupling of the local residues to the global dynamics of the protein [Wong et al. 2004]. Some examples are described in "Applications."

18.3 APPLICATIONS

In the following we discuss the use of NMA for studying biological molecules and while NMA is based on the harmonic approximation of the potential energy surface, it can have a significant impact for structural biology [Ma 2005].

18.3.1 RTB Approach to Study Large Biological Systems

The RTB method reduces the size of the matrix to be diagonalized, and minimizes the computation time, enabling NMA on very large systems. While coarse-graining the protein as a collection of rigid blocks leads to approximate normal modes, it has been shown that the low-frequency normal modes, those responsible for large conformational changes, are well conserved. For example, studies on several proteins have shown that atomic fluctuations (as calculated by an equation over the 30 lowest-frequency normal modes) derived from RTB had a correlation with atomic fluctuation calculated with a full Cartesian approach of at least 0.88 to a maximum of 0.96 in some cases. In addition, it was shown that the conformational change observed experimentally could be reproduced with very high accuracy.

Different levels of coarse-graining can be implemented with the RTB method. Blocks can be made of one to several residues or secondary structure elements such as helices. The size of block does not need to be equal. Table 18.1 shows the comparison of the correlation between atomic fluctuations obtained with the RTB approach and standard approach using different partitioning. Clearly a good correlation with exact calculation and experimentally observed conformational change can be obtained. To obtain a good approximation of the low-frequency normal mode it is better to consider the largest number of blocks that can be handled computationally.

As we mentioned, the coarse-graining introduced with RTB has a limited effect on the eigenvectors that are obtained from NMA, however such approximation affects the frequency of the normal modes. Those frequencies are shifted to higher values, which indicates that the potential energy surface is stiffer in the subspace considered within the framework of the RTB approximation. Interestingly, from frequencies below 40 cm^{-1}, a linear relationship (factor $d_p = 1.7$) between exact and RTB frequency was observed for a set of several proteins. This result suggests that in the

TABLE 18.1
(1) Comparison of the Correlation Between Atomic Fluctuations Obtained With the RTB Approach and Standard Approach. (2) Comparison of Normal Mode Vector Obtained With the RTB Method or the Standard Approach with Experimentally Observed Conformational Change

			Overlap with Conformational Change Mode Number[c]				
Nb of Blocks	**Nb of Residues per Block**	**Correlation with Standard Approach**	**1**	**2**	**3**	**4**	**5**
RTB approach							
218	1	0.93	**0.58**	0.37	0.21	0.23	0.37
109	2	0.93	**0.60**	0.41	0.14	0.26	0.23
73	3	0.92	**0.53**	0.48	0.12	0.29	0.26
44	5	0.88	**0.60**	0.44	0.27	0.02	0.12
40	1–14[a]	0.87					
40	1–14[b]	0.89					
37	6	0.86					
Standard approach: Cartesian diagonalization							
n/a	n/a	1	0.32	0.48	0.25	**0.53**	0.17

[a] Each secondary structure element is put in a block.
[b] Blocks of same length as in [a], but randomly distributed along the polypeptidic chain.
[c] For the definition of overlap, see page 277.

case of low-frequency normal modes, the slight deformations experienced by the residues during these motions are a consequence of local constraints, nearly identical in average. We should note that by increasing the number of residues per block, the linear relationship is lost for frequencies higher than 10 cm^{-1}. d_p also increases as block size increases.

The RTB method has been applied to proteins in varieties of size by using larger blocks for larger systems. In order to study even larger systems, the RTB approach is often combined with the elastic network model to further coarse-grain the system from the full atomic description. As the elastic network model does not require full atomic information, only the C_α atoms of the system can be considered. In such cases, the minimum number of residues per block is three C_α, since three atoms are needed to define the rotation and translation of a block. Using such approaches, systems as large as ribosomes [Tama et al. 2003], myosins [Tama et al. 2005], and viruses [Tama and Brooks 2002, 2005] can be studied in a very short amount of time.

In the following we discuss applications of coarse-grained C_α/RTB NMA to viruses. The large majority of viruses display icosahedral symmetry. This requires 60 structural units (the icosahedral asymmetric unit) to complete a shell. However, very few viruses contain only 60 copies of the capsid protein. Most of the icosahedral viruses display quasi symmetry; that is, they have 60 × T identical subunits in the shell, where T is the triangulation number [Caspar and Klug 1962], which reflects the selection rules for distributing capsomers (hexamers and pentamers) on a surface lattice. Large conformational changes have been observed among viruses displaying different quasi symmetry. Therefore, our studies had a special emphasis on the general relationship between the dynamics and the structure of the virus capsid [Tama and Brooks 2005].

Normal modes were computed using a C_α coarse-grained model, with each protein treated as a rigid block in the RTB approach (Figure 18.1a and Figure 18.1b) for three viral capsids: Cowpea Clorotic Mosaic Virus (CCMV, T = 3), Nudaurelia Capensis Virus (NωV, T = 4), and bacteriophage HK97 (HK 97, T = 7).

For CCMV and NωV, the overall shapes of the two conformational states are rather similar; that is, they both have an apparent round shape [Canady et al. 2000; Speir et al. 1995]. For the T = 7, HK97, virus particle, the conformational change is quite complex [Conway et al. 2001]. The morphology of the virus particles changes upon expansion from an apparent round shape to a more polyhedral shape (Figure 18.1c and Figure 18.1d). Each normal mode $\mathbf{a}_n$ was compared to the experimentally observed conformational change $\Delta\mathbf{r} = \mathbf{r}_1 - \mathbf{r}_2$ ($\mathbf{r}_1$ and $\mathbf{r}_2$ representing the two known conformational states of each virus) by measuring the overlap, $I_n = \mathbf{a}_n \cdot \Delta\mathbf{r} / |\mathbf{a}_n||\Delta\mathbf{r}|$. An overlap value of one indicates that the direction of the conformational change, $\Delta\mathbf{r}$, and the direction given by the normal mode are identical. The contribution of normal modes to the displacement between the two virus states is given by the cumulative sum of the squares of the overlap between the modes and the vector difference. For the three viruses considered in this study, the two lowest-frequency normal modes that adhere to the symmetry of the system can account for more than 90% of the conformational change.

Differences in the mechanical properties of the viral capsids appear in the difference of the modes obtained from NMA. For native CCMV and native NωV, we observe one predominant mode (overlap ~0.95), which accounts for more than 90% of the conformational change. In each case, this mode is the first nondegenerate normal mode (i.e., the motion fully adheres to the icosahedral symmetry of the system). In both case these modes consist in a radial expansion of the capsid. For HK97, the T = 7 quasi symmetry virus, the nature of the modes is quite different from what is observed for CCMV and NωV. The first nondegenerate normal mode has a lower overlap than those observed for the other two viruses. This mode itself describes only 65% of the conformational change, and we observe that the second nondegenerate normal mode still has a quite high overlap (~0.55), which is sufficient in combination with the first one to provide a nearly complete description of the conformational change (more than 90% of the overall expansion and shape transition of the capsid).

Another unique feature of HK97 is that the protein itself is flexible; the E-loop is different between the two virus conformations, while no structural difference is present in the proteins for CCMV and NωV. In the results shown above for HK97, each subunit is considered as a rigid body and the flexibility of the loop is not considered. However, such flexibility has only a small effect for the global mechanical property of the entire capsid. The analysis using RTB in which the loop is allowed to move as an individual rigid body block unit showed no difference in the low-frequency modes that are relevant to the capsid conformational transition. Comparison of the two sets of modes indicates that they are similar by more than 97%. One the other hand, NMA on the single protein of the virus capsid gives a few modes that can accommodate the displacement of the E-loop. These indicate that the internal flexibility of each subunit and the mechanical property of the whole capsid are determined from the structural properties at different scales. Thus, the internal flexibility of the subunit would need to be considered if atomic detail of conformational change is of interest. However, for studying the global mechanical property of the capsid, such subunit flexibility has little effect on the lowest-frequency normal modes.

The conformational change of HK97 involves a unique shape transformation, in particular the icosahedral facet changes to become almost planar. This characteristic transition appears in NMA as the difference in the character of the lowest-frequency mode. For HK97, a pair of low-frequency normal modes is necessary to produce this nonuniform conformational transition, while for NωV and CCMV one single normal mode provides the nearly uniform overall translation associated with the conformational transition. It is surprising that even though the shape of HK97 prohead II is apparently spherical, as with the other viruses, the nature of the normal modes that describe its conformational change is totally different. While the shape of the system is important in predicting its mechanical properties, this study indicates that the underlying network of interactions makes it possible for certain types of viruses to achieve the necessary change in shape.

18.3.2 Strain Energy Analysis

18.3.2.1 The Linear Elastic Model

Adenylate kinase is a system in which a few normal modes successfully describe its conformational transitions. Using strain energy analysis, not only structural insights but also energetics of the conformational transitions can be discussed.

Adenylate kinase is open without substrates and closed with substrates. The low-frequency modes of the open form were determined using the elastic network model. Among the normal modes, the normal mode #1 (the lowest frequency mode) is most relevant, having the highest overlap with the conformational change. By generating some deformed structures following mode #1, a representative conformational transition pathway from the open form to the closed form can be generated.

Calculating the strain energy for each of the structures along the transition pathway, the strain energy surface of the conformational deformation from the open to the closed form was obtained. If the potential of the elastic network model was perfectly harmonic, this would give a quadratic function of the displacement from the open form with a curvature dependent on the normal-mode frequency. However the actual energy surface of the conformational transition agrees with this ideal harmonic description only in the vicinity of the energy minimum structure, being larger than expected from the standard NMA. This is because a single normal mode does not accurately represent the conformational changes during the deformation. The conformational change of adenylate kinase resembles the bending of a rod. Obviously, when bending a rod through a finite angle, a single normal mode only indicates the initial direction of the transformation [Miyashita, Onuchic, and Wolynes 2003].

18.3.2.2 Nonlinear Elastic Models

To overcome this limitation and extend the model to include the bending nonlinearity, iterative NMA can be used. We used the combinations of one or three normal modes that are most relevant to conformational change (high overlap modes). The obtained pathways are only representatives of the possible pathways. Using only low-frequency normal modes for deformation, we generate conformational transitions that resemble the energy minimum pathway of the potential energy surface. For adenylate kinase, following the pathway with only one mode, the LID domain closure precedes the NMP bind domain closure. On the pathway with three modes, both domains close simultaneously. Detailed analysis of conformational transition energy surface shows that the closure of the LID domain is energetically easier than the closure of the NMP bind domain.

18.3.2.3 Strain Energy is Localized

The energy surface of the reactant state was calculated as strain energy along the resulting nonlinear path from the open to closed form. The same procedure was performed from the closed to open form to obtain the energy surface of the ligand bound state. From the intersection of two energy surfaces, the model predicts the activation energy barrier of 20–25 kcal/mol depending on the energy difference of the two states (Figure 18.2). This value is high; higher than the stability of adenylate kinase. Distribution of the strain energy within the molecule needs to be closely examined to explain this puzzling result.

The strain intensity for all residues was determined for each structure along the nonlinear conformational change using Equation 18.17. The analysis shows that the strain in the molecule when it is deformed along the reaction path is not uniform. Figure 18.2 shows the spatial pattern of residue strain as the molecule is deformed. When the RMSD to the open form becomes larger than 3 Å, some residues become particularly highly strained. In particular the T1 region and helix $\alpha 6$, and helix $\alpha 7$ are under high strain. There is a clear correlation between the high strain energy regions along the nonlinear conformational conversion path and what have been called hinges.

The observation that particular contiguous residues are under high strain led us to hypothesize that these special regions may "crack"; that is, unfold partially, during the conformational change. To include this possibility, we considered a cracking model, in which a residue is allowed to unfold if the strain energy of the residue is higher than the difference of local folding energy and the structural entropy to be gained by locally unfolding. As shown in Figure 18.2, partial unfolding gives a lower energy barrier. The extent of cracking strongly depends on the choice of a local stability threshold.

This work provides new insights into the structure–function relationship. Many structural studies assume that the function of a protein occurs entirely within the confines of structures that are completely folded, probably involving rigid body motions of preexisting fragments of structure. However, functional motions of protein do not need to occur smoothly within the folded state itself but may involve "cracking" in which part of the molecule may unfold and refold. Many aspects of motions in biomolecules may resemble macroscopic machines and be described as hinge motions, however, unlike macroscopic machines, the functioning of biological machines may involve catastrophic events such as cracking and subsequent reassembly [Miyashita, Onuchic, and Wolynes 2003].

18.3.3 Flexible Fitting of Atomic Structures into Low-Resolution Electron Density Maps

The third application presented in this chapter describes the use of iterative coarse-grained NMA to interpret low-resolution structural data as obtained from cryoelectron microscopy (cryo-EM). Large biomolecular assemblies are difficult to study by X-ray crystallography and therefore electron

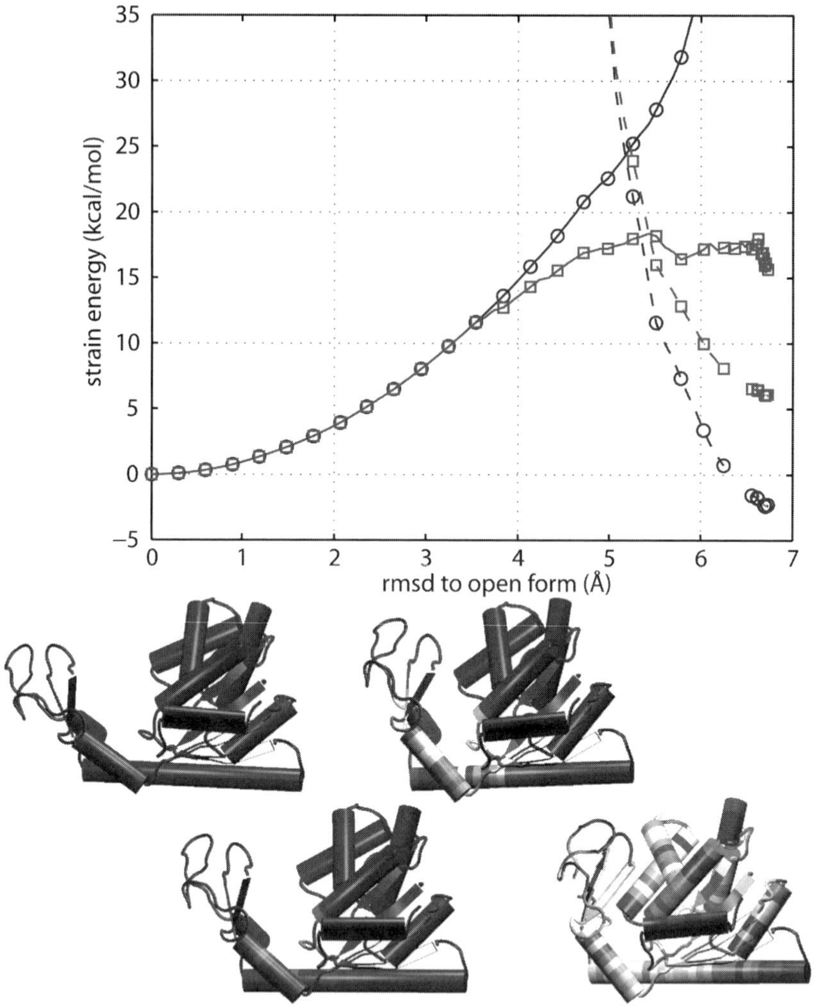

FIGURE 18.2 (See color insert following page 238.) Energy surface of adenylate kinase conformational transition calculated from the nonlinear elastic model. Energies of the ligand-unbound state are shown by solid lines, and the ligand bound state are shown by broken lines. Energy profiles without cracking are marked • and with cracking are marked ⊠. The cracking effect lowers the energy barrier. The energy surface of the ligand bound states with low surface tension has higher energy than without cracking, since some residues are partially unfolded. Actual conformational change would involve a refolding process; that is, shift from the energy profile with cracking to the energy profile without cracking. The structures along the conformational change pathway are also shown (open form, transition state and closed form from left side). The residues are colored according to the strain energy calculated for the ligand-unbound state; that is, the open form has no strain and the closed form has the largest strain; blue corresponding to no strain and red residues high strain. At the transition state, the strain energy is very localized and it could cause cracking or partial unfolding.

microscopy plays an important role, providing a growing number of structures of such large assemblies, although at low resolution. In particular, large conformational changes of macromolecular complexes have been characterized by cryo-EM [Saibil 2000]. Nevertheless, despite the amount of available low-resolution structural information, few theoretical methods have been developed to assist the interpretation of inferred dynamical transitions from known X-ray structures [Wriggers and Birmanns 2001; Fabiola and Chapman 2005; Topf and Sali 2005; Volkmann and Hanein 1999].

The fitting of a known X-ray structure into low-resolution structural data from electron microscopy or other lower-resolution methods can provide a deeper understanding of the low-resolution data. However, since biological systems are highly dynamic, in some cases the conformation of the known X-ray structure does not correspond to the conformation of the low-resolution map. Thus, it is necessary to take into account the flexibility of the system during the fitting. Computationally, low-frequency distortions from NMA can be used as search directions in structure refinement protocols. A protocol following this idea was implemented into the iterative NMA flexible fitting (NMFF) package [Tama, Miyashita, and Brooks 2004a, 2004b]. In the following we describe the procedure.

For the fitting of a high-resolution structure into low-resolution structural data, one wants to maximize the correlation coefficient between the experimental and model map:

$$\mathrm{c.c.} = \frac{\sum_{ijk} \rho^{\mathrm{exp}}(i,j,k)\rho^{\mathrm{sim}}(i,j,k)}{\sqrt{\sum_{ijk} \rho^{\mathrm{exp}}(i,j,k)^2 \sum_{ijk} \rho^{\mathrm{sim}}(i,j,k)^2}}, \tag{18.18}$$

where $\rho^{\mathrm{exp}}(i,j,k)$ is the target experimental density map and $\rho^{\mathrm{sim}}(i,j,k)$ is a given map simulated from the deformed X-ray structure.

To choose the normal mode that increases the correlation coefficient and to determine the modes to be followed, one can calculate the gradient of the correlation coefficient $F_n = \partial \mathrm{c.c.} / \partial q_n$ for each mode $\mathbf{a}_n$. The gradient can be derived analytically. The absolute value of the gradient indicates how each mode will increase the correlation coefficient, while the sign determines the direction that needs to be followed along each mode. The magnitude of the displacement can be set to be proportional to the gradient $q_n = \lambda F_n$ where λ is a parameter that is chosen to make suitable displacements that expedite convergence while minimizing local distortions. This refinement process is equivalent to a steepest descent/ascent optimization of the correlation coefficient in the normal mode coordinate space.

During the maximization process, we also use a Newton-Raphson-type algorithm near a maximum of the correlation coefficient. In this case, one needs to calculate the second derivative, or Hessian, of the correlation coefficient. From this, the approximate best amplitude to displace the structure along each mode is given as $\mathbf{q} = \boldsymbol{V}^{-1}\boldsymbol{F}$ where $\boldsymbol{F} = \{F_k\}$, $\boldsymbol{V} = \{\partial^2 \ \mathrm{c.c.} / \partial q_n \partial q_m\}$ is the Hessian, and $\mathbf{q}$ is the amplitude of displacement.

The NMFF procedure combines steepest ascent and Newton-Raphson techniques to optimize displacements of the X-ray structures along the 20 lowest-frequency normal modes to achieve high correlation between the experimental cryo-EM map and the modeled structure. At each step, the structure is displaced along the five best modes. The procedure is performed until convergence of the correlation coefficient (see Figure 18.3a). Using simulated data of several proteins for which large conformational changes are known, it was shown that NMFF provides an accurate and fast method for the flexible fitting of high-resolution structure into a low-resolution density map determined by electron microscopy [Tama, Miyashita, and Brooks 2004a].

This methodology was then successfully applied to experimental cryo-EM maps with different degrees of complexity [Tama, Miyashita, and Brooks 2004b]. In each case there were experimental data suggesting that a large conformational change was needed to fit the known x-ray structure of the biomolecule or molecular complex into the cryo-EM map. Conformational change of a single protein, elongation factor G (EF-G), bound to the ribosome was elucidated using this method. Figure 18.3b shows that after initial rigid body fitting some regions of the density remain unaccounted for. Figures 18.3c illustrates the structure after flexible fitting, using coarse-grained C_α atom NMA as implemented in the NMFF algorithm. Large rearrangements between domain motions as obtained from NMA lead to a significant improvement of the fit to the density. The NMFF procedure was also shown to be successful in flexibly fitting complex macromolecular protein assemblies such

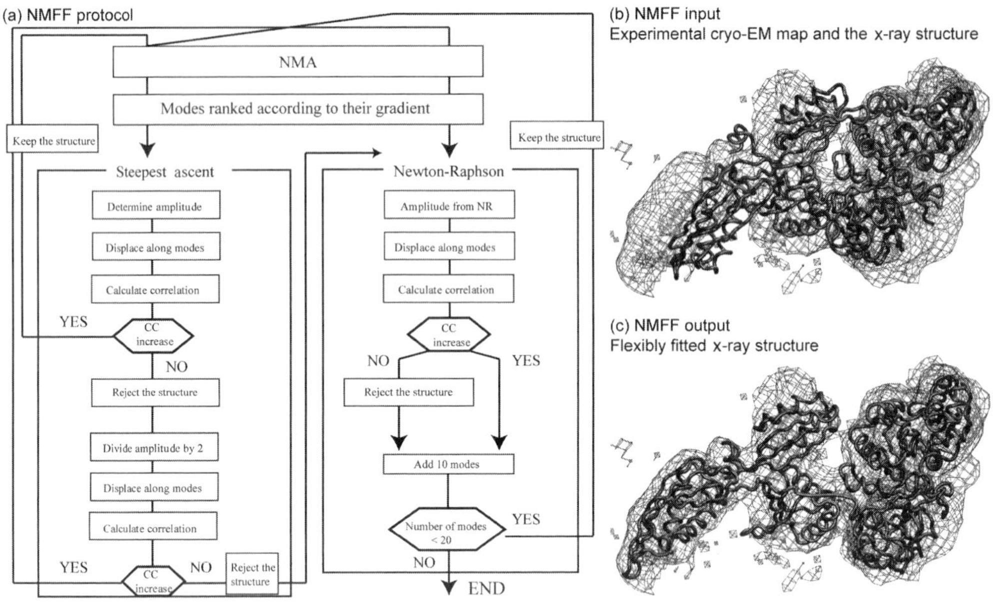

FIGURE 18.3 (a) Schematic diagram of the NMFF protocol. Flexible fitting of EF-G into the experimental cryo-EM map at 11.8 Å resolution: (b) structure rigidly fitted into the experimental EM map, (c) the final flexibly fitted structure using iterative coarse-grained C_α normal mode analysis as implemented in NMFF.

as the *Escherichia coli* RNA polymerase [Darst et al. 2002], anthrax complex [Tama et al. 2006], GroEL [Falke et al. 2005], and viruses [Sherman et al. 2006].

18.4 CONCLUSION

Demand for new computational techniques to study the dynamics of biological systems never end. As experimental techniques for structure determination develop, we gain access to larger and important biological systems. The standard molecular mechanics method requires very large computer resources to be used for those systems and thus its application has been limited. There are two main reasons why the standard molecular mechanics method is time consuming. First, it describes the dynamics of all atoms in the system. Second, it follows the time development of the system with the precision of femtoseconds. For large macromolecules, those are more precise than necessary, and thus, coarse-graining approaches are required.

NMA is an approach to answer the second problem. In this approach, the energy surface is approximated (coarse-grained) as harmonic and the dynamics are described as a set of harmonic oscillators, rather than following the time development of the system by solving the equation of motions. Due to its harmonic approximation, NMA itself cannot fully describe nonlinear motions often seen in biological molecules, however an iterative approach enables us to describe certain nonlinear motions.

NMA also benefits from new coarse-grained models for biological molecules, such as the RTB method and elastic network model. A limitation of NMA applications has been the amount of memory required for the diagonalization process (related to the first problem). New multiscale coarse-grained models enable us to use only a subset of atoms to represent the shape of the molecule, while providing an accurate description of the dynamics. Thus the applicability of NMA is further enhanced to large systems such as viruses and ribosomes. It also speeds up the calculation for regular size systems, enabling us to perform NMA iteratively without significant computational cost, and also to combine with algorithms for fitting electron microscopy data.

As experimental techniques develop, structural information for larger systems at higher resolution will become available. Coarse-grained NMA will continue to be a powerful tool to examine the dynamics of those systems. For those applications, it would be beneficial to have descriptions of large-scale dynamics of the whole molecule in atomic detail; that is, to cover the whole spectrum of motions within the system—from large to small scale. This would require the development of new multiscale models that could connect, within a unique framework, a fine-level atomic description of the motions with a large-scale coarse-grained description.

ACKNOWLEDGMENTS

We are grateful to Charles L. Brooks III, José N. Onuchic, Yves-Henri Sanejouand, and Peter G. Wolynes, who we have had the chance to work with on these projects.

REFERENCES

Bahar, I., A. R. Atilgan, and B. Erman. 1997. Direct evaluation of thermal fluctuations in proteins using a single-parameter harmonic potential. *Fold. Des.* 2 (3): 173–81.

Brooks, B. R., and M. Karplus. 1983. Harmonic dynamics of proteins: Normal mode and fluctuations in bovine pancreatic trypsin inhibitor. *Proc. Natl. Acad. Sci. U.S.A.* 80: 6571–75.

Canady, M. A., M. Tihova, T. N. Hanzlik, J. E. Johnson, and M. Yeager. 2000. Large conformational changes in the maturation of a simple RNA virus, Nudaurelia capensis ω virus (N ω V). *J. Mol. Biol.* 299 (3): 573–84.

Caspar, D. L. D., and A. Klug. 1962. Physical principles in the construction of regular viruses. *Cold Spring Harbor Symp. Quant. Biol.* 27: 1–24.

Conway, J. F., W. R. Wikoff, N. Cheng, R. L. Duda, R. W. Hendrix, J. E. Johnson, and A. C. Steven. 2001. Virus maturation involving large subunit rotations and local refolding. *Science* 292 (5517): 744–48.

Darst, S. A., N. Opalka, P. Chacon, A. Polyakov, C. Richter, G. Y. Zhang, and W. Wriggers. 2002. Conformational flexibility of bacterial RNA polymerase. *Proc. Natl. Acad. Sci. U.S.A.* 99 (7): 4296–4301.

Doruker, P., R. L. Jernigan, and I. Bahar. 2002. Dynamics of large proteins through hierarchical levels of coarse-grained structures. *J. Comput. Chem.* 23 (1): 119–27.

Durand, P., G. Trinquier, and Y. H. Sanejouand. 1994. New approach for determining low-frequency normal-modes in macromolecules. *Biopolymers* 34 (6): 759–71.

Fabiola, F., and M. S. Chapman. 2005. Fitting of high-resolution structures into electron microscopy reconstruction images. *Structure* 13 (3): 389–400.

Falke, S., F. Tama, C. L. Brooks, E. P. Gogol, and M. T. Fisher. 2005. The 13 Å structure of a chaperonin GroEL-protein substrate complex by cryo-electron microscopy. *J. Mol. Biol.* 348 (1): 219–30.

Freddolino, P. L., A. S. Arkhipov, S. B. Larson, A. McPherson, and K. Schulten. 2006. Molecular dynamics simulations of the complete satellite tobacco mosaic virus. *Structure* 14 (3): 437–49.

Go, N., T. Noguti, and T. Nishikawa. 1983. Dynamics of a small globular proteins in terms of low-frequency vibrational modes. *Proc. Natl. Acad. Sci. U.S.A.* 80: 3696–3700.

Goldstein, H. 1950. *Classical Mechanics*. Reading, MA: Addison-Wesley.

Guilbert, C., D. Perahia, and L. Mouawad. 1995. A method to explore transition paths in macromolecules: Applications to hemoglobin and phosphoglycerate kinase. *Comput. Phys. Commun.* 91 (1–3): 263–73.

Hinsen, K. 1998. Analysis of domain motions by approximate normal mode calculations. *Proteins* 33 (3): 417–29.

Karplus, M., and J. A. McCammon. 2002. Molecular dynamics simulations of biomolecules. *Nat. Struct. Biol.* 9 (9): 646–52.

Kundu, S., J. S. Melton, D. C. Sorensen, and G. N. Phillips. 2002. Dynamics of proteins in crystals: Comparison of experiment with simple models. *Biophys. J.* 83 (2): 723–32.

Ma, J. P. 2005. Usefulness and limitations of normal mode analysis in modeling dynamics of biomolecular complexes. *Structure* 13 (3): 373–80.

Maragakis, P., and M. Karplus. 2005. Large amplitude conformational change in proteins explored with a plastic network model: Adenylate kinase. *J. Mol. Biol.* 352 (4): 807–22.

Ming, D., Y. F. Kong, M. A. Lambert, Z. Huang, and J. P. Ma. 2002. How to describe protein motion without amino acid sequence and atomic coordinates. *Proc. Natl. Acad. Sci. U.S.A.* 99 (13): 8620–25.

Miyashita, O., J. N. Onuchic, and P. G. Wolynes. 2003. Nonlinear elasticity, proteinquakes, and the energy landscapes of functional transitions in proteins. *Proc. Natl. Acad. Sci. U.S.A.* 100 (22): 12570–75.

Miyashita, O., P. G. Wolynes, and J. N. Onuchic. 2005. Simple energy landscape model for the kinetics of functional transitions in proteins. *J. Phys. Chem. B* 109 (5): 1959–69.

Mouawad, L., and D. Perahia. 1996. Motions in hemoglobin studied by normal mode analysis and energy minimization: Evidence for the existence of tertiary T-like, quaternary R-like intermediate structures. *J. Mol. Biol.* 258 (2): 393–410.

Saibil, H. R. 2000. Conformational changes studied by cryo-electron microscopy. *Nat. Struct. Biol.* 7 (9): 711–14.

Schlitter, J., M. Engels, P. Kruger, E. Jacoby, and A. Wollmer. 1993. Targeted molecular-dynamics simulation of conformational change: Application to the T–R transition in insulin. *Mol. Simul.* 10 (2–6): 291–308.

Shepherd, C. M., I. A. Borelli, G. Lander, P. Natarajan, V. Siddavanahalli, C. Bajaj, J. E. Johnson, C. L. Brooks, and V. S. Reddy. 2006. VIPERdb: A relational database for structural virology. *Nucleic Acids Res.* 34: D386–89.

Sherman, M. B., R. H. Guenther, F. Tama, T. L. Sit, C. L. Brooks, A. M. Mikhailov, E. V. Orlova, T. S. Baker, and S. A. Lommel. 2006. Removal of divalent cations induces structural transitions in red clover necrotic mosaic virus, revealing a potential mechanism for RNA release. *J. Virol.* 80 (21): 10395–406.

Speir, J. A., S. Munshi, G. J. Wang, T. S. Baker, and J. E. Johnson. 1995. Structures of the native and swollen forms of cowpea chlorotic mottle virus determined by X-ray crystallography and cryoelectron microscopy. *Structure* 3 (1): 63–78.

Tama, F., and C. L. Brooks, III. 2002. The mechanism and pathway of pH induced swelling in cowpea chlorotic mottle virus. *J. Mol. Biol.* 318: 733–47.

Tama, F., and C. L. Brooks III. 2005. Diversity and identity of mechanical properties of icosahedral viral capsids studied with elastic network normal mode analysis. *J. Mol. Biol.* 345 (2): 299–314.

Tama, F., M. Feig, J. Liu, C. L. Brooks III, and K. A. Taylor. 2005. The requirement for mechanical coupling between head and S2 domains in smooth muscle myosin ATPase regulation and its implications for dimeric motor function. *J. Mol. Biol.* 345 (4): 837–54.

Tama, F., F. X. Gadea, O. Marques, and Y. H. Sanejouand. 2000. Building-block approach for determining low-frequency normal modes of macromolecules. *Proteins: Struct. Funct. Genet.* 41 (1): 1–7.

Tama, F., O. Miyashita, and C. L. Brooks III. 2004a. Flexible multi-scale fitting of atomic structures into low-resolution electron density maps with elastic network normal mode analysis. *J. Mol. Biol.* 337: 985–99.

Tama, F., O. Miyashita, and C. L. Brooks III. 2004b. NMFF: Flexible high-resolution annotation of low-resolution experimental data from cryo-EM maps using normal mode analysis. *J. Struct. Biol.* 147: 315–26.

Tama, F., G. Ren, C. L. Brooks, 3rd, and A. K. Mitra. 2006. Model of the toxic complex of anthrax: Responsive conformational changes in both the lethal factor and the protective antigen heptamer. *Protein Sci.* 15 (9): 2190–2200.

Tama, F., and Y. H. Sanejouand. 2001. Conformational change of proteins arising from normal mode calculations. *Protein Eng.* 14 (1): 1–6.

Tama, F., M. Valle, J. Frank, and C. L. Brooks, III. 2003. Dynamic reorganization of the functionally active ribosome explored by normal mode analysis and cryo-electron microscopy. *Proc. Natl. Acad. Sci. U.S.A.* 100 (16): 9319–23.

Tama, F., W. Wriggers, and C. L. Brooks, III. 2002. Exploring global distortions of biological macromolecules and assemblies from low-resolution structural information and elastic network theory. *J. Mol. Biol.* 321 (2): 297–305.

Tirion, M. M. 1996. Large amplitude elastic motions in proteins from a single-parameter, atomic analysis. *Phys. Rev. Lett.* 77 (9): 1905–1908.

Topf, M., and A. Sali. 2005. Combining electron microscopy and comparative protein structure modeling. *Curr. Opin. Struct. Biol.* 15 (5): 578–85.

Tozzini, V. 2005. Coarse-grained models for proteins. *Curr. Opin. Struct. Biol.* 15 (2): 144–50.

Volkmann, N., and D. Hanein. 1999. Quantitative fitting of atomic models into observed densities derived by electron microscopy. *J. Struct. Biol.* 125 (2–3): 176–84.

Wong, L., S. Lieser, B. Chie-Leon, O. Miyashita, B. Aubol, J. Shaffer, J. N. Onuchic, P. A. Jennings, V. L. Woods, and J. A. Adams. 2004. Dynamic coupling between the SH2 domain and active site of the COOH terminal Src kinase, Csk. *J. Mol. Biol.* 341 (1): 93–106.

Wriggers, W., and S. Birmanns. 2001. Using Situs for flexible and rigid-body fitting of multiresolution single-molecule data. *J. Struct. Biol.* 133 (2–3): 193–202.

Xu, C. Y., D. Tobi, and I. Bahar. 2003. Allosteric changes in protein structure computed by a simple mechanical model: Hemoglobin T<->R2 transition. *J. Mol. Biol.* 333 (1): 153–68.

19 One-Bead Coarse-Grained Models for Proteins

Valentina Tozzini
NEST-CNR-INFM Scuola Normale Superiore

J. Andrew McCammon
Department of Chemistry and Biochemistry, Center for Theoretical Biological Physics, Howard Hughes Medical Institute, University of California

CONTENTS

19.1 INTRODUCTION

Most biomolecular processes involve macromolecular aggregates (on the size scale of 10–100 nm or more, including the cell membranes) and occur on a time scale of microseconds to milliseconds (or even hours to days, including folding and amyloid aggregation, for instance). Computer simulations based on atomic force fields are not yet able to reach these scales, since they are currently restricted in most cases to systems comprising fewer than a million atoms for times of less than 1 μs. In consideration of these facts, the idea of simplifying the description of a macromolecular system by including groups of atoms in a single interaction center (coarse-graining, CG) in order to reduce the number of internal degrees of freedom and, with them, the computational cost, is quite simple and somewhat natural, even considering the hierarchical structural organization of the proteins and nucleic acids.

CG can be done at many different levels.[1] In general, the coarser the description, the larger the saving in computational cost. The simulation speed-up occurs not only because the degrees of freedom are fewer, but also because the highest vibrational frequencies of the system—usually pertaining to the eliminated local degrees of freedom—are eliminated, allowing increases in the time step of the simulation. This elimination of degrees of freedom typically causes the potential energy surface of the system to be smoothed, so that the simulated dynamics is unphysically accelerated, because the system does not lose time in exploring many local minima. This allows a more efficient and faster phase space sampling, an additional acceleration factor that can be useful when only the thermodynamics is being studied, but must be corrected when a realistic dynamics of the system is under evaluation. This aspect is addressed in detail elsewhere in this book (Chapter 3) and in Ref. 2. Here this issue will be marginally addressed in the context of implicit representation of the solvent, which can be considered an extreme example of CG.

In this chapter, we focus on the so-called one-bead models for proteins, representing a whole amino acid with a single interaction center (bead). These are sometimes called minimalist models for proteins, although in the literature a vast variety of coarser models are available (see Ref. 3 and elsewhere in this book), even including multiscale approaches.[4] There are several advantages in using one-bead models. First, among all the CG models, they are the coarsest models for which the internal conformational variables needed to describe changes in the secondary structure are still definable. The subsequent levels of coarsening (e.g., n amino acids or a whole helix or strand represented as a single rigid body) must have the secondary structure already built in. This is an important issue, because many important biological processes occurring in or between proteins involve a transition in the secondary structure as the triggering step.[5] Conversely, with respect to the less coarse models, the one-bead simplification allows a significant saving of computer time and a rather simple implementation of the model, but still retains the possibility of including amino-acid-dependent terms that can allow a certain degree of predictivity. Additionally, the one-bead CG level corresponds to the resolution of the structural data obtained with electron cryomicroscopy, a technique widely used to resolve the structure of large biomolecular aggregates. Actually, this experimental technique and the one-bead CG models can take reciprocal advantage and are often used synergistically (see Chapters 17 and 18 of this volume and Refs. 6 and 7).

The disadvantages with respect to less coarse CG models (in particular four- and six-bead models) are mainly related to the accuracy in the description of the local interactions. While isotropic medium- and long-range interactions (hydrophobicity and electrostatics) are relatively easy to parameterize, it turns out to be quite difficult to include highly specific and strongly directional interactions (such as hydrogen bonds, and side-chain conformational flexibility effects) in the very few parameters available in the one-bead representation in a way that is both accurate and predictive.

Attention will be devoted to the different strategies adopted for the parameterization. These characterize the several different one-bead models that were made available in the literature starting from the 1970s, and determine their level of accuracy-predictivity. It turns out that, generally, predictivity (transferability) and accuracy are competing factors in one-bead models, unless one accepts extremely complex parameterizations. A possible route to the ultimate goal of optimizing the predictivity-accuracy/simplicity balance is indicated in this chapter, and intermediate steps to reach this goal are reported.

19.2 METHODS

19.2.1 The All-Atom to One-Bead Mapping

The backbone conformation description in one-bead models differs substantially from that in the all-atom models: once the *trans* or *cis* conformation of the peptide bond is fixed (see Figure 19.1a, $\omega = 180°$ or $0°$), in the all-atom representation the backbone conformation is entirely determined by the dihedral conformational angles (ϕ, ψ). Conversely, in the CG representation these internal

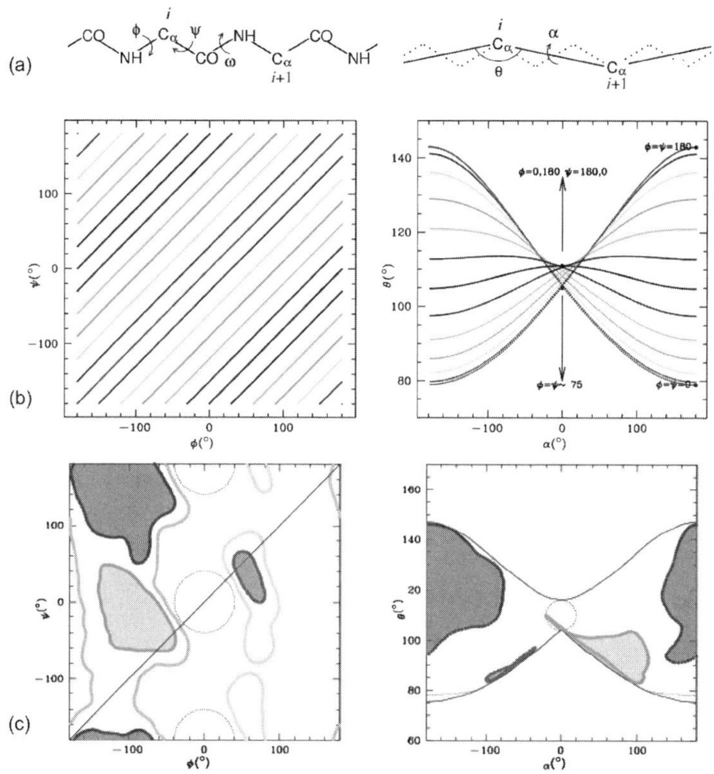

FIGURE 19.1 The all-atom to one-bead mapping (for the *trans* peptide bond configuration, which is understood in the discussion). (a) All-atom (left) and CG (right) representations of the protein backbone. (b) Schematic representation of the $(\phi, \psi) \rightarrow (\alpha, \theta)$ mapping. The parallel lines in the (ϕ, ψ) plane are mapped into lines in the (α, θ) plane with corresponding gray shades. (c) Mapping of the generic Ramachandran plot into the (α, θ) plot. Color coding: light gray filled region: right-handed helices; dark gray border with dark gray filling: strands or extended structures; medium gray with medium gray border: left-handed helices.

variables are no longer explicitly represented, and the backbone conformation is determined by the (θ, α) couple, the pseudobond angle between three subsequent C_α, and the pseudodihedral between four subsequent C_α (see Figure 19.1a). The C_α–C_α is a pseudobond because it does not correspond to any real chemical bond, but rather crosses the peptide bond. Its equilibrium length depends on the *cis* or *trans* conformation of the peptide bond (here we restrict the discussion to the *trans* case).

A mapping between the (ϕ, ψ) and the (θ, α) variables can be analytically derived,[8] as is illustrated in Figure 19.1b: the whole (ϕ, ψ) plane is mapped onto a butterfly-shaped region of the (θ, α) plane. The mapping is not one-to-one: corresponding regions of the upper triangular and lower triangular half of the (ϕ, ψ) plane are mapped onto the same (θ, α) region. In spite of this, the typical secondary structure regions that are separated in the Ramachandran plot remain separate also in the (θ, α) plot (see Figure 19.1c), indicating that (θ, α) still distinguishes and describes the different secondary structures. The (θ, α) plot contains the same information as the Ramachandran plot: the shaded regions indicate combinations of the internal variables whose probability is higher. These depend on the secondary structure: the right-handed helices are characterized by $\theta \approx 90°$ and $\alpha \approx 40$–$90°$; left-handed helices have the same value of θ and opposite values of α; the strand-extended regions have $\theta \approx 120$–$150°$ and $\alpha \approx 140$ to $-140°$ (crossing 180°). The capability of any one-bead model to reproduce the secondary structures can be validated on this plot, as the all-atom models of proteins are often validated on the Ramachandran plot.

19.2.2 The Parameterization and a First Classification of the One-Bead Models

The force field expression for a generic one-bead model can be written as

$$U = U_{\mathrm{b}} + U_{\theta\alpha} + U_{\mathrm{nb}} \tag{19.1}$$

that is, the sum of a (pseudo)bond term, a conformational term, and a nonbonded interaction term. Different one-bead models differ in the functional form of the terms and in the strategy used for their parameterization.

19.2.2.1 The Biased Models

The simplest one-bead models described in the literature are the harmonic networks and the Go models (see also Chapters 11, 17 and 18 of this volume for a detailed description and Ref. 1 for a recent review). In the harmonic network models, U_{b} is a constraint or an elastic potential maintaining the C_αs at the known equilibrium distance for the *trans* peptide bond configuration (3.8 Å), the conformational term $U_{\theta\alpha}$ is absent, and $U_{\mathrm{nb}} = \sum_{i \neq j} 1/2k(r_{ij} - r_{0,ij})^2$ is a sum of elastic terms between all the *i,j* couples whose distance in a reference structure is less than a cutoff distance r_{c}; $r_{0,ij}$ is taken from the reference structure, and k and r_{c} are fitted to the experimental thermal fluctuations. In spite of their extreme simplicity, the network models are able to reproduce correctly the fundamental slow motions of large systems. In fact, these are mainly determined by the topology and 3D shape of the structure that is an input of the model. Furthermore, they are extremely adaptable: they can be applied to different levels of CG and are amenable to a multiscale approach. Generalizations of the network models involve using distance-dependent elastic constants $k(r_{0,ij})$[4] or amino-acid-type dependent elastic constants,[9] and/or anharmonic terms for U_{nb} (Morse or Lennard-Jones-like potentials), and parameterizing the force constants through more sophisticated methods (fluctuation matching to all-atom simulations,[4] or the Boltzmann inversion, described further on). These add some transferability and predictivity to the model for the fluctuations and vibrational modes, but still the bias towards the reference structure limits the exploration of the conformational space to the local minimum.

The Go-like models, specifically tailored for protein folding, are formally similar to the network models. U_{b} is usually a constraint or a harmonic term. When present, the conformational potential is decoupled in the bond angle and dihedral part

$$U_{\theta\alpha} = U_{\theta} + U_{\alpha}, \tag{19.2}$$

where U_θ is harmonic and U_α is a cosine sum, both biased towards the reference structure, which in this case is the folded native structure. U_{nb} is also completely biased: the interactions within r_{c} (called the native contacts) are represented as Lennard-Jones terms with equilibrium distances taken from the native structure, while the others are represented with repulsive terms. The main difference between Go-like models and the anharmonic networks is in the philosophy of the parameterization: in Go models the force constants (k_{b}, k_θ, k_α, $\varepsilon_{\mathrm{native}}$, $\varepsilon_{\mathrm{nonnative}}$) are tuned to match thermodynamic properties (folding temperature, free-energy barriers). These models reproduce a minimally frustrated folding: the folding funnel, representing the free-energy landscape of the protein during folding, is smooth in this model, while it is more rugged with local minima in the real protein or in standard all-atom representation. Nevertheless, interesting information about the folding intermediates can be obtained.

Generalization of the Go-like models involves increasing frustration in different ways: using *i,j*-dependent $\varepsilon_{\mathrm{native}}$ values fitted on a set of native structures of similar proteins (e.g., mutants),[10] using different levels of coarsening, or using more sophisticated U_{nb} to include desolvation effects.[11]

Recently, some steps towards an increasing predictivity/transferability were taken, by using amino-acid-type dependent (and structure-independent) ε_{native} and r_0 parameterized on statistical sets of experimental structures,[12] while maintaining the bias in the conformational term. More details are given in Chapter 11.

19.2.2.2 Unbiased Models with Empirical Parameterization

Apart from some very recent generalizations of Go models, the parameterization of network and Go models is characterized by a strong bias towards a reference structure. This, on the one hand, ensures good accuracy of the structure and its correct topology, which, as we have seen, already determines many of the most important properties of a protein. But on the other hand, this largely limits the predictivity of the model: no ab initio prediction can be made on folding if the folded structure is unknown, or more generally on structures far from the reference structure. For this, a model that is independent of any reference structure would be necessary.

A paradigmatic one-bead model for such an approach is that of Sorenson and Head-Gordon.[13] The functional form is given by Equation 19.1 and Equation 19.2 with

$$U_\theta = \sum_{\text{angles}} \frac{1}{2} k_\theta (\theta - \theta_0)^2,$$

$$U_\alpha = \sum_{\text{dihedrals}} A[1+\cos\alpha] + B[1-\cos\alpha] + C[1+\cos 3\alpha] + D[1+\cos(\alpha + \pi/4)],$$

$$U_{\text{nb}} = \sum_{\text{nonbonded}} 4\varepsilon_H S_1[(\sigma/r)^{12} - S_2(\sigma/r)^6]. \qquad (19.3)$$

σ and ε_H are the units of length and energy of the system, respectively; θ_0 is fixed at an intermediate value between that typical of helices and that for sheets (105°) and k_θ is fixed at $20\varepsilon_H$ independent of the sequence. Some a priori knowledge on the secondary structure is introduced in U_α, where the parameters A, B, C, and D assume different values depending on the secondary structure (helix, strand, turn). However, this term could also be parameterized based on sequence, assuming the known secondary structure propensities for each amino acid. U_{nb}, conversely, is a directly sequence-based term: S_1 and S_2 assume values for a repulsive or attractive U_{nb} depending on the classes to which the two amino acids in interaction belong. These classes can be hydrophobic, hydrophilic, or neutral: interactions between hydrophobic amino acids are attractive, while all the others are repulsive with different strengths. Similar models were previously proposed, including separate helix stabilization, electrostatics, and solvent effects terms.[14,15]

This model has, in principle, a high degree of transferability-predictivity, since no knowledge is necessary on the structure of the protein. However, the predicted structures are not expected to be very accurate. This is mainly due to two intrinsic features of the model: U_θ is independent of the secondary structure or amino acid type, and U_{nb} does not contain amino-acid-specific terms to reproduce the side-chain shape effect and the hydrogen-bond propensity. Both of these facts contribute to a poor description of the local structure. Conversely, all the responsibility of reproducing a correct local environment is imputed to the four parameters of U_α.

Possible ways to circumvent these problems are to adopt multiple beads for the side chain, and/or for the backbone, and/or more complex parametric forms including angle-dependent and correlated conformational terms. These routes are explored elsewhere in this volume (see Chapters 8 and 10) and usually lead to large increases in the number of parameters and, in general, to considerable increases in the complexity of the model.

19.2.2.3 Statistics-Based Parameterization: The Boltzmann Inversion and the Thermodynamic Consistency

A possible way to improve the accuracy while preserving the predictivity and simplicity is to statistically optimize the relatively few parameters present in the one-bead potential. This involves (i) considering an amino-acid-specific parameterization, (ii) choosing appropriate functional forms for the terms, and (ii) fitting them based on some objective criterion, instead of using empirical, though reasonable, rules.

One way is to use the so-called Boltzmann inversion method, based on the definition of the free energy and its internal-variable-dependent version:

$$F = -kT \ln \int dQ_1 \cdots dQ_N \exp[-U(Q_1, \ldots, Q_N)/kT] = -kT \ln[Z],$$

$$F(Q_1) = -kT \ln \int dQ_2 \cdots dQ_N \exp[-U(Q_1, Q_2, \ldots, Q_N)/kT] = -kT \ln[P_J(Q_J)/P_0],$$

where Z is the partition function, $U(Q_1, \ldots, Q_N)$ is... the potential energy, $P_J(Q_J)$ is the probability distribution of the internal variable Q_1 and P_0 is a normalization factor that is equal to the inverse of Z. If the internal variables can be chosen in such a way that the potential is completely and exactly separable in a sum of single variable terms $U(Q_1, \ldots, Q_N) = U_1(Q_1) + \cdots + U_N(Q_N)$, then $F(Q_1)$ coincides with $U_1(Q_1)$ for any internal variable. Otherwise the relationship

$$W_I(Q_I) = -kT \ln[P_I(Q_I)/P_0] \tag{19.4}$$

is called the Boltzmann inversion, and defines the so-called potential of mean force (PMF). At this level, it is important to mention the thermodynamic consistency: the free energy of the system should be the same whether evaluated on the all-atom or CG set of internal variables. Given the above definitions, this is already included in the Boltzmann inversion method: if $P_I(Q_I)$ is accurately evaluated, $W_I(Q_I)$ will be the thermodynamically consistent PMF of the internal variable Q_I, which can be chosen to be any variable, for instance a CG variable. As we have seen, $W_I(Q_I)$ and the "true" CG potential term $U_I(Q_I)$ coincide only if there is no correlation between the internal variables, otherwise $W_I(Q_I)$ can only be considered a rough approximation for $U_I(Q_I)$. It was shown by several authors (see for instance Refs. 2 and 16 and Chapter 3) that $W_I(Q_I)$ is generally more structured and harder than $U_I(Q_I)$, because of the influence of the other correlated degrees of freedom. Thus in general $U_I(Q_I)$ could be evaluated with an iterative procedure consisting in numerically evaluating $W_I(Q_I)$ on a statistical set of configurations of the system, fitting the parameters of $U_I(Q_I)$ on it, reevaluating $P_I(Q_I)$ and $W_I(Q_I)$ on a simulation performed with $U_I(Q_I)$, and repeating the iteration.[16] At convergence, the probability distribution functions of the CG and all-atom reference system match (*distribution function matching*). It was also shown[2] that this procedure is especially needed for the nonbonded potential terms, while for the bond, bond angle, and dihedral terms the direct Boltzmann inverse is already a rather good approximation of the "true" CG potential.

The concepts of PMF and Boltzmann inversion are illustrated in Figure 19.2, on the internal variables θ and α: the probability distribution of the two variables (basically, the projections of the (θ, α) map onto the θ and α axes) for glycine are built on a set of diverse proteins, and then Boltzmann inverted. $W(\alpha)$ has a multiwell structure that is quite similar to that proposed by the Head-Gordon model. $P(\alpha)$ distributions for other amino acids have the maxima at $\pm 60°$ more or less predominant depending on the helical propensity, and there can be a more or less strong asymmetry introduced by the amino-acid chirality. $W(\theta)$ has a deeper minimum at $\sim 90°$, corresponding to the helical structures and a shallower minimum at ~ 130–$140°$, corresponding to the extended

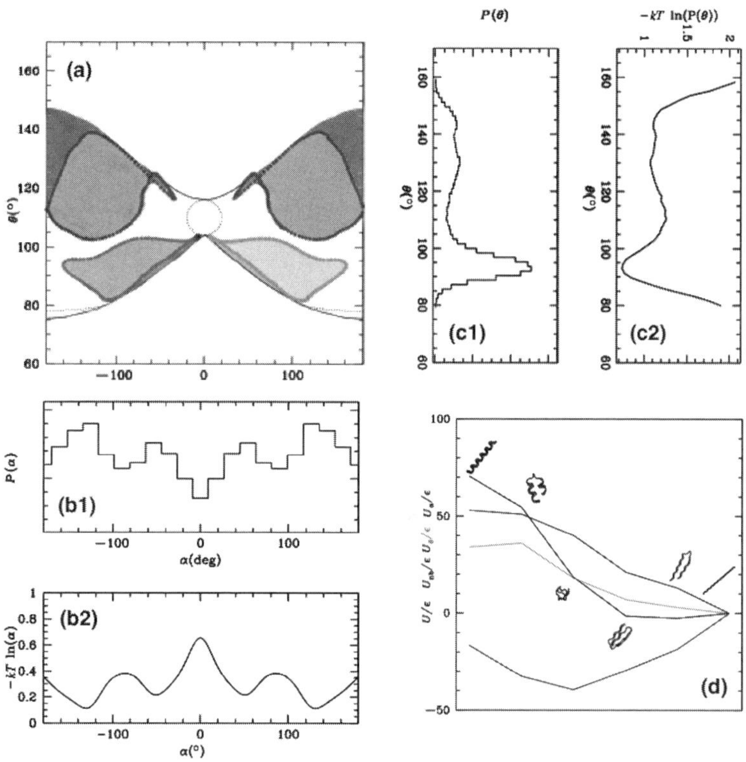

FIGURE 19.2 (a) The (θ, α) map for glycine is reported with the same gray-shade code as in Figure 19.1. In this case the map is symmetric with respect to $\alpha = 0$ because the glycine is not chiral. The graphs represent the probability distributions $P(\theta)$ and $P(\alpha)$ (c1 and b1) and their corresponding potential of mean force obtained by Boltzmann inversion (c2 and b2). (d) A plot of the relative values of the internal energy terms in the different secondary structures, as represented.

structures. Probability distributions for other amino acids have the helical minimum basically unvaried, while the location of the extended structure's minimum (and its relative energy) depends on the amino acid.[17]

It is important to observe that, while the $U(\alpha)$ term is usually quite accurately represented in the CG models, neither the Head-Gordon potential nor any of the previously mentioned one-bead models include a double peak $U(\theta)$: all of them use a single-well harmonic $U(\theta)$ with equilibrium position located somewhere in between the two peaks. Conversely, a double-well quartic potential was included in Ref. 17 and in Ref. 8, where it was shown to be important to reproduce the correct local geometry and dynamical behavior. Amino-acid-dependent $U(\theta)$ can be parameterized[17] based on the amino-acid-dependent probability distributions and on the amino-acid helical propensities, and the secondary structure information can be included into it more easily than in the dihedral term. It can be shown that,[8] properly tuning the parameters of the $U_{\theta\alpha}$ (helix versus extended structures relative energy) with respect to the nonbonded interaction energy, one can stabilize one or the other of the main secondary structures (helix, strand, random coil, see Figure 19.2d). These secondary structure tendencies could be included in an amino-acid-specific way into the potential terms, to have a model that can predict the secondary structure based on the sequence.

The accuracy of the parameterization derived with this method is also determined by the statistical set chosen.[18] If this is a set of experimental structures (usually a set of diverse proteins), the parameters might be affected by crystal-packing artifacts and do not explore all the possible conformations of the molecules. On the other hand, if the statistical set is extracted by an all-atom simulation, the crystallographic constraints are removed but the structures are not diverse (usually they are extracted from simulations on the same proteins) and in any case a limited region of the phase space is explored.

19.2.2.4 The Force-Matching Method: The Mechanical Consistency

An alternative method to derive a coarse-grained potential is the force-matching method. This is described in detail for instance in Ref. 19 and also in this book (Chapter 3). It consists in matching the forces of the CG model to those from all-atom simulation trajectories via a least-square fitting procedure. Numerical CG potentials are derived, which can be subsequently fitted with analytical functions. This method ensures high accuracy and the mechanical consistency of the CG model by definition, and it can clearly be used in multiscale approaches. It was recently applied to 2–4-bead models for polyalanine[2] (only the nonbonded interaction part, while the $U_{\theta\alpha}$ was derived with the Boltzmann inversion). This method has the positive side effect of directly showing the difference between $U(r)$ and the Boltzmann inverse of the pair distribution function $g(r)$, the PMF $W(r)$; as already pointed out, the PMF has more and more pronounced wells than $U(r)$. However, $U(r)$ has a rather complex form that is not particularly prone to a simple analytical parameterization. The main drawback of the force-matching method is that it is very demanding from a computational point of view. Additionally, there is still a dependence of the $U(r)$ evaluation on the conformational space sampling that is limited by the length of the all-atom simulation used for the fit. The transferability/predictivity of potentials derived based on these methods critically depend on which conformations are explored during the all-atom simulation.

19.2.3 The Parameterization of the Nonbonded Interactions: A Critical Issue

Whatever the method to obtain the effective potential is, the parameterization of the nonbonded interactions for one-bead models remains the most critical issue for CG models. The reason is basically due to the fact that in the nonbonded interaction potential for one-bead models several effects different in nature must be included. Consider the C_α–C_α pair distribution function for a generic set of proteins reported in Figure 19.3. The sharp peak at 3.8 Å corresponds to the 1–2 C_α–C_α pseudobond distance, already included in the U_b term, thus it must be subtracted to obtain the nonbonded distribution. A similar adjustment is made for the 1–3 and 1–4 C_α–C_α terms (already included in $U_{\theta\alpha}$). What is left is the $g(r)$ for the nonbonded interactions (gray histogram, Figure 19.3a). Once it is inverted to obtain the

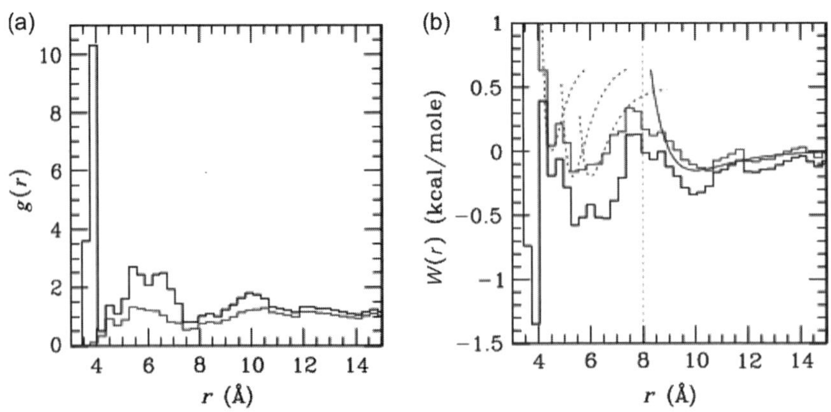

FIGURE 19.3 (a) The pair distribution function of a generic set of proteins (black histogram) and its nonbonded part (gray). (b) The potential of mean force (black histogram) and its nonbonded part (gray histogram). Dotted and continuous gray lines are Morse fitting functions.

nonbonded PMF, two minima are clearly visible that are separated by a barrier at about 8 Å. Noticeably, these are present also in the $U_{nb}(r)$ obtained with Boltzmann procedure iteration or with force matching, thus they are not an artifact of $W(r)$. The first minimum < 8 Å is very structured and includes short-range nonbonded interactions; that is, hydrogen bonds and Van der Waals packing effects. It is important to accurately reproduce this part, because it mainly determines the secondary structure of the protein. However, it is extremely difficult to parameterize with an isotropic—though possibly amino-acid-type dependent—potential, due to the intrinsic directionality of the hydrogen bond and to the fact that the shape of the amino acid is not isotropic itself. This is why this part of the potential looks more like a sum of wells. The second minimum > 8 Å is smoother and could be approximated with an amino-acid-dependent potential. This includes mainly hydrophobicity and electrostatics.

This problem of the "double-nature" of the nonbonded interactions in CG models can only be partially solved adding a separate bead for the side chain and additional beads for the backbone, as in the multiple bead models the double-well structure of the $U_{nb}(r)$ are still present,[2] although less pronounced. Thus, remaining within the one-bead model, a possibility is to parameterize more accurate amino-acid-type dependent $U_{nb}(r)$, separating it into local and nonlocal parts:

$$U_{nb}(r) = U_{nb\text{-}loc}(r) + U_{nb\text{-}nonloc}(r), \tag{19.5}$$

the first including anisotropic potentials and the second including isotropic terms. Some very preliminary steps towards this completely unbiased and accurate/predictive model were made in Ref. 8. The minimal polypeptide model in Ref. 8 shows secondary structure transitions and quite accurate structures, but only for oligopeptides, and an amino-acid-based optimized parameterization is in progress.

A possible way to get around this difficulty is to preserve a local bias in the model, as will be shown in detail in the "Results" section of this chapter.

19.2.3.1 The Inclusion of Electrostatics: Solvent Effects

In the parameterization of $U_{nb\text{-}nonloc}(r)$ it should be borne in mind that in order to maintain the advantage of saving computational cost, the solvent is treated implicitly. This term, however, is in principle easier to parameterize, since it can be treated as isotropic, and basically includes only two effects: the hydrophobicity and the electrostatics. The two effects cannot be separated if the parameterization is based on the $g(r)$, however the first one can be conveniently represented as a Morse-like potential whose parameters depend on the hydrophobicity value of the amino-acid-interacting couple. This problem was addressed by several authors (see for instance Ref. 20). Conversely, a proper treatment of electrostatics based on the Boltzmann inversion is difficult, because it involves inverting $g(r)$ at medium and long range, where it is less well determined. An attractive direction of work in the future will be the development of hybrid models that combine the one-bead CG elements with implicit solvation models that have been developed for arbitrary types of molecular dynamics and molecular modeling simulations.[21] For long-ranged electrostatic interactions, these include new developments that use finite element or boundary element methods to solve the Poisson–Boltzmann equation very efficiently.[22] Apolar interactions have traditionally been accounted for separately from the electrostatic interactions, by means of effective surface tension models.[21] Newer solvation models are becoming available that allow for a more accurate, coupled treatment of the polar and apolar interactions.[23,24]

To conclude this section, in Figure 19.4 a schematic representation of the classification of the mentioned one-bead models for proteins is presented (highlighted in dark gray). The models are placed according to their predictivity-transferability and accuracy. As already remarked, the more transferable models are also in general less accurate in reproducing local structures. While the ultimate goal in one-bead model parameterization is to have good accuracy together with high predictivity, in the following section a compromise model will be presented, having both good

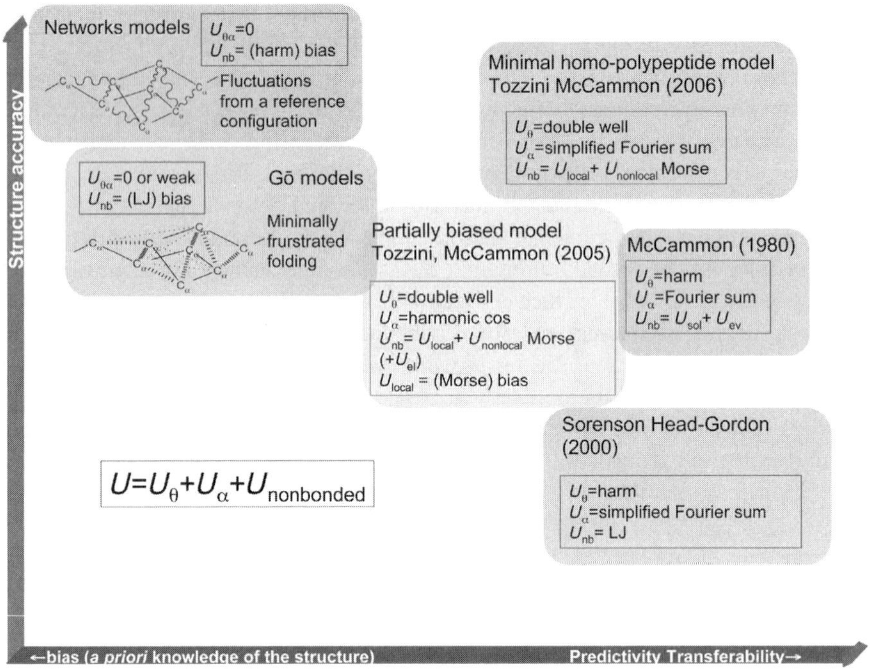

FIGURE 19.4 A qualitative predictivity-versus-accuracy diagram of the one-bead models. The main characteristics of the models are given.

accuracy and predictivity, which can be considered an intermediate temporary step on the path toward developing a truly unbiased model for proteins

19.3 RESULTS: INTERMEDIATE STEPS TOWARDS A COMPLETELY UNBIASED ONE-BEAD MODEL

As we have seen in the previous paragraph, the main problem in the parameterization of the one-bead models is the representation of local nonbonded interactions, which are various in nature, highly nonisotropic, and very specific. Always having in mind that the final goal should be a sequence-based parameterization of these interactions, we report a possible way to get around this problem. The idea is to keep a local bias towards a known reference structure.[17,25] In practice, the nonbonded potential is separated into two parts, as in Equation 19.5, the cutoff between the two being conveniently located at $\sim$8 Å, and both of them are represented by Morse potentials $v_M(r)=\varepsilon\{[\exp(-\alpha(r-r_0))-1]^2-1\}$. The $U_{\text{nb-loc}}(r)$ bears a bias: $r_0=r_{0,ij}$, where $r_{0,ij}$ is taken from a reference structure. Additionally, the dissociation energy is made exponentially decreasing with the equilibrium distance, $\varepsilon(r_{0,ij})=A\exp(-\lambda r_{0,ij})$, to account for the decreasing of the bond strength as the equilibrium distance increases. $U_{\text{nb-nonloc}}(r)$ has the same functional form but r_0 and ε are independent of the reference structure, and matched with those of the local part. The parameters A and λ were determined based on the iterative Boltzmann inversion technique. The conformational terms (double well for the bond angle and a simplified cosine form for the dihedral) are unbiased and parameterized by amino-acid type.

This model was applied to the HIV-1 protease, a key enzyme in the HIV replication cycle. HIV-1pr binds to the viral poly-proteins and cleaves them into functional pieces. The enzyme mechanism is thought to involve the opening of two beta-hairpin structures that protect the active site, called the flaps, but the opening frequency is on the micro-millisecond time scale. Thus this was a good test case, since the multi-microsecond time scale is feasible with one-bead models, while not reachable with all-atom simulations. The presence of the local bias towards the crystallographic apo-HIV-1-pr structure maintains a very good accuracy of the secondary structure of the protein. However, this bias has also proven to be weak enough to ensure the possibility of large fluctuations from the reference structure: the protease flap can open, leaving the active site completely exposed to the solvent (see Figure 19.5). Indeed, the comparison of the simulation with a recently crystallized semiopen structure shows very good accuracy (see Figure 19.5a). Since no bias towards this structure was included in the model, this is an a posteriori indicator of the predictivity of the model. Additionally, the flap opening was allowed by the particularly accurate form of the double-well bond angle term: the flap tip needs to curl for the flap to open. This local conformational transition involves three subsequent C_αs and it does not occur with a harmonic potential.

Multiple microsecond simulations were performed, showing that the flap opening fraction depends on the temperature and follows a sigmoid curve that is typical of phase transitions: the

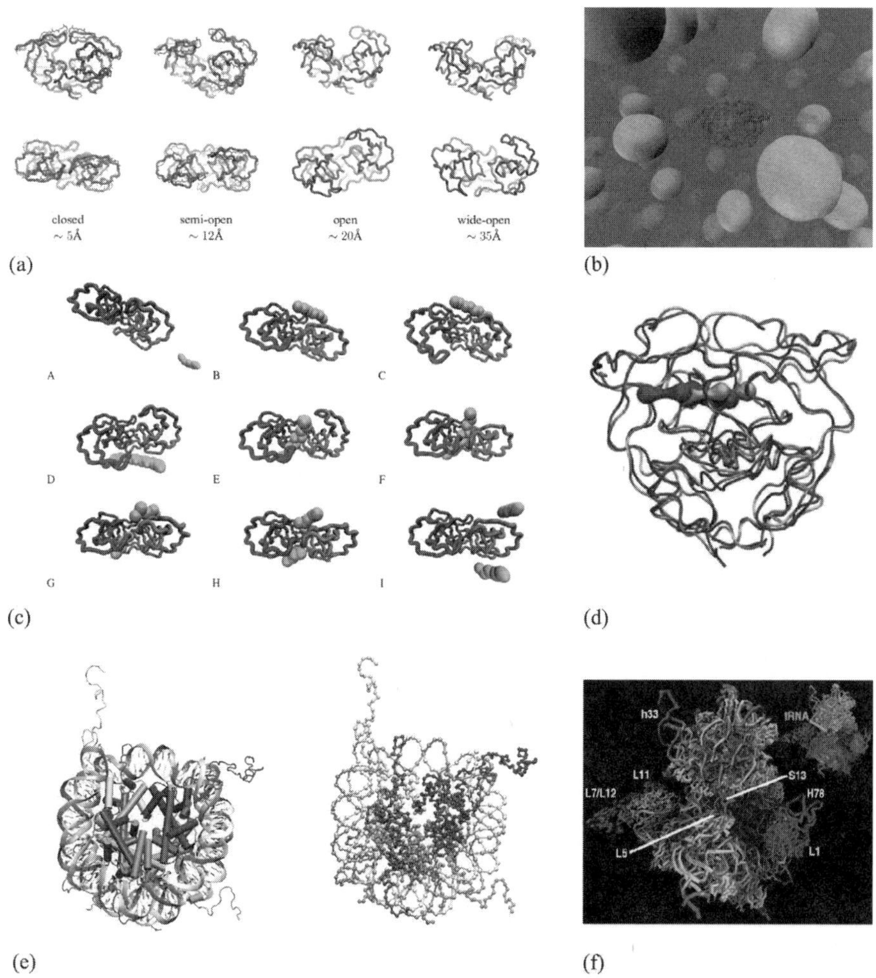

FIGURE 19.5 (See color insert following page 238.) (a) Snapshots from the free protease simulation, showing the steps of the flap opening. For the first two steps the experimental structures are available (in blue in the color figure) that superimpose very well on the simulated structures (in red). (b) A snapshot of the simulation in the presence of crowders. (c) Substrate approach (A, B), interaction with the flaps (C, D), substrate adjustment and flap closing (E, F), cleavage and release (G–I). (d) The ligand binding with closed flaps (for small ligands). (e) Coarse-graining of the nucleosome: the all-atom cartoon representation (left) and the one-bead model (right). (f) The one-bead model of the S70 bacterial ribosome.

model describes a system stable in the closed state, but very near to the transition, as one would expect considering the mechanism. This is also in agreement with experimental association rates, if one assumes that the substrate association is mainly triggered by the flap opening. The flap-opening frequency, conversely, depends on the damping constant in the Langevin dynamics (or on the hydrodynamic radius in the Brownian dynamics). For physically reasonable values it reaches the microsecond time scale, indicating again a good accuracy of the model, even for what concerns the characteristic times, and demonstrating how the CG model can be combined with a stochastic approach in order to reproduce the correct dynamics. The study of the correlations between the flap opening and other principal modes has revealed the location of a possible allosteric inhibition site.[27] The influence of the crowder molecules on the flap-opening dynamics was studied, by representing the crowders as large soft spheres, showing that at certain concentrations the presence of crowders can hinder the flap opening[26] (see Figure 19.5b).

The model was also applied to the ligand binding[27,28] and substrate binding-cleavage dynamics.[27,29] The effect of mutations on the binding affinity was evaluated and found in agreement with the experiment. It was also shown that while small ligands can enter HIV-1pr from the sides with partial opening of the flaps, the opening must be complete for the substrate to enter (Figure 19.5c,d). The initial approach of the substrate is by free diffusion. Subsequently, the substrate explores different possible approach angles, interacts with the flaps, and modifies their dynamics, favoring the open state. When the flaps open, if the substrate is in the correct orientation it enters and correctly positions into the active site. The flaps close and the HIV-1pr–substrate complex is stable in the closed conformation. When the substrate is cleaved, the products are released without flap opening. This is the first simulation of the entire process of capture, cleavage, and release of this kind of system.

In conclusion to this section, the local bias retained in the model has the positive effect of correctly reproducing the local hydrogen-bond network and shape effects of the side chains, without precluding the possibility to explore conformations very far from the reference one, thanks to the other unbiased terms of the potential. In other words, this is a good compromise between completely biased models (Go, networks) and completely unbiased models. In addition, depending on the cases, the bias can be made weaker or stronger, by tuning the cutoff between the local and nonlocal part of the potential and/or biasing other term of the potential. For instance, in Ref. 7 the bias is stronger to increase the stability of the structural features of the system, the S70 bacterial ribosome. This is a huge system (about 9000 residues) that was simulated around the microsecond time scale with the one-bead approach, revealing the slow motions responsible for the translocation process (see Figure 19.5f). In Ref. 18, in a model for the nucleosome (about 1300 residues, see Figure 19.5e), the bias is intermediate, and a very accurate parameterization leads to a particularly good comparison between CG and all-atom simulations on the nanosecond time scale. This model is designed to simulate the nucleosome unwrapping preliminary to transcription, replication, or chromosome condensation phases.[18]

19.4 CONCLUDING REMARKS

In this chapter we have described some of the most representative one-bead CG models available in the literature. We focused on the one-bead models because they have several advantages with respect to other CG models. Their resolution matches with that of cryoelectron microscopy, which is a fortunate circumstance that allows CG modeling and experiments to synergistically give a realistic and accurate view of a system's structure and internal dynamics. Additionally, this level of coarsening allows simulating the maximum sizes and time scales while preserving the possibility of explicitly describing complex structural transitions. This requires a preliminary study of the mapping between the all-atom and one-bead internal variables describing the backbone conformation, so that the available secondary structure information in the Ramachandran map can be used also in the one-bead representation. Finally, but maybe most obviously, the one-bead models are the simplest to implement and the most "natural" from the point of view of the hierarchical structural organization of proteins, since the amino acid is the basic unit of the proteins.

The one-bead models were reviewed and classified according to the functional forms and parameterization philosophy of their force fields. These determine the accuracy and transferability-predictivity of the models, which are usually competing factors in these approaches. This is basically due to the fact that, if one wants to preserve the intrinsic simplicity of the model, it is very difficult to include many complex interactions that occur between amino acids in the relatively few parameters. However to have an accurate and yet transferable and predictive one-bead force field is the ultimate goal, and some steps forward have been taken, as described in this chapter. In particular, we indicate possible ways to exploit at best the potential of the conformational term (bond angles and dihedrals) of the force field, to properly introduce solvent effects into the electrostatic term and to treat intermediate-range polar and apolar interactions.

However, the most critical issue is the parameterization of the short-range nonbonded interactions, which must include many highly specific physicochemical effects in a few parameters. Good and simple recipes for the parameterization of this term are not yet available in the general case, and work is in progress. However, less general, "intermediate" models were presented. Those are not yet completely independent of some a priori knowledge of the system, yet are able to include a high degree of accuracy together with predictivity and, as shown in the chapter, have proven to be capable of simulating very slow processes (such as the HIV-1 protease substrate capture) occurring in very large systems (nucleosomes and ribosomes).

The ultimate goal remains to reduce to zero, if possible, the necessary a priori knowledge of the system and to accurately predict structures and internal dynamics, a hard task that hopefully will stimulate researchers from different fields, including biochemistry, biophysics, bioinformatics, and mathematics.

ACKNOWLEDGMENTS

Work in VT's group is supported in part by "INFM-CNR parallel computing initiative 2005–2006" and by IIT. VT also wishes to thank Karine Voltz and Joanna Trylska for useful discussions and for having provided material for figures. Work in JAM's group is supported in part by the NIH, NSF, HHMI, CTBP, NBCR, and Accelrys.

REFERENCES

1. Tozzini, V. 2005. Coarse-grained models for proteins. *Curr. Opin. Struct. Biol.* 15:144–50.
2. Izvekov, S., and Voth, G. A. 2006. Modeling real dynamics in the coarse-grained representation of condensed phase systems. *J. Chem. Phys.* 125:151101.
3. Reynwar, B. J., Illya, G., Harmadaris, V. A., Müller, M. M., Kremer K., and Deserno M. 2007. Aggregation and vesiculation of membrane proteins by curvature-mediated interactions. *Nature* 447:461–64.
4. Chu, J.-W., and Voth, G. A. 2006. Coarse-grained modeling of the actin filament derived from atomistic-scale simulations. *Biophys. J.* 90:1572–82.
5. Eghiaian, F. 2005. Structuring the puzzle of prion propagation. *Curr. Opin. Struct. Biol.* 15:724–30.
6. Tama, F., Valle, M., Frank, J., and Brooks, C. L., III. 2003. Dynamic reorganization of the functionally active ribosome explored by normal mode analysis and cryo-electron microscopy. *Proc. Natl. Acad. Sci. U.S.A.* 100:9319–23.
7. Trylska, J., Tozzini, V., and McCammon, J. A. 2005. Exploring global motions and correlations in the ribosome. *Biophys. J.* 89:1455–63.
8. Tozzini, V., Rocchia, W., and McCammon, J. A. 2006. Mapping all-atom models onto one-bead coarse-grained models: General properties and applications to a minimal polypeptide model. *J. Chem. Theor. Comp.* 2:667–73.
9. Hamacher, K., and McCammon, J. A. 2006. Computing the amino-acid specificity of fluctuations in biomolecular systems. *J. Chem. Theory Comput.* 2:873–78.

10. Matysiak, S., and Clementi, C. 2004. Optimal combination of theory and experiment for the characterization of the protein folding landscape of S6: How far can a minimalist model go? *J. Mol. Biol.* 343:235–48.
11. Liu, Z., and Chan, H. S. 2005. Desolvation is a likely origin of robust enthalpic barriers to protein folding. *J. Mol. Biol.* 349:872–89.
12. Das, P., Matysiak, S., and Clementi, C. 2005. Balancing energy and entropy: A minimalist model for the characterization of protein folding landscapes. *Proc. Natl. Acad. Sci. U.S.A.* 102:10141–46.
13. Sorenson, J. M., and Head-Gordon, T. 2000. Matching simulation with experiment: A new simplified model for simulating protein folding. *J. Comput. Chem.* 7:469–81.
14. McCammon, J. A., and Northup, S. H. 1980. Helix-coil transition in a simple polypeptide model. *Biopolymers* 19:2033–45.
15. Wade, R. C., Davis, M. E., Luty, B. A., Madura, J. D., and McCammon, J. A. Gating of the active site of triose phosphate isomerase: Brownian dynamics simulations of flexible peptide loops in the enzyme. *Biophys. J.* 64:9–15.
16. Reith, D., Pütz, M., and Müller-Plathe, F. 2003. Deriving effective mesoscale potentials for atomistic simulations. *J. Comput. Chem.* 24:1624–36.
17. Tozzini, V., and McCammon, J. A. 2005. A coarse-grained model for the dynamics of flap opening in HIV-1 protease. *Chem. Phys. Lett.* 413:123–28.
18. Voltz, K., Trylska, J., Tozzini, V., Kurkal-Siebert, K., Langowsky, J., and Smith, J. 2008. Coarse-grained force field for the nucleosome from self-consistent multiscaling. *J. Comput. Chem.* 29:1429–39.
19. Izvencov, S., and Voth, G. A. 2005. A multiscale coarse-graining method for biomolecular systems. *J. Phys. Chem. B.* 109:2469–73.
20. Levitt, M. 1976. A simplified representation of protein conformations for rapid simulation of protein folding. *J. Mol. Biol.* 104:59–107.
21. Adcock, S. A., and McCammon, J. A. 2006. Molecular dynamics: A survey of methods for simulating the activity of proteins. *Chem. Rev.* 106:1589–1615.
22. Lu, B., Cheng, X., Huang, J., and McCammon, J. A. 2006. An order N algorithm for computation of electrostatic interactions in biomolecular systems. *Proc. Natl. Acad. Sci. U.S.A.* 59:19314–15.
23. Dzubiella, J., Swanson, J. M. J., and McCammon, J. A. 2006. Coupling hydrophobic, dispersion, and electrostatic contributions in continuum solvent models. *Phys. Rev. Lett.* 96:087802.
24. Dzubiella, J., Swanson, J. M. J., and McCammon, J. A. 2006. Coupling nonpolar and polar solvation free energies in implicit solvent models. *J. Chem. Phys.* 124:084905.
25. Tozzini, V., Trylska, J., Chang, C. E., and McCammon, J. A. 2007. Flap opening dynamics in HIV-1 protease explored with a coarse-grained model. *J. Struct. Biol.* 157:606–15.
26. Minh, D. D. L., Chang, C. E., Trylska, J., Tozzini, V., and McCammon, J. A. 2006. The influence of macromolecular crowding on HIV-1 protease internal dynamics. *J. Am. Chem. Soc.* 128:6006–6007.
27. Chang, C. E., Shen, T., Trylska, J., Tozzini, V., and McCammon, J. A. 2006. Gated binding of ligands to HIV-1 protease: Brownian dynamics simulations in a coarse-grained model. *Biophys. J.* 90:3880–85.
28. Chang, C. E., Trylska, J., Tozzini, V., McCammon, J. A. 2007. Binding pathways of ligands to HIV-1 protease: Coarse-grained and atomistic simulations. *Chem. Biol. Drug Des.* 65:5–13.
29. Trylska, J., Tozzini, V., Chang, C.-E., and McCammon, J. A. 2007. HIV-1 protease substrate binding and product release pathway explored with a coarse-grained molecular dynamics. *Biophys. J.* 92:4179–87.

20 Application of Residue-Based and Shape-Based Coarse-Graining to Biomolecular Simulations

Peter L. Freddolino and Amy Y. Shih
Center for Biophysics and Computational Biology,
University of Illinois at Urbana-Champaign

Anton Arkhipov, Ying Ying, Zhongzhou Chen, and Klaus Schulten
Department of Physics, University of Illinois at Urbana-Champaign

CONTENTS

20.1 INTRODUCTION

A vast array of problems currently addressed by computer simulations, including biological systems, involve the analysis of properties on long time and length scales derived from simulations on relatively short time and length scales [Katsoulakis, Majda, and Vlachos 2003]. Although these techniques can provide a great deal of insight into the processes under study, traditional simulations of this type are limited in scope by their computational costs, which impose an upper limit on the time scale that can be studied (currently in the nanosecond range, for biological systems [Sastry et al. 2005]). This limitation has lead to the development of a wide variety of techniques attempting to provide longer time and length scale information than traditional (usually atomistic) simulations, many of which fall into the category of coarse-graining. In the broadest possible sense, the term "coarse-graining" (CG) can be used to refer to any simulation technique in which a simulated system is simplified by clustering several subcomponents of it into one component, thus effectively reducing the computational complexity by removing both degrees of freedom and interactions from the system. The fundamental assumption behind such techniques is that by eliminating insignificant degrees of freedom, one will be able to obtain physically correct data on the properties of a system over longer time scales than would otherwise be achievable [Schütte et al. 1999].

A wide variety of CG methods for biological systems currently exist, ranging in some sense from united-atom models to elastic network models. We focus on the principles and applications of two classes of biological CG, namely residue-based and shape-based CG. Residue-based CG is a broad family of methods in which clusters of 10–20 covalently bonded atoms are represented by one bead; it is a fairly natural and common method for CG when a speedup of 1–2 orders of magnitude over all-atom simulations is required. Shape-based CG is a method recently developed in our group that uses a neural network algorithm to assign CG beads to domains of a protein, efficiently reproducing the shape of the protein with a minimal number of particles. Interactions between beads are then parameterized from all-atom simulations of the bead components. In this chapter we present a summary of both methods, along with exemplary applications of residue-based CG to two lipid-protein systems involving large-scale conformational changes, and of shape-based CG to the mechanical properties of multiprotein systems.

20.2 RESIDUE-BASED COARSE-GRAINING

The most natural (and frequently used) method for coarse-graining a biological system is to assign sections of each biological molecule (or monomer, in the case of a biopolymer) with similar chemical properties and spatial location to a "bead," and then treat the CG system as an ensemble of beads. This type of description is henceforth referred to as "residue-based coarse-graining." For example, in one possible description of a protein each amino acid residue would be represented by two beads, one representing the backbone atoms and a second (different for each residue type) representing the side-chain atoms [Shih et al. 2006, 2007b].

While in principle similar to the united-atom models common in the early stages of molecular dynamics (MD) [Leach 1996], modern residue-based CG methods are generally geared toward much longer time scales, and are thus coarser. The strategy of making a cluster of connected heavy atoms the unit particle, rather than atoms or heavy atoms, permits a longer time-step and thereby yields a larger reduction in computational effort than united-atom models, but obviously carries a commensurate loss of detail. Recent interest in residue-based CG has emerged in the field of lipid simulations, where several groups have developed CG lipid models either by attempting to reconstruct the forces observed in all-atom MD [Shelley et al. 2001; Stevens, Hoh, and Woolf 2003; Stevens 2004; Nielsen et al. 2004; Nielsen and McCammon 2003] or by using a created potential with parameters tuned to match experimental thermodynamic data [Marrink and Mark 2002, 2003, 2004; Marrink, de Vries, and Mark 2004; Marrink, Risselada, and Mark 2005; Baron et al. 2007]. In both of these cases, the CG process maps approximately 10 atoms to one coarse-grained particle ("bead"), and the resultant CG model reproduced both the physical properties and (to the extent that they are experimentally known) assembly mechanisms of bilayers, micelles, and other lipid aggregates on microsecond time scales. Similar efforts have recently been extended to proteins, including simulation of protein-lipid assemblies [Shih et al. 2006; Bond and Sansom 2006] and protein folding [Das, Matysiak, and Clementi 2005].

20.2.1 Interaction Potentials for Residue-Based CG

In the broadest sense, the force fields used in residue-based CG models tend to fall into one of two categories, either being derived phenomenologically or through MD-based parameterization. The former approach, exemplified by the lipid-water force fields of Marrink and co-workers [Marrink and Mark 2003, 2004; Marrink, de Vries, and Mark 2004; Marrink, Risselada, and Mark 2005] and by the more recent MARTINI force field [Marrink et al. 2007], involves partitioning clusters of atoms into abstract "types" based on their physical properties (for example, polarity and ability to hydrogen bond); the interactions between beads are then parameterized to reproduce experimental data such as partition energies [Marrink, de Vries, and Mark 2004]. The latter approach is a direct analogue of parameterization of all-atom MD models from quantum mechanical calculations; here, all-atom simulations are performed on some system including the CG beads whose interactions are to be parameterized, and the results are used to construct an effective potential between the beads. Both approaches have been successfully applied to a number of systems, but potentials derived from all-atom MD simulations carry the added benefit of improved miscibility of all-atom and CG components, which is likely to become increasingly important as mixed all-atom/CG simulations [Shi, Izvekov, and Voth 2006; Praprotnik, Site, and Kremer 2005, 2006; Lyman, Ytreberg, and Zuckerman 2006] become more common.

MD-based parameterization can be carried out in a variety of ways, depending on the scope and intended use of the parameter set in question. Given an all-atom simulation including the components whose interactions are to be parameterized, an effective interaction potential between CG beads can be constructed by attempting to match the forces present between the beads in the all-atom description as a function of distance [Izvekov and Voth 2005a, 2005b, 2006; Shi, Izvekov, and Voth 2006] or through a process such as Boltzmann inversion [Reith, Pütz, and Müller-Plathe 2003; Tozzini and McCammon 2005], which is described in more detail in the following sections. Note that although the example given below is for shape-based CG, the same techniques can be applied to determine interactions for residue-based CG models.

Both in the case of MD-based and phenomenological parameterization, the resulting potentials may either be fitted to an existing potential form (for example, the Lennard–Jones potential for nonbonded interactions) or used directly (for example, in the form of an energy/force lookup table). While making use of an existing potential form has long been preferred because it allows the use of existing MD packages without further modification, the use of tabulated potentials allows more control over the exact potential form being used, and is increasingly supported in common MD packages such as DL-POLY and NAMD.

20.2.2 Reverse Coarse-Graining and Resolution Switching

Coarse-grained MD simulations have proven quite useful for obtaining data on the behavior of systems, where the relevant time or length scales (or both) are inaccessible to all-atom MD. However, even heavier use of CG simulations could be made if CG could be used as an accelerator, with atomic detail either maintained in regions of interest or recoverable from snapshots in the CG trajectory. Recent progress has been made along both these fronts, in the form of mixed CG-all-atom simulations [Shi, Izvekov, and Voth 2006] and simulations involving dynamic switching of components between CG and all-atom descriptions [Praprotnik, Site, and Kremer 2006; Lyman, Ytreberg, and Zuckerman 2006]. The primary new challenges faced in either of these cases lie in deriving accurate potentials for interactions between CG and all-atom components, and in effectively mapping CG conformations to all-atom conformations. The latter challenge is particularly significant both because any given conformation of CG particles can be taken to represent an ensemble of conformations of the corresponding all-atom system (any set of states where the centers of mass of the component atoms for each bead correspond to the CG bead positions), and because switching to the all-atom system will almost certainly cause a change in the energy of the system due to the introduction of new interactions.

Early efforts in switching of scales have focused on building a method allowing true mixed-scale dynamics, either by allowing particles to transition between all-atom and CG representations while passing through a specific region in space [Praprotnik, Site, and Kremer 2006] or by allowing exchange between low-resolution and high-resolution replicas of a system being simulated in parallel [Lyman, Ytreberg, and Zuckerman 2006]. Outgrowths of these methods will likely be quite useful in the future, although both face the difficulty that deterministically mapping a given CG conformation to an all-atom conformation may be insufficient for more complex beads (such as beads representing an amino acid side-chain or significant fraction thereof) and that the free-energy discontinuities experienced during scale-switching may become prohibitively high if a poor initial all-atom conformation is chosen during exchange.

In some cases where a CG model is used to accelerate sampling, there is no need to repeatedly switch between CG and all-atom descriptions; it is sufficient to sample the conformational space of the system using the CG model and then analyze the results in terms of a consistent all-atom model. This is the case, for example, in the studies of nanodiscs presented below, where all-atom conformations had to be extracted from various snapshots of the CG simulation for comparison with experimental data. In this case, it proved sufficient to reverse coarse-grain the system by superimposing the all-atom components of the system on the CG structure such that the center of mass of each cluster of atoms is located on the corresponding CG bead, and then minimizing and annealing the resulting all-atom structure with the center of mass of each atom cluster constrained to the bead location. This can be conceptually interpreted as sampling the conformational space of the all-atom structure in the region consistent with the CG structure being converted. While this method is far too time-consuming to use when rapid switching of all-atom and CG representations is desired, and does not preserve the dynamic or thermodynamic properties of the CG system, it is sufficient for recovering an all-atom snapshot from a CG simulation, and some conformational sampling scheme similar to that used here is likely to become necessary in resolution exchange for cases where mapping the CG conformation to an all-atom conformation is nontrivial.

20.2.3 Application to Nanodiscs and HDL

High-density lipoproteins (HDL) are lipid–protein particles that function in the body to remove cholesterol from peripheral tissues and return them to the liver for processing. These particles, which occur in a wide variety of shapes and sizes in vivo, are known to play an important role in protecting the body from heart disease [Wang and Briggs 2004]. HDL particles are known to be composed of a disc-shaped patch of membrane enclosed by two or more copies of apolipoprotein A-I (ApoA-I). In addition to their medical importance, a truncated form of the protein component of HDL particles has recently been used to assemble homogeneous protein–lipid particles known as nanodiscs [Bayburt, Grinkova, and Sligar 2002; Sligar 2003], which can incorporate membrane proteins and thus be used to study them in an environment more realistic than micelles or liposomes [Seddon, Curnow, and Booth 2004; Davydov et al. 2005; Baas, Denisov, and Sligar 2004; Duan et al. 2004; Civjan et al. 2003; Boldog et al. 2006; Shih et al. 2005].

The conditions needed to cause nanodiscs to assemble around a protein, however, are very dependent on the protein itself, and different conditions are required to efficiently incorporate different proteins [Denisov et al. 2004; Bayburt, Grinkova, and Sligar 2006; Boldog et al. 2006]. Obtaining information on the structure and assembly of nanodiscs would thus be useful in the rational design of nanodisc assembly protocols, and would additionally provide data on HDL assembly and characteristics. Unfortunately, no high-resolution structure has been obtained for a complete HDL particle or nanodisc, although a consensus double-belt model is emerging for the general layout of the proteins and lipids in the particle [Koppaka et al. 1999; Panagotopulos et al. 2001; Li et al. 2000; Tricerri et al. 2001; Silva et al. 2005; Li et al. 2006; Gorshkova et al. 2006].

Unfortunately, nanodisc assembly takes place on a time scale of microseconds to milliseconds, far longer than can be treated using all-atom MD simulations. The nature of the type of data sought—relatively coarse data on important stages of nanodisc assembly and factors affecting it—is in principle appropriate for a residue-based CG model. In addition, the fact that hydrophobic interactions and the properties of a lipid patch are the primary features likely to drive the simulation meant that the bulk of the force field in this case could be taken from the lipid–water model of Marrink and coworkers [Marrink, de Vries, and Mark 2004], a phenomenological model which had shown excellent results in the assembly and physical properties of micelles and bilayers. For the protein component of the system, the bead types of Marrink's force field were assigned to protein components according to their properties, with each amino acid residue represented by a backbone bead (the same type for each residue) and a side-chain bead [Shih et al. 2006]. A very similar model was proposed by Bond and coworkers in their simulations of the bacterial membrane protein OmpA [Bond and Sansom 2006]. The use of a CG model on the nanodisc provides a factor of 500 speedup compared with all-atom simulations, due to the use of 50 fs time-steps and reduction in number of particles by a factor of 10 [Shih et al. 2006].

Simulation of the components of a single nanodisc beginning from a random mixture with water, over a period of 10 μs, revealed a complete pathway for the assembly of nanodiscs from their components, as shown in Figure 20.1. Further simulations from other starting points showed both similar assembly pathways and mechanisms [Shih et al. 2006, 2007b; Shih et al. 2007a]. Analysis of the energetics of assembly illustrated that it occurs as a three-step process. First, nucleation of assembly occurs as the lipids form pseudomicelles, which are roughly spherical in shape; at this point, the hydrophobic face of the Apo A-I proteins (each of which contains a set of amphipathic α-helices) binds to the pseudomicelle in a random conformation. After this initial aggregation, the proteins reorient along the surface to bring themselves into more favorable contact with each other, eventually forming a series of salt bridges that force the double belt orientation to form.

Although no high-resolution structural data on formed nanodiscs or HDL are available, the assembly mechanism and final structure obtained from CG simulations could still be compared to low-resolution information from SAXS studies [Shih et al. 2007c]. Theoretical SAXS curves can

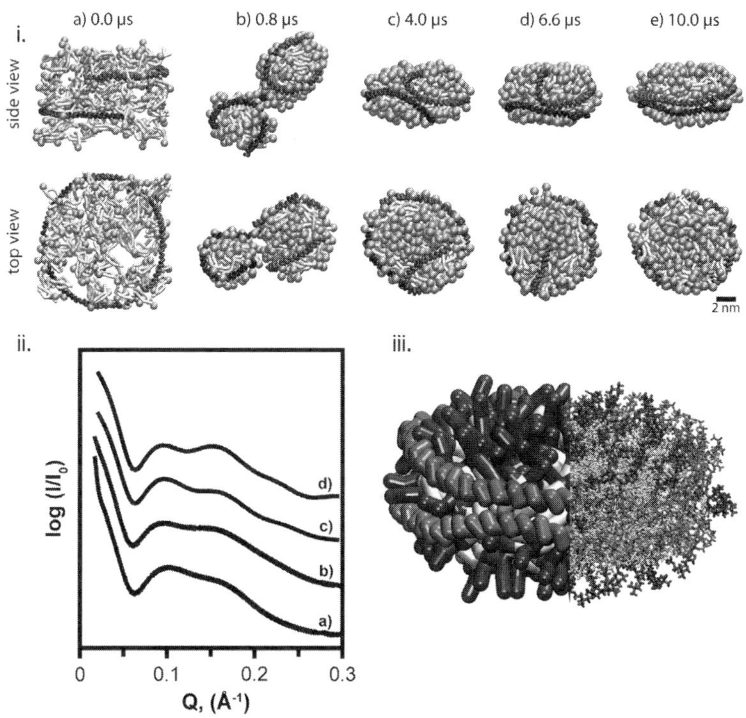

FIGURE 20.1 (i) Snapshots from an assembly simulation in which 160 DPPC lipids and two Apo A-I proteins were assembled from a random mixture over 10 μs. CG water is present in all cases but omitted from images for clarity. (ii) Comparison of SAXS curves between experimental results for DPPC nanodiscs (a), DMPC nanodiscs (b), an ideal all-atom model of a double-belt nanodisc (c), and the final structure from a 10 μs CG assembly simulation (d). Note that the curves are separated vertically for clarity. (iii) Example of a CG conformation (left) mapped onto a corresponding all-atom conformation (right).

be calculated from an all-atom structure using the program CRYSOL [Svergun, Barberato, and Koch 1995]; however, obtaining a SAXS curve from CG simulations first requires reverse CG of CG snapshots. Because there was no need to significantly continue the simulations after reverse CG in this case, a fairly simple scheme was used, in which the centers of mass of the all-atom components of each bead were aligned with this bead, and then the system annealed with the center of mass of the components of each bead constrained, allowing the structure to relax while remaining consistent with the CG snapshot. A comparison of the SAXS curve obtained from the assembled CG nanodisc with experimental results is shown in panel (ii) of Figure 20.1, and a time course of the SAXS curve observed during the CG assembly process in panel (iii) of Figure 20.1. The excellent agreement between experimental and theoretical results illustrates both the success of the CG model in reproducing the nanodisc assembly process and structure, and the utility of even fairly simple reverse CG methods.

20.2.4 Application to the BAR Domain

BAR domains constitute an ubiquitous type of protein, found in many organisms and performing the function of driving the formation of tubulated and vesiculated membrane structures inside cells [Sakamuro et al. 1996]. BAR domains contain a conserved protein motif and are involved in a variety of cellular processes including fission of synaptic vesicles, endocytosis, and apoptosis [Ren et al. 2006]. Structurally, BAR domains form crescent-shaped dimers (see Figure 20.2) with a high density of positively charged residues on their concave face. The shape and charge distribution suggest that BAR domains induce membrane curvature by binding to negatively charged lipids [Peter et al. 2004; Blood and Voth 2006]. However, the common molecular mechanism underlying membrane sculpting by BAR domains remains largely unknown.

Recently, all-atom simulations [Blood and Voth 2006] have demonstrated that a single BAR domain induces membrane curvature. The all-atom study required a simulation of up to 700,000 atoms on the time scales of 50 ns. The next demanding question after the discovery of the membrane bending by a single BAR domain is how multiple BAR domains work together to bend membranes. All-atom simulations of this process are too challenging at present, since one would have to consider millions of atoms in each simulation. However, the residue-based CG method appears to be

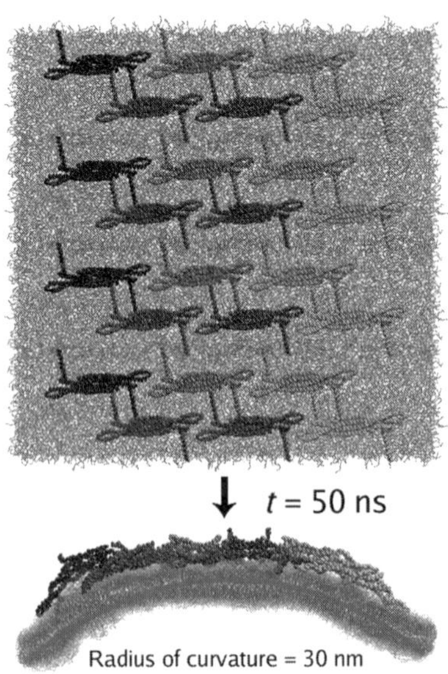

FIGURE 20.2 Membrane curvature induced by BAR domains. Upper panel: top view of the initial arrangement (four periodic cells along the vertical axis); lower panel: side view after 50 ns.

a good option for this application, and, thus, we have performed CG simulations of systems with multiple BAR domains, in order to determine how the cooperative interaction of the latter with the membrane induces global membrane curvature.

The residue-based CG model [Shih et al. 2007b, 2006] described above is ideally suited to describe the membrane remodeling by BAR domains since it has demonstrated its power before on the tasks where lipids assemble, disassemble, and reshape membranes [Shih et al. 2007b, 2006; Marrink, de Vries, and Mark 2004]. The only difficulty is that the residue-based protein CG model has not been developed to work for proteins of arbitrary shapes. In particular, the model has not been designed to maintain tertiary structure of proteins, which is determined by the protection of hydrophobic side groups in the protein amino acid sequence from solvent (well described by the residue-based CG model), but also, to a large extent, by atomic level interactions that the residue-based CG model does not capture. Indeed, when the model was applied to the BAR domain, the tertiary structure was not preserved. Accordingly, we added harmonic bonds and angles connecting protein beads that conserve protein shape and flexibility. A minimal set of bonds and angles was selected for this purpose. The strength of these bonds and angles was chosen to reproduce the tertiary structure flexibility as observed in the all-atom simulations. As a result, the protein was not heavily constrained, but the tertiary structure (the BAR domain's crescent shape) was maintained well. This feature has been implemented through a NAMD [Phillips et al. 2005] functionality that allows one to add extra bonded interactions to simulations.

In our previous residue-based CG simulations [Shih et al. 2007b, 2006; Marrink, de Vries, and Mark 2004], a relative dielectric constant ε of 20 was employed. In the case of the BAR domain simulations we chose $\varepsilon = 1$. Such a low ε-value is necessary for membrane curvature to be induced by BAR domains, which is driven by short-range electrostatics, when charged groups from the protein's concave surface interact at close range with charged lipid heads. Interactions at larger distances should be screened by water, requiring, in principle, higher values of ε. However, the electrostatic interactions at large distances appear to be relatively weak in the present case such that $\varepsilon = 1$ has no adverse effect on long-range electrostatics in case of the BAR domain simulations.

The rather rough CG model of the BAR domain and lipid membrane, described above, has been applied to study the behavior of multiple BAR domains [Arkhipov, Yin, and Schulten 2008], as shown in Figure 20.2. The all-atom simulations with a single BAR domain [Blood and Voth 2006], from other groups as well as our own, have been reproduced well by the residue-based CG simulations (not shown), in terms of both membrane curvature and protein structure. Six BAR domains interacting with a patch of membrane were then simulated. Two rows of three BAR domains each were placed in parallel (shifted with respect to each other) on top of a planar membrane, composed of electrostatically neutral DOPC lipids mixed with negatively charged DOPS lipids (30% DOPS). BAR domains produced a global bending mode [Arkhipov, Yin, and Schulten 2008], exhibiting a radius of curvature of 30 nm within 50 ns (comparable to experimental values for the curvature [Peter et al. 2004]). This result suggests how BAR domains quickly generate membrane curvature, as possibly occurs in cells during the formation of subcellular membrane structures [Ren et al. 2006].

20.3 SHAPE-BASED COARSE-GRAINING

The shape-based CG [Arkhipov, Freddolino, and Schulten 2006; Arkhipov et al. 2006] method offers a higher degree of CG than the residue-based method, but at the price that the biopolymers described are restricted in their motion to elastic vibration around a given shape. The method is available through the molecular visualization software VMD [Humphrey, Dalke, and Schulten 1996].

20.3.1 Selection of Bead Arrangement and Potentials

Biomolecules, and proteins in particular, assume a variety of shapes, often featuring both compact domains and elongated tails, the compact regions and tails often being equally important. To our knowledge, all existing CG methods assign CG beads to represent a fixed group of atoms, but this is not efficient for the CG of molecules with complex shapes, because with such an approach either the tails are misrepresented or too many CG beads are used for the compact domains. With shape-based CG, one addresses the task of representing shapes with as few CG beads as possible by so-called topology-conserving maps [Martinetz and Schulten 1994].

Consider a molecule consisting of Na atoms with coordinates $\mathbf{r}_n$ and masses m_n, $n = 1, 2, \ldots, N_a$. One seeks to reproduce the shape of the molecule with N CG beads. The mass distribution $p_n = m_n/M$ $(M = \sum_n m_n)$ is used as a target probability distribution for the evolving map. CG beads are assigned their initial positions randomly; then, the beads are considered as nodes of a network [Martinetz and Schulten 1994], on which S adaptation steps are performed. At each step the following procedures are carried out. First, the nth atom is chosen randomly, according to the probability distribution p_n; its coordinates $\mathbf{r}_n = \mathbf{v}$ are used to adapt the neural network (see Equation 20.1). Second, for each CG bead i $(i = 1, 2, \ldots, N)$, one determines the number k_i of CG beads j, obeying the condition $|\mathbf{v} - \mathbf{R}_j| < |\mathbf{v} - \mathbf{R}_i|$, where $\mathbf{R}_j$ is the position of the jth bead. Third, positions of the beads are updated $(i = 1, 2, \ldots, N)$, according to the rule

$$\mathbf{R}^{new}{}_i = \mathbf{R}^{old}{}_i + \xi e^{-k_i/\lambda}(\mathbf{v} - \mathbf{R}^{old}{}_i). \tag{20.1}$$

Parameters ξ and λ are adapted at each step according to the functional form $f_s = f_0(f_S/f_0)^{s/S,}$ where s is the current step, $\lambda_0 = 0.2N$, $\lambda_S = 0.01$, $\xi_0 = 0.3$, and $\xi_S = 0.05$. We use $S = 200N$; typical adaptation steps are shown in Figure 20.3. Once beads are placed, an all-atom "domain" is found for each bead (the domain includes all atoms closer to this bead than to any other bead). The total mass and charge of a domain is assigned to the respective bead. Since the shape of a molecule is reproduced by this

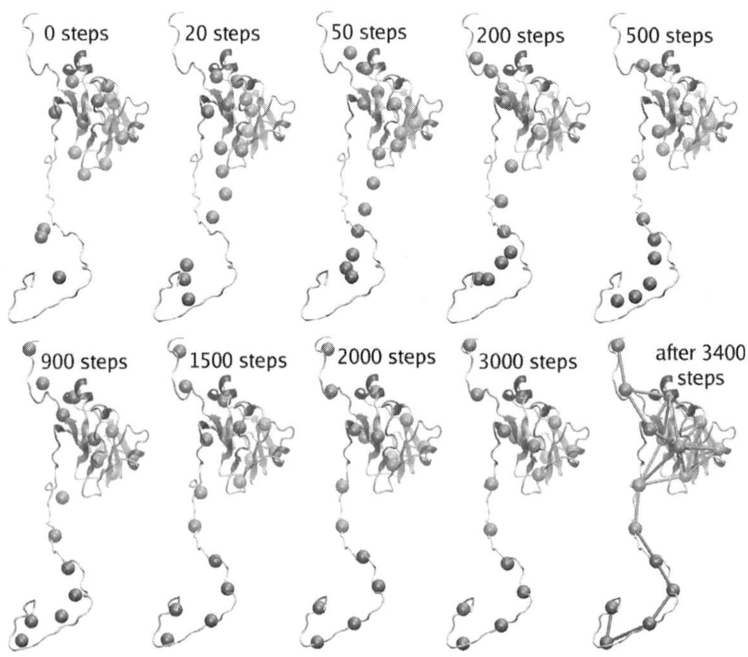

FIGURE 20.3 Shape-based coarse-graining algorithm assigning CG beads. The CG beads (spheres) are the nodes of the network; their positions are updated throughout the learning steps (3400 steps for 17 beads in this example). As a result, the shape of a protein (here, the capsid unit protein of the brome mosaic virus) is reproduced with a small number of beads (chosen prior to starting the algorithm). After the assignment converged, the beads are connected by bonds. The algorithm is of a neural network type described in Martinetz and Schulten (1994).

CG model, the method is termed shape-based CG. The molecular graphics program VMD [Humphrey, Dalke, and Schulten 1996], through its shape-based CG plug-in, can also build CG models from volumetric data, such as density maps obtained from cryoelectron microscopy.

Currently, two ways of establishing bonds between CG beads are implemented. In one case, a bond is established if the distance between two beads is below a cutoff distance (chosen by the researcher). Another possibility is to establish a bond between two CG beads if their respective all-atom domains are connected by protein or nucleic backbone trace; in the latter case, the topology of the molecular polymeric chain is reproduced better.

Interactions between beads are described by a CHARMM-like force field [MacKerell et al. 1998]; that is, bonded interactions are represented by harmonic bond and angle potentials (no dihedral potentials). The nonbonded potentials include 6–12 Lennard-Jones (LJ) and Coulomb terms:

$$V = \sum_{bonds\ i} \frac{K_i}{2}(R_i - L_i)^2 + \sum_{angles\ k} \frac{M_k}{2}(\theta_k - \Theta_k)^2 + \sum_{m,n} 4E_{mn}\left[\left(\frac{\sigma_{mn}}{r_{mn}}\right)^{12} - \left(\frac{\sigma_{mn}}{r_{mn}}\right)^{6}\right] + \sum_{m,n} \frac{q_m q_n}{4\pi\varepsilon\varepsilon_0 r_{mn}}, \tag{20.2}$$

where R_i and θ_k are the distance and angle for bond i and angle k, K_i and M_k are the force constants, L_i and Θ_k are the equilibrium bond length and angle; r_{mn} is the distance between beads m and n, E_{mn} and σ_{mn} are the LJ parameters, q_m is the charge of the mth bead, and the sum over m and n runs over all pairs of CG beads. The constant ε_0 is the vacuum dielectric permittivity; ε is a relative dielectric constant.

Bonded parameters K_i, L_i, etc., can be extracted from all-atom MD simulations of the considered system. For each CG bond and angle, one follows the distances between the centers of mass of corresponding atomic domains; CG force-field parameters are chosen so that in the CG simulation of a protein unit, the mean distances (angles) and respective root mean square deviations (rmsd) reproduce those found in an all-atom simulation. This procedure can be illustrated by the simple example of a one-dimensional harmonic oscillator, with a particle moving along the x coordinate in the potential $V(x) = f(x - x_0)^2/2$. With the system in equilibrium at temperature T, the average position $\langle x \rangle$ is equal to x_0, and the rmsd is given by $(k_B T/f)^{1/2}$ (k_B is the Boltzmann constant). Using an MD simulation, one can compute $\langle x \rangle$ and the rmsd, thus obtaining x_0 and f.

In all-atom simulations, LJ radius σ_{mn} for a pair m,n is usually approximated by $\sigma_{mn} = (\sigma_m + \sigma_n)/2$, where σ_m is the LJ radius of the mth atom. We use the same approach for CG beads; σ_m for the mth bead is calculated as the radius of gyration of its all-atom domain, increased by 2 Å (an average LJ radius of an atom in the CHARMM force field). The LJ well depth σ_{mn} is set to a uniform value for all pairs m-n; usually, we used $E_{mn} = 4$ kcal/mol. This choice for σ_{mn} and E_{mn} was supported by all-atom simulations of pairs of protein segments about 500 atoms each (roughly representing a single CG bead in one of our applications). Several such simulations were performed, for about 10 ns each. The effective potential of interaction between two segments was obtained for every pair using the Boltzmann inversion method [Reith, Pütz, and Müller-Plathe 2003; Tozzini and McCammon 2005]: assuming that the distribution of the distance between the segments x is given by $\rho(x) = e^{-V(x)/k_B T}$, where $V(x)$ is the potential, one computes $\rho(x)$ from the simulation and finds the potential as $V(x) = -k_B T \ln[\rho(x)] + \text{const}$. The potentials computed from all-atom simulations were similar to a LJ potential in shape, and for each pair the well depth was about 4 kcal/mol; the LJ radius was well represented using the procedure (radius of gyration + 2 Å) described above [Arkhipov et al. 2006].

An effect of the solvent is modeled implicitly, by reproducing three basic features of water, namely, viscosity, fluctuations due to Brownian motion, and dielectric permittivity. The relative dielectric constant ε is set to 80 everywhere (the experimental value for liquid water). Frictional and fluctuating forces are introduced through the Langevin equation that describes the time evolution of the CG system for each bead

$$m\frac{\partial^2 \mathbf{r}}{\partial t^2} = \mathbf{F} - m\,\gamma\,\frac{\partial \mathbf{r}}{\partial t} + \chi\,\psi(t). \tag{20.3}$$

Here, $\mathbf{r}$ is the position of the bead, $\mathbf{F}$ is the force acting on the bead from other beads in the system, γ is a damping coefficient, $\psi(t)$ is a univariate Gaussian random process, and χ is related to the frictional forces through the fluctuation-dissipation theorem, $\chi = (2\gamma k_B T/m)^{1/2}$, with m being the bead's mass. With $\mathbf{F} = 0$, Equation 20.3 describes free diffusion, where γ is related to the diffusion constant D, $D = k_B T/(m\gamma)$. In principle, γ can be computed from all-atom simulations by calculating D for the molecule under study (although the force fields used in such simulations might not be good enough to reproduce the water viscosity), but a much better approach is to use an experimental value of D if available, for example, D for a molecule of similar size. Contrary to the extraction of D from all-atom simulation, which is often difficult due to insufficient sampling, γ can be easily tuned in CG simulations to give the appropriate value of D for a given molecule, since one achieves sampling for the center of mass displacements much faster in CG simulations than in all-atom simulations. Based on estimates from the all-atom simulations and experimental data for various proteins, the appropriate values of γ for 500 atoms per CG bead should be in the range 3–15 ps^{-1}.

The dynamics of the CG system is realized through MD simulations using NAMD [Phillips et al. 2005]. For the case of 500 atoms per CG bead the CG allows one to simulate systems 500 times larger than possible in all-atom representation. As water often accounts for 80% of atoms in biomolecular simulations, and since the solvent is treated implicitly, the real gain is even higher, typically 2000–3000 times. Due to slower motions of CG beads in comparison with atoms, one can use a time-step of 500 fs to integrate the equations of motion, instead of the 1 fs time-step common for all-atom simulations. As a result, the shape-based CG with a typical ratio of 500 atoms per bead allows one to simulate dynamics of micrometer-sized objects on time scales of 100 μs using just one to three processors, while all-atom simulations even with 1000 processors are limited now to ~20 nm in size and 100 ns in time. Of course, this gain comes at the price of limited resolution.

20.4 APPLICATION TO STRUCTURAL DYNAMICS OF VIRUSES

Shape-based CG was successfully applied to study the structural dynamics of viruses. A virus [Levine 1991; Flint et al. 2004] is a macromolecular complex, normally 10–100 nm across, consisting of a genome enclosed in a protein coat (capsid); usually, the capsid is a symmetric assembly, often an icosahedron, formed by multiple copies of a few proteins. Other accessory molecules can be contained inside the capsid; additional proteins and a lipid bilayer envelope are also found on the surface of some viruses. The viral replication cycle starts with the delivery of the viral genome into a host cell, a step usually involving capsid disintegration. Then, the host cell replicates the viral genome and produces viral proteins, often at the cost of reducing the cell's normal functionality. Finally, the newly produced parts of the virus assemble into viral particles and leave the host cell, which is usually destroyed as a result. Outside of the host cell a viral particle has to be stable and relatively rigid to protect the genome, but it also has to become unstable when virulence factors need to be released into the host cell. In order to determine the stability of viral capsids and transitions between stable and unstable structures, we performed MD simulations of several viruses, both in all-atom [Freddolino et al. 2006] and CG representations [Arkhipov, Freddolino, and Schulten 2006].

Employing the shape-based CG method [Arkhipov, Freddolino, and Schulten 2006], we were able to study large viral capsids (up to 75 nm in diameter, see Figure 20.4) on 1.5–25 μs time scales. Most of the simulations were performed on a single processor, but parallel simulations on up to 48 processors were also carried out; the latter exhibited good parallel scaling similar to that of all-atom simulations with NAMD [Phillips et al. 2005].

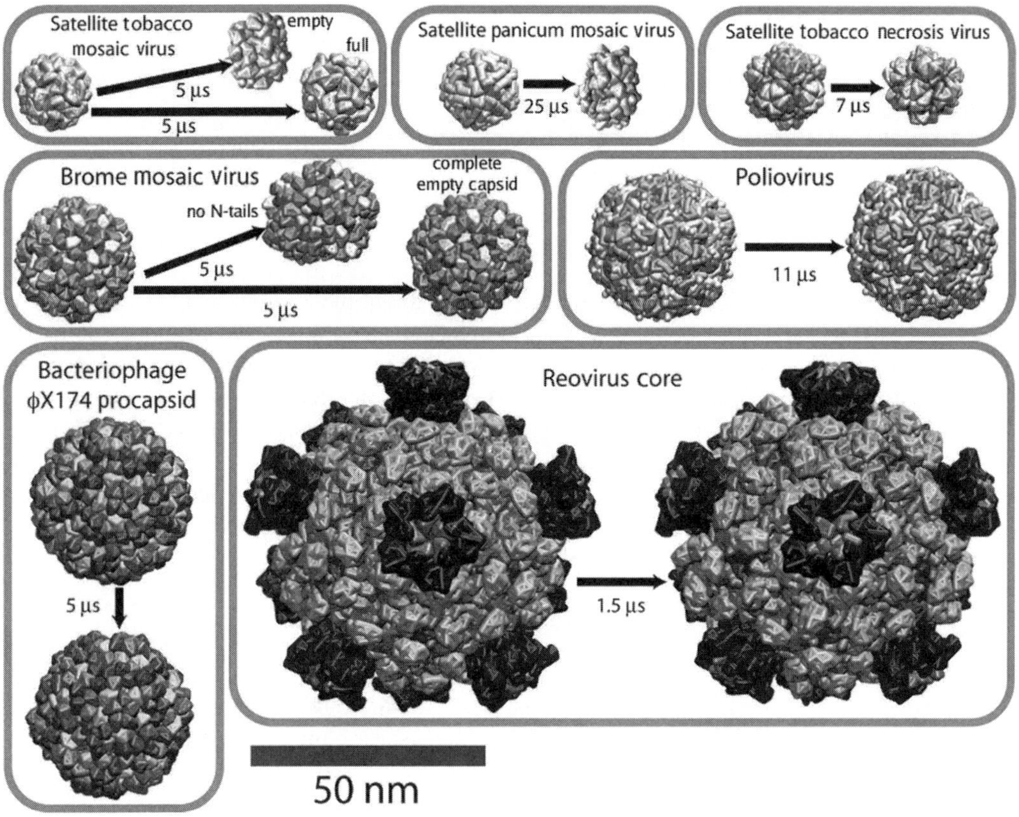

FIGURE 20.4 CG simulations of viral capsids. The initial and final structures for each simulation are shown (all particles are drawn to scale). The ratio of 200 atoms per CG particle is used. All capsids are simulated without gene content; that is, empty, except in the case of the satellite tobacco mosaic virus, in which case both empty and full capsids were simulated. From Arkhipov, Freddolino, and Schulten (2006).

First [Arkhipov, Freddolino, and Schulten 2006], we performed CG simulations of satellite tobacco mosaic virus (STMV), found to be in good agreement with previous all-atom simulations [Freddolino et al. 2006]. STMV is one of the smallest and simplest viruses, only 17 nm in diameter (Figure 20.4), yet, to describe it using all-atom simulations required dealing with a one-million-atom system. MD simulations on the complete STMV showed that it is perfectly stable on a time scale of 10 ns. The STMV capsid without genome, in contrast, was unstable, showing a remarkable collapse over the first 5–10 ns of simulation. The CG simulation of STMV reproduced the patterns and time scales of the collapse observed for the STMV capsid in all-atom simulations. For both complete STMV and the capsid alone, several other quantities computed in CG simulations, such as the average capsid radius, were within a few angstroms from those in the all-atom study.

CG simulations of capsids of several more viruses were then carried out (Figure 20.4), of the satellite panicum mosaic virus (SPMV), the satellite tobacco necrosis virus (STNV), the brome mosaic virus (BMV), the poliovirus, the bacteriophage ϕX174, and reovirus. In CG simulations, the empty capsids of STMV, SPMV, and STNV collapsed. The reovirus core, the bacteriophage ϕX174 procapsid, and the poliovirus capsid were stable, and indeed, it is known experimentally that these are stable even without their respective genetic material. For BMV, empty capsids have been observed experimentally, while a cleavage of the N-terminal tails of the unit proteins makes the capsid unstable [Lucas, Larson, and McPherson 2002]. In agreement with these observations, the BMV capsid was stable in our simulations, although very flexible, but when the N-terminal tails were removed, the capsid collapsed.

Thus, results of CG simulations agree with all-atom studies and experimental data, where available. The simulations also provide new quantitative information about viral dynamics. Perhaps the main finding in this regard is that some of the capsids (STMV, SPMV, and STNV) cannot maintain their structural integrity in the absence of the genome. This suggests a specific self-assembly pathway for these viruses: it must be the RNA, and not the protein, which nucleates assembly of the complete virus. Apparently, the RNA forms a spherical particle, and then capsid proteins attach to its surface. It is known for some viruses that they assemble "capsid first" [Flint et al. 2004], the genome being pulled into the preformed capsid. Our simulations and emerging experimental evidence [Lucas, Larson, and McPherson 2002; Kuznetsov et al. 2005] suggest that this might be different for some viruses. Related to what determines the stability, we found that the stability and flexibility of viral capsids are closely correlated with the strength of interactions between capsid subunits. Larger capsids, such as the reovirus core, have proteins that intricately intertwine with each other, featuring even a "thread and needle" arrangement. For STMV, SPMV, and STNV, unit proteins only touch each other by the edges. With more contacts between the protein units, a capsid has more hydrogen bonds and salt bridges per unit area (reflected in the CG model by generalized nonbonded LJ and Coulomb forces), and the frictional force between capsid faces a rise. These factors enhance capsid stability. Our simulations suggest that viruses like STMV, SPMV, and STNV have relatively few contacts between the capsid subunits and only their genomes render the capsids stable.

20.4.1 Application to the Bacterial Flagellum

The shape-based CG method has recently been applied to study the molecular basis of bacterial swimming. Many types of bacteria propel themselves through liquid media using whip-like structures known as flagella. The bacterial flagellum is a huge (several micrometers long, 20 nm wide), multiprotein assembly built of three domains: a basal body, fixed in the cell body below the outer membrane and acting as a motor; a filament, which grows out of the cell, making up the bulk of the length of the flagellum and interacting with solvent to propel the bacterium; and a hook, connecting basal body and filament and acting as a joint transmitting the torque from the former to the latter. Depending on the direction of the torque applied by the basal body, the filament assumes different helical shapes. Under counterclockwise rotation (as viewed from the exterior of the cell), several flagella form a single helical bundle which propels the cell along a straight line (running mode) [Berg 2000]. Under clockwise rotation, the individual flagella dissociate from the bundle and form separate right-handed helices, causing the cell to tumble. Varying the duration of running and tumbling, bacteria can move up or down a gradient of an attractant or repellent by a biased random walk.

One of the unresolved questions about the flagellum is how the reversal of torque applied by the motor results in a switching between the helical shapes of the flagellar filament. This switching is a result of polymorphic transitions in the filament, when individual protein units slide against each other [Samatey et al. 2001], but its molecular mechanism remains poorly understood. Trying to answer this question, we performed CG MD studies of the flagellar filament [Arkhipov et al. 2006], which is formed by thousands of copies of a single protein, flagellin. Flagellin was coarse-grained with 500 atoms per CG bead, as shown in Figure 20.5. Segments of the filament (1100 flagellin units, or 0.5 μm long) were rotated clockwise and counterclockwise, with a constant rotation speed one turn in 10 μs applied to 33 protein units at the bottom of the segment. The simulations covered 30 μs each.

The filament is built by the helical arrangement of flagellin units, 11 per turn. A thread of units each separated by one turn is called a "protofilament" (see Figure 20.5); 11 protofilaments comprise the filament. In the CG simulations, the filament segments remained stable when rotated, but protofilaments rearranged dramatically (though it must be noticed that the torque applied to the model flagellum exceeded by far the one arising under native conditions). In the straight filament, which was the starting structure, the protofilaments form a right-handed helix with large helical period. When the torque is applied counterclockwise (as viewed from the base to the tip), the protofilaments remain arranged in right-handed helices, but the pitch of the helices rises; when the torque is opposite, the

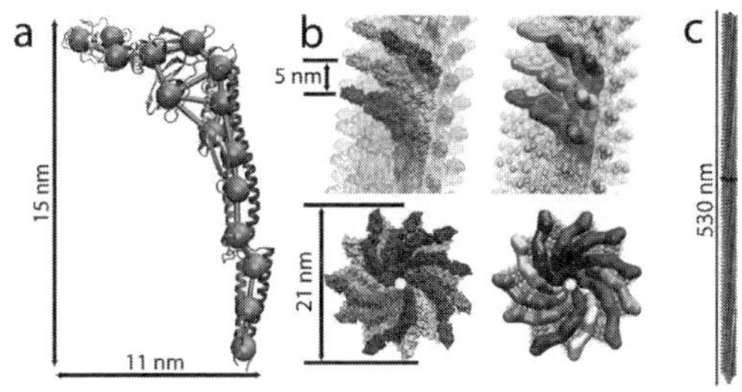

FIGURE 20.5 (See color insert following page 238.) CG of the flagellar filament. Unit proteins are represented by 15 CG beads (a). In (b), the flagellar filament viewed from the side and from the top is shown in all-atom (left) and CG (right) representations. A filament segment (1100 monomers) is shown in CG representation in (c). A single helix turn of 11 unit proteins is highlighted in black.

helices become left-handed. The filament also forms a helix as a whole. For the rotation corresponding to the running mode, the filament forms a left-handed helix, whereas for the tumbling mode it becomes a right-handed helix. The same difference in handedness between these helices is found in living bacteria [Turner, Ryu, and Berg 2000].

Running and tumbling modes of bacterial swimming are determined by structural transitions in the flagellar filament, depending on the direction of the applied torque. Clearly, interactions between protein units play an important role in enabling this transition. However, flagella act in solvent (water), and, curiously, the role of the solvent had not been analyzed much before. The effect of solvent was taken into account using Equation 20.3 [Arkhipov et al. 2006]. It was found that without friction due to solvent, flagella rotate as a rigid body; that is, the mutual positions of monomers are frozen, both for running and tumbling mode. With the solvent's friction present, the protofilaments rearrange as explained above, in agreement with structural changes in the flagellum suggested by experimental studies. Thus, the solvent (friction) plays a crucial role in the switching between the arrangements of protofilaments and, consequently, in producing supercoiling along the entire filament, or running and tumbling modes of motion.

20.5 FUTURE APPLICATIONS OF COARSE-GRAINING

Due to growing interest in large biomolecules and systems biology, coarse-grained simulations have grown increasingly common over the past few years as a means of accessing time and size scales that cannot be reached with all-atom MD. Recent advances such as more reliable force fields for residue-based CG [Marrink et al. 2007; Zhou et al. 2007], mixed CG and all-atom simulations [Shi, Izvekov, and Voth 2006; Praprotnik, Site, and Kremer 2006], and low-resolution shape-based CG models [Arkhipov et al. 2006; Arkhipov, Freddolino, and Schulten 2006] have improved the accuracy, flexibility, and potential scope of CG simulations. Since, however, coarse-grained simulations will never offer the same level of accuracy as all-atom simulations, it seems likely that CG simulations will naturally evolve in directions allowing closer links to atomistic descriptions. Both the aforementioned techniques of dynamic changes of scale and mixing CG and all-atom descriptions serve as useful and distinct models for how this can be accomplished, with the former using CG as an accelerant to improve sampling and then using all-atom simulations to flesh out the details of the sampled states, and the latter allowing less important parts of a system (such as bulk solvent) to be treated with a lower resolution than the regions of interest.

The utility of further development and application of these techniques can be illustrated, for example, for the case of the bacterial flagellum. Coarse-grained simulations have been used to investigate both the large-scale behavior of the flagellar filament during supercoiling [Arkhipov et al. 2006] and solvent dynamics around the supercoiled flagellum [Gebremichael, Ayton, and Voth 2006]; at the same time, large-scale all-atom simulations have offered a potential atomic-scale mechanism for differential supercoiling [Kitao et al. 2006]. The remaining challenge for theory is to fully link the CG and atomistic descriptions to provide a coherent and fully testable model for filament supercoiling; the most likely path for developing such a model is to use rotation of a shape-based CG filament to develop an ensemble of conformations at different points along the flagellum, which can then be simulated and perturbed in an all-atom representation to understand what interactions and structural transitions are important for the supercoiling process. A similar scale-switching approach could be applied to other systems, including viral capsids (allowing the study of assembly intermediates obtained from shape-based CG).

The shape-based CG methods should be further developed in a few important directions. Our present shape-based CG methodology [Arkhipov et al. 2006; Arkhipov, Freddolino, and Schulten 2006] allows one to simulate proteins. Despite initial successes, the protein model remains relatively rough and needs to be further refined, in particular with respect to the interaction potentials employed. These potentials can be improved using systematic all-atom parameterizing simulations for target systems. The same is true for the solvent model, which should be further developed along the lines of a true implicit solvent model, such as the generalized Born approach [Dominy and Brooks, 1999; Bashford and Case 2000; Mongan, Case, and McCammon 2004]. The CG method should also be extended to biomolecules other than proteins; to that end, we have recently started the development of a shape-based CG membrane model [Arkhipov, Yin, and Schulten 2008]. In this model, each leaflet of a lipid bilayer is represented by a collection of two-bead "molecules" (two beads connected by a spring), held together by nonbonded interactions tuned to mimic the bilayer stability, thickness, and area per lipid. This approach is similar to previous attempts of CG membrane simulations, such as by Reynwar et al. (2007). However, in our model each two-bead "molecule" represents a patch of a leaflet (not necessarily an integer number of lipid molecules), rather than a single lipid. Using the model, we have been able to simulate bilayer self-assembly and reproduce the results of all-atom and residue-based CG simulations of BAR domains (see above); much larger BAR domain simulations using the new model are under way. The shape-based CG model describing proteins and lipids will be very useful for simulations of subcellular processes, where multiple proteins interact with each other and with cellular membranes on long time scales.

Future residue-based CG simulations of nanodiscs will continue to further our understanding of HDL assembly and maturation, as well as aiding in the use of synthetic nanodiscs as protein scaffolds. HDL particles acting in vivo absorb esterified cholesterol for transport [Wang and Briggs 2004]; understanding the structural transitions involved in this process will be a key step in the overall goal of characterizing HDL function. This absorption process can be studied through residue-based CG simulations designed to observe how the structure of a nanodisc adjusts to the presence of esterified cholesterol. Ongoing simulations of nanodiscs will also be used to refine reverse CG methods for residue-based CG models to move from the snapshot-only reversal described above to a thermodynamically correct method for changing from all-atom to residue-based CG models.

The continued development and application of CG, along with ongoing improvements in generally available computational resources, promises to enable biomolecular simulations to treat many systems which were previously inaccessible. The increasing application of all-atom and CG simulations to the same system should greatly increase the impact of CG by allowing the CG method to be thoroughly tested or replaced by all-atom calculations when desired. CG simulations will be useful for understanding the behavior of cell-scale systems over millisecond time scales, and their role will increase with continuing improvements to CG potentials.

REFERENCES

Arkhipov, A., P. L. Freddolino, K. Imada, K. Namba, and K. Schulten. 2006. Coarse-grained molecular dynamics simulations of a rotating bacterial flagellum. *Biophys. J.* 91:4589–97.

Arkhipov, A., P. L. Freddolino, and K. Schulten. 2006. Stability and dynamics of virus capsids described by coarse-grained modeling. *Structure* 14:1767–77.

Arkhipov, A., Y. Yin, and K. Schulten. 2008. Four-scale description of membrane sculpting by BAR domains. *Biophys. J.* In press.

Baas, B. J., I. G. Denisov, and S. G. Sligar. 2004. Homotropic cooperativity of monomeric cytochrome P450 3A4 in a nanoscale native bilayer environment. *Arch. Biochem. Biophys.* 430:218–28.

Baron, R., D. Trzesniak, A. H. de Vries, A. Elsener, S. J. Marrink, and W. F. van Gunsteren. 2007. Comparison of thermodynamic properties of coarse-grained and atomic-level simulation models. *Chem. Phys. Chem.* 8:452–61.

Bashford, D., and D. A. Case. 2000. Generalized Born models of macromolecular solvation effects. *Annu. Rev. Phys. Chem.* 51:129–52.

Bayburt, T. H., Y. V. Grinkova, and S. G. Sligar. 2002. Self-assembly of discoidal phospholipid bilayer nanoparticles with membrane scaffold proteins. *Nano Lett.* 2:853–56.

———. 2006. Assembly of single bacteriorhodopsin trimers in bilayer nanodiscs. *Arch. Biochem. Biophys.* 450:215–22.

Berg, H. C. 2000. Motile behavior of bacteria. *Phys. Today* 53:24–29.

Blood, P. D., and G. A. Voth. 2006. Direct observation of Bin/amphiphysin/Rvs (BAR) domain-induced membrane curvature by means of molecular dynamics simulations. *Proc. Natl. Acad. Sci. U.S.A.* 103:15068–72.

Boldog, T., S. Grimme, M. Li, S. G. Sligar, and G. L. Hazelbauer. 2006. Nanodiscs separate chemoreceptor oligomeric states and reveal their signaling properties. *Proc. Natl. Acad. Sci. U.S.A.* 103:11509–14.

Bond, P. J., and M. S. P. Sansom. 2006. Insertion and assembly of membrane proteins via simulation. *J. Am. Chem. Soc.* 128:2697–704.

Civjan, N. R., T. H. Bayburt, M. A. Schuler, and S. G. Sligar. 2003. Direct solubilization of heterologously expressed membrane proteins by incorporation into nanoscale lipid bilayers. *Biotechniques* 35:556–60, 562–63.

Das, P., S. Matysiak, and C. Clementi. 2005. Balancing energy and entropy: A minimalist model for the characterization of protein folding landscapes. *Proc. Natl. Acad. Sci. U.S.A.* 102 (29):10141–46.

Davydov, D. R., H. Fernando, B. J. Baas, S. G. Sligar, and J. R. Halpert. 2005. Kinetics of dithionite-dependent reduction of cytochrome P450 3A4: Heterogeneity of the enzyme caused by its oligomerization. *Biochemistry* 44:13902–13.

Denisov, I. G., Y. V. Grinkova, A. A. Lazarides, and S. G. Sligar. 2004. Directed self-assembly of monodisperse phospholipid bilayer nanodiscs with controlled size. *J. Am. Chem. Soc.* 126:3477–87.

Dominy, B. N., and C. L. Brooks, III. 1999. Development of a generalized Born model parametrization for proteins and nucleic acids. *J. Phys. Chem. B.* 103:3765–73.

Duan, H., N. R. Civjan, S. G. Sligar, and M. A. Schuler. 2004. Coincorporation of heterologously expressed Arabidopsis cytochrome P450 and P450 reductase into soluble nanoscale lipid bilayers. *Arch. Biochem. Biophys.* 424:141–53.

Flint, S. J., L. W. Enquist, V. R. Racaniello, and A. M. Skalka. 2004. *Principles of virology.* 2nd ed. Washington, DC: ASM Press.

Freddolino, P. L., A. S. Arkhipov, S. B. Larson, A. McPherson, and K. Schulten. 2006. Molecular dynamics simulations of the complete satellite tobacco mosaic virus. *Structure* 14:437–49.

Gebremichael, Y., G. S. Ayton, and G. A. Voth. 2006. Mesoscopic modeling of bacterial flagellar microhydrodynamics. *Biophys. J.* 91:3640–52.

Gorshkova, I. N., T. Liu, H. Y. Kan, A. Chroni, V. I. Zannis, and D. Atkinson. 2006. Structure and stability of apolipoprotein a-I in solution and in discoidal high-density lipoprotein probed by double charge ablation and deletion mutation. *Biochemistry* 45:1242–54.

Humphrey, W., A. Dalke, and K. Schulten. 1996. VMD: Visual molecular dynamics. *J. Mol. Graphics* 14:33–38.

Izvekov, S., and G. A. Voth. 2005a. A multiscale coarse-graining method for biomolecular systems. *J. Phys. Chem. B.* 109 (7):2469–73.

———. 2005b. Multiscale coarse graining of liquid-state systems. *J. Chem. Phys.* 123:134105.

———. 2006. Multiscale coarse-graining of mixed phospholipid/cholesterol bilayers. *J. Chem. Theory Comput.* 2:637–48.

Katsoulakis, M. A., A. J. Majda, and D. G. Vlachos. 2003. Coarse-grained stochastic processes for microscopic lattice systems. *Proc. Natl. Acad. Sci. U.S.A.* 100 (3):782–87.

Kitao, A., K. Yonekura, S. Maki-Yonekura, F. A. Samatey, K. Imada, K. Namba, and N. Go. 2006. Switch interactions control energy frustration and multiple flagellar filament structures. *Proc. Natl. Acad. Sci. U.S.A.* 103 (13):4894–99.

Koppaka, V., L. Silvestro, J. A. Engler, C. G. Brouillette, and P. H. Axelsen. 1999. The structure of human lipoprotein A-I. Evidence for the "belt" model. *J. Biol. Chem.* 274:14541–44.

Kuznetsov, Y. G., S. Daijogo, J. Zhou, B. L. Semler, and A. McPherson. 2005. Atomic force microscopy analysis of icosahedral virus RNA. *J. Mol. Biol.* 347:41–52.

Leach, A. R. 1996. *Molecular modelling, principles and applications*. Essex: Addison Wesley Longman Limited, Harlow, England.

Levine, A. J. 1991. *Viruses*. Scientific American Library.

Li, H., D. S. Lyles, M. J. Thomas, W. Pan, and M. G. Sorci-Thomas. 2000. Structural determination of lipid-bound ApoA-I using fluorescence resonance energy transfer. *J. Biol. Chem.* 275:37048–54.

Li, Y., A. Z. Kijac, S. G. Sligar, and C. M. Rienstra. 2006. Structural analysis of nanoscale self-assembled discoidal lipid bilayers by solid-state NMR spectroscopy. *Biophys. J.* 91:3819–28.

Lucas, R. W., S. B. Larson, and A. McPherson. 2002. The crystallographic structure of brome mosaic virus. *J. Mol. Biol.* 317:95–108.

Lyman, E., F. M. Ytreberg, and D. M. Zuckerman. 2006. Resolution exchange simulation. *Phys. Rev. Lett.* 96 (2):028105.

MacKerell, A. D., Jr., D. Bashford, M. Bellott, R. L. Dunbrack, Jr., J. Evanseck, M. J. Field, S. Fischer, J. Gao, H. Guo, S. Ha, D. Joseph, L. Kuchnir, K. Kuczera, F. T. K. Lau, C. Mattos, S. Michnick, T. Ngo, D. T. Nguyen, B. Prodhom, I. W. E. Reiher, B. Roux, M. Schlenkrich, J. Smith, R. Stote, J. Straub, M. Watanabe, J. Wiorkiewicz-Kuczera, D. Yin, and M. Karplus. 1998. All-atom empirical potential for molecular modeling and dynamics studies of proteins. *J. Phys. Chem. B* 102:3586–616.

Marrink, S. J., A. H. de Vries, and A. E. Mark. 2004. Coarse grained model for semiquantitative lipid simulations. *J. Phys. Chem. B* 108:750–60.

Marrink, S. J., and A. E. Mark. 2002. Molecular dynamics simulations of mixed micelles modelling human bile. *Biochemistry* 41:5375–82.

Marrink, S. J., and A. E. Mark. 2003. Molecular dynamics simulation of the formation, structure, and dynamics of small phospholipid vesicles. *J. Am. Chem. Soc.* 125:15233–42.

———. 2004. Molecular view of hexagonal phase formation in phospholipid membranes. *Biophys. J.* 87:3894–900.

Marrink, S. J., J. Risselada, and A. E. Mark. 2005. Simulation of gel phase formation and melting in lipid bilayers using a coarse grained model. *Chem. Phys. of Lipids* 135 (2):223–44.

Marrink, S. J., H. J. Risselada, S. Yefimov, D. P. Tieleman, and A. H. de Vries. 2007. The Martini forcefield: Coarse grained model for biomolecular simulations. *J. Phys. Chem. B* 111:7812–24.

Martinetz, T., and K. Schulten. 1994. Topology representing networks. *Neural Netw.* 7 (3):507–22.

Mongan, J., D. A. Case, and J. A. McCammon. 2004. Constant pH molecular dynamics in generalized Born implicit solvent. *J. Comp. Chem.* 25:2038–48.

Nielsen, J. E., and J. A. McCammon. 2003. On the evaluation and optimization of protein X-ray structures for pKa calculations. *Protein Sci.* 12:313–26.

Nielsen, S. O., C. F. Lopez, G. Srinivas, and M. L. Klein. 2004. Coarse grain models and the computer simulation of soft materials. *J. Phys.: Condens. Matter* 16:R481–512.

Panagotopulos, S. E., E. M. Horace, J. N. Maiorano, and W. S. Davidson. 2001. Apolipoprotein A-I adopts a belt-like orientation in reconstituted high density lipoproteins. *J. Biol. Chem.* 276:42965–70.

Peter, B. J., H. M. Kent, I. G. Mills, Y. Vallis, P. Johnathon, G. Butler, P. R. Evans, and H. T. McMahon. 2004. BAR domains as sensors of membrane curvature: The amphiphysin BAR structure. *Science* 303:495–99.

Phillips, J. C., R. Braun, W. Wang, J. Gumbart, E. Tajkhorshid, E. Villa, C. Chipot, R. D. Skeel, L. Kale, and K. Schulten. 2005. Scalable molecular dynamics with NAMD. *J. Comp. Chem.* 26:1781–802.

Praprotnik, M., L. D. Site, and K. Kremer. 2005. Adaptive resolution molecular dynamics simulation: Changing the degrees of freedom on the fly. *J. Chem. Phys.* 123:224106.

———. 2006. Adaptive resolution scheme for efficient hybrid atomistic-mesoscale molecular dynamics simulations of dense liquids. *Phys. Rev. E* 73:066701.

Reith, D., M. Pütz, and F. Müller-Plathe. 2003. Deriving effective mesoscale potentials from atomistic simulations. *J. Comp. Chem.* 24:1624–36.

Ren, G., P. Vajjhala, J. S. Lee, B. Winsor, and A. L. Munn. 2006. The BAR domain proteins: Molding membranes in fission, fusion, and phagy. *Microbiol. Mol. Biol. Rev.* 70:37–120.

Reynwar, B. J., G. Illya, V. A. Harmandaris, M. M. Müller, K. Kremer, and M. Deserno. 2007. Aggregation and vesiculation of membrane proteins by curvature-mediated interactions. Nature 447:461–64.

Sakamuro, D., K. J. Elliott, R. Wechsler-Reya, and G. C. Prendergast. 1996. BIN1 is a novel myc-interacting protein with features of a tumour suppressor. *Nat. Genet.* 14:69–77.

Samatey, F. A., K. Imada, S. Nagashima, F. Vonderviszt, T. Kumasaka, M. Yamamoto, and K. Namba. 2001. Structure of the bacterial flagellar protofilament and implications for a switch for supercoiling. *Nature* 410:331–37.

Sastry, K., D. D. Johnson, D. E. Goldberg, and P. Bellon. 2005. Genetic programming for multitimescale modeling. *Phys. Rev. B* 72 (8):085438.

Schütte, Ch., A. Fischer, W. Hiosinga, and P. Deuflhard. 1999. A direct approach to conformational dynamics based on hybrid Monte Carlo. *J. Comput. Phys.* 151:146–68.

Seddon, A. M., P. Curnow, and P. J. Booth. 2004. Membrane proteins, lipids and detergents: Not just a soap opera. *Biochim. Biophys. Acta* 1666:105–17.

Shelley, J. C., M. Y. Shelley, R. C. Reeder, S. Bandyopadhyay, P. B. Moore, and M. L. Klein. 2001. Simulations of phospholipids using a coarse grain model. *J. Phys. Chem. B* 105:9785–92.

Shi, Q., S. Izvekov, and G. A. Voth. 2006. Mixed atomistic and coarse-grained molecular dynamics: Simulation of a membrane-bound ion channel. *J. Phys. Chem. B* 110 (31):15045–48.

Shih, A. Y., A. Arkhipov, P. L. Freddolino, and K. Schulten. 2006. Coarse grained protein–lipid model with application to lipoprotein particles. *J. Phys. Chem. B* 110:3674–84.

Shih, A. Y., A. Arkhipov, P. L. Freddolino, S. G. Sligar, and K. Schulten. 2007a. Assembly of lipids and proteins into lipoprotein particles. *J. Phys. Chem. B* 111:11095–104.

Shih, A. Y., I. G. Denisov, J. C. Phillips, S. G. Sligar, and K. Schulten. 2005. Molecular dynamics simulations of discoidal bilayers assembled from truncated human lipoproteins. *Biophys. J.* 88:548–56.

Shih, A. Y., P. L. Freddolino, A. Arkhipov, and K. Schulten. 2007b. Assembly of lipoprotein particles revealed by coarse-grained molecular dynamics simulations. *J. Struct. Biol.* 157:579–92.

Shih, A. Y., P. L. Freddolino, S. G. Sligar, and K. Schulten. 2007c. Disassembly of nanodiscs with cholate. *Nano Lett.* 7:1692–96.

Silva, R. A. G. D., G. M. Hilliard, L. Li, J. P. Segrest, and W. S. Davidson. 2005. A mass spectrometric determination of the conformation of dimeric apolipoprotein A-I in discoidal high density lipoproteins. *Biochemistry* 44:8600–607.

Sligar, S. G. 2003. Finding a single-molecule solution for membrane proteins. *Biochem. Biophys. Res. Commun.* 312:115–19.

Stevens, M. J. 2004. Coarse-grained simulations of lipid bilayers. *J. Chem. Phys.* 121:11942–48.

Stevens, M. J., J. H. Hoh, and T. B. Woolf. 2003. Insights into the molecular mechanism of membrane fusion from simulations: Evidence for the association of splayer tails. *Phys. Rev. Lett.* 91:188102.

Svergun, D. I., C. Barberato, and M. H. J. Koch. 1995. CRYSOL: A program to evaluate X-ray solution scattering of biological macromolecules from atomic coordinates. *J. Appl. Cryst.* 28:768–73.

Tozzini, V., and A. McCammon. 2005. A coarse grained model for the dynamics of flap opening in HIV-1 protease. *Chem. Phys. Lett.* 413:123–28.

Tricerri, M. A., A. K. Behling Agree, S. A. Sanchez, J. Bronski, and A. Jonas. 2001. Arrangement of apolipoprotein A-I in reconstituted high-density lipoprotein disks: An alternative model based on fluorescence resonance energy transfer experiments. *Biochemistry* 40:5065–74.

Turner, L., W. S. Ryu, and H. C. Berg. 2000. Real-time imaging of fluorescent flagellar filaments. *J. Bacteriol.* 182 (10):2793–801.

Wang, M., and M. R. Briggs. 2004. HDL: The metabolism, function, and therapeutic importance. *Chem. Rev.* 104:119–37.

Zhou, J., I. F. Thorpe, S. Izvekov, and G. A. Voth. 2007. Coarse-grained peptide modeling using a systematic multiscale approach. *Biophys. J.* 92 (12):4289–303.

21 Coarse-Graining Protein Mechanics

Richard Lavery
Institute de Biologie et Chimie des Protéines, Université de Lyon

Sophie Sacquin-Mora
Laboratoire de Biochimie Théorique, Institut de Biologie Physico-Chimique

CONTENTS

21.1 INTRODUCTION

Almost 50 years after the first protein structures were solved [1,2], structural databases now contain tens of thousands of structures, which have been extensively analyzed and classified. Despite these efforts, we still have relatively little understanding of how structure is related to the mechanical and dynamical properties of proteins, which are nevertheless indissociable features of protein function. This situation is beginning to change because of progress in both experimental and theoretical approaches.

Experimentally, both of the methods for determining high-resolution structures, X-ray crystallography and NMR spectroscopy, also provide some information on protein flexibility. First, it is possible to compare structures resolved with or without interacting species, or, in the case of enzymes, to capture intermediate conformational states using unreactive substrate analogs. Both methods can also provide finer data on the positional fluctuations of individual residues within proteins in terms of Debye–Waller temperature factors or order parameters.

A new route to mechanical probing has recently arisen with the development of single-molecule experiments [3,4], which enable a protein to be pulled apart, either by tethers on its N- and C-termini or, in "triangulation" experiments, between other residue pairs [5,6]. The latter approach has convincingly demonstrating that, not surprisingly, proteins respond differently depending on the direction of the applied forces.

Theoretically, a number of different approaches have been applied to analyzing protein flexibility. First amongst these are all-atom molecular dynamics simulations, taking into account the surrounding solvent (generally represented by explicit solvent molecules, but also, potentially, with simpler continuum representations). Such simulations are generally limited to the nanosecond time scale and are expensive in terms of computer resources. They are thus generally limited to studying specific cases, although this situation is changing today [7]. Dynamic trajectories can be analyzed to understand which parts of a protein are the most mobile, how domains move with respect to one another within multidomain structures or how much and how fast individual amino acid side chains can change their conformational substates [8,9]. Trajectories can also be biased in a number of ways to mimic external forces acting on proteins and thus to model single-molecule experiments (albeit on a very different and much faster time scale) [10] or environmental forces such as membrane tension [11].

Simpler methods, notably those based on elastic network models [12,13], can also provide valuable data on protein deformations, despite the fact that these models generally ignore the difference between individual amino acid residues and are guided only by the proximity of residues within the 3D structure of the protein. Thus, the so-called Gaussian network model (GNM), which extracts normal modes from an elastic network protein representation, has been shown to provide useful information on the slow, large-amplitude, collective motions which characterize domain movements, allosteric effects, and enzyme activity [14]. Elastic network models can also be used to calculate the atomic fluctuations. These can be converted to temperature factors (also termed *B*-factors), which generally show good overall correlations with those measured crystallographically. It has recently been shown that this correlation can be further improved by taking crystal-packing effects into account [15]. Good correlations have also been found with the conformational fluctuations represented by the multiple structures compatible with NMR data [16]. It has also been found that elastic network models are capable of reproducing the anisotropy of protein fluctuations to a surprisingly good extent [17].

Other coarse-grain approaches to protein flexibility include graph-theoretic models based on the concept of tensegrity (which determines the residual degrees of freedom in a mechanically linked system) [18]. These, along with elastic network approaches, have also become the basis of a variety of multiscale coarse-grain models [19–21].

We started to become interested in protein mechanics as a result of our earlier work on the mechanics of DNA [22,23] and the associated base-sequence-dependent mechanical properties for understanding protein–DNA recognition [24,25]. From the beginning of our studies, we were interested in defining mechanical properties on the residue level since this seemed to be the easiest way of making comparisons with data on biological function, the impact of point mutations, differences between homologous proteins and so on.

We were unsatisfied with the possibility of using temperature factors to answer these questions, notably because of the work of Halle [26], which showed convincingly that temperature factors basically reflect only local structure, and, in particular, local atomic packing densities. We consequently looked for a new measure. Although one obvious approach was to copy the single-molecule triangulation experiments cited above and test the resistance of all residue–residue (or atom-atom) vectors, this method has the disadvantage that it does not easily yield properties that can be associated with individual residues. Tests on the ease of displacing residues with respect to the center of mass of the protein also turned out to be unsatisfactory because observed flexibility could again be attributed either to the probed residue or to the center of mass (for example, because of the movement of a flexible region on the distal side of the protein with respect to the probed residue) [27].

We finally found that testing the displacement of each residue with respect to the rest of the protein structure gave the most interesting results. This involved asking how much energy was necessary to change the mean distance d_i from residue i to all other residues $j \neq i$ in an N residue protein:

$$d_i = \frac{1}{N-1} \sum_{j=1, j \neq i}^{N} \left| r_i - r_j \right|.$$

Note that the position of each residue r_i can be characterized by a single atom, such as C_α. The mean distance d_i can alternatively be obtained by averaging over the mean distances for each atom in a given residue. If the mean distance was successively decreased and increased, we obtained an energy versus mean distance plot. For distance changes of the order of a few tenths of angstroms, these plots turned out to be virtually quadratic and could thus be characterized by the second derivative at the energy minimum, or, in other words, an effective force constant (hereafter denoted k_i) for displacing a residue i within the whole protein structure. Note that d_i is a scalar quantity. Changes in d_i leave all residues free to move in their energetically optimal directions.

The studies we have subsequently carried out on a variety of proteins [27–29] show that the associated force constants are a very interesting guide to protein mechanics. They reveal the extent of the mechanical heterogeneity induced by the complex 3D shapes of proteins and suggest that this heterogeneity plays a significant role in preparing proteins for their biological functions. We have notably found that mechanical properties seem to be very useful in identifying active sites, which in turn provides valuable information for determining protein function [30], a major problem in our postgenomic era [31]. This chapter summarizes the approaches that we have used to obtain residue-by-residue force constants, gives an example of their application to a specific protein, and speculates on future developments.

21.2 METHODOLOGY

Our earliest studies in this field used all-atom protein representations and a conventional AMBER force field [32] combined with a generalized Born continuum solvent model [33]. An internal coordinate minimization program based on JUMNA [34] was used to relax the protein structure and then to perturb the C_α position of each residue in turn by constraining the mean distance to all other C_αs to increase or decrease. This approach was naturally slow since it typically required four energy minimizations (± 0.1 Å, ± 0.2 Å) for each residue. We thus looked for ways of speeding up the calculation.

This was achieved in two steps. Firstly, we noted that rather than physically constraining each residue to move within the overall protein structure, we could simply analyze the fluctuations of the mean distance d_i from each residue (to the rest of the structure) occurring naturally within a molecular dynamics simulation [27]:

$$k_i = \frac{3k_\mathrm{B}T}{\left\langle \left(d_i - \langle d_i \rangle\right)^2 \right\rangle},$$

where d_i is the mean distance defined above, $\langle\, \rangle$ denotes the average over the simulation, k_B is the Boltzmann constant and T is the temperature of the simulation. This implied that the N-residue force constants could be obtained from a single dynamic trajectory rather than from $4N+1$ minimizations. The results obtained in this way were very similar to those derived from constrained energy minimization. However, since all-atom dynamics simulations generally require an explicit solvent representation to avoid deforming the initial protein structure, the resulting computational cost was still high.

We consequently turned to simpler elastic network models [12–14] to gain time. Although we made initial trials with one point per residue models, where each amino acid gives rise to a single node in the elastic network (positioned on the C_α atom), it was clear that a more refined model which could distinguish between the various types of amino acid would be necessary if we wanted to study the impact of sequence mutations. We consequently adopted the model proposed by Zacharias [35,36] which has two or three points per amino acid and has already proved effective in protein-protein docking studies. In this model, each amino acid has one pseudoatom at the C_α position. Small side chains (excepting glycine) have a second pseudoatom at the geometric center of the heavy atoms of the side chain, while larger side chains (Arg, Gln, Glu, His, Lys, Met, Trp, Tyr) have a pseudoatom at the center of the C_β-C_γ bond and a third pseudoatom at the geometrical center of the heavy atoms of the side-chain atoms beyond C_γ [35].

With this coarse-grain protein representation, the force field was also simplified to a set of quadratic springs placed between all pseudoatoms lying below a chosen cutoff distance. We chose a distance of 9 Å. All springs had identical force constants of 0.6 kcal mol^{-1} Å^{-2} (note that changing this value simply acts as an overall scale factor on the final results). With this type of representation, it is appropriate to replace Newtonian dynamics with stochastic Brownian dynamics (BD), which ignores inertial effects and treats solvent only through random forces and hydrodynamic drag. Full details of the BD simulation protocol we use can be found in one of our earlier publication [28].

21.3 RESULTS AND DISCUSSION

21.3.1 Force Constant "Spectra"

We have chosen to illustrate our force constant calculations using a soluble enolase [37]. The structure of this dimeric protein, PDB 2AL1 [38], has been solved to a resolution of 1.5 Å in the presence of its substrates, 2-phospho-D-glycerate (2-PGA) and phosphoenol-pyruvate (PEP), and two magnesium ions. Figure 21.1a shows a cartoon version of this α/β-fold protein with its two monomers colored dark and light gray. Each monomer consists of two domains and the substrates, in this case 2-PGA (black), are bound within the C-terminal domains. The substrate-binding pocket shown in Figure 21.1b involves residues Ser-39, His-159, Glu-168, Glu-211, Lys-345, His-373, and Lys-396, with Lys-345 and Glu-211 serving as acid/base catalysts in the interconversion of 2-PGA and PEP [37]. Note that Ser-39 has been excluded from Figure 21.1b for clarity. The coordination of the two magnesium ions in the enolase (black spheres) also involves residues Ser-39, Asp-246, Glu-295, and Asp-320 [39].

The force constants calculated for this protein, by analyzing the fluctuations from a BD simulation on a 2–3-point representation, are shown in Figure 21.2. The inhibitor was not represented by elastic network points and consequently has no impact on the force constant calculation. The reader

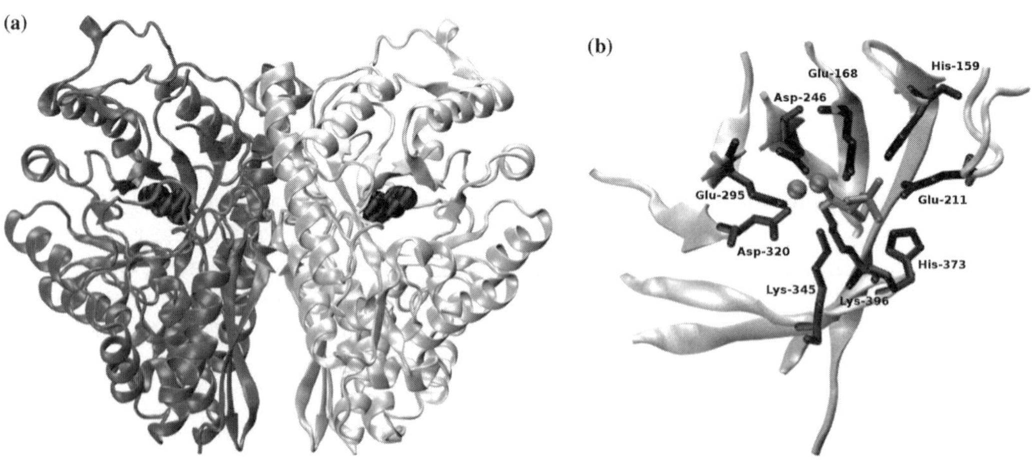

FIGURE 21.1 (a) Cartoon representation of a yeast enolase dimer, PDB 2AL1 [37]. The two monomeric units are colored in light and dark gray and the 2-PGA substrates are shown in black. All the molecular graphics in this article were prepared using VMD [50]. (b) Simplified representation of the active site and β-barrel of enolase (2AL1). Catalytic and magnesium-binding residues are in black, and the two magnesium ions and the 2-PGA substrate are in dark gray. Ser-39 has been omitted for clarity.

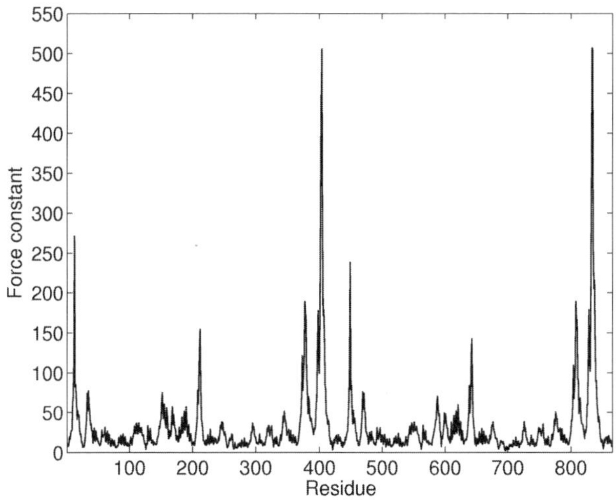

FIGURE 21.2 Force constant plot for enolase. The residues are numbered consecutively and the two monomeric units follow one another along the abscissa. Force constants in Figure 21.2, Figure 21.3, and Figure 21.5 are in units of kcal mol^{-1} Å $^{-2}$.

is referred to our earlier publications, which show that very similar results are obtained whether the force constants are calculated by energy minimization or BD simulations and also that bound ligands generally have very little effect on the results [28,29]. Note that the residues have been numbered consecutively in the force plot shown in Figure 21.2.

The first striking observation concerning these results is that the force constants are highly variable and often change sharply from one residue to the next. Here the values range from 3 to 507 kcal mol^{-1} Å^{-2} with a standard deviation of 48 around an average of 32 kcal mol^{-1} Å^{-2} (note: 1 kcal mol^{-1} Å^{-2} = 0.07 nN Å^{-1}). In Figure 21.2, the force constants for the two monomers follow one another, giving rise to the horizontally repeating pattern. Figure 21.3a shows the results for the first monomer in more detail. It can be seen that the largest force constants occur for residues in the core of the dimer. Their location is illustrated graphically by dark shading in Figure 21.4a for the residues in the right-hand monomer. (Note that these results can be seen better in the color version of Figure 21.4, where high force constant residues are shown in green.) It can be seen that the highest force constants occur for residues at the junction between the two monomers. In contrast, except for Glu-211 and His-373, no residues with high force constants are found in the active site pocket, as can been seen in Figure 21.3a, where the circles and triangles indicate the values corresponding respectively to the active site and the magnesium-ion-binding residues cited above.

We have found that this behavior is common to most multidomain proteins and reflects the fact that domain movements leave the residues at the junctions virtually undisturbed [28,29]. This leads to high force constants in our approach, since the mainly rotational movements of the domains do not modify the distances of other residues to these hinge points. Similar findings have been observed with normal mode analyses of elastic network models [40,41]. To avoid this effect dominating the force constant spectra, we have developed a so-called domain separation approach. This consists of calculating force constants for changing the mean distance for a given residue with respect to the subset of other residues belonging to the same domain. Note that this change does not influence the elastic network representation, which still includes all residues from all domains.

The results of this procedure are shown for a single monomer in the plot in Figure 21.3b and illustrated graphically in Figure 21.4b. It is now seen that the residues with the highest force constants (black, or green in the color version of the figure) lie near the center of the C-terminal domain, shown for the right-hand monomer in Figure 21.4b, and close to the substrate-binding site. Five of

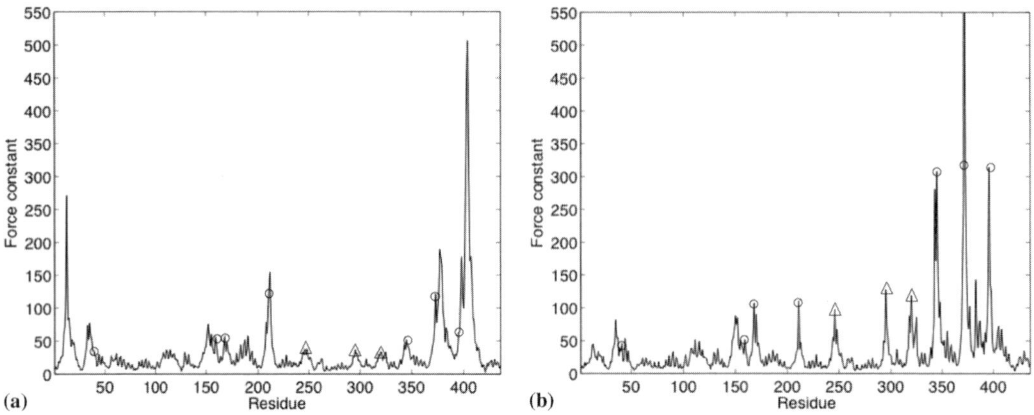

FIGURE 21.3 (a) Force constant plot for the first monomer of enolase. Circles indicate the active site residues (from left to right: Ser-39, His-159, Glu-168, Glu-211, Lys-345, His-373, and Lys-396) and triangles indicate residues binding the magnesium ions (from left to right: Asp-246, Glu-295, and Asp-320). (b) Force constant plot for the first monomer of enolase after domain separation. Circles indicate the active site residues (from left to right: Ser-39, His-159, Glu-168, Glu-211, Lys-345, His-373, and Lys-396) and triangles indicate residues binding the magnesium ions (from left to right: Asp-246, Glu-295, and Asp-320).

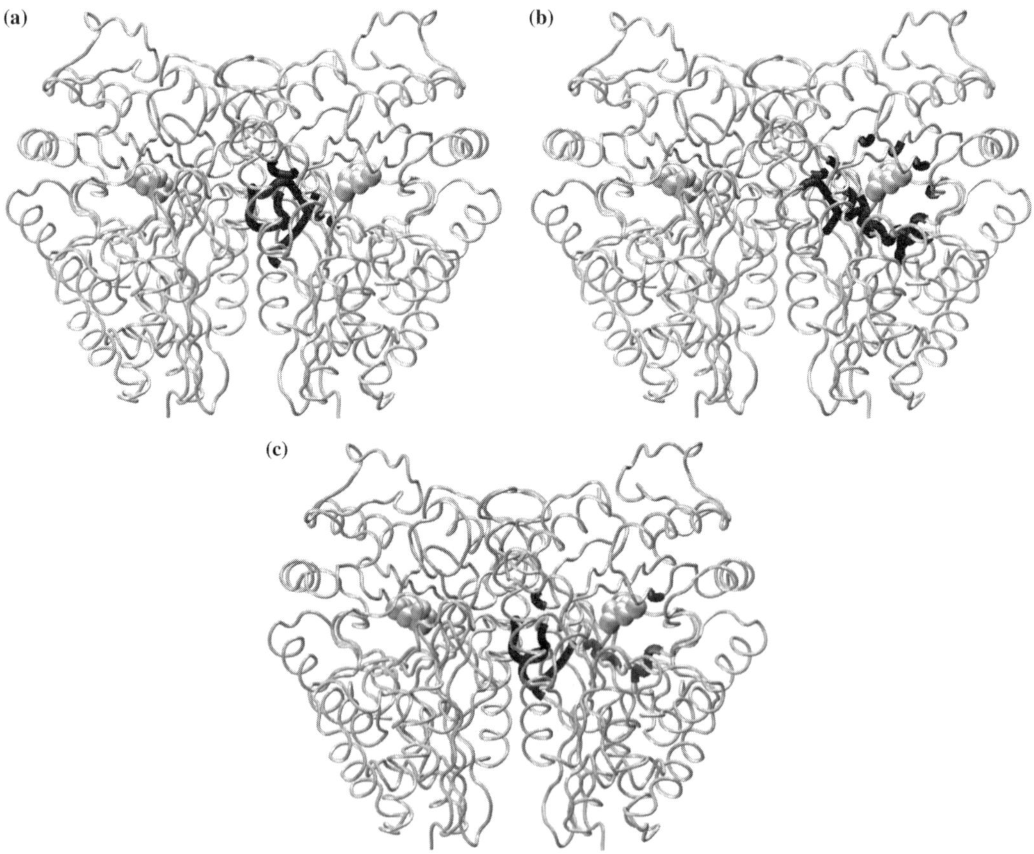

FIGURE 21.4 (See color insert following page 238.) (a) Backbone diagram of enolase. Residues with high force constants within the right-hand monomer are shown in black. (b) Following domain separation, residues with high force constants within the right-hand domain are shown in black. (c) Mechanical changes in passing from the monomeric to the dimeric form of enolase. Residues with significantly increased force constants are shown in black and those with significantly decreased force constants in gray (changes are only shown for the right-hand domain).

the seven key active-site residues, indicated by circles in Figure 21.3b, now lie within force constant peaks and, in particular, the catalytic residues Lys-345, His-373, and Lys-396 represent three of only four residues having force constants above 300 kcal mol^{-1} $Å^{-2}$ within the monomer. It is interesting to note that, after domain separation, rigidity peaks corresponding to the magnesium-binding residues also become visible. As shown by the triangles in Figure 21.3b, all three of these residues (Asp-246, Glu-295, and Asp-320) are now in force constant peaks.

21.3.2 Locating Active Sites

The example of enolase illustrates the general behavior of mechanical properties of enzymes. In a recent study, we looked at a group of almost 100 enzymes containing proteins belonging to all the main enzymatic families [29]. In the vast majority of the cases studied, the active sites of these residues, as defined in the Catalytic Site Atlas database [42] or in an earlier elastic network study [43], turned out to be amongst the most strongly fixed residues within the protein structures. During this study, bound ligands or inhibitors where again ignored and the domain separation approach was applied to proteins with nonsymmetric domains and more than one active site.

Since the range of force constants varies with the size of each protein (being in general larger for larger proteins), we also normalized their values by converting them to Z-scores, that is, units of standard variation $\sigma(k)$ with respect to the mean $\langle k \rangle$:

$$k' = \frac{k - \langle k \rangle}{\sigma(k)},$$

where both $\sigma(k)$ and $\langle k \rangle$ are calculated protein by protein. Using these values, it turns out that active site residues are generally associated with force constants well above the mean. By setting a cutoff at $k' = 0$, the residues with force constants above the average represent only 28% of the total set (the overall distribution is highly skewed to lower values). This set is very highly enriched in active site residues, containing 78% of all such residues and only 25% of other residues. Consequently, rigidity within the overall protein structure seems to be a good guide to catalytic activity.

This is a somewhat surprising result, given that active site residues are generally assumed to be amongst the most flexible, flexibility being necessary for them to carry out their catalytic functions [44]. However, the reverse has already been found by an analysis of temperature factors [43,45,46] and by looking at the residue fluctuations associated with the low-frequency normal modes representing collective motions [43]. These results are in line with our present findings.

21.3.3 Conformational Versus Mechanical Changes

We have used the increased resolution of the multipoint Zacharias amino acid representation to compare the impact of conformational change within a given protein on its mechanical properties. In our study of hemoproteins [28], we were able to detect differences in the rigidity profile of the active and inactive forms of cytochrome c peroxidase, which correlated well with the known role of the active site residues in this enzyme's function. Here we compare the flexibility of yeast enolase in its active form, complexed with two magnesium ions, PDB 2AL1, and in an inactive form, complexed with one calcium ion, PDB 5ENL [47]. These two structures are very similar to one another, with an average C_α RMSD of 1.2 Å. The main conformational change involves an important opening movement of the backbone loop between residues 36 and 44. The average C_α RMSD of these amino acids is 7.4 Å. The variations in residue rigidity when changing from the active to the inactive form of enolase are shown in the upper curve of Figure 21.5. Except for His-373, all the residues involved in substrate and magnesium binding show a decrease in their force constants, thus suggesting a globally more flexible catalytic site in the inactive form of the protein.

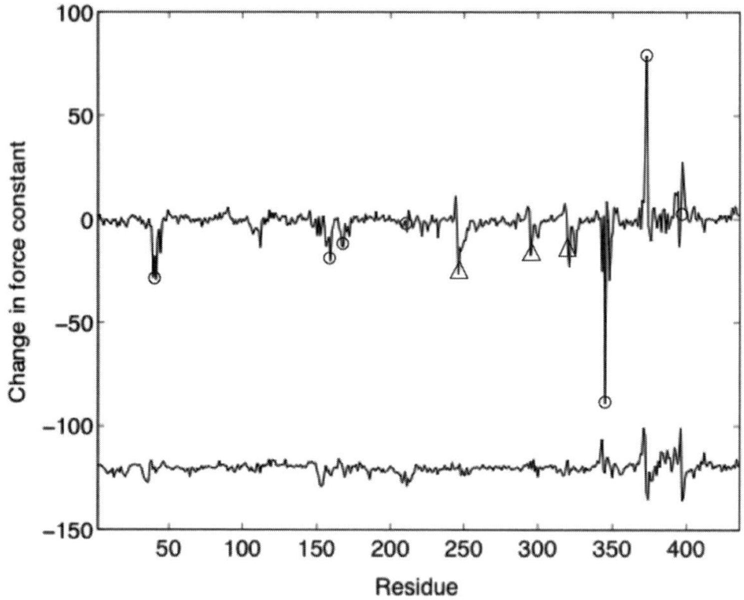

FIGURE 21.5 Changes in the force constants when passing from the active to the inactive form of enolase. Upper curve: Force constants calculated using the Zacharias reduced multipoint amino acid representation. Lower curve: Force constants calculated using a single-point-per-residue representation (with a vertical offset of −120 kcal mol^{-1} $Å^{-2}$ for clarity).

It is worth noting that these more detailed studies of protein mechanics require the improved resolution of the multipoint Zacharias representation. This is clearly shown in the lower plot in Figure 21.5, which was obtained using a one-point-per-residue protein representation. As seen, this cruder representation which ignores the size and conformation of the amino acid side chains shows little structure and does not single out any particular behavior for the active site or ion-binding residues.

21.3.4 Architectural Fingerprints in the Force Constant Spectra

Most of the proteins studied to date show high force constants for a number of residues other than those in the active sites. In some case, these residues are simply close to the active site residues and presumably play a role in maintaining its overall rigidity. However, in other cases, the residues are far from the active site. One such example, seen in our study of hemoproteins [28], involved two pairs of highly conserved residues at the junction between two α-helices within proteins of the cytochrome *c* family. These residues have been identified as playing key roles in the folding of such proteins [48]. Another very preliminary study of cytochrome *c* (see the supplementary material of Ref. [28]) suggested that there might be some correlation between the folding units ("foldons") identified by hydrogen exchange experiments [49] and our calculated force constants, the groups of highly rigid residues along the primary sequence being generally associated with early folding units. This suggests that mechanical properties may reflect to some extent the protein-folding pathways. More data are however needed to test this hypothesis.

We have also observed that, in some cases, high force constants are a signature of the overall protein structure, as in the case of residues lying within each β-strand within β-barrel domains [28]. This behavior is seen in our enolase test case where the active site is located at the top of an eight-stranded β-barrel. The barrel fold of this protein is reflected in the force constant spectra as the series of peaks starting at residue Asn-152. Obviously, much remains to be studied in this area. One possibility is that such "architectural fingerprints" can be defined for each family of protein folds and then removed from the overall force constant spectra, making it still easier to detect active site residues.

Finally, it is also possible to study the build-up of mechanical properties by taking a protein apart at the monomer or domain levels. This is illustrated for our example of enolase in Figure 21.6. In this case we have calculated the change in force constants (after normalization by conversion to Z-scores, see Methods) in passing from a single monomer to the full dimeric structure. Note that here, in contrast to the domain separation technique, we are actually changing the elastic network representation being studied (monomer or dimer). The plot in Figure 21.6 shows that moving from a monomer to a dimer does not simply lead to a general increase in force constants, since both increases and decreases are seen. The location of the changes are illustrated in Figure 21.4c where it is observed that force constants understandably increase at the junction between the two monomers, but, more surprisingly, decrease in the C-terminal domain, in a region not far from the substrate-binding site. We have seen complex, and not easily predictable, changes such as this in other proteins, both as the result of conformational changes or as a result of point mutations.

21.4 CONCLUSIONS

The complex structures of proteins appear to lead to equally complex mechanical properties. The coarse-graining approach described here makes it possible to analyze such properties on a residue-by-residue basis. The results suggest that proteins are very heterogeneous in mechanical terms and that active sites, and possibly other functionally important residues, have unusual properties, generally being associated with above average force constants. While a single-point-per-residue representation captures the main features of a protein's mechanical properties, a finer representation, taking side-chain size and orientation into account, is necessary for analyzing the effects of point mutations or small conformational changes. We have also shown that the fluctuations of our mean-distance function observed using BD simulations enable residue force constants to be calculated quickly, while giving results very close to those obtained with all-atom minimization or molecular dynamics approaches.

Although more work clearly remains to be done to understand how mechanical heterogeneity is actually generated and is related to the structural classes of proteins, this property seems well worth studying in a systemic way.

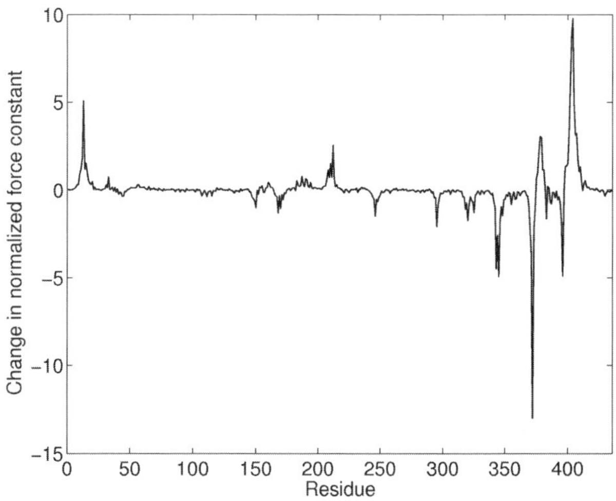

FIGURE 21.6 Changes in normalized force constants (units of standard deviation) in passing from a single monomer to the dimeric form of enolase.

REFERENCES

1. Kendrew, J. C., G. Bodo, H. M. Dintzis, R. G. Parrish, H. Wyckoff, and D. C. Phillips. 1958. A three-dimensional model of the myoglobin molecule obtained by x-ray analysis. *Nature* 181:662–66.
2. Perutz, M. F. 1960. Structure of hemoglobin. *Brookhaven Symp. Biol.* 13:165–83.
3. Bustamante, C. 2004. Of torques, forces and protein machines. *Protein Sci.* 13:3061–65.
4. Lavery, R., A. Lebrun, J.-F. Allemand, D. Bensimon, and V. Croquette. 2002. Structure and mechanics of single biomolecules: Experiment and simulation. *J. Phys. Condens. Matter* 14:R383–414.
5. Dietz, H., and M. Rief. 2006. Protein structure by mechanical triangulation. *Proc. Natl. Acad. Sci. U.S.A.* 103:1244–47.
6. Dietz, H., F. Berkemeier, M. Bertz, and M. Rief. 2006. Anisotropic deformation response of single protein molecules. *Proc. Natl. Acad. Sci. U.S.A.* 103:12724–28.
7. Rueda, M., C. Ferrer-Costa, T. Meyer, A. Perez, J. Camps, A. Hospital, J. L. Gelpi, and M. Orozco. 2007. A consensus view of protein dynamics. *Proc. Natl. Acad. Sci. U.S.A.* 104:796–801.
8. Norberg, J., and L. Nilsson. 2003. Advances in biomolecular simulations: Methodology and recent applications. *Q. Rev. Biophys.* 36:257–306.
9. Karplus, M., and J. Kuriyan. 2005. Molecular dynamics and protein function. *Proc. Natl. Acad. Sci. U.S.A.* 102:6679–85.
10. Gao, M., D. Craig, O. Lequin, I. D. Campbell, V. Vogel and K. Schulten. 2003. Structure and functional significance of mechanically unfolded fibronectin type III1 intermediates. *Proc. Natl. Acad. Sci. U.S.A.* 100:14784–89.
11. Gullingsrud, J., D. Kosztin, and K. Schulten. 2001. Structural determinants of MscL gating studied by molecular dynamics. *Biophys. J.* 80:2074–81.
12. Tirion, M. M. 1996. Large amplitude elastic motions in proteins from a single-parameter, atomic analysis. *Phys. Rev. Lett.* 77:1905–1908.
13. Tozzini, V. 2005. Coarse-grained models for proteins. *Curr. Opin. Struct. Biol.* 15:144–50.
14. Chennubhotla, C., A. J. Rader, L. W. Yang, and I. Bahar. 2005. Elastic network models for understanding biomolecular machinery: From enzymes to supramolecular assemblies. *Phys. Biol.* 2:S173–80.
15. Song, G., and R.L. Jernigan. 2007. vGNM: A better model for understanding the dynamics of proteins in crystals. *J. Mol. Biol.* 369:880–93.
16. Yang, L.-W., E. Eyal, C. Chennubhotla, J. G. Jee, A. M. Gronenborn, and I. Bahar. 2007. Insights into equilibrium dynamics of proteins from comparison of NMR and X-ray data with computational procedures. *Structure* 15:741–49.
17. Kondrashov, D. A., Q. A. Cui, and G. N. Phillips Jr. 2006. Optimization and evaluation of a coarse-grained model of protein motion using X-ray crystal data. *Biophys. J.* 91:2760–67.
18. Jacobs, D. J., A. J. Rader, L. A. Kuhn, and M. F. Thorpe. 2001. Protein flexibility predictions using graph theory. *Proteins* 44:150–65.
19. Kurkcuoglu, O., R. L. Jernigan, and P. Doruker. 2005. Collective dynamics of large proteins from mixed coarse-grained elastic network model. *QSAR Comb. Sci.* 24:443–48.
20. Aqeel, A., and H. Gohlke. 2006. Multi-scale modeling of macromolecular conformational changes combining concepts from rigidity and elastic network theory. *Proteins* 63:1038–51.
21. Zhao, Y., D. Stoffler, and M. Sanner. 2006. Hierarchical and multi-resolution representation of protein flexibility. *Bioinformatics* 22:2768–74.
22. Cluzel, P., A. Lebrun, C. Heller, R. Lavery, J. L. Viovy, D. Chatenay, and F. Caron. 1996. DNA: An extensible molecule. *Science* 271:792–94.
23. Allemand, J. F., D. Bensimon, R. Lavery, and V. Croquette. 1998. Stretched and overwound DNA forms a Pauling-like structure with exposed bases. *Proc. Natl. Acad. Sci. U.S.A.* 95:14152–57.
24. Lebrun, A., Z. Shakked, and R. Lavery. 1997. Local DNA stretching mimics the distortion caused by the TATA box-binding protein. *Proc. Natl. Acad. Sci. U.S.A.* 94:2993–98.
25. Paillard, G., and R. Lavery. 2004. Analyzing protein-DNA recognition mechanisms. *Structure* 12:113–22.
26. Halle, B. 2002. Flexibility and packing in proteins. *Proc. Natl. Acad. Sci. U.S.A.* 99:1274–79.
27. Navizet, I., F. Cailliez, and R. Lavery. 2004. Probing protein mechanics: Residue-level properties and their use in defining domains. *Biophys. J.* 87:1426–35.
28. Sacquin-Mora, S., and R. Lavery. 2006. Investigating the local flexibility of functional residues in hemoproteins. *Biophys. J.* 90:2706–17.
29. Sacquin-Mora, S., E. Laforet, and R. Lavery. 2007. Locating the active sites of enzymes using mechanical properties. *Proteins* 67:350–59.

30. Glaser, F., R. J. Morris, R. J. Najmanovich, R. A. Laskowski, and J. M. Thornton. 2006. A method for localizing ligand binding pockets in protein structures. *Proteins* 62:479–88.
31. Soro, S., and A. Tramontano. 2005. The prediction of protein function at CASP6. *Proteins* 61 (Suppl. 7): 201–13.
32. Wang, J. M., P. Cieplak, and P. A. Kollman. 2000. How well does a restrained electrostatic potential (RESP) model perform in calculating conformational energies of organic and biological molecules? *J. Comput. Chem.* 21:1049–74.
33. Tsui, V., and D. A. Case. 2000. Molecular dynamics simulations of nucleic acids with a generalized born solvation model. *J. Am. Chem. Soc.* 122:2489–98.
34. Lavery, R., K. Zakrzewska, and H. Sklenar. 1995. JUMNA (junction minimization of nucleic-acids). *Comp. Phys. Commun.* 91:135–58.
35. Zacharias, M. 2003. Protein-protein docking with a reduced protein model accounting for side-chain flexibility. *Protein Sci.* 12:1271–82.
36. Bastard, K., C. Prevost, and M. Zacharias. 2006. Accounting for loop flexibility during protein–protein docking. *Proteins* 62:956–69.
37. Sims, P. A., A. L. Menefee, T. M. Larsen, S. O. Mansoorabadi, and G. H. Reed. 2006. Structure and catalytic properties of an engineered heterodimer of enolase composed of one active and one inactive subunit. *J. Mol. Biol.* 355:422–31.
38. Berman, H. M., J. Westbrook, Z. Feng, G. Gilliland, T. N. Bhat, H. Weissig, I. N. Shindyalov, and P. E. Bourne. 2000. The protein data bank. *Nucleic Acids Res.* 28:235–42.
39. Larsen, T. M., J. E. Wedekind, I. Rayment, and G. H. Reed. 1996. A carboxylate oxygen of the substrate bridges the magnesium ions at the active site of enolase: Structure of the yeast enzyme complexed with the equilibrium mixture of 2-phosphoglycerate and phosphoenolpyruvate at 1.8 Å resolution. *Biochemistry* 35:4349–58.
40. Isin, B., P. Doruker, and I. Bahar. 2002. Functional motions of influenza virus hemagglutinin: A structure-based analytical approach. *Biophys. J.* 82:569–81.
41. Bahar, I., and R. L. Jernigan. 1999. Cooperative fluctuations and subunit communication in tryptophan synthase. *Biochemistry* 38:3478–90.
42. Porter, C. T., G. J. Bartlett, and J. M. Thornton. 2004. The catalytic site atlas: A resource of catalytic sites and residues identified in enzymes using structural data. *Nucleic Acids Res.* 32:D129–33.
43. Yang, L. W., and I. Bahar. 2005. Coupling between catalytic site and collective dynamics: A requirement for mechanochemical activity of enzymes. *Structure* 13:893–904.
44. Daniel, R. M., R. V. Dunn, J. L. Finney, and J. C. Smith. 2003. The role of dynamics in enzyme activity. *Annu. Rev. Biophys. Biomol. Struct.* 32:69–92.
45. Bartlett, G. J., C. T. Porter, N. Borkakoti, and J. M. Thornton. 2002. Analysis of catalytic residues in enzyme active sites. *J. Mol. Biol.* 324:105–21.
46. Yuan, Z., J. Zhao, and Z. X. Wang. 2003. Flexibility analysis of enzyme active sites by crystallographic temperature factors. *Protein Eng.* 16:109–14.
47. Lebioda, L., B. Stec, J. M. Brewer, and E. Tykarska. 1991. Inhibition of enolase: The crystal structures of enolase-Ca^{2+}-2-phosphoglycerate and enolase-Zn^{2+}-phosphoglycolate complexes at 2.2 Å resolution. *Biochemistry* 30:2823–27.
48. Ptitsyn, O. B. 1998. Protein folding and protein evolution: Common folding nucleus in different subfamilies of c-type cytochromes? *J. Mol. Biol.* 278:655–66.
49. Krishna, M. M. G., Y. Lin, L. Mayne, and S. W. Englander. 2003. Intimate view of a kinetic protein folding intermediate: Residue-resolved structure, interactions, stability, folding and unfolding rates, homogeneity. *J. Mol. Biol.* 334:501–13.
50. Humphrey, W., A. Dalke, and K. Schulten. 1996. VMD: Visual molecular dynamics. *J. Mol. Graph.* 14:33–38, 27–28.

22 Self-Assembly of Surfactants in Bulk Phases and at Interfaces Using Coarse-Grain Models

Wataru Shinoda
Research Institute of Computational Science, National Institute of Advanced Industrial Science and Technology

Russell DeVane, and Michael L. Klein
The Laboratory for Research on the Structure of Matter, University of Pennsylvania

CONTENTS

22.1 INTRODUCTION

The amphiphilic nature of surfactant molecules leads to their aggregation and self-assembly into a variety of morphologies when exposed to solvents. The observed morphology depends on a number of variables including the molecular structure of the specific surfactant, its solvophilicity, the concentration of the surfactant, the solvent properties and finally the thermodynamic conditions. Understanding such a complex interplay of variables at the atomic level is a natural goal of molecular simulations using high-performance computing resources. However, even with generous access to multiterascale machines, this goal is particularly challenging due to both the temporal and spatial scales involved. Simply put, the study of surfactant self-assembly is beyond the capabilities of current computational resources if one desires an all-atom representation. To overcome this difficulty, two approaches are commonly adopted: (1) use of enhanced sampling techniques and (2) simplified molecular representation of the surfactant molecules; that is, coarse-graining. With the relentless increase in available computer resources, some of the issues that arise in the investigation of complex phenomena will likely be resolved via currently available and recently enhanced sampling techniques. However, for the foreseeable future many aspects of the time-scale problem are likely to persist and remain beyond the scope of all-atom simulations.

Coarse-graining (CG) models reduce computational demand by reducing the number of degrees of freedom for the molecules (i.e., number of atomic sites) that comprise the system of interest. Of course, with this reduction in the description of the system comes a reduction in the level of chemical detail that is retained. An early example of a coarse-grain approach is the molecular dynamics (MD) simulations of the folding of a small protein by Levitt and Warshel [1]. The polymer community has also adopted CG models with considerable success [2,3]. More recently the study of surfactants by Smit and co-workers [4] inspired our first-generation CG model for lipid bilayers [5]. Ultimately, any coarse-grain approach requires a selection of "key" molecular properties or attributes to be retained in advance of model development (parameterization). The inherent limitation of a typical CG model is illustrated by the work of Siepmann et al. [6,7], who presented a new approach to constructing an intermolecular potential, called the TraPPE force field, in which liquid–gas phase equilibrium data were used as a target property to be reproduced by the model. To parameterize and test the force field, simulations were performed using the configurational-bias Monte Carlo techniques in the Gibbs ensemble. In a series of alkane models they changed the resolution from a united-atom (UA) to an all-atom (AA) description [6,7]. Both models perfectly reproduced the phase equilibrium diagram. However, a comparison of UA and AA models for several thermodynamic quantities at ambient conditions revealed deficiencies in the UA model. This simply implies that a reduction in the number of degrees of freedom yields a model with less adjustability and consequently a model with a more modest scope of applicability. This degradation is inevitable even with minimal coarse-graining, for example, AA to UA, and in general it is impossible to reproduce all of the properties that are obtained from the (original) AA model. Thus, CG models should be designed for a more specific purpose than finer-grained AA models.

The primary motivation of CG modeling is a reduction of computational overhead, thus allowing larger system and time scales to be accessed and explored. Accordingly, it is necessary to strike a balance between the complexity of the model and the increase in computational efficiency such that the level of accuracy required to provide insight into the behavior of the system of interest is retained while still providing computational efficiency. That is to say, even with a reduction in the number of degrees of freedom (interaction sites), it is possible to maintain a high level of accuracy in the force field model by using a more complex description of the intermolecular interactions [8]. However this typically comes at the expense of an increase in computational overhead, which in turn could easily offset all gains made by reducing the description of the system. Thus, if one wishes to reduce the computational cost significantly compared with an AA model, one is forced to focus on a selection of the properties to be reproduced in the CG model while keeping the force field as simple as possible to meet that goal. With this target in mind, the question arises as to what experimental properties should be retained and how to preserve those in the CG model. There is no unique answer, as will be evident by the fact that readers will find several different approaches in the other chapters of this book [5,9–16]. Nonetheless, herein we will outline a systematic procedure to build a CG model for surfactant systems.

As mentioned above, surfactant solutions exhibit a variety of morphologies depending on the thermodynamic conditions. These morphologies are mainly determined by the interfacial properties so that the surface/interfacial tension is one of the key properties that characterize the system. Thus, surface tension and density are used as target properties to fix the parameters for the nonbonded interaction of pure solvents. The specific functional form is selected to refine the structural data and compressibility (of water). Interfacial tension is used to parameterize the interaction between phase-separated fluids, while solvation free energy is employed for the interaction between soluble fluids. The systems used in the parameterization are examined at both the AA and CG levels. By keeping the number of unknown parameters smaller than the number of target properties at each fitting step, it is possible to find suitable parameters straightforwardly and unambiguously. The extensive use of many molecular systems for fitting is essential for a systematic parameterization.

As a result of this parameterization approach, we have several favorable features in the CG model. For example, the model guarantees the correct molecular partitioning and is applicable to systems having an air/solution interface. The former is guaranteed by requiring the model to predict the correct solvation (or transfer) free energy, and the latter is a result of using the surface tension and density as target values in the parameterization. Polyethylene glycol (PEG) surfactant solutions will be presented here to exemplify our strategy to build a CG model.

22.2 COARSE-GRAINED SURFACTANT MODEL

The initial step of the approach is to systematically map the system into groups of atoms that will each be represented by a CG site (see Figure 22.1). The atomic groups needed to construct a CG PEG/water system with our level of mapping (roughly three to four heavy atoms with associated hydrogens per CG site) are W, CT, CM, CT2, EO, EOT, and OA, which represent $(H_2O)_3$, CH_3–CH_2–CH_2–, –CH_2–CH_2–CH_2–, CH_3–CH_2–, –CH_2–O–CH_2–, CH_3–O–CH_2–, and HO–CH_2–, respectively. The CG water, W, is special because the site represents three molecules, while the other CG particle corresponds just to a segment of a single molecule. With just seven CG sites, there are 28 pair-interactions that have to be determined. For ease of implementation, a Lennard–Jones (LJ) function is used for the nonbonded interactions:

$$U_{\mathrm{LJ}}(r_{ij}) = B\varepsilon_{ij}\left\{\left(\frac{\sigma_{ij}}{r_{ij}}\right)^m - \left(\frac{\sigma_{ij}}{r_{ij}}\right)^n\right\}.$$

Several pairs of the repulsive and attractive parameters, m and n, were tested to search for a suitable functional form to give the best structural and thermodynamic properties. Ultimately, the values of (m,n) chosen were (12,4) and (9,6). The choice depends on the type of interaction with the nonbonded interactions involving "W" modeled with the LJ12-4 function, while all others employ the LJ9-6 functional form. The prefactor B, which is chosen such that $U_{\mathrm{LJ}}(\sigma) = 0$ and $\min(U_{\mathrm{LJ}}) = \varepsilon$ is given by $3\sqrt{3}/2$ and 27/4 for LJ12-4 and LJ9-6, respectively. The long-range force is simply truncated at 15 Å so that the cutoff distance should affect the calculated system properties. Note that here we have a nonionic system and use no ionic particles. For ionic systems, it may be necessary to employ alternative methods to handle the long-range interactions.

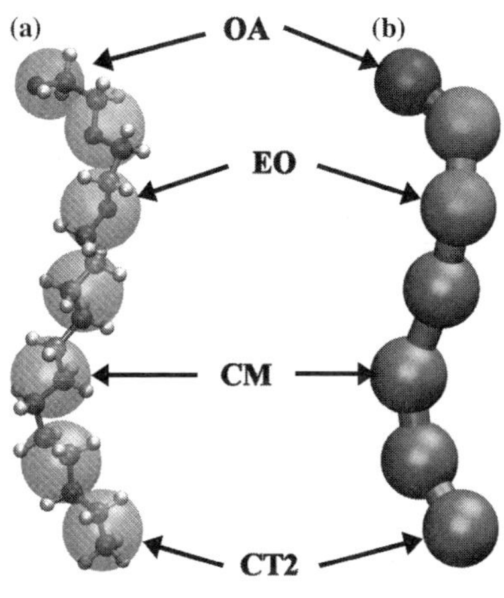

FIGURE 22.1 (See color insert following page 238.) Atomistic (a) and coarse-grained (b) representation of $C_{12}E_2$ molecule. The atomic groups, (HO–CH_2–), (–CH_2–O–CH_2–), (–CH_2–CH_2–CH_2–), and (CH_3–CH_2–), are referred to as OA, EO, CM, and CT2 segments, respectively.

For the bonded interactions, we employ simple harmonic potentials for 1-2 bond stretching and 1-2-3 angle bending given by

$$U_{\text{stretching}}(r_{ij}) = k_{\text{b}}(r_{ij} - r_0)^2,$$

$$U_{\text{bending}}(\theta_{ijk}) = k_{\theta}(\theta_{ijk} - \theta_0)^2.$$

Here the force constants and zero-force distance and angle are fitted to reproduce the corresponding distribution functions from AA-MD trajectories. Although we sometimes find bimodal probability distributions in AA results, the above functions are used and the CG parameters are fit to give the average and dispersion of the AA distribution. In our experience, this simplification does not give significant error in the assembled morphologies. The bonded interactions exceeding three bodies, for example, torsions and dihedrals, are not treated with an internal potential. However, these CG sites do interact via the nonbonded pair potential with no scaling of the potential introduced.

Importantly, a target temperature at 30°C was selected for the parameterization presented herein, although it is possible to select any arbitrary temperature, with the only constraint being that the target molecules be in the liquid state in order to use the condensed phase surface tension and density data. The transferability of the CG model to a different temperature is not expected in principle, though a test with the CG water model showed good transferability with respect to surface tension and density within the liquid temperature range of water [9].

To optimize the CG force field, we have carried out a series of MD simulations for the systems shown below. The methods used for the simulations are briefly summarized here. The CHARMM PARAM27 force field was used for all AA-MD simulations except for the PEG headgroups [17]. The interaction parameters for the PEG headgroup were taken from Ref [18]. The van der Waals interactions were truncated at 12 Å by applying the standard CHARMM smoothing function for the tail region of 10–12 Å, while the Coulomb interaction was calculated using the Ewald or particle mesh Ewald method [19]. The SHAKE/RATTLE (ROLL) method was used to fix the bond lengths involving hydrogen atoms and allowed the use of a 2 fs time-step [20]. For CG-MD, two time-step sizes were used to solve the equations of motion by employing the rRESPA algorithm [20,21]; for updating long-range nonbonded forces (0.6–1.5 nm) a 10 fs time-step was used and 2 fs was used for updating short-range nonbonded and bonded forces. Those can be extended to 40 and 5 fs, respectively, without changing the system properties.

22.2.1 Parameter Fitting for Pure Solvents

For pure solvents, surface tension and density data were used to fix the LJ parameters, σ and ε. These parameters were fit by a trial-and-error approach. To do this efficiently, we employed the following technique. First, a cubic simulation box is prepared with the edge length of approximately 40 Å and the proper target density. To fix the pressure, short NVT-MD runs (typically 100 ps) are performed while adjusting the LJ parameters. After selecting the parameters to give zero pressure, the simulation box is elongated in the z-direction to 400 Å to create a system with a liquid/vacuum interface. Again, NVT-MD simulations are carried out on the elongated box to measure the surface tension, which was calculated by

$$\gamma = \frac{L_z}{2}\left\{P_{zz} - \frac{P_{xx} + P_{yy}}{2}\right\}.$$

Here, the factor of 1/2 is included to account for the two interfaces in the simulation box, and P_{ij} is the ij component of the averaged pressure tensor. To achieve the convergence of surface tension with a precision of 1 dyne/cm, 5–10 ns MD simulations are usually needed. Finally, to confirm the system density, NPT-MD is also carried out for 1 ns on the cubic simulation box.

TABLE 22.1
Comparison of CG-MD and Experiments for Surface Tension, γ (dyne/cm), and Density, ρ (g/cm³) at 303 K

System	Molecular structure	Interaction[a]	MD γ	MD ρ	Exp[b] γ	Exp[b] ρ
Water	W	W-W	70.8	0.9949	71.20	0.9957
Hexane	CT-CT	CT-CT	17.5	0.6498	17.43	0.6518
Nonane	CT-CM-CT		22.3	0.7129	21.94	0.7114
Dodecane	CT-$(CM)_2$-CT	CM-CT, CM-CM	24.5	0.7422	24.48	0.7415
Pentadecane	CT-$(CM)_3$-CT		25.9	0.7603	26.23	0.7616
Octadecane	CT-$(CM)_4$-CT		27.6	0.7726	27.53	0.7722
Heptane	CT2-CM-CT2	CT2-CT2, CT2-CM	19.4	0.6791	19.27	0.6773
Decane	CT2-$(CM)_2$-CT2		22.3	0.7239	22.92	0.7247
Dimethoxyethane	EOT-EOT	EOT-EOT	19.9	0.8617	19.45	0.8593
Diethylene glycol dimethyl ether	EOT-EO-EOT	EOT-EO, EO-EO	25.6	0.9374	28.60	0.9372
Triethylene glycol dimethyl ether	EOT-$(EO)_2$-EOT		29.8	0.9804	27.83	0.9735
Tetraethylene glycol dimethyl ether	EOT-$(EO)_3$-EOT		31.7	1.0060	32.88	1.0010
Ethylene glycol	OA-OA	OA-OA	50.2	1.1060	49.01	1.1070
Diethylene glycol	OA-EO-OA		44.8	1.0990	48.86	1.1100
Triethylene glycol	OA-$(EO)_2$-OA	EO-OA	45.2	1.1150	45.80	1.1180
Tetraethylene glycol	OA-$(EO)_3$-OA		45.0	1.1200	43.53	1.1170
Diethylene glycol di-n-butyl ether	CT-$(EO)_3$-CT	EO-CT	26.3	0.8767	26.07	0.8774
Dipropylether	CT2-EO-CT2	EO-CT2	19.1	0.7379	19.46	0.7366
Di-n-hexyl-ether	CT2-CM-EO-CM-CT2	EO-CM	24.9	0.7858	24.91	0.7860
1-Propanol	CT2-OA	CT2-OA	22.8	0.7943	23.80	0.7950
1-Hexanol	CT2-CM-OA	CM-OA	25.7	0.8121	25.48	0.8123

[a] Interaction column gives the CG particle pair parameterized using the system.
[b] Experimental data are taken from Ref. [22].

This approach was used to parameterize pure solvents; that is, water, alkanes, and ethylene glycols, which are listed in Table 22.1. The LJ12-4 function was used for the CG water model in order to maintain a liquid state from 0 to 100°C while simultaneously optimizing the model with respect to compressibility and interfacial properties (at the alkane–water interface) and obtaining the correct transfer free energy of alkane from its bulk to water (see the next subsection). The choice of LJ9-6 for chained molecules was made in order to preserve structural detail as much as possible. Figure 22.2 plots the radial distribution functions for a triethylene glycol dimethyl ether (EOT–EO–EO–EOT) system. Although a slightly higher first peak is observed for EOT–EOT with the CG model, the overall structure agrees reasonably well with the AA results. It is worth reiterating the agreement that is achieved with the simple interaction functions used here and point out that we have only slight degradation of the structural properties when compared to the tabulated potentials based on the inverse Boltzmann method. It should be noted that, as shown in Table 22.1, the model is transferable to chains of various length. This is achieved by making use of segments of various lengths; that is, CT, CM, and CT2, that

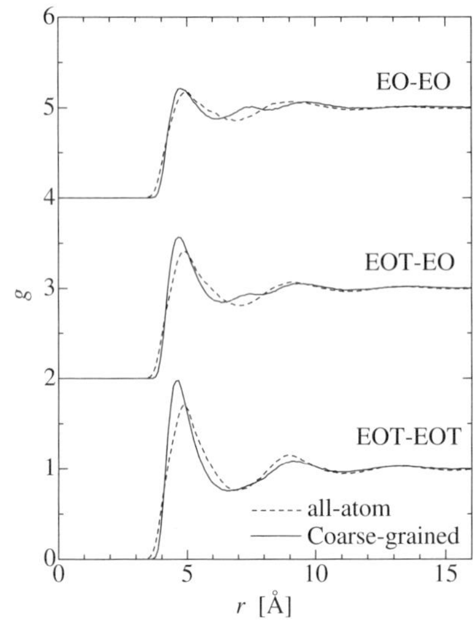

FIGURE 22.2 Radial distribution functions from AA and CG simulations of triethylene glycol dimethyl ether (EOT–EO–EO–EOT).

can be assembled into alkanes with a variety of lengths, all of which give reasonable surface tension, density, and pair-distribution functions.

22.2.2 Parameters for Immiscible Solvents

Next we discuss the parameterization of the alkane-type CG sites (CT, CM, and CT2) and water. To fix the LJ parameter, ε, between alkane sites and water, the interfacial tension was used as a target property to be reproduced. As for σ, which represents a contact distance between the particles, the arithmetic average between the alkane particle and W was used. The LJ12-4 function was used for water–alkane interactions in order to produce a more attenuated interfacial width. The broadening of the interfacial width is usually observed with CG models simply due to the larger size of the CG particles [12]. Choosing a steeper function (more strongly repulsive term) gives better agreement in the interfacial width compared with the AA model [9]. Following this parameterization approach, the model was systematically built to have the correct interfacial tension for a series of alkane chains with water (Table 22.2).

It should be noted here that our CG model reproduces the experimental transfer free energy accurately. We have carried out a series of steered MD simulations [24] which involve dragging an *n*-hexane molecule from the bulk *n*-hexane region to the bulk water region along the interface normal. The free energy cost for the transfer was calculated using Jarzynski's theory based on 15 sets of steered MD calculations [25]. The transfer free energy is estimated to be ~8 kcal/mol, which is in good agreement with the experimental value, 7.74 kcal/mol, as shown in Figure 22.3 [26]. We also confirm the convergence of the free-energy profile by measuring the work with the reverse operation; that is, dragging a *n*-hexane molecule from the bulk water region to the bulk *n*-hexane region. We should emphasize that the accurate transfer free energy is not just a coincidence but a result of extensive exploration of suitable interaction function and parameters.

22.2.3 Parameters for Solutes

For the interaction between miscible pairs, for example, PEG/water, values for the solvation free energy are used for fitting the LJ parameter, ε. Although the combination rule for σ can be used, a different approach is taken to estimate the σ value in this case because the effective size of a CG site in water will change depending on the hydrophilicity and may be different from that in bulk

TABLE 22.2
Comparison of CG-MD and Experiment for Interfacial Tension at Alkane/Water Interface at 303 K

		Interfacial tension (dyne/cm)	
Mixture	**Interaction**	**MD**	**Exp[a]**
Water/hexane	CT-W	50.0	49.96
Water/nonane		51.9	51.21
Water/dodecane	CM-W	52.9	52.14
Water/pentadecane		52.9	–
Water/heptane	CT2-W	50.1	50.30

[a] Experimental data are taken from Ref. [23].

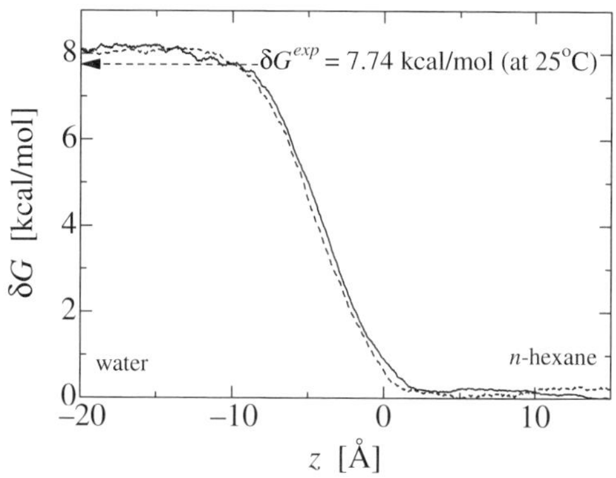

FIGURE 22.3 Free-energy profile of n-hexane molecule across the interface between n-hexane ($z > 0$ Å) and water ($z < 0$ Å). The solid line denotes the work needed to drag a hexane molecule from the bulk hexane region to the bulk water region and the dotted line gives the work along the inverse pathway.

solution. To estimate the effective size of a CG site, we use a potential of mean force (PMF) analysis of an AA-MD trajectory of a single molecule (or fragment of molecule) corresponding to the CG site in bulk water. Details of this procedure are given in the previous publication [9]. After σ is fixed, a series of free-energy calculations are needed to find a suitable ε to reproduce the experimental hydration free energy. This approach is generally useful for a variety of molecules as long as the experimental hydration free energy data are available. Thus, a systematic parameterization for a series of CG segments is feasible. We choose the parameters for the OA–W interaction with this protocol using the experimental hydration free energy for ethylene glycol.

All parameters have been fixed except for the EO–W interaction. Since no experimental hydration free-energy data are available to parameterize this interaction, structural data of the lamellar phase of the $C_{12}E_2$/water system are used [27]. As mentioned above, σ is estimated from the effective size of the EO segments in water from AA-MD simulations. Since the lamella spacing and the molecular area of $C_{12}E_2$ are available from X-ray diffraction measurements, ε is fixed using these quantities. A series of NP_nAT-MD simulations of the lamellar systems at the surfactant composition of 67 wt% have been carried out with the cross-sectional area fixed to have a experimental molecular area of 30 Å^2. With these simulations, ε is fit to give zero surface tension. After the parameterization, 10 ns-NPT-MD simulations of the $C_{12}E_2$/water system were performed to assess the membrane properties. The average molecular area shows perfect agreement with the experimental value, while the lamellar spacing, 48.1 Å, is slightly overestimated compared with the experiment (47.3 Å). Figure 22.4 plots the number density for each CG segment along the bilayer normal and

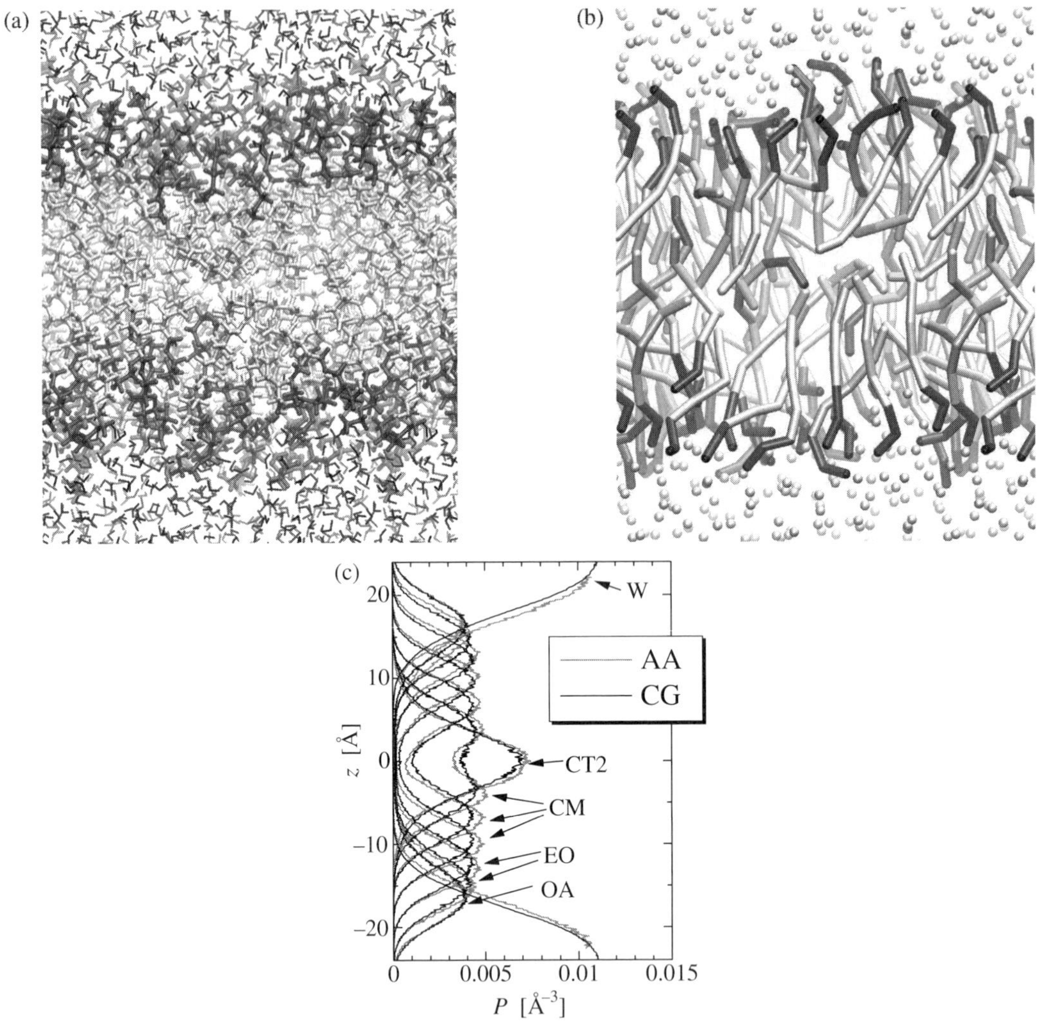

FIGURE 22.4 A snapshot of $C_{12}E_2$ lamellar system with (a) AA and (b) CG models. Thick lines denote $C_{12}E_2$ molecules with headgroup in dark gray. Water is depicted by solid line and white particle in AA and CG, respectively. (c) The density profile of each component of the $C_{12}E_2$ lamellar system along the bilayer normal is shown for the AA and CG simulation.

compares it with the equivalent measurement from the AA-MD simulations. Considering the fact that no structural details of the surfactant, other than the molecular area, were used in the parameterization, the agreement is remarkable.

Due to the fact that the AA force field is not guaranteed to predict the correct surface tension, it was necessary to perform the AA simulations using the fixed area ensemble. For example, TIP3P water, which is the most widely used water model, gives a surface tension of about 52.7 dyne/cm, while it should be 72.8 dyne/cm at ambient temperature [28]. In addition, it was reported that the CHARMM force field overestimates the surface tension of the DPPC bilayer system; consequently, a long timescale MD simulation will eventually generate a gel-like bilayer in the NPT ensemble even at liquid-crystal conditions [29]. Thus, the surface properties are subtle and can be difficult to reproduce even with a widely adopted AA description of the system. This point helps to highlight the advantages of a model guaranteed to properly reproduce experimental properties.

22.3 SELECTED APPLICATIONS

22.3.1 Lamellar Phase Formation

One of the goals of developing a CG force field is to be able to investigate nonequilibrium molecular processes that take place on time scales not accessible by AA-MD. A self-organized mesostructure of amphiphiles is one such example. We demonstrate here an application of our new CG potential to observe the self-assembly process of $C_{12}E_2$ molecules in water into the lamellar phase. The initial configuration was made with 1296 $C_{12}E_2$ molecules and 3456 W particles randomly packed into the cubic simulation box with an edge of approximately 100 Å. Lamellar formation occurred in a 10 ns-CG-MD run (Figure 22.5). The simulation time does not correspond to the physical time straightforwardly, because of much simplified energy surface results from the coarse-graining procedure. Although we do not have a sophisticated measure for "real" time in the CG simulations, a comparison of the diffusion coefficients of surfactant molecules calculated for CG and AA models, respectively, suggests that the physical time is longer than the simulation time by at least two orders of magnitude.

The initial stage of the structural reorganization was a local rearrangement to reduce the contact area between hydrophilic and hydrophobic components. This process proceeded in a short time period, <0.5 ns, to bring about a large energy gain. A bicontinuous three-dimensional structure was observed at ~ 1 ns with the hydrophobic core almost completely covered with hydrophilic components. After 4 ns, a lamellar phase appeared though the two bilayers were still interconnected, showing a spiraled continuous membrane. Then the interconnection of the two bilayers gradually thinned. This process took place relatively slowly, allowing a symmetric partitioning of surfactant molecules into two adjacent bilayers. The resultant four layers (two bilayers) of the surfactants consisted of 325, 327, 323, and 321 $C_{12}E_2$ molecules, respectively. Even at 7 ns, a water channel penetration of the membrane was observed in each bilayer, and these two channels disappeared simultaneously at roughly 8.3 ns. A similar long-lasting water pore, during the self-assembling process of bilayer-forming lipids, has been observed in an AA-MD simulation previously [30], though the system may have been too small to observe the interconnection we observed between adjacent bilayers. In the present case, the water channels always coexisted with

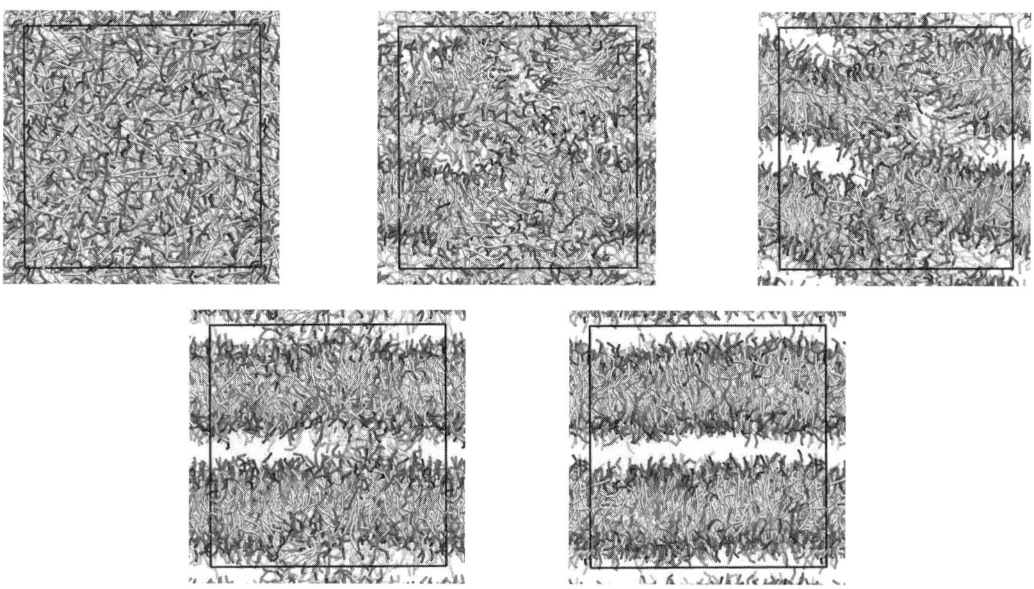

FIGURE 22.5 Snapshots of lamellar formation of the $C_{12}E_2$/water system from a random initial configuration. The time stamps are 0, 1, 4, 7, and 10 ns from left to right. Water is not shown for clarity.

a surfactant tube interconnecting two bilayers. After the 10 ns NVT-MD simulation, another 10 ns MD simulation was carried out in the NPT ensemble to obtain fully relaxed bilayers. Again, the average molecular area for a $C_{12}E_2$ molecule, 30.1 Å^2, was confirmed, showing excellent agreement with the experimental value [27].

22.3.2 Monolayer at the Air/Water Interface

To further test the transferability of the model, simulations of a monolayer of $C_{12}E_{12}$ surfactant molecules at air/water interface were performed. Starting with a slab of water containing 22,249 W particles, 144 $C_{12}E_{12}$ molecules were placed at each of two air/water interfaces. The simulation box was $101.823 \times 101.823 \times 500$ Å, with the monolayer normal taken along the z-axis. The area per $C_{12}E_{12}$ molecule was set to 72 Å^2, which corresponds to the value at the critical micelle concentration (CMC) suggested by neutron reflection experiments [31]. The thickness of water slab was approximately 200 Å, which was large enough to ignore the interaction between the two interfaces. During the MD run for 25 ns, all $C_{12}E_{12}$ molecules remained at the interface with no desorption from the interface observed. The surface tension was calculated as 39 dyne/cm, which is in excellent agreement with the reported experimental value of 38.5 dyne/cm at the CMC. This is a rather surprising agreement, considering that only the surface tension data of small molecules were used for parameterization and that the $C_{12}E_{12}$ molecule has a long and flexible headgroup chain adopting a variety of different conformations at the air/water interface. As can be seen in the snapshot shown in Figure 22.6, some of the $C_{12}E_{12}$ molecules project their headgroup section straight into water while a non-negligible number project their headgroup along the interface. The computed probability density of headgroup chains along the normal to the interface actually shows a peak near the interface. A comparison of the computed density profile with one derived from analysis of neutron reflectivity data, assuming a Gaussian distribution of the components, is also given in Figure 22.6. Applying Gaussian fits to the computed headgroup and hydrocarbon chain distributions, the effective thickness, σ, of each of these distributions was found to be 22 and 7.5 Å, respectively, while the neutron reflection data suggest values of 21.5 ± 4 and 15 ± 3 Å, respectively. The thickness of the headgroup region shows an excellent agreement, though a significant discrepancy is found in the thickness of

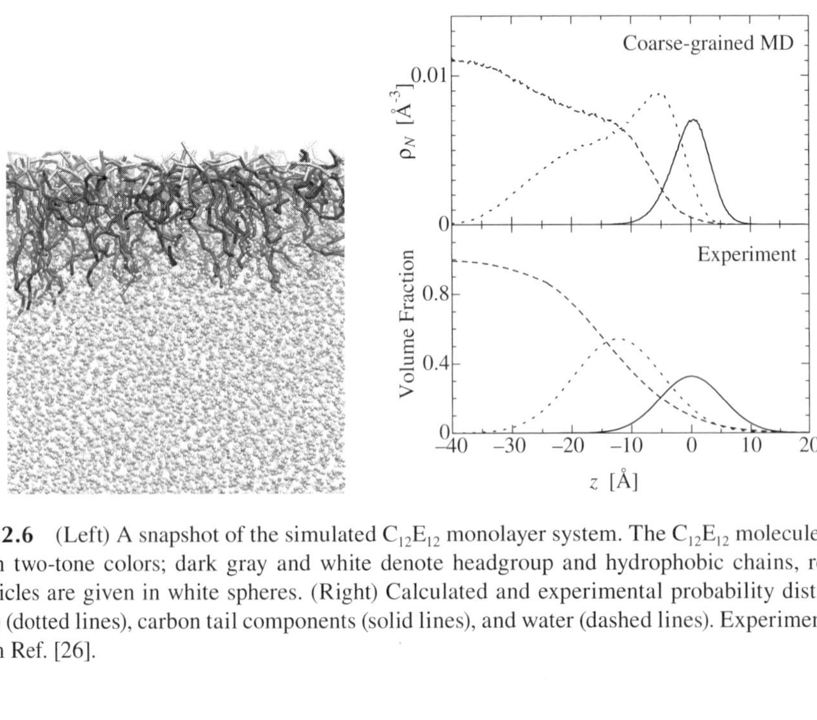

FIGURE 22.6 (Left) A snapshot of the simulated $C_{12}E_{12}$ monolayer system. The $C_{12}E_{12}$ molecule is depicted by lines in two-tone colors; dark gray and white denote headgroup and hydrophobic chains, respectively; water particles are given in white spheres. (Right) Calculated and experimental probability distributions of headgroup (dotted lines), carbon tail components (solid lines), and water (dashed lines). Experimental data are taken from Ref. [26].

the hydrocarbon chain distribution. Figure 22.6 shows a snapshot of the $C_{12}E_{12}$ monolayer. One sees that most of the hydrophobic chains are lying on the surface, which is reasonable for the area per surfactant of 72 Å^2 [31].

According to the neutron reflection data for a series of PEG chains, the chain length of the headgroup does not greatly affect the hydrocarbon thickness; for example, the thickness for $C_{12}E_2$ is reported as 17 ± 2 Å, which is in the error bar of that for $C_{12}E_{12}$ [32]. However, the experimental area per $C_{12}E_2$ is 33 Å^2, which is less than half of that of the $C_{12}E_{12}$ monolayer. A short MD run of a $C_{12}E_2$ monolayer suggests that in this case the hydrophobic chains mostly stand straight along the interface normal. From the molecular viewpoint, it is hard to imagine having a similar chain alignment along the interface normal with such a large change in the area per molecule. Although further study will be needed to confirm the results, it should be emphasized that we now have a molecular model that yields the correct surface tension and density. This should be useful for interpreting the experimental data of long-chained surfactant monolayer systems from the molecular viewpoint.

22.4 FUTURE PERSPECTIVES

The driving motivation for the development of CG surfactant models is to push the limits of the available computational resources well beyond what is possible with AA approaches. This naturally requires the use of cutting edge hardware and software. Timings on current supercomputers indicate that with the present CG models it is possible to simulate systems containing over 1 million CG particles, which corresponds to AA systems of approximately 10 million atoms. Moreover, with a few thousand processors it is possible to generate trajectories for up to 100 ns/day on such large systems [33,34]. Such simulations are thus effectively approaching the millisecond regime, and as such, are likely to be of relevance to laboratory experiments.

A typical example of such a system is shown in Figure 22.7 where the simulation included 807,360 CG particles composed of 61,696 CG $C_{12}E_6$ molecules and 555,264 CG water particles. To put this simulation in perspective, it is instructive to point out that an AA representation of this system would require nearly 5,000,000 atoms to represent the solvent alone. The simulation was set up by starting with an equilibrated system containing 50 wt% $C_{12}E_6$ in the hexagonal

FIGURE 22.7 (See color insert following page 238.) Shown here is a snapshot of a simulation box containing 807,360 CG particles composed of $C_{12}E_6$ and water in the lamellar phase after 470 ns of MD simulation started from the hexagonal phase. The hydrophobic tails are shown in blue, the headgroups are shown in green and red, and the water is shown in white. The unit cell is shown in yellow.

phase and removing the water to obtain a system composed of 80 wt% $C_{12}E_6$. This composition corresponds to the lamellar phase in the phase diagram of the binary mixture of $C_{12}E_6$ and water. The figure shows the system after 470 ns of simulation time where it has reached the lamellar phase although with some minor defects still present. The model is also currently being applied to the transition from a random configuration to the hexagonal phase. However, the simulation time scale for the latter transition is much longer and is expected to be In the order of 1 μs of MD, corresponding to roughly the millisecond 'real' time domain. Further analysis of the results of these large-scale CG-MD simulations is ongoing and will be the subject of forthcoming publications.

22.5 CONCLUSIONS

A straightforward approach to deriving CG interaction parameters has been presented herein for PEG surfactant solutions. A key characteristic of the resulting model is the transferability of parameters among PEG surfactants with arbitrary chain length. This was demonstrated in the above applications and is a direct result of the systematic scheme used to parameterize the force field model. Furthermore, parameterization against key experimental quantities helps to assure that the models are applicable to systems having an air/water interface, which is a significant advantage of this model.

As has been mentioned above, the CG description of the system necessarily requires a reduction in the capabilities compared to an AA model. Further, with the complexity of such systems as those studied herein there are reasons to assume that the details of an AA simulation would occasionally be necessary. Thus, it is natural to assume that both AA and CG descriptions will continue to play equally important roles in future MD simulation studies. Using the CG simulations to more efficiently explore the phase space associated with a system and the AA simulations to provide atomic-level detail offers a logical, synergistic approach to computational investigation of complex systems.

Another approach that has recently gained attention involves the mapping of systems from AA to CG representations and vice versa. This allows atomically detailed systems to be used as starting configurations that are then explored using CG simulations and finally toggling back to the AA representation to evaluate the effectiveness of the CG trajectory. A CG model designed for a specific molecule or atomic group should be more useful for such approaches. This suggestion is backed up by the reasonable agreement of the comparison of the pair distribution functions of the CG system with the AA simulations that were reported above. Pushing this further are multiscale approaches that model the system of interest simultaneously with both AA and CG representations, for example, surfactant molecules represented at the AA level with the solvent represented at the CG level [35–37]. It is likely that such approaches will require extremely refined CG models to properly integrate with the AA systems.

In closing, we have presented a systematic approach to the parameterization of CG models with applications to surfactant systems and have demonstrated the efficacy of this procedure. This method is currently being extended to other systems including phospholipids and proteins, which represent a very rich and active area of research.

ACKNOWLEDGMENTS

The authors thank Dr. Axel Kohlmeyer for his invaluable help in obtaining results on the national supercomputers at PSC, SDSC, and NCSA at UIUC and Dr. Kelly Anderson for his continuing interest in the problem of surfactant self-assembly. The research described herein was supported in part by the National Science Foundation and Procter & Gamble. RD was supported by a National Science Foundation Biological Informatics Postdoctoral Fellowship.

REFERENCES

1. Levitt, M., and Warshel, A. 1975. Computer-simulation of protein folding. *Nature* 253:694–98.
2. Eisenriegler, E., Kremer, K., and Binder, K. 1982. Adsorption of polymer-chains at surfaces: Scaling and Monte-Carlo analyses. *J. Chem. Phys.* 77:6296–320.
3. Kremer, K., and Grest, G. S. 1990. Dynamics of entangled linear polymer melts: A molecular-dynamics simulation. *J. Chem. Phys.* 92:5057–86.
4. Smit, B., Hilbers, P. A. J., Esselink, K., Rupert, L. A. M., Vanos, N. M., and Schlijper, A. G. 1990. Computer-simulations of a water oil interface in the presence of micelles. *Nature* 348:624–25.
5. Shelley, J. C., Shelly, M. Y., Reeder, R. C., Bandyopadhyay, S., and Klein. M. L. 2001. A coarse grain model for phospholipid simulations. *J. Phys. Chem. B* 105:4464–70.
6. Martin, M. G., and Siepmann, J. I. 1998. Transferable potentials for phase equilibria. 1. United-atom description of n-alkanes. *J. Phys. Chem. B* 102:2569–77.
7. Chen, B., and Siepmann, J. I. 1999. Transferable potentials for phase equilibria. 3. Explicit-hydrogen description of normal alkanes. *J. Phys. Chem. B* 103:5370–79.
8. Stone, A. J., and Price, S. L. 1988. Some new ideas in the theory of intermolecular forces: Anisotropic atom-atom potentials. *J. Phys. Chem.* 92:3325–35.
9. Shinoda, W., DeVane, R., and Klein, M. L. 2007. Multi-property fitting and parameterization of a coarse grained model for aqueous surfactants. *Mol. Simul.* 33:27–36.
10. Nielsen, S. O., Lopez, C. F., Srinivas, G., and Klein, M. L. 2003. A coarse grain model for n-alkanes parameterized from surface tension data. *J. Chem. Phys.* 119:7043–49.
11. Srinivas, G., Shelley, J. C., Nielsen, S. O., Discher, D. E., and Klein, M. L. 2004. Simulation of diblock copolymer self-assembly, using a coarse-grain model. *J. Phys. Chem. B* 108:8153–60.
12. Marrink, S. J., de Vries, A. H., and Mark, A. E. 2004. Coarse grained model for semiquantitative lipid simulations. *J. Phys. Chem. B* 108:750–60.
13. Marrink, S. J., Risselada, H. J., Yefimov, S., Tieleman, D. P., and de Vries, A. H. 2007. The MARTINI force field: Coarse grained model for biomolecular simulations. *J. Phys. Chem. B* 111:7812–24.
14. Izvekov, S., and Voth, G. A. 2005. A multiscale coarse-graining method for biomolecular systems. *J. Phys. Chem. B* 109:2469–73.
15. Izvekov, S., and Voth, G. A. 2005. Multiscale coarse graining of liquid-state systems. *J. Chem. Phys.* 123:134105.
16. Lyubartsev, A. P. 2005. Multiscale modeling of lipids and lipid bilayers. *Eur. Biophys. J. Biophys. Lett.* 35:53–61.
17. MacKerell, A. D., Bashford, D., Bellott, M., Dunbrack, R. L., Evanseck, J. D., Field, M. J., Fischer, S., Gao, J., Guo, H., Ha, S., Joseph-McCarthy, D., Kuchnir, L., Kuczera, K., Lau, F. T. K., Mattos, C., Michnick, S., Ngo, T., Nguyen, D. T., Prodhom, B., Reiher, W. E., Roux, B., Schlenkrich, M., Smith, J. C., Stote, R., Straub, J., Watanabe, M., Wiorkiewicz-Kuczera, J., Yin, D., and Karplus, M. 1998. All-atom empirical potential for molecular modeling and dynamics studies of proteins. *J. Phys. Chem. B* 102:3586–616.
18. Shinoda, K., Shinoda, W., Baba, T., and Mikami, M. 2004. Comparative molecular dynamics study of ether- and ester-linked phospholipid bilayers. *J. Chem. Phys.* 121:9648–54.
19. Frenkel, D., and Berend, S. 2002. *Understanding Molecular Simulation*. 2nd ed. San Diego: Academic Press.
20. Martyna, G. J., Tuckerman, M. E., Tobias, D. J., and Klein, M. L. 1996. Explicit reversible integrators for extended systems dynamics. *Mol. Phys.* 87:1117–57.
21. Tuckerman, M. E., Berne, B. J., and Martyna, G. J. 1992. Reversible multiple time scale molecular-dynamics. *J. Chem. Phys.* 97:1990–2001.
22. Yaws, C. L. 1999. *Chemical Properties Handbook*, McGraw-Hill. http://www.knovel.com/knovel2/Toc.jsp?BookID=49.
23. Zeppieri, S., Rodriguez, J., and de Ramos, A. L. L. 2001. Interfacial tension of alkane plus water systems. *J. Chem. Eng. Data* 46:1086–88.
24. Park, S., and Schulten, K. 2004. Calculating potentials of mean force from steered molecular dynamics simulations. *J. Chem. Phys.* 120:5946–61.
25. Jarzynski, C. 1997. Nonequilibrium equality for free energy differences. *Phys. Rev. Lett.* 78:2690–93.
26. White, S. H., and Wimley, W. C. 1999. Membrane protein folding and stability: Physical principles. *Ann. Rev. Biophys. Biomol. Struct.* 28:319–65.

27. Funari, S. S., and Rapp, G. 1997. X-ray studies on the C(12)EO(2)/water system. *J. Phys. Chem. B* 101:732–39.
28. Khurana, E., Nielsen, S. O., and Klein, M. L. 2006. Gemini surfactants at the air/water interface: A fully atomistic molecular dynamics study. *J. Phys. Chem. B* 110:22136–42.
29. Jensen, M. O., Mouritsen, O. G., and Peters, G. H. 2004. Simulations of a membrane-anchored peptide: Structure, dynamics, and influence on bilayer properties. *Biophys. J.* 86:3556–75.
30. Marrink, S. J., Lindahl, E., Edholm, O., and Mark, A. E. 2001. Simulation of the spontaneous aggregation of phospholipids into bilayers. *J. Am. Chem. Soc.* 123:8638–39.
31. Lu, J. R., Su, T. J., Li, Z. X., Thomas, R. K., Staples, E. J., Tucker, I., and Penfold, J. 1997. Structure of monolayers of monododecyl dodecaethylene glycol at the air–water interface studied by neutron reflection. *J. Phys. Chem. B.* 101:10332–39.
32. Lu, J. R., Thomas, R. K., and Penfold, J. 2000. Surfactant layers at the air/water interface: Structure and composition. *Adv. Colloid Interf. Sci.* 84:143–304.
33. Plimpton, S. J. 1995. Fast parallel algorithms for short-range molecular dynamics. *J. Comp. Phys.* 117:1–19. http://lammps.sandia.gov.
34. The benchmark simulations included up to 1,003,500 CG particles in a bilayer assembly employing the LAMMPS MD simulation code from Sandia National Laboratories. Simulations were performed on several supercomputers including the Cray XT3 (Big Ben) at the Pittsburgh Supercomputing Center, the Intel 64 Linux Cluster (ABE) at the National Center for Supercomputing Applications and the IBM Blue Gene machine at San Diego Supercomputing Center. Timings reported here were for 2000 processor simulations utilizing multiple timestep integration methods.
35. Shi, Q., Izvekov, S., and Voth, G. A. 2006. Mixed atomistic and coarse-grained molecular dynamics: Simulation of a membrane-bound ion channel. *J. Phys. Chem. B* 110:15045–48.
36. Ensing, B., Nielsen, S. O., Moore, P. B., Klein, M. L., and Parrinello, M. 2007. Energy conservation in adaptive hybrid atomistic/coarse-grain molecular dynamics. *J. Chem. Theory Comput.* 3:1100–105.
37. Praprotnik, M., Delle Site, L., and Kremer, K. 2005. Adaptive resolution molecular-dynamics simulation: Changing the degrees of freedom on the fly. *J. Chem. Phys.* 123:224106.

23 Coarse-Grained Simulations of Polyelectrolytes

Mark J. Stevens
Sandia National Laboratories

CONTENTS

23.1 INTRODUCTION

A polyelectrolyte is a polymer that has charged monomers of one sign. Because of the variety of ways charges can be placed on a polymer, there is a wide range of polyelectrolytes with a corresponding range of applications. Polyelectrolytes thus comprise one of the most important classes of molecules. DNA is the most well known example, and some of its properties depend on the charged polymeric structure [Williams 2000; Schiessel 2003]. Polyacrylic acid is a common industrial polyelectrolyte, which is the main ingredient in disposable diapers. One of the common uses for synthetic polyelectrolytes is stabilization of colloids [Napper 1983]. In this case the polyelectrolytes are grafted onto the surface. One of the main polyelectrolytes in experimental studies is sodium polystyrene sulfonate (NaPSS). Just about every monomer is charged in NaPSS, which makes it a strong polyelectrolyte. Polyacrylic acid is an example of a weak polyelectrolyte, which has relatively long separation of charges on the backbone.

A main characteristic of polyelectrolytes is their water solubility. Not surprisingly, most biopolymers are charged. If both positive and negative charges are present the polymer is a polyampholyte. Proteins are heteropolymer polyampholytes. In cases where one sign is more prevalent, the polyampholyte may effectively be a polyelectrolyte. For example, biopolymers such as F-actin, which is a polymer composed of monomers of the protein G-actin, are viewed as polyelectrolytes since the monomer has a net charge and the F-actin is homopolymer. There are a variety of such biopolymers composed of protein monomers.

Even though there was tremendous interest in polyelectrolytes and the theory of neutral polymer systems was well developed [de Gennes 1979], polyelectrolytes remained for a long time poorly understood. The complications of theoretically handling Coulomb interactions limited progress in the field. The simulation methods described here to treat polyelectrolytes lead to significant progress pertaining to some of the fundamental questions about polyelectrolytes in solution.

The starting point for modeling polyelectrolytes is the well-developed coarse-grained models for neutral polymers [Kremer and Grest 1990, 1995]. The main addition is the inclusion of charges on the polymer and ions in solution. Much of the coarse-grained perspective is derived from neutral polymer simulations. However, polyelectrolytes are different structurally, and the intuition concerning their properties does not follow from neutral polymers. The basic polymer conformation of polyelectrolytes is not a random walk and the presence of charges introduces another length scale into the basic polymer model, which can have fundamental consequences. Thus, while polyelectrolytes have their roots in neutral polymers, the practical issue of performing and understanding the simulations is quite distinct.

The fundamental model of a polyelectrolyte chain in solution includes a chain of N monomers with bond length b, and distance between charges a. If all monomers are charged, then a and b are equal. One can also include the fraction of charged monomers $f = b/a$ as a parameter. The total charge in the polymer is fN. The contour length of the chain is $L = (N-1)b$. The solvent is typically treated as a continuum dielectric with dielectric constant ε. The nature of the Coulomb interaction adds major issues to be dealt with and the simulations are much more time consuming than neutral polymer simulations. The computational aspects of the Coulomb interaction will be discussed below.

This basic model is used in analytic theories as well as simulations. As with these theories, the goal of coarse-grained simulations is to treat a class of polyelectrolytes. The detailed properties of a particular polyelectrolyte that depend on its monomer chemistry are out of range for such models. Much of the polymer behavior depends on long length scale structure and has universal dependences, which is well described by coarse-grained models. Within the coarse-grained picture of polyelectrolytes, going beyond mean field theory has been difficult [Barrat and Joanny 1996]. Simulations treat the model Hamiltonian without approximation and a means to handle the complications of polyelectrolyte systems.

For a linear polyelectrolyte we can write the basic Hamiltonian as

$$H = \sum_{i=1,N-1} k_{\mathrm{b}}(\mathbf{r}_i - \mathbf{r}_{i+1})^2 + \sum_{i=2,N-1} k_{\mathrm{a}}(\theta_i - \theta_{i,0})^2 + k_{\mathrm{B}}T\sum_{i \neq j} \frac{\ell_{\mathrm{B}} z_i z_j}{\left|\mathbf{r}_i - \mathbf{r}_j\right|}, \tag{23.1}$$

where k_{b} is the spring constant for the bond potential, k_{a} is the spring constant for the three-body angle potential, θ_i is the angle between three consecutive monomers with the ith monomer being the central one, $\theta_{i,0}$ is the equilibrium angle (usually 180° for coarse-grained models), z_i is the valence of the ith monomer, $\ell_{\mathrm{B}} = e^2/\varepsilon k_{\mathrm{B}}T$ is the Bjerrum length with e the electron charge, k_{B} the Boltzmann constant, and T the temperature. The Bjerrum length is a critical parameter that defines the strength of the Coulomb interaction with respect to the thermal interactions (i.e., entropy). For separations less than ℓ_{B}, the Coulomb interaction between ions is stronger than the thermal interactions. At room temperature in water, $\ell_{\mathrm{B}} = 7.1$ Å. A strong polyelectrolyte has $\ell_{\mathrm{B}}/a > 1$. Examples include DNA and NaPSS. A weak polyelectrolyte such as polyacrylic acid has $\ell_{\mathrm{B}}/a \ll 1$.

More care and awareness must be applied in treatment of charged systems, because many of the rules that apply to short-ranged interactions do not apply to Coulomb interactions. For most of the Coulomb pair interactions in Equation 23.1, the energies are less than $k_{\mathrm{B}}T$, but the Coulomb interactions can sum up to be much greater than $k_{\mathrm{B}}T$. Consider a straight configuration of a polyelectrolyte and the monomer in the middle. The Coulomb energy for this monomer is

$$U = 2\sum_{j=1}^{N/2} \frac{e^2}{\varepsilon a j} = 2k_{\mathrm{B}}T\frac{\ell_{\mathrm{B}}}{a}\sum_{j=1}^{N/2}\frac{1}{j}, \tag{23.2}$$

which diverges as $N \to \infty$. This divergence shows the long-ranged nature of the Coulomb interaction. The Coulomb energy of the total system will be finite, but the order of the summation is important, as the sum is conditionally convergent. In physical terms, the nature of the screening by the ions in solution is important. The screening can yield a net interaction that is short ranged, and that can be treated effectively in ways similar to other short-ranged interactions. For example, when the Debye–Hückel (DH) approximation is valid the Yukawa potential can be substituted for the $1/r$ potential. On the other hand, many of the interesting phenomena occur for strong Coulomb interactions, which can demand explicit long-ranged evaluations. Fortunately, in the last decade the development and availability of fast Coulomb codes has greatly reduced the computational cost of treating such systems, and the long-ranged interaction can be treated at minimal extra cost in most cases. These codes use particle-mesh methods, which are discussed in the Methods section.

The starting point for treating the Coulomb interactions in theoretical works is the DH approximation for the electrostatic interactions. Briefly, for a system containing added salt, the approximation is as follows. The Poisson equation in a uniform dielectric with constant ε, is

$$\nabla^2\phi = \frac{4\pi e}{\varepsilon}\rho(r) = \frac{4\pi e}{\varepsilon}\sum_\alpha z_\alpha \rho_\alpha(r), \tag{23.3}$$

where ϕ is the electrostatic potential, ρ_α is the number density of mobile ion species α, and z_α is the valence of species α. Using the Boltzmann distribution for the ion densities and expanding, one finds

$$\rho(r) \approx \sum_\alpha \rho_\alpha e^{-z_\alpha e\beta\phi}, \tag{23.4}$$

where $\beta = 1/k_B T$. The Poisson-Boltzmann (PB) approximation is a combination of Equation 23.2 and Equation 23.3 and is a mean-field approximation. The nonlinear PB equations can be solved only for selected geometries such as charged lines and cylinders [Lifson and Katchalsky 1954]. Linearizing Equation 23.3 yields the DH approximation:

$$\nabla^2\phi \approx \frac{4\pi e}{\varepsilon}\sum_\alpha z_\alpha\rho_\alpha(1 - z_\alpha e\beta\phi) = -\frac{4\pi e^2}{\varepsilon k_B T}\sum_\alpha z_\alpha^2\rho_\alpha\,\phi = -\kappa^2\phi, \tag{23.5}$$

where Debye length is

$$\ell_D = \kappa^{-1} = 4\pi\ell_B\sum_\alpha z_\alpha^2\rho_\alpha. \tag{23.6}$$

This can be a rather severe approximation, especially for a fully charged polyelectrolyte and for r near the chain. The solution for the DH equation is

$$\phi(r) = \phi_0\frac{e^{-r/\ell_D}}{r}, \tag{23.7}$$

which is the screened Coulomb or Yukawa potential. Interactions beyond one or two Debye lengths can be neglected.

Manning determined important physical aspects of polyelectrolytes from the solution of the DH equations for the simplest model of a polyelectrolyte, namely a charged line [Manning 1969]. One of the key concepts to arise from these calculations and from the work of Oosawa (1971), is the idea of counterion condensation. The solution of the DH equations in terms of the Manning parameter

$\xi = \ell_B/a$ has a singularity at $\xi = 1$. The physical interpretation of the singularity is that for $\xi > 1$, a sufficient number of counterions condense onto the chain neutralizing some of the monomer charge and effectively change a such that the renormalized ξ is 1. Oosawa simultaneously pointed out that for such strong polyelectrolytes there are two types of counterions: free and condensed. The condensed counterions are localized (trapped) near the polymer chain by the strong Coulomb interactions.

There is a simple, charged model system that has a complete solution particularly for the strong coupling regime. This system is the one-component plasma (OCP), which consists of charged, point particles in a uniform neutralizing background [Brush, Sahlin, and Teller 1966; Stringfellow, DeWitt, and Slattery 1990]. The thermodynamics for the OCP are all a function of just one parameter, $\Gamma = \ell_B/a$. Here a is the average spacing between the charged particles defined in terms of the volume per particle or number density as $V/N = 1/\rho = 4\pi a^3/3$, which is similar to the definition of a for polyelectrolyte chains. Also, Γ is similar to the Manning parameter. The pressure as a function of Γ is shown in Figure 23.1. The plot shows the full OCP pressure [Stringfellow, DeWitt, and Slattery 1990] and the pressure in the PB approximation, which is the ideal gas pressure. This plot provides a basis for understanding some of the most interesting behavior of polyelectrolyte systems. At low Γ the PB approximation is accurate. In this regime, entropy dominates the interaction, which is the regime of validity for the PB approximation. As Γ approaches 1, the PB begins to break down and the pressures diverge. At larger Γ the OCP pressure exhibits some fundamental differences. First, there is a mechanical instability where $dP/dV < 0$ for $\Gamma > 3.09$. This is in the regime where Coulomb interactions dominate. The system wants to collapse to reduce the Coulomb energy. Consider the case of a crystal with NaCl structure of positive and negative point charges and lattice spacing a. Decreasing a will lower the Coulomb energy since this among other things brings the nearest neighbor $\pm$ pairs closer together. For point particles, the decreasing a will reduce the energy ultimately to $-\infty$, since there is no steric repulsion to limit the contraction. Thus, as in the plot the pressure becomes negative at sufficiently large Γ. We will see that these negative pressures do occur in more realistic polyelectrolyte systems. As in the OCP system, their origin is in the Coulomb interaction being stronger than entropy. An important point to keep in mind is that the OCP is a fluid in the range shown in Figure 23.1. The solid phase (Wigner crystal) does not form until very large values of Γ (~ 170) [Stringfellow, DeWitt, and Slattery 1990]. In the fluid phase, some degree of charge ordering does occur and is related to the structural origins of the instability. However, the degree of ordering is that of a liquid (small peaks in correlation functions) and not a solid (delta function peaks).

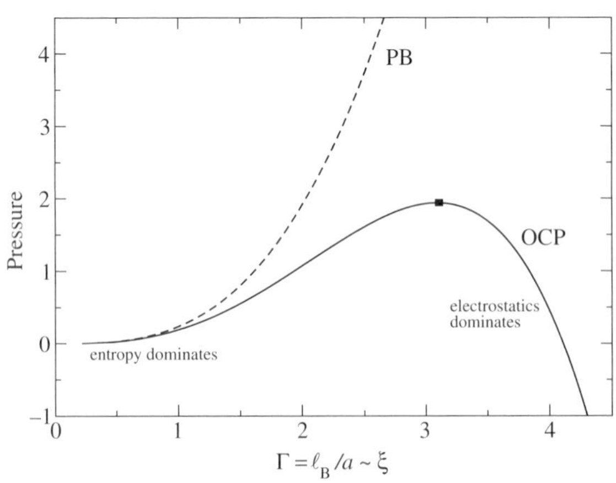

FIGURE 23.1 The pressure (solid line) of the one-component plasma (OCP) is given as a function of the ratio of the Bjerrum length ℓ_B and the average interparticle spacing a. The pressure in the Poisson–Boltzmann (PB) approximation is shown by the dashed line. The square point denotes the instability point where $dP/dV = 0$. (From Stevens, M.J. and Robbins, M.O., *Europhys. Lett.*, 12, 81, 1990. With permission.)

One of the hallmarks of polymer theory are scaling theories. In the Flory scaling argument for polyelectrolytes, the free energy F is the sum of the chain entropy and Coulomb energy.

$$F = k_B T \frac{R^2}{Nb^2} + \frac{(efN)^2}{\varepsilon R}. \tag{23.8}$$

Minimizing with respect to the end-to-end distance R yields

$$R \sim Nbf^{2/3}\left(\frac{\ell_B}{b}\right)^{1/3}, \tag{23.9}$$

which gives a Flory exponent of $\nu = 1 (R \sim N^{\nu})$. This is very different from neutral polymers, which have $\nu = 1/2$ for ideal chains and $\nu = 3/5$ in a good solvent. More details of scaling theory are given in the references [Odijk 1979; Dobrynin, Colby, and Rubinstein 1995].

23.2 METHODS

Coarse-grained models have been applied to polymers for a long time, especially flexible, neutral polymers [Binder 1995]. The coil diameter for flexible, neutral polymers is of the order 100 Å, which is much larger than the bond length (~ 1–2 Å) or the Kuhn length (~ 10 Å), the length that the intramolecular interactions keep the polymer locally stiff and straight. From the perspective of understanding the conformation of the polymer on the coil diameter scale, the local details of the structure on the atomic scale are secondary. The understanding of the physics of neutral polymers has come from realizing that the fundamental conformation is the random walk, and this conformation determines much of the physical properties. That is, that the dependence of the properties (e.g., viscosity as a function of concentration) is primarily due to the coarse-grained, random walk structure. Many physical properties can be scaled so that plots of data for different polymers coincide. The absolute magnitudes do depend on the chemical detail.

Coarse-grained models are thus inherent in polymer physics. Analytic theories are based on coarse-grained models. Simulations have traditionally been a means by which calculations of the polymer properties can be performed without the further approximations that are necessary in most analytic calculations and that often limit the range of validity of the calculation. This is particularly true for polyelectrolytes.

As noted above, the model of polyelectrolytes is an extension of the successful methods used for neutral polymers [Kremer and Grest 1995]. The basic model of the polymer is a bead-spring chain that can treat flexible polyelectrolytes like RNA and NaPSS and semiflexible polyelectrolytes like DNA and actin. The systems studied are composed of N_p bead-chain polymers of N monomers and N_c counterions. All particles are monovalent, and since the system is neutral, the total number of monomers, $N = N_p N$, equals the number of counterions. In this chapter the focus will be on salt-free solutions as the extension to include salt is straightforward. Added salt primarily increases the screening without adding additional physical phenomena [Stevens and Plimpton 1998]. The counterion and monomer number densities are the same ($\rho_m = \rho_c$), and we drop the subscript and use ρ as either density.

The interaction between beads is the Lennard-Jones (LJ) potential:

$$U_{LJ}(r) = \begin{cases} 4\varepsilon\left[\left(\frac{\sigma}{r}\right)^{12} - \left(\frac{\sigma}{r}\right)^{6} - \left(\frac{\sigma}{r_c}\right)^{12} + \left(\frac{\sigma}{r_c}\right)^{6}\right]; & r \le r_c \\ 0; \quad r > r_c. \end{cases} \tag{23.10}$$

For polyelectrolytes with a good solvent backbone, the cutoff is chosen to be $r_c = 2^{1/6}\sigma$, which yields a purely repulsive interaction. Treating poor solvent condition can be done by including an attractive part of the LJ interaction [Micka, Holm, and Kremer 1999]. However, the interaction of the backbone with the water may require treatment using implicit solvent models used in protein simulations [Reddy and Yethiraj 2006].

The monomers of a chain that represent several atoms are connected by a 'bond' potential. Here, we consider only the case where each monomer is charged; that is, $b = a$. For work treating poor solvent chains, generally not all the beads are charged ($b < a$). The attractive part of the bond potential (FENE) is given by

$$U_{\text{FENE}} = -\frac{1}{2}k_b R_0^2 \ln\left(1 - \frac{r^2}{R_0^2}\right), \tag{23.11}$$

with k_b being the spring constant and R_0 the maximum extent of the bond. The FENE bond potential has a singularity at $r = R_0$, which prevents the bond length from becoming larger than R_0. The repulsive part of the LJ potential is combined with the FENE potential to yield the total bond potential. A key physical characteristic of polymers is that the chains cannot cross. This requires the bond potential to prevent bonded beads separating enough to allow chains to cross. The FENE potential inherently achieves this. A harmonic bond potential does not limit the bond length and may be problematic, although in many cases a sufficiently strong harmonic bond potential will work fine. For systems with entanglements that put large stresses locally at the point where two chains intersect, one must be more careful and the FENE bond potential is preferred. From a computational point of view the cost of the FENE potential, while larger than a harmonic potential, is negligible overall, since the computational cost is dominated by the nonbond interactions. In either case, the spring constant is chosen primarily to maintain the polymer connectivity and is much weaker than a chemical bond length, which allows time-steps equal to that used for the LJ potential.

Particularly, biopolymers have an intrinsic stiffness due to the intramolecular bonding of the polymer. We modify the angle bending potential in Equation 23.1 by including the quartic term

$$U_{\text{angle}}(\theta) = k_{a2}(\theta - \theta_0)^2 + k_{a4}(\theta - \theta_0)^4, \tag{23.12}$$

where k_a are the bending constants, θ is the angle between three consecutive monomers on the chain, and θ_0 is the equilibrium angle, which is typically 180°. The quartic term is included in Equation 23.9 to make sharp bends prohibitively expensive [Stevens 2001].

The persistence length L_p is the quantity used to define the choice of the k_a, since L_p is a measured quantity. The persistence length is conceptually the length over which the chain is straight. For separations $s < L_p$ the tangent vectors of the chain are parallel. For larger separations the tangent vectors become uncorrelated. The definition of L_p is

$$\left\langle (t(s) - t(0))^2 \right\rangle = e^{-2L_p s}, \tag{23.13}$$

where $t(s)$ is the tangent vector at position s along the chain [Doi and Edwards 1986].

As noted above, the Coulomb interactions are long ranged and require special treatment. Not only must all ion pair interactions within the simulation be calculated, but also the interactions with the images. This summation is the Ewald sum, which treats the Coulomb energy for a periodic system with boundary dielectric ε_m at a radius much larger than the cell dimensions [Allen and Tildesley 1987]. The Ewald sum splits the calculation into two parts, a real space sum and a reciprocal space sum, such that each part converges "rapidly". A parameter G is used to control the convergence, or equivalently, the number of terms in the two sums required to achieve the desired accuracy. The real space sum for a system of N_{tot} total charged particles of valence z_i is

$$U_r = \sum_{i=1}^{N_{\text{tot}}-1}\sum_{j>i}^{N_{\text{tot}}} \frac{z_i z_j}{r_{ij}} \operatorname{erfc}\left(G\frac{r_{ij}}{\sqrt{2}}\right) - \frac{G}{\sqrt{2\pi}}\sum_{i=1}^{N_{\text{tot}}} z_i^2 + \frac{4\pi}{(1+2\varepsilon_{\text{m}})V}\mathbf{M}^2, \tag{23.14}$$

where $\mathbf{M}$ is the simulation cell's total dipole moment, and the sums only involve particles within the simulation cell, no image particles (the full Ewald sum includes these). The volume V is the simulation cell volume. The complementary error function limits the range of the first term. The last two terms are generally not used in simulations. The second term is a constant and thus neglectable. In most cases, the system has no net dipole moment and the last term is zero. However, this may not always be the case. An interesting example of the system dipole moment being relevant is in some dipolar systems [Wei and Patey 1992]. The reciprocal space contribution to the total system energy is

$$U_k = \frac{1}{\pi V}\sum_{i=1}^{N_{\text{tot}}-1}\sum_{j>i}^{N_{\text{tot}}}\sum_{\mathbf{k}\neq\mathbf{0}} z_i z_j \frac{4\pi^2}{k^2}\exp\left(-\frac{k^2}{2G^2}\right)\cos(\mathbf{k}\cdot\mathbf{r}_{ij}), \tag{23.15}$$

where the sum over $\mathbf{k}$ is over the reciprocal lattice vectors of the simulation cell lattice. The exponential limits the range in k-space of this sum. The double sum over i and j can be made into a single sum using trigonometric sum rules. However, this applies to the full total energy and not to forces on individual particles.

Using cutoff methods instead of variants of the Ewald sum to evaluate the Coulomb interaction can lead to gross errors in some cases. For example, the solidification of the OCP is off by an order of magnitude when calculated using the minimum image cutoff [Brush, Sahlin, and Teller 1966]. One of the subtle aspects of the Coulomb interaction is that the energy can often be calculated relatively accurately and sometimes even radial distribution functions are not so bad, but the orientational correlation functions are poor [Schreiber and Steinhauser 1992]. Given the speed of present particle-mesh algorithms, it is best to use them and to know that the calculation (with right parameters) is accurate.

The basic idea of fast particle-mesh calculations of the Coulomb interactions is to calculate the k-space sums using fast Fourier transforms (FFTs). The great advantage is that algorithm scales as $N \log N$, where N is the number of charged particles in the simulation. In addition the algorithm is parallelizable [Plimpton, Pollock, and Stevens 1997]. Furthermore, the crossover where the particle-mesh algorithms are faster than standard Ewald methods is a small number of particles (~ 100). The basic algorithm is to interpolate the charges to a 3D mesh; solve Poisson's equation on the mesh using FFTs; and interpolate back the electric fields to the atoms from which the forces are calculated. There are various particle-mesh methods available [Hockney and Eastwood 1988; Darden, York, and Pederson 1993; Pollock and Glosli 1996; Deserno and Holm 1998]. The advantages both computationally and physically for using one of these are so significant that standard cutoff methods are not worth considering. A discussion of the parallel implementation of particle-mesh methods can be found in the references [Plimpton 1995; Pollock and Glosli 1996; Plimpton, Pollock, and Stevens 1997]. In addition, the LAMMPS molecular dynamics code is open source and available online [Plimpton].

A comment concerning the nature of dielectric screening is worthwhile, since the issue comes up particularly when comparing with analytic calculations. The models discussed here treat the solvent, typically water, as a uniform dielectric medium. The temperature dependence of this approximation is subtle and often neglected. The dielectric constant is temperature dependent. In thermodynamics the relevant coupling parameter is ℓ_{B}, because the Boltzmann weighting involves $U/k_{\text{B}}T$. However, for water the temperature dependence of the Bjerrum length is small (15%) over the range from 0 to 100°C, because the temperature dependence of the ε is canceled by the $k_{\text{B}}T$ in ℓ_{B}.

As an example of mapping the coarse-grained model to a real polyelectrolyte, we consider the NaPSS system. In NaPSS, every other carbon atom in the backbone has a sulfonate group, which is typically charged. While not all the sulfonate groups are charged, to a good approximation we can consider them as charged. The distance between charges along the backbone is then $a = 2.5$ Å. Using the FENE bond potential with $k_b = 7\varepsilon/\sigma^2$ and $R_0 = 2\sigma$, the average bond length is 1.1σ. Equating the values of a defines the LJ unit as $\sigma = 2.2$ Å. We also would have $\ell_B = 7.1$ Å $= 3.2\sigma$. To treat a polyelectrolyte with a fraction f of charged monomers, we equate $a/f = 1.1\sigma$. For NaPSS, $f = 0.29$ yields $\sigma = 8.6$ Å and $\ell_B = 0.83\sigma$.

23.2.1 Dynamics

The dynamics of the system are performed at constant temperature $T = 1.2\varepsilon$ using the Langevin thermostat [Schneider and Stoll 1978]. The dynamical equations of motion with random noise term $\mathbf{W}$ are

$$m\ddot{\mathbf{r}}_i = \mathbf{F}_i - m\Gamma\dot{\mathbf{r}}_i + \mathbf{W}_i(t), \tag{23.16}$$

where $\mathbf{r}_i$ and $\mathbf{F}_i$ are the ith particle's position and force, respectively, and Γ is the damping constant such that

$$\langle \mathbf{W}_i(t)\mathbf{W}_j(t') \rangle = 6k_B T m \Gamma \delta_{ij} \delta(t - t'). \tag{23.17}$$

The two additional terms to Newton's equation couple the system to a heat bath maintaining a constant, average temperature. To thermostat the polymer beads we use $\Gamma = 1\tau^{-1}$. The time-step is 0.015 τ. Typically about 3×10^5 time-steps are used for $N = 32$ systems and 8×10^5 time-steps are used for $N = 64$. However, some circumstances such as multivalent ions require longer simulations [Stevens 2001].

23.3 APPLICATIONS

23.3.1 Polyelectrolytes in Salt-Free Solution

In polymer physics the structure of the single polymer in the low-density, noninteracting limit is the basis for the more complex calculations of the polymer structure as a function of concentration. While this perspective remains true in polyelectrolytes, the dilute limit for polyelectrolytes is not as trivial as for neutral polymers. For neutral polymers, the single-chain structure is the same for all concentrations below the overlap concentration c*, since there is no chain–chain interactions below c*. In polyelectrolyte systems there is an interaction due to the long-ranged Coulomb potential and to the screening by counterions. Thus, to know that one has reached the dilute limit structure, simulations have to be performed as a function of concentration (even if only one chain is treated in the simulation). For such simulations, we consider flexible polyelectrolyte chains of length $N = 16$, 32, and 64 in salt-free solution. Since the chains are flexible, there is no angle term in the potential. The interactions are just electrostatic and bond forces. In these simulations all the beads are singly charged and $\ell_B = 0.83\sigma$, which corresponds to $a = 8.6$ Å.

To characterize the structure of the single chain, we calculate the ration $r = R^2/R_G^2$, where R is the average end-to-end distance of the polyelectrolyte chain and R_G is the average radius of gyration. For a rod the ratio is 12 and for an ideal chain the ratio is 6. Thus, this ratio encompasses the two limits one expects for polyelectrolytes. Figure 23.2 shows the plot of r as a function of monomer density and chain length. At low densities the value of r obtains a limiting value that depends on the

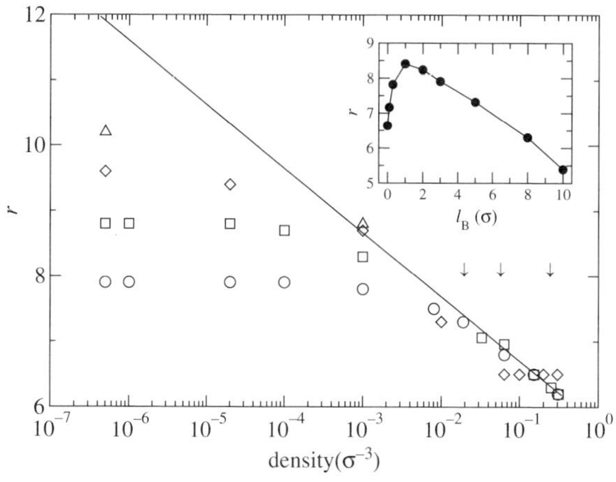

FIGURE 23.2 The ratio $r = R^2/R_G^2$ is plotted as a function of the monomer density for salt-free solutions at chain lengths of $N = 16$ (circles), 32 (squares), 64 (diamonds), and 128 (triangles). The arrows denote the overlap density for $N = 16$, 32, and 64 going from high to low density. The straight line is a guide to the eye for the part of the curves that is independent of N. Inset: The ratio r is plotted as a function of Bjerrum length for $N = 32$ at $\rho = 0.001\sigma^{-3}$. (Modified from Stevens, M.J. and Kremer, K., *J. Chem. Phys.*, 103, 1669, 1995. With permission.)

chain length. The increase in r with chain length is consistent with the Coulomb interaction being long ranged and longer chains having a larger net Coulomb repulsion among the monomers. The low-density limit of r is much greater than the neutral chain values, yet all the chain lengths have r below the ideal rod limit. There are still fluctuations within the chain structure; the chain entropy is not zero [Stevens and Kremer 1995]. To obtain the $r = 12$ rod-like structure requires much larger N *and* lower densities. The rod-like limit is a double limit in N and ρ.

The overlap densities ρ^* for $N = 16$, 32, and 64 are marked by arrows in the figure. For $\rho < \rho^*$, the value of r is still increasing. As noted above, this behavior is different from neutral polymers and different from early theoretical work [de Gennes et al. 1976]. The screening of the monomer repulsion by the counterions is substantial at concentrations near ρ^*. As the density decreases, this screening decreases and r increases. The saturation limit occurs when the local concentration of counterions becomes negligible. The density at which a single counterion will occupy the volume of the chain assuming uniform counterion density is a good approximation for the saturation density.

Experimentally, the single-chain structure factor is the measurable quantity. For polyelectrolytes measuring the dilute limit structure factor is very difficult for reasons apparent from the discussion of Figure 23.2. To obtain a structure factor that is independent of concentration requires going to very low concentrations—orders of magnitude lower than in neutral polymers—which greatly reduces the signal. This is a case where the simulations are much easier to perform than the experiments. The single-chain structure factor is

$$S(\mathbf{q}) = \frac{1}{N}\left|\sum_{j=1}^{N} \exp(i\mathbf{q}\cdot\mathbf{r}_j)\right|^2, \tag{23.18}$$

where the normalization is $S(0) = N$. The spherically averaged quantity $S(q)$ is calculated for $2\pi/b < q < R$. This range of q corresponds to structure on length scales between the bond length and the end-to-end distance, R. The concentration dependence is in the slope in the range $-1 < \log q\sigma < 0$, which corresponds to the lengths between $b < r < L$. The slope is related to the Flory exponent ν, which defines the scaling relations $R \sim N^{\nu}$ and $S(q) \sim q^{-1/\nu}$. For the lowest densities the ν is near 1.0, which is the rod-like limit [Stevens and Kremer 1993, 1995]. A finer examination gives $\nu = 0.93$, which is consistent with the data for the ratio r. As the density increases, the screening of the monomer repulsion increases and consequently ν decreases, reaching the neutral value of 5/3 at the highest densities. Thus, the single-chain polyelectrolyte structure as a function of concentration possesses the range of conformations from almost rod-like to self-avoiding random walk.

Validation of the simulation results by comparison to experimental data is essential. There were two measurable quantities that could be compared with the simulation data at the time of the original work. The osmotic pressure of the polyelectrolyte solution shows a density dependence. The data from several groups are presented in a paper by Wang and Bloomfield (1990). They found that the osmotic pressure scaled as a function of concentration $P \sim c^{\alpha}$ with $\alpha \approx 1$ at low concentrations and $\alpha = 9/4$ at high concentrations, which is the neutral limit. The simulations reproduced these measured results and provided a refinement due to the strength of the Coulomb interactions [Stevens and Kremer 1995].

The other measured quantity is the peak in the monomer–monomer structure factor. While the single-chain structure factor is difficult to measure, the total monomer–monomer structure factor is measurable and also shows a density dependence. The position of the peak in the structure factor scales as $c^{1/3}$ at low concentrations and has a chain length dependent crossover to $c^{1/2}$ at high concentrations [Kaji et al. 1988]. This result was reproduced by the simulations [Stevens and Kremer 1995]. In addition the relation between the crossover point and the overlap concentration could be directly calculated.

23.3.1.1 Counterion Condensation and the Strength of Coulomb Interactions

Counterion condensation is an important physical characteristic of polyelectrolyte systems. As noted above, for strong polyelectrolytes the total charge that resides in the chain is so large that some counterions are captured by the chain much like a nucleus binds electrons. In the same vein, the effective net charge is reduced by the condensed counterions. The Debye–Huckel approximation breaks down when the interactions are strong enough to yield counterion condensation. With simulations we can perform calculations without any approximation. A result that is very indicative of the nature of charged interactions and has broad implications is the counterion condensation and its connection with chain conformation at a dilute concentration as a function of varying ℓ_B/a [Gonzales-Mozuelos and Cruz 1995; Stevens and Kremer 1995].

The inset of Figure 23.2 shows the ratio r calculated for $N = 32$ at $\rho = 0.001\sigma^{-3}$ for varying values of ℓ_B [Stevens and Kremer 1995]. In the main plot of Figure 23.2, ℓ_B is 0.83σ. As ℓ_B decreases from this value in the main plot, r decreases, which is as expected due to the reduced Coulomb repulsion between the monomers yielding a more coiled structure. The ratio r goes toward the correct, neutral limit ($\sim$6.3) as $\ell_B \rightarrow 0$. Very interesting behavior occurs for $\ell_B > 1\sigma$. Instead of r continuing to increase with larger ℓ_B and stronger Coulomb interactions, r decreases. The reason lies in the strong interaction regime of the OCP pressure discussed above. For $\ell_B > 1\sigma$ the Coulomb interaction begins to dominate the entropy. Thus, the counterion attraction to the polyelectrolyte chains becomes strong enough that counterions are captured by the chain, and the number of condensed counterions increases with ℓ_B. Figure 23.3 shows a set of single-chain images with counterions within 2σ. The individual chains were chosen such that their eigenvalues of the radius of gyration tensor match the average values of the simulation for the given ℓ_B. The chains are shown oriented such that the largest eigenvector of the R_G tensor is along the width of the page and the second largest is along the height of the page. The set of images shows variation in the size of the average configuration and the increase of counterion condensation as a function of ℓ_B. The dual nature of the Coulomb interactions is evident from these images. While the Coulomb repulsion between monomers on the chain will yield a more rod-like conformation, the attraction of the counterions screens the monomer repulsion and shrinks the chain. In fact for the largest ℓ_B in the inset figure, the chain size is smaller than the neutral chain size ($r < 6$).

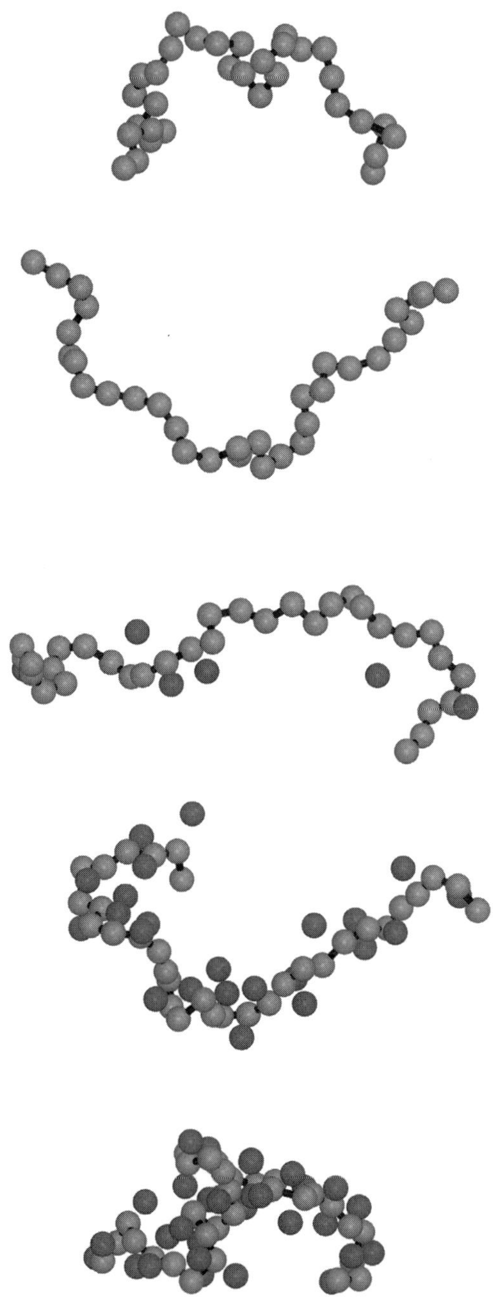

FIGURE 23.3 Images of single-chain conformations for (top to bottom) $\ell_B = 0.0$, 0.3, 1.0, 5.0, and 10.0σ for system in Figure 23.2. The chains have been oriented such that their largest eigenvector of R_G is along the width of the paper and the second largest eigenvector is along the height of the page. The light gray spheres are polyelectrolyte monomers and the dark gray spheres are counterions within 2σ of the chain.

In the simulations the variation of ℓ_B was done by changing the dielectric constant. This is certainly possible, but there is a limit of physically realizable values. This raises the question of what is the relevant range of ℓ_B/a (the ratio is the relevant quantity; a was kept fixed in the discussion above). Nature gives us a guide. We have treated monovalent ions in the discussion to this point.

In the expression for OCP Γ parameter, the valence z enters as z^2. Thus, a trivalent ion can push the value of ℓ_B/a by a factor of 10! This brings us to the next section.

23.3.1.2 DNA Condensation

DNA is one of the prototypical polyelectrolytes and is one of the most highly charged polymers with a charge every 1.7 Å along the axis. Thus, DNA is well into the counterion condensation regime with $\xi = 4.2$. A fundamental issue is packing DNA into cells. The contour length of the DNA can be larger than the cell. The simplest case is packing of DNA into viral capsids. For example, the λ bacteriophage has a capsid diameter of 60 nm. The λ bacteriophage's DNA has a contour length of 16 μm and can be coiled up within the capsid. How is the electrostatic repulsion between the highly charged DNA overcome in order to pack the DNA into the capsid? Moreover, like many biopolymers, double-stranded DNA is intrinsically stiff (due to the double-stranded structure), with $L_p = 500$ Å. Bending DNA must overcome both the Coulomb interaction and the intrinsic mechanical stiffness of the polymer.

We saw above that flexible polyelectrolytes can collapse for large values of ℓ_B/a or equivalently strong Coulomb interactions can dominate entropic interactions. With respect to DNA, the value of ℓ_B/a is fixed. Recall in Figure 23.3 that the counterions become more attracted to the polyelectrolyte with increasing ℓ_B/a and the strong Coulomb interactions yield the more compact structure as the system tries to obtain a charge-ordered structure. The strength of the Coulomb interaction between the DNA and counterions can be increased by increasing the valence of the counterion. In fact, it is well known that DNA will pack into toroidal structures in the presence of counterions with valences $z \geq 3$ [Kleinschmidt et al. 1962; Widom and Baldwin 1980]. This effect is purely electrostatic in that it does not depend on the chemical structure of the counterion [Widom and Baldwin 1980]. This behavior of DNA packing into condensed structures is called DNA condensation.

We can examine DNA condensation with coarse-grained polyelectrolyte simulations using the model described above. We now include the angle potential to produce an intrinsic mechanical stiffness in the polymer. We treat the DNA as a bead-spring polymer (i.e., no double strand) with every bead charged $b = a = 1.7$ Å. The persistence length of DNA is prohibitively long to treat even in the coarse-grained simulations. However, the issue is what happens to a semiflexible polyelectrolyte with $L \gg L_p \gg a$ in the presence of counterions of different valence. We can perform simulations with this constraint. Simulations were performed with $N = 256$, $k_{a2} = 5$ ε/rad^2 and $k_{a4} = 200$ ε/rad^4. The bead diameters were chosen to be 4 Å, which corresponds to a typical ionic diameter with $\sigma = 1.5$ Å [Stevens 2001].

First, the effect of divalent counterions was examined to see whether condensation can occur. Starting from random conformations, simulations performed with divalent ions do not form any condensed structures. This does not demonstrate that condensation does not occur with divalent ions, because there is always the issue of whether condensation would occur if the simulation were run longer; that is, that the nucleation event occurs on a time scale longer than that simulated. The result does show that there is a barrier to condensation of the polyelectrolyte. To treat this computational issue, simulations were performed starting with initial conformations near the toroid structure to determine if the structure is stable with divalent ions. The initial polyelectrolyte conformation was a spiral. The counterions are placed on a separate, translated spiral such that they are between successive arcs of the polymer's spiral. The energy of the single conformation with counterions was calculated for varying spiral radii and pitches. For the above force-field parameters the minimum energy conformation was found and used as the initial state. In this minimum energy state, one turn of the spiral has 40 beads and the pitch is $2 \times 2^{1/6} d$, where d is the bead diameter. This value of the pitch puts the counterions and charged monomers as close as possible without overlap of the LJ spheres. The spiral structure should be able to evolve easily into a toroidal structure, which is a slightly more condensed structure with charge ordering in three dimensions.

Figure 23.4a shows the conformations of the eight polyelectrolytes in the simulations with divalent counterions after 5×10^6 time-steps starting from an initial spiral conformation. Clearly,

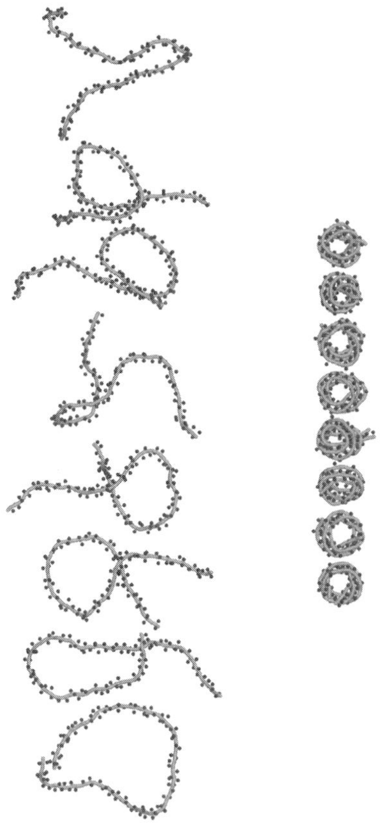

FIGURE 23.4 Images of $N=256$ chains (light gray) with counterions (dark gray) showing chain conformations. Each chain is oriented as in Figure 23.3 and placed on the figure individually. (a) Divalent counterion case showing that chains do not form toroidal structures starting from a spiral initial conformation. (b) Tetravalent counterions form toroidal structures. (Adapted from Stevens, M.J., *Biophys. J.*, 80, 130, 2001. With permission.)

the chains unwind from the spiral structure and the toroidal structure is not stable for the divalent system. Some counterions are delocalized, and as a whole the counterions are not fully screening the monomeric charges. On average, 116 out of 128 counterions per chain condense to within $2d$ of the polyelectrolytes. Each chain in combination with these counterions has a net negative charge. The simulations show that this net charge results in a net repulsion within the molecule and an extended structure. For divalent ions, not enough of the counterion and chain entropy can be overcome by Coulomb interactions to yield DNA condensation.

For the same parameter set but with tetravalent counterions, toroidal structures form and are stable. Figure 23.4b shows the eight conformations that evolved to be toroids. Even starting from random polymer conformations, condensed structures form for $z=3$ and 4. (Depending on the angle bend potential, kinked rod structures as well as toroids can form [Stevens 2001].) In general, for $z=4$, all the counterions condense to the chains. While the counterions are condensed, they still move about in the volume near the polymer. In other words, the counterions are bound to the polyelectrolyte, not to individual monomers. As such, they do not lose all their entropy in becoming condensed. These results show the competition between entropy, particularly of the counterions, and the Coulomb free energy. Condensing the counterions reduces their entropy. This can occur only if the Coulomb free energy of condensing the counterions compensates for the entropy loss.

Thus, the Coulomb coupling strength must be large enough to achieve this compensation. Also, divalent counterions have more entropic costs, since there are more of them than larger valence counterions. In condensation of single, semiflexible polyelectrolytes such as DNA, this competition requires $z \geq 3$ in agreement with experimental data [Bloomfield 1996].

23.3.1.3 Bundle Formation

The competition between entropy and Coulomb interactions is further elucidated by the formation of bundles in stiff polyelectrolytes. DNA condensation is the collapse of a single polyelectrolyte whose length is greater than its persistence length. A set of polyelectrolyte chains can collapse as a group. Of particular interest is the case when $L > L_p$. There are a variety of very stiff biopolymers that fall in this class, for example, F-actin, fd virus, short DNA [Tang and Janmey 1996; Tang et al. 2002]. These polyelectrolytes are also highly charged and will form bundles in the presence of multivalent ions. The basic principle is the same; the Coulomb interaction dominates entropy and the system forms a charge-ordered structure.

Simulations use the same basic model as the DNA model above, using just the harmonic angle bending potential. The chains have $N = 8$–64 monomers. The spring constant $k_a = 60\varepsilon/\sigma^2$ is large enough to make $L_p \geq L$. The system density is chosen below the onset of liquid crystal phase, but not so dilute that the chains do not interact. The systems start with the chains and counterions randomly placed without overlap. Simulations were performed with monovalent and divalent counterions. Figure 23.5 shows the interchain monomer–monomer radial distribution function $g_{mm}(r)$ as a function of N for divalent ions and for monovalent ions with $N = 32$. For the monovalent ions the bundles do not form, which also verifies that the system would not form a liquid crystal phase in the neutral case. There is a correlation hole in $g_{mm}(r)$ for the monovalent ion showing that the monomers on separate chains do not get close and form bundles. In contrast, for $z = 2$ the correlation function has a peak at $r = 2\sigma$ for all N, which grows with chain length. The peak occurs at this location because two parallel chains with counterions packed between them have a separation of 2σ. This is an indication of the charge-ordered structure that exists within these Coulomb-dominated systems. The growth in the peak is due to the stronger ordering of the chains within the bundle for larger N. This is in part due to the stronger total electrostatic interactions with the longer chains (with larger total charge).

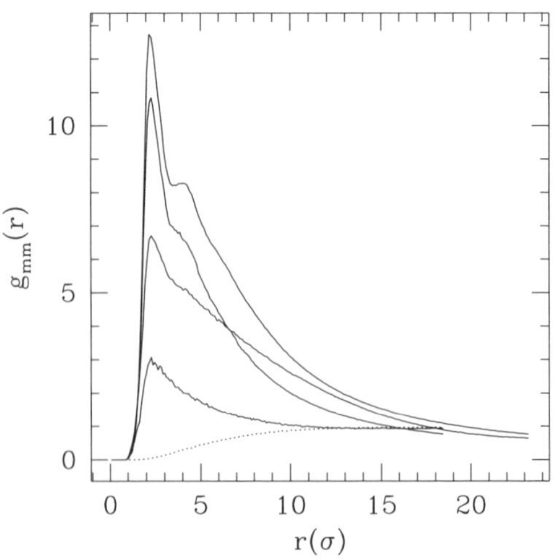

FIGURE 23.5 The monomer–monomer radial distribution function for the stiff polyelectrolytes at $\rho = 0.01\sigma^{-3}$. The solid lines are for divalent counterions. From top to bottom at the peak position $r = 2\sigma$, the lines are for $N = 64$, 32, 16, and 8. The dotted line is for monovalent counterions and $N = 32$. (From Stevens, M.J., *Phys. Rev. Lett.*, 82, 101, 1999. With permission.)

Examination of the counterion dynamics reveals the connection between the necessary valence to form condensed structures and the competition between entropic and Coulomb interactions. The divalent counterions are localized to the whole bundle. In comparison to the DNA condensation, the counterion entropy is larger in the bundle, because the counterions occupy a larger volume. For this reason, only divalent counterions are needed to form bundles. In general, the greater the loss of entropy in the system, the greater the Coulomb strength must be to compensate. The DNA condensation is an example where there is self-attraction within a macromolecule due to charge ordering and bundle formation is attraction between like-charged macromolecules again due to charge ordering in strongly coupled Coulomb systems [Stevens and Robbins 1990].

23.3.2 Grafted Polyelectrolytes

Grafted polyelectrolyte systems are a common application of polyelectrolytes. A main use of synthetic, grafted polyelectrolytes is stabilization of colloidal suspensions [Napper 1983]. A more recent technological example is in DNA microarrays, where DNA is grafted to a surface. Recent articles explore much of the progress on grafted polyelectrolyte topics [Netz and Andelman 2003; Ruhe et al. 2004; Naji, Seidel, and Netz 2006].

In the last few years, simulations of the basic polyelectrolyte brush systems have been performed [Csajka and Seidel 2000; Hehmeyer and Stevens 2005; Kumar and Seidel 2005, 2007]. The model is an extension of the model described above for polyelectrolytes in solution. For grafted polyelectrolytes one end of each chain is bound to a flat surface. The geometry consists of a system periodic in x and y. The substrate is typically a repulsive wall at $z=0$ that is treated as a z-dependent potential,

$$U_{\text{wall}}(z) = U_{\text{LJ}}(z). \tag{23.19}$$

Typically the cutoff is chosen so that the interaction is purely repulsive. The polyelectrolyte chains are bound to the surface with an area per chain A.

As one of the applications of grafted polyelectrolytes is the repulsion between two surfaces coated with the chains, simulations of two apposed surfaces is of fundamental interest and has been a focus of analytic work [Pincus 1991]. To treat such systems, the basic geometry described above is doubled with the chains grafted on the inside of opposite walls with separation D. Each wall has $N_p = 16$ chains arranged in a triangular lattice. In all LJ pair interactions, the bead diameter is set to be $d=4$ Å, the value used in primitive model electrolytes. The polyelectrolytes are treated as flexible (no angle potential). One of the interesting results is the density profile as a function of surface separation distance D. Figure 23.6 shows the density profile for $A=77.4\sigma^2$, $N=32$, and separations defined by $\delta = D/L$, the ratio of the separation to the chain contour length. For separations larger than the contour length, the chain profiles naturally do not overlap. As the separation shrinks, the peak in the density shifts from close to the substrate to the middle of the system. At these short separations, the density of the system is large and the screening is strong. In addition, much like in the solution as the concentration goes from well below overlap toward overlap, the polyelectrolyte chains contract as they avoid each other and the counterion screening increases. Overall, the chains have conformations more like neutral coils at small δ.

23.4 FUTURE DIRECTIONS

There are many future directions for coarse-grained modeling of polyelectrolytes, because of the variety of polyelectrolytes. Work to date has focused on the simplest cases. Future work will include more complex polyelectrolytes and systems of polyelectrolytes with other molecules. Two examples are given below that will likely have a big impact beyond just the polyelectrolyte field.

Recently is has become understood that a large fraction (30% eukaryotic genome) of proteins are not natively folded. These "unstructured proteins" are polyampholytes strictly speaking, but

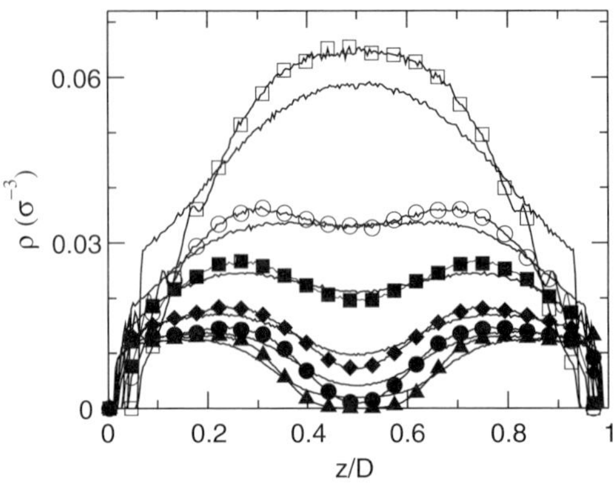

FIGURE 23.6 Density profiles of apposing polyelectrolyte brushes. Monomer density is indicated by a solid line with different point types to indicate the gap width. The series are for gap widths of $\delta = 1.42$ (triangles), 1.14 (circles), 0.85 (diamonds), 0.57 (squares), 0.43 (open circles), and 0.28 (open squares). Each monomer density profile is paired with a counterion density profile that is indicated by a solid line. (From Hehmeyer, O. and Stevens, M.J., *J. Chem. Phys.*, 122, 134909, 2005. With permission.)

they typically have a net charge and can behave on large length scales as polyelectrolytes. In some cases unstructured proteins only partially unfolded; often the tail segments are unstructured. The unstructured proteins are biologically functional. Many bind to nuclei acids; thus, a polyelectrolyte-polyelectrolyte binding occurs in this case.

An interesting example of the function of unstructured proteins is the neurofilament fibers (NF) and microtubule associating proteins (MAPs) [Bright, Woolf, and Hoh 2001; Weathers et al. 2004]. There are three NF polymers, called NF-L, NF-M, and NF-H for low, medium, and high molecular weight. The NF-M and NF-H have long tails that are unstructured. The NF-L, which has the same beginning amino acid sequence as NF-M and NF-H, forms a coiled-coil structure that binds together into 10 nm wide fibers. The long unstructured tails of the NF-M and NF-H form a polyelectrolyte brush bound to the fiber. In a related manner MAPs form brushes in conjunction with microtubules. Part of the MAP is folded and binds to the microtubule. The unstructured part extends from the microtubule and forms the polymer brush. Functionally these polymer brushes provide mechanical stability for neural axons. The polymer brush core (coiled coils of NF and the microtubules) are oriented along the axis of the axon tube. The systems of NF and MAP-microtubules form a polymer brush that, much like colloids with grafted polymers, forms a liquid structure [Brown and Hoh 1997; Mukhopadhyay and Hoh 2001]. The grafted polyelectrolyte repulsion provides a mechanical stability for the axon. This mechanical structure is quite flexible. A rigid, bonded network does not exist. This is just one example of unstructured proteins possessing interesting physical properties. This class of polymers is very broad and most likely contains a large number of interesting physical phenomena to study. A strong basis has been provided for coarse-grained modeling by work using support vector machine algorithms to characterize unstructured proteins based on reduced amino acid groups, which found that only four groups are necessary [Weathers et al. 2004].

The combination of polyelectrolytes and other charged molecules is an area of growing interest. An example is the complexes formed by cationic lipid bilayers and DNA as well as other biopolymers. One application of such systems is packing of DNA as a delivery mechanism for gene therapy. The complexes self-assemble to form hierarchical structures. At one level there is the self-assembly of the lipids into a bilayer. The bilayers and the DNA form another level, which can have different structures depending on the lipid compositions. Lamellar and hexagonal phases have been observed [Wong et al. 2000; Liang, Harries, and Wong 2005]. The basic mechanism for forming the complexes is believed to be electrostatic interactions. With the advent of coarse-grained models for lipid molecules, [reference Chapters 2 and 3] it is now possible to simulate these complexes. The connection of the fundamental interactions and the complex structure can be investigated in such systems.

Because of the richness of possibilities of putting charges on polymers and mixing different polyelectrolytes, lipids, nanoparticles, etc., this is a very exciting research area. Over the last decade, major progress has been made in some of the basic polyelectrolyte systems. Future research is now possible on more complex assemblies that will naturally possess more complex properties.

REFERENCES

Allen, M. P., and D. J. Tildesley. 1987. *Computer Simulation of Liquids*. New York: Oxford University.

Barrat, J. L., and J. F. Joanny. 1996. Theory of polyelectrolyte solutions. *Adv. Chem. Phys.* 94:1–66.

Binder, K., ed. 1995. *Monte Carlo and Molecular Dynamics Simulations in Polymer Science*. New York: Oxford.

Bloomfield, V. A. 1996. DNA condensation. *Curr. Op. in Str. Biol.* 6:334–41.

Bright, J. N., T. B. Woolf, and J. H. Hoh. 2001. Predicting properties of intrinsically unstructured proteins. *Prog. Biophys. Mol. Biol.* 76:131–73.

Brown, H. G., and J. H. Hoh. 1997. Entropic exclusion by neurofilament sidearms: A mechanism for maintaining interfilament spacing. *Biochemistry* 36 (49):15035–40.

Brush, S. G., H. L. Sahlin, and E. Teller. 1966. Monte Carlo study of a one-component plasma. I. *J. Chem. Phys.* 45:2102.

Csajka, F. S., and C. Seidel. 2000. Strongly charged polyelectrolyte brushes: A molecular dynamics study. *Macromolecules* 33:2728–39.

Darden, T., D. York, and L. Pederson. 1993. Particle mesh Ewald: An N log(N) method for Ewald sums in large systems. *J. Chem. Phys.* 98:10089.

de Gennes, P. G. 1979. *Scaling Concepts in Polymer Physics*. Ithaca, NY: Cornell University.

de Gennes, P. G., P. Pincus, R. M. Valesco, and F. Brochard. 1976. Remarks on polyelectrolyte conformation. *J. Physique* 37:1461.

Deserno, M., and C. Holm. 1998. How to mesh up Ewald sums. I. A theoretical and numerical comparison of various particle mesh routines. *J. Chem. Phys.* 109:7678–793.

Dobrynin, A. V., R. H. Colby, and M. Rubinstein. 1995. Scaling theory of polyelectrolyte solutions. *Macromolecules* 28:1859–71.

Doi, M., and S. F. Edwards. 1986. *The Theory of Polymer Dynamics*. New York: Oxford University Press.

Gonzales-Mozuelos, P., and M. O. de la Cruz. 1995. Ion condensation in salt-free dilute polyelectrolyte solutions. *J. Chem. Phys.* 103:3145–57.

Grest, G. S., and K. Kremer. 1986. Molecular dynamics simulation of polymers in the presence of a heat bath. *Phys. Rev. A* 33:3628–31.

Hehmeyer, O. J., and M. J. Stevens. 2005. Molecular dynamics simulations of grafted polyelectrolytes on two apposing walls. *J. Chem. Phys.* 122:134909.

Hockney, R. W., and J. W. Eastwood. 1988. *Computer Simulation Using Particles*. New York: Adam Hilger.

Kaji, K., H. Urakawa, T. Kanaya, and R. Kitamaru. 1988. Phase diagram of polyelectrolyte solutions. *J. Physique* 49:993.

Kleinschmidt, A. K., D. Lang, D. Jacherts, and R. K. Zahn. 1962. Darstellung und Längenmessungen des gesamten desoxyribonucleinsäure: Inhaltes von T_2-Bakteriophagen. *Biochim. Biophys. Acta* 61:857–64.

Kremer, K., and G. S. Grest. 1990. Dynamics of entangled linear polymer melts: A molecular-dynamics simulation. *J. Chem. Phys.* 92:5057–86.

———. 1995. Entanglement effects in polymer melts and networks. In *Monte Carlo and Molecular Dynamics Simulations in Polymer Science*, ed. K. Binder, 194–262. New York: Oxford.

Kumar, N. A., and C. Seidel. 2005. Polyelectrolyte brushes with added salt. *Macromolecules* 38:9341–50.

———. 2007. Interaction between two polyelectrolyte brushes. *Phys. Rev. E* 76:020801.

Liang, H., D. Harries, and G. C. L. Wong. 2005. Polymorphism of DNA-anionic liposome complexes reveals hierarchy of ion-mediated interactions. *Proc. Natl. Acad. Sci. U.S.A.* 102:11173–78.

Lifson, S., and A. Katchalsky. 1954. The electrostatic free energy of polyelectrolyte solutions. I. Fully stretched macromolecules. *J. Polym. Sci.* 13:43.

Manning, G. 1969. Limiting laws and counterion condensation in polyelectrolyte solutions I. Colligative properties. *J. Chem. Phys.* 51:924–33.

Micka, U., C. Holm, and K. Kremer.1999. Strongly charged, flexible polyelectrolytes in poor solvents: Molecular dynamics simulations. *Langmuir* 15:4033–44.

Mukhopadhyay, R., and J. H. Hoh. 2001. AFM force measurements on microtubule-associated proteins: The projection domain exerts a long-range repulsive force. *FEBS Lett.* 505 (3):374–78.
Naji, A., C. Seidel, and R. R. Netz. 2006. Theoretical approaches to neutral and charged polymer brushes. *Adv. Polym. Sci.* 198:149–83.
Napper, H. 1983. *Polymeric Stabilisation of Colloidal Dispersions.* London: Academic Press.
Netz, R. R., and D. Andelman. 2003. Neutral and charged polymers at interfaces. *Phys. Rep.* 380:1.
Odijk, T. 1979. Possible scaling relations for semidilute polyelectrolyte solutions. *Macromolecules* 12:688.
Oosawa, F. 1971. *Polyelectrolytes.* New York: Marcel Dekker.
Pincus, P. 1991. Colloid stabilization with grafted polyelectrolytes. *Macromolecules* 24:2912–19.
Plimpton, S. J. n.d. LAMMPS. Code may be downloaded at lammps.sandia.gov.
———. 1995. Fast parallel algorithms for short-range molecular dynamics. *J. Comput. Phys.* 117:1–19.
Plimpton, S. J., E. L. Pollock, and M. Stevens. 1997. Particle-mesh Ewald and rRESPA for parallel molecular dynamics simulations. In *Eighth SIAM Conference on Parallel Processing for Scientific Computing*, Minneapolis, MN, SIAM.
Pollock, E. L., and J. Glosli. 1996. Comments on P3M, FMM and the Ewald method for large periodic coulombic systems. *Comput. Phys. Commun.* 95:93.
Reddy, G., and A. Yethiraj. 2006. Implicit and Explicit Solvent Models for the simulations of dilute polymer solutions. *Macromolecules* 39:8536–42.
Ruhe, J., M. Ballauff, et al. 2004. Polyelectrolyte brushes. In *Polyelectrolytes with Defined Molecular Architecture I.* ed. M. Schmidt, 79–150. Berlin: Springer.
Schiessel, H. 2003. The physics of chromatin. *J. Phys.: Condens. Matter* 15 (19): R699–774.
Schneider, T., and E. Stoll. 1978. Molecular dynamics study of a three-dimensional one-component model for distortive transitions. *Phys. Rev. B* 17:1302–22.
Schreiber, H., and O. Steinhauser. 1992. Molecular dynamics studies of solvated polypeptides: Why the cut-off scheme does not work. *Chem. Phys.* 168:75–89.
Stevens, M. J. 2001. Simple simulations of DNA condensation. *Biophys. J.* 80:130–39.
Stevens, M. J., and K. Kremer. 1993. Form factor of salt-free linear polyelectrolytes. *Macromolecules* 26:4717–19.
———. 1995. The nature of flexible linear polyelectrolytes in salt free solution: A molecular dynamics study. *J. Chem. Phys.* 103:1669–90.
Stevens, M. J., and S. J. Plimpton. 1998. The effect of added salt on polyelectrolyte structure. *Euro. Phys. J. B* 2:341.
Stevens, M. J., and M. O. Robbins. 1990. Density functional theory of ionic screening: When do like charges attract. *Europhys. Lett.* 12:81.
Stringfellow, G. S., H. E. DeWitt, and W. L. Slattery. 1990. Equation of state of the one-component plasma derived from precision Monte Carlo calculations. *Phys. Rev. A* 41:1105.
Tang, J. X., and P. Janmey. 1996. The polyelectrolyte nature of F-actin and the mechanism of actin bundle formation. *J. Biol. Chem.* 271:8556–63.
Tang, J. X., P. Janmey, A. Lyubartsev, and L. Nordenskiold. 2002. Metal ion-induced lateral aggregation of filamentous viruses fd and M13. *Biophys. J.* 83:566–81.
Wang, L., and V. A. Bloomfield. 1990. Osmotic pressure of polyelectrolytes without added salt. *Macromolecules* 23:804.
Weathers, E. A., M. E. Paulaitis, T. B. Woolf, and J. H. Hoh. 2004. Reduced amino acid alphabet is sufficient to accurately recognize intrinsically disordered protein. *FEBS Lett.* 576:348–52.
Wei, D., and G. N. Patey. 1992. Ferroelectric liquid-crystal and solid phases formed by strongly interacting dipolar soft spheres. *Phys. Rev. A* 46:7783–92.
Widom, J., and R. L. Baldwin. 1980. Cation-induced toroidal condensation of DNA. *J. Mol. Biol.* 144:431.
Williams, L. D. 2000. Electrostatic mechanisms of DNA deformation. *Annu. Rev. Biophys. Biomol. Struct.* 29:497–521.
Wong, G. C. L., J. X. Tang, A. Lin, Y. Li, P. A. Janmey, and C. R. Safinya. 2000. Hierarchical self-assembly of F-actin and cationic lipid complexes: Stacked three-layer tubule networks. *Science* 288:2035–39.

24 Monte Carlo Simulations of a Coarse-Grain Model for Block Copolymer Systems

F.A. Detcheverry
Department of Chemical and Biological Engineering, University of Wisconsin-Madison

K.Ch. Daoulas and M. Müller
Institut für Theoretische Physik, Georg-August Universität

P.F. Nealey and J.J. de Pablo
Department of Chemical and Biological Engineering, University of Wisconsin-Madison

CONTENTS

24.1 INTRODUCTION

Polymeric systems are characterized by a wide spectrum of length scales that range from short chemical bonds (Å) to chain dimensions (10 nm) and macroscopic behavior. The corresponding time scales associated with motions on such length scales are even broader; bond vibrations occur on the scale of picoseconds (10^{-13} s) and, depending on molecular weight, temperature and density, chain relaxation and morphology formation, can occur over seconds, minutes, or hours. Multiple length scales are inherently linked through the connectivity of the chain molecules, and different levels of description are therefore coupled and cannot be treated independently. Molecular dynamics simulations using atomistic force fields are unable to access the time scales necessary to achieve chain relaxation for polymeric systems of intermediate or high molecular weights. Advanced Monte Carlo (MC) methods have been developed for the equilibration of dense polymeric systems with long chains. Nevertheless, the size of the systems that can be efficiently simulated is still limited by the performance of present-day computers [Binder 1995; Kotelyanskii and Theodorou 2004].

In order to study polymeric systems, particularly their ability to self-assemble over tens or hundreds of nanometers, it is necessary to reduce the number of degrees of freedom. In doing so, it is essential that one preserves a number of relevant key features that give rise to the characteristic behavior on mesoscopic and macroscopic length scales. This coarse-graining procedure leads to a hierarchy in degrees of freedom with increasing length scales: for polymers, we start with atoms, continue with monomers (groups of 10–100 atoms), and then have polymer chains (soft fluid) [Murat and Kremer 1998; Louis et al. 2000; Eurich and Maass 2001; Yatsenko et al. 2004; Pierleoni et al. 2006], or one integrates out the microscopic degrees of freedom of the molecular conformations and describes the system via local, spatially varying densities. The latter are the central object in self-consistent field (SCF) theoretic treatments [Fredrickson 2006]. On the largest length scale, phenomenological treatments of polymeric systems in terms of Ginzburg-Landau functionals utilize an even coarser description that ignores much of the spatially extended molecular architecture and is mainly based upon symmetry considerations. At each level of coarse-graining some information is irreversibly lost: for instance, in the case of dynamics of polymeric melts, the reptation motion of individual molecules in highly entangled melts can no longer be captured when polymer chains are represented as collections of just a few beads, or as simple ellipsoids that can overlap with each other.

There are two broad classes of coarse-grain models that are particularly relevant to our discussion. In the first of these approaches—denoted "systematic coarse-graining", the model retains, even at the coarse-grain level, the specificity of the polymer under consideration. Such systematically coarse-grained models are useful to relate the macroscopic properties of a polymeric material to the chemical structure of the individual chains. The degrees of freedom are often effective segments comprised of a small number of atoms whose characteristics (interaction parameters) are adjusted to match results obtained at a fully atomistic level. A coarse-grain model of that type provides access to properties that arise over longer time and length scales than those accessible to a fully atomistic model. In a process known as fine graining or reverse mapping, the details of the atomistic model can be reintroduced. On the other hand, minimally coarse-grained models only retain relevant features common to a class of systems; they assume that universal properties emerge on large length scales (an example is provided by the Gaussian nature of chain molecules in a polymeric melt). Contrary to "systematically coarse-grained" models, whose predictions are absolute quantities for a specific material, minimally coarse-grained models predict the mesoscopic and large-scale properties of a class of materials. Their predictions can be quantitatively related to a specific material by matching a small number of coarse-grain parameters or invariants (e.g., the end-to-end distance or the Flory-Huggins parameter) that define the strength of the relevant interactions in the specific material and the minimally coarse-grained model.

In this chapter we discuss a minimally coarse-grained model for a polymer melt. It is based on models introduced in the context of SCF theory [Muller and Schmid 2005; Matsen and Schick 1994; Fredrickson 2006; Edwards 1965; Helfand and Tagami 1972; Hong and Noolandi 1981], but it is viewed from a different perspective theory since the fundamental degrees of freedom are not the local densities but the positions of polymer segments. This particle-based approach describes chain conformations explicitly. Perhaps more importantly, it facilitates description of complex chain architecture and nonpolymeric objects such as nanoparticles. We do not invoke a saddle-point approximation that is utilized in SCF theory, but we study the exact statistical mechanics of the particle-based Hamiltonian via MC simulations. The underlying idea has been explored previously in the context of polymer brushes [Laradji, Guo, and Zuckermann 1994; Soga, Zuckermann, and

Guo 1996; Soga, Guo, and Zuckermann 1995] and in recent studies of polymeric melt [Daoulas and Muller 2006; Kang et al. 2008; Detcheverry et al. 2008]. Here we examine various aspects of the proposed approach that turn out to have important consequences for the results, and we illustrate its promise by presenting applications to mixtures of copolymers and nanoparticles—systems whose description in the context of SCF remains particularly challenging.

After describing the method, and briefly discussing its relation to other approaches, we illustrate a few possible applications, such as the prediction of diblock copolymer morphologies and self-assembly in mixtures of block-copolymer/nanoparticles. We end with a few concluding remarks regarding the general applicability of the approach outlined in this work.

24.2 METHOD

24.2.1 Model and Coarse-Grain Parameters

For simplicity, the model is presented in the context of a diblock copolymer melt; extensions to multicomponent systems, including multiblock systems or copolymer–homopolymer blends, are straightforward. Consider n copolymer molecules in a volume V at temperature T. The polymer chains are assumed to be Gaussian and are represented by a bead-spring model. The chain contour is discretized with N beads; $\mathbf{r}_i(s)$ denotes the position of the sth bead in the ith chain. For an isolated, noninteracting chain, the probability of adopting a given conformation $\mathbf{r}(s)$ is given by:

$$P[\mathbf{r}(s)] = \exp\left[-\frac{H_b[\mathbf{r}(s)]}{k_B T}\right],$$

where k_B is the Boltzmann constant. The bonded interactions $H_b\left[\mathbf{r}(s)\right]$ between the beads correspond to harmonic springs and are given by:

$$\frac{H_b[\mathbf{r}(s)]}{k_B T} = \frac{3}{2}\sum_{s=1}^{N-1}\frac{[\mathbf{r}(s+1)-\mathbf{r}(s)]^2}{b^2}, \tag{24.1}$$

where b^2 is the mean squared bond length. Nonbonded interactions among the effective segments are taken into account through an interaction functional $F[\phi_A, \phi_B]$ that is comprised of enthalpic and entropic contributions due to the coarse-graining procedure. It depends on the local, normalized bead densities $\phi_A(\mathbf{r})$ and $\phi_B(\mathbf{r})$. In this work, $F[\phi_A, \phi_B]$ is given by the simple choice:

$$F[\phi_A, \phi_B] = \rho_0 \int_V d^3\mathbf{r}\left[\chi\phi_A\phi_B + \frac{\kappa}{2}(1-\phi_A-\phi_B)^2\right], \tag{24.2}$$

where $\rho_0 = nN/V$ is the average bead number density. The first term in the sum accounts for the incompatibility between beads of different type, the strength of which is quantified by the Flory–Huggins parameter χ. The second term enforces the finite compressibility of the melt, which is inversely proportional to κ. The so-called Helfand quadratic approximation [Helfand and Tagami 1972] does not aim to describe the liquid-like structure of the polymeric melt; rather, it is the simplest form that penalizes fluctuations of the local densities away from their average value, thereby enforcing near incompressibility of the melt on long length scales.

The resulting Hamiltonian is given by:

$$\frac{H[\{\mathbf{r}_i(s)\}]}{k_B T} = \frac{3}{2}\sum_{i=1}^{n}\sum_{s=1}^{N-1}\frac{N-1}{R_e^2}[\mathbf{r}_i(s+1)-\mathbf{r}_i(s)]^2 + \sqrt{\bar{N}}\int_V \frac{d^3\mathbf{r}}{R_e^3}\left[\chi N\phi_A\phi_B + \frac{\kappa N}{2}(1-\phi_A-\phi_B)^2\right]. \tag{24.3}$$

It incorporates the three relevant ingredients necessary to describe the physics of diblock copolymers on mesoscopic length scales: the chain structure and connectivity, the incompatibility between unlike molecules, and the small but finite compressibility of the melt that stems from the excluded volume of the beads. It should be apparent from Equation 24.3 that only a few coarse-grain parameters emerge from this model. The first is the mean squared end-to-end distance $R_e^2 = (N-1)b^2$ for a noninteracting chain; R_e sets the length scale for the coarse-grain representation. The chain contour discretization N (number of beads in the chain) is not a physical parameter in itself; only the products χN and κN are meaningful. The last parameter, $\bar{N} = (\rho_0 R_e^3/N)^2$, controls the strength of fluctuations; $\bar{N}$ is referred to as the invariant degree of polymerization because in a dense melt, $R_e \sim \sqrt{N}$ and $\bar{N} \sim N$; $\sqrt{\bar{N}}$ is a dimensionless density that measures the number of chains found in the typical volume of a single chain (estimated as R_e^3), and it also provides a rough estimate of the number of chains that a given molecule interacts with.

24.2.2 Definition of Local Densities

To completely describe the model, we need to specify how the local, normalized densities $\phi_A(\mathbf{r})$ and $\phi_B(\mathbf{r})$ are defined in terms of the bead positions $\{\mathbf{r}_i(s)\}$. Note that the densities are not given by the microscopic expression $\phi_A(\mathbf{r}) = \sum_{i \in \text{A-bead}} \delta(\mathbf{r} - \mathbf{r}_i)$, as in liquid state theory. Instead, as in SCF theory, they are defined after a coarse-graining procedure that introduces a microscopic cut-off and results in a continuous scalar field. There are at least two ways to define such densities from the bead positions. The first is to associate to each bead a cloud density, such as a Gaussian instead of the δ-function in the microscopic expression [Laradji, Guo, and Zuckermann 1994]. The local densities are then unambiguously defined and the width of the Gaussian sets the microscopic cut-off. The second possibility is to use a particle-to-mesh technique [Soga, Zuckermann, and Guo 1996]. In the simplest scheme (zeroth-order interpolation), a regular, cubic mesh is introduced and from the number n_A^k of A beads in the cell k, the local, normalized density is given by:

$$\phi_A^k = n_A^k / n_{\text{cell}}, \tag{24.4}$$

where n_{cell} is the average number of beads in a cell (see Figure 24.1). In this case, the grid spacing sets the microscopic cut-off. Alternatively, one can use other assignment functions between the particle positions and the grid.

No matter which technique is chosen, cloud-density or particle-to-mesh, the definition of local densities requires the introduction of a new discretization parameter—the microscopic cutoff ΔL. Physically, ΔL corresponds to the range of interaction between beads. Its choice must meet several constraints. On the one hand, ΔL cannot be too small: if, for example, ΔL were much smaller than the mean distance between neighboring beads, the beads would barely interact with each other. In the following we use $10 \le n_{\text{cell}}$, which enforces a minimal value for the grid spacing $\Delta L = (n_{\text{cell}}/\rho_0)^{1/3}$. On the other hand, the grid spacing cannot be too large if one aims to spatially resolve the inhomogeneous density distribution. If the grid spacing was much larger than the smallest length scale over which the density exhibits significant variations (e.g., the width of an interface between A-rich and B-rich domains), one would observe an explicit dependence of the results on the grid spacing. Following those two constraints, ΔL is chosen to be the smallest distance over which it remains meaningful to define the local densities. The use of the cloud density is more computationally demanding than the particle-to-mesh technique; therefore, in what follows, we restrict our discussion to the particle-to-mesh approach with a zeroth-order interpolation (Equation 24.4). Higher-order interpolations [Deserno and Holm 1998] are straightforward to implement, but require longer computation times. To avoid any artifact associated with a fixed grid, the position of the grid is randomly chosen at each MC step.

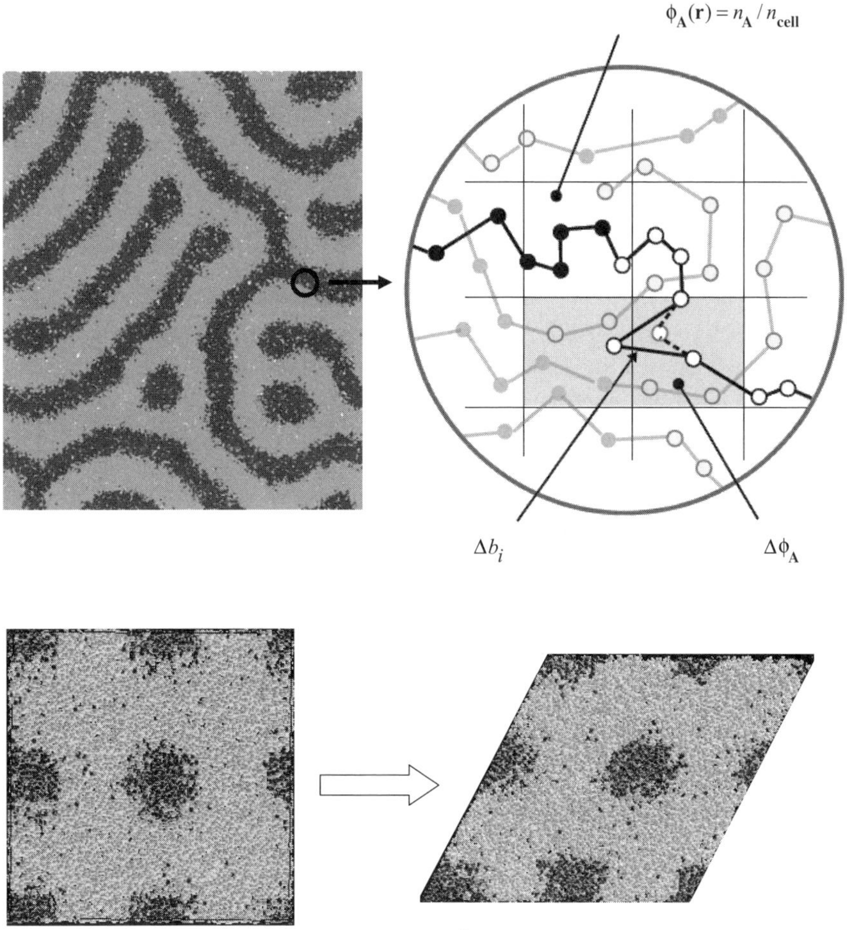

FIGURE 24.1 Top: illustration of the coarse-grain MC simulations proposed in this work. Local densities are defined by using a regular grid and by counting the number of beads in each cell (more accurate interpolation schemes can be used). When a MC move is proposed, for one bead or more, the difference in energy comes from the change in bond lengths (Δb_i) and the change in local densities ($\Delta\phi_{A,B}$); these changes can be computed efficiently. Bottom: in the variable cell shape method, the simulation box changes its shape and size to accommodate the natural symmetry and periodicity of the mesophase.

24.2.3 MC Simulations

The equilibrium properties of the model defined above are determined by MC simulations. A realization of the system consists of many molecules interacting as described above. Distinct configurations are sampled according to their Boltzmann weight. A MC move consists in choosing at random a chain molecule or a subset of beads, proposing trial positions, and determining whether the trial positions should be accepted on the basis of the energy change. This difference in energy ΔE between the old and the trial configuration stems from changes in the bond lengths (bonded interactions) and changes in the local densities (nonbonded interactions). The move is then accepted according to the Metropolis criterion; that is, with probability: $p_{acc} = \min(1, \exp(-\Delta E / k_B T))$. The simplest MC move is the random displacement of a single bead; other types of move include reptation of individual or multiple beads, translation of entire chains, switching the order of blocks while keeping the same chain conformation, and deleting an entire chain and randomly rebuilding it at a different position. Drastic, global moves are particularly helpful for rapidly reaching the equilibrium morphology of the system.

We now briefly discuss the present method in relation to several other approaches. In Equation 24.3, the basic variables are the bead positions; our approach is thus particle-based, and it describes the conformation of individual chains explicitly. Equation 24.3 has also been taken as the starting point for SCF theory [Fredrickson 2006]; note, however, that in SCF theory the local densities $\phi_A(\mathbf{r})$ and $\phi_B(\mathbf{r})$ are the fundamental variables and, as in other field-theoretic techniques, the configurational degrees of freedom of the chains have been integrated out. This tacitly assumes that local chain conformations are always in equilibrium with the density distribution. The SCF theory neglects fluctuations: the saddle point approximation involved in SCF theory selects from all possible density distributions only the one that minimizes the functional $F[\phi_A,\phi_B]$. This treatment becomes valid in the limit $\bar{N} \to \infty$; it is important to emphasize that much of our current understanding of block copolymer behavior has been generated in the context of SCF theory. In the limit $\bar{N} \to \infty$, our proposed MC simulations recover the SCF solution; note, however, that they do not rely on the saddle-point approximation invoked in SCF theory [Fredrickson, Ganesan, and Drolet 2002; Fredrickson 2006]. As such, its results could be viewed as an exact solution of the Hamiltonian of Equation 24.3. The specific form of the Hamiltonian in Equation 24.3 is essential for efficient solution of the SCF equations. The Gaussian nature of the chain is required to integrate out the conformational degrees of freedom via the solution of a modified diffusion equation, and the quadratic approximation allows the decoupling of interacting fields via a Hubbard–Stratonovitch transformation. In contrast, virtually any kind of bonded interactions (bond length potentials, angular and torsional contributions, or chain branching) or interaction functional can be used in the MC simulations outlined above, without a significant computational overhead.

From a different standpoint, our coarse-grain approach can also be viewed as a traditional MC simulation of a model defined by an unusual kind of interaction potential [Daoulas and Muller 2006]. The interaction between two beads depends not only on their relative positions, but also involves the grid. In the scheme presented here, only those beads that are found within the same cell interact with each other. The interaction is therefore discontinuous and anisotropic, and it is not translationally invariant. As we show later in this work, this simple approach suffices to capture the block copolymer properties on long length scales while drastically reducing the computational demands of energy calculations (vis-à-vis those encountered when conventional pairwise additive interactions are employed). In contrast, keeping track of the local densities in our simulations is relatively straightforward, and the computational time remains strictly proportional to the number of beads.

The coarse-grain approach described here is closely related to single chain in mean field (SCMF) simulations. In SCMF simulations [Daoulas et al. 2006; Muller and Smith 2005; Stoykovich et al. 2005; Daoulas et al. 2006], the free-energy functional is expressed as

$$\frac{F[\phi_A,\phi_B]}{k_B T} = \frac{\rho_0 \int d^3\mathbf{r}[\phi_A w_A + \phi_B w_B]}{2}. \tag{24.5}$$

As in SCF theory, the fields w_A and w_B are defined from the local densities, for instance $w_A = \chi\phi_B - \kappa(1-\phi_A-\phi_B)$, but they fluctuate in time instead of being self-consistently determined static quantities. SCMF simulations consist of the following two steps that are repeated until sufficient statistics are generated: (1) perform a short MC simulation of the chains placed in the given external fields $w_A(\mathbf{r})$ and $w_B(\mathbf{r})$, with the nonbonded energy given by the Equation 24.5, (2) update the fields from the instantaneous value of the local densities. Due to a temporary decoupling between the field value and the chain conformations, the energy associated with a MC move is only an approximate form of the exact expression given by Equation 24.3. As discussed by Daoulas and Muller (2006), this "quasi-instantaneous approximation" is controlled by a small parameter ε which plays a role similar to that of the Ginzburg parameter in SCF theory. This parameter ε depends on the discretization of the chain contour N and space ΔL and therefore it can be made small even if the Ginzburg parameter is large and fluctuations are important to capture the physics. The quasi-instantaneous approximation becomes accurate if the external fields mimic the instantaneous fluctuating interactions of a chain with its fluctuating environment. To this end, the fields have to be frequently updated and the density should not change significantly between updates. The main advantage of SCMF simulations over MC simulations is that they can be parallelized in a straightforward manner since, for a given value of the fields, the chains evolve independently from one another. On the other hand, SCMF simulations are less appropriate for dilute systems, the use of global MC moves that facilitate rapid equilibration is somewhat restricted, and extension to systems such as polymer nanocomposites is more demanding.

24.2.4 Choice of Parameters

We now explain our choice of parameters. From the melt density and the molecular weight of the diblock used in a specific experiment, one can deduce $\bar{N}$; a typical order of magnitude for a polymer melt is 10^2. The product χN is determined by taking the value of χ extracted in experiments and, for N, the degree of polymerization (number of monomers). The inverse of κ can be related to the isothermal compressibility through $1/\kappa = -(\rho_0 k_B T / V)(\partial V/\partial p)|_T$, where p is the pressure. In accord with previous studies we utilize $\kappa N = 50$, a value which is high enough to prevent fluctuations of the total densities on length scales larger than a fraction of R_e. On the practical side, higher values are difficult to consider because they increase the equilibration time: most MC moves (particularly global moves) induce a local density fluctuation and are rejected in the nearly incompressible limit.

The choice of N is dictated by the properties one wants to study. For instance, an accurate description of the narrow width and the detailed density profile at a hard surface or a liquid–vapor interface would require high N ($>10^3$). The high number of degrees of freedom per chain would then imply that only small systems (a few R_e^3) could easily be accessible with common computational resources. The spirit of the coarse-grain model is to study properties on length scales set by R_e, such as the morphologies formed by the copolymer. The simulation of large systems (many R_e^3) favors a choice of N as low as possible, while still faithfully describing the chain architecture. We found that $N = 32$ provides a good compromise between those two requirements. Note, however, that triblock and other multiblock copolymers might require higher N, since each block must be represented by a sufficient number of beads. Taking $\bar{N} = 128^2$, $N = 32$, and $n_{\text{cell}} \approx 15$ results in a grid spacing $\Delta L \approx 0.15 R_e$. With these parameters, systems containing more than a million beads can be simulated on a single processor machine.

24.2.5 Stress Tensor and Variable Cell Shape Method

Within the mean-field approximation the internal stress tensor [Doi and Edwards 1986; Maurits, Zvelindovsky, and Fraaije 1998; Tyler and Morse 2003; Barrat, Fredrickson, and Sides 2005] for vanishing interaction range is given by:

$$\frac{\sigma_{\alpha\beta}}{k_B T / V} = -nN\delta_{\alpha\beta} + 3\sum_{i=1}^{n}\sum_{s=1}^{N-1}\frac{N-1}{R_e^2}\left\langle b_{i,\alpha}(s) b_{i,\beta}(s)\right\rangle, \tag{24.6}$$

where $b_i(s) = \mathbf{r}_i(s+1) - \mathbf{r}_i(s)$ is the bond vector joining two adjacent beads. This approximation for the internal stress tensor can be computed from a given configuration and averaged over several MC steps.

When the block copolymer forms ordered microphases, such as lamellae or cylinders, the size and shape of the simulation box are bound to influence the geometric properties. In particular, it is important to avoid finite-size effects in order to determine the true periodicity of the microphase. A first possibility is to use a large cell calculation, with a simulation box as large as possible, thereby minimizing the influence of the finite-size constraints. The long computational times required by large system sizes are not the only difficulty; it is also desirable to obtain a defect-free microphase (i.e., perfect, long-range order of the domains), which is often a challenge. This is why unit cell calculations are usually more efficient. Assuming a particular symmetry for the microphase (that can be deduced from a large cell calculation), the size and shape of the simulation box are then relaxed in order to minimize the free energy (SCF) or relieve the internal stress. Such variable cell shape methods were originally introduced by Parinello, Ray and Ramhan [Parrinello and Rahman 1981; Ray and Rahman 1984] and used, for example, to describe crystalline structures in solids. More recently, variable shape methods have been implemented within the context of SCF theory [Barrat, Fredrickson, and Sides 2005; Tyler and Morse 2003].

Here we briefly describe how this technique can be applied to compute the natural periodicity of the microphase. In the following we utilize the notation of Barrat, Fredrickson, and Sides (2005). The geometry of the simulation box, assumed to be a parallelepiped (but not necessarily orthorhombic), is specified by three vectors $\mathbf{h}_1$, $\mathbf{h}_2$, and $\mathbf{h}_3$ that constitute the box sides. Let $\mathbf{H}$ be the matrix obtained by concatenating these three vectors: $\mathbf{H} = [\mathbf{h}_1, \mathbf{h}_2, \mathbf{h}_3]$. The $\mathbf{H}$ matrix evolves during a MC simulation according to the following equation:

$$\frac{d\mathbf{H}}{dt} = -\lambda \mathbf{D}[\mathbf{H}^{-1}\Sigma\mathbf{H}^{T-1}]. \tag{24.7}$$

In the context of our MC approach, the time t corresponds to the number of MC steps. $\mathbf{H}^{-1}$ is the inverse of $\mathbf{H}$ and $\mathbf{H}^{T-1}$ denotes its transpose; $\mathbf{D}$ is a matrix operator defined as $\mathbf{DA} = \mathbf{A} - (1/3)Tr(\mathbf{A})\mathbf{I}$. This evolution equation drives a change in the box shape and dimensions until the system reaches a stress-free configuration, that is $\sigma = 0$, while keeping the volume of the box constant. In practice, the box shape is updated after a given number of MC steps; the amplitude of the shape changes can be tuned by the parameter λ. Because the box relaxation towards the natural periodicity of an ordered microphase can be slow, it is sometimes faster to compute the stress-strain relationship. For instance, lamellae can initially be formed with a nonequilibrium lamellar spacing imposed by the box dimensions and periodic boundary conditions; the stress tensor is then easily computed as a function of the lamellar spacing. The natural periodicity is reached when the stress tensor is isotropic.

24.2.6 Soft Nanoparticles

The model defined so far can describe bulk systems of pure copolymers (or blends and other multicomponent polymer systems). We now introduce an approximate but simple approach to include in our model nonpolymeric objects such as nanoparticles. Such nanoparticles often consist of a solid, metallic core to which short polymer chains are grafted so as to facilitate dispersion in the polymer melt [Mackay et al. 2006]. Nanoparticles have two effects on the surrounding chains: (1) they enforce an excluded volume and (2) the brush coating might exhibit a preferential interaction with one block of the copolymer. In the following, we propose to describe a nanoparticle as a cluster of beads, all attached together to form a rigid object of spherical shape. The density of beads inside the sphere is chosen equal to the average density outside so that the compressibility constraint partially prevents the chains from penetrating the nanoparticle. The interaction between the nanoparticle and the neighboring chains is controlled by changing its composition; that is, the proportion of A and B beads forming the nanoparticle. When dispersed in a diblock copolymer melt, an A-like nanoparticle interacts preferentially with the A block; this situation corresponds to a nanoparticle covered with chains that are chemically identical to one of the copolymer blocks. Taking for each bead a random position inside the sphere, a nanoparticle made of A and B beads in equal proportion would be non-selective, having no preference for either A or B domains. Alternatively, the nanoparticle beads could be of type C and, in that case, two additional parameters, χ_{AC} and χ_{BC}, would be introduced to describe the nanoparticle-block interactions. Extensions to nonspherical nanoparticles (e.g., rods) or to systems having more elaborated brush structures do no present any additional difficulties. In the following, we consider the case of Janus-like nanoparticles, which are spherical but are coated with two hemispheres having a different brush. Such nanoparticles have been recently studied by Kramer and co-workers [Kim et al. 2007], and it is therefore of interest to consider whether the model proposed here can describe some of their main experimental observations. The simplest case is to have one hemisphere entirely made of A beads, and the other entirely made of B beads.

This model of nanoparticle is a crude representation that does not focus on the effect of an isolated nanoparticle. In particular, polymer chains are not strictly prevented from entering the nanoparticle (hence "soft nanoparticle"). Rather, the model is designed to reproduce the collective behavior arising when many nanoparticles are dispersed in block copolymer melt, and their interplay with the copolymer morphology. A more accurate model of nanoparticle would treat them as a potential that explicitly interacts with the polymer beads. The degrees of freedom associated with each nanoparticle include its position, and its orientation when anisotropic. Therefore, the only MC moves that are needed are translation and rotation; they are treated in the same way as the MC moves for the polymer.

24.3 APPLICATIONS

24.3.1 Equilibrium Morphologies

We begin this section by examining the capability of the proposed coarse-grain model and method to predict the morphology of block copolymers. We consider the simple case of a linear diblock copolymer in the bulk but more complex situations could be addressed, including thin films, patterned substrates, linear or star triblocks. Figure 24.1 illustrates the variable cell shape method. The simulation box, initially cubic, deforms to accommodate the hexagonal cylindrical phase and adjusts its size to match the natural periodicity. Our coarse-grain description is primarily designed to describe properties on the length scale R_e. Nevertheless it is interesting to consider the validity of our predictions on small length scales (small fractions of R_e). Figure 24.2 shows the average density profiles of A and B beads in the lamellar phase of a symmetric diblock copolymer. For comparison, the result of a one-dimensional SCF calculation is also included. The equilibrium lamellar spacings found in both cases are very close to each other: $L_o = 1.80R_e$ in MC simulations and $L_o = 1.83R_e$ in SCF calculations. The corresponding density profiles are almost similar to each other when plotted in units of the lamellar spacing. Compared to the one-dimensional mean-field calculation, the variations in total density observed in three-dimensional simulations are less pronounced and the density profile is not as steep, thereby resulting in a slightly wider interface. Considering the rather low contour and space discretization employed here ($N = 32$ and $\Delta L \approx 0.15R_e$) the MC method is surprisingly accurate. Also note that using higher discretization N (and lower ΔL) or a more accurate interpolation scheme to compute local densities from bead positions does not seem to yield significant changes in the MC profiles, suggesting that the main differences with SCF can be ascribed to interface fluctuations. These capillary waves in the three-dimensional MC simulations are expected to broaden the width of the interface.

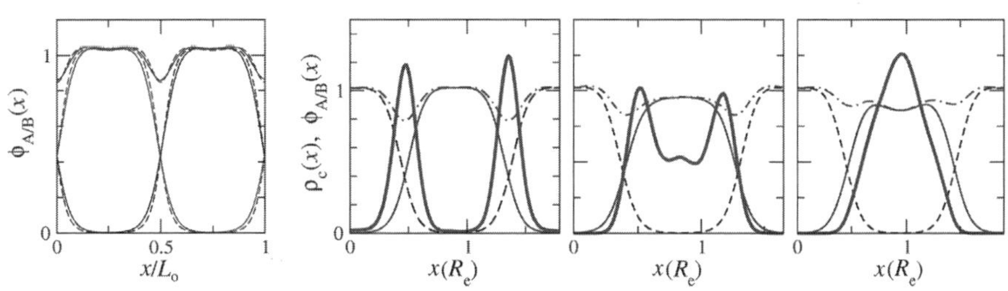

FIGURE 24.2 Left: density profiles of A and B beads and total density in the lamellar phase of a symmetric diblock copolymer, computed with MC simulations (solid lines) and SCF theory (dashed lines). The unit length is the lamellar spacing. Parameters: $\chi N = 36.7$, $\kappa N = 50$, $\sqrt{\bar{N}} = 128$ ($N = 32$, $\Delta L = 0.15R_e$) Right: distribution of nanoparticles in the lamellar phase of a symmetric diblock copolymer. The nanoparticle composition (fraction of A beads) is 0.5, 0.8, and 1, from left to right. Each graph shows the distribution of nanoparticle centers (solid line), the density profile of A beads (continuous line), B beads (dashed line), and the total density profile (dashed-dotted line). Parameters: $R_p = 0.16R_e$, $\phi_p = 0.05$, $\chi N = 40$, $\kappa N = 50$, $\sqrt{\bar{N}} = 128$ ($N = 32$, $\Delta L = 0.19R_e$).

24.3.2 Qualitative Description of the Dynamics

In addition to equilibrium properties, MC simulations can provide an approximate but reasonable account of the dynamics on long length scales at a qualitative level. In order to do so, it is important to avoid "drastic" nonlocal MC moves, and use instead MC moves that mimic those encountered in real polymeric systems and, in particular, give rise to a diffusive relaxation of the local densities. In what follows, only two types of MC moves are used: the random displacement of a single bead which, when used alone, would lead to Rouse-type dynamics, and the slithering-snake move, which mimics the reptation of the chains in the "tube" created by the topological constraints imposed by neighboring chains. Since the simulated chains can cross each other, all entanglement effects are neglected and the dynamics cannot be realistic at the level of an individual chain. However, when the collective, global motion of many chains is required, such as during the formation of an ordered microphase or during the relaxation of a structural defect, this MC dynamics might become qualitatively correct on large time and length scales because it captures the diffusive relaxation kinetics of composition fluctuations. In this case, the time scale can be identified by matching the single-chain diffusion coefficient in the simulation and the experimental system.

Those assumptions are less restrictive than those involved in most dynamics approaches within the context of SCF theory. In such approaches, the time evolution is driven by the spatial variation of a local chemical potential and chain conformations are assumed to be fast variables that adjust instantaneously to the slow variables (local densities and fields). Kinetic coefficients must be introduced that relate the time evolution of the slow variables to gradients of chemical potentials. Exact expressions for those Onsager coefficients are not available for inhomogeneous systems, and assumptions on the chain structure must be made to obtain approximate parameters.

Figure 24.3 provides an example of simulated dynamics in a thin film of a symmetric diblock copolymer confined between two hard walls (with periodic boundary conditions in the other directions), so as to represent a thin film laid over a neutral substrate. The simulation starts with the chain positions and conformations chosen randomly (in experiments this would correspond to a quench from a high temperature). Very rapidly, the copolymer forms lamellae perpendicular to the substrate. The ordering remains only local, however, and the lamellae form the characteristic fingerprint pattern that is seen in experiments. From a series of snapshots one can analyze the type of defects that are formed, and the mechanisms by which they annihilate each other and disappear. While the system shown in Figure 24.3 has been simulated using a single-processor machine, the use of SCMF simulations on a parallel computer cluster could permit study of the relaxation of defects on larger length scales and over longer times [Edwards et al. 2005]. The phenomenological approach that has been used so far to study such phenomena generally relies on a Landau expansion

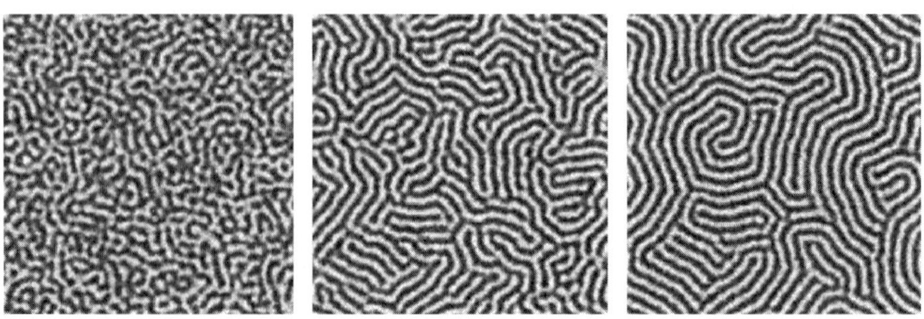

FIGURE 24.3 Simulation of a symmetric diblock copolymer confined between two hard walls, starting from a random initial configuration. The figures provide top-down snapshots of three configurations. The MC moves employed for these calculations were local. From left to right, configurations correspond to 50, 500, and 6000 MC steps, respectively. The natural lamellar spacing is $L_o = 1.53R_e$ and the system size is $40 \times 40 \times 1.53R_e^3$. Parameters: $\chi N = 18$, $\kappa N = 50$, $\sqrt{\bar{N}} = 128$ ($N = 32$, $\Delta L = 0.19R_e$).

of the local order parameter (difference between A and B local densities), namely a coarser representation where information about the chain conformations is lost.

24.3.3 Nanoparticle-Induced Phase Transition (Soft Nanoparticles)

Incorporating nanoparticles into diblock copolymers is of interest for design of new functional materials [Bockstaller, Mickiewicz, and Thomas 2005]. Experiments with a low loading of nanoparticles [Chiu et al. 2005; Kim et al. 2006; Bockstaller et al. 2003] have shown that both the nanoparticle

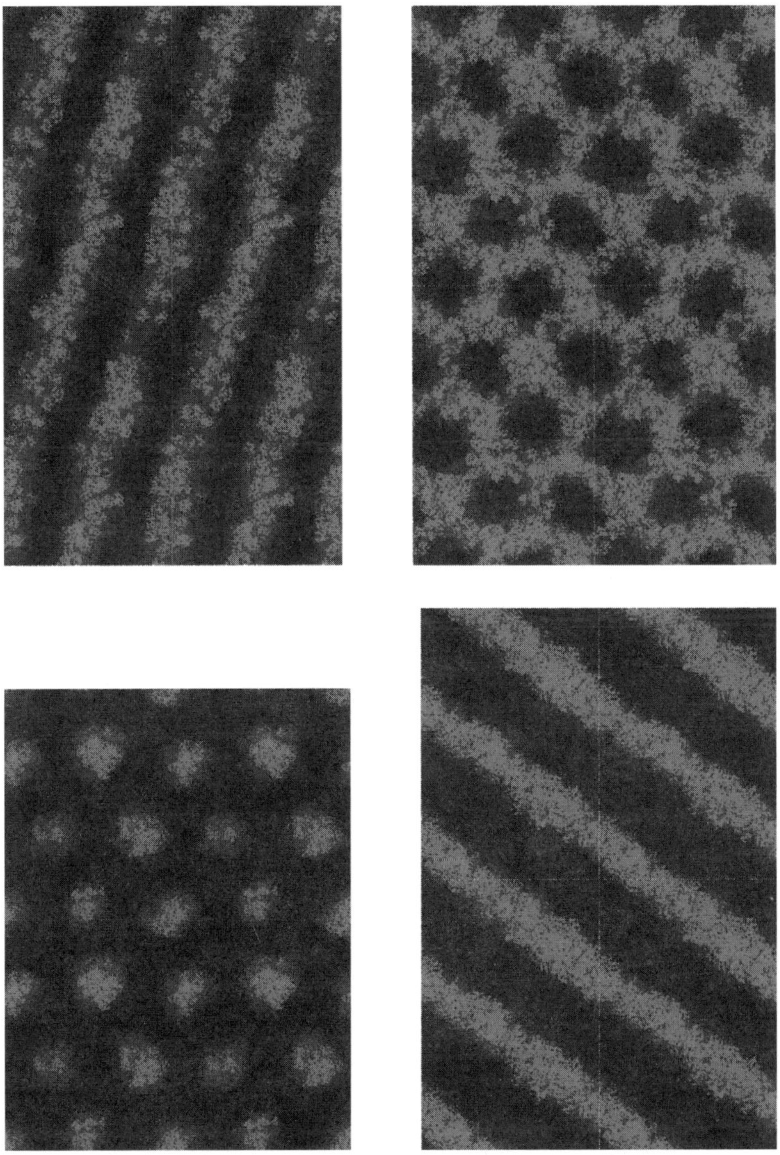

FIGURE 24.4 Change in morphology induced by a high loading of selective (A-like) nanoparticles. In these cross-sections, the A beads are shown in dark gray, the B beads in black, and the nanoparticle beads in light gray. The system has been replicated once in each direction. Top row: symmetric copolymer ($f = 0.5$) with $\phi_p = 0.2$ and $\phi_p = 0.4$ (from left to right). Bottom row: asymmetric copolymer ($f = 0.25$) with $\phi_p = 0.1$ and $\phi_p = 0.3$. Parameters: $R_p = 0.16R_e$, $\chi N = 25$, $\kappa N = 50$, $\sqrt{\bar{N}} = 128$ ($N = 32$, $\Delta L = 0.19R_e$).

size and the type of brush that covers it are important in determining the location of the nanoparticle, be it in specific domains or at the interface between them. At high loadings of nanoparticles [Sides et al. 2006; Kim et al. 2005], it is possible to induce a change in morphology. For instance, the lamellar morphology of symmetric diblock can be coaxed into forming a hexagonal morphology when the volume fraction of nanoparticles exceeds a critical threshold.

Considerable effort has been directed towards predicting the morphology of nanoparticle/block copolymer mixtures. In their initial studies, Balazs and coworkers combined the SCF theory for the polymer with a density functional theory for hard spheres to describe the nanoparticles (SCF-DFT method). A variety of systems, including nanoparticles in lamellae, bulk or confined, were considered by these authors [Thompson et al. 2001; Lee et al. 2002; Thompson et al. 2002; Lee, Shou, and Balazs 2003a, 2003b; Lin et al. 2005; Smith, Tyagi, and Balazs 2005; Balazs, Emrick, and Russell 2006]. One limitation of that approach is that the coupling between the nanoparticle and the melt is only approximate, since the correlations between nanoparticles are assumed to reproduce the structure of a hard sphere fluid. The hybrid particle-field method (HPF), recently introduced [Sides et al. 2006], does not involve such an approximation; the nanoparticle positions remain explicit degrees of freedom, and a Brownian dynamics technique is used to describe their evolution in time and space. This approach has been shown to reproduce experimentally observed nanoparticle-induced changes in morphology. However, both the SCF-DFT and HPF methods have so far been restricted to two-dimensional systems. MC simulations of many-body models for nanoparticle/diblock mixtures [Schultz, Hall, and Genzer 2005; Pryamitsyn and Ganesan 2006] are fully three-dimensional, but they remain computationally intensive, particularly for large systems. Recently, we have introduced a more tractable approach [Kang et al. 2008] that maintains a full coupling between the nanoparticles and the polymer chains.

In the interest of brevity, here we only outline some results obtained with the simple 'soft nanoparticle' model presented above. The nanoparticle radius (including the solid core and the brush) is chosen in the range $R_p = 0.16 - 0.21 R_e$, thus in a range intermediate between the protein limit ($R_p \ll R_e$) and the colloid limit ($R_p \ll R_e$). Computing the pair correlation function $g(r)$ between the nanoparticle center and the polymer beads shows that the local density of polymer beads does not vanish inside the nanoparticle; however, it is significantly reduced. In the worst case considered here (A-like nanoparticle with $R_p = 0.16$ placed in an A homopolymer melt), $g(r = 0) \approx 0.4$; larger nanoparticles lead to a stronger exclusion of the polymer chains. The fact that chains overlap with the nanoparticle can be expected, as the compressibility constraint is enforced only at the scale of the grid spacing, which is not much smaller than the nanoparticle's radius. A higher discretization N (and therefore a smaller grid spacing) or a lower compressibility (higher κ) would improve exclusion effects, but it would also lead to longer computational times. The properties we focus on here are the collective effects induced by a high volume fraction of nanoparticles, not the influence of an isolated nanoparticle on the neighboring chains.

The first property we consider is the location of nanoparticles in the diblock. Figure 24.2 shows the density profile of spherically symmetric nanoparticles dispersed in the lamellar phase of a symmetric diblock, for various compositions of the nanoparticle (i.e., fraction of A beads). As expected, A-like nanoparticles are found exclusively in the A domains (with a preference for the center). On the other hand, nonselective (neutral) nanoparticles are found at the interface between domains, where they screen contacts between A and B blocks and reduce the penalty associated with the decrease in total density. Given that the nanoparticles are rather small and soft, the entropic penalty they impose on the chains by restricting the possible conformations is not dominant here, thus enthalpic factors are expected to be most important. As shown in Figure 24.5, Janus-like nanoparticles are found only at the interface, with each hemisphere located in its respective domain.

When dispersed at high-volume fractions, neutral nanoparticles tend to aggregate. Therefore, in what follows only the cases of selective and Janus-like nanoparticles are considered. Figure 24.6 shows the predicted morphology when nanoparticles are dispersed in a symmetric block copolymer with a volume fraction ϕ_p ranging from 0.1 to 0.4. The simulation box has fixed dimensions $L_x \times L_y \times L_z = 40 \times 40 \times 1.53\, R_e^3$ (here $L_0 = 1.53\ R_e$). Choosing a small thickness L_z

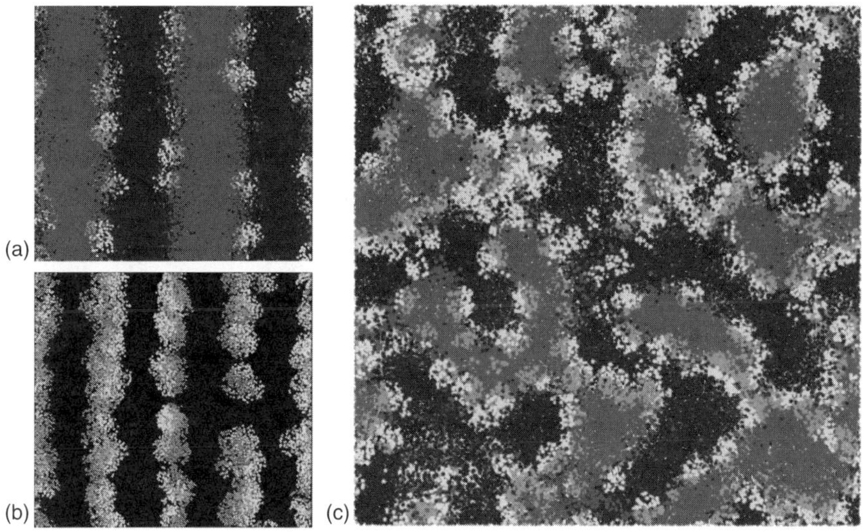

FIGURE 24.5 Morphology obtained with Janus-like nanoparticles dispersed in a symmetric AB block copolymer. The nanoparticle radius is $R_p = 0.21R_e$. In these cross-sections, the A beads are shown in dark gray, the B beads in black. The A-like hemisphere of the nanoparticle is light gray, the B-like hemisphere white. (a) $\phi_p = 0.05$ yields a lamellar morphology. (b) Same as previously with only nanoparticles shown. (c) The morphology observed with $\phi_p = 0.35$ suggests a bicontinuous phase. The system size is $7\times7\times6.9R_e^3$. Parameters: $\chi N = 40$, $\kappa N = 50$, $\sqrt{\bar{N}} = 128$ ($N = 32$, $\Delta L = 0.19R_e$).

favors the ordering in the x–y plane and helps to identify the morphology. Note that the system is indeed three-dimensional and the nanoparticle are spheres, not disks or rods. As ϕ_p increases, the A lamellae form more T-junctions and rings until the B domain is finally fragmented into isolated cylinders. Even on a local scale, the hexagonal order is barely visible, since the cylinders widely vary in radius, but the change in morphology is clear.

To better characterize the morphology, we used a smaller simulation box of variable shape, as shown in Figure 24.4. All simulation boxes have converged towards different stable dimensions to better accommodate the periodicity of the microphase. Figure 24.4 shows a second example of a nanoparticle-induced change in morphology, where the cylindrical phase of an asymmetric block copolymer is converted into a lamellar phase. The mechanism driving the transition is the same: the A domains are swollen by nanoparticles, and deform until the initial morphology becomes unstable. Note that replacing the nanoparticles with a homopolymer we would not reproduce the same transition. Instead, depending on the ratio between the molecular weights of homopolymer and the diblock copolymer, the lamellar morphology would be conserved, but with A domains swollen by the homopolymer and a larger lamellar spacing. Alternatively, a microemulsion could be formed or the homopolymers could macroscopically phase-separate from the diblock domains. The results presented here do not provide an estimate of the critical volume fraction at which the transition occurs, but the window is compatible with that observed in experiments.

A high loading of Janus-like nanoparticles can also induce a change in morphology: for instance, above a critical volume fraction, the cylindrical phase of an asymmetric copolymer is replaced by a lamellar phase (not shown). The mechanism is different from that observed for selective nanoparticles: instead of swelling their preferred domain, Janus-like nanoparticles decrease the interface tension between domains (this difference is already reflected at low loading in the lamellar phase: selective nanoparticles lead to an increase of lamellar spacing, and Janus-like nanoparticles to a decrease). They also might modify the spontaneous curvature and bending rigidity of the interface

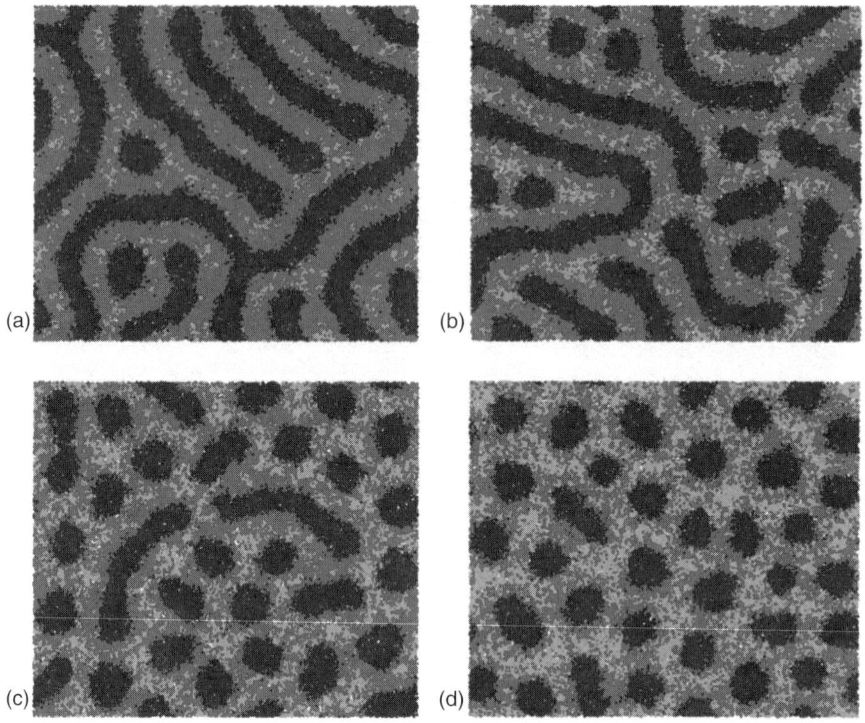

FIGURE 24.6 (See color insert following page 238.) Morphology of nanoparticle/copolymer mixtures. In these cross-sections, the A beads are shown in red, the B beads in blue, and the nanoparticle beads in green. As the nanoparticle volume fraction increases, the morphology changes from lamellar to cylindrical. The diblock copolymer is symmetric; the nanoparticles are A-selective and have a radius $R_p = 0.16R_e$ and the volume fraction is $\phi_p = 0.1, 0.2, 0.3$, and 0.4 in (a), (b), (c), and (d) respectively. Parameters: $\chi N = 40$, $\kappa N = 50$, $\sqrt{\bar{N}} = 128$ ($N = 32$, $\Delta L = 0.19R_e$).

[Pryamitsyn and Ganesan 2006]. As the loading increases, it becomes favorable to increase the amount of interface between domains, which is a possible driving force for the transition. Figure 24.5c shows a mixture of symmetric copolymer and nanoparticles. Even at the local scale, the nature of the morphology is unclear; it does not seem to be lamellar or cylindrical. Besides, any cross-section of the system shows A and B domains interpenetrating each other, and separated by an interface packed with nanoparticles. This suggests the possibility of a bicontinuous phase, in agreement with experimental observations [Kim et al. 2007].

24.4 CONCLUSION

The SCF theory has been instrumental in understanding the properties of block copolymers. However, there are systems of considerable fundamental and technological interest, including complex multiblock materials and nanoparticle/copolymer mixtures, that continue to pose challenges for traditional SCF treatments. We have presented in this chapter an alternative numerical framework for description of polymeric systems that exhibits several attractive features. Since it is a particle-based method, it treats the conformations of the chains in an explicit manner; it is therefore relatively straightforward to describe polymeric molecules of arbitrary architecture. When a rough description of the system suffices, nonpolymeric objects such as functionalized nanoparticles can be represented as a rigid cluster of beads, as was shown here. If a more accurate description is needed, nanoparticles can be represented through a potential energy function that interacts explicitly with the polymer beads. Such an approach has been used to predict the spatial distribution of nanoparticles for nanoparticle/copolymer thin films on nanopatterned substrates [Kang et al. 2008; Detcheverry et al. 2008]. The predictions of our MC simulations are not restricted to equilibrium properties but can be extended, at least at the qualitative level, to the dynamics. Since our approach does not rely on a saddle-point approximation, it should be able to describe fluctuations effects; such fluctuations, however, must still be characterized and it remains to be seen if the MC simulations proposed here provide a simpler alternative to field-theoretic methods. This will require a better understanding of the conditions under which discretization effects are negligible. In contrast to SCF theory, the MC simulations described in this work do not directly provide the free energy of the system; it is therefore difficult to determine and trace precise phase boundaries. Methods that permit efficient calculation of the chemical potential or free energy of the system must be developed for systematic studies of phase behavior.

ACKNOWLEDGMENTS

This research was supported by the National Science Foundation through the Nanoscale Science and Engineering Center. Support from the Semiconductor Research Corporation is also gratefully acknowledged.

REFERENCES

Balazs, A. C., T. Emrick, and T. P. Russell. 2006. Nanoparticle polymer composites: Where two small worlds meet. *Science* 314 (5802):1107–10.

Barrat, J. L., G. H. Fredrickson, and S. W. Sides. 2005. Introducing variable cell shape methods in field theory simulations of polymers. *J. Phys. Chem. B* 109 (14):6694–6700.

Binder, K. 1995. *Monte Carlo and Molecular Dynamics Simulations in Polymer Science*. New York: Oxford University Press.

Bockstaller, M. R., Y. Lapetnikov, S. Margel, and E. L. Thomas. 2003. Size-selective organization of enthalpic compatibilized nanocrystals in ternary block copolymer/particle mixtures. *J. Am. Chem. Soc.* 125 (18):5276–77.

Bockstaller, M. R., R. A. Mickiewicz, and E. L. Thomas. 2005. Block copolymer nanocomposites: Perspectives for tailored functional materials. *Adv. Mater.* 17 (11):1331–49.

Chiu, J. J., B. J. Kim, E. J. Kramer, and D. J. Pine. 2005. Control of nanoparticle location in block copolymers. *J. Am. Chem. Soc.* 127 (14):5036–37.

Daoulas, K. C., and M. Muller. 2006. Single chain in mean field simulations: Quasi-instantaneous field approximation and quantitative comparison with Monte Carlo simulations. *J. Chem. Phys.* 125 (18):18.

Daoulas, K. C., M. Muller, J. J. de Pablo, P. F. Nealey, and G. D. Smith. 2006. Morphology of multi-component polymer systems: Single chain in mean field simulation studies. *Soft Matter* 2 (7):573–83.

Daoulas, K. C., M. Muller, M. P. Stoykovich, S. M. Park, Y. J. Papakonstantopoulos, J. J. de Pablo, P. F. Nealey, and H. H. Solak. 2006. Fabrication of complex three-dimensional nanostructures from self-assembling block copolymer materials on two-dimensional chemically patterned templates with mismatched symmetry. *Phys. Rev. Lett.* 96 (3):4.

Deserno, M., and C. Holm. 1998. How to mesh up Ewald sums. I. A theoretical and numerical comparison of various particle mesh routines. *J. Chem. Phys.* 109 (18):7678–93.

Detcheverry, F. A., H. Kang, K. Ch. Daoulas, M. Muller, P. F. Nealey, and J. J. de Pablo. 2008. Monte Carlo simulations of a coarse grain model for block copolymers and nanocomposites. To appear in *Macromolecules*.

Doi, M., and S. F. Edwards. 1986. *The Theory of Polymer Dynamics*. Oxford: Oxford University Press.

Edwards, E. W., M. P. Stoykovich, M. Muller, H. H. Solak, J. J. De Pablo, and P. F. Nealey. 2005. Mechanism and kinetics of ordering in diblock copolymer thin films on chemically nanopatterned substrates. *J. Polym. Sci. Part B: Polym. Phys.* 43 (23):3444–59.

Edwards, S. F. 1965. Statistical mechanics of polymers with excluded volume. *Proc. Phys. Soc. London* 85 (546P):613.

Eurich, F., and P. Maass. 2001. Soft ellipsoid model for Gaussian polymer chains. *J. Chem. Phys.* 114 (17):7655–68.

Fredrickson, G.H. 2006. *The Equilibrium Theory of Inhomogeneous Polymers*. Oxford: Clarendon Press.

Fredrickson, G. H., V. Ganesan, and F. Drolet. 2002. Field-theoretic computer simulation methods for polymers and complex fluids. *Macromolecules* 35 (1):16–39.

Helfand, E., and Y. Tagami. 1972. Theory of interfaces between immiscible polymers. II. *J. Chem. Phys.* 56 (7):3592.

Hong, K. M., and J. Noolandi. 1981. Theory of inhomogeneous multicomponent polymer systems. *Macromolecules* 14 (3):727–36.

Kang, H., F. A. Detcheverry, A. N. Mangham, M. P. Stoykovich, K. Ch. Daoulas, R. J. Hamers, M. Müller, J. J. de Pablo, and P. F. Nealey. 2008. Hierarchical assembly of Nanoparticle superstructures from block copolymer-nanoparticle composites. *Phys. Rev. Lett.* 100:148303.

Kim, B. J., J. Bang, C. J. Hawker, and E. J. Kramer. 2006. Effect of areal chain density on the location of polymer-modified gold nanoparticles in a block copolymer template. *Macromolecules* 39 (12):4108–14.

Kim, B. J., J. J. Chiu, G. R. Yi, D. J. Pine, and E. J. Kramer. 2005. Nanoparticle-induced phase transitions in diblock-copolymer films. *Adv. Mater.* 17 (21):2618.

Kim, B.J., G. H. Fredrickson, C. J. Hawker, and E. J. Kramer. 2007. Nanoparticle surfactants as a route to bicontinuous block copolymer morphologies. *Langmuir* 23 (14):7804.

Kotelyanskii, M., and D. N. Theodorou. 2004. *Simulation Methods for Polymers*. New York: Dekker.

Laradji, M., H. Guo, and M. J. Zuckermann. 1994. Off-lattice Monte-Carlo simulations of polymer brushes in good solvents. *Phys. Rev. E* 49 (4):3199–3206.

Lee, J. Y., Z. Shou, and A. C. Balazs. 2003a. Modeling the self-assembly of copolymer-nanoparticle mixtures confined between solid surfaces. *Phys. Rev. Lett.* 91 (13).

Lee, J. Y., Z. Y. Shou, and A. C. Balazs. 2003b. Predicting the morphologies of confined copolymer/nanoparticle mixtures. *Macromolecules* 36 (20):7730–39.

Lee, J. Y., R. B. Thompson, D. Jasnow, and A. C. Balazs. 2002. Entropically driven formation of hierarchically ordered nanocomposites. *Phys. Rev. Lett.* 89 (15).

Lin, Y., A. Boker, J. B. He, K. Sill, H. Q. Xiang, C. Abetz, X. F. Li, J. Wang, T. Emrick, S. Long, Q. Wang, A. Balazs, and T. P. Russell. 2005. Self-directed self-assembly of nanoparticle/copolymer mixtures. *Nature* 434 (7029):55–59.

Louis, A. A., P. G. Bolhuis, J. P. Hansen, and E. J. Meijer. 2000. Can polymer coils be modeled as "soft colloids"? *Phys. Rev. Lett.* 85 (12):2522–25.

Mackay, M. E., A. Tuteja, P. M. Duxbury, C. J. Hawker, B. Van Horn, Z. B. Guan, G. H. Chen, and R. S. Krishnan. 2006. General strategies for nanoparticle dispersion. *Science* 311 (5768):1740–43.

Matsen, M. W., and M. Schick. 1994. Stable and unstable phases of a diblock copolymer melt. *Phys. Rev. Lett.* 72 (16):2660–63.

Maurits, N. M., A. V. Zvelindovsky, and J. G. E. M Fraaije. 1998. Equation of state and stress tensor in inhomogeneous compressible copolymer melts: Dynamic mean-field density functional approach. *J. Chem. Phys.* 108 (6):2638–50.

Muller, M., and F. Schmid. 2005. Incorporating fluctuations and dynamics in self-consistent field theories for polymer blends. *Adv. Polym. Sci.* 185:1–58.

Muller, M., and G. D. Smith. 2005. Phase separation in binary mixtures containing polymers: A quantitative comparison of single-chain-in-mean-field simulations and computer simulations of the corresponding multichain systems. *J. Polym. Sci. Part B: Polym. Phys.* 43 (8):934–58.

Murat, M., and K. Kremer. 1998. From many monomers to many polymers: Soft ellipsoid model for polymer melts and mixtures. *J. Chem. Phys.* 108 (10):4340–48.

Parrinello, M., and A. Rahman. 1981. Polymorphic transitions in single-crystals: A new molecular dynamics method. *J. Appl. Phys.* 52 (12):7182–90.

Pierleoni, C., C. Addison, J. P. Hansen, and V. Krakoviack. 2006. Multiscale coarse graining of diblock copolymer self-assembly: From monomers to ordered micelles. *Phys. Rev. Lett.* 96 (12):4.

Pryamitsyn, V., and V. Ganesan. 2006. Strong segregation theory of block copolymer–nanoparticle composites. *Macromolecules* 39 (24):8499–8510.

Ray, J. R., and A. Rahman. 1984. Statistical ensemble and molecular-dynamics studies of anisotropic solids. *J. Chem. Phys.* 80 (9):4423–28.

Schultz, A. J., C. K. Hall, and J. Genzer. 2005. Computer simulation of block copolymer/nanoparticle composites. *Macromolecules* 38 (7):3007–16.

Sides, S. W., B. J. Kim, E. J. Kramer, and G. H. Fredrickson. 2006. Hybrid particle-field simulations of polymer nanocomposites. *Phys. Rev. Lett.* 96 (25):250601.

Smith, K. A., S. Tyagi, and A. C. Balazs. 2005. Healing surface defects with nanoparticle-filled polymer coatings: Effect of particle geometry. *Macromolecules* 38 (24):10138–47.
Soga, K. G., H. Guo, and M. J. Zuckermann. 1995. Polymer brushes in a poor solvent. *Europhys. Lett.* 29 (7):531–36.
Soga, K. G., M. J. Zuckermann, and H. Guo. 1996. Binary polymer brush in a solvent. *Macromolecules* 29 (6):1998–2005.
Stoykovich, M. P., M. Muller, S. O. Kim, H. H. Solak, E. W. Edwards, J. J. de Pablo, and P. F. Nealey. 2005. Directed assembly of block copolymer blends into nonregular device-oriented structures. *Science* 308 (5727):1442–46.
Thompson, R. B., V. V. Ginzburg, M. W. Matsen, and A. C. Balazs. 2001. Predicting the mesophases of copolymer-nanoparticle composites. *Science* 292 (5526):2469–72.
Thompson, R. B., V. V. Ginzburg, M. W. Matsen, and A. C. Balazs. 2002. Block copolymer-directed assembly of nanoparticles: Forming mesoscopically ordered hybrid materials. *Macromolecules* 35 (3):1060–71.
Tyler, C. A., and D. C. Morse. 2003. Stress in self-consistent-field theory. *Macromolecules* 36 (21):8184–88.
Yatsenko, G., E. J. Sambriski, M. A. Nemirovskaya, and M. Guenza. 2004. Analytical soft-core potentials for macromolecular fluids and mixtures. *Phys. Rev. Lett.* 93 (25):4.

25 Structure-Based Coarse- and Fine-Graining in Soft Matter Simulations

Nico F.A. van der Vegt, Christine Peter, and Kurt Kremer
Max Planck Institute for Polymer Research

CONTENTS

25.1 INTRODUCTION

Many physical phenomena in biology, chemistry, and materials science involve processes occurring on atomistic length and time scales, which affect structural and dynamical properties on mesoscopic scales exceeding far beyond atomistic ones. Because it is infeasible (and most often undesirable) to run computer simulations of very large systems with atomically detailed models, mesoscale (coarse-grained) models are being developed through which structural relaxations can be studied at large length scales, allowing for full system equilibration on mesoscopic time scales [1–5]. Ideally, coarse-grained (CG) models stay reasonably close to the chemical structure of the material so that inverse-mapping (reintroduction of chemical details) procedures can be employed and atomically detailed processes can be studied in various windows of the CG trajectory where "something interesting happens." Only with that possibility at hand, the corresponding CG models can be used to describe chemically realistic systems over a wide range of length and times scales in a hierarchical, sequential set of simulations at multiple resolution levels, or in *a single, multiscale* simulation where the level of resolution can be changed at will, locally or adaptively (in the course of a simulation).

Linking chemical structure to properties and behavior of materials on different time and length scales can be achieved only if the various (high- and low-) resolution models involved are structurally consistent. Ideally, the structural agreement should hold down to the smallest possible length scale, which is the dimension of a CG unit. It is important to realize that, depending on the extent of coarse-graining, many all-atom (AT) states correspond to one CG configuration. Although a one-to-one correspondence between AT and CG configurations therefore does not exist, it is crucial that the conformational ensemble obtained with a CG model corresponds to that of the all-atom system, with the latter being analyzed in terms of the CG degrees of freedom. If we limit ourselves to the classical (non-quantum mechanical) case it means that the CG model must be parameterized such that the statistical weights of CG configurations are obtained from a (Boltzmann) weighted average over all corresponding AT states. Although for many systems we are still far from achieving this goal, it makes clear that quantum mechanical (QM), classical atomistic (AT) and coarse-grained (CG) mesoscopic models should ideally be developed such that "scale-hopping" [1,6–8] is possible in both forward and backward directions. It is the purpose of this chapter to discuss some of these issues and provide examples of CG models and multiscale modeling methods recently developed in our lab. We will emphasize *structure-based coarse-graining* for reasons following from the goal to allow for structure-based scale hopping as outlined above. In doing so, we follow a coarse-graining prescription *without* using ad hoc input in order to get the desired properties right. Alternative coarse-graining approaches (described elsewhere in this book) will not be discussed. Also, approaches that go much further and map the whole chain to one ellipsoidal [9] particle or just a soft sphere [10] are not considered here.

Figure 25.1 shows the systems that are discussed in this chapter. It includes bisphenol-A polycarbonate (BPA-PC) [11,12], polystyrene (PS) [13,14], and the liquid crystalline (LC) azobenzene derivative 8AB8 [15]. The CG representations are superimposed onto the chemical structures illustrating the typical level of coarse-graining. In Section 25.2 we shall discuss the coarse-graining and inverse-mapping procedures employed. In Section 25.3, several aspects of the CG models representing the above molecules are being discussed in terms of the structure (melt structure, chain conformations, LC order) and dynamics they predict. In this section, we focus on recent developments (what can be done nowadays with structure-based coarse-graining approaches and where possible pitfalls are that need to be avoided), inverse-mapped atomistic structures, and issues concerning the time-mapping procedure. In Section 25.4, an outlook to future developments and recent extensions to an adaptive scheme is being presented.

25.2 METHODS

25.2.1 GENERAL CONCEPT

The following section will be organized along the sequence of steps in deriving a CG model: first, we have to formulate a mapping scheme that relates the coordinates in the atomistic description with the centers of the CG particles. Second, one has to decide on a strategy concerning bonded and nonbonded interactions. In the coarse-graining procedure used by us, nonbonded and bonded interactions are strictly separated and derived sequentially. Such a clear separation makes the possibility

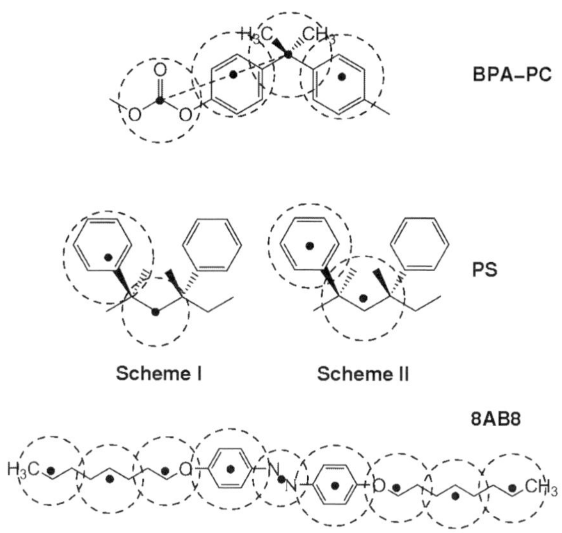

FIGURE 25.1 Atomistic and coarse-grained models of bisphenol-A-polycarbonate (BPA-PC), polystyrene (PS), and 4,4′-dioctyloxyazobenzene (8AB8). The CG mapping points are indicated with black dots. The corresponding CG superatoms, centered on the CG mapping points, are represented by the dashed spheres. For PS, two mapping schemes are shown. For BPA-PC, mapping points on the carbonate, phenyl, and isopropylidene groups are connected through a single CG bond.

to transfer potentials more likely, and allows us to distinguish between effects due to inter- and intramolecular potentials. Consequently, we describe separately how bond stretching, bond angle bending, and dihedral torsion potentials in the CG scheme are derived based on an atomistically detailed simulation of the isolated molecule in vacuo. Next, nonbonded interaction potentials between CG beads are derived based on the liquid structure of polymer melts or low-molecular-weight fluids (i.e., fragments of the target molecule or chain). These interaction potentials are subsequently used to generate well-equilibrated mesoscale structures and long-time trajectories of the system of interest. A last step, which also belongs to the coarse-graining procedure in the sense that it is a crucial link between the atomistic and the CG level of resolution, is the procedure of reintroducing atomistic details into a CG simulation trajectory ("back-mapping" or "inverse mapping").

25.2.2 Mapping Scheme

The mapping scheme relates the atomistic coordinates of a structure to the bead positions in the CG model. (Our models usually rely on CG centers with spherically isotropic potentials.) It is clear that there is no unique way to map a given set of atoms onto a coarser description. However, depending on the specific system and on the properties of the system that one wants to see reflected on the coarse level, one can define criteria to determine mapping points. Examples for such criteria are requirements to keep the ability to account for stereoregularity of chain molecules (e.g., PS [13,14]), or to capture certain geometry changes. For example, for azobenzene-containing LCs (8AB8) [15] one needs a clear distinction between the *cis* and *trans* geometry of the AB unit if one wants to investigate photoinduced phase transitions. There are other criteria that make a certain CG model more or less appealing, for example, in the PS example, a mapping was chosen which avoids "branching off" dangling side groups; that is, all CG beads are linearly connected in the chain, which saves complicated torsion and angle potentials [13,14].

When discussing the computational efficiency of a specific mapping scheme, one has to take several aspects into account. Trivially one would assume that fewer CG beads per molecule result in higher computational efficiency. In addition to a reduction in number of degrees of freedom

(DOFs), there is a speedup of the dynamics of the system due to a reduced molecular friction (larger beads, smoother potentials) of the CG model. In the case of chain molecules, there is, however, another aspect that should be kept in mind. Chain dynamics is faster if the envelope of the beads of the chain is tube-like, preventing optimised sphere packing and subsequent cage formation with corresponding higher friction [16]. A measure for this commensurability is given by the ratio of mean bond length and bead diameter. This criterion was used to explain why, for BPA-PC, a mapping scheme of Figure 25.1 with more beads is computationally more efficient than another one where the phenyl rings were included in somewhat larger spheres at the carbonate and isopropylidene units [11].

Another criterion that needs to be accounted for when devising a mapping scheme relates to the statistical correlations of internal DOFs. The mapping should be chosen such that these correlations are as weak as possible so that the intramolecular (bonded) potentials can be separated into bond stretching, bond angle bending, and torsion terms, as outlined in the next subsection.

25.2.3 Bonded Interaction Potentials

First of all, the determination of interaction potentials for the CG model is based on the assumption that the total potential energy U^{CG} can be separated into bonded/covalent ($U_{\mathrm{B}}^{\mathrm{CG}}$) and nonbonded ($U_{\mathrm{NB}}^{\mathrm{CG}}$) contributions [1]:

$$U^{\mathrm{CG}} = \sum U_{\mathrm{B}}^{\mathrm{CG}} + \sum U_{\mathrm{NB}}^{\mathrm{CG}}. \tag{25.1}$$

Intramolecular bonded/covalent interactions of the CG model are determined by sampling the distributions of (CG) conformational DOFs based on an atomically detailed simulation (Monte Carlo or molecular dynamics (MD) using a stochastic thermostat to ensure proper equilibration) of an isolated molecule in vacuo. These conformational distributions are in general characterized by CG bond lengths $\{r\}$, bond angles $\{\theta\}$, and torsions $\{\phi\}$; that is, $P^{\mathrm{CG}}(r,\theta,\phi,T)$ and are clearly temperature dependent (for simplicity we assume here that there is only one kind of bond, bond angle, or torsion). If one assumes that the different CG internal DOFs are uncorrelated, $P^{\mathrm{CG}}(r,\theta,\phi,T)$ factorizes into independent probability distributions of bond length, angle, and torsional DOFs:

$$P^{\mathrm{CG}}(r,\theta,\phi,T) = P^{\mathrm{CG}}(r,T)\,P^{\mathrm{CG}}(\theta,T)\,P^{\mathrm{CG}}(\phi,T). \tag{25.2}$$

This assumption has to be carefully checked (it is not uncommon that CG DOFs are correlated, for example that certain combinations of CG bonds, angles, and torsions are "forbidden" in the distributions obtained from the "real" atomistic chain), and is an important test of the suitability of a mapping scheme [14], because a mapping scheme that requires complex multiparameter potentials is computationally rather inefficient.

The individual probability distributions $P^{\mathrm{CG}}(r,T)$, $P^{\mathrm{CG}}(\theta,T)$, and $P^{\mathrm{CG}}(\phi,T)$ are then Boltzmann inverted to obtain the corresponding potentials and—through taking the derivatives—the forces

$$U^{\mathrm{CG}}(r,T) = -k_{\mathrm{B}}T\ln[P^{\mathrm{CG}}(r,T)/r^2] + C_r, \tag{25.3}$$

$$U^{\mathrm{CG}}(\theta,T) = -k_{\mathrm{B}}T\ln[P^{\mathrm{CG}}(\theta,T)/\sin\theta] + C_\theta, \tag{25.4}$$

$$U^{\mathrm{CG}}(\phi,T) = -k_{\mathrm{B}}T\ln P^{\mathrm{CG}}(\phi,T) + C_\phi. \tag{25.5}$$

When deriving potentials from bond and angle distributions one has to account for the respective volume elements r^2 and $\sin\theta$. Using the inverted distributions as potentials means that these potentials are in fact potentials of mean force. Ergo they are free energies and consequently temperature dependent. As mentioned before, this temperature dependence originates not only from the prefactor k_BT, but from the distributions P themselves. Strictly speaking they can only be applied at the temperature (state point) they were derived at.

The approach outlined in this section is in contrast to other approaches, where the CG internal DOFs are determined based on the distributions obtained from an atomistic simulation of the liquid phase [3]. In the latter case one obtains potentials for bonded and nonbonded interactions simultaneously from the same liquid simulation; consequently they are potentially interdependent; that is, there is no clear separation between covalent and nonbonded interaction potentials. We achieve this separation by deriving CG bond length, bond angle, and torsional distributions from the atomically detailed conformations sampled by a single (chain) molecule in vacuo. In the atomistic simulation performed to generate the distributions of CG intramolecular DOFs, the inclusion of nonbonded interactions has to be taken with care to avoid "double counting" of interactions. This means that long-range intrachain nonbonded interactions (beyond the distance between CG beads which are explicitly covered via bonded interaction potentials, for example, beyond the distance of three CG bonds if torsion potentials are used) should be excluded when the single chains are sampled. Instead these long-range interactions should be treated equivalently to CG intermolecular nonbonded interactions.

25.2.4 Nonbonded Interaction Potentials

The general principle when deriving nonbonded interaction potentials is to reproduce structural properties; that is, radial distribution functions of (low-molecular-weight) liquids or polymer melts (experimentally known or obtained from atomistic simulations). Similarly to the above case of bonded interaction functions, one has two principal options: either (1) to use analytical potentials, in which case one would optimize the parameters of a chosen analytical function to reproduce the structure of the atomistic melt/liquid as accurately as possible (or to account for the excluded volume interaction only, in which case no further optimization is being done, see BPA-PC [1,11]); or (2) one would use numerically derived tabulated potentials, which are designed such that the CG liquid reproduces the atomistic liquid structure, when the latter is analyzed in terms of the overlaid CG structure the microstate corresponds to.

In the first case, analytical potentials of various types can be used: the "normal" Lennard–Jones 12-6 potential is frequently used; it has, however, been proven to be in many cases too steeply repulsive; that is, too "hard," for CG particles, which are rather large and soft. In that case, softer Lennard–Jones-type (e.g., 9-6 or 7-6) [14], Buckingham or Morse potentials [15] are employed. These potentials are usually made purely repulsive in the spirit of the WCA potential [17] by shifting upwards and truncating in the minimum. In order to search in parameter space to optimize these analytical potentials to reproduce a given liquid or melt structure, a simplex algorithm can be used [18,19].

Concerning the second option to generate numerically a tabulated potential that closely reproduces a given melt structure; that is, a given radial distribution function $g(r)$, the iterative Boltzmann inversion method has been developed [20,21]. This method relies on an initial guess for a nonbonded potential $U^{\mathrm{CG}}_{\mathrm{NB},0}$. Usually the Boltzmann inverse of the target $g_{\mathrm{target}}(r)$; that is, the potential of mean force,

$$U^{\mathrm{CG}}_{\mathrm{NB},0} = -k_BT \ln g_{\mathrm{target}}(r), \tag{25.6}$$

is used, with which one then generates a CG simulation trajectory of the liquid. The resulting structure will not match the target structure since, due to multibody interactions, the potential of mean force is a good estimate for the pair potential only at very high dilution. However, using the iteration scheme

$$U^{CG}_{NB,i+1} = U^{CG}_{NB,i} + k_B T \ln\left[\frac{g_i(r)}{g_{target}(r)}\right], \tag{25.7}$$

the original guess can be self-consistently refined until the desired structure is obtained. There can be limits to this approach because it is not always clear whether the chosen CG mapping scheme can converge to an optimal fit.

For complex molecules with a large number of different CG beads or more importantly in the case of molecules that form complex or anisotropic liquid or melt structures, for example, liquid crystals, the procedure to determine nonbonded interaction functions is more complicated. In these cases it is advantageous to split the target molecule into fragments so that the nonbonded interactions between different bead types can be determined based on the structure of isotropic liquids of these fragment molecules. One principal problem that arises if one uses smaller fragments to generate nonbonded interaction potentials for larger molecules is that different conformations may contribute to the structure of the liquid of the fragment molecules differently than in the (polymeric) melt [22]. One example where such an effect may play a role is in the parameterization of phenyl rings based on the structure of liquid benzene: in that case the relative population of parallel and perpendicular arrangements of two phenyl rings that are part of longer chain molecules potentially differs from the arrangements in liquid benzene for steric reasons. Despite these potential problems, the procedure to parameterize CG nonbonded interactions based on small molecules is promising to generate CG parameters for complex molecules and it also allows reuse of certain CG potentials for reoccurring building blocks (such as alkyl or phenyl groups), which aims at some sort of building block or LEGO set of molecule fragments for CG simulations. Of course, this approach needs to be carefully tested and the transferability of the potentials generated from these fragments to (slightly) different conditions needs to be carefully evaluated (as will be further discussed in the Examples section).

25.2.5 Coarse-Grained Simulations: Equilibration of Mesoscale Structures

Even with the dynamic speedup gained by CG models, it is not trivial to obtain well-equilibrated structures of mesoscale (polymeric) systems. In particular for long-chain molecules (beyond a few entanglement lengths), branched polymers, or polymers at interfaces, brute force MD algorithms that follow the slow dynamics of the system will not easily lead to complete equilibration of the chains. Besides, criteria are needed to judge whether a melt structure is really equilibrated since local monomer packing and the statistics of end-to-end distances or radii of gyration are not sufficient. Auhl et al. [23] describe such criteria and investigate various methods to generate well-equilibrated polymer melts using MD simulations. Based on such CG structures and simulation trajectories it is in the next step possible to reintroduce atomistic coordinates and to obtain equilibrated atomistic structures on the mesoscale or long-time atomistic trajectories.

25.2.6 Reintroduction of Atomistic Details ("Inverse Mapping")

Inverse mapping; that is, reintroduction of atomistic detail, requires finding a set of atomistic coordinates that corresponds to a given CG structure. In general there is no unique solution to that problem since each CG structure corresponds to many all-atom configurations. Therefore, one needs to find one representative all-atom structure, with the correct statistical weight of those DOFs that are not resolved in the CG description. Several slightly different strategies to reintroduce atomistic detail into a CG structure have been presented [2,3,12,13,15,24].

If the (polymer) chain consists of reasonably rigid (all-atom) fragments, it is sufficient to fit these rigid all-atom units onto the corresponding CG chain segment coordinates. The atomistic fragments can be taken from a pool of structures that correctly reflect the statistical weight of those DOFs (certain torsions, ring flips, etc.) that are not resolved in the CG description and that relax too slowly to be properly equilibrated in a short equilibration run of the resulting atomistic structure.

If the CG molecule/polymer chain consists of very flexible units, for example, alkyl tails, and in particular if the CG structure consists of small molecules (8AB8, a low-molecular-weight LC), where even in a very short equilibration step, the atomistic structure significantly diffuses away from the CG coordinates, a slightly different strategy was employed: atomistic coordinates were inserted into the CG structure using fragments for the rigid units and random atomistic positions for the flexible units (with the constraint that the atomistic coordinates have to satisfy the "mapping" condition; that is, the atomistic coordinates have to correspond to the CG structure if one applies the mapping scheme). The resulting structure was then relaxed (energy minimized and equilibrated by MD simulations), while restraining the atomistic coordinates to CG mapping points. This results in a perfectly equilibrated structure that (almost, depending on the strength of the restraining potential) exactly reproduces the CG structure.

25.3 EXAMPLES

In this section, we discuss, on the basis of the three examples shown in Figure 25.1 (and Figure 25.6), various aspects of structure-based coarse-graining focusing on recent developments, inverse-mapped atomistic structures and dynamics. In Section 25.3.1 ("Structure") we discuss experimental validation of inverse-mapped BPA-PC and PS melt structures and the prospects that open up due to the resulting well equilibrated long-time/large-scale atomistic trajectories; we illustrate the consequences of the choice of a CG mapping scheme using the example of PS, and we show the application of the present coarse-graining approach to LC molecules. In Section 25.3.2 ("Dynamics") we discuss how, by application of CG models, the corresponding time scales are modified. In that context we compare BPA-PC chain dynamics in all-atom and CG molecular liquids as well as diffusion of low-molecular-weight additives in CG PS melts.

25.3.1 STRUCTURE

25.3.1.1 Inverse-Mapped BPA-PC Melts

Although many aspects of, for example, polymer dynamics, overall chain conformations, or LC order can be well described with CG resolution, for many other questions chemical details need to be reintroduced by inverse-mapping methods described in the previous section. This we illustrate here by discussing aspects of packing of BPA-PC polymeric liquids [2,12] and the evaluation of interactions (chemical potentials) of small molecules inside polymeric microstructures [25].

To check the quality of BPA-PC melt structures, we calculated neutron scattering functions of the (reintroduced) all-atom melts. Figure 25.2a shows the coherent neutron scattering function for a melt containing 100 chains of $N = 20$ chemical repeat units at two temperatures [12,26]. The simulated functions are compared with experiments obtained at $T = 1.5$ K [26] and consequently most probably a slightly higher density. The peak at 0.6 Å^{-1} corresponds to the intrachain sequential carbonate-carbonate distance of about 11 Å and not to interchain correlations. This could be concluded from the simulations, where the n-scattering functions were calculated. For the "computer samples" one can vary the atomic scattering lengths in the analysis and delete or create scattering contrast for any correlation at will. The main peak (amorphous halo) corresponds to the typical interchain (packing) distance. The agreement between the experimental data and the simulations is close to perfect. The discrepancies are due to the higher temperature of the simulated melts, which causes the amorphous halo to broaden and to shift to slightly larger distances and the peak corresponding to intrachain

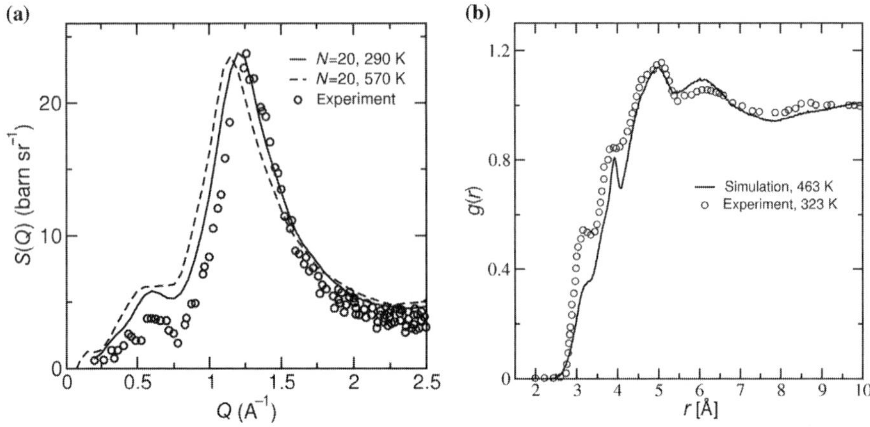

FIGURE 25.2 (a) Coherent neutron scattering function of two BPA-PC melts (290 and 570 K) [12] in comparison with experiments of a sample, which was cooled down and kept at a temperature of 1.5 K [26]. The solid and dashed curves were obtained by inverse mapping of chemical details for a system containing 100 chains of 20 repeat units each. (b) Radial distribution function of a simulated atactic polystyrene melt obtained by inverse mapping of chemical details [13] in comparison with the experimental RDF obtained from X-ray diffraction [27]. All atom-atom correlations are included except those between atoms within phenyl rings and atoms along the backbone separated by less than three chemical bonds.

carbonate–carbonate correlations to wash out. A comparison of simulated scattering curves with experimental data for partially deuterated BPA-PC samples was also made [2,12], which further supported the overall agreement with experiments [26].

A similar comparison was made for a PS melt. Figure 25.2b shows the total radial distribution function obtained after reintroducing chemical details together with experimental data obtained by wide-angle X-ray diffraction measurements [27]. In both the simulation and the experimental data, intramolecular correlations due to 1-2 and 1-3 bonded neighbors (along the backbone) as well as all intraring correlations have been removed in order to emphasize the features deriving from the packing of nonbonded segments. Despite differences in temperature and chain length of the simulated and experimental samples, the overall agreement is very good. Moreover, in our analysis of the simulation trajectories we employed a united-atom model. Because of that, we assumed Q-independent atomic scattering functions taking the carbon nuclear positions as scattering centers. This assumption gives rise to a stronger developed peak in the simulated data slightly below 4 Å in comparison with the X-ray experiment.

As a second example we mention a significant advantage of using inverse-mapped polymer microstructures in studying permeation of small molecules (so-called 'penetrants'). The first application using this approach was a computational study of phenol in BPA-PC [28]. The phenol diffusion process revealed a strong coupling between size and shape fluctuations of the pore space and the hopping of the penetrant. The pore structure was also analyzed in terms of the positronium annihilation time [29]. The resulting lifetime distribution functions compared very well to those from experiments, again supporting the overall consistency of the approach. In addition to diffusion, the penetrant solubility or excess chemical potential inside the polymer microstructure is also of interest. With currently available methods, penetrant excess chemical potentials can *only* be computed with sufficient statistical accuracy for fairly small penetrants. These are usually pure substances, such as gases under ambient conditions. A polymeric simulation box with a typical linear dimension of 4–5 nm is usually large enough to contain a statistically meaningful number of pre-existing, empty cavities, which can host a small molecule without significantly modifying the matrix. Thus standard methods, such as test-particle insertion techniques, can be used to obtain reliable data. However, calculations of excess chemical potentials of larger penetrants, with equally high statistical reliability, are extremely cumbersome for several reasons. Most importantly, larger penetrants (e.g., phenol, propane, chloroform) occupy larger cavities, which in microstructures with the above-mentioned linear dimensions occur very infrequently, albeit contributing significantly to the excess chemical potential. This problem can be resolved only if a large number of statistically uncorrelated microstructures can be generated at small computational expense. Obviously, reinserted all-atom microstructures generated from CG mesoscale simulations can be used to resolve this problem. Based on large systems generated in this way, we currently explore an alternative, nonequilibrium free-energy sampling technique, in order to resolve insertion problems usually encountered with large molecules in dense systems [25].

25.3.1.2 Two Mapping Schemes for Polystyrene

As discussed in Section 25.2, CG intramolecular potentials are developed assuming that the CG bond length, bond angle, and dihedral angle have no interdependencies. The validity of this assumption depends however on how we choose the CG mapping points. Figure 25.1 shows two CG representations for PS [14]. In the first scheme (I), the PS repeat unit is represented by a CG bead (type "*A*") localized on the methylene position, and another CG bead (type "*B*") is localized on the mass center position of the remaining atoms. *A* and *B* beads are connected by CG bonds giving rise to bond angles θ_{ABA} and θ_{BAB}, and dihedral torsions φ_{ABAB} and φ_{BABA}. In the second scheme (II), bead *A* is positioned at the center of mass defined by the methylene group and the two adjacent CH-groups (taking however the half-masses rather than the full CH masses in defining the CG bead mass center). Bead *B* corresponds to the phenyl group. The *A* and *B* beads in scheme (II) are also connected by CG bonds, giving rise to the same number of DOF (see Figure 25.3a). We note that the corresponding intramolecular potentials depend on the chain stereoregularity (i.e., the type of dyad [13]), hence the model can in principle be used in simulations of atactic, isotactic, and syndiotactic PS.

The PS conformation on the left-hand side in Figure 25.3a is based on CG mapping scheme (I) and is shown to illustrate how the (θ,φ) CG angles are correlated. If the *A* bead on the left end of the picture is being rotated along the indicated CG bond, the adjacent *B* bead will also be rotated because these two beads are directly connected through two underlying chemical bonds. This causes variations of φ_{ABAB} and θ_{BAB} to be correlated. Whether at all and to which degree such correlations lead to erroneous conformational sampling in the CG simulations depends on the mapping scheme and needs to be tested to assess the quality of a mapping scheme.

Figure 25.3b shows energy diagrams (defined as $-\ln[P^{CG}(\theta,\varphi)/\sin\theta]$) in a contour map representation for the racemic PS dyad [14]. The bond bending angle θ corresponds to BAB and the dihedral angle φ to ABAB. The diagrams presented in the left part of this figure are obtained from simulations of a single united atom chain and diagrams on the right were obtained with the corresponding CG models. The upper panel corresponds to mapping scheme (I) and the lower panel to mapping scheme (II) (see Figure 25.3a). From the contour maps obtained with the CG models, the (θ,φ) correlation discussed above is lost to some extent. For example, CG scheme (I) has an energy minimum at $\theta \approx 150°$ (upper panel, left), which is about 3 k_BT deeper than the minimum at $\theta \approx 100°$. Therefore, CG model (I) predominantly samples $\theta \approx 150°$, independent from the torsion angle φ, which causes the energy basin at $(\theta,\varphi) \approx (100°, 240°)$ observed with the united-atom model (upper panel, left) to shift to a region $(\theta,\varphi) \approx (150°, 240°)$ (upper panel, right) hardly ever sampled by the united-atom model.

With mapping scheme (I), the CG model also samples parts in (θ,φ)-space not at all accessible by the united-atom model (e.g., (80°, 300°) or (80°, 30°)). These 'forbidden' regions include conformations with excluded volume violations of CG 1-4 interaction sites (methylene units partly overlapping with phenyl groups). These overlaps can be avoided by introducing a special 1-4 nonbonded interaction in the CG model [13]. Noteworthy, CG model (II) clearly performs much better in this respect. Because special 1-4 nonbonded terms are not needed [14], it is also more consistent with

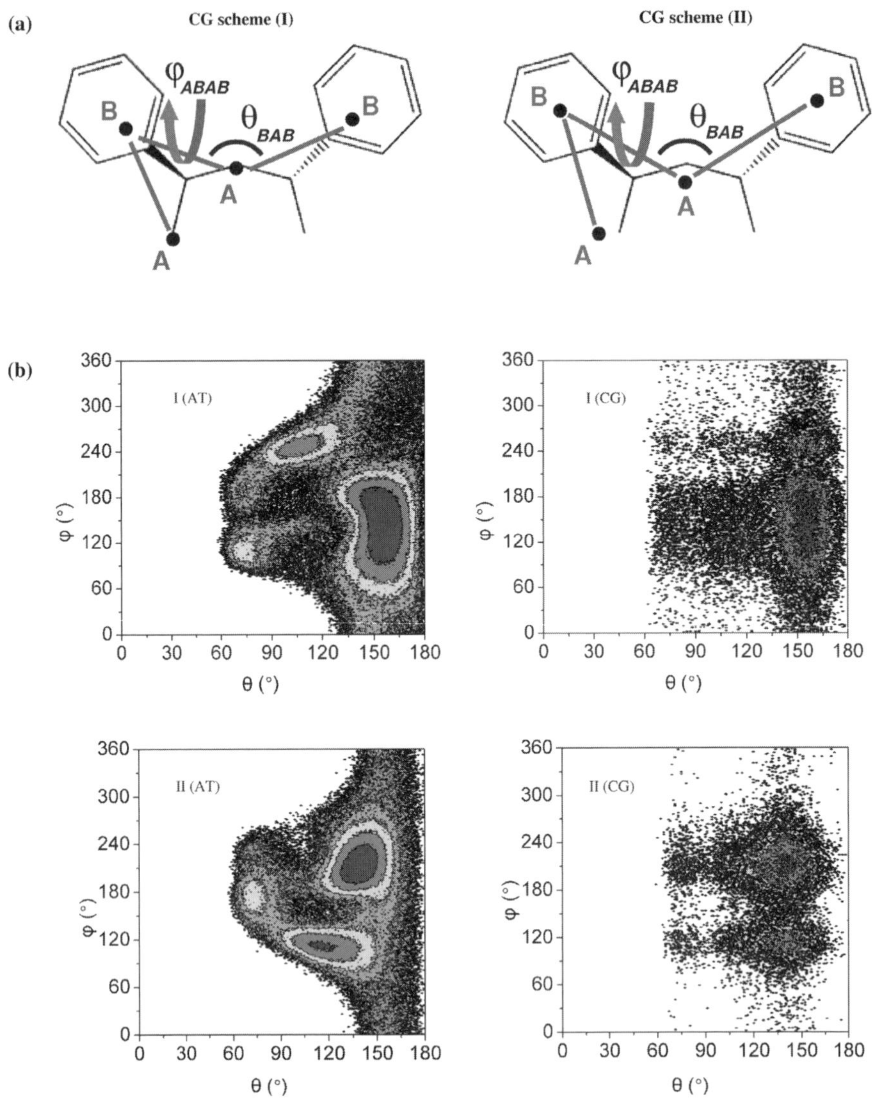

FIGURE 25.3 (a) PS conformation with CG mapping points based on schemes (I) and (II) (cf. Figure 25.1). The CG mapping points are indicated with black dots, CG bonds are indicated by thick gray lines. (b) (θ,φ)-energy surfaces: I (AT), obtained by sampling the atomistic model, analyzed in terms of CG scheme (I); I (CG), obtained by sampling with the CG model, scheme (I); II (AT), obtained by sampling the atomistic model, analyzed in terms of CG scheme (II); II (CG), obtained by sampling with the CG model, scheme (II).

the general CG strategy outlined in the previous sections. In addition, there are certain advantages when studying dynamical properties compared to CG model (I).

It is very important to be aware of correlations of internal DOFs in CG simulations, even though artifacts introduced by decoupling the bond-angle bending and dihedral torsion potentials in CG models have so far been shown to affect neither the overall chain conformations nor the ability to successfully perform the inverse mapping in polymer modeling [13,14]. This is potentially more problematic in CG models for biomolecules. Here a similar decoupling of the bonded potentials is likely to be more tedious because specific (θ,φ)-combinations may turn out to be needed for discriminating turns, helices, sheets, etc. which will be a significant criterion to distinguish "good" and "bad" mapping schemes [30].

25.3.1.3 Azobenzene-Based Mesogens

In the previous two examples the coarse-graining procedure (Section 25.2) was applied to polymeric systems, where the behavior of the melt is very much determined through chain connectivity and excluded volume interactions of the polymeric beads. Consequently, it often is not essential to introduce attractive (nonbonded/intermolecular) interactions in order to correctly predict melt structure and dynamics on the mesoscale. It is however very interesting to explore how far the above coarse-graining scheme carries if one tries to apply it to systems where attractive nonbonded interactions are likely to be more important than in amorphous polymers; that is, where the balance of attractions between different chemical units plays a possible role in structure formation. Biopolymers, liquid crystals, and in general self-assembling systems are examples where this can be of importance. The compound 8AB8 (see Figure 25.1) is a LC compound that contains azobenzene as a mesogen and forms a thermotropic nematic phase (and a monotropic smectic). This system is used to study how the coarse-graining approach can be adapted to LC systems. It is of particular interest to build a CG model that is close to an atomistic description not only in order to obtain as much chemical accuracy as possible but also because a close link between the coarse (mesoscale) and the atomistic level is important for multiscale simulation purposes. The reason for this is that azobenzene is a photoswitchable mesogen; that is, it undergoes a *trans/cis* photoisomerization, which goes along with a drastic shape change: in its *trans* form it is rod-shaped and functions as a mesogen; in its *cis* form, it is bent and does not induce a mesophase. Therefore, with 8AB8 a photoinduced nematic-to-isotropic phase transition is observed. This LC phase change and the photoisomerization mechanism are interdependent since on the one hand the LC phase change obviously depends on the degree of *trans/cis* isomerization, and on the other hand it is believed that the photoisomerization mechanism depends on the (anisotropic) environment or the mechanical pulling of the tails that are attached to the azobenzene group. Therefore, the LC-photoswitching of azobenzene compounds is a true multiscale problem, since the photoisomerization mechanism can be studied using quantum-mechanical (QM) simulation techniques, whereas investigations of the LC phase change requires much longer length and time scales that can only be achieved by mesoscale (CG) techniques. In this constellation it is important to be able to switch between the levels of resolution, where the atomistic description can function as a link; that is, the coarse model needs to be built on the atomistic description, and the inverse mapping from the CG to the all-atom level is essential to link to QM calculations of the transition.

Ref. 15 describes how a CG model for 8AB8 was developed using the CG techniques developed for polymers. It was shown how intramolecular (bonded) potentials were obtained from simulations of an all-atom single 8AB8 molecule, and how intermolecular potentials were developed based on all-atom simulations of isotropic liquids of fragments of the 8AB8 molecule. The isotropic liquids that were used in the parameterization process were liquid benzene, liquid azobenzene (in its *trans* and in its *cis* form), liquid octadecane, and various mixtures of these compounds. Based on the structure of these liquids (radial distribution functions), nonbonded interaction potentials were determined, both using analytical potential functions and the iterative Boltzmann inversion method as detailed in the Methods section (for the case of octadecane see Figure 25.4a). The resulting interaction functions were then used for liquid (*trans*) 8AB8, where we tried to reproduce the experimentally observed LC phase behavior. In particular we aimed at obtaining a stable nematic phase. One could observe that the use of (soft) analytical potentials that are purely repulsive (in the spirit of the previous coarse-graining examples of polymeric systems) did not yield the correct mesophase behavior of 8AB8; in fact no long-range ordering was observed for the model chosen (see Figure 25.4b), even with a rather wide scan of temperatures and pressures. With potentials generated with the iterative Boltzmann inversion method; that is, numerical (tabulated) potentials which are also partly attractive, it is however possible to generate nematic-like (and smectic) phase of 8AB8. Thus, for the given molecule; that is, the given size and shape of the mesogen and the given molecular flexibility of the alkoxy tails, it seems to be important to account for attractions

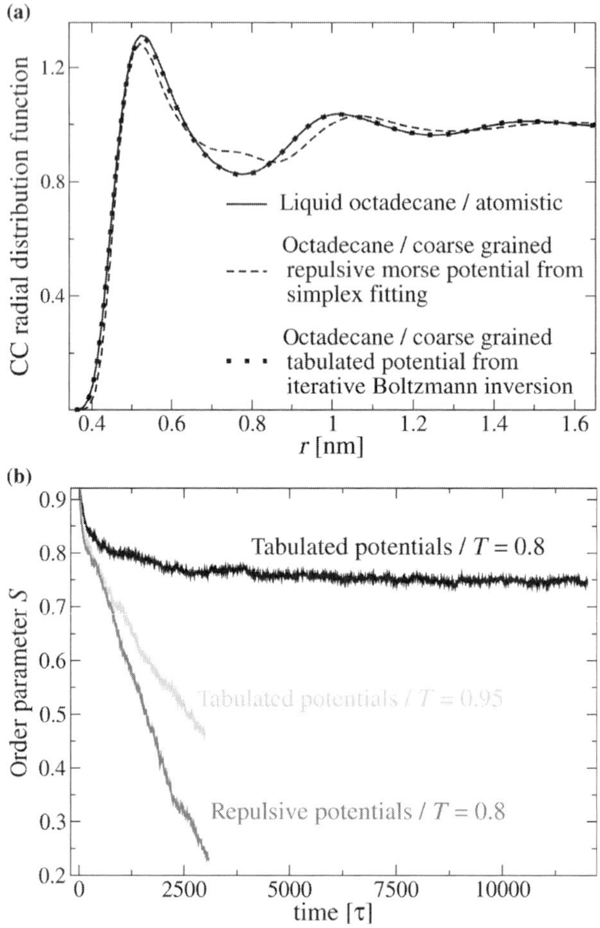

FIGURE 25.4 (a) Structure-based derivation of nonbonded interaction potentials: carbon–carbon radial distribution functions (RDF) of CG centers in an octadecane liquid at 400 K. Thin straight line: RDF obtained from atomistic simulation, mapped onto CG centers. Thin dashed line: RDF obtained in CG simulation after optimizing a purely repulsive Morse potential to reproduce the atomistic structure as well as possible. Fat dotted line: RDF obtained in CG simulation after determining a numerical potential through iterative Boltzmann inversion so that the atomistic structure is reproduced. (b) Order parameter of 8AB8 system in coarse-grained simulations (initial setup fully ordered: four smectic layers). Black and light gray lines: simulations with potentials obtained through iterative Boltzmann inversion (partly attractive). Black line: the system remains ordered at $T = 0.8$ (corresponds to 320 K) (nematic-like structures are observed). Light gray line: the system becomes isotropic at $T = 0.95$ (corresponds to 380 K). Dark gray line: Simulation with purely repulsive Morse potentials—the system becomes disordered (at a wide range of temperatures and densities).

between the different beads in the CG model in order to reproduce the ordered phase of 8AB8. A snapshot of a structure that shows the alignment of the 8AB8 molecules in a nematic-like phase can be seen in the Color Figure 25.6 in the center of the book. This structure was generated by MD simulations using the CG model, the atomistic coordinates that are also shown in the figure were obtained using the inverse-mapping procedure as outlined above (restraining the atom coordinates during equilibration such that the "mapping criterion" is satisfied and the CG structure is therefore preserved). It shows that the structure-based coarse-graining approach originally developed in the polymer framework can be extended to LC systems, where mesoscale (with both large length and long time scales) simulations are essential to probe phase behavior and to generate well equilibrated mesostructures. With the given approach the mesoscale simulations also maintain an important link to the chemical structure, and through the inverse-mapping procedure it is possible to obtain atomistic coordinates of the system.

In the course of the parameterization process of the nonbonded interactions, we also performed preliminary tests on the transferability of these fragment-based potentials. We tested, for example, the applicability of potentials derived for pure liquids on mixtures of various compositions and of potentials derived for liquid benzene on liquid *trans* or *cis* azobenzene. Overall, the transfer of the nonbonded potentials worked surprisingly well; the limitations are more thoroughly discussed in Ref. 15, and these investigations will also be extended in the future.

25.3.2 Dynamics

Within CG models length scales are usually well defined through the construction of the coarse-graining itself. In most dynamic CG simulations reported in the literature little attention is paid however to the corresponding "coarse-graining" of the time unit. From polymer simulations of both simple continuum as well as lattice models it is known that such simulations reproduce the essential generic features of polymer dynamics; that is, the crossover from the Rouse to the entangled reptation regime, qualitatively and to a certain extent quantitatively [31,32]. While such previous studies concern motion distances on scales well above a typical monomer extension and provide quantitative information on characteristic time ratios, this still leaves a number of open questions. These refer to the predictive quantitative modeling of diffusion, viscosity, rates, and correlation times, etc. of dynamic events as well as to the question of minimal time and length scales CG simulations apply to. Particle mass, size, and energy scale, which are all well defined within a CG model, of course trivially fix a time scale, too, and it is indeed this time scale that is most often reported in MD simulations of CG systems. However, it does not usually correspond to the true physical time scale, because part of the friction experienced by a (sub)molecule (in the AT representation) is lost in the CG representation, causing the CG system to evolve faster. (Note that this is in principle also the case for atomistic simulations that make use of so-called united atoms where aliphatic hydrogen atoms are incorporated into the carbon atoms.) In other words, the fluctuating random forces of atomic DOFs, which are integrated out in the CG model, contribute to a "background friction" that must be considered in order to obtain a realistic time scale in the CG dynamics simulation. In their study of CG blob dynamics in polyethylene melts, Padding and Briels [33] employed effective potentials, frictions, and random forces all derived from detailed MD simulations. Izvekov and Voth [34] proposed a closely related recipe within the coarse-graining framework of force matching. Alternatively, CG dynamic quantities can in some cases be mapped directly onto the corresponding quantity obtained from detailed MD simulations or from experiments. For example, a diffusion coefficient D^{CG} in units $[m^2/\tau]$ can be mapped onto the diffusion coefficient D^{AT} in units $[m^2/s]$ providing the time units of the CG simulation $\tau = x$ [sec]. Alternatively, the CG mean squared displacement curve can be superimposed with the atomistic curve at (for atomistic simulations) long times [35]. This approach was used to study entangled polycarbonate (BPA-PC) melts of up to 20 entanglement lengths. The CG simulations provided truly quantitative information on the different measures of the entanglement molecular weight (from displacements, scattering functions, modulus and topological analysis) and the ratios of the different crossover times.

25.3.2.1 Long-Time Atomistic BPA-PC Trajectories Obtained by Inverse Mapping

All CG mapping schemes shown in Figure 25.1 stay close to the atomistic structure of the molecules. Therefore, the dynamics of the CG system is expected to follow quite closely that of the atomistic system down to small length and time scales. Moreover, due to significant dynamic speedup, the CG systems can be simulated up to times that exceed far beyond what is possible in brute force detailed atomistic simulations, allowing for in silico experiments looking at exactly the same quantities as in experiments. The idea is to reintroduce atomic details in long-time CG trajectories of the system (BPA-PC for the present case) and measure dynamic relaxations on time scales that altogether cover at least nine decades and overlap the experimental regime probed, for example, with spectroscopic techniques. Here we only discuss the dynamic chain scattering function $S(Q, t)$ as obtained in neutron spin echo experiments:

$$S(Q,t) = \frac{1}{n}\left\langle \sum_{i,j} l_i l_j \exp[i\mathbf{Q}\cdot(\mathbf{r}_i(t) - \mathbf{r}_j(0))] \right\rangle_{|\mathbf{Q}|} . \tag{25.8}$$

The double sum runs otver all n atoms in the chain. The term $\mathbf{r}_i$ is the position of atom i and l_i is the neutron scattering length of atom i. The index $|\mathbf{Q}|$ indicates spherical averaging. For nonentangled melts on time scales above the local fast oscillations and above the persistence length of the polymer the Rouse model predicts $S(Q,t)/S(Q,0) \propto \exp(-WQ^2t^{1/2})$, where W is related to the effective bead friction. The onset of this universal behavior is typically small compared to the diffusion time and chain extension. For larger times the overall diffusion takes over; that is, $S(Q,t)/S(Q,0) \propto \exp(-DQ^2t)$. In the case of entangled polymers, $S(Q,t)$ displays a qualitatively different behavior due to the tube-like confinement of the monomer motion. On intermediate time scales the scatterer "sees" a smeared-out monomer density in the tube of diameter d_T leading to an analog of a Debye Waller factor with, in the simplest approximation $S(Q,t)/S(Q,0) = 1 - Q^2d_T^2/36$. CG and atomistic MD simulations of BPA-PC melts were performed with $N=5$ up to $N=120$ repeat units [35] and used to analyze this property. The entanglement molecular weight of BPA-PC (1200–1400 g/mol) corresponds to $N_e \approx 5$–6 repeat units. Based on performing a time mapping by superimposing repeat unit mean squared displacements of the CG and atomistic systems for $N=5$ and $N=20$ for long times, a time unit is obtained. While the intrinsic time unit of the CG model (determined through conversion of Lennard–Jones reduced units, assuming the same mass for all beads) is $\tau \approx 1.7$ ps, the physical time unit of the underlying BPA-PC is much larger, namely $\tau = 30$ ps at the temperature studied here ($T=570$ K) [35]. Note that the typical time-step in a CG dynamic simulation is 0.01 τ, thus roughly 0.3 ps. For $N=20$ the atomistic simulations only covered a bead motion up to about the monomer size. This time mapping unit was used in Figure 25.5a, which shows $S(Q,t)/S(Q,0)$ for a $N=5$ and $N=20$ BPA-PC melt [12]. For each chain length two independent sets of data are shown; the first has been obtained after reinsertion of chemical details in long-time CG trajectories (symbols); the second has been obtained from separate detailed, all-atom simulations (lines). Data

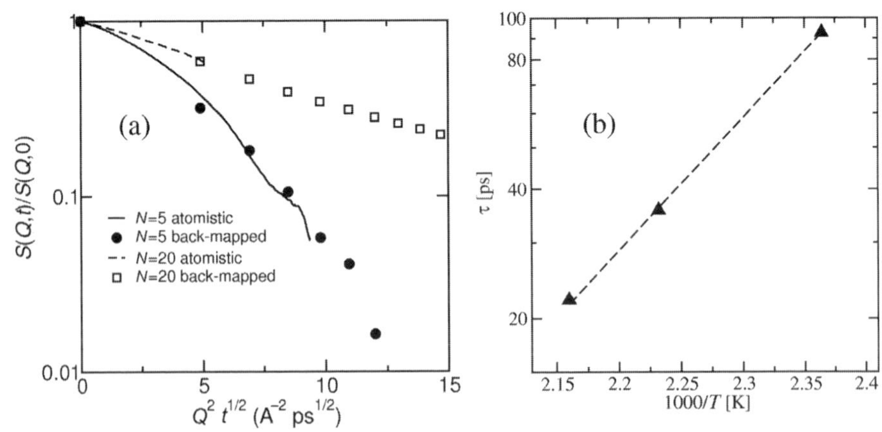

FIGURE 25.5 (a) Dynamic scattering function $S(Q,t)/S(Q,0)$ of BPA-PC chains in the melt (570 K) as measured by n-spin echo experiments versus the scaled time $Q^2t^{1/2}$ for $Q=0.2$ Å^{-1} [12]. Data obtained by original atomistic simulations are shown by the solid and dashed line; data obtained from inverse mapped conformations are shown by the symbols. (b) Arrhenius representation of the time mapping constant for the ethylbenzene motions in PS melt [36].

are presented for $Q=0.2$ Å^{-1}, which covers the typical chain extension. A remarkable agreement is observed by the data obtained based on the CG trajectory and the all-atom simulations. This perfect agreement of trajectories illustrates that the CG dynamic trajectories are physically meaningful down to very small length and time scales. It also shows that with such a time mapping of CG and atomistic simulations absolute data for long time and large scale dynamic quantities can be obtained without calibrating simulation timescales using experimental data. Based on the above time-mapping and inverse-mapping methods, the largest all-atom system simulated consisted of 200 BPA-PC chains of $N=120$ (corresponding to roughly 800,000 atoms in a box with a linear dimension of 100 nm) up to 4×10^{-5} sec.

25.3.2.2 Dynamic Speedup: Additive Molecules in a Long-Chain Polystyrene Melt

The above route to determining the physical time scale in a CG simulation has been applied to several systems. To better understand the physical origin of the dynamic speedup in comparison with all-atom models and real-life experimental systems, we discuss in this section an example of a simulation study of the dynamics of CG ethylbenzene (EB) molecules dissolved in a CG PS microstructure. A physical time scale was obtained by mapping the simulated EB diffusion coefficients onto the corresponding experimental data obtained by pulse field gradient NMR [36]. The time

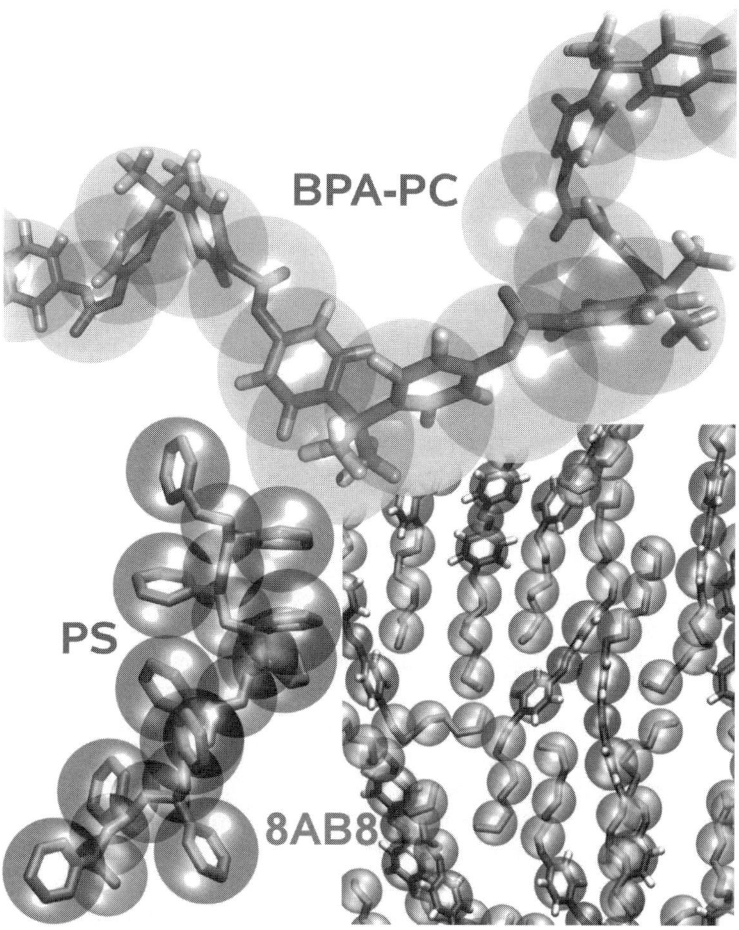

FIGURE 25.6 (See color insert following page 238.) Snapshots of selected molecules from CG simulations of BPA-PC, PS, and 8AB8 indicating both CG centers and atomistic coordinates obtained through inverse mapping.

conversion unit $\tau = D^{CG}/D^{exp}$ (expressed in picoseconds) is presented in Figure 25.5b on a logarithmic scale versus the inverse temperature. The key observation is that τ depends exponentially on the temperature; that is, $\tau = \tau_0 \exp(-A/T)$, where the constant A is positive. This observation originates from the fact that energy barriers for EB diffusional motions are lower in the CG system where interparticle potentials are softer and more smoothly varying with distance. The time mapping $\tau(T)$ between the real and the CG system therefore follows an Arrhenius dependency with an "activation energy" $k_B A$ describing an average reduction of energy barriers in the CG system. It should be noted that D^{CG} and D^{exp} do not follow an Arrhenius dependency. Because the time scale for migration of the relatively large EB molecules is coupled to chain rearrangements of the PS matrix, it is important that the CG model is capable of reproducing the non-Arrhenius (Vogel-Fulcher) type temperature dependence of structural relaxation of the melt.

25.4 SOME RECENT DEVELOPMENTS AND FUTURE PERSPECTIVES

25.4.1 Adaptive Resolution MD

In many systems formation (e.g., self-assembly) and dynamics of large-scale structures and conformations cannot be decoupled from local, chemical processes and specific intermolecular interactions. To perform computer simulations for those cases, dual-scale resolution schemes can be used [37–42]. One can however even go beyond using molecular models with fixed (single or dual) resolution and allow for a dynamic change of molecular resolution by changing the number of molecular DOF on-the-fly during the course of an MD simulation. Recently, such an adaptive resolution scheme (AdResS) has been introduced in which molecules can freely exchange between a high-resolution and low-resolution region [43–45]. A key ingredient in this new method is a transition region in which a weighting function is applied that mixes the high-resolution and low-resolution pair *forces* thereby slowly modifying the resolution of the molecules that move through [46]. The 'latent heat' associated with increasing or decreasing the number of molecular DOF is supplied or removed by a properly chosen thermostat. By these means thermodynamic equilibrium is maintained throughout the system. This method, which so far has been used for liquid water [44] and a polymer-solvent system [45], is of great interest in a much wider variety of systems. An example could be an active site on a protein where the biological function requires an explicit description of solvent molecules. It would clearly be beneficial if far away from the active site the system could be described at lower resolution to avoid spending 99% of computer time on moving water molecules around in regions not of primary interest.

25.4.2 Surface Interactions of Biomolecules

Interactions of biomolecules with metal and inorganic surfaces are becoming increasingly important in nanobiotechnology. Typical questions involve how the functionality of a bio/inorganic hybrid device depends on the conformation of adsorbed biomolecules and how conformations are affected by the nature of the surface interactions involved. Multiscale modeling techniques that bridge between quantum, classical atomistic, and CG model descriptions are needed to approach such issues. Recently, initial steps have been made to bridge between the quantum and classical atomistic levels by performing a quantum-classical modeling of statistical conformations and interactions of amino acids and water molecules with metal surfaces [47]. This work has provided a recipe for treating surface interactions of amino acid residues in a classical-level description through an interactive quantum-classical modeling approach that can in principle be applied to larger organic molecules. Further progress will rely on the development of dual-resolution or adaptive resolution models that can be used to describe the system (solute and solvent) at high resolution close to the surface, combined with a description at lower resolution far away from the surface.

25.4.3 Nonbonded Interactions

Although the iterative Boltzmann inversion method (Equation 25.7) provides nonbonded interaction potentials for CG models, it is based on radial distributions functions, which usually do not precisely define the system. In addition this can lead to very complicated and long-range potentials, which reduce the efficiency of the CG simulation significantly. Ideally one should aim to put as little as possible prior information into the model because that unavoidably leads to CG potentials that lack transferability and thus predictive potential. With respect to CG models for biopolymers in solution (e.g., oligopeptides) one ideally develops the CG force field in a way that distinguishes the bonded and nonbonded parts of the interaction potentials (in analogy to the method described above). Whereas for the bonded part, lessons learned from polymer coarse-graining could be applied, for the nonbonded part important challenges remain. Current developments include empirical parameterization against thermodynamic data [48] and force-matching approaches [49,50]. Alternative to these approaches, intermolecular pair potentials of mean force obtained from atomistic MD simulations can be used. Based on this approach, CG potentials for aqueous electrolytes were recently reported [51,52]. This method has been extended to a wide range of electrolytes including, for example, alkylammonium salts for which a realistic description of the ion pairing and dissociation equilibrium requires accounting for aspects of hydrophobicity that—in addition to standard electrostatics—gives rise to an additional attraction between the ions [53].

25.4.4 Perspectives

Questions related to the specific systems discussed in this chapter lead automatically to another, almost philosophical aspect, namely—how specific is specific? In polymer physics one knows which properties are universal and which are chemistry specific. The systems considered there, however, are, in the end, very simple systems, where the above question is rather simple to answer. In problems related to structure formation, self-assembly, and surface interactions in synthetic and biological systems, specific interactions are operating. In these cases, it is far less understood which (chemistry) specific details should be kept in CG models (and which can safely be ignored). Moreover, it is not clear at what length scales the various CG modeling approaches described throughout this book merge and equally well describe these types of systems. Especially for biological molecules or complex structures employed in organic electronics, we, however, are still far away from such an understanding.

ACKNOWLEDGMENTS

We wish to acknowledge Berk Hess and Vagelis Harmandaris for providing data and figures. We wish to thank Berk Hess, Vagelis Harmandaris, Pim Schravendijk, Matej Praprotnik, and Luigi Delle Site for many stimulating discussions and fruitful collaborations. CP acknowledges financial support from the Volkswagen Foundation. Most atomistic simulations were carried out using the Gromacs simulation package [54]; CG simulations were mainly performed with the ESPResSo suit of programs [55].

REFERENCES

1. Tschöp, W., Kremer, K., Batoulis, J., Bürger, T., and Hahn, O. 1998. Simulation of polymer melts. I. Coarse-graining procedure for polycarbonates. *Acta Polym*. 49:61–74.
2. Tschöp, W., Kremer, K., Hahn, O., Batoulis, J., and Bürger, T. 1998. Simulation of polymer melts. II. From coarse-grained models back to atomistic description. *Acta Polym*. 49:75–79.
3. Müller-Plathe, F. 2002. Coarse-graining in polymer simulation: From the atomistic to the mesoscopic scale and back. *ChemPhysChem* 3:754–69.
4. Müller, M., Katsov, K., and Schick, M. 2006. Biological and synthetic membranes: What can be learned from a coarse-grained description? *Phys. Rep*. 434:113–76.

5. Ayton, G. S., Noid, W. G., and Voth, G. A. 2007. Multiscale modeling of biomolecular systems: In serial and in parallel. *Curr. Opin. Struct. Biol.* 17:192–98.
6. Kremer, K. 2000. Computer simulations in soft matter science. In *Soft and Fragile Matter, Nonequilibrium Dynamics, Metastability and Flow*, ed. M. E. Cates and M. R. Evans, 145–84. Bristol: Institute of Physics.
7. Baschnagel, J., Binder, K., Doruker, P., Gusev, A. A., Hahn, O., Kremer, K., Mattice, W. L., Müller-Plathe, F., Murat, M., Paul, W., Santos, S., Suter, U. W., and Tries, V. 2000. Bridging the gap between atomistic and coarse-grained models of polymers: Status and perspectives. *Adv. Polym. Sci.* 152:41–156.
8. Müller-Plathe, F. 2003. Scale-hopping in computer simulations of polymers. *Soft Mater.* 1:1–31.
9. Murat, M., and Kremer, K. 1998. From many monomers to many polymers: Soft ellipsoid model for polymer melts and mixtures. *J. Chem. Phys.* 108:4340–48.
10. Bolhuis, P. G., Louis, A. A., Hansen, J. P., and Meijer, E. J. 2001. Accurate effective pair potentials for polymer solutions. *J. Chem. Phys.* 114:4296–311.
11. Abrams, K., and Kremer, K. 2003. Combined coarse-grained and atomistic simulation of liquid bisphenol A-polycarbonate: Liquid packing and intramolecular structure. *Macromolecules* 36:260–67.
12. Hess, B., León, S., Van der Vegt, N., and Kremer, K. 2006. Long time atomistic polymer trajectories from coarse grained simulations: Bisphenol-A polycarbonate. *Soft Mater.* 2:409–14.
13. Harmandaris, V. A., Adhikari, N. P., Van der Vegt, N. F. A., and Kremer, K. 2006. Hierarchical modeling of polystyrene: From atomistic to coarse-grained simulations. *Macromolecules* 39:6708–19.
14. Harmandaris, V. A., Reith, D., Van der Vegt, N. F. A., and Kremer, K. 2007. Comparison between coarse-graining models for polymer systems: Two mapping schemes for polystyrene. *Macromol. Chem. Phys.* 208:2109–20.
15. Peter, C., Delle Site, L., and Kremer, K. 2008. Classical simulations from the atomistic to the mesoscale and back: Coarse graining an azobenzene liquid crystal. *Soft Matter* 4:859–69.
16. Abrams, C. F., and Kremer, K. 2002. Effects of excluded volume and bond length on the dynamics of dense bead-spring polymer melts. *J. Chem. Phys.* 116:3162–65.
17. Weeks, J. D., Chandler, D., and Andersen, H. C. 1971. Role of repulsive forces in determining equilibrium structure of simple liquids. *J. Chem. Phys.* 54:5237–47.
18. Meyer, H., Biermann, O., Faller, R., Reith, D., and Müller-Plathe, F. 2000. Coarse graining of nonbonded inter-particle potentials using automatic simplex optimization to fit structural properties. *J. Chem. Phys.* 113:6264–75.
19. Press, W. H., Teukolsky, S. A., Vetterling, W. T., and Flannery, B. P. 1992. *Numerical Recipes in C. The Art of Scientific Computing*. Cambridge: Cambridge University Press.
20. Lyubartsev, A. P., and Laaksonen, A. 1995. Calculation of effective interaction potentials from radial-distribution functions: A reverse Monte-Carlo approach. *Phys. Rev. E* 52:3730–37.
21. Reith, D., Pütz, M., and Müller-Plathe, F. 2003. Deriving effective mesoscale potentials from atomistic simulations. *J. Comp. Chem.* 24:1624–36.
22. McCoy, J. D. and Curro, J. G. 1998. Mapping of explicit atom onto united atom potentials. *Macromolecules* 31:9362–68.
23. Auhl, R., Everaers, R., Grest, G. S., Kremer, K., and Plimpton, S. J. 2003. Equilibration of long chain polymer melts in computer simulations. *J. Chem. Phys.* 119:12718–28.
24. Santangelo, G., Di Matteo, A., Müller-Plathe F., and Milano, G. 2007. From mesoscale back to atomistic models: A fast reverse-mapping procedure for vinyl polymer chains. *J. Phys. Chem. B* 111:2765–73.
25. Hess, B., Peter, C., Özal, T. A., Van der Vegt, N. F. A. 2008. Fast-growth thermodynamic integration: Calculating excess chemical potentials of additive molecules in polymer microstructures. *Macromolecules* 41:2283–89.
26. Eilhard, J., Zirkel, A., Tschop, W., Hahn, O., Kremer, K., Scharpf, O., Richter, D., and Buchenau, U. 1999. Spatial correlations in polycarbonates: Neutron scattering and simulation. *J. Chem. Phys.* 110:1819–30.
27. Londono, J. D., Habenschuss, A., Curro, J. G., and Rajasekaran, J. J. 1996. Short-range order in some polymer melts from X-ray diffraction. *J. Polym. Sci. B* 34:3055–61.
28. Hahn, O., Mooney, D. A., Müller-Plathe, F., and Kremer, K. 1999. A new mechanism for penetrant diffusion in amorphous polymers: Molecular dynamics simulations of phenol diffusion in bisphenol-A-polycarbonate. *J. Chem. Phys.* 111:6061–68.
29. Schmitz, H. 1999. Computersimulation von positronium-annihilation in polymeren. PhD thesis, University of Mainz, Germany.
30. Tozzini, V., Rocchia, W., and McCammon, J. A. 2006. Mapping all-atom models onto one-bead coarse-grained models: General properties and applications to a minimal polypeptide model. *J. Chem. Theory Comput.* 2:667–73.

31. Kremer, K., and Grest, G. S. 1990. Dynamics of entangled linear polymer melts: A molecular-dynamics simulation. *J. Chem. Phys.* 92:5057–86.
32. Kremer, K. 2006. Polymer dynamics: Long time simulations and topological constraints. In *Computer Simulations in Condensed Matter: From Materials to Chemical Biology*, vol. 2. ed. M. Ferrario, G. Cicotti, and K. Binder, 341–78. Lect. Notes. Phys., vol. 704. Berlin, Heidelberg: Springer.
33. Padding, J. T., and Briels, W. J. 2002. Time and length scales of polymer melts studied by coarse-grained molecular dynamics simulations. *J. Chem. Phys.* 117:925–43.
34. Izvekov, S., and Voth, G. A. 2006. Modeling real dynamics in the coarse-grained representation of condensed phase systems. *J. Chem. Phys.* 125:151101.
35. León, S., Van der Vegt, N., Delle Site, L., and Kremer, K. 2005. Bisphenol A polycarbonate: Entanglement analysis from coarse-grained MD simulations. *Macromolecules* 38:8078–92.
36. Harmandaris, V. A., Adhikari, N. P., Van der Vegt, N. F. A., Kremer, K., Mann, B. A., Voelkel, R., Weiss, H., and Liew, C. 2007. Ethylbenzene diffusion in polystyrene: United atom atomistic/coarse grained simulations and experiments. *Macromolecules* 40:7026–35.
37. Chun, H. M., Padilla, C. E., Chin, D. N., Watanabe, M., Karlov, V. I., Alper, H. E., Soosaar, K., Blair, K. B., Becker, O. M., Caves, L. S. D., Nagle, R., Haney, D. N., and Farmer, B. 2000. MBO(N)D: A multibody method for long-time molecular dynamics simulations. *J. Comput. Chem.* 21:159–84.
38. Malevanets, A., and Kapral, R. 2000. Solute molecular dynamics in a mesoscale solvent. *J. Chem. Phys.* 112:7260–69.
39. Abrams, C. F., Delle Site, L., and Kremer, K. 2003. Dual-resolution coarse-grained simulation of the bisphenol-A-polycarbonate/nickel interface. *Phys. Rev. E* 67:021807.
40. Villa, E., Balaeff, A., Mahadevan, L., and Schulten, K. 2004. Multiscale method for simulating protein-DNA complexes. *Multiscale Model. Simul.* 2:527–53.
41. Delle Site, L., Leon, S., and Kremer, K. 2004. BPA-PC on a Ni(111) surface: The interplay between adsorption energy and conformational entropy for different chain-end modifications. *J. Am. Chem. Soc.* 126:2944–55.
42. Schravendijk, P., Van der Vegt, N., Delle Site, L., and Kremer, K. 2005. Dual-scale modeling of benzene adsorption onto Ni(111) and Au(111) surfaces in explicit water. *Chemphyschem* 6:1866–71.
43. Praprotnik, M., Delle Site, L., and Kremer, K. 2005. Adaptive resolution molecular-dynamics simulation: Changing the degrees of freedom on the fly. *J. Chem. Phys.* 123:224106.
44. Praprotnik, M., Matysiak, S., Delle Site, L., Kremer, K, and Clementi, C. 2007. Adaptive resolution simulation of liquid water. *J. Phys. Condens. Mater* 19:292201.
45. Praprotnik, M., Delle Site, L., and Kremer, K. 2007. A macromolecule in a solvent: Adaptive resolution molecular dynamics simulation. *J. Chem. Phys.* 126:134902.
46. Praprotnik, M., Kremer, K., and Delle Site, L. 2007. Fractional dimensions of phase space variables: A tool for varying the degrees of freedom of a system in a multiscale treatment. *J. Phys. A: Math. Theor.* 40:F281–88.
47. Schravendijk, P., Ghiringhelli, L., Delle Site, L., and Van der Vegt, N. F. A. 2007. Interaction of hydrated amino acids with metal surfaces: A multiscale modeling description. *J. Phys. Chem. C* 111:2631–42.
48. Marrink, S.-J., de Vries, A. H., and Mark, A. E. 2004. Coarse grained model for semiquantitative lipid simulations. *J. Phys. Chem. B* 108:750–60.
49. Izvekov, S., Parrinello, M., Burnham, C. J., and Voth, G. A. 2004. Effective force fields for condensed phase systems from ab initio molecular dynamics simulation: A new method for force-matching. *J. Chem. Phys.* 120:10896–913.
50. Izvekov, S., and Voth, G. A. 2005. A multiscale coarse-graining method for biomolecular systems. *J. Phys. Chem. B* 109:2469–73.
51. Hess, B., Holm, C., and Van der Vegt, N. F. A. 2006. Modeling multibody effects in ionic solutions with a concentration dependent dielectric permittivity. *Phys. Rev. Lett.* 96:147801.
52. Hess, B., Holm, C., and Van der Vegt, N. F. A. 2006. Osmotic coefficients of atomistic NaCl (aq) force fields. *J. Chem. Phys.* 124:164509.
53. Hess, B., and Van der Vegt, N. F. A. 2007. Solvent-averaged potentials for alkali-, earth alkali- and alkylammonium halide aqueous solutions. *J. Chem. Phys.* 127:234508.
54. Van der Spoel, D., Lindahl, E., Hess, B., Groenhof, G., Mark, A. E., and Berendsen, H. J. C. 2005. GROMACS: Fast, flexible, and free. *J. Comput. Chem.* 26:1701–18.
55. Limbach, H.-J., Arnold, A., Mann, B. A., and Holm, C. 2006. ESPResSo: An extensible simulation package for research on soft matter systems. *Comput. Phys. Commun.* 174:704–27.

26 From Atomistic Modeling of Macromolecules Toward Equations of State for Polymer Solutions and Melts: How Important Is the Accurate Description of the Local Structure?

Kurt Binder, Wolfgang Paul, Peter Virnau, and Leonid Yelash
Institut für Physik, Johannes Gutenberg-Universität Mainz

Marcus Müller
Institut für Theoretische Physik, Georg-August-Universität Göttingen

Luis González MacDowell
Departamento de Quimica Fisica, Universidad Compluteuse de Madrid

CONTENTS

26.1 INTRODUCTION

For designing the properties of polymeric materials one often uses multicomponent systems (polymer blends, copolymers of various architectures, etc.), and in the process of making them solvents play a key role. This is particularly true when nucleation processes are considered. Structure formation processes may occur, which start out at the nanometer scale but create nontrivial structures on mesoscopic scales up to 100 μm. A good example of high industrial relevance is the creation of polymeric foam materials (by using polystyrene in supercritical carbon dioxide as a solvent for instance).

Clearly, a detailed theoretical understanding of these processes and the resulting structure–property relationships is a challenging problem, also in its own right, as a problem of the statistical thermodynamics and physical chemistry of condensed matter. Due to the complexity of this problem, any approach exclusively relying on analytical theory will be extremely limited, and developing approaches based on computer simulation is highly desirable. However, due to the range of length scales involved and the multiscale character of the problem, a straightforward chemically realistic all-atom approach is unfeasible. In addition, there is the problem that methods based on classical molecular dynamics need force fields that often contain parameters of doubtful accuracy, in particular with respect to intermolecular nonbonded interactions, which are often modeled in an ad hoc manner by Lennard–Jones parameters fitted to some experimental data. For a recent critical assessment of force fields, see Smith (2005).

In view of these problems, it has been a very attractive and longstanding idea [Baschnagel et al. 1991, 1992; Batoulis et al. 1991; Paul et al. 1991; Paul and Pistoor 1994; Tries et al. 1997; Tschöp et al. 1998a, 1998b; Baschnagel et al. 2000; Hahn, delle Site, and Kremer 2001; Müller-Plathe 2002, 2003; Milano and Müller-Plathe 2005; Theodorou 2006; Bedrov, Ayyagari, and Smith 2006] to provide an explicit connection between a chemically realistic atomistic model and coarse-grained models, which describe only certain degrees of freedom on the mesoscopic scale. In fact, there is a wealth of coarse-grained models, both lattice models such as the simple self-avoiding walk model [Kremer and Binder 1988; Sokal 1995] and the bond fluctuation model [Carmesin and Kremer 1988; Deutsch and Binder 1991; Paul et al. 1991], and off-lattice models such as various types of bead-spring models [Grest and Kremer 1986; Kremer and Grest 1990; Gerroff et al. 1993; Milchev, Paul, Binder 1993; Bennemann et al. 1998; Milchev and Binder 2002]).

While a large variety of simulation methods exists for these models [Baumgärtner 1984, 1992; Binder 1995; Baschnagel, Wittmer, and Meyer 2004; Kotelyanskii and Theodorou 2004], in most cases studies lack any connection to specific systems exhibiting chemical detail, and rather address "universal" properties of polymers [de Gennes 1979].

First attempts to create such a connection have focused on an intramolecular mapping procedure from atomistic models of polycarbonate [Paul et al. 1991] or polyethylene [Baschnagel et al. 1991, 1992; Paul and Pistoor 1994; Tries et al. 1997] to the bond fluctuation model. These studies are based on the idea that $n \approx 3-5$ successive chemical carbon–carbon bonds along the backbone of the chain are mapped into one bond of the bond fluctuation model (recall that the length of the bonds in this model may vary from 2 to $\sqrt{10}$ lattice spacings). The intrachain potentials of the atomistic model (potentials for the lengths of the chemical bonds and the angles between them, as well as the torsional potential) are then used to construct the distribution $P_n(\ell)$ of the length ℓ of an effective segment of the atomistic model containing n bonds, as well as the distribution $P_n(\theta)$ of the angle θ between two such (subsequent) effective segments. These distributions are then used to fit suitable effective potentials $U(\ell)$ and $V(\theta)$ controlling the length ℓ of the bonds in the bond fluctuation model and the angle θ between two such subsequent lattice bonds. In this way it is possible, for instance, to obtain the temperature dependence of the characteristic ratio C_N for polyethylene (see Figure 26.1). In the regime where real polyethylene is chemically stable and hence C_N can be measured, the simulation results are in reasonable agreement with experimental data. For describing the dynamics, one needs to use a measure of the local mobility of the real chain determined by the barriers of the torsional potential to construct a hopping rate for the effective monomers of the lattice model. With the derivation of a time rescaling factor, which relates the time unit of the Monte Carlo simulation (1 Monte Carlo step per effective monomer) to the physical time, a self-consistent coarse-grained description of the statics and dynamics of the considered polymer melt (polyethylene, polycarbonate, etc.) is obtained [Paul et al. 1991; Tries et al. 1997]. Although this approach is surprisingly successful with respect to the prediction of glass transition temperatures [Paul et al. 1991], many problems remain: (i) The lattice structure limits the accuracy with which

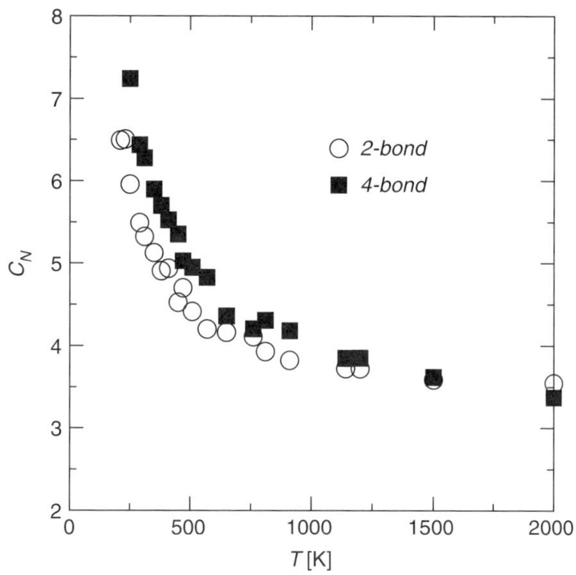

FIGURE 26.1 Characteristic ratio of polyethylene plotted vs. temperature, for $N = 20$ effective monomers. Two versions of the mapping procedure are shown: the two-bond approximation uses properties of two successive lattice bonds for the optimization procedure of the potential, while the four-bond approximation is believed to yield better results, but is more cumbersome to use. (From Tries et al., *J. Chem. Phys.* 1997, 106, 738–48, Copyright American Institute of Physics.)

structural properties can be predicted. (ii) Apart from excluded volume interactions (since each lattice site can be occupied only once) no intermolecular interactions are accounted for, and it is not at all straightforward to include them in a quantitatively meaningful manner. (iii) Due to the use of a discrete lattice model, only the *NVT* ensemble (both the volume V and the particle number N are fixed) can be straightforwardly simulated. However, from the point of view of experiments, a *NpT* ensemble, p being the pressure, would be preferable. (iv) Both the effective interactions and the effective monomeric jump rate are clearly state-dependent (i.e., depend both on temperature T and density $\rho = N/V$).

Clearly, drawbacks (i) and (ii) can be mitigated by using off-lattice bead-spring-type models, onto which a mapping of the atomistic model is performed [Tschöp et al. 1998a, 1998b; Hahn, delle Site, and Kremer 2001; Reith, Meyer, and Müller-Plathe 2001; Müller-Plathe 2002, 2003; Milano and Müller-Plathe 2005]. Typically, these models involve a chain of spherically symmetric effective monomers bound together by stiff springs to model chain connectivity, a purely repulsive intermolecular potential (like the repulsive part of a Lennard–Jones-like potential, see Reith, Meyer, and Müller-Plathe (2001)), and a bond-angle potential. The latter is derived from the atomistic model in a rather direct and elegant way, from the angular distribution of the effective bonds, applying a Boltzmann inversion procedure [Tschöp et al. 1998a, 1998b; Müller-Plathe 2002, 2003]. While it is clearly an advantage that on the level of the coarse-grained model one no longer has to deal with a torsional potential, it must be noted that the angular potential is strongly state dependent and often rather complicated. For example, in the case of poly(vinyl alcohol) studied by Reith, Meyer, and Müller-Plathe (2001) the angular potential has a complicated shape with three minima.

Due to the lack of intermolecular attractive potentials, the models of Tschöp et al. (1998a, 1998b) and Reith, Meyer, and Müller-Plathe (2001) are unsuitable to include solvents. Only in more recent work [Reith, Pütz, and Müller-Plathe 2003; Milano and Müller-Plathe 2005; Bedrov, Ayyagari, and Smith 2006] intermolecular attractive interactions are extracted from Boltzmann inversion procedures as well. However, these effective potentials are strongly state dependent again. In addition it is doubtful to what extent effective interactions that are always assumed to be of a pairwise form are accurate at all. The atomistic foundation of effective potentials for the mesoscale modeling of complex binary fluids is a fundamental problem of statistical mechanics [Silbermann et al. 2006]. For instance, in the case of colloid–polymer mixtures it is well known that even in the framework of very simplified models, such as the Asakura-Oosawa (AO) model where the polymer–polymer interaction is ideal gas-like, integrating out the polymers one creates multibody interactions among the colloids, and not just pairwise interactions, that become important for a polymer to colloid size ratio exceeding about 15% [Dijkstra, Brader, and Evans 1999]. Similar nonpairwise contributions to effective potentials between effective monomers (and solvent molecules) must be expected when one integrates out degrees of freedom of an atomistic model of a polymer plus solvent system as well.

Thus it is clear that the task of systematically integrating out short-wavelength degrees of freedom to construct a coarse-grained model which contains only degrees of freedom on the nanoscale or even mesoscale but nevertheless provides a very accurate description of structure and dynamics is very difficult if at all feasible. Therefore, we pursue a more modest approach in the present chapter: we no longer require our coarse-grained model to accurately describe the local geometric structure of the polymer chains, nor their dynamics faithfully, but we focus on thermodynamic properties. In particular, we ask what is the minimal coarse-grained model for polymer solutions and melts that is required to describe their equation of state with sufficient accuracy?

In fact, the theoretical modeling of the equation of state of polymer solutions, melts, and blends has been a central topic of polymer science since the work of Flory (1941, 1953) and Huggins (1941). It now is well known, however, that the predictive power of these descriptions, which are based on simple lattice models and their generalizations [Sanchez and Lacombe 1978], is somewhat limited [Binder 1994]. In the dilute and semidilute regime, the (osmotic) pressure exhibits universal behavior which can be described by scaling considerations [blob picture, de Gennes 1979] or renormalization group theory [Des Cloizeaux and Jannink 1990]. In this regime, minimal models are well suited to investigate the equation of state and have made significant contributions. In a dense melt, however, the pressure is dictated by the packing of the fluid of segments and the equation of state is expected to sensitively depend on the nonuniversal details of the chemical structure. It is this technically important regime of dense polymer melts that we focus on in the chapter. At present, state-of-the art analytical theories of equation of state of polymeric systems rely mostly on liquid-state theories known as "statistical associating fluid theory" (SAFT) [Chapman et al. 1989] and their various generalizations [see e.g., Müller and Gubbins 2001; Economou 2002 for reviews]. Using a "reference fluid" of unconnected monomers as a starting point, one treats the chain connectivity in the framework of a thermodynamic perturbation theory for chain molecules (TPT1). This perturbative treatment prevents the approach from capturing the power-law dependencies that characterize the semidilute regime, but it is justifiable in a dense melt. Particularly popular is the so-called perturbed chain-SAFT (PC-SAFT) method [Gross and Sadowski 2001, 2002], although it has recently been shown that this approach suffers from artificial multiple criticality in the predicted phase diagrams [Yelash et al. 2005a, 2005b]. It is based upon a hard-chain reference system, with attractive interactions being accounted for by a perturbation approach [Barker and Henderson 1967], and free parameters adjusted to experimental data. However, in view of the problems with PC-SAFT mentioned above [Yelash et al. 2005a, 2005b], an alternative approach [MacDowell et al. 2000, 2002] based on SAFT seems preferable: unlike PC-SAFT, which is based on a repulsive hard-sphere potential (with a temperature-dependent diameter, derived from the potential of Chen and Kreglewski (1977)) augmented by an attractive square well interaction, a Lennard–Jones fluid is utilized as a reference system, which is analytically describable within the mean spherical approximation (MSA). The extension to chain molecules is referred to as TPT1-MSA in the literature [MacDowell et al. 2000, 2002]. It is essentially a liquid-state theory based on the same type of coarse-grained bead-spring models that are commonly used in many computer simulations [Bennemann et al. 1998; Müller and MacDowell 2003; Binder, Baschnagel, and Paul 2003]. However, this model differs from the coarse-grained models resulting from mapping procedures based on atomistic models in one very important aspect: it completely lacks an effective bond-angle potential!

However, the PC-SAFT approach [Gross and Sadowski 2001, 2002] also lacks such a bond-angle potential, and moreover provides a poor description of both intramolecular and intermolecular pair correlation functions between effective monomers, since the steps of the potential lead to corresponding jumps in the correlation functions. This point is exemplified in Figure 26.2 and Figure 26.3, where Monte Carlo simulations of the Lennard–Jones bead-spring chains [Yelash et al. 2006] are compared with corresponding results of Chen-Kreglewski chains, both with 29 beads/molecule, and results obtained from a real coarse-graining of a united-atom (UA) model of polybutadiene [Krushev 2002]. More details on these simulations will be given in Section 26.2. One can see that the Chen-Kreglewski chains provide a rather poor representation of the data derived from the UA model, while the Lennard–Jones chains perform somewhat better. However, it is known from the literature [Gross and Sadowski 2001, 2002] that PC-SAFT does provide a rather

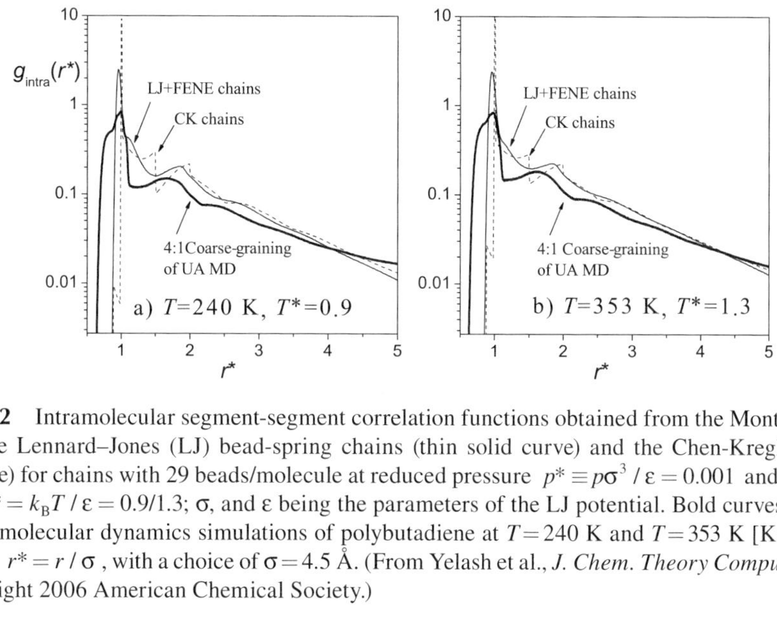

FIGURE 26.2 Intramolecular segment-segment correlation functions obtained from the Monte Carlo simulations of the Lennard–Jones (LJ) bead-spring chains (thin solid curve) and the Chen-Kreglewski chains (dashed curve) for chains with 29 beads/molecule at reduced pressure $p^* \equiv p\sigma^3 / \varepsilon = 0.001$ and reduced temperatures $T^* = k_B T / \varepsilon = 0.9/1.3$; σ, and ε being the parameters of the LJ potential. Bold curves are from the united-atom molecular dynamics simulations of polybutadiene at $T = 240$ K and $T = 353$ K [Krushev 2002]. The distance $r^* = r / \sigma$, with a choice of $\sigma = 4.5$ Å. (From Yelash et al., *J. Chem. Theory Comput.* 2, 588–597, 2006. Copyright 2006 American Chemical Society.)

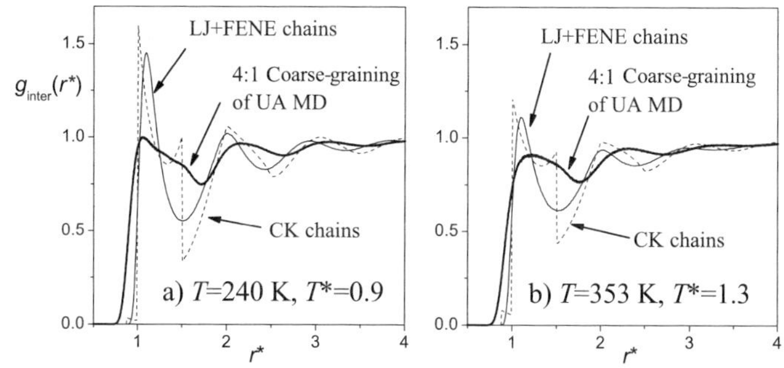

FIGURE 26.3 Intermolecular pair correlation functions obtained from the Monte Carlo simulations of the bead-spring chains (thin solid curves) and Chen-Kreglewski chains (dashed curves) for the same systems as in Figure 26.2. For explanations of the simulated model see Figure 26.2 and Section 26.2. (From Yelash et al., *J. Chem. Theory Comput.* 2, 588–597, 2006. Copyright 2006 American Chemical Society.)

reasonable fit of a large body of equation of state data for a huge variety of polymer melts, solutions, and blends. In fact, in the case referred to in Figure 26.2 and Figure 26.3, one also finds that equation of state data of polybutadiene are described by PC-SAFT by a fit of fair quality (Figure 26.4), though some systematic deviations are noticeable, which arise from a spurious liquid–liquid unmixing predicted to occur by PC-SAFT at higher densities [Yelash et al. 2005a, 2005b], while the fit based on TPT1-MSA is quite perfect.

Figure 26.4 thus suggests that equation of state data of polymeric systems can be described by a simple bead-spring model of the polymer, with state-independent parameters for the intermolecular Lennard–Jones interaction, over a wide range of temperatures and pressures, although the description of both intra- and intermolecular structure provided by the model (Figure 26.2 and Figure 26.3) is only in qualitative accord with the corresponding description based on an atomistic model. Quantitative distinctions can be seen clearly, and with respect to distributions of effective bond angles there is even qualitative disagreement [Yelash et al. 2006]. This can be expected, however, because our model does not include any effective bond-angle potential.

Thus the concept followed in the present chapter is the idea that a much cruder model is sufficient, if the only goal of the modeling is the description of the equation of state at fairly elevated temperatures where the system is fluid, rather than describing structure and dynamics on nanoscopic scales. It is clear that for the latter goal a description in terms of simple potentials that are independent of temperature and pressure over a wide range of these variables cannot be expected: for example, as one can see from Figure 26.1 for alkane melts, the effective chain stiffness depends considerably on temperature. While in the melt the mean-square end-to-end distance $\langle R^2 \rangle$ of a chain with N carbon–carbon bonds along the backbone varies as $\langle R^2 \rangle = C_N \ell_{cc}^2 N$, where $\ell_{cc} \approx 1.53$ Å is the length of a carbon–carbon covalent bond and C_N the characteristic ratio shown in Figure 26.1, a rather different behavior applies for low pressures and densities where the vapor–liquid transition of the alkane chains occurs: in the vapor phase, the chains form collapsed globules for temperatures below the vapor–liquid critical point [de Gennes 1979], while far above the critical point they form swollen coils, with $\langle R^2 \rangle \propto N^{2\nu}$ with $\nu \approx 0.59$. Analogous changes occur in the single-chain structure when we consider the polymer–solvent equilibrium, where below the theta temperature of the solution [Flory 1953; de Gennes 1979] a demixing occurs in a solvent-rich and a polymer-rich phase. Both structure and dynamics of the macromolecules in these various phases that are of interest will depend very much on the thermodynamic state of the system, and there would be little hope to describe the system accurately with state-independent

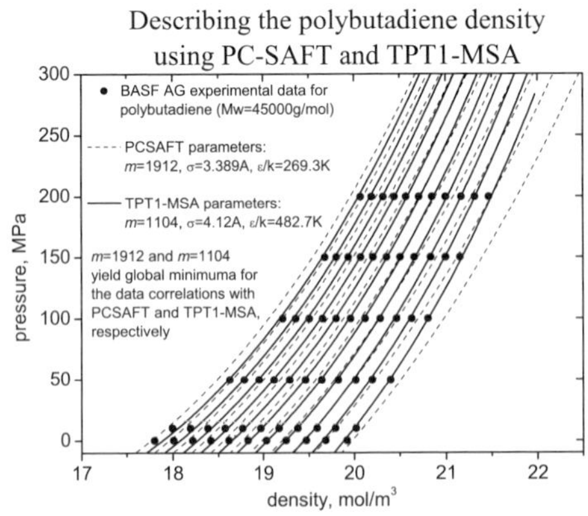

FIGURE 26.4 A comparison between experimental data for polybutadiene melts in the temperature range from 299 to 461 K (symbols) and calculations using PC-SAFT (dashed curves) and TPT1-MSA (solid curves) models. At high pressure, the PC-SAFT calculation predicts a much too large density as a result of the vicinity of the spurious "liquid–liquid" phase separation predicted by PC-SAFT, as discussed in detail by Yelash et al. (2005a, 2005b), from which papers the data for polybutadiene reanalyzed here are taken.

potentials under all these various conditions. However, for many applications this is not necessary, and one just wishes to describe the macroscopic thermodynamic properties of a polymer melt or polymer solution with reasonable accuracy.

In the present chapter, we discuss such a description where the polymer is modeled by a simple bead-spring-type chain, and the solvent is modeled by spherical particles, interacting with each other and the effective monomers of the macromolecule. The effective Lennard–Jones potentials are suitably chosen with state-independent parameters. We suggest that a preferable choice of these parameters is made such that the critical points of the vapor–liquid phase diagrams of the solvent and polymer are correctly reproduced. Then we test to what extent the solution phase diagram can be predicted. This is a very nontrivial test, since in binary fluid mixtures a large variety of phase diagrams can be realized [Scott and van Konynenburg 1970]. In addition, the approach to use the critical points to fix the parameters of the coarse-grained models implies that analytical theories such as the variants of SAFT, including TPT1-MSA, should not be used to fix these parameters: all these theories describe criticality in terms of a mean-field-type approximation, similar to the van der Waals equation. The mean-field character of these theories implies that the extent over which liquid–vapor or liquid–liquid phase separation occurs in the parameter space of the model (temperature T, pressure p, mole fraction x in a binary system) is overestimated significantly (and the shape of the coexistence curves is described by mean-field exponents rather than those of the Ising model universality class; see Binder et al. (2005) for a more detailed discussion of this issue). Thus, it is important to use computer simulation methods for the prediction of the phase diagrams of these coarse-grained models and the resulting adjustment of their parameters to critical point data of the real systems to be modeled.

In the next section we summarize this methodology in more detail, while in the third section we present applications to alkanes and carbon dioxide as a solvent, while the fourth section gives some concluding remarks and an outlook to unsolved problems.

26.2 METHODS

Having in mind that we wish to present solvent particles (such as CO_2 molecules, for instance) as spherical particles, and a macromolecule as a bead-spring chain without bond angle or torsional potentials, the question arises how many carbon atoms along the backbone of the polymer should be integrated into one effective unit of the coarse-grained chain. Of course, there is neither a rigorous nor a general answer to this question. In the mapping of polyethylene to the bond fluctuation model it was found that $n=5$ CH_2 groups was a useful choice [Tries et al. 1997]. However, varying n systematically from $n=2$ to $n=16$ for polybutadiene it was found that $n=4$ was the optimum choice [Yelash et al. 2006]. But with respect to the solvent–polymer mixing thermodynamics, it is also important to roughly preserve the geometrical size ratio between the solvent molecule and the effective polymer segment, which determines the intermolecular packing [Virnau et al. 2002, 2004a]. Having in mind an application to the system hexadecane ($C_{16}H_{34}$) plus CO_2, it was decided that the most plausible choice was to replace the 15 covalent C-C bonds by four effective beads in the bead-spring model; that is, we work with $N=5$ effective beads. This means literally that $n=3.2$ CH_2 groups correspond to one effective segment.

The reader may be bewildered by this choice for n, which is noninteger. However, since we disregard here the geometric structure of the polymer, this is not at all a problem. Note that in analytical models such as PC-SAFT even the number of effective beads N is treated as noninteger in the fitting to experimental data [Gross and Sadowski 2001, 2002]. For Monte Carlo simulations, however, N must be integer, while a non-integer n is no problem at all for the theory.

A comment also deserves to be made on why a short polymer such as $C_{16}H_{34}$ and not a much larger macromolecule was chosen. The answer is that for $C_{16}H_{34}$ experimental data on the properties of the vapor–liquid critical point of the pure polymer are still available. For much longer alkanes, such data do not exist, since the critical temperature T_c would be so high that the polymer is no longer chemically stable. Of course, it is an interesting question to what extent the effective Lennard–Jones parameters extracted for $C_{16}H_{34}$ can be used for a reliable modeling of other alkanes as well. Note that in our description no account is made for the fact that the two chemical end groups (CH_3) differ from the interior chemical monomers (CH_2). We shall return to this important question of the transferability of a coarse-grained model description to a chemically similar system in the last section of this chapter.

For the nonbonded interaction between the effective monomers, we use a truncated and shifted Lennard–Jones potential:

$$V_{LJ}(r) = \begin{cases} 4\varepsilon_{pp}[(\sigma_{pp}/r)^{12} - (\sigma_{pp}/r)^6 + 127/16384], & \text{for } r < r_c, \\ 0, & \text{for } r \geq r_c \end{cases} \tag{26.1}$$

where the cutoff r_c is twice the distance of the potential minimum from the origin, $r_c = 2\cdot\sqrt[6]{2}\sigma_{pp}$. The additive constant in Equation 26.1 is chosen such that $V_{LJ}(r)$ is continuous at r_c.

Effective monomers along a chain also interact with this potential, and in addition are bonded together via FENE (finitely extensible nonlinear elastic) springs [Kremer and Grest 1990]:

$$V_{FENE}(r) = -33.75\varepsilon_{pp}\ln\left[1-(r/R_{pp})^z\right], \tag{26.2}$$

with $R_{pp} = 1.5\sigma_{pp}$. The solvent particles were described in Virnau et al. (2002, 2004b) by exactly the same type of potential as Equation 26.1, but with different parameters, namely σ_{ss} and ε_{ss}.

With current Monte Carlo techniques, which will be briefly characterized below, it is nowadays possible to predict critical temperatures and densities T_c, ρ_c of models such as those introduced above with a relative accuracy of a few parts in a thousand (or better). Thus, ε_{ss}, σ_{ss} have been adjusted such that the experimental T_c and ρ_c of the solvent are reproduced, and ε_{pp}, σ_{pp} are chosen such that the experimental T_c and ρ_c of hexadecane are reproduced. This yields $\sigma_{pp} = 4.52 \times 10^{-10}$ m, $\varepsilon_{pp} = 5.79 \times 10^{-21}$ J, while $\sigma_{ss} = 0.816\sigma_{pp}$ and $\varepsilon_{ss} = 0.726\varepsilon_{pp}$. Given these values, our model for each of these materials no longer exhibits any adjustable parameter whatsoever.

In view of this fact, it is rather remarkable that for both materials a rather good description of phase coexistence simultaneously in the temperature–density plane and in the pressure–temperature plane is obtained (Figure 26.5) [Virnau et al. 2002]. For CO_2, one notes a slight systematic discrepancy on the liquid branch of the coexistence curve in the (T,ρ) plane. This discrepancy is mostly due

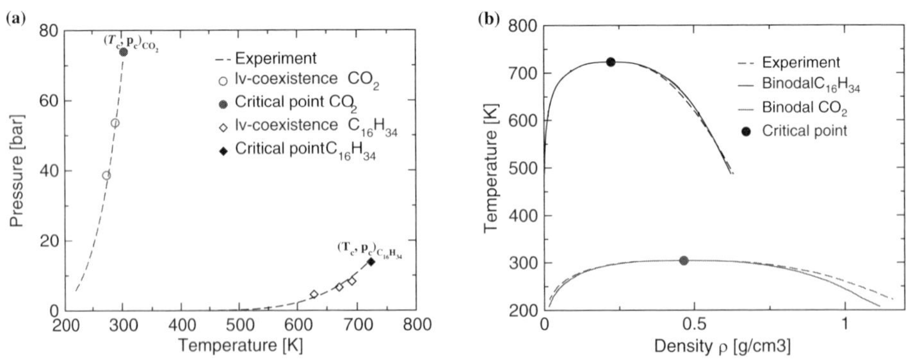

FIGURE 26.5 (a) Phase diagrams of pure CO_2 (lower two curves) and pure $C_{16}H_{34}$ (upper two curves) in the temperature–density plane. (b) Same as (a) but in the pressure–temperature plane. (From Virnau et al., *Comput. Phys. Comm.* 147, 378, 2002. Copyright 2002 Elsevier.)

to the neglect of the quadrupole moment, which is rather large for the CO_2 molecule. If one takes the quadrupole–quadrupole interaction between CO_2 molecules into account, using the experimental value of the quadrupole moment as a further input to the model, the agreement between the model results and experiment is improved significantly [Mognetti et al. 2008]. A similar improvement also occurs with respect to the description of the temperature dependence of the interfacial tension between the coexisting phases [Mognetti et al. 2008]. Note that no further parameter is available to be fitted for the interfacial tension, and hence the fact that it can be predicted so accurately [Virnau et al. 2002, 2004a; Mognetti et al. 2008] is very remarkable.

In the following we summarize the methodic aspects relevant for the construction of phase diagrams such as shown in Figure 26.5 from Monte Carlo simulations.

A key ingredient is the sampling of the density distribution function $P_L(\rho)$ using $L \times L \times L$ boxes with periodic boundary conditions in the grand-canonical μVT ensemble [Virnau et al. 2002, 2004a; Landau and Binder 2005]. Varying the chemical potential μ for $T < T_c$, $P_L(\rho)$ exhibits a single maximum (of approximately Gaussian shape) deep in the one-phase region, but it adopts a double peak shape when μ is close to μ_{coex}, the chemical potential for which two-phase coexistence occurs. When μ varies through μ_{coex}, the weights of the two peaks (one centered near the density ρ_υ of the vapor phase, the other centered near the density ρ_ℓ of the liquid phase) change gradually, and μ_{coex} can actually be located with high precision when the weights of both peaks are equal [Binder and Landau 1984; Borgs and Kotecky 1990].

Getting accurate data for the weights of both peaks of $P_L(\rho)$ for μ near μ_{coex} is not at all straightforward, however, since there is often a pronounced hysteresis since the two states with densities near ρ_υ and ρ_ℓ are separated by a high free-energy barrier in phase space (due to the interfacial free-energy cost of a mixed-phase configuration). This difficulty can be overcome by suitable biased sampling methods, such as "successive umbrella sampling" [Virnau and Müller 2004]. Another difficulty is that the acceptance rate for inserting a particle in a rather dense configuration (a move that is necessary in the grand-canonical ensemble simulation) may be negligibly small. This problem constrains the applicability of the μVT simulation approach to rather short polymer chains and not very low temperatures. Even then the particle insertions and deletions require the implementation of configurational bias Monte Carlo methods [Laso, de Pablo, and Suter 1992; Siepmann and Frenkel 1992; Siepmann, Karaborni, and Smit 1993]. In addition, the chain configurations in between the configurational bias moves are relaxed by local monomer displacements and slithering snake movements [Binder 1995; Kotelyanskii and Theodorou 2004].

From the methods mentioned above, one obtains $\mu_{coex}(T)$ and the associated estimates for the coexisting liquid and vapor densities, $\rho_\upsilon(T)$ and $\rho_\ell(T)$, as well as the coexistence diameter $\rho_d(T) = (\rho_\upsilon(T) + \rho_\ell(T))/2$. It must be stressed, however, that the "naïve" estimates of $\rho_\upsilon(T)$ and $\rho_\ell(T)$ extracted from the peak positions of $P_L(\rho)$ are not at all reliable estimates of bulk behavior near the critical temperature, due to pronounced finite size effects [Landau and Binder 2005]. Applying finite size scaling methods [Binder 1992; Wilding 1996], a reliable extrapolation of such Monte Carlo data for finite box linear dimensions L to the thermodynamic limit ($L \to \infty$) is, however, possible, and such techniques were in fact used by Virnau et al. (2002, 2004a) to obtain the results shown in Figure 26.5.

Having studied both the phase behavior of both the pure solvent and of the pure polymer melt, the next step is the study of the phase behavior of the polymer solution, of course. First of all, the interaction between the solvent molecules and the effective monomers needs to be specified. A simple and widely used approximation relies on the Lorentz–Berthelot mixing rules for the Lennard–Jones parameters ε_{sp}, σ_{sp} for this solvent–polymer mixture [Maitland et al. 1987]:

$$\sigma_{sp} = (\sigma_{ss} + \sigma_{pp})/2\,, \ \varepsilon_{sp} = \sqrt{\varepsilon_{ss}\varepsilon_{pp}}\,. \tag{26.3}$$

Since it is well known that in many cases of interest Equation 26.3 is not accurate enough, a parameter ξ is commonly introduced, describing deviations from the Lorentz–Berthelot mixing rule for the energy parameters:

$$\varepsilon_{sp} = \xi\sqrt{\varepsilon_{ss}\varepsilon_{pp}}\,. \tag{26.4}$$

The simulation in the grand-canonical ensemble then amounts to the variation of two chemical potentials μ_s for the solvent particles and μ_p for the polymers, respectively, and a distribution function involving, correspondingly, two densities ρ_s, ρ_p is recorded $P_L(\rho_s,\rho_p)$. This task is practically feasible when suitable reweighting methods are applied [see Virnau et al. 2002]. A correct description of the order parameter for the mixtures would in principle require a linear combination of the densities of the polymer and the solvent particle. In most cases, however, it is sufficient to consider a single density because one of the two usually exhibits only Gaussian fluctuations. This corresponds to a projection of the joint probability distribution $P_L(\rho_s,\rho_p)$ onto either the polymer or the solvent axis. Methods for determining the probability weight can still be applied with a one-dimensional weight function. Gaussian fluctuations in the second density do not constitute a barrier and need not be considered. Therefore, a single scalar order parameter (e.g., the polymer density) characterizes the phase transition, which then belongs to the Ising model universality class as the pure systems do. Thus, one can apply the same finite-size scaling techniques as for the pure systems.

26.3 APPLICATIONS

In this section, we describe rather briefly the application of the concepts sketched in the previous section to the mixture of CO_2 and $C_{16}H_{34}$. Note that no adjustable parameters whatsoever are any longer available for the models of the pure systems, after we have requested that their vapor–liquid critical temperatures and densities should coincide with their experimental counterparts. However, no a priori information is available on the parameter ξ in Equation 26.4, describing the deviation from the Lorentz–Berthelot mixing rule. Thus, rather arbitrarily three choices were tried: $\xi = 1$, $\xi = 0.9$, and $\xi = 0.886$ [Virnau et al. 2002, 2004a,b].

Figure 26.6a shows the projection of the critical line of the vapor–liquid transition of the mixed system onto the (T,p) plane [Binder et al. 2005]. Along the critical line the molar fraction x of CO_2 quickly rises as $T_c(x)$ decreases from its maximum value $T_c(0)$ for pure hexadecane. For $\xi = 1$ one can clearly see that x monotonously rises to $x = 1$ (Figure 26.6b) and the critical line $p_c(T,x)$ just connects smoothly the critical points of both pure substances. This is the simplest case among all possible scenarios of binary mixture phase diagrams, namely the "type I" diagram in the classification scheme of Scott and van Konynenburg (1970). It is well known, however, that the real hexadecane + carbon dioxide system does not belong to this class, but rather it belongs to "type III" in this classification. This implies that $p_c(T,x)$ does not decrease smoothly toward $p_c(T_c^{CO_2}, x = 1)$ [Schneider et al. 1967] as x increases towards unity. Instead the critical line $p_c(T,x)$ reaches a minimum value at some $x < 1$, and this minimum value p_c^{min} exceeds the critical pressure of pure carbon dioxide. For temperatures less than the associated temperature T_{min} of this minimum the curve $p_c(T,x)$ rises sharply.

It was empirically found [Virnau et al. 2004a] that a choice $\xi = 0.886$ for the parameter that characterizes the deviation from the Lorentz–Berthelot rule corresponds rather clearly to the behavior of the real material. However, we add two caveats: first of all, even for this rather simple system (both CO_2 and $C_{16}H_{34}$ are chemically very stable molecules, cheap and easy to handle in the laboratory) there is still a significant uncertainty about the phase diagram, as the discrepancy between the data reported by Schneider et al. (1967) and by Amon, Martin, and Kobayashi (1986), that we have included in Figure 26.6a, shows. This scarcity of accurate experimental data on the phase behavior of polymer solutions as a function of temperature, pressure, and molar fraction of solvent is an

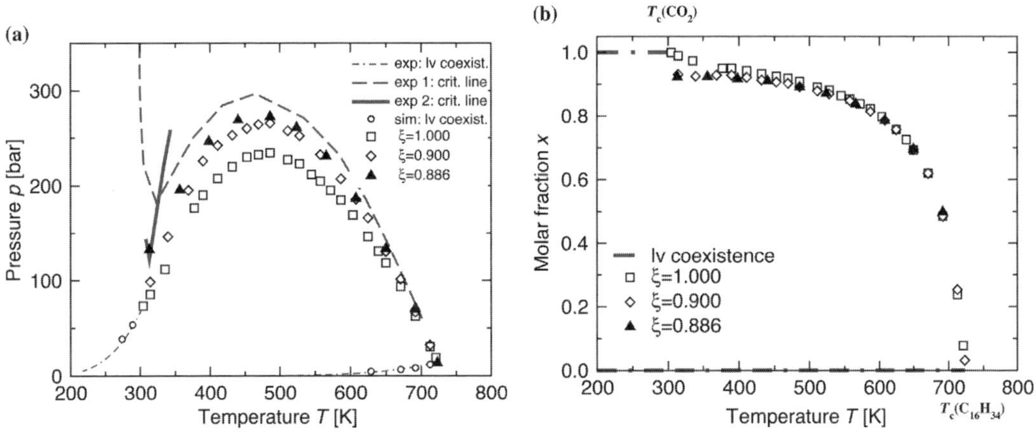

FIGURE 26.6 (a) Phase diagram of the model for the hexadecane-carbon dioxide mixture as a function of temperature and pressure for three different trial values of the parameter ξ. Squares correspond to $\xi = 1$, diamonds to $\xi = 0.9$, and triangles to $\xi = 0.886$. The simulation results for the liquid-vapor coexistence of the pure components are shown by circles. Thick lines mark two experimental observations of the critical lines in hexadecane and CO_2 from Schneider et al. (1967) and Amon et al. (1986), respectively (from Binder et al. (2005)). (b) Molar fraction x of CO_2 along the critical line plotted as function of the critical temperature, for the same systems as in (a). (From Binder et al., *Adv. Polym. Sci.* 173, 1–110, 2005. Copyright 2005 Springer.)

even more acute problem for less common materials, of course (in particular for solvents which are highly poisonous or chemically reactive or even explosive). Secondly, the physical significance of the parameter ξ is open to doubt; its existence has no first-principles theoretical justification at all. The need to use such a parameter ξ may rather indicate that the description of the pure materials may be too crude in certain respects. Indeed, including quadrupolar interactions in the description of carbon dioxide not only gives a much more accurate account of the properties of pure CO_2 but also seems to provide a significant improvement of the description of the mixture behavior. Mognetti et al. (2008) demonstrated that such a model with $\xi = 1$ yields a phase diagram that almost coincides with results such as those shown in Figure 26.6 for $\xi = 0.9$. Clearly, it would be a significant improvement of the theoretical modeling of mixture phase behavior if Equation 26.3 would hold strictly, and no need to fit such a ξ-parameter would arise.

Of course, the theoretical modeling is not at all restricted to a prediction of the vapor–liquid and liquid–liquid demixing critical lines, but one can also study two-phase coexistence very nicely. As an example, Figure 26.7 presents an isothermal slice of the phase diagram at $T = 486$ K [Virnau et al. 2004b]. Here, corresponding results from the TPT1-MSA approach are included (assuming exactly the same interactions). One sees that the coexistence curves are in very good agreement, apart from the (expected) discrepancies close to the critical point. Even three-phase coexistence along the triple line where solvent vapor plus solvent liquid plus a dense polymer-rich phase coexist could be studied (in the $\rho_s - \rho_p$ plane then three peaks grow, corresponding to the three coexisting phases; see Virnau et al. (2004a)). Thus, the simulations of such coarse-grained models can predict their phase behavior in impressive detail.

26.4 CONCLUDING REMARKS

In this chapter we have discussed an approach devoted to deriving a coarse-grained model of polymer plus solvent systems which is able to describe the equation of state of these systems with reasonable accuracy, even though no attempt is made to reproduce intra- and intermolecular correlations reliably. Note that this endeavor is a formidable task, since the interference of liquid–vapor and liquid–liquid phase separation in these systems leads to a very rich variety of phase diagrams in the

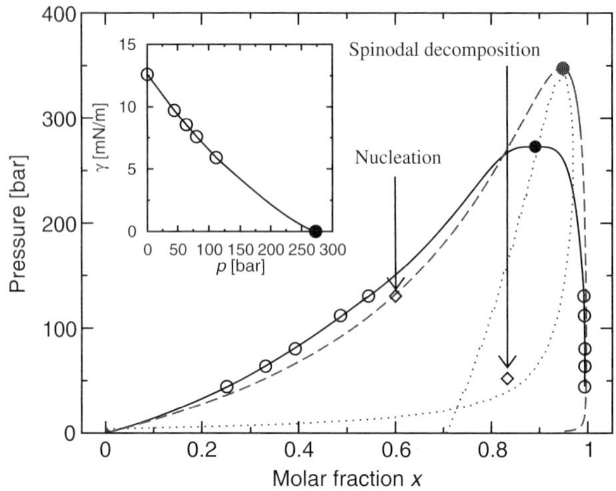

FIGURE 26.7 Isothermal slice of the phase diagram of CO_2–$C_{16}H_{34}$ at $T = 486$ K as obtained from Monte Carlo simulation (thick solid line and open symbols) and the TPT1-MSA approach (long-dashed line). The spinodals obtained from the TPT1-MSA equation of state are indicated as short-dashed lines. The arrows indicate the study of possible pressure quench experiments. The inset presents the interfacial tension between the coexisting phases as a function of pressure. (From Virnau et al., *New J. Phys.* 6, 7, 2004. Copyright 2004 Institute of Physics.)

space of the three relevant thermodynamic control parameters: temperature, pressure, and molar fraction. Using the example of the system hexadecane plus carbon dioxide solvent as a test case, and implementing the idea to fix Lennard–Jones parameters of the pure materials in terms of their critical temperatures and densities, a surprisingly accurate description of surprisingly many physical quantities of interest (coexistence curves, associated pressure at phase coexistence in the *pT*-plane, interface tension between coexisting phases) is obtained. Unfortunately, it is less clear how one should determine the exact interaction potential between the polymer and the solvent. The simple Lorentz–Berthelot mixing rule does not seem to be accurate enough. However, with a slight modification of this mixing rule also a rich variety of useful predictions for the full binary system can be obtained. Note that in spite of the fact that short alkanes at low temperatures are rather stiff, with a persistence length (manifested in a characteristic ratio C_N much larger than one) that distinctly grows as the temperature is lowered, we have used a fully flexible bead-spring model (similar to the way the common analytical equations of state such as PC-SAFT and TPT1-MSA, etc. do, although some of these analytical methods suffer from other problems). This observation leads to one of the main messages of this chapter, namely the suggestion that for a description of the equation of state of polymer plus solvent systems the variable local stiffness of the polymer chains is less important. To a first approximation bond angle potentials for the coarse-grained models can be disregarded. As a consequence, an accurate description of local intra- and intermolecular structure of the polymer solution or melt is no longer obtained. However, this does not seem to matter too much for the equation of state.

Of course, one should not overemphasize this conclusion: when one deals with rather stiff short chains, the possibility of nematic order in the polymer solution arises, and this new phase changes the phase diagram significantly. Such nematic order in polymer solutions is clearly beyond the realm of the present model. Thus, it would be very interesting to extend the present approach by including a bond-angle potential and apply it to such a solution of stiff chains. Then one could also make contact with the traditional mapping approaches, where via Boltzmann inversion from an atomistic model a bond angle potential on the coarse-grained scales inevitably comes into play.

Another important extension is the inclusion of electrostatic interactions. While it is clearly a long way to go from the present approach where the solvent molecules are described as Lennard–Jones-type point particles towards systems, such as biopolymers or synthetic polyelectrolytes in aqueous solution, a very desirable first step is the inclusion of dipole or quadrupole moments of the molecules. Current work has shown [Mognetti et al. 2008] that even for carbon dioxide the consideration of the quadrupolar interactions leads to a very significant improvement in agreement between the model results and the experimental data. Considering mixtures, deviations from the Lorentz–Berthelot rule are much reduced. Thus, the work reviewed in this chapter is only a small first step. The trend, however, is promising.

ACKNOWLEDGMENTS

Early stages of the research reviewed here were supported by the German Federal Ministry of Education and Research (BMBF), Bayer AG, and BASF AG. We thank J. Baschnagel, K. Kremer, and F. Müller-Plathe for many useful discussions, and V. Tries for the fruitful collaboration that led to Figure 26.1.

REFERENCES

Amon, C. T., Martin, R. J., and Kobayashi, R. 1986. Application of a generalized multiproperty apparatus to measure phase equilibrium and vapor phase densities of supercritical carbon dioxide in n-hexadecane systems up to 26 MPa. *Fluid Phase Equil.* 31:89–104.

Barker, J. A., and Henderson, D. 1967. Perturbation theory and equation of state for fluids. II. A successful theory of liquids. *J. Chem. Phys.* 47:2856–61.

Baschnagel, J., Binder, K., Doruker, P., Gusev, A. A., Hahn, O., Kremer, K., Mattice, W. L., Müller-Plathe, F., Murat, M., Paul, R., Santos, S., Suter, U. W., and Tries, V. 2000. Bridging the gap between atomistic and coarse-grained models of polymers: Status and perspectives. *Adv. Polym. Sci.* 152:41–156.

Baschnagel, J., Binder, K., Paul, W., Laso, M., Suter, U. W., Batoulis, I., Jilge, W., and Bürger, T. 1991. On the construction of coarse-grained models for linear flexible polymer chains: Distribution functions for groups of consecutive monomers. *J. Chem. Phys.* 95:6014–25.

Baschnagel, J., Qin, K., Paul, W., and Binder, K. 1992. Monte Carlo simulation of models for single polyethylene coils. *Macromolecules* 25:3117–24.

Baschnagel, J., Wittmer, J. P., and Meyer, H. 2004. Monte Carlo simulation of polymers: Coarse-grained models. In *Computational Soft Matter: From Synthetic Polymers to Proteins*, ed. N. Attig, K. Binder, H. Grubmüller, and K. Kremer, 83–140. Juelich: John von Neumann Institute for Computing (NIC).

Batoulis, J., Binder, K., Gentile, F. T., Heermann, D. W., Jilge, W., Kremer, K., Laso, M., Ludovice, P. J., Morbitzer, L., Paul, W., Pittel, B., Plaetschke, R., Reuter, K., Sommer, K., Suter, U. W., Timmermann, R., and Weymans, G. 1991. Correlation between primary chemical structure and property phenomena in polycondensates. *Adv. Mater.* 3:590–99.

Baumgärtner, A. 1984. Simulations of polymer models. In *Applications of the Monte Carlo Method in Statistical Physics*. ed. K. Binder, 145–80. Berlin: Springer.

———. 1992. Simulations of macromolecules. In *The Monte Carlo Method in Condensed Matter Physics*, ed. K. Binder, 285–316. Berlin: Springer.

Bedrov, D., Ayyagari, C., and Smith, G. D. 2006. Multiscale modeling of poly(ethylene oxide)-poly(propylene oxide)-poly(ethylene oxide) triblock copolymer micelles in aqueous solution. *J. Chem. Theory Comput.* 2:598–606.

Bennemann, C., Paul, W., Binder, K., and Dünweg, B. 1998. Molecular-dynamics simulations of the thermal glass transition in polymer melts: α-Relaxation behavior. *Phys. Rev. E* 57:843–57.

Binder, K. 1992. In *Computational Methods in Field Theory*, ed. C. B. Lang and H. Gausterer, 59–125. Berlin: Springer.

———. 1994. Phase transitions in polymer blends and block-copolymer melts: Some recent developments. *Adv. Polymer Sci.* 112:181–99.

———. 1995. ed. *Monte Carlo and Molecular Dynamics Simulations in Polymer Science*. New York: Oxford University Press.

Binder, K., Baschnagel, J., and Paul, W. 2003. Glass transition of polymer melts: Test of theoretical concepts by computer simulation. *Prog. Polym. Sci.* 18:115–72.

Binder, K., and Landau, D. P. 1984. Finite-size scaling at first-order phase transitions. *Phys. Rev. B* 30:1477–85.

Binder, K., Müller, M., Virnau, P., and MacDowell, L. G. 2005. Polymer+solvent systems: Phase diagrams, interface free energies, and nucleation. *Adv. Polymer Sci.* 173:1–110.

Borgs, C., and Kotecky, R. 1990. A rigorous theory of finite-size scaling at first-order phase transitions. *J. Stat. Phys.* 61:79–119.

Carmesin, I., and Kremer, K. 1988. The bond fluctuation method: A new effective algorithm for the dynamics of polymers in all spatial dimensions. *Macromolecules* 21:2819–23.

Chapman, W. G., Gubbins, K. E., Jackson, G., and Radosz, M. 1989. SAFT: Equation-of-state solution model for associating fluids. *Fluid Phase Equilibria* 52:31–38.

Chen, S. S., and Kreglewski, A. 1977. Applications of the augmented van der Waals theory of fluids. I. Pure fluids. *Ber. Bunsenges. Phys. Chem.* 81:1048–49.

Des Cloizeaux, J., and Jannink, G. 1990. *Polymers in Solution: Their Modeling and Structure.* Oxford: Oxford University Press.

Deutsch, H.-P., and Binder, K. 1991. Interdiffusion and self-diffusion in polymer mixtures: A Monte Carlo study. *J. Chem. Phys.* 94:2292–2304.

Dijkstra, M., Brader, J. M., and Evans, R. 1999. Phase behaviour and structure of model colloid–polymer mixtures. *J. Phys.: Condens. Matter* 11:10079–106.

Economou, I. G. 2002. Statistical associating fluid theory: A successful model for the calculation of thermodynamic and phase equilibrium properties of complex fluid mixtures. *Ind. Eng. Chem. Res.* 41:953–62.

Flory, P. J. 1941. Thermodynamics of high polymer solutions. *J. Chem. Phys.* 9:660–61.

———. 1953. *Principles of Polymer Chemistry.* Ithaca: Cornell University Press.

de Gennes, P. G. 1979. *Scaling Concepts in Polymer Physics.* Ithaca: Cornell University Press.

Gerroff, I., Milchev, A., Binder, K., and Paul, W. 1993. A new off-lattice Monte Carlo model for polymers: A comparison of static and dynamic properties with the bond-fluctuation model and application to random media. *J. Chem. Phys.* 98:6526–39.

Grest, G. S., and Kremer, K. 1986. Molecular dynamics simulation for polymers in the presence of a heat bath. *Phys. Rev. A* 33:3628–31.

Gross, J., and Sadowski, G. 2001. Perturbed-chain SAFT: An equation of state based on a perturbation theory for chain molecules. *Ind. Eng. Chem. Res.* 40:1244–60.

———. 2002. Modeling polymer systems using the perturbed-chain statistical associating fluid theory equation of state. *Ind. Eng. Chem. Res.* 41:1084–93.

Hahn, O., delle Site, L., and Kremer, K. 2001. Simulation of polymer melts: From spherical to ellipsoidal beads. *Macromol. Theory Simul.* 10:288–303.

Huggins, M. J. 1941. Solutions of long chain compounds. *J. Chem. Phys.* 9:440.

Kotelyanskii, M. J., and Theodorou, D. Y. eds. 2004. *Simulation Method for Polymers.* New York: Marcel Dekker.

Kremer, K., and Binder, K. 1988. Monte Carlo simulation of lattice models for macromolecules. *Computer Phys. Rep.* 7:259–310.

Kremer, K., and Grest, G. S. 1990. Dynamics of entangled linear polymer melts: A molecular-dynamics simulation. *J. Chem. Phys.* 92:5057–86.

Krushev, S. 2002. Computersimulationen zur Dynamik und Statik von Polybutadienschmelzen. Ph.D. Thesis (unpublished), Universität Mainz.

Landau, D. P., and Binder, K. 2005. *A Guide to Monte Carlo Simulation in Statistical Physics.* 2nd ed. Cambridge: Cambridge University Press.

Laso, M., de Pablo, J. J., and Suter, U. W. 1992. Simulation of phase equilibria for chain molecules. *J. Chem. Phys.* 97:2817–19.

MacDowell, L. G., Müller, M., Vega, C., and Binder, K. 2000. Equation of state and critical behavior of polymer models: A quantitative comparison between Wertheim's thermodynamic perturbation theory and computer simulations. *J. Chem. Phys.* 113:419–33.

MacDowell, L. G., Virnau, P., Müller, M., and Binder, K. 2002. Critical lines and phase coexistence of polymer solutions: A quantitative comparison between Wertheim's thermodynamic perturbation theory and computer simulations. *J. Chem. Phys.* 117:6360–71.

Maitland, G. C., Rigby, M., Smith, E. B., and Wakeham, W. A. 1987. *Intermolecular Forces.* Oxford: Clarendon.

Milano, G., and Müller-Plathe, F. 2005. Mapping atomistic simulations to mesoscopic models: A systematic coarse-graining procedure for vinyl polymer chains. *J. Phys. Chem. B* 109:18609–19.
Milchev, A., and Binder, K. 2002. Off-lattice Monte Carlo methods for coarse-grained models of polymeric materials and selected applications. *J. Computer-Aided Mater. Des.* 9:33–74.
Milchev, A., Paul, W., Binder, K. 1993. Off-lattice Monte Carlo simulation of dilute and concentrated polymer solutions under theta conditions. *J. Chem. Phys.* 99:4786–98.
Mognetti, B. M., Yelash, L., Virnau, P., Paul, W., Binder, K., Müller, M., MacDowell, L. G. (2008). Efficient prediction of thermodynamic properties of quadrupolar fluids from simulation of a coarse-grained model: The case of carbon dioxide. *J. Chem. Phys.* 128:104501, 1–13.
Müller, E. A., and Gubbins, K. E. 2001. Molecular based equations of state for associating fluids: A review of SAFT and related approaches. *Ind. Eng. Chem. Res.* 40:2198–2211.
Müller, M., and MacDowell, L. G. 2003. Wetting of polymer liquids: Monte Carlo simulations and self-consistent field calculations. *J. Phys.: Condens. Matter* 15:R609–53.
Müller-Plathe, F. 2002. Coarse-graining in polymer simulation: From the atomistic to the mesoscopic scale and back. *Chem. Phys. Chem.* 3:754–69.
———. 2003. Scale-hopping in computer simulations of polymers. *Soft Mater.* 1:1–31.
Paul, W., Binder, K., Heermann, D. W., and Kremer, K. 1991. Crossover scaling in semidilute polymer solutions: A Monte Carlo test. *J. Phys. (Paris) II* 1:37–60.
Paul, W., Binder, K., Kremer, K., and Heermann, D. W. 1991. Structure–property correlation of polymers, a Monte Carlo approach. *Macromolecules* 24:6332–34.
Paul, W., and Pistoor, N. 1994. A mapping of realistic onto abstract polymer models and an application to two bisphenol polycarbonates. *Macromolecules* 27:1249–55.
Reith, D., Meyer, H., and Müller-Plathe, F. 2001. Mapping atomistic to coarse-grained polymer models using automatic simplex optimization to fit structural properties. *Macromolecules* 34:2335–45.
Reith, D., Pütz, M., and Müller-Plathe, F. 2003. Deriving effective mesoscale potentials from atomistic simulations. *J. Comput. Chem.* 24:1624–36.
Sanchez, I. C., and Lacombe, R. H. 1978. Statistical thermodynamics of polymer solutions. *Macromolecules* 11:1145–56.
Schneider, G., Alwani, Z., Heim, W., Horvath, E., and Franck, E. U. 1967. Phase equilibriums and critical phenomena in binary mixed systems to 1500 bars. Carbon dioxide with n-octane, n-undecane, n-tridecane, and n-hexadecane. *Chem. Ingr. Tech.* 39:649–56.
Scott, R. L., and van Konynenburg, P. H. 1970. Van der Waals and related models for hydrocarbon mixtures. *Discuss. Faraday Soc.* 49:87–97.
Siepmann, J. I., and Frenkel, D. 1992. Configurational bias Monte Carlo: A new sampling scheme for flexible chains. *Mol. Phys.* 75:59–70.
Siepmann, J. I., Karaborni, S., and Smit, B. 1993. Vapor–liquid equilibria of model alkanes. *J. Am. Chem. Soc.* 115:6454–55.
Silbermann, J. R. Klapp, S. H. K., Schoen, M., Channamsetty, N., Bock, H., and Gubbins, K. E. 2006. Mesoscale modeling of complex binary fluid mixtures: Towards an atomistic foundation of effective potentials. *J. Chem. Phys.* 124:074105.
Smith, G. D. 2005. Atomistic potentials for polymers and organic materials. In *Handbook of Materials Modeling*, ed. S. Yip, 2561–71. Berlin: Springer.
Sokal, A. D. 1995. Monte Carlo methods for the self-avoiding walk. In *Monte Carlo and Molecular Dynamics Simulations in Polymer Science*, ed. K. Binder, 47–124. New York: Oxford University Press.
Theodorou, D. N. 2006. Equilibration and coarse-graining methods for polymers. In *Computer Simulations in Condensed Matter: From Materials to Chemical Biology*, vol. 2. ed. F. Ferrario, G. Ciccotti, and K. Binder, 419–48. Berlin: Springer
Tries, V., Paul, W., Baschnagel, J., and Binder, K. 1997. Modeling polyethylene with the bond fluctuation model. *J. Chem. Phys.* 106:738–48.
Tschöp, W., Kremer, K., Batoulis, J., Bürger, T., and Hahn, O. 1998a. Simulation of polymer melts I: Coarse graining procedure for polycarbonates. *Acta Polym.* 49:61–74.
———. 1998b. Simulation of polymer melts {II}: From coarse grained models back to atomistic description. *Acta Polym.* 49:75–79.
Virnau, P., and Müller, M. 2004. Calculation of free energy through successive umbrella sampling. *J. Chem. Phys.* 120:10925–30.
Virnau, P., Müller, M., MacDowell, L. G., and Binder, K. 2002. Phase diagrams of hexadecane–CO_2 mixtures from histogram-reweighting Monte Carlo. *Comput. Phys. Comm.* 147:378–81.

———. 2004a. Phase behavior of *n*-alkanes in supercritical solution: A Monte Carlo study. *J. Chem. Phys.* 121:2169–79.

———. 2004b. Phase separation kinetics in compressible polymer solutions: Computer simulation of the early stages. *New J. Phys.* 6:7.

Wilding, N. B. 1996. Critical phenomena in simple and complex fluids. In *Annual Reviews of Computational Physics*, vol. 4, ed. D. Stauffer, 37–73. Singapore: World Scientific.

Yelash, L., Müller, M., Paul, W., and Binder, K. 2005a. Artificial multiple criticality and phase equilibria: An investigation of the PC-SAFT approach. *Phys. Chem. Chem. Phys.* 7:3728–32.

———. 2005b. A global investigation of phase equilibria using the perturbed-chain statistical-associating-fluid-theory approach. *J. Chem. Phys.* 123:014908.

———. 2006. How well can coarse-grained models of real polymers describe their structure? The case of polybutadiene. *J. Chem. Theory Comput.* 2:588–97.

27 Effective Interaction Potentials for Coarse-Grained Simulations of Polymer-Tethered Nanoparticle Self-Assembly in Solution

Elaine R. Chan
Semiconductor Electronics Division,
Electronics and Electrical Engineering Laboratory,
National Institute of Standards and Technology

Alberto Striolo
School of Chemical, Biological and Materials Engineering, The University of Oklahoma

Clare McCabe, Peter T. Cummings
Department of Chemical Engineering, Vanderbilt University

Sharon C. Glotzer
Department of Chemical Engineering and Department of Materials Science and Engineering, University of Michigan

CONTENTS

27.1 INTRODUCTION

Self-assembly is a highly promising route for constructing new and enhanced nanoparticle-based materials and devices with unique properties. However, fabrication of these nanoscale materials and devices requires knowledge of the processes that occur during self-assembly at the relevant length and time scales. Theory and simulation are useful tools for probing self-assembly in nanoscale systems because they allow access to pertinent length and time scales and enable exploration of the vast parameter space efficiently and systematically. The development and application of multiscale modeling and simulation techniques are increasingly desirable for investigating assemblies of molecular nanoparticles having various geometries and/or functionalized with appropriate substituents.

Polyhedral oligomeric silsesquioxane (POSS) molecules with the formula $(RSiO_{1.5})_8$ [Lichtenhan 1995] is one example of such nanoparticles. These molecules resemble cubes with silicon atoms at the corners and oxygen atoms interspersed between them (Figure 27.1). The silicon atoms can be functionalized with nonreactive organic substituents R to render the molecules compatible with

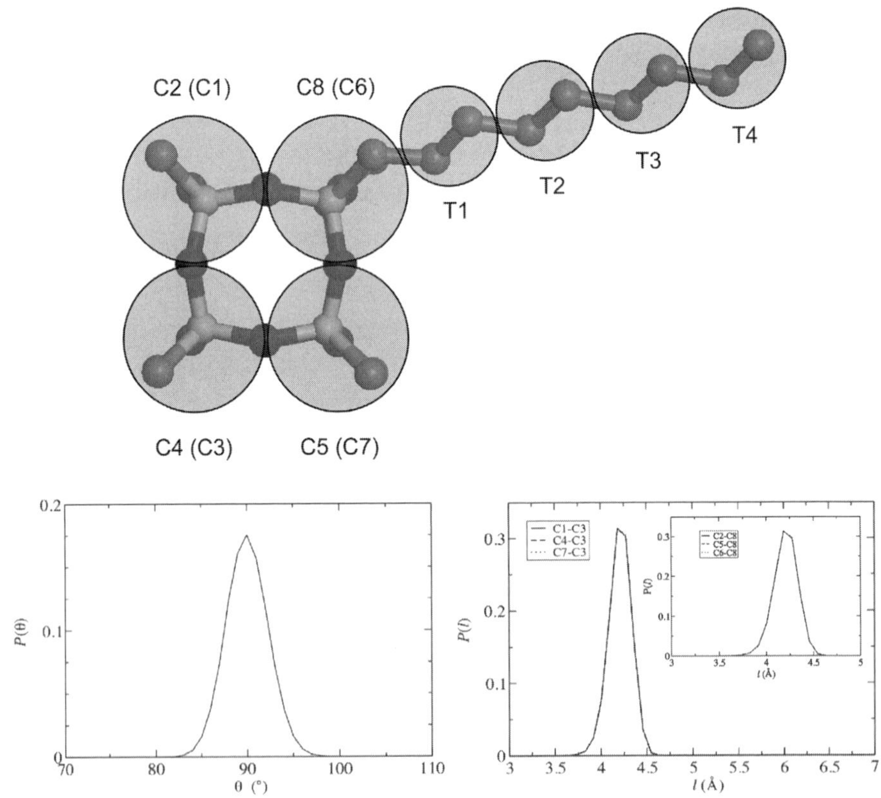

FIGURE 27.1 (Top) Mapping of the CG tethered POSS molecule onto its atomistic counterpart. CG bead labels in parentheses denote beads in the background (not shown). (Bottom) C3–C7–C5 bond angle probability distribution (left) and bond length probability distributions for CG cube bead pairs (right) computed from AA simulations at $T = 400$ K.

polymers and surfaces, or with reactive functional groups R that provide sites for polymerization, grafting, and surface bonding. POSS molecules are therefore attractive candidates for engineering precursor structures or assemblies to construct hybrid organic/inorganic nanostructured materials with enhanced properties. In particular, previous experiments have demonstrated that POSS molecules functionalized with polymer tethers can be synthesized, and that POSS/polymer pendant copolymers self-assemble into lamellar, cylindrical, and micellar structures in solution or melt states [Knischka et al. 1999; Kim and Mather 2002; Kim, Keum, and Chujo 2003; Cardoen and Coughlin 2004]. In conjunction with these experiments, molecular simulations have been performed to predict the types of structures that can arise from self-assembly of polymer-tethered POSS in solution when concentration and temperature are varied [Chan et al. 2005; Zhang, Chan, and Glotzer 2005; Chan, Ho, and Glotzer 2006]. These simulations utilized a minimal model of tethered POSS that was developed on the basis of structural and energetic insights from quantum mechanical calculations. To investigate self-assembly phenomena at the mesoscale, hundreds and thousands of minimal model molecules were considered simultaneously. Such simulations are presently computationally prohibitive at the explicit atom level because they involve hundreds of thousands of atoms. The inclusion of atomistic detail limits the possible simulation times compared to that achievable in mesoscale simulations, and thus self-assembled structures that may form on longer time scales may not be observed. Despite these limitations, progress has been made [McCabe et al. 2004]. It has been demonstrated that standard force fields are sufficiently accurate to describe systems of POSS monomers at the explicit atom level [Ionescu et al. 2006; Li et al. 2007]. Detailed all-atom (AA) molecular dynamics simulations have been conducted for POSS monomers dissolved in common organic solvents and provide insights on effective POSS–POSS interactions in solution under varying temperatures and solvent compositions [Striolo, McCabe, and Cummings 2005a, 2005b; Striolo et al. 2007]. Other groups have also reported additional AA simulation studies of systems containing POSS monomers [Bharadwaj, Berry, and Farmer 2000; Capaldi, Rutledge, and Boyce 2005; Capaldi, Boyce, and Rutledge 2006; Patel, Mohanraj, and Pittman 2006; Qi, Durandurdu, and Kieffer 2007; Zhou and Kieffer 2007; Zhou et al. 2007]. However, it remains generally unclear how to relate the parameters in minimal models of POSS monomers to the properties of these systems obtained from AA simulations.

To accurately examine self-assembly of POSS monomers into bulk structures at long length and time scales, it is necessary to develop mapping schemes that relate coarse-grained (CG) models to their underlying AA representations. Presented herein is the development of a CG force field for accurately simulating monotethered POSS molecule self-assembly in an organic solvent. The force field consists of effective solvent-mediated interaction potentials that implicitly account for POSS-solvent molecule interactions. Hence, the solvent molecules do not need to be explicitly accounted for in the CG simulations, resulting in a reduced number of particles. Our effort builds upon recent results obtained for systems of linear molecules such as polymer melts [Müller-Plathe 2002; Ashbaugh et al. 2005; Milano and Müller-Plathe 2005] and phospholipids in water [Shelley et al. 2001; Lyubartsev 2005]. We extend those methods here to coarse-grain cubic molecules such as POSS monomers.

Coarse-graining approaches aim to improve the computational efficiency of a simulation by reducing the number of degrees of freedom in the system in a systematic fashion [Baschnagel et al. 2000; Kremer and Müller-Plathe 2001; Glotzer and Paul 2002; Kremer and Müller-Plathe 2002; Müller-Plathe 2002, 2003; Nielsen et al. 2004; Lu and Kaxiras 2005]. These methods reduce the central processing unit (CPU) time by two to four orders of magnitude compared to the corresponding AA simulations. Currently, CG methodologies typically involve two steps: (1) mapping a detailed atomistic or molecular representation onto a CG representation, and (2) deriving the equivalent CG interaction potentials. The approach utilized in this work is to map specific groups of atoms onto CG particles and derive CG numerical effective potentials that sufficiently reproduce at the mesocale structural properties observed in the AA simulations. The mapping scheme preserves important molecular details, such as connectivity, in the CG representation as well as relevant physical properties, such as intermolecular packing, which should be captured in the mesoscale simulations. With regard to the development of the effective potentials, two methodologies are often employed, namely, analytical potentials with tunable parameters or numerical potentials in tabulated form. Although analytical potentials are desirable because they can be parameterized according to experimental data or quantum mechanical calculations, the processes available to obtain the correct parameter values can be time-consuming, and in some cases, the data necessary for parameterization are unavailable. Hence, most current CG models utilize solely numerical potentials or combinations of numerical and simple analytical potentials to describe complex interactions.

CG numerical potentials can be derived by requiring that the mesoscale simulations reproduce specific intra- and intermolecular probability distribution functions computed from the underlying AA simulations [Lyubartsev and Laaksonen 1995; Soper 1996; Tschöp et al. 1998; Eilhard et al. 1999; Lyubartsev et al. 2003; Reith, Putz, and Müller-Plathe 2003; Lyubartsev 2005]. These structural-based coarse-graining schemes require iterative numerical methods and are attractive because they can be automated. However, one caveat of the method is that the resulting effective potentials lack transferability across thermodynamic state space, as the CG Hamiltonians are only parameterized to reproduce structural correlations correctly [Ashbaugh et al. 2005]. It has been suggested that such transferability could be obtained if the effect of enthalpy and entropy are decoupled and the CG force fields account for the decoupling [Baron et al. 2006, 2007]. Another drawback is the nonuniqueness of the derived effective potentials; that is, different effective potentials exist that can each reproduce the target distribution functions from the AA simulations.

The coarse-graining approach undertaken in the following examples is a structural-based one where effective numerical potentials are derived that reproduce in the CG simulations target structures in the underlying AA simulations. These target structural features are expected to influence the local intermolecular packing within self-assembled structures of polymer-tethered POSS molecules, and consequently the formation of specific types of bulk structures at longer length and time scales. In addition to obtaining the CG force field for simulating POSS molecule self-assembly, particular aspects of the coarse-graining approach, including nonuniqueness of the effective potentials and variations on the numerical iteration algorithm, are examined. The work presented herein is adapted from previous publications [Chan 2006; Chan et al. 2007; Striolo et al. 2007], which the reader can refer to for additional details and discussion.

27.2 COARSE-GRAINING METHODOLOGY

27.2.1 Physical Mapping of the Coarse-Grained Model

We have developed a CG model of a POSS molecule functionalized with a single nonyl tether on one corner and nonreactive methyl groups on the remaining seven silicon atoms (Figure 27.1). The hydrocarbon substituents render the molecule soluble in chemically similar and common solvents such as hexane. Because the silsesquioxane core is symmetric, one starting point is to model the cage as a rigid cube with interaction sites on the corners, as in our previous minimal model [Chan et al. 2005]. Each of the resulting eight cube corner beads thus represents one silicon atom, the neighboring oxygen atoms, and the methyl (or methylene in the case of the nonyl tether) substituent attached to the silicon atom. The bead interaction sites are at the centers of the silicon-carbon bonds that connect each substituent to the cage. The beads are connected by rigid bonds.

To examine the physical appropriateness of this CG model of the silsesquioxane cage, the bond length and bond angle probability distributions are compared with those computed in AA simulations of nonyl-tethered POSS molecules dissolved in hexane [Chan et al. 2007; Striolo et al. 2007]. Figure 27.1 shows one example of an AA simulated bond angle distribution that is sharply peaked at about 90°, thereby indicating that the grouping of atoms on the silsesquioxane cage is commensurate with a rigid cube model having eight corner sites. The distances l between the centers of the silicon–carbon bonds that correspond to the interaction sites in the CG rigid cube model exhibit peaks centered at $l = 4.2$ Å. Mapping the AA simulation results to the CG model establishes a length scale in the CG simulations by specifying the edge of the CG cube equal to this value. Each cube corner bead in the model is thus assigned a diameter of $\sigma_c = 4.2$ Å.

To model the nonyl tether, two methylene groups are assigned to each CG tether bead. Although this mapping is on a finer scale compared to previous CG models of hydrocarbon chains that employ groupings of three or more methylene groups per CG bead [Baschnagel et al. 1991; Marrink, de Vries, and Mark 2004; Ashbaugh et al. 2005; Depa and Maranas 2005], it is chosen in order to facilitate future efforts to bridge length and time scales in polymer-tethered POSS self-assembly via reverse mapping schemes where the CG model is mapped back onto its explicit atom counterpart. Note that the end tether bead actually represents a CH_2–CH_3 group in the model, and it is assumed that the behavior and physical properties of this group are not significantly different from those of a CH_2–CH_2 group. The interaction sites for the CG tether beads occur along the center of the corresponding carbon–carbon bond in the AA molecule. The bond-length distributions between pairs of tether bead sites are computed from AA simulations [Striolo et al. 2007]. On the basis of these results [Chan et al. 2007], we assign a diameter of $\sigma_t = 2.5$ Å to each bead in the CG tether.

27.2.2 Derivation of Solvent-Mediated Effective Potentials

27.2.2.1 Approach

We seek to reproduce in mesoscale simulations a select set of target structural quantities computed from the underlying AA simulations that correspond to the bead interaction sites in the CG model. These quantities are the intramolecular bond length and bond angle probability distributions and the intermolecular radial distribution function (RDF) between cube corner beads on different molecules.

The algorithm used to derive the effective potentials is a numerical iteration scheme that produces effective potentials via the following equation [Lyubartsev and Laaksonen 1995; Soper 1996; Reith, Putz, and Müller-Plathe 2003; Ashbaugh et al. 2005]:

$$U_{i+1}(x) = U_i(x) + \alpha k_B T \ln\left[\frac{P_i(x)}{P_{\text{target}}(x)}\right] \quad i = 0, 1, 2, \ldots, \tag{27.1}$$

where i is the iteration step number, k_B is Boltzmann's constant, T is the temperature, x is the independent variable, and $P(x)$ is a probability distribution function, such as a RDF, bond length probability distribution, or bond angle probability distribution. The algorithm updates trial effective potentials $U_i(x)$ at each iteration step by adding a correction term based on the deviation between the trial CG-simulated probability distribution function and target AA-simulated distribution function. The term α is an arbitrary number that scales the magnitude of the correction term to ensure algorithm stability and convergence. A CG effective potential that reproduces the desired structural features in the underlying AA simulations is obtained when the trial CG-simulated and target AA-simulated distribution functions are sufficiently close according to some prescribed tolerance value.

It is important to emphasize that Equation 27.1 has no theoretical basis [Chan et al. 2007] and is employed here with the understanding that it is simply one of many possible numerical algorithms that yield CG effective potentials that satisfactorily reproduce the target distribution functions in the AA simulations. Briefly, the concept of structural-based coarse-graining is motivated by the proof that there is a unique mapping between the RDF and the intermolecular potential for simple pairwise additive and spherically symmetric potentials at a given thermodynamic state point [Henderson 1974]. The relationship between the potential of mean force (PMF) and the RDF at infinite dilution for molecular centers of mass is given by the following equation [McQuarrie 2000]:

$$U_{\mathrm{PMF}}(r) = -k_{\mathrm{B}}T \ln[g(r)]. \tag{27.2}$$

The PMF is precisely equal to the intermolecular pair potential between two point particles. It is strictly applicable to particles or molecules described as single interaction sites and is invalid for molecules treated as collections of multiple interaction sites or beads, such as polymer chains and the CG-tethered POSS molecules of interest here. These types of molecules exhibit orientational correlations that are not accounted for in Equation 27.2, as explained further in the Appendix of Chan et al. (2007). Instead, Equation 27.1 is merely a convenient algorithm to use here, as it satisfies the boundary condition that the trial CG effective potentials converge when the CG-simulated distribution functions match the target AA-simulated ones. We explore this point further in Section 27.2.2.2.

To generate the initial guesses ($i=0$) for the CG effective potentials, the target RDF, bond length probability distribution $P(l)$, and bond angle probability distribution $P(\theta)$ computed from the AA simulations are Boltzmann inverted using the equations below, respectively. Note these choices for the initial guesses are rather arbitrary, as discussed further in Section 27.2.2.2 and in the Appendix of Chan et al. (2007).

$$U_0(r) = -k_{\mathrm{B}}T \ln[g_{\mathrm{target}}(r)], \tag{27.3}$$

$$U_0(l) = -k_{\mathrm{B}}T \ln[P_{\mathrm{target}}(l)], \tag{27.4}$$

$$U_0(\theta) = -k_{\mathrm{B}}T \ln\left[\frac{P_{\mathrm{target}}(\theta)}{\sin\theta}\right]. \tag{27.5}$$

To assess convergence of the derived effective potentials, during each iteration step the following merit functions [Müller-Plathe 2002; Reith, Putz, and Müller-Plathe 2003] are computed for the intermolecular RDF between cube corner beads, intramolecular bond length probability distributions, and intramolecular bond angle probability distributions, respectively.

$$f_{\mathrm{merit,RDF}} = \int w(r)[g_i(r) - g_{\mathrm{target}}(r)]^2\, dr, \tag{27.6}$$

$$f_{\mathrm{merit,bond}} = \int w(l)[P_i(l) - P_{\mathrm{target}}(l)]^2\, dl, \tag{27.7}$$

$$f_{\mathrm{merit,angle}} = \int [P_i(\alpha) - P_{\mathrm{target}}(\alpha)]^2\, d\alpha. \tag{27.8}$$

Optional non-negative weighting functions $w(r) = \exp(-r/\sigma_c)$ and $w(l) = \exp(-l/\sigma_t)$ are also utilized to penalize deviations between the distribution functions in the CG and AA simulations at small separation distances.

On the basis of RDFs computed from the AA simulations for the CG tether bead sites [Striolo et al. 2007], a purely repulsive soft-sphere potential [Leach 2001] is used to capture the intermolecular excluded volume interactions between tether beads.

$$U(r) = \frac{27\varepsilon}{4}\left[\left(\frac{\sigma}{r}\right)^9 - \left(\frac{\sigma}{r}\right)^6\right] + \varepsilon \quad r \le r_c, \qquad (27.9)$$

$$U(r) = 0 \quad r > r_c.$$

In this expression, $r_c = (3/2)^{1/3}$ and $\varepsilon = k_B T$. The choice of this potential is not expected to significantly affect the resulting CG probability distribution functions involving the tether beads or self-assembly of the molecules.

27.2.2.2 Alternative routes

As the effective potentials obtained using the approach discussed in Section 27.2.2.1 are nonunique, one means to evaluate their accuracy is to derive them from different types of initial guesses using the same numerical iteration algorithm and compare the results. This exercise is helpful for corroborating an effective potential in cases where different initial guesses yield the same result or for assessing the best effective potential if different potentials result. The intermolecular POSS cube corner bead effective potentials are first obtained by deriving them using initial guesses generated by Equation 27.3; that is, Boltzmann inversions of the target AA-simulated RDFs. As there is no theoretical basis for using this expression to generate the initial guesses [Chan et al. 2007], we next derive effective potentials using a different initial guess; that is, the purely repulsive Weeks–Chandler–Andersen (WCA) [Allen and Tildesley 1987] interaction potential:

$$U(r) = 4\varepsilon\left[\left(\frac{\sigma_c}{r}\right)^{12} - \left(\frac{\sigma_c}{r}\right)^6\right] + \varepsilon \quad r \le r_c, \qquad (27.10)$$

$$U(r) = 0 \quad r > r_c,$$

where $r_c = 2^{1/6}\sigma_c$ and $\varepsilon = k_B T$.

We also compare the effective potentials derived from a different numerical equation since the iterative scheme of Equation 27.1 has no theoretical basis [Chan et al. 2007]. Equation 27.1 is a successful algorithm for deriving CG effective potentials because the logarithmic term is able to change sign ($+/-$) accordingly so that the updated effective potential produces a CG distribution function that is in better agreement with the AA target distribution function. Thus, this property of the correction term functions as one criterion for devising alternative numerical algorithms that are equally or potentially superior to Equation 27.1. A simple correction term that takes the linear difference between the RDFs computed in the CG and AA simulations satisfies both the above criterion and the boundary condition that the effective potential converges when the CG and target RDFs are equal. We thus propose the following numerical equation for deriving effective potentials:

$$U_{i+1}(r) = U_i(r) + \alpha k_B T[g_i(r) - g_{\text{target}}(r)], \qquad (27.11)$$

where α is an arbitrary number used to scale the magnitude of the correction term. We compare below the effective potentials generated by Equation 27.1 and Equation 27.11 from identical initial guesses. The speed of each algorithm is also examined.

27.2.2.3 Simulation Details

Brownian dynamics, a stochastic molecular dynamics simulation method that samples the canonical ensemble, is utilized to conduct the CG simulations. Additional details on this method are presented elsewhere [van Gunsteren, Berendsen, and Rullmann 1981; Chan et al. 2005; Zhang, Chan, and Glotzer 2005; Chan, Ho, and Glotzer 2006]. Systems containing $Nb = 5$ and $Nb = 20$ CG nonyl-tethered POSS molecules ($N = 40$ and 240 total particles, respectively) are simulated at overall density $\rho = 0.75$ g/cm^3 and temperatures $T = 300$ and 400 K. The simulations employ cubic boxes and periodic boundary conditions. The equations of motion are integrated using the leap-frog algorithm, and the rigid-body motion of the cubes is captured using the method of quaternions [Allen and Tildesley 1987]. Each system is first relaxed athermally to generate initial configurations. Self-assembly of the molecules over time is monitored by inspecting simulation snapshots of configurations. These configurations are subsequently compared to those in the corresponding AA molecular dynamics simulations having the same number of molecules and at the same temperature and density. AA simulations of $Nb = 20$ nonyl-tethered POSS molecules dissolved in 987 hexane solvent molecules ($N = 6642$ total atoms) are performed using the DL_POLY [Smith and Forester 1996] simulation package. The Frischknecht-Curro force field [Frischknecht and Curro 2003] is employed to describe the POSS cage, and the TRAPPE force field [Martin and Siepmann 1998] is used to describe the nonyl tether and hexane solvent. Further details of these simulations are reported in Striolo et al. (2007).

27.3 COARSE-GRAINED POTENTIALS FOR BARE POSS MOLECULES

Initially, an intermolecular CG effective potential that captures the interactions between cube corner beads is derived. This is a logical starting point since the addition of a single hydrocarbon tether on one corner of the silsesquioxane cage has little impact on cage behavior [Li et al. 2007]. Hence, the tether should have minimal impact on the intermolecular interactions between nonreactive "bare," or nontethered, POSS monomers.

AA molecular dynamics simulations of $Nb = 5$ octamethyl functionalized POSS monomers dissolved in hexane have been previously performed at overall density $\rho = 0.75$ g/cm^3 and temperatures $T = 300$ and 400 K [Striolo et al. 2007]. Figure 27.2 displays target RDFs computed from these simulations that characterize the local structure among the CG cube corner bead sites from the underlying atomistic molecules. The RDFs exhibit pronounced peaks that occur primarily at integer values of the cube edge length. This behavior in the RDF was also previously observed in simulations of POSS monomers dissolved in hexadecane [Striolo, McCabe, and Cummings 2005a]. The tails in the RDFs at large separation distances fall below unity at both temperatures because of a combination of three factors: (1) small system size effects ($Nb = 5$ molecules or $N = 40$ particles), which are corrected by multiplying the RDF by the correction factor $N/(N-1)$ [Barker and Henderson 1971; McQuarrie 2000]; (2) not accounting for the close proximity of the cube corner beads that are rigidly bound together when normalizing the RDF; and (3) nonuniform clustering of the POSS monomers throughout the simulation box.

The CG effective pair potentials derived on the basis of these RDFs and the initial guesses used in the iteration algorithm are also shown in Figure 27.2. The interaction potential cutoff value used in the CG simulations is $r_c = 28$ Å. Small correction steps ($\alpha = 0.01$–0.1) are required during numerical iteration to ensure algorithm stability and convergence of the potentials, most likely because explicit solvent molecules are absent in the CG model. Previous applications of Equation 27.1 to derive effective potentials for polymer melts report success with larger parameter values $\alpha = 0.2$ [Ashbaugh et al. 2005] and $\alpha = 1$ [Reith, Putz, and Müller-Plathe 2003]. At $T = 300$ K, the effective potential consists of an alternating series of attractive wells and repulsive peaks that correspond to the peaks and valleys in the target RDF computed from the AA simulations,

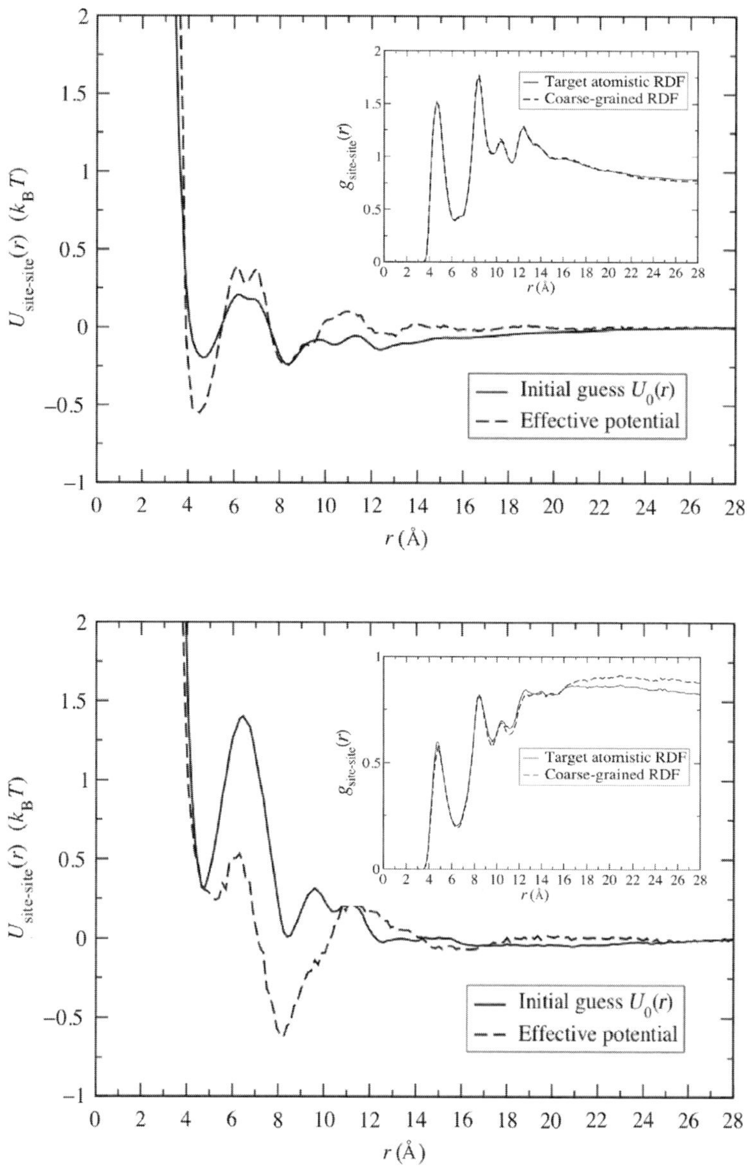

FIGURE 27.2 Site-site CG effective potentials of bare POSS molecules at $T = 300$ K (top) and $T = 400$ K (bottom). The corresponding intermolecular radial distribution functions are shown in the insets.

respectively. This relationship between the shapes of the effective potential and the target RDF is absent at a higher temperature, $T = 400$ K. The effective pair potential here exhibits a steep attractive well at $r = 8.3$ Å followed by broader attractive wells and repulsive peaks compared to those observed at lower temperature. The latter behavior indicates loss of long-range structure with increasing temperature.

The RDFs produced in the CG simulations by the effective potentials are shown in Figure 27.2. The agreement between the CG and AA target RDFs at $T = 300$ K is excellent and the merit function value is $f_{\text{merit,RDF}} \approx 10^{-5}$ when the iteration algorithm reaches convergence [Reith, Putz, and Müller-Plathe 2003]. The agreement between the two RDFs at $T = 400$ K is good, as indicated by $f_{\text{merit,RDF}} \approx 10^{-4}$ when convergence is attained.

27.4 COARSE-GRAINED POTENTIALS FOR MONOTETHERED POSS MOLECULES

We next build upon the model developed thus far for bare POSS cubes by considering the interactions introduced when a nonyl tether is attached to one cube corner (Figure 27.1). Effective potentials are derived to capture the bond stretching and bending interactions now present in this CG monotethered POSS molecule. Because the POSS cages are treated as rigid cubes, only four bonded interactions are considered between the following pairs of beads: C8–T1, T1–T2, T2–T3, and T3–T4 (see Figure 27.1). Four bending interactions due to the angles defined by the bead triplets C6–C8–T1, C8–T1–T2, T1–T2–T3, and T2–T3–T4 are included in the model. Dihedral interactions are not incorporated to maintain model simplicity.

An example of simulated probability distributions for the effective bond stretching and bending interactions, along with the corresponding effective potentials, is presented in Figure 27.3. The CG and target distributions for the C8–T1 bond match closely with $f_{\text{merit,bond}} \approx 10^{-5}$ when the iteration algorithm converges. The two distributions for the tether bonds T1–T2, T2–T3, and T3–T4 are in good agreement with $f_{\text{merit,bond}} \approx 10^{-4}$. The bond bending distributions display multiple peaks that can probably be attributed to dihedral transitions along the alkyl chain that are captured in the fine level of coarse-graining adopted here. The corresponding effective potentials exhibit peaks and valleys that mirror the shape and relative magnitude of these features in the target distribution functions. There is excellent agreement between the CG and target bond bending distributions for each of the four angles treated in the model. The merit function values are $f_{\text{merit,angle}} < 10^{-6}$ when convergence is reached in each case.

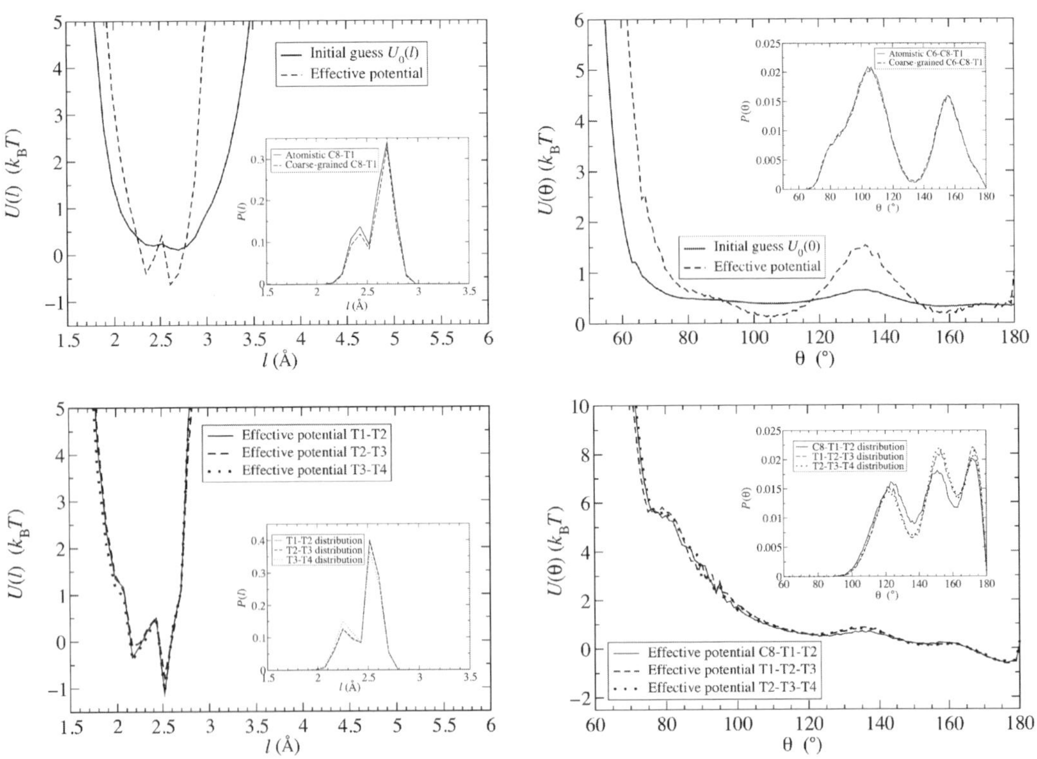

FIGURE 27.3 Effective bond stretching (left column) and bond bending (right column) potentials for CG monotethered POSS molecules at $T = 400$ K. The corresponding probability distribution functions are shown in the insets.

27.5 COARSE-GRAINED FORCE FIELD EVALUATION AND VALIDATION

27.5.1 Varying Initial Guesses for the Effective Potentials

Because the CG effective potentials obtained from the structural-based coarse-graining approach adopted here are nonunique, it is instructive to examine how a different initial guess affects the resulting pair potentials. As an example, bare cube effective potentials are derived using Equation 27.1 with the WCA potential as the initial guess and $\alpha = 0.01$. At $T = 300$ K, the effective potentials obtained using this initial guess and that generated by Equation 27.3 are nearly identical, as shown in Figure 27.4. However, at $T = 400$ K the effective potential derived from the WCA initial guess is significantly different from the one obtained using the initial guess of Equation 27.3. Figure 27.4 shows that there is good agreement between the cube corner bead RDFs and the target RDFs for

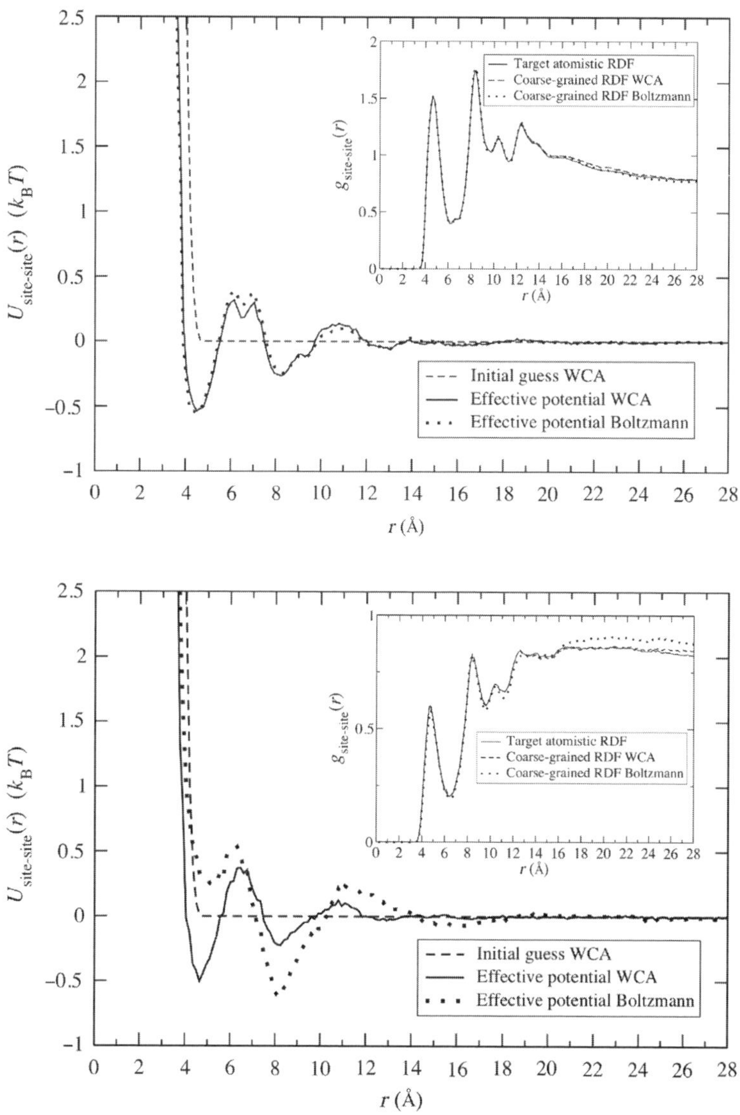

FIGURE 27.4 Site-site CG effective potentials for bare POSS molecules derived from different initial guesses at $T = 300$ K (top) and $T = 400$ K (bottom). The corresponding intermolecular radial distribution functions are shown in the insets.

the effective potentials derived from both types of initial guesses at both temperatures. Furthermore, there is better agreement between these RDFs over the entire range of separation distances at $T = 400$ K for the effective potential obtained from the WCA initial guess, as indicated by smaller merit function values $f_{\text{merit,RDF}} \approx 10^{-5}$ in this case. The WCA initial guess may yield a more accurate effective potential because it lacks the multiple peaks and valleys that are present in the initial guess generated by Equation 27.3. These latter features tend to impede the evolution of the initial guess towards the optimal effective potential.

27.5.2 Varying Numerical Iteration Algorithms

Another means to assess the accuracy of the effective potentials is to use a different numerical equation with identical initial guesses. This approach can yield an additional measure of accuracy if two different numerical equations produce the same potential. As an example, bare POSS cube effective pair potentials derived using Equation 27.1 and Equation 27.11 are examined. In both cases, the WCA potential is used as the initial guess and $\alpha = 0.01$. Figure 27.5 displays the effective potentials obtained from both numerical equations. The overall shapes of the potentials are identical, but the linear algorithm generates attractive wells and repulsive peaks of slightly different magnitude at separation distances $r < 9$ Å. This result may arise because the two equations update the potentials differently when either the AA or CG RDF is zero. Because the logarithmic correction term in Equation 27.1 requires that the RDFs be nonzero over the entire range of separation distances, the trial effective potentials are extrapolated at small separation distances where the target AA RDF is zero during the iteration process. The linear correction term in Equation 27.11, however, exists for all RDF values and the effective potentials are derived without additional approximations or measures. Figure 27.5 demonstrates that the shape of the extrapolated potential matches that of the corresponding portion derived via Equation 27.11, but there are slight differences in magnitude. Nevertheless, the results indicate that the effective potentials overall are essentially independent of the numerical equation used to obtain them. The RDFs generated by the CG effective potentials derived from Equation 27.11 closely match those from the underlying AA simulations.

The number of iteration steps required for the effective potential to converge is examined to compare the speed of Equation 27.1 and Equation 27.11. Our findings indicate that slightly less time is required to reach convergence with Equation 27.11. At $T = 300$ K, convergence is attained with Equation 27.11 after 232 iterations, while Equation 27.1 requires 243 iterations. At $T = 400$ K, convergence is attained after 218 iterations when Equation 27.11 is employed, while the algorithm which employs Equation 27.1 requires 229 iterations. Approximately 17–18 min CPU time is needed to complete each iteration step on a single 2 GHz Apple G5 processor.

27.5.3 Validation from Atomistic Simulations

To corroborate the solvent-mediated effective potentials for simulating monotethered POSS self-assembly accurately, CG simulations of small systems of these molecules ($Nb = 20$) are conducted. The self-assembled structures and local intermolecular packing of cubes are compared in both CG and AA simulations.

Small aggregates containing two to three POSS cubes are observed in the AA and CG simulations of bare POSS molecules. The cubes exhibit face–face, corner–corner, corner–edge, and edge–edge local packings within the aggregates (Figure 27.6). These local packings of the silsesquioxane cages were also previously found in AA simulations of octamethyl functionalized POSS monomers dissolved in hexadecane solution [Striolo, McCabe, and Cummings 2005a]. The face–face packing of cubes appears to be the most predominant one in both the CG and AA simulations and accounts for the large peaks in the RDFs that occur at integer values of the CG cube edge length. The variety of local cube packings captured in the CG model here denotes an improvement in accuracy over the minimal model developed previously [Chan et al. 2005]. In simulations of monotethered POSS

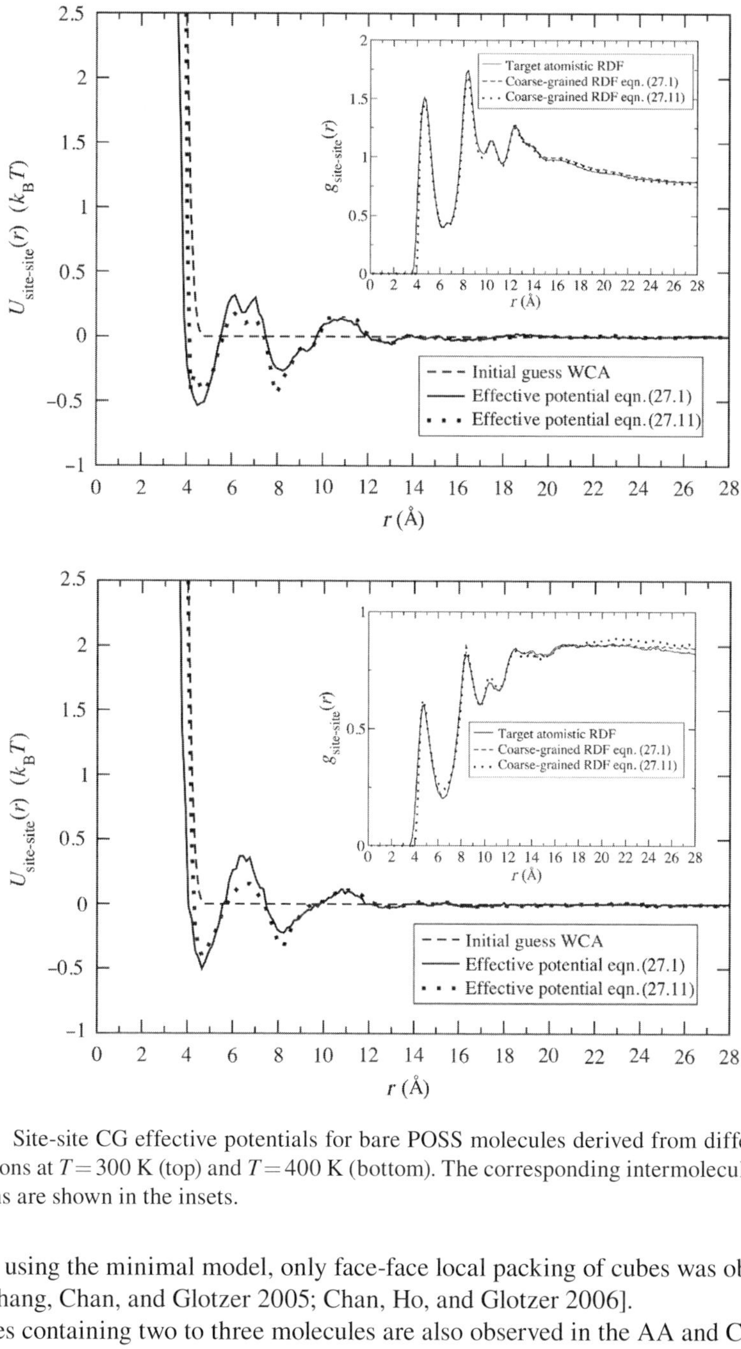

FIGURE 27.5 Site-site CG effective potentials for bare POSS molecules derived from different numerical iteration equations at $T = 300$ K (top) and $T = 400$ K (bottom). The corresponding intermolecular radial distribution functions are shown in the insets.

self-assembly using the minimal model, only face-face local packing of cubes was observed [Chan et al. 2005; Zhang, Chan, and Glotzer 2005; Chan, Ho, and Glotzer 2006].

Aggregates containing two to three molecules are also observed in the AA and CG simulations of monotethered POSS monomers. The intermolecular packings of cubes within the aggregates are similar to those seen in simulations of bare POSS monomers (see Figure 27.6). The hydrocarbon tethers in both the atomistic and CG simulations point outwards away from the aggregates cores comprised of the cages. This behavior is consistent with the expectation that the alkyl tethers prefer to interact with the chemically identical hexane solvent molecules. The tethers appear to sample nearly perfectly linear conformations at low temperature and various bent conformations at higher temperature. The conformational behavior of the tethers is assessed quantitatively by computing the intramolecular distance distribution functions for nonbonded bead pairs C8–T4 and C6–T4. Effective

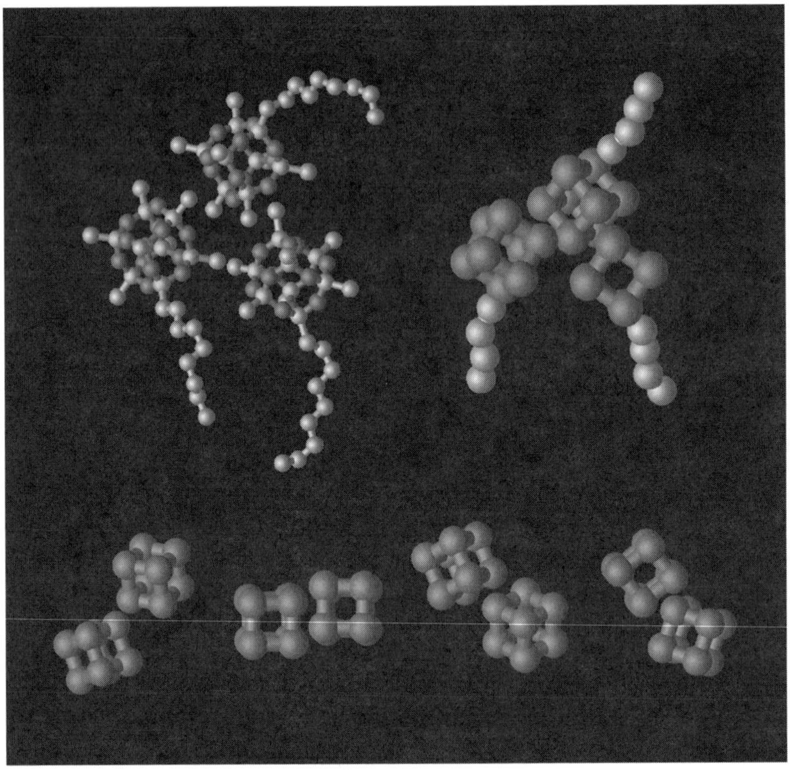

FIGURE 27.6 (See color insert following page 238.) (Top) Self-assembled aggregates observed in all-atom and coarse-grained simulations of polyhedral oligomeric silsesquioxane (POSS) nanoparticles functionalized with a single nonyl tether and dissolved in hexane solution. Solvent molecules have been omitted from the atomistic simulation snapshot for clarity. (Bottom) Simulated local packings of coarse-grained POSS nanoparticles functionalized solely with methyl groups. Left to right: corner-corner, face-face, edge-edge, and corner-edge orientations.

potentials for these bead pairs were not derived here because the effective bond stretching and bond bending potentials derived in Section 27.4 should be sufficient, in principle, to capture these intramolecular nonbonded distance distributions. The distributions computed in the CG and underlying AA simulations are in good overall agreement [Chan et al. 2007].

The CG model developed here affords a savings of roughly two orders of magnitude in computing time compared to the AA simulations. The increased computational efficiency of the CG simulations is largely due to the reduced number of particles as a result of treating the solvent molecules implicitly. The AA simulations required approximately 420 CPU · h on a single 2 GHz Apple G5 processor to simulate 5 ns, while the corresponding CG simulations required only 3.5 CPU · h.

27.6 CONCLUSIONS AND OUTLOOK

We have presented the development of a CG model and force field for accurately simulating self-assembly of nonyl-tethered POSS molecules dissolved in hexane solvent. First, a one-to-one mapping of groups of atoms on the atomistic molecule onto CG beads in the mesoscale model was performed. Next, solvent-mediated effective pair potentials were derived from a structural-based coarse-graining numerical iteration scheme. The potentials sufficiently reproduced target probability distribution functions for intramolecular bond lengths, intramolecular bond angles, and intermolecular bare cube packing that were computed from explicit atom simulations.

The CG force field was evaluated by examining bare POSS cube effective potentials obtained from different initial guesses and numerical equations. Initially, we found that different initial guesses could reproduce essentially the same potential using Equation 27.1 at low temperature. However, at higher temperature substantially different potentials were obtained from two different initial guesses, thereby demonstrating the nonunique character of the potentials obtained from the approach utilized here. In this case, merit functions that indicate the level of agreement between the simulated AA and CG probability distribution functions were computed. Second, effective potentials obtained from two numerical equations with different correction terms and from the same initial guess were compared. Both numerical algorithms yielded essentially identical potentials. Furthermore, the alternative algorithm proposed here with a linear correction term required a smaller number of iteration steps to reach convergence.

Regarding the self-assembly behavior captured by the CG force field, both the AA and CG simulations of monotethered POSS molecules exhibited aggregation into small clusters with the cages in the aggregate cores and the tethers pointing outward. Various types of local packings of the silsesquioxane cages and tether conformations were observed in the AA simulations and sufficiently captured in the CG model. The CG force field afforded a savings of about two orders of magnitude in computing time.

The CG model developed here can be utilized subsequently in mesoscale simulations to predict the types of self-assembled structures or phases that arise in systems containing several thousand tethered POSS monomers dissolved in hexane. These results can then be compared to those of the corresponding experimental studies. The coarse-graining approach employed here can also be extended to develop CG force fields for POSS molecules with different functionalities and in a variety of solvents. One can envision developing solvent-mediated CG effective potentials that capture the interactions important for elucidating the self-assembly behavior of POSS molecules functionalized with tethers of varying chemical composition, number, and location on the silsesquioxane cage. These force fields can then be used to accurately simulate and quickly predict self-assembled structures to guide and motivate ongoing experimental efforts. Advances in reverse mapping techniques, where individual atoms from the atomistic representation are reinserted back onto the CG beads in the mesoscale model, could enable predictions of other useful properties of the self-assembled systems via AA simulations and quantum mechanical calculations in the near future.

Remarkable advances in the development of methods that attempt to bridge the length and time scales accessible in theory, simulation, and experiment have been accomplished within the last decade, and it is anticipated that many more significant breakthroughs will be forthcoming. In conjunction with these developments, ongoing work in the area of explicit atom force field development will be essential for successfully deriving CG force fields that enable one to span longer length and time scales pertinent to self-assembly processes of interest. The active development and application of multiscale modeling and simulation methodologies are expected to play an important role in furthering our understanding of the fundamental principles governing nanoparticle self-assembly. This information should prove valuable for guiding the construction of novel nanoscale materials with enhanced properties.

ACKNOWLEDGMENTS

The authors are grateful to the National Science Foundation for financial support under grant number DMR-0103399. ERC thanks the National Academies for support through a National Research Council Research Associateship award. AS acknowledges the Vice President for Research at The University of Oklahoma for financial support through a Junior Faculty Research Program Award. The work presented here represents part of the PhD dissertation of ERC at the University of Michigan, and we thank the University of Michigan Center for Advanced Computing for computer cluster support. The AA molecular dynamics simulations were mainly conducted on the VAMPIRE cluster at Vanderbilt University.

Commercial names of software and equipment are identified only to specify the simulation procedure. Such identification is not intended to imply recommendation or endorsement by NIST, nor does it imply that they are the best available for the purpose.

REFERENCES

Allen, M. P., and Tildesley, D. J. 1987. *Computer Simulation of Liquids*. Oxford: Clarendon Press.

Ashbaugh, H. S., Patel, H. A., Kumar, S. K., and Garde, S. 2005. Mesoscale model of polymer melt structure: Self-consistent mapping of molecular correlations to coarse-grained potentials. *J. Chem. Phys.* 122:104908.

Barker, J. A., and Henderson, D. 1971. Monte Carlo values for the radial distribution function of a system of fluid hard spheres. *Mol. Phys.* 21:187–91.

Baron, R., de Vries, A. H., Hunenberger, P. H., and van Gunsteren, W. F. 2006. Comparison of atomic-level and coarse-grained models for liquid hydrocarbons from molecular dynamics configurational entropy estimates. *J. Phys. Chem. B* 110:8464–73.

Baron, R., Trzesniak, D., de Vries, A. H. et al. 2007. Comparison of thermodynamic properties of coarse-grained and atomic-level simulation models. *ChemPhysChem* 8:452–61.

Baschnagel, J., Binder, K., Doruker, P. et al. 2000. Bridging the gap between atomistic and coarse-grained models of polymers: Status and perspectives. *Adv. Polymer Sci.* 152:41–156.

Baschnagel, J., Binder, K., Paul, W. et al. 1991. On the construction of coarse-grained models for linear flexible polymer chains: Distribution functions for groups of consecutive monomers. *J. Chem. Phys.* 95:6014–25.

Bharadwaj, R. K., Berry, R. J., and Farmer, B. L. 2000. Molecular dynamics simulation study of norbornene-POSS polymers. *Polymer* 41:7209–21.

Capaldi, F. M., Boyce, M. C., and Rutledge, G. C. 2006. The mechanical properties of crystalline cyclopentyl polyhedral oligomeric silsesquioxane. *J. Chem. Phys.* 124:214709.

Capaldi, F. M., Rutledge, G. C., and Boyce, M. C. 2005. Structure and dynamics of blends of polyhedral oligomeric silsesquioxanes and polyethylene by atomistic simulation. *Macromolecules* 38:6700–709.

Cardoen, G., and Coughlin, E. B. 2004. Hemi-telechelic polystyrene-POSS copolymers as model systems for the study of well-defined inorganic/organic hybrid materials. *Macromolecules* 37:5123–26.

Chan, E. R. 2006. Multiscale modeling and computer simulation of polyhedral oligomeric silsesquioxane assemblies. PhD diss., University of Michigan.

Chan, E. R., Ho, L. C., and Glotzer, S. C. 2006. Computer simulations of block copolymer tethered nanoparticle self-assembly. *J. Chem. Phys.* 125:064905.

Chan, E. R., Striolo, A., McCabe, C., Cummings, P. T., and Glotzer, S. C. 2007. Coarse-grained force field for simulating polymer-tethered silsesquioxane self-assembly in solution. *J. Chem. Phys.* 127:114102.

Chan, E. R., Zhang, X., Lee, C.-Y., Neurock, M., and Glotzer, S. C. 2005. Simulations of tetra-tethered organic/inorganic nanocube-polymer assemblies. *Macromolecules* 38:6168–80.

Depa, P. K., and Maranas, J. K. 2005. Speed up of dynamic observables in coarse-grained molecular-dynamics simulations of unentangled polymers. *J. Chem. Phys.* 123:094901.

Eilhard, J., Zirkel, A., Tschöp, W. et al. 1999. Spatial correlations in polycarbonates: Neutron scattering and simulation. *J. Chem. Phys.* 110:1819–30.

Frischknecht, A. L., and Curro, J. G. 2003. Improved united atom force field for poly(dimethylsiloxane). *Macromolecules* 36:2122–29.

Glotzer, S. C., and Paul, W. 2002. Molecular and mesoscale simulation methods for polymer materials. *Annu. Rev. Mater. Res.* 32:401–36.

Henderson, R. L. 1974. A uniqueness theorem for fluid pair correlation functions. *Phys. Lett.* 49A:197–98.

Ionescu, T. C., Qi, F., McCabe, C. et al. 2006. Evaluation of force fields for molecular simulation of polyhedral oligomeric silsesquioxanes. *J. Phys. Chem. B.* 110:2502–10.

Kim, B. S., and Mather, P. T. 2002. Amphiphilic telechelics incorporating polyhedral oligosilsesquioxane: 1. Synthesis and characterization. *Macromolecules* 35:8378–84.

Kim, K. M., Keum, D. K., and Chujo, Y. 2003. Organic–inorganic polymer hybrids using polyoxazoline initiated by functionalized silsesquioxane. *Macromolecules* 36:867–75.

Knischka, R., Dietsche, F., Hanselmann, R. et al. 1999. Silsesquioxane-based amphiphiles. *Langmuir* 15:4752–56.

Kremer, K., and Müller-Plathe, F. 2001. Mültiscale problems in polymer science: Simulation approaches. *MRS Bull.* 26:205–10.

______. 2002. Multiscale simulation in polymer science. *Mol. Simulat.* 28:729–50.

Leach, A. R. 2001. *Molecular Modelling: Principles and Applications*. 2nd ed. Harlow: Prentice Hall.

Li, H.-C., Lee, C.-Y., McCabe, C., Striolo, A., and Neurock, M. 2007. Ab initio analysis of the structural properties of alkyl-substituted polyhedral oligomeric silsesquioxanes. *J. Phys. Chem. A* 111:3577–84.

Lichtenhan, J. D. 1995. Polyhedral oligomeric silsesquioxanes: Building blocks for silsesquioxane-based polymers and hybrid materials. *Comments Inorg. Chem.* 17:115–30.

Lu, G., and Kaxiras, E. 2005. Overview of multiscale simulations of materials. In *Handbook of Theoretical and Computational Nanotechnology*, ed. M. Rieth and W. Schommers, vol. 10. Stevenson Ranch, CA: ASP.

Lyubartsev, A. P. 2005. Multiscale modeling of lipids and lipid bilayers. *Eur. Biophys. J. Biophys. Lett.* 35:53–61.

Lyubartsev, A. P., Karttunen, M., Vattulainen, I., and Laaksonen, A. 2003. On coarse-graining by the inverse Monte Carlo method: Dissipative particle dynamics simulations made to a precise tool in soft matter modeling. *Soft Mater.* 1:121–37.

Lyubartsev, A. P., and Laaksonen, A. 1995. Calculation of effective interaction potentials from radial distribution functions: A reverse Monte Carlo approach. *Phys. Rev. E* 52:3730–37.

Marrink, S. J., de Vries, A. H., and Mark, A. E. 2004. Coarse grained model for semiquantitative lipid simulations. *J. Phys. Chem. B* 108:750–60.

Martin, M. G., and Siepmann, J. I. 1998. Transferable potentials for phase equilibria. 1. United-atom description of n-alkanes. *J. Phys. Chem. B* 102:2569–77.

McCabe, C., Glotzer, S. C., Kieffer, J., Neurock, M., and Cummings, P. T. 2004. Multiscale simulation of the synthesis, assembly and properties of nanostructured organic/inorganic hybrid materials. *J. Comput. Theor. Nanosci.* 1:265–79.

McQuarrie, D. A. 2000. *Statistical Mechanics*. Sausalito: University Science Books.

Milano, G., and Müller-Plathe, F. 2005. Mapping atomistic simulations to mesoscopic models: A systematic coarse-graining procedure for vinyl polymer chains. *J. Phys. Chem. B* 109:18609–19.

Müller-Plathe, F. 2002. Coarse-graining in polymer simulation: From the atomistic to the mesoscopic scale and back. *ChemPhysChem* 3:754–69.

———. 2003. Scale-hopping in computer simulations of polymers. *Soft Mater.* 1:1–31.

Nielsen, S. O., Lopez, C. F., Srinivas, G., and Klein, M. L. 2004. Coarse grain models and the computer simulation of soft materials. *J. Phys. Condensed Matter* 16:R481–512.

Patel, R. R., Mohanraj, R., and Pittman, C. U. 2006. Properties of polystyrene and polymethyl methacrylate copolymers of polyhedral oligomeric silsesquioxanes: A molecular dynamics study. *J. Polym. Sci. B Polym. Phys.* 44:234–48.

Qi, F., Durandurdu, M., and Kieffer, J. 2007. Structure and properties of acene-functionalized polyhedral oligomeric silsesquioxanes, preprint.

Reith, D., Putz, M., and Müller-Plathe, F. 2003. Deriving effective mesoscale potentials from atomistic simulations. *J. Comput. Chem.* 24:1624–36.

Shelley, J. C., Shelley, M. Y., Reeder, R. C., Bandyopadhyay, S., and Klein, M. L. 2001. A coarse grain model for phospholipid simulations. *J. Phys. Chem. B* 105:4464–70.

Smith, W., and Forester, T. R. 1996. DL_POLY_2.0: A general-purpose parallel molecular dynamics simulation package. *J. Mol. Graph.* 14:136–41.

Soper, A. K. 1996. Empirical potential Monte Carlo simulation of fluid structure. *Chem. Phys.* 202:295–306.

Striolo, A., McCabe, C., Cummings, P. T., Chan, E. R., and Glotzer, S. C. 2007. Aggregation of POSS monomers in liquid hexane: A molecular simulation study. *J. Phys. Chem. B* 111:12248–56.

Striolo, A., McCabe, C., Cummings, P. T., Chan, E. R., and Glotzer, S. C. 2007. Aggregation of POSS monomers in liquid hexane: A molecular simulation study. *J. Phys. Chem. B*. 111:12248–56.

———. 2005b. Thermodynamic and transport properties of polyhedral oligomeric silsesquioxanes in poly(dimethylsiloxane). *J. Phys. Chem. B* 109:14300–307.

Tschöp, W., Kremer, K., Batoulis, J., Bürger, T., and Hahn, O. 1998. Simulation of polymer melts. I. Coarse-graining procedure for polycarbonates. *Acta Polym.* 49:61–74.

van Gunsteren, W. F., Berendsen, H. J. C., and Rullmann, J. A. C. 1981. Stochastic dynamics for molecules with constraints: Brownian dynamics of n-alkanes. *Mol. Phys.* 44:69–95.

Zhang, X., Chan, E. R., and Glotzer, S. C. 2005. Self-assembled morphologies of monotethered polyhedral oligomeric silsesquioxane nanocubes from computer simulation. *J. Chem. Phys.* 123:184718.

Zhou, J. H. and Kieffer, J. 2008. Molecular dynamics simulations of monofunctionalized polyhedral oligomeric silsesquioxane $C_6H_{13}(H_7Si_8O_{12})$. *J. Phys. Chem. C*. 112:3473–81.

Zhou, J. H. and Kieffer, J., Lee, C.-Y., and Neurock, M. 2007. Charge-transfer function for a reactive force field, preprint.

28 Coarse-Graining in Time: From Microscopics to Macroscopics

Angela Violi
Department of Mechanical Engineering, University of Michigan

CONTENTS

28.1 INTRODUCTION

Natural phenomena occur on a variety of length scales, which not only roughly define scientific disciplines (physics and chemistry for the very small, biology for intermediate, geology for very long times and large length scales, and cosmology for the largest), but also delineate subfields within a given discipline, due to the very different experimental methods and theoretical models applicable at each scale [1]. In some cases, these models form a multiscale hierarchy in which the parameters used in the larger scale can be measured or calculated using modeling carried out on a smaller scale.

In a large class of problems, however, the length scales cannot be separated in this way because the coupling between them is strong and "bi-directional," and microscopic phenomena are not only driven by macroscopic forces but also change these macroscopic forces. In this case, there is a feedback loop between the microscopic and macroscopic scales. For example, many bioassemblies explicitly contain both micro- and macroscale dimensions in their structure, and the proper modeling of transtemporal and transspatial structures involves crossing this microscopic/macroscopic barrier to incorporate an adequate degree of information transfer between these disparate scales [2].

One of the challenges in carrying out realistic nonequilibrium simulations of complex systems is that many important processes occur on a time scale that is much longer than the typical atomic vibrational time scale. For example, because of the small time step required to integrate the equations of motion, the standard method for dynamical simulations at the atomistic level, Molecular Dynamics (MD), is generally limited to nanoseconds. However, the relevant thermally activated processes typically take place over time scales of microseconds or even longer.

Although MD simulations provide a powerful tool for investigating complex systems, their substantial computational costs limit their application to investigations on time scales that are less than microseconds and length scales that are significantly less than micrometers [3,4]. Such atomistic

MD simulations are often inadequate to model a variety of processes such as protein folding [5,6], signal transduction [7] for the biology realm, or multiscale materials modeling that aims to incorporate information about materials structures and processes from various length and time scales as necessary.

An alternative to all-atom representations is to employ spatial coarse-grained (CG) models which reduce computational costs by recasting the system in a reduced degree of resolution, while, at the same time, retaining the ability to predict the properties of interest (for recent reviews, see, for example, Refs 8–10). The impetus for CG molecular modeling and simulation primarily derives from the need to bridge the atomistic and mesoscopic scales that typically are separated by two to three orders of magnitude in length and time.

Although CG models have been successfully used to achieve fast simulations for large systems, the lack of detailed atomistic information in CG simulations still limits the systems and processes that can be studied using these models. In addition, the CG approach may be further complicated due to the fact that CG potentials are expected to have less transferability compared to atomistic potentials. This is because an effective interaction between the structural units intended for coarse-graining is defined by the average structure (e.g., average orientations, distances) within the complexes formed by those units in a particular phase [11]. Therefore, the structural properties and thus the CG potentials even for the same phase can, in principle, be sensitive to variations in temperature and other thermodynamic conditions. A poor transferability of a CG potential will undermine the reliability of simulations in which the same CG potential was used to simulate different system phases. Ideally, a CG potential should be fit (or refit) to a system under the same thermodynamic conditions at which the system is intended to be simulated.

As an alternative to the spatial coarse-graining techniques, in this chapter we present a novel method to describe the evolution of complex systems by coarse-graining in time. This new multiscale computational approach describes the evolution of reactive systems (diffusion processes and chemical reactions) over long time scales while preserving an all-atom description of the system. Diffusion processes and chemical reactions can be incorporated in the time evolution of an atomic-scale system and the information can be bridged in a multiscale fashion to elucidate long time-scale changes.

28.2 A NOVEL MULTISCALE COMPUTATIONAL APPROACH: THE AMPI CODE

The novel multiscale approach is embedded in the Atomistic Model for Particle Inception (AMPI) code that combines the MD methodology with kinetic Monte Carlo (KMC) to allow the extension of the accessible time scales by orders of magnitude relative to direct MD, while retaining full atomistic details of the systems of interest [12].

Below we first describe the characteristics of the individual modules (KMC and MD) and subsequently their coupling and integration within the AMPI code. A specific application to describe the evolution of a nanosystem composed of thousands of atoms over a time scale of seconds is then reported in the last section of this chapter.

28.2.1 The Kinetic Monte Carlo Method

Monte Carlo (MC) methods, which first emerged in the late 1940s, refer to a broad class of algorithms that solve problems through the use of random numbers [13,14]. One of the famous MC methods is called kinetic Monte Carlo (KMC) and it was developed for evolving systems dynamically from state to state [15]. The underlying principle in all KMC algorithms is the random selection of a process based on the transition probabilities of all processes, execution of the selected process, and updating the time clock and the transition probabilities [16].

The starting point for a stochastic description of a system is the underlying master equation given by [17]:

$$\frac{dP(\underline{\sigma})}{dt} = \sum_{\underline{\sigma}'} G(\underline{\sigma}' \to \underline{\sigma})P(\underline{\sigma}') - \sum_{\underline{\sigma}'} G(\underline{\sigma} \to \underline{\sigma}')P(\underline{\sigma}), \tag{28.1}$$

with $\underline{\sigma}' \neq \underline{\sigma}$, which gives the evolution of the probability density function $P(\underline{\sigma})$ of observing a state $\underline{\sigma}$; $G(\underline{\sigma} \to \underline{\sigma}')$ is an element of the transition matrix for the transition from state $\underline{\sigma}$ to state $\underline{\sigma}'$ [18]. Atoms or molecules vibrate around locations separated by large free-energy barriers and occasionally jump from one location to a nearby one. The momentum degrees of freedom of all atoms/molecules and thermal vibrations are integrated out to compute the microscopic rates used in a stochastic description of a system. The resulting microscopic state variable $\underline{\sigma}$ is a function of only spatial and time coordinates. The atomic jump results in a change in $\underline{\sigma}$ once the jump has occurred. An equivalent difference-differential equation is:

$$d\sigma_i = \sum_i \Gamma_{ij}^{+}(\underline{\sigma})\, dt - \sum_i \Gamma_{ij}^{-}(\underline{\sigma})\, dt, \tag{28.2}$$

which gives the evolution of σ_i in terms of the transition probability $\Gamma_{ij}^{+}(\underline{\sigma}) / \Gamma_{ij}^{-}(\underline{\sigma})$ of all processes j that lead to a particle addition/removal at site i [19].

Equation 28.2 cannot be solved analytically due to the large number of dimensions present in most systems. If the total number of processes that can occur at site i is N_p, then $\{\varepsilon_{ij}\}_{j=1}^{N_p}$ is the participation indices of site i and $\varepsilon_{ij} = 1$ if site i participates in process j, otherwise $\varepsilon_{ij} = 0$. The transition probabilities $\{\Gamma_{ij}\}_{j=1}^{N_p}$ depend on the participation index of a site in a process and are given by $\Gamma_{ij} = \Gamma_j \varepsilon_{ij}$ with $i = 1,\ldots, N_L$, with N_L being the total number of sites and $j = 1,\ldots, N_p$. The product is equal to Γ_j if site i participates in process j or otherwise is 0.

MC simulations follow stochastic dynamics of the master equation and with appropriate parameterization of the transition probabilities they provide continuous time information. The MC method that employs transition probabilities is usually known as the KMC method.

Recently, Chatterjee and Vlachos published a thorough overview of the KMC methods and we refer the reader to their paper [20] and citations therein for detailed information on the microscopic and accelerated KMC methodologies.

The algorithm employed in the multiscale AMPI code for spatial simulations is the direct KMC method [16,21] and it involves the following two steps: (a) it computes a priori the microscopic rates of all processes and then (b) it selects a process and a site using two random numbers at each event.

The jth process at the ith site is selected with a probability

$$p_{ij} = \frac{\Gamma_{ij}}{\sum_{j=1}^{N_p} \sum_{i=1}^{N_L} \Gamma_{ij}} = \frac{\Gamma_{ij}}{\sum_{i=1}^{N_L} \Gamma_{i,\text{tot}}} = \frac{\Gamma_{ij}}{\Gamma_{\text{tot}}}. \tag{28.3}$$

Here $\Gamma_{ij} = \Gamma_j \varepsilon_{ij}$ is the transition probability of the jth process at site i, $\Gamma_{i,\text{tot}}$ is the total transition probability of all processes at site i, and Γ_{tot} is the total transition probability over the entire system. All Γ_{ij} and $\Gamma_{i,\text{tot}}$ are computed prior to selecting an event.

(a) The (i,j) pair is selected using a uniform distribution random number $\xi_1 \in (0,1)$. according to:

$$\sum_{j=1}^{J-1} \sum_{i=1}^{I} \frac{\Gamma_{ij}}{\Gamma_{\text{tot}}} < \xi_1 < \sum_{j=1}^{J} \sum_{i=1}^{I} \frac{\Gamma_{ij}}{\Gamma_{\text{tot}}}. \tag{28.4}$$

(b) The time is advanced via an increment given from the exponential distribution:

$$\Delta t = -\frac{\ln \xi_2}{\Gamma_{\text{tot}}}, \tag{28.5}$$

where $\xi_2 \in (0,1)$ is another uniform distribution random number. The average time step is the inverse of the total transition probability and can be written as:

$$\Delta t = \frac{1}{\Gamma_{\text{tot}}} = \frac{1}{\sum_{j=1}^{N_p}\sum_{i=1}^{N_L} \varepsilon_{ij}\Gamma_j} = \frac{1}{\sum_{j=1}^{N_p} n_j\Gamma_j} = \frac{1}{N_L\sum_{j=1}^{N_p} \varphi_j\Gamma_j}, \tag{28.6}$$

where n_j is the number of sites of the growing structure that can participate in process j and φ_j is the fraction of sites that can participate in process j.

An important point here is that the time step is inversely proportional to the size of the growing structure. As the size of the structure grows, not only updating and searching per event take more time, but also the real time reached is much smaller and additional KMC events are needed to reach the same real time.

The search and update steps are at the heart of any KMC simulation and they are computationally intensive. The reaction network is usually comprised of hundreds to thousands of elementary reaction steps and effective algorithms for search and update are critical for accessing reasonably large time scales on large systems. The n-fold method of Bortz et al. [22] is an efficient search spatial KMC method and it is an example of hash-table search [23,24]. A hash table, or a hash map, is a data structure that associates keys with values. The primary operation it supports efficiently is a lookup: given a key (e.g., a person's name), find the corresponding value (e.g., that person's telephone number). It works by transforming the key using a hash function into a hash, a number that is used to index into an array to locate the desired location ("bucket") where the values should be. Hash tables support the efficient addition of new entries, and the time spent searching for the required data is independent of the number of items stored (i.e., O(1)).

For KMC simulations, the hashing function categorizes processes according to their transition probability. Each site in the system allows only a small number of possible reactions, $\varepsilon_{ij} = 1$. All sites with $\varepsilon_{ij} = 1$ are grouped into a class, jth, where n_j is the number of sites belonging to the jth class. First a class is selected and then a site from this class is randomly chosen. The jth class is selected with a probability:

$$p_j^{\text{class}} = \frac{\sum_{i=1}^{N_L} \Gamma_j\varepsilon_{ij}}{\sum_{k=1}^{N_p}\sum_{i=1}^{N_L} \Gamma_k\varepsilon_{ik}} = \frac{\Gamma_j\sum_{i=1}^{N_L} \varepsilon_{ij}}{\sum_{i=1}^{N_L} \Gamma_k\sum_{i=1}^{N_L} \varepsilon_{ik}} = \frac{n_j\Gamma_j}{\sum_{i=1}^{N_L} n_k\Gamma_k} = \frac{n_j\Gamma_j}{\Gamma_{\text{tot}}}, \tag{28.7}$$

using a uniform distribution random number $\xi_1 \in (0,1)$.

Once a class is selected, a site from this class (m) is randomly chosen using another random number.

The computational cost of the selection now scales linearly with N_p and the computational savings are of order O(N_L). The low storage requirements given by the number of processes N_p and the low CPU requirements given by O(1) are the main advantages of the n-fold method.

The update of the transition probability matrix is done by employing the global update method where all the interparticle interactions and all the process transition probabilities are reevaluated after a process is executed.

In summary, in the KMC method, the system starts in some state and equations of motion are never directly evolved, but a set of escape pathways connect a state to neighboring states. The key requirement is to specify these escape paths and the rate constant for each one. Given this set of rates, a stochastic algorithm is employed to advance the time and to choose a neighboring state to which the system is moved. The rate constants for the possible moves out of each state might be approximated by a few additive parameters, or computed more accurately using transition state theory (TST) to make a rate catalog [13], perhaps even using first-principles calculations [15].

28.2.2 Molecular Dynamics

Molecular Dynamics has a long history and has evolved into an important and widely used theoretical tool that allows researchers in physics, chemistry, and biology to model the microscopic dynamical behavior of many different types of systems, such as gas, liquids, surfaces, solids, and clusters. MD methodology is founded upon the basic principles of classical mechanics and can provide a window into the microscopic dynamical behavior of the individual atoms of a system.

The goal of this section is to summarize the main concepts behind MD and its use. For additional information we refer the reader to other references [18,25].

In the broadest sense, MD is concerned with molecular motion and conformational transitions and local vibrations are the usual subjects of MD. MD alters the intramolecular degrees of freedom in a stepwise fashion, analogous to energy minimization. The individual steps in energy minimization are merely directed at establishing a downhill direction to a minimum. The steps in MD, on the other hand, meaningfully represent the changes in atomic position over time (i.e., velocity).

Newton's equation is used in the MD formalism to simulate atomic motion. The rate and direction of motion (velocity) are governed by the forces that the atoms of the system exert on each other as described by Newton's equation. Knowledge of the atomic forces and masses can then be used to solve for the positions of each atom along a series of extremely small time steps (in the order of femtoseconds $= 10^{-15}$ s).

Dynamical properties, such as absorption spectra, rate constants, and transport properties, as well as equilibrium properties, such as thermodynamic quantities, structure, and free energies along reaction paths, can be calculated using MD [26].

Integrating the classical equation of motion forward in time, the behavior of the system emerges requiring no input from the user. Complicated and surprising events may occur. If the potential gives an accurate description of the atomic forces of the system and assuming both that quantum dynamical effects are not important and that electron–phonon coupling (non-Born–Oppenheimer) effects are negligible unless atoms are moving extremely fast, then the dynamical evolution will be an accurate representation of the real physical system. This explains the popularity of the MD method.

A serious limitation, however, is that accurate integration requires time steps short enough ($\sim 10^{-15}$ s) to resolve the atomic vibrations. The total simulation time is typically limited to less than one microsecond, while important processes such as diffusion, annihilation, and nanoparticle assembly often take place on much longer time scales.

To summarize, MD simulations follow the dynamical evolution of a system and generate a sequence of points in phase space as a function of time corresponding to different conformations of the system and their momenta. This requires choosing an interatomic potential for the atoms and a set of boundary conditions.

28.2.3 The AMPI Code: Synergy between KMC and MD

The AMPI code combines the strengths of KMC and MD methodologies in a unique way. The KMC method is very effective for conformational changes that jump to a completely different area of phase space, whereas MD is useful for exploration of the local phase space. The two procedures are placed on an equal footing and the code alternates between MD and KMC steps during the simulation.

This feature represents the peculiarity of the newly proposed methodology; the two approaches have been widely used [27] but have never been integrated in this way. The novelty of this code lies in the synthesis of methods that consent to describe reacting systems over long time scales while retaining fully atomistic detail of the system itself. In fact, a common problem in theoretical chemistry, condensed matter physics, and materials science is the calculation of the time evolution of an atomic-scale system where, for example, chemical reactions and/or diffusion occur causing changes in the atom configurations.

The system analyzed continuously undergoes a KMC step followed by a MD run. The KMC method is an extremely efficient way to simulate processes involving activated events, but in order to carry out a realistic simulation, it is important to know in advance all the relevant atomic processes along with their corresponding rates. Since many important processes may involve complicated concerted motions, it may not be possible to "guess" them a priori, and thus they will not be included in the KMC rate table. In some systems, it is very difficult to guess where the state-to-state evolution might lead. The underlying mechanisms may be too numerous, too complicated and have an interplay whose consequences are unpredictable [28]. As a result, in KMC simulations only a relatively limited catalog of atomic-scale processes is typically included (along with their corresponding rates), thus limiting the accuracy of the simulations.

In recent years, it has become clear that these mechanisms can be surprisingly complex, often involving many atoms even in relatively simple systems [15]. Whether these unexpected mechanisms are important to the long-time dynamical evolution depends on the system, but their very existence provides a clear motivation to develop methods that make no a priori assumptions about the mechanisms.

The AMPI code overcomes this limitation of the KMC methodology by adding MD runs to the time evolution of the system. During the MD module, the behavior of the system emerges naturally without intuition or further input from the user, and complicated and surprising events can occur reflecting the correct dynamical evolution of the system for a specific potential and boundary conditions. In the example reported in the next section, we clearly show the occurrence of rearrangement reactions during the MD run that were not taken into account during the KMC step [12,29].

AMPI Algorithm

The algorithm used in the AMPI code includes the following steps:

1. The code reads a set of inputs, such as reaction rates for the KMC module and MD-related quantities.
2. It counts the number of sites available for reactions in the system by identifying each atom as belonging to a specific reaction class and a rate catalog of possible reactions at a specific time t is constructed.
3. After classifying the reactions as either accepted or rejected, the code creates a list of the reaction rates for the accepted reactions along with possible sites and reaction rates. A reaction is then selected based on the reaction probability that is proportional to its reaction rate.
4. The code randomly selects the site for the reaction to occur, assuming that all possible sites for this reaction have an equal probability.

5. A KMC step is performed and the structure of the growing species is modified according to the selected reaction. The clock is then incremented in a way that is consistent with the average time for escape from that state, which can be determined easily from the rate constants for the possible escape paths in the list.
6. Finally, the code relaxes the newly formed structure using a MD run allowing for relaxation of the structure towards thermal equilibrium.

At this point, a new KMC step is performed and the sequence of steps 1–6 is repeated. Before performing the next KMC step, the global update method is used in order to reevaluate all interparticle interactions and all process transition probabilities after a process is executed.

Temporal Acceleration

While time acceleration with the KMC method is still achieved, further acceleration is obtained via temporal coarse-graining, for example, executing multiple processes at once using the τ-leap method. The τ-leap method is an approximate stochastic simulation method introduced by Gillespie [30] for modeling reaction networks in spatially uniform systems. There are several variations of the original method, and numerical simulations of simple reaction networks and complex biological systems have shown that the method captures the probability density function for the time-dependent species population for small τ [31–34].

An important assumption in the τ-leap method—the leap condition [30]—requires τ to be sufficiently small so that the change in the population for all chemical species is small. In the AMPI code, however, by grouping sites with the same transition probability into the same group or class, hence using the *n*-fold method, a large number of sites in each class are present. Each class then becomes a lumped reaction and the τ-leaping can be applied to the classes without violating the leap condition.

28.3 NANOPARTICLE FORMATION IN HIGH-TEMPERATURE REGIMES: AN APPLICATION OF THE AMPI CODE

The nanoscale is an exciting regime for practitioners of atomistic simulation techniques because in the last few years there has been a crossing of the length scales accessible to simulation and experiment. The size of a nano feature that can be directly synthesized and probed experimentally has been decreasing, while simulation sizes are continuously increasing.

As an application of the AMPI code, we report a study on the formation of carbonaceous nanoparticles with an average diameter of 50 nm, using as building blocks organic species of 0.5 nm. This scenario is representative of many interesting processes, such as new material designs, natural and biological materials with nanostructured hierarchical designs, energy processes, and environmental issues, such as particulate formation.

In this section, we will focus on the formation of anthropogenic nanoparticles in the environment that are mainly produced by combustion sources. These combustion-generated particles are present not only in a very large amount, but are produced, at the smallest scale, in the form of clusters with nanometric dimensions [35]. Given that such great quantities of carbonaceous nanoparticles are being introduced into the atmosphere, the question naturally arises as to their environmental fate. But the most direct and serious risk is the direct absorption of these particles into the living systems of humans and animals through the process of respiration, especially in more urban environments [36,37].

In high-temperature environments, particle formation from small gas-phase species is a fascinating multiscale problem of nanoparticle growth, both in length scale and in time scale. Starting from simple gas-phase molecules, such as ethylene and acetylene, polycyclic aromatic hydrocarbons, including naphthalene and pyrene, are formed. The system then undergoes a nucleation process and the first particles are formed by reactions with the gas-phase species. Coagulation and agglomeration of these nanoparticles lead to particle size of the order of hundreds of nanometers in diameter [38].

The processes involved in the formation of particles exhibit a wide range of time scales, spanning pico- or nanoseconds for intramolecular processes, to milliseconds for intermolecular reactions. At the same time, the length scale also undergoes significant changes, going from a few angstroms for small polycyclic aromatic hydrocarbons to hundreds of nanometers for particle aggregates.

When the nanoparticles are able to reach a critical size (which is not always the case), they begin to aggregate on two separate length scales. The assembly on the smallest of these scales seems to be governed by the strongest interparticle interactions, so interesting patterns reminiscent of a spinodal decomposition process become evident. These smaller assemblies will, in turn, aggregate on an even larger length scale through a process that resembles colloidal nucleation. Thus, the molecular science of the nanoparticles bridges with the statistical mechanics of nonequilibrium self-assembly and nucleation over two disparate length and time scales in a truly fascinating way.

Within this context, the AMPI is used to understand the physical and chemical processes involved in the formation of nanoparticles in different conditions of temperature, pressure, and species concentrations starting from small aromatic compounds. The initial system, used in this example, is composed of 300 gas-phase species at 1500 K and 1 atm. The goal is to understand how one of these species, for example, benzene, evolves into a three-dimensional structure of thousands of atoms over long time scales. The three main modules are: (a) gas-phase species that can react with the growing species, (b) reaction sites on the growing particle, and (c) reaction rates. For the gas-phase input, the user defines the temperature, and the concentrations of the species that contribute to the formation of the nanoparticles. Some examples of common gas-phase species for the high temperature environment are H, OH, H_2, and aliphatic and aromatic species C_xH_y, with x and y varying from 1 to 24.

Once the species are defined, it is important to list the possible reactions that can occur over time and describe them in terms of reaction sites; that is, atoms where a particular reaction can occur. The reaction site module governs the definition and counting algorithm of reaction sites, which are capable of undergoing modification (e.g., addition reactions). Some examples of reaction sites included in the AMPI code are: (a) atom on a five- or six-membered ring, (b) sp^3 carbon, (c) HCCCCR and HCCCCCR chains prone to close to form five- and six-membered rings where R is a radical carbon, and (d) HCCCCCH and HCCCCCCH to form five- and six-membered rings via the route of dehydrogenation. Figure 28.1 shows a schematic of the sites for the abstraction of H from a five-membered ring, a six-membered ring, and HCCCCCH and HCCCCCCH that are chains prone to close to form six- and five-membered rings for cyclodehydrogenation reactions.

Proximity conditions are also placed on radical sites. In the site-counting procedure, a radical carbon is considered as a possible site for termination or addition by a gas-phase species only if the gas-phase species of interest can be accommodated.

The third input to the AMPI code is the list of the rate constants that are extracted from smaller length and time scale simulation tools, such as density functional theory (DFT) [39–41], transition state methods, transition state theory (TST) [42], and MD [43]. The reactions included in the AMPI code can be broadly classified into three categories.

Ring-Closure Reactions

Ring-closure reactions involve the formation of nascent rings through dehydrogenation mechanisms. Two hydrogen atoms are lost and a new ring is created by the formation of a new transannular C–C bond. We recently reported new reaction pathways for the cyclodehydrogenation of PAH to form cyclopentafused aromatic hydrocarbons [44], and the computed reaction rates have been included in the AMPI code.

Abstraction Reactions

An example of an abstraction reaction can be expressed as: $R_i{-}H + H = R_i + H_2$, where $R_i{-}H$ is a saturated site on the growing structure, and R_i is its radical. Usually this initiation step is represented by H abstraction reactions, which activate the aromatic structures.

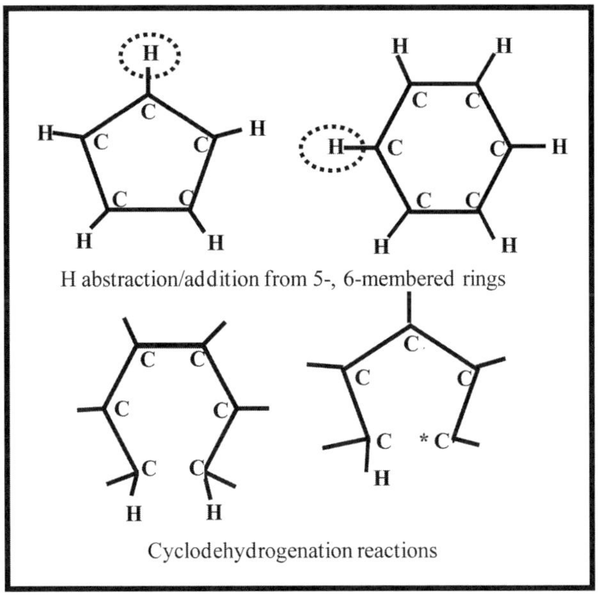

FIGURE 28.1 Examples of reaction site in the AMPI code.

Addition Reactions

Addition reactions are divided into two categories: termination and elimination reactions. In a termination reaction, a reactive intermediate is rendered inactive, thus ending the reaction sequence.

In an elimination reaction, two groups are removed from a molecule in either a one- or two-step mechanism. Either the degree of saturation of the compound diminishes or the valence of an atom in the molecule decreases by two (this is known as reductive elimination).

$$R_i + R_j - H = R_i - R_j + H$$

Here, the radical R_i is added to the neutral site $R_j - H$ on the growing structure, removing a hydrogen atom [45,46].

As mentioned in the previous section, MD simulations within the AMPI code allow for relaxation of the growing species towards thermal equilibrium and possible rearrangements of the structure. Within the MD module, the potential used to describe hydrocarbon interactions is the Adaptive Intermolecular Reactive Bond Order (AIREBO) [47], based on the reactive empirical bond-order potential of Brenner [48,49]. The potential describes the covalent bonding interactions largely as in Brenner's Reactive Empirical Bond Order potential [50]. In addition, there are terms in the energy to account for torsional energies of single-bond rotations, as well as the nonbonded van der Waals interactions. The torsional energy is modeled as a pairwise interaction between atoms. The advantage of this formulation is that the symmetry of the dihedral angle potential is dictated by the symmetry of the molecule, and can change as reactions occur. The van der Waals interaction is modeled using the Lennard–Jones 12-6 potential. The thermodynamic ensemble realized for the MD module is the (N,V,T) canonical ensemble, where the number of particles, the volume, and the temperature are fixed to prescribed values. The integral thermostat method was selected for our MD simulations in the AMPI code. The integral thermostat method (which is sometimes called the extended system method or the Nose–Hoover algorithm [51,52]) introduces additional degrees of freedom into the system's Hamiltonian, for which equation of motion can be derived. These equations for the additional degrees of freedom are integrated together with "usual" equations for spatial coordinates and momenta.

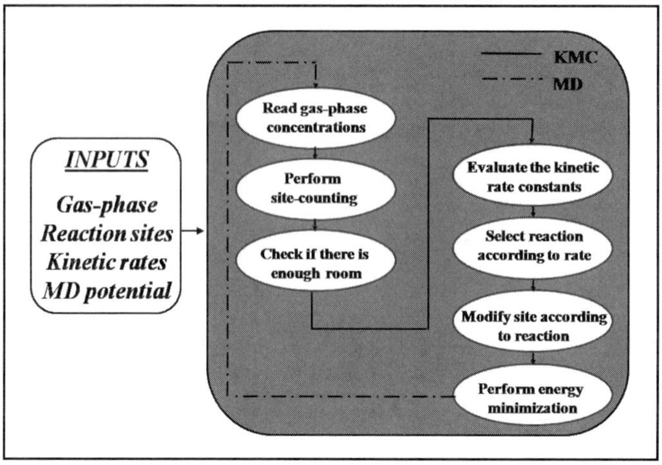

FIGURE 28.2 Flowchart for the AMPI methodology as applied to nanoparticle formation.

The idea of the method proposed by Nose [53] was to reduce the effect of an external system, acting as heat reservoir, to an additional degree of freedom. This heat reservoir controls the temperature of the given system; that is, the temperature fluctuates around the target value. Actually, the thermal interaction between the heat reservoir and the system results in exchange of the kinetic energy between them [54]. The predictor corrector method [18] is used to integrate the equations of motion.

Bonding pairs in the reaction sites described above are defined according to the formalism of Brenner. Hence, chains and rings are identified through efficient nested loops directed by neighbor list pointers. Additional proximity and geometric conditions are placed on the definition of the HC..CH chains to ensure that false-positive sites are not identified. Figure 28.2 shows the flowchart of the AMPI methodology.

After reading the various inputs, including temperature and species concentrations, the AMPI code classifies the sites on the growing species as belonging to a specific class. Then the code checks for steric hindrance, and the reactions available at that time are classified as either accepted or rejected. The reaction rates for the accepted reactions [$k = AT^n \exp(-E_a/RT)$] are then computed and a list is created along with possible sites and reaction rates. The code randomly selects a reaction based on the reaction probability that is proportional to its reaction rate and randomly chooses the site for the reaction to occur.

The reaction rates among the compounds present in the system are specified as probabilities and the surface configuration over time is then given by a master equation, describing the time evolution of the probability distribution of system configurations.

The code modifies the structure of the growing particle according to the selected reaction. For example, if abstraction occurs, the code removes the selected atom(s) from the specified sites; if addition is chosen, the fragment is added to the selected site.

Finally, the code relaxes the newly formed structure using a MD run, allowing for relaxation of the system towards thermal equilibrium and possible rearrangement reactions. The sequence is then repeated and a new KMC step is performed.

The CPU requirements of a global update method for computing only the interactions are $O(N_{\text{interac}}{*}N_L)$ for two-body potentials. The CPU requirements for updating the transition probabilities are $O(N_P{*}N_L)$ where N interact is the average number of interacting neighboring sites of a single site.

For these simulations the coarse time step has been chosen as that of the microscopic KMC simulation multiplied by an acceleration factor f ($f > 1$) [32] $\tau = f/\Gamma_{\text{tot}}$. The acceleration depends on the size of the time increment τ.

Figure 28.3 shows snapshots of nanoparticles obtained using benzene as initial seed at a temperature of 1500 K.

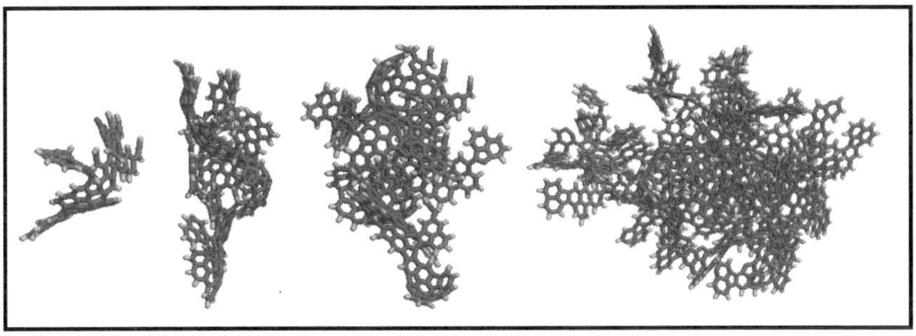

FIGURE 28.3 (See color insert following page 238.) Nanoparticles formed with the AMPI code.

The particles depicted in Figure 28.3 evolve from a few carbon atoms (benzene was used as initial seed) to 3000 carbon atoms reaching a diameter of ~ 8 nm. Table 28.1 reports the sequence of reactions performed by the code during one trajectory together with the corresponding KMC time.

Every time the nanoparticle undergoes bond cleavage, a new fragment is formed. The AMPI code assigns a reaction rate to this fragment and adds it to the list of species that can interact with the growing particle. The rate is assigned based on similarity.

The AMPI code has been used to describe particle formation up to 50 nm in high-temperature regimes. Figure 28.4 shows the time evolution of H/C, and mass and diameter of nanoparticles produced in this environment at 1500 K (dashed curves) and 2000 K. Specifically for the runs carried out at 1600 K, 0.02 s was reached to produce structures of ~ 3000 atoms (~ 12 nm) and 0.1 s for structures of ~ 50 nm.

At higher temperature (solid curves), rearrangement reactions such as cage closure reactions, the breaking of C–H bonds, and intramolecular rearrangements have large activation energies and are favored [19]. This environment leads to the formation of particles with different chemical and physical characteristics, for example, for the same mass the H/C ratio is lower at 2000 K than 1600 K.

At any point in time, the AMPI code can be used to describe the chemical and physical characteristics of the growing structure. Properties such as chemical composition, radical concentration, functional group distribution, surface reactivity, morphology, porosity, pore size distribution, density, etc. can be determined. This ability makes the AMPI code unique.

The unique capability of the AMPI code has been very successful in reproducing physical and chemical characteristics of nanoparticles in different environments [12,55,56]. Nanoparticles have been characterized in terms of chemical structure/components and relationships between structure and pathways, structure and properties, and structure and reactivity population of active sites have been addressed. Results have been reported showing the comparison between computed properties and experimental data in terms of H/C trends, particle morphology, depolarization ratio, and free radical concentration [12,17,57–60].

The AMPI code has been very successful in identifying reaction mechanisms for the formation of nanoparticles. For example, the AMPI code can identify the role of accessibility to reaction sites on the relative rate of reactions of different molecules. During the site-counting procedure, the code identifies every atom with a local environment that fits the definition of each site listed as input to the code. At the same time, it also checks if there is sufficient space next to the site for a gas-phase species to penetrate and react. In this way, an atom is considered as a possible site for termination or addition by a gas-phase species only if the gas-phase species of interest can be accommodated. The code goes over all atoms in the cluster, including all the possible sites, even the "hidden" ones that may be located inside the structure. Once the site is found, the code loops over species that may be added to that specific site. The spherical volume around each adduct atom is computed, and if the space around the site is large enough for a specific species to fit in, then the site is accepted, otherwise it is rejected.

TABLE 28.1
List of the reactions picked by the KMC module of the AMPI code for 2 picoseconds of simulation

Time (s)	Reactions
2.1E-7	H abstraction from six-membered ring by H atom
2.6E-7	C_6H_6 addition to radical on six-membered ring
4.0E-7	H abstraction from six-membered ring by H atom
6.1E-7	C_6H_5 addition to radical on six-membered ring
6.5E-7	Dehydrogenation reaction to form six-membered ring—site HCCCCCCR
7.5E-7	H addition to radical on six-membered ring
7.9E-7	H abstraction from five-membered ring
8.1E-7	H addition to six-membered ring
8.4E-7	C_2H_2 abstraction
9.5E-5	H addition to radical on six-membered ring
1.15E-6	Dehydrogenation reaction to form five-membered ring—site HCCCCCC*
1.3E-6	H abstraction by CH_3
1.4E-6	H addition to five-membered ring
1.6E-6	CH_3 abstraction
1.7E-6	C_6H_6 addition to radical on aliphatic C
1.8E-6	H addition to radical on six-membered ring
1.9E-6	CH_3 addition
2.1E-6	H addition to radical on five-membered ring

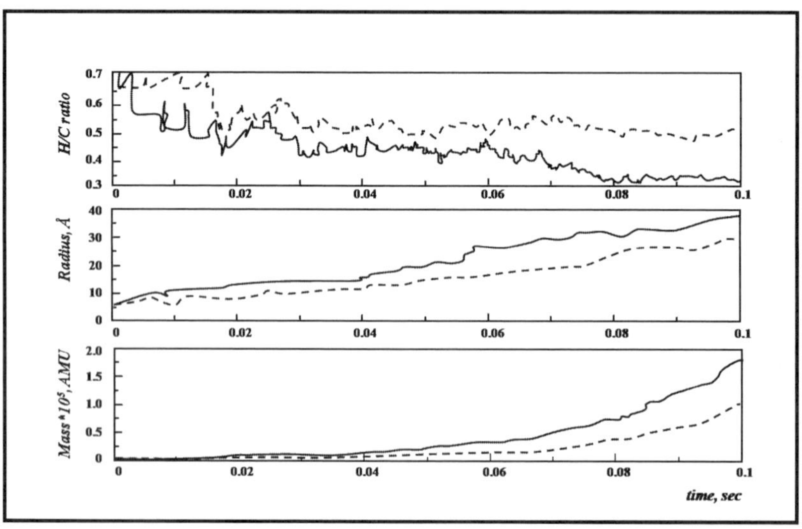

FIGURE 28.4 Some properties of the nanoparticles computed with the AMPI code at 1600 K (dashed line) and 2000 K.

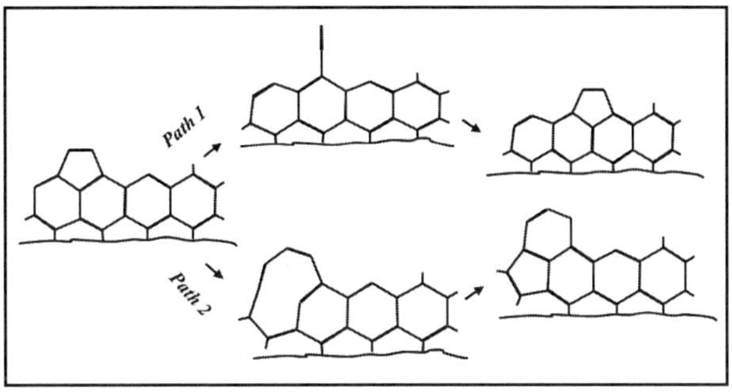

FIGURE 28.5 Intramolecular rearrangements during the MD simulation: (*Path 1*) 5-membered ring migration, and (*Path 2*) 5- and 6-membered ring interconversion.

It was found that the accessibility of the sites on the growing seed molecule decreases with increases in the curvature of the seed particle and size of the reacting molecule. This result can be used to identify new reaction mechanisms for the formation of nanoparticles of specified morphology.

For these systems, the presence of the MD module in the AMPI code is determinant for a correct representation of the system. Rearrangement reactions are identified by MD leading to structural changes of the particle. Figure 28.5 shows examples of two reaction pathways identified during the MD module. Five-member ring migration on the surface of the particle (*Path 1*) and five-, six-membered ring interconversion (*Path 2*) are very difficult to guess in advance, but at the same time they need to be part of the structural evolution for a realistic description. The collision of two migrating rings could lead to the formation of a defect on the surface that prevents any further growth, reducing the surface reactivity of the particle [61].

28.4 CONCLUDING REMARKS

In this chapter, we presented a new multiscale computational methodology for coarse-graining in time. The unique and powerful tool, embedded in the AMPI code, utilizes information on the chemistry occurring at the atomistic level and bridges it in a multiscale fashion to describe long time-scale changes.

The code combines the strengths of MD and KMC and integrates them in a novel way. The synergy of the methods leads to a powerful computational approach that can describe the evolution of reactive systems over long time scales (seconds). The KMC method is employed to describe conformational changes that jump to a completely different area of phase space, and MD is used for exploration of the local phase space and rearrangement reactions. The time duration between MC events can be arbitrarily long (depending on the kinetics, model, etc.), while in MD we require time steps that are a small fraction of the atomic vibrational period. The combination of the two techniques spans two time and equilibrium scales: MD allows for relaxation as well as processes very far from equilibrium, while the KMC module allows much larger time-scale changes to the system, provided that the system is at equilibrium.

An additional reason to combine the two methods resides in the limited catalog of atomic-scale processes that can be included in the KMC module, since some reaction mechanisms can be very complex and difficult to predict a priori. The AMPI code overcomes this limitation by adding a MD run after each KMC step. In this way, possible reactions that were not taken into account during the KMC run can naturally emerge during the MD module.

An illustrative application of the overall approach is reported for the formation of environmental particles generated from high-temperature sources. Specifically, the formation of carbonaceous nanoparticles with an average diameter of 50 nm is described in a detailed way using initial species of 0.5 nm in size. The technique relies on a closed feedback loop between KMC and MD in which chemical reactions or diffusion processes are included over time scales of the order of seconds.

This approach allows for the examination of the transtemporal and at the same time transspatial phenomena that can occur in many real systems, bridging the microscale to the macroscale.

Detailed chemical and physical properties of the systems investigated can be obtained using the AMPI code. For the formation of carbonaceous nanoparticles, the AMPI code provides the time evolution of properties such as chemical composition, morphology, density, porosity, etc.

The AMPI code represents a unique tool for studying complex phenomena happening on different time scales. The methodology is pretty general and it can be applied to different fields, such as biological assembly or atomistic processes in nanostructured materials, predicting material strength, damage, and fracture. For example, AMPI can be used to describe changes in the structure, morphology, and surface properties of materials in chemically aggressive environments. In this case, the chemistry occurring at the atomistic level needs to be integrated into coarser models to describe gradual changes in material properties. The KMC steps can be used to describe slow chemical and/or large-scale structural transitions in the system. As inputs to the KMC/MD procedure, reaction rates and mechanisms can be derived from a combination of ab initio reaction path calculations and large-scale MD simulations. The basic KMC/MD algorithm can also be combined with CG representations of the system to significantly expand the length scale of the long time simulations to the key mesoscopic regime where the complex material morphology is manifest. In turn, these simulations can be bridged to continuum representations to predict properties of the systems of interest in a more meaningful and realistic fashion.

ACKNOWLEDGMENTS

This research has been supported by the National Science Foundation (Nanoscale Interdisciplinary Research Team grant EEC-0304433 and CTS 0553764) and by the University of Utah Center for the Simulation of Accidental Fires and Explosions (C-SAFE), funded by the Department of Energy, Lawrence Livermore National Laboratory, under subcontract B341493.

REFERENCES

1. Csanyi G, Albaret T, Moras G, Payne MC, De Vita A. 2005. Multiscale hybrid simulation methods for material systems. *Journal of Physics: Condensed Matter* 17 (27): R691–703.
2. Wang CY, Zhang X. 2006. Multiscale modeling and related hybrid approaches. *Current Opinion in Solid State & Materials Science* 10 (1): 2–14.
3. Muller-Plathe F. 2002. Coarse-graining in polymer simulation: From the atomistic to the mesoscopic scale and back. *Chemphyschem* 3 (9): 754–69.
4. Duan Y, Kollman PA. 1998. Pathways to a protein folding intermediate observed in a 1-microsecond simulation in aqueous solution. *Science* 282 (5389): 740–44.
5. Karplus M, McCammon JA. 2002. Molecular dynamics simulations of biomolecules. *Nature Structural and Molecular Biology* 9 (10): 646–652.
6. Daggett V. 2006. Alpha-sheet: The toxic conformer in amyloid diseases? *Accounts of Chemical Research* 39 (9): 594–602.
7. Wulfing C, Sjaastad MD, Davis MM. 1998. Visualizing the dynamics of T cell activation: Intracellular adhesion molecule 1 migrates rapidly to the T cell/B cell interface and acts to sustain calcium levels. *Proceedings of the National Academy of Sciences of the United States of America* 95 (11): 6302–6307.
8. Nielsen SO, Lopez CF, Srinivas G, Klein ML. 2004. Coarse-grain models and the computer simulation of soft materials. *Journal of Physics: Condensed Matter* 16 (15): R481–512.
9. Tozzini V. 2005. Coarse-grained models for proteins. *Current Opinion in Structural Biology* 15 (2): 144–50.
10. Chu JW, Ayton GS, Izvekov S, Voth GA. 2007. Emerging methods for multiscale simulation of biomolecular systems. *Molecular Physics* 105 (2–3): 167–75.

11. Marrink SJ, Mark AE. 2003. Molecular dynamics simulation of the formation, structure, and dynamics of small phospholipid vesicles. *Journal of the American Chemical Society* 125 (49): 15233–42.
12. Violi A. 2004. Modeling of soot particle inception in aromatic and aliphatic premixed flames. *Combustion and Flame* 139 (4): 279–87.
13. Metropolis N, Rosenbluth AW, Rosenbluth MN, Teller AH, Teller E. 1953. Equation of state calculations by fast computing machines. *Journal of Chemical Physics* 21 (6): 1087–92.
14. Metropolis N, Ulam S. 1949. The Monte Carlo method. *Journal of the American Statistical Association* 44 (247): 335–41.
15. Voter AF. 2006. Introduction to the Kinetic Monte Carlo method. In *Radiation Effects in Solids*, ed. EA Skeak. Dordrecht, The Netherlands: Springer, NATO Publishing Unit.
16. Gillespie DT. 1976. General method for numerically simulating stochastic time evolution of coupled chemical-reactions. *Journal of Computational Physics* 22 (4): 403–34.
17. Violi A, Kubota A, Truong TN, Pitz WJ, Westbrook CK, Sarofim AF. 2003. A fully, integrated kinetic Monte Carlo/molecular dynamics approach for the simulation of soot precursor growth. *Proceedings of the Combustion Institute* 29: 2343–49.
18. Allen MP, Tildesley DJ. 1989. *Computer Simulation of Liquids*. Oxford: Oxford Science Publications.
19. Ghez R. 1988. *A Primer of Diffusion Problems*. New York: John Wiley & Sons.
20. Chatterjee A., Vlachos DG. 2007. An overview of spatial microscopic and accelerated kinetic Monte Carlo methods. *Journal of Computer-Aided Materials Design* 14 (2): 253–308.
21. Gillespie DT. 1977. Exact stochastic simulation of coupled chemical-reactions. *Journal of Physical Chemistry* 81 (25): 2340–61.
22. Bortz AB, Kalos MH, Lebowitz JL. 1975. New algorithm for Monte-Carlo simulation of Ising spin systems. *Journal of Computational Physics* 17 (1): 10–18.
23. Litwin W, Neimat MA, Schneider DA. 1996. LH*: A scalable, distributed data structure. *ACM Transactions on Database Systems* 21 (4): 480–525.
24. Thomas H, Cormen CEL, Ronald L. 1990. *Rivest. Introduction to Algorithms*. Cambridge, MA: The MIT Press.
25. Frenkel DSB. 1996. *Understanding Molecular Simulation: From Algorithms to Applications*. San Diego, CA: Academic Press.
26. Tuckerman ME, Martyna GJ. 2000. Understanding modern molecular dynamics: Techniques and applications. *Journal of Physical Chemistry B* 104 (2): 159–78.
27. McDermott JB, Libanati C, Lamarca C, Klein MT. 1990. Quantitative use of model-compound information: Monte-Carlo simulation of the reactions of complex macromolecules. *Industrial & Engineering Chemistry Research* 29 (1): 22–29.
28. Voter AF, Montalenti F, Germann TC. 2002. Extending the time scale in atomistic simulation of materials. *Annual Review of Materials Research* 32:321–46.
29. Violi A, Voth GA. 2005. A multi-scale computational approach for nanoparticle growth in combustion environments. In *High Performance Computing and Communications*. Berlin/Heidelberg: Springer.
30. Gillespie DT. 2001. Approximate accelerated stochastic simulation of chemically reacting systems. *Journal of Chemical Physics* 115 (4): 1716–33.
31. Burrage K, Tian T. 2004. Implicit stochastic Runge-Kutta methods for stochastic differential equations. *Bit Numerical Mathematics* 44 (1): 21–39.
32. Chatterjee AVD, Katsoulakis MA. 2005. Binomial distribution based tau-leap accelerated stochastic simulation. *Journal of Chemical Physics* 122 (2): 024112.
33. Chatterjee AMK, Edwards JS, Vlachos DG. 2005. Time accelerated Monte Carlo simulations of biological networks using the binomial tau-leap method. *Bioinformatics* 21 (9): 2136–37.
34. Rathinam MPL, Cao Y, Gillespie DT. 2003. Stiffness in stochastic chemically reacting systems: The implicit tau-leaping method. *Journal of Chemical Physics* 119 (24): 12784–94.
35. Kittelson DB. 1998. Engines and nanoparticles: A review. *Journal of Aerosol Science* 29 (5–6): 575–88.
36. Semmler-Behnke M, Takenaka S, Fertsch S, Wenk A, Seitz J, Mayer P, et al. 2007. Efficient elimination of inhaled nanoparticles from the alveolar region: Evidence for interstitial uptake and subsequent reentrainment onto airway epithelium. *Environmental Health Perspectives* 115 (5): 728–33.
37. Oberdorster G. 2006. Ultrafine particles in lung injury and inflammation. *Free Radical Biology and Medicine* 41: Suppl. 1.
38. Homann KH, Wagner HG. 1968. Chemistry of carbon formation in flames. *Proceedings of the Royal Society of London. Series A. Mathematical and Physical Sciences* 307 (1489): 141–52.

39. Hansen EW, Neurock M. 2000. First-principles-based Monte Carlo simulation of ethylene hydrogenation kinetics on Pd. *Journal of Catalysis* 196 (2): 241–52.
40. Neurock M, Hansen EW. 1998. First-principles-based molecular simulations of heterogeneous catalytic surface chemistry. *Journal of Computational Chemistry and Engineering* 22:S1045–60.
41. Violi A, Truong TN, Sarofim AF. 2004. Kinetics of hydrogen abstraction reactions from polycyclic aromatic hydrocarbons by H atoms. *Journal of Physical Chemistry A* 108 (22): 4846–52.
42. Truhlar DG, Garrett BC, Klippenstein SJ. 1996. Current status of transition-state theory. *Journal of Physical Chemistry* 100 (31): 12771–800.
43. Car R, Parrinello M. 1985. Unified approach for molecular-dynamics and density-functional theory. *Physical Review Letters* 55 (22): 2471–74.
44. Violi A. 2005. Cyclodehydrogenation reactions to cyclopentafused polycyclic aromatic hydrocarbons. *Journal of Physical Chemistry A* 109 (34): 7781–87.
45. Violi A, Sarofim AF, Truong TN. 2001. Quantum mechanical study of molecular weight growth process by combination of aromatic molecules. *Combustion and Flame* 126 (1–2): 1506–15.
46. D'Anna A, Violi A, D'Alessio A. 2000. Modeling the rich combustion of aliphatic hydrocarbons. *Combustion and Flame* 121 (3): 418–29.
47. Stuart SJ, Tutein AB, Harrison JA. 2000. A reactive potential for hydrocarbons with intermolecular interactions. *Journal of Chemical Physics* 112 (14): 6472–86.
48. Brenner DW. 1990. Empirical potential for hydrocarbons for use in simulating the chemical vapor-deposition of diamond films. *Physical Review B* 42 (15): 9458–71.
49. Brenner DW, Harrison JA, White CT, Colton RJ. 1991. Molecular-dynamics simulations of the nanometer-scale mechanical-properties of compressed Buckminsterfullerene. *Thin Solid Films* 206 (1–2): 220–23.
50. Brenner DW, Shenderova OA, Harrison JA, Stuart SJ, Ni B, Sinnott SB. 2002. A second-generation reactive empirical bond order (REBO) potential energy expression for hydrocarbons. *Journal of Physics: Condensed Matter* 14 (4): 783–802.
51. Branka AC, Wojciechowski KW. 2000. Generalization of Nose and Nose-Hoover isothermal dynamics. *Physical Review E* 62 (3): 3281–92.
52. Hoover WG. 2007. Nose-Hoover nonequilibrium dynamics and statistical mechanics. *Molecular Simulation* 33 (1–2): 13–19.
53. Nose S. 1984. A molecular-dynamics method for simulations in the canonical ensemble. *Molecular Physics* 52 (2): 255–68.
54. Hoover WG. 1985. Canonical dynamics: Equilibrium phase-space distributions. *Physical Review A* 31 (3): 1695–97.
55. Violi A, Venkatnathan A. 2006. Combustion-generated nanoparticles produced in a benzene flame: A multiscale approach. *Journal of Chemical Physics* 125 (5): 054302
56. Chung SH, Violi A. 2007. Insights on the nanoparticle formation process in counterflow diffusion flames. *Carbon* 45:2400–10.
57. Violi A, Sarofim AF, Voth GA. 2004. Kinetic Monte Carlo molecular dynamics approach to model soot inception. *Combustion Science and Technology* 176 (5–6): 991–1005.
58. Violi A, Sarofim AF, Truong TN. 2002. Mechanistic pathways to explain H/C ratio of soot precursors. *Combustion Science and Technology* 174 (11–2): 205–22.
59. Violi A, Izvekov S. 2007. Soot primary particle formation from multiscale coarse-grained molecular dynamics simulation. *Proceedings of the Combustion Institute* 31 (1): 529–37.
60. Fiedler SL, Izvekov S, Violi A. 2007. The effect of temperature on nanoparticle clustering. *Carbon* 45 (9): 1786–94.
61. Violi A, Voth GA, Sarofim AF. 2005. The relative roles of acetylene and aromatic precursors during soot particle inception. *Proceedings of the Combustion Institute* 30: 1343–51.

Index

N

O

P

R

S

T

U

V

W

X

Y